COMPUTATIONAL TECHNIQUES AND APPLICATIONS

COMPUTATIONAL TECHNIQUES AND APPLICATIONS: CTAC-89

Proceedings of the Computational Techniques and Applications Conference held at Griffith University, Australia 10–12 July 1989

Edited by

W. L. Hogarth
Division of Australian Environmental Studies
Griffith University
Queensland, Australia

B. J. Noye
Department of Applied Mathematics
The University of Adelaide
South Australia, Australia

⊙ HEMISPHERE PUBLISHING CORPORATION
A member of the Taylor & Francis Group

New York   Washington   Philadelphia   London

COMPUTATIONAL TECHNIQUES AND APPLICATIONS: CTAC-89

1 2 3 4 5 6 7 8 9 0   E B E B   9 8 7 6 5 4 3 2 1 0

Cover design by Sharon DePass.
A CIP catalog record for this book is available from the British Library.

Library of Congress Cataloging-in-Publication Data

International Conference on Computational Techniques and Applications (4th : 1989 : Griffith University, Australia).
    Computational techniques and applications : proceedings of the 4th International Computational Techniques and Applications Conference, held at Griffith University, Australia, 10–12 July 1989 / edited by W. L. Hogarth, B. J. Noye.
        p. cm.

1. Mathematical analysis—Congresses.   I. Hogarth, W. L. (William L.), date.   II. Noye, John, date.   III. Title.
QA299.6.I56   1989
515—dc20   89-78234
ISBN 1-56032-047-8   CIP

Contents

INVITED PAPERS

ALGORITHMS AND SUPERCOMPUTING

BOUNDARY INTEGRALS AND ELEMENTS

COMBINED METHODS AND COMPARISONS

FINITE DIFFERENCE METHODS AND APPLICATIONS

FINITE ELEMENT METHODS AND APPLICATIONS

INDUSTRIAL MATHEMATICS

NETWORKS

NUMERICAL METHODS

OPTIMIZATION

GENERAL

Preface

This volume is the proceedings of the fourth International Conference on Computational Techniques and Applications (CTAC-89) which was held at Griffith University, Brisbane, Australia from 10 to 12 July 1989. It is the fourth in the CTAC series with preceding conferences being held at the University of Sydney, Australia (CTAC-83), University of Melbourne, Australia (CTAC-85) and the University of Sydney, Australia (CTAC-87). The Computational Mathematics Group of the Applied Mathematics Division of the Australian Mathematical Society is responsible for overseeing the organization of these biennial conferences in the CTAC series.

There were nine scheduled invited and review speakers who gave one hour lectures. They included Professor Heinz Engl, University of Linz, Austria, "Inverse Problems in Industry: Case Studies"; Associate Professor Janusz Filipiak, Cracow University, Poland, "Computational Problems in Telecommunnications"; Dr. Nick Fisher, Division of Mathematics and Statistics, CSIRO, Sydney, Australia, "On the Application of Some Computer Intensive Statistical Methods in Earth Sciences"; Professor Jim Liggett, Cornell University, Ithaca, United States of America, "Boundary Integral Equation Calculations for Flow in Fractured and Heterogenous Media"; Dr. Carlos Marino, Cray Research, Massachusetts, United States of America, "Impact of Cray Supercomputers in Science and Engineering Simulations"; Dr. Gerard Meurant, CEA Centre d'Etudes de Limeil-Valenton, France, "The Conjugate Gradient Method on Vector and Parallel Supercomputers"; Associate Professor Bruce Murtagh, University of New South Wales, Australia, "Nonlinear Integer Programming with Applications in Manufacturing and Process Engineering", and Professor Val Pinczewski, University of New South Wales, Australia, "Computational Methods in Petroleum Reservoir Simulation." Professor David Evans, the ninth speaker, was unable to attend. Professor Graham Carey, University of Texas, Austin, United States of America, kindly gave an address on "Adaptive Grids and Supercomputers" to the conference, in place of Professor Evans. This paper appears in the contributed paper section of the proceedings.

In addition, 90 research papers of 25 minutes duration were presented. These papers could be broadly categorized into the following groups: Algorithms and Supercomputers, Boundary Integrals and Elements, Combined Methods and Comparisons, Finite Differences and Applications, Finite Elements and Applications, Industrial Mathematics, Networks, Numerical Analysis, Optimization and General. The papers presented in this volume have been referred by the editorial board, the members of which are listed separately.

A special mention must be made of the work undertaken by the committee which organized the conference so efficiently. Thanks also goes to the various organizations which gave assistance financially and otherwise—they are listed separately.

Finally, we express our appreciation to Hemisphere Publishers for their assistance in the publications of these proceedings.

Bill Hogarth
John Noye

Conference Organizing Committee

Convenor: Dr. W. L. HOGARTH
Division of Australian Environmental Studies
Griffith University
Brisbane, Queensland, Australia

Treasurer: Dr. A. J. O'CONNOR
Division of Science and Technology
Griffith University
Brisbane, Queensland, Australia

Committee: Dr. R. ANDERSSEN
Division of Mathematics and Statistics
CSIRO
Canberra, Australian Capital Territory, Australia

Dr. J. BARRY
ANSTO
Lucas Heights, New South Wales, Australia

Dr. J. BELWARD
Department of Mathematics
University of Queensland
St. Lucia, Queensland, Australia

Professor L. BERRY
School of Information Technology and Computing
Bond University
Gold Coast, Queensland, Australia

Dr. R. BRADDOCK
Division of Australian Environmental Studies
Griffith University
Brisbane, Queensland, Australia

Dr. F. DE HOOG
Division of Mathematics and Statistics
CSIRO
Canberra, Australian National Territory, Australia

Dr. P. DIAMOND
Department of Mathematics
University of Queensland
St. Lucia, Queensland, Australia

Dr. J. HOLT
Department of Mathematics
University of Queensland
St. Lucia, Queensland, Australia

Associate Professor J. MEEK
Department of Civil Engineering
University of Queensland
St. Lucia, Queensland, Australia

Associate Professor J. NOYE
Department of Applied Mathematics
University of Adelaide
Adelaide, South Australia, Australia

Dr. J. PATTERSON
Department of Civil Engineering
University of Western Australia
Perth, Western Australia, Australia

Dr. G. SANDER
Division of Australian Environmental Studies
Griffith University
Brisbane, Queensland, Australia

Dr. J. WATSON
Department of Mining
University of New South Wales
Kensington, New South Wales, Australia

International Steering Group

The Computational Mathematics Group
Division of Applied Mathematics
Australian Mathematical Society

The biennial meeting of the Computational Mathematics Group of the Division of Applied Mathematics of the Australian Mathematical Society took place on 11 July during CTAC-89. At this meeting, the committee of the Computational Mathematics Group for 1989–1991 was elected. It consists of:

Convenor: Dr. C. A. J. FLETCHER
Department of Mechanical Engineering
University of Sydney
Sydney, New South Wales, Australia

Treasurer: Mr. B. BENJAMIN
Department of Mathematics and Computer Studies
South Australian Institute of Technology
The Levels, South Australia, Australia

Secretary: Dr. J. BARRY
ANSTO
Lucas Heights, New South Wales, Australia

Committee: Dr. R. ANDERSSEN
Division of Mathematics and Statistics
CSIRO
Canberra, Australian Capital Territory, Australia

Dr. M. DAVIDSON
Division of Mineral and Process Engineering
CSIRO
Lucas Heights, New South Wales, Australia

Dr. R. MAY
Department of Mathematics
Royal Melbourne Institute of Technology
Box 2476V
Melbourne, Victoria, Australia

Association Professor B. J. NOYE
Department of Applied Mathematics
University of Adelaide
Adelaide, South Australia, Australia

The major responsibility of this group is the organization of the biennial international conferences on Computational Techniques and Applications which are intended to foster communication between the users of computational mathematics who work in a wide range of disciplines.

The next such conference (CTAC-91) will be held in Adelaide, Australia, in July 1991. The Director will be Basil Benjamin, Department of Mathematics and Computer Studies, South Australian Institute of Technology, The Levels, South Australian 5095. All inquiries concerning CTAC-91 should be addressed to him.

THE AUSTRALIAN APPLIED MATHEMATICS DIVISION STUDENT PRIZE

A prize for the best student lecture at the CTAC conference series is now sponsored by the Applied Mathematics Division of the Australian Mathematical Society.

There were 20 student papers presented at the conference. The recipient of the prize in 1989 is indicated below:

Conference   **Recipient**
CTAC-89   *R. G. Brookes,* University of Canterbury, New Zealand

Acknowledgements

The assistance of the following organizations is gratefully acknowledged:

Ansett Airlines of Australia

Applied Mathematics Division, Australian Mathematical Society

The British Council

Centre for Mathematical Analysis, Australian National University, Canberra

Cray Research Australia

Division of Australian Environmental Studies, Griffith University, Brisbane, Australia

National Bank of Australia

Queensland Division of the Australian Mathematical Society

List of Participants

CTAC-89: International Conference on Computational Techniques and Applications, Griffith University, Brisbane, Australia
10 - 12 July 1989

ABRAMSON, Dr D., CSIRO, RMIT, GPO Box 2476V, Melbourne VIC 3001, Australia
ANDERSSEN, Dr R., CSIRO, Division of Mathematics and Statistics, GPO Box 1965, Canberra ACT 2601, Australia
ANDREW, Dr A., Department of Mathematics, La Trobe University, Bundoora VIC 3083, Australia
ARMFIELD, Dr S. Centre for Water Research, University of Western Australia, Nedlands, WA 6009, Australia
AUCHMUTY, Prof. G., Department of Mathematics, University of Houston, Houston, Texas 77204, USA
BARRY, Dr J., ANSTO, Private Mail Bag 1, Menai NSW 2234, Australia
BARTON, Dr N., CSIRO Division of Mathematics and Statistics, PO Box 218, Lindfield NSW 2070, Australia
BASHIR, Dr M., Department of Mathematics, University of Bahrain, PO Box 32038, Bahrain
BASU, Dr A., Department of Mechanical Engineering, University of Wollongong, PO Box 1144, Wollongong NSW 2500, Australia
BECKER, Ms A., Department of Mathematics, Monash University, Clayton, VIC 3168, Australia
BEER, Dr G., CSIRO Division of Geomechanics, Meiers Rd, St Lucia, QLD 4067, Australia
BELWARD, Dr J., Department of Mathematics, University of Queensland, St Lucia QLD 4067, Australia
BENJAMIN, Mr B., School of Mathematics, South Australia Institute of Technology, The Levels SA 5095, Australia
BENNETT, Prof. J., PO Box 22, Balgowlah NSW 2093, Australia
BERRY, Prof. L., Bond University, Private Bag 10, Gold Coast Mail Centre QLD 4215, Australia
BEST, Mr J., Department of Mathematics, University of Wollongong, PO Box 1144, Wollongong 2500
BEVERLY, Mr C., Department of Mechanical Engineering, University of Sydney, Sydney NSW 2006, Australia
BILLINGHURST, Mr D., Comalco Research Centre, 15 Edgars Rd, Thomastown VIC 3074, Australia
BIRTWISTLE, Mr D., Department of Mathematics, Queensland University of Technology, GPO Box 2434, Brisbane QLD 4001, Australia
BRADDOCK, Dr R., Division of Australian Environmental Studies, Griffith University, Nathan QLD 4111, Australia
BROOKES, Mr R., Department of Mathematics, University of Canterbury, Christchurch, New Zealand
BROUGHAN, Dr K., Department of Mathematics, Waikato University, Hamilton, New Zealand
CAREY, Prof. G., Department of Aerospace Engineering, University of Texas, Austin, Texas 78712-1085, USA
CARTER, Dr J., Department of Civil Engineering, University of Sydney NSW 2006, Australia

CHANG, Mr P., CSIRO, RMIT, GPO Box 2476V, VIC 3001, Australia
CHEN, Prof. T.C., Department of Computer Science, Chinese University, Hong Kong
CHU, Dr E., Department of Mathematics, Monash University, Clayton VIC 3053, Australia
COHEN, Dr H., Department of Mathematics, La Trobe University, Bundoora VIC 3083, Australia
COHEN, Mr R., Department of Mechanical Engineering, University of Sydney NSW 2006, Australia
COLGAN, Mr L., Department of Mathematics, South Australia Institute of Technology, The Levels SA 5095, Australia
COLLINGS, Dr I., Department of Computing and Mathematics, Deakin University, Geelong VIC 3217, Australia
DANIELS, Dr W., Department of Mechanical Engineering, University of Queensland, St Lucia QLD 4067, Australia
DAVIDSON, Dr M., CSIRO Division of Mineral and Process Engineering, Lucas Heights NSW 2234, Australia
DAVIES, Dr G., Research School of Earth Sciences, Australian National University, GPO Box 4, Canberra ACT 2601, Australia
DEWAR, Dr R., Department of Theoretical Physics, Australian National University, Canberra ACT 2601, Australia
DIETACHMAYER, Dr G., Bureau of Meteorology Research, GPO Box 1289K, Melbourne VIC 3001, Australia
DOW, Mr M., Computer Services Section, Australian National University, GPO Box 4, Canberra ACT 2601, Australia
ENGL, Prof. H., Institut fur Mathematik, Johannes Kepler University, Linz, A - 4040, Austria
FILIPIAK, Dr J., Department of Applied Mathematics, University of Adelaide, GPO Box 488, Adelaide SA 5001, Australia
FISHER, Dr N., CSIRO Division of Mathematics and Statistics, PO Box 218, Lindfield NSW 2070, Australia
FLEMING, Dr A., Telecom Research Laboratories, 770 Blackburn Rd, Clayton VIC 3168, Australia
FLETCHER, Dr C., Department of Mechanical Engineering, University of Sydney NSW 2006, Australia
FORBES, Dr L., Department of Mathematics, University of Queensland, St Lucia QLD 4067, Australia
FORBES, Mr M., 5 Trephina Close, Riverhills QLD 4074, Australia
FREUND, Mr N., 1 Cope Place, Bulli NSW 2516, Australia
GOLLEY, Dr B., Department of Civil Engineering Dept, AFDA, Campbell ACT 2601, Australia
GOTTSCHALL, Dr N., Department of Mechanical Engineering, University of Queensland, St Lucia QLD 4067, Australia
GUO, Prof. B., Shanghai University of Science and Technology, Shanghai City, China
HA, Dr J., CSIRO, Division of Mathematics and Statistics, Private Bag 10, Clayton VIC 3168, Australia
HAKIM, Dr L., Department of Mechanical Engineering, Chisholm Institute of Technology, PO Box 197, Caulfield East VIC 3145, Australia
HARITOS, Dr N., Department of Civil Engineering, University of Melbourne, Parkville VIC 3052, Australia
HARMAN, Dr C., Darling Downs Institute of Advanced Education, Toowoomba QLD 4350, Australia
HELFGOTT, Dr A., 9 Bundarra Road, Marino Rocks SA 5049, Australia
HOGARTH, Dr W., Division of Australian Environmental Studies, Griffith University, Nathan QLD 4111, Australia

HOLT, Dr J., Department of Mathematics, University of Queensland, St Lucia QLD 4067, Australia

ILIC, Mr M., School of Mathematics, Queensland University of Technology, GPO Box 2434, Brisbane QLD 4001, Australia

INAYAT-HUSSAIN, Dr A., BHP Melbourne Research Laboratories, PO Box 264, Clayton VIC 3168, Australia

ISAACS, Dr L., Department of Civil Engineering, University of Queensland, St Lucia QLD 4067, Australia

ITADANI, Dr Y., Nakayama Propeller Company, 688-1 Jodo Kitagata, Okayama, Japan

JENKINS, Dr D., CSIRO Division of Mathematics and Statistics, Lindfield, PO Box 218, Lindfield NSW 2070, Australia

JOHNSON, Dr G., School of Mathematics, Macquarie University, NSW 2109, Australia

KAGAN, Mr M., Centre for Petroleum Engineering, University of New South Wales, Kensington NSW 2033, Australia

KAHN, Mrs M., Computer Sciences Laboratory, Research School of Physical Sciences, Australian National University, GPO Box 4, Canberra ACT 2600, Australia

KELLY, Dr D., Department of Mechanical Engineering, University of New South Wales, Sydney NSW 2006, Australia

KHOSLA, Prof. P., Department of Aerospace Engineering, University of Cincinatti, Cincinatti, Ohio 45221, USA

KILBY, Mr P., Department of Mathematics, University of Queensland, St Lucia QLD 4067, Australia

KITCHEN, Mr A., 99 Victoria Rd, West Ryde, Sydney NSW 2114, Australia

KOJOVIC, Dr A., Julis Kruttschnidt Mineral Researh Centre, Isles RD, Indooroopilly Qld 4068, Australia

KUCERA, Dr A., Department of Mathematics, La Trobe University, Bundoora VIC 3083, Australia

LAM, Dr Y., Department of Mechanical Engineering, Monash University, Clayton VIC 3168, Australia

LEONARD, Prof. B., Department of Mechanical Engineering, University of Akron, Akron , Ohio 44325, USA

LIGGETT, Prof. J., Department of Civil and Environmental Engineering, Cornell University, Cornell, N.Y., USA

LIN, Ms W., Department of Civil Engineering, University of Queensland, St Lucia QLD 4067, Australia

LOPEZ, Dr J., Aeronautical Research Laboratories, PO Box 4431, Melbourne VIC 3001, Australia

LOWE, Mr S., Department of Mathematics, Wollongong University, PO Box 1144, Wollongong NSW 2500, Australia

LUCAS, Mr S., Department of Mechanical Engineering, University of Sydney NSW 2006, Australia

MACK, Dr A., School of Mechanical Engineering, University of Technology, PO Box 123, Broadway NSW 2007, Australia

MAGDY, Dr A., Department of Electrical Engineering, University of Wollongong, PO Box 1144, Wollongong NSW 2500, Australia

MANI, Dr S., Division of Australian Environmental Studies, Griffith University, Nathan QLD 4111, Australia

MARINO, Dr C., Cray Research Inc, Minneapolis, Minnesota, USA

MAY, Dr R., Department of Mathematics, RMIT, Box 2476V, Melbourne VIC 3001, Australia

MCINNES, Dr A., Department of Mathematics, University of Canterbury, Christchurch, New Zealand

MCLEAN, Dr W., School of Mathematics, University of New South Wales, PO Box 1, Kensington NSW 2030, Australia

MCLEAN, Prof. K., Department of Electrical Engineering, University of Technology, Lae, Papua New Guinea

MEEK, Dr J., Department of Civil Engineering, University of Queensland, St Lucia QLD 4067, Australia

MEURANT, Dr G., Department of Applied Mathematics, BP 27, 94195, Villeneuve St Georges Cedex, France

MOONEY, Dr J., CSIRO Division of Information Technology, GPO Box 664, Canberra ACT 2601

MORROW, Dr R., CSIRO Division of Applied Physics, Box 218, Lindfield NSW 2070, Australia

MURATA, Dr S., Department of Mechanical Engineering, Kyoto Institute of Technology, Matsugasaki, Sakyo-ku, Kyoto 606, Japan

MURTAGH, Prof. B., PO Box 196, Collaroy NSW 2097, Australia

MYERSCOUGH, Dr M., School of Chemistry, Macquarie University, Sydney NSW 2109, Australia

NGUYEN, Dr T., CSIRO Division of Mineral and Process Engineering, Clayton VIC 3168, Australia

NIXON, Mr J., 61 The Esplanade, Henley Beach South, SA 5022, Australia

NOYE, Assoc Prof J., Department of Applied Mathematics, University of Adelaide, SA 5000, Australia

O'CONNOR, Dr A., Division of Science and Technology, Griffith University, Nathan QLD 4111, Australia

OWEN, Mr P., BHP Melbourne Research Laboratories, PO Box 264, Clayton VIC 3168, Australia

PAGE, Dr M., Department of Mathematics, Monash University, Clayton VIC 3168, Australia

PETROLITO, Dr J., Department of Civil Engineering, AFDA, Northcott Drive, Campbell ACT 2600, Australia

PETTIT, Dr A., CSIRO, Cunningham Laboratory, St Lucia QLD 4067, Australia

PINCZEWSKI, Prof. W., Centre for Petroleum Engineering, University of New South Wales, PO Box 1, Kensington NSW 2033, Australia

PLETZER, Mr A., Department of Theoretical Physics, Australian National University, GPO Box 1, Canberra ACT 2601, Australia

PRICE, Dr P., Advanced Computing Support Group, CSIRO, Canberra, GPO Box 664, Canberra ACT 2601, Australia

PRICE, Mr C., Department of Mathematics, University of Canterbury, Christchurch, New Zealand

QUY, Dr N., Research and Technology Centre, Coated Products Division, BHP Steel, PO Box 77, Port Kembla NSW 2505, Australia

READ, Dr W., Department of Mathematics, James Cook University, Townsville QLD 4811, Australia

ROBERTS, Dr A., Department of Applied Mathematics, University of Adelaide SA 5000, Australia

SANDER, Dr G., Division of Australian Environmental Studies, Griffith University, Nathan QLD 4111, Australia

SANDLAND, Dr R., CSIRO Division of Mathematics and Statistics, PO Box 218, Lindfield NSW 2070, Australia

SARDANA, Dr V., School of Civil Engineering, University of New South Wales, PO Box 1, Kensington NSW 2033, Australia

SHEPARD, Mr L., DSTO, WSRL, PO Box 1700, Salisbury SA 5108, Australia

SIEW, Miss W., Department of Mechanical Engineering, Monash University, Clayton VIC 3168, Australia

SPENCER, Mr S., CSIRO Division of Mathematics and Statistics, Box 218, Lindfield NSW 2070, Australia

STEVENS, Mr M., 44 Cudmore Terrace, Henley Beach SA 5022, Australia

STEVENSON, Mr M., Centre for Petroleum Engineeering, University of New South Wales, PO Box 1, Kensington, NSW 2033, Australia

STOKES, Dr N., CSIRO, Division of Mathematics and Statistics, Private Bag 10, Clayton VIC 3168, Australia

SUGDEN, Mr S., Bond University, Private Bag 10, Gold Coast Mail Centre QLD 4215, Australia

SUTER, Dr D., Department of Computer Science, La Trobe University, Bundoora VIC 3083, Australia

SWEET, Dr D., Department of Computer Science, James Cook University, Townsville QLD 4810, Australia

TAN, Dr A., School of Mechanical Engineering, Queensland University of Technology , GPO Box 2434, Brisbane 4001, Australia

TEIPEL, Prof I., Institute for Mechanics, University of Hanover, Appelstrasse 11, 3000 Hanover, West Germany

THOMPSON, Dr A., CSIRO Division of Geomechanics, PO Box 437, Nedlands WA 6009, Australia

TICKLE, Dr K., Department of Mathematics and Computing, Capricornia Institute, Rockhampton QLD 4700, Australia

TIEU, Dr A., Department of Civil Engineering, University of Wollongong, PO Box 1144, Wollongong NSW 2500, Australia

TIPPER, Dr J., Department of Geology, Australian National University, GPO Box 4, Canberra ACT 2601, Australia

TORDESILLAS, Miss A., Department of Mathematics, University of Wollongong, Box 1144, Wollongong NSW 2500, Australia

TSAO, Prof. N-K., Department of Computer Science, Wayne State University, Detroit, Michigan 48202, USA

TURNER, Mr I., 12 Tarnook Drive, Ferny Hills QLD 4055, Australia

TYNDALL, Ms M., Department of Mathematics, Monash University, Clayton VIC 3168, Australia

VAN KEER, Dr R., Department of Mathematics, State University of Ghent, Sint Peitermieniestraat 39, 9000 Ghent, Belgium

WATSON, Dr K., 55 Copeland Road, Beecroft NSW 2119, Australia

WATSON, Dr J., Department of Mining Engineering, University of New South Wales , PO Box 1, Kensington NSW 2033, Australia

WATTS, Dr A., Department of Mathematics, University of Queensland, St Lucia QLD 4067, Australia

WENDTLAND, Dr B., Materials Research Laboratory, DSTO, PO Box 50, Ascot Vale VIC 3032, Australia

WHITEN, Dr W., Julius Kruttschnidt Mineral Research Centre, Isles Rd, Indooroopilly QLD 4068, Australia

WICKS, Dr T., Boeing Computer Corporation, Seattle, USA

WIEDERMANN, Dr A., Institute for Mechanics, University of Hanover, Appelstrasse 11, D-3000, Hanover, West Germany

WU, Mr Y.H., Department of Civil Engineering, University of Wollongong, PO Box 1144, Wollongong NSW 2500, Australia

YASSIN, Prof. A., Chemical and Natural Resources Engineering, University of Technology, Johor, Malaysia

YOUNG, Prof. D-L., Department of Civil Engineering, National Taiwan University, Taipei, Taiwan 10764

ZHENG, Mr R., Department of Mechanical Engineering, University of Sydney, Sydney NSW 2006, Australia

ZHU, Mr Z., Department of Mechanical Engineering, University of Sydney, Sydney NSW 2006, Australia

INVITED PAPERS

On an Inverse Problem from Magnetostatics

H. W. ENGL and A. NEUBAUER
Institut für Mathematik
Johannes-Kepler-Universität
A-4040 Linz, Austria

INTRODUCTION

The problem to be described stems from a current industrial cooperation with a large European company, so that details about the precise technological background are confidential. The problem can be thought of as a nondestructive–testing problem of the following type:

A large block of material with known magnetic permeability μ_0 contains a collection of comparably small enclosures of cylindrical shape with a known magnetic permeability $\mu_1 \neq \mu_0$. A stationary magnetic field is generated by use of a permanent magnet located outside the large block of material. The magnetic field is scattered by the enclosures. The aim is to determine the locations and precise shapes of the enclosures (using the knowledge that they are cylinders) from measurements of the scattered field. Figure 1 shows a typical situation. In this Figure, S_0 is the permanent magnet, which is assumed to be a rectangular parallelepiped with faces parallel to the coordinate planes. The $S_i (i = 1, \ldots, n;$ in Fig.1: $n = 2)$ are the cylindrical enclosures. The relevant parameters are:

a_M, b_M, h_M:	dimensions of the magnet
(x_M, y_M, z_M):	coordinates of the center of one face of the magnet
n :	number of cylindrical enclosures
d_i, l_i:	diameter and length of S_i
$u_i = (ux_i, uy_i, uz_i)$:	unit vector in direction of the axis of S_i
$Q_i = (Qx_i, Qy_i, Qz_i)$:	center of S_i .

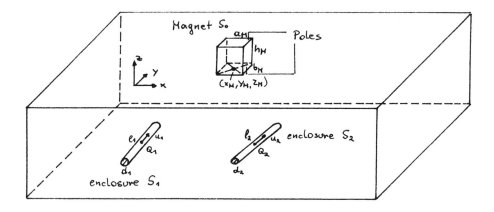

FIGURE 1. A typical situation

We deal with the following two problems:

Direct Problem: Given the location and the magnetization of S_0 and the locations and shapes of the $S_i(i = 1, \ldots, n)$, compute the total magnetic field $H = H^i + H^s$, where H^i is the "incident field" generated by S_0 if no enclosures $S_i(i = 1, \ldots, n)$ are present; H^s is called the "scattered field".

Inverse Problem: Given the location and the magnetization of S_0 (and hence the incident field H^i), determine the S_i from measurements of certain components of the field H (which give the corresponding components of the scattered field H^s).

We will now derive a mathematical model for the Direct Problem and describe its numerical solution. Then we will mention known theoretical results for the Inverse Problem (or at least for problems closely related to our Inverse Problem) and show that although they suggest that the Inverse Problem is uniquely solvable, the accuracy requirements for measuring H to get any useful information about the S_i are higher than practically feasible.

THE MATHEMATICAL MODEL FOR THE DIRECT PROBLEM

We assume throughout that the magnetic induction B is related to the magnetic field H in a linear way, i.e.,

$$B = \mu H, \tag{1}$$

where μ is the magnetic permeability. We assume that $\mu = \mu_0 > 0$ in $\mathbb{R}^3 \setminus \bigcup_{i=1}^n \bar{S}_i$ (i.e., the magnetic permeabilities in the magnet and in the large block containing the enclosures equal the permeability of air) and $0 < \mu = \mu_1 \neq \mu_0$ in $\bigcup_{i=1}^n S_i$. Of course, we assume that

4

$$\bar{S}_i \cap \bar{S}_j = \emptyset \quad \text{for} \quad i \neq j \in \{0, \dots, n\}. \tag{2}$$

The $S_i (i \in \{0, \dots, n\})$ are assumed to be open and connected. The permanent magnet S_0 is assumed to be unifomly magnetized in negative z–direction, i.e., the magnetization is assumed to have the form

$$M = (0, 0, -d_M) \tag{3}$$

with a known positive constant d_M. This assumption was fulfilled in our practical problem, but is not needed in this Section.

With these assumptions, Maxwell's equations for describing our configuration read as follows:

$$B = \mu_0(H + M) \quad \text{in } S_0 \tag{4}$$

$$B = \mu_0 H \quad \text{in } \mathbb{R}^3 \setminus \bigcup_{i=0}^{n} \bar{S}_i \tag{5}$$

$$B = \mu_1 H \quad \text{in } \bigcup_{i=1}^{n} \bar{S}_i \tag{6}$$

$$\text{div } B = 0 \quad \text{in } \mathbb{R}^3 \setminus \bigcup_{i=0}^{n} \partial S_i \tag{7}$$

$$\text{rot } B = 0 \quad \text{in } \mathbb{R}^3 \setminus \bigcup_{i=0}^{n} \partial S_i \tag{8}$$

$$(n_i, B)_+ = (n_i, B)_- \quad \text{on } \partial S_i \text{ for } i \in \{0, \dots, n\} \tag{9}$$

$$[n_i, H]_+ = [n_i, H]_- \quad \text{on } \partial S_i \text{ for } i \in \{0, \dots, n\} \tag{10}$$

$$H(x) = o(1) \quad \text{as } \| x \| \to +\infty$$

$$\text{uniformly in all directions.} \tag{11}$$

In (9) and (10), we denote by (,) and [,] the scalar and vector products, respectively. The subscripts $+$ and $-$ denote limits taken from the exterior or the interior of S_i, respectively; n_i is the unit outward normal to S_i, which is assumed to exist almost everywhere.

We now derive a representation of solutions to (4) – (11) via a boundary integral equation. Let

$$\gamma(x, y) := \frac{1}{4\pi \| x - y \|} \quad (x \neq y \in \mathbb{R}^3), \tag{12}$$

where $\| \, . \, \|$ denotes the Euclidian norm in $\mathbb{R}^3$, be a fundamental solution of Laplace's equation. Because of (10), it follows from Cauchy's integral formula for harmonic vector fields (Martensen (1968)) that

$$H(x) = -\text{grad}_x \int_{\partial V} \varphi(y)\gamma(x, y)dS(y) \quad (x \in \mathbb{R}^3 \setminus \partial V) \tag{13}$$

with

$$\varphi(y) := (n(y), H(y))_+ - (n(y), H(y))_- \quad (y \in \partial V). \tag{14}$$

Here, $\partial V := \bigcup_{i=0}^{n} \partial S_i$ and $n(y)$ denotes the unit outward normal to ∂V, i.e., $n(y) = n_i(y)$ if $y \in \partial S_i \subseteq \partial V$ in the notation of (9).

While the transmission conditions (10) are already contained in the representation (13), the conditions (9) have to be taken into account; for this, we use the potential theoretic jump relations (cf. Colton – Kreß (1983)). We define a single layer potential with a density Ψ defined on ∂V via

$$(E\Psi)(x) := \int_{\partial V} \Psi(y)\gamma(x,y)dS(y) \quad (x \in \mathbb{R}^3). \tag{15}$$

Then

$$(\operatorname{grad} E\Psi)_\pm(x) = \int_{\partial V} \Psi(y)\operatorname{grad}_x\gamma(x,y)dS(y) \mp \frac{1}{2}n(x)\Psi(x) \quad (x \in \partial V) \tag{16}$$

holds, if Ψ and ∂V are sufficiently smooth, as will always be assumed: E.g., $\partial V = \bigcup_{i=0}^{n} \partial S_i$ consisting of C^2–curves ∂S_i and Ψ being continuous suffices.

Combining (13) and (16), we obtain

$$H(x)_\pm = -\int_{\partial V} \varphi(y)\operatorname{grad}_x\gamma(x,y)dS(y) \pm \frac{1}{2}n(x)\varphi(x) \quad (x \in \partial V) \tag{17}$$

with φ as in (14). We now insert (17) into (9) and obtain for $x \in \partial S_0$:

$$
\begin{aligned}
\varphi(x) &= (n(x), n(x))\varphi(x) = (n(x), H(x))_+ - (n(x), H(x))_- = \\
&= \frac{1}{\mu_0}(n(x), B(x))_+ - [\frac{1}{\mu_0}(n(x), B(x))_- - (n(x), M)_-] = \\
&= (n(x), M)_-
\end{aligned}
$$

because of (4), (5), and (9). Hence, we have

$$\varphi(x) = (n(x), M)_- \quad (x \in \partial S_0) \tag{18}$$

with M as in (3).

Now, let $x \in \partial S_i$ $(i \in \{1, \ldots, n\})$. From (5), (6) and (9) we obtain

$$(n(x), H(x))_- = \chi(n(x), H(x))_+ \tag{19}$$

with

$$\chi := \frac{\mu_0}{\mu_1}. \tag{20}$$

Combining (19) with (17) yields

$$\frac{1+\chi}{2}\varphi(x) + (1-\chi)\int_{\partial V} \varphi(y)\frac{\partial}{\partial n(x)}\gamma(x,y)dS(y) = 0 \quad (x \in \bigcup_{i=1}^{n} S_i). \tag{21}$$

We now use (18) and obtain from (21) the following integral equation for φ:

$$\frac{\mu_0 + \mu_1}{2(\mu_1 - \mu_0)}\varphi(x) + \sum_{i=1}^{n}\int_{\partial S_i} \varphi(y)\frac{\partial}{\partial n(x)}\gamma(x,y)dS(y) = \tag{22}$$

$$= -\int_{\partial S_0} (n(y), M)_-\frac{\partial}{\partial n(x)}\gamma(x,y)dS(y) \quad (x \in \partial S_j, j \in \{1, \ldots, n\}).$$

Note that in (22), the differentation inside the integral takes place with respect to the independent variable, the direction $n(x)$ is the normal to ∂S_j, if $x \in S_j$, not to the curve of integration ∂S_i (unless $i = j$); (22) should actually be considered as a system of integral equations for the unknowns $\varphi|_{S_j}(j \in \{1, \ldots, n\})$.

It follows with standard methods of potential theory (cf. the proof of Theorem 3.2 in Engl – Kreß (1981)) that any solution of (22) generates via (13) a solution to (4) – (11). Hence, we have under our smoothness assumptions:

Theorem 1: *If H solves (4)–(11), then φ defined by (14) will solve the system of integral equations (22). Conversely, if φ solves (22), then H defined by (13) will solve (4)–(11).*

Thus, we have reformulated the problem of calculating the magnetic field as a system of integral equations for the jump of its normal components across the boundaries of the enclosures $\partial S_i(i = \{1, \ldots, n\})$. This is crucial for a numerical solution, since now we have a two–dimensional problem over the bounded domain $\bigcup_{i=1}^{n} \partial S_i$, while the original problem was three–dimensional and had an unbounded domain.

Another possibility of deriving a boundary integral equation would be to consider $[n(y), B(y)]_+ - [n(y), B(y)]_-$ as unknown and proceed as in Theorem 4.2 of Engl – Kreß (1981). We did not pursue this approach.

Of course, the incident field can be computed directly via (18) and (13).

Under our smoothness assumptions, the following can be shown:

Theorem 2: *The system of integral equations (22) and hence the Direct Problem (4)–(11) each have a unique solution.*

Proof: by standard potential theoretic methods. Alternatively, the result follows as a special case of Theorem 3.4 from Engl – Kreß (1981).

We finally remark that in an analogous way, an equivalent integral equation like (22) can also be derived if the S_j have different magnetic permeabilities $\mu_j(j \in \{1, \ldots, n\})$. The factor multiplying $\varphi(x)$ in (22) has then to be replaced by $\frac{\mu_0 + \mu_j}{2(\mu_j - \mu_0)}$.

Note that if some of the μ_i tend to $+\infty$, the corresponding limiting problem will have multiple solutions, as can be seen from Section 3 of Engl – Kreß (1981). Thus, numerical difficulties have to be expected for large permeabilities μ_i.

We record for later use that also the function $u := E\varphi$ defined by (14) and (15) solves a boundary value problem, namely

$$\triangle u = 0 \qquad \text{in } \mathbb{R}^3 \setminus (\bigcup_{i=0}^{n} \partial S_i \cup S_0) \tag{23}$$

$$\triangle u = \text{div } M \qquad \text{in } S_0 \tag{24}$$

$$u_+ = u_- \qquad \text{on } \partial S_i \ (i \in \{0, \ldots, n\}) \tag{25}$$

$$\mu_0 \left(\frac{\partial u}{\partial n}\right)_+ = \mu_1 \left(\frac{\partial u}{\partial n}\right)_- \qquad \text{on } \partial S_i \ (i \in \{1, \ldots, n\}) \tag{26}$$

$$\left(\frac{\partial u}{\partial n}\right)_+ = \left(\frac{\partial u}{\partial n}\right)_- - (M, n)_- \qquad \text{on } \partial S_0 \tag{27}$$

together with the usual boundary conditions at infinity. Note that because of (13), u is a scalar magnetic potential for H (except on $\bigcup_{i=0}^{n} \partial S_i$).

7

In principle, it is clear from the preceding Section how to proceed for solving the Direct Problem:

For computing the incident field, one uses (13) with $\partial V = \partial S_0$ and φ as in (18). For computing the total field, one solves the system of integral equations (22) and then uses (13).

We now report about our experience with solving the system (22) for the specific configuration assumed above, where the $S_i (i \in \{1, \ldots, n\})$ are actually cylinders, whose mantles we denote by M_i and whose top and bottom are denoted by D_{i1} and D_{i2}, respectively. We regard these sets as parameterized surfaces over $[0, 1]^2$ via

$$
\left.
\begin{aligned}
D_{i1} &= \Psi_{i1} \; ([0, 1]^2), \\[2mm]
D_{i2} &= \Psi_{i2} \; ([0, 1]^2), \\[2mm]
M_i &= \Psi_{i3} \; ([0, 1]^2),
\end{aligned}
\right\}
\tag{28}
$$

$$
\text{with}
\left.
\begin{aligned}
\Psi_{i1}(\alpha, r) &:= Q_i - \tfrac{l_i}{2} u_i & + \tfrac{d_i}{2} r (\sin 2\pi\alpha . v_{1,i} + \cos 2\pi\alpha . v_{2,i}), \\[2mm]
\Psi_{i2}(\alpha, r) &:= Q_i + \tfrac{l_i}{2} u_i & + \tfrac{d_i}{2} r (\sin 2\pi\alpha . v_{1,i} + \cos 2\pi\alpha . v_{2,i}), \\[2mm]
\Psi_{i3}(\alpha, r) &:= Q_i + l_i (r - \tfrac{1}{2}) u_i & + \tfrac{d_i}{2} (\sin 2\pi\alpha . v_{1,i} + \cos 2\pi\alpha . v_{2,i}),
\end{aligned}
\right\}
\tag{29}
$$

where

$$
\left.
v_{1,i} := \frac{1}{\sqrt{uy_i{}^2 + uz_i{}^2}}
\begin{pmatrix} 0 \\ -uz_i \\ uy_i \end{pmatrix},
\qquad
v_{2,i} := \frac{1}{\sqrt{uy_i{}^2 + uz_i{}^2}}
\begin{pmatrix} uy_i{}^2 + uz_i{}^2 \\ -uy_i . ux_i \\ -uz_i . ux_i \end{pmatrix}
\right\}
\tag{30}
$$

$$
\text{if} \quad |uz_i| > \tfrac{1}{2}
$$

and

$$
\left.
v_{1,i} := \frac{1}{\sqrt{ux_i{}^2 + uy_i{}^2}}
\begin{pmatrix} -uy_i \\ ux_i \\ 0 \end{pmatrix},
\qquad
v_{2,i} := \frac{1}{\sqrt{ux_i{}^2 + uy_i{}^2}}
\begin{pmatrix} -ux_i . uz_i \\ -uy_i . uz_i \\ ux_i{}^2 + uy_i{}^2 \end{pmatrix}
\right\}
\tag{31}
$$

otherwise.

These transformations using cylindrical coordinates hold for $i \in \{1, \ldots, n\}$. The poles of the permanent magnet, i.e., the faces of the parallelepiped S_0 orthogonal to the direction of magnetization, are parameterized over $[0,1]^2$ via

$$
\Psi_0(\alpha, r) := \begin{pmatrix} x_M - \frac{a_M}{2} + \alpha a_M \\[2mm] y_M - \frac{b_M}{2} + r b_M \\[2mm] z_M \end{pmatrix}, \tag{32}
$$

$$
\check{\Psi}_0(\alpha, r) := \Psi_0(\alpha, r) + (0, 0, h_M)^T. \tag{33}
$$

We now introduce functions f_{jk} ($j \in \{1, \ldots, n\}, k \in \{1,2,3\}$) on $[0,1]^2$ via

$$
f_{jk} := \varphi \circ \Psi_{jk} \tag{34}
$$

and

$$
\rho := \frac{\mu_1 + \mu_0}{2(\mu_1 - \mu_0)}. \tag{35}
$$

Then the system (22) can be reformulated equivalently as the following system of integral equations for the unknown functions f_{jk}:

$$
\left.
\begin{aligned}
& \sum_{i=1}^{n} \sum_{l=1}^{3} \int_0^1 \int_0^1 f_{il}(\alpha, r) g_{jkil}(\beta, s, \alpha, r).p_{il}(r)\,d\alpha\,dr + \rho_j f_{jk}(\beta, s) \\[2mm]
& = \frac{a_M.b_M.d_M}{4\pi} \int_0^1 \int_0^1 (\bar{g}_{jk}(\beta, s, \alpha, r) - g_{jk}(\beta, s, \alpha, r))\,d\alpha\,dr, \\[2mm]
& j = 1, \ldots, n; \quad k = 1, 2, 3; \quad (\beta, s) \in [0,1]^2,
\end{aligned}
\right\} \tag{36}
$$

where

$$
\left.
\begin{aligned}
& p_{il}(r) := \begin{cases} r.\frac{d_i^2}{8}, & l = 1, 2, \\[3mm] d_i.\frac{l_i}{4}, & l = 3, \end{cases} \\[6mm]
& g_{jkil}(\beta, s, \alpha, r) := \frac{(\Psi_{il}(\alpha, r) - \Psi_{jk}(\beta, s), n_{jk}(\beta))}{\| \Psi_{il}(\alpha, r) - \Psi_{jk}(\beta, s) \|^3}, \\[4mm]
& \overset{(-)}{g}_{jk}(\beta, s, \alpha, r) := \frac{(\overset{(-)}{\Psi}_o(\alpha, r) - \Psi_{jk}(\beta, s), n_{jk}(\beta))}{\| \overset{(-)}{\Psi}_o(\alpha, r) - \Psi_{jk}(\beta, s) \|^3}, \\[4mm]
& n_{jk}(\beta) := \begin{cases} -u_j, & k = 1, \\[3mm] u_j, & k = 2, \\[3mm] \sin 2\pi\beta.v_{1,j} + \cos 2\pi\beta.v_{2,j}, & k = 3. \end{cases}
\end{aligned}
\right\} \tag{37}
$$

9

To see this, note that

$$\frac{\partial}{\partial n(x)}\gamma(x,y) = \frac{(y-x,n(x))}{4\pi\,\|\,y-x\,\|^3} \tag{38}$$

holds.

It is also important to note that in the preceding Section, we assumed the boundaries ∂S_i to be smoother than they actually are at points where the mantles of the cylinders S_i meet the top and bottom. In our case, the integral equation (22) and its equivalent form (36) holds at these points with ρ replaced by a different value. Since in our numerical scheme, we will never use the values of the unknown functions exactly at these corners, we can as well work with (22) and (36) in their original forms. However, due to the corners in ∂S_i, the solution φ of (22) will not be continuous, but will have singularities in the "corners" of ∂S_i. This can be seen e.g. from Sloan – Spence (1988) taking into account the representation (14) for φ. These singularities cause major numerical problems. This is only true for computing the function φ. Both computations and physical experience show that if one smoothes out the corners in ∂S_i, so that the functions $\varphi\,|_{\partial S_i}$ become continuous, this makes no relevant difference for computing H via (13) away from the enclosures S_i. However, since by transforming (22) into (36), we used the fact that the S_i are cylinders, we have to live with the singularities in φ; smoothing ∂S_i in a C^2–way first would have made the transformations much more complicated.

There are basically two ways of coping with these singularities. Either one uses a basis for approximating φ (or the f_{jk}) that also contains functions with singularities, or one uses graded meshes. For the first approach, one has to know the type of the singularity. While for the two–dimensional case, one has information of that type (cf. Jawson – Symm (1977)), this seems not to be the case in three dimensions. Thus we used graded meshes in the following Galerkin method:

We approximated the functions f_{jk} by piecewise polynomials and obtained a system of linear equations by requiring that the L^2–inner products of both sides of (36) with these Ansatz functions coincide. We first used piecewise linear polynomials as basis functions, then piecewise constant functions. The first choice gave worse results than the second choice for the following reasons:

First, because of the continuity of piecewise linear functions, the error that comes from trying to approximate the singularities in the corners propagates away from the corners more strongly than when using (discontinuous) piecewise constant functions. Second, the Galerkin approach involves the calculation of quadruple integrals (see (42)). With piecewise constant functions, we managed to evaluate at least some of these integrals in closed form, thus reducing the error in computing these integrals considerably. For piecewise linear basis functions, the effort for evaluating these integrals explicitly would be overwhelming. We had to use numerical quadrature exclusively there and observed errors of up to 7 % for integrals over basis functions with support close to the corners, although we used Gaußian quadrature with as many points as we could justify taking the resulting computation time into account. Note that these integrals enter into the matrix for the linear system of equations (42) determining the coefficients.

As mentioned in the preceding Section, (22) loses its unique solvability as $\mu_1 \to +\infty$. Due to the large errors in the matrix elements when using piecewise linear Ansatz functions, this led to severe numerical problems already for $\mu \approx 100$, while the code with piecewise constant basis functions performed well for much larger values of μ. In that code, we approximate the functions f_{jk} by

$$
\left.
\begin{array}{l}
f_{jk}(\alpha, r) \approx \sum\limits_{g=1}^{mW} \sum\limits_{h=1}^{mr} c_{gh,jk} \cdot v_g(\alpha) \cdot \bar{v}_h(r) \;, \quad k = 1, 2 \\[4mm]
f_{j3}(\alpha, r) \approx \sum\limits_{g=1}^{mW} \sum\limits_{h=1}^{ml} c_{gh,j3} \cdot v_g(\alpha) \cdot \tilde{v}_h(r) \;.
\end{array}
\right\}
\tag{39}
$$

Here, mW, mr, and ml denote the number of grid points for discretizing the angular variable, the radius of the top and the bottom and the length of the cylinder, respectively; furthermore,

$$
\left.
\begin{array}{l}
v_g(\alpha) \;\; := \;\; \begin{cases} 1 & , \quad \alpha \in [\alpha_{g-1}, \alpha_g] \\[3mm] 0 & , \quad \text{otherwise} \end{cases} \\[8mm]
\bar{v}_h(r) \;\; := \;\; \begin{cases} 1 & , \quad r \in [fr_{h-1}, fr_h] \\[3mm] 0 & , \quad \text{otherwise} \end{cases} \\[8mm]
\tilde{v}_h(r) \;\; := \;\; \begin{cases} 1 & , \quad r \in [fl_{h-1}, fl_h] \\[3mm] 0 & , \quad \text{otherwise.} \end{cases}
\end{array}
\right\}
\tag{40}
$$

The grid points were chosen as follows:

$$
\left.
\begin{array}{lll}
\alpha_g & := \;\; \dfrac{g}{mW} & , \quad g = 0, 1, \ldots, mW \\[4mm]
fr_h & := \;\; \sqrt{\sin(\frac{\pi}{2} \cdot \frac{h}{mr})} & , \quad h = 0, 1, \ldots, mr \\[4mm]
fl_h & := \;\; \frac{1}{2}(1 - \cos(\pi \cdot \frac{h}{ml})) & , \quad h = 0, 1, \ldots, ml \;,
\end{array}
\right\}
\tag{41}
$$

i.e., the mesh was graded with respect to radius and length of the cylinder in a way that more mesh points lie close to the "corners" where singularities of the solution are to be expected.

The Galerkin approach now leads to the following system of equations for the unknown coefficients $\{c_{gh,jk}\}$ in the approximation (39):

$$\sum_{i=1}^{n} \left[\frac{d_i^2}{8} \sum_{l=1}^{2} \sum_{p=1}^{mW} \sum_{q=1}^{mr} c_{pq,il} \int_0^1 \int_0^1 \int_0^1 \int_0^1 r \cdot g_{jkil}(\beta, s, \alpha, r) \cdot v_p(\alpha) \bar{v}_q(r) v_g(\beta) \hat{v}_h(s) \, d\alpha dr d\beta ds \; + \right.$$

$$\left. + \frac{d_i l_i}{4} \sum_{p=1}^{mW} \sum_{q=1}^{ml} c_{pq,i3} \int_0^1 \int_0^1 \int_0^1 \int_0^1 g_{jki3}(\beta, s, \alpha, r) v_p(\alpha) \bar{v}_q(r) v_g(\beta) \cdot \hat{v}_h(s) \, d\alpha dr d\beta ds \right] \qquad (42)$$

$$+ \, p \cdot c_{gh,jk} \cdot (\hat{f}_h - \hat{f}_{h-1}) \frac{1}{mW} \; =$$

$$= \frac{a_M \cdot b_M \cdot d_M}{4\pi} \int_0^1 \int_0^1 \int_0^1 \int_0^1 (\bar{g}_{jk}(\beta, s, \alpha, r) - g_{jk}(\beta, s, \alpha, r)) v_g(\beta) \hat{v}_h(s) \, d\alpha dr d\beta ds,$$

with

$$\hat{v}_h := \begin{cases} \bar{v}_h & , \quad k = 1, 2 \quad , \\[2mm] \tilde{v}_h & , \quad k = 3 \quad , \end{cases}$$

$$\hat{f}_h := \begin{cases} fr_h & , \quad k = 1, 2 \quad , \\[2mm] fl_h & , \quad k = 3 \quad , \end{cases}$$

and

$$j = 1, \ldots, n; \quad k = 1, 2, 3;$$

$$g = 1, \ldots, mW; h = \begin{cases} 1, \ldots, mr \;, \quad k = 1, 2 \\[2mm] 1, \ldots, ml \;, \quad k = 3 \quad . \end{cases}$$

In the quadruple integrals in (42), the order of integrations can be interchanged. The integrals with respect to r and s can be explicitely evaluated (see Neubauer (1989)). The necessary computations are lengthy and have been verified by use of the computer algebra system REDUCE.

The remaining double integrals were evaluated numerically using Gaußian quadrature with 4 Gauß points per subinterval. The integrals on the right-hand side of (42) were evaluated by using Gaußian quadrature with 4 knots on 3 subintervals of [0,1].

The resulting system has $N := n * mW * (2mr + ml)$ unknowns and can efficiently be solved by the Gaußian algorithm with column pivoting; it has no sparsity structure.

The number of operations for setting up the matrix in (42) is proportional to N^2, while solving it is (in the absence of sparsity) proportional to N^3. Thus, for large N, solving (42) is the most time consuming task. For smaller N, setting up the matrix needs most of the CPU–time.

To validate our code, we tried to find an example with an analytically known solution without success. Thus, we used an example from Sowerby (1974) for comparison, where the enclosure is a ball instead of a cylinder. As incident field, we used a uniform field

TABLE 1. Uniform incident field

z	ball	cylinder,Case A	cylinder,Case B	cylinder,Case C
5	$-0.17143 * 10^{-2}$	$-0.10292 * 10^{-2}$	$-0.15042 * 10^{-2}$	$-0.16634 * 10^{-2}$
10	$-0.21429 * 10^{-3}$	$-0.13282 * 10^{-3}$	$-0.18876 * 10^{-3}$	$-0.20881 * 10^{-3}$
15	$-0.63492 * 10^{-4}$	$-0.39753 * 10^{-4}$	$-0.55970 * 10^{-4}$	$-0.61917 * 10^{-4}$
CPU–time	—	0.04 min.	0.14 min.	2.49 min.

$H^i = (0, 0, -1)$ (which is not generated by a permanent magnet). Then we compared the z−component of the scattered field for a ball with radius $\sqrt[3]{\frac{3}{16}}$ centered at the origin (computed analytically) with the corresponding quantity for one cylindrical enclosure with the same volume as the ball with the data $Q_1 = (0, 0, 0), u_1 = (0, 1, 0), d_1 = l_1 = 1$, $\mu_0 = 1, \mu_1 = 5$, computed via our code). We report the results at points $(0, 0, z)$ for three different discretizations, namely

$Case\ A$: mr = mW = ml = 3,   hence N=   27

$Case\ B$: mr = 3, mW = ml = 6,   hence N=   72

$Case\ C$: mr = 6, mW = ml = 12,   hence N=   288

in Table 1. The CPU–time is given for a BASF 7/78.
While the discretization in Case A is clearly too coarse, Cases B and C give reasonable results. Complete coincidence with the results for the ball cannot be expected anyway. Already this example shows that the scattered field is by orders of magnitude smaller than the incident field. However, a uniform magnetic field as incident field is unrealistic. In the following example we used the setting shown in Figure 1 with the following data: $a_M = 4, b_M = 2.5, h_M = 8, d_M = 400, x_M = y_M = 0, z_M = 5; n = 1, u_1 = (0, 1, 0)$, $Q_1 = (0, 0, \ 5), d_1 - 1, l_1 - 5, \mu_0 = 1, \mu_1 = 5$.

We computed the z−component of the incident and of the scattered field at points $(x, 0, 0)$ with x between 0 and 20 (in increments of 0.5). The CPU–times on a MicroVAX 3500 depending on the discretization parameters are shown in Table 2.
For the intermediate discretization, the results for the scattered field were good (compared to the finest discretization) only away from the region where the scattered field changes its direction. We give a sample from the results obtained with the finest discretization in Table 3.
The results show that the scattered field in this configuration is by three orders of magnitude smaller than the incident field.
More details about the numerical solution of the Direct Problem are reported in Neubauer (1989).

TABLE 3. Incident vs. scattered field

TABLE 2. CPU–time

mr	mW	ml	CPU–time in minutes
3	6	6	1:17
6	12	12	10:12
12	24	24	179:06

x	z–component of incident field	z–component of scattered field
0	$0.96312 * 10^{+1}$	$0.12748 * 10^{-1}$
3	$0.62225 * 10^{+1}$	$0.50389 * 10^{-2}$
6	$0.21021 * 10^{+1}$	$0.41310 * 10^{-3}$
9	$0.47721 * 10^{+1}$	$-0.24772 * 10^{-3}$
12	$-0.21581 * 10^{-2}$	$-0.23581 * 10^{-3}$
15	$-0.12025 * 10^{+0}$	$-0.16613 * 10^{-3}$

CONCLUSIONS: THEORY VS. PRACTICE

So far, we have only discussed the Direct Problem. We now turn to our original objective, namely solving the Inverse Problem of determining the enclosures S_i from measurements of the scattered field.

There are numerous results in the literature on closely related inverse problems, namely (essentially) on determining either the S_i in the exterior boundary value problem (23), (25), (26) from measurements on the boundary of a set enclosing all S_i, or on determining a spatially varying μ from the weak form of (23), (25), (26) from such boundary measurements. See e.g. Prilepko (1971), Kohn – Vogelius (1985), Sylvester – Uhlmann (1988), Rundell (1988) (although there are doubts about the correctness of a proof there, see Jodlbauer (1989)), Friedman – Isakov (1988), Isakov – Powell (1989). These results indicate that in our Inverse Problem, the enclosures should be uniquely determined from measurements of the scattered field (for one incident field) around the whole configuration. Aside from the fact that in the practical application where this problem comes from, one cannot measure on a closed surface around the configuration, the numerical results reported in in the preceding Section indicate that the scattered field is so small compared to the incident field that it cannot be measured reliably enough by conventional techniques. Thus, our Inverse Problem cannot be solved in practice with this approach. However, we are currently investigating a different approach based on this work which might (hopefully) lead to a practically feasible method for solving the Inverse Problem.

ACKNOWLEDGEMENT

The authors acknowledge partial support from the Austrian Fonds zur Förderung der wissenschaftlichen Forschung (project S32/03). Travel to this conference was supported by the Centre for Mathematical Analysis, Australian National University, Canberra.

REFERENCES

1. Colton, D., Kreß, R., *Integral Equation Methods in Scattering Theory*, Wiley, New York, 1983.

2. Engl, H.W., Kreß, R., A singular perturbation problem for linear operators with an application to electrostatic and magnetostatic boundary and transmission problems, *Math.Meth.in the Appl.Sc.*, vol. 3, pp. 249–274, 1981.

3. Friedman, A., Isakov, V., On the uniqueness in the inverse conductivity problem with one measurement, *Indiana Univ.Math.J.*, forthcoming.

4. Isakov, V., Powell, J., On uniqueness in the inverse conductivity problem with one measurement for cylinders and disks, *Inverse Problems*, forthcoming.

5. Jawson, M., Symm, G., *Integral Equation Methods in Potential Theory and Elastostatics*, Academic Press, New York, 1977.

6. Jodlbauer, H., *Herleitung und Untersuchung eines mathematischen Modells zum Problem der Identifizierung von Armierungseisen in Beton*, Diploma Thesis, Johannes–Kepler–Universität Linz, 1989.

7. Kohn, R., Vogelius, M., Determining conductivity by boundary measurements II: Interior results, *Comm.Pure and Appl.Math.*, vol. 38, pp. 643–667, 1985.

8. Martensen, E., *Potentialtheorie*, Teubner, Stuttgart, 1968.

9. Neubauer, A., On the numerical solution of a 3–D magnetostatic scattering problem by boundary integral equation methods, *Proc.Internat.Symp.and Team Workshop on 3–D Electromagn. Field Analysis, Okayama (Japan)*, forthcoming.

10. Prilepko, A., On the stability and uniqueness of a solution of inverse problems of generalized potentials of a simple layer, *Sibir.Math.Journ.*, vol. 12, pp. 594–601, 1971.

11. Rundell, W., Some inverse problems for elliptic equations, *Applic.Anal.*, vol. 28, pp. 67–78, 1988.

12. Sloan, I., Spence, A., The Galerkin method for integral equations of the first kind with logarithmic kernel: Theory, *IMA Journ.Num.Anal.*, vol. 8, pp. 105–122, 1988.

13. Sowerby, L., *Vector Field Theory with Applications*, Longman, London 1974.

14. Sylvester, J., Uhlmann, G., Inverse boundary value problems at the boundary — Continuous dependence, *Comm.Pure and Appl.Math.*, vol. 41, pp. 197–219, 1988.

Computational Problems
in Telecommunications

JANUSZ FILIPIAK
The Teletraffic Research Centre,
The University of Adelaide
G.P.O. Box 498
Adelaide 5001, South Australia

INTRODUCTION

Analysis and design of communication systems require that new mathematical methods be developed. At present, efforts of communication engineers are oriented toward increasing the efficiency of transmission systems by statistical multiplexing of calls, bursts and packets. Statistical multiplexing is crucial for design of future integrated networks and remains in the center of interest of many researchers. In this paper we are concentrating on presentation of the work relevant to that problem. Discrete and continuous time descriptions of the multiplexing process are presented. It is indicated that performance indices of practical systems cannot be obtained in closed form. Numerical computations are needed. Computational problems are discussed in detail.

Multiplexing process is described by the following variables:

$U_3(t)$ = number of calls in progress at time t,
$U_2(t)$ = number of bursts in progress at time t,
$U_1(t)$ = number of packet arrivals during time interval $(t - \Delta t, t)$,
$X(t)$ = number of packets in the system at time t.

A number of packets arriving to the buffer is determined by events in the burst and call layers. When the number of calls or bursts in progress increases an intensity of packet arrivals also increases. In other words, the state of the process $U_1(t)$ is conditioned on the state of the process $U_2(t)$, and possibly, on the state of the process $U_3(t)$.

Usually, an attempt is made to disregard the complex process structure. This is done by constructing a Markov renewal process in the packet layer. In that way the upper layers of the arrival process can be neglected and the one dimensional analysis performed.

Such approach is followed by Sriram and Whitt (1986) Heffes and Lucantoni (1986), and more recently, by Ide (1988). Heffes and Lucantoni model packet arrivals as a Markov modulated Poisson process. Sriram and Whitt investigate the dependence among successive interarrival times in the aggregated packet arrival process and use the Queueing Network Analyser to investigate performance of packet multiplexers for voice and data.

Different models of packet arrival process are considered by Daigle and Langford (1986). They indicate that the type of model applied depends on the parameter range investigated. A survey of methods which can be used to examine individual traffic streams in superpositions was given by Holtzman (1987). The switched Poisson process, the special case of Markov

---

J. Filipiak is on leave from Cracow University, Poland.

modulated Poisson process, was applied to performance evaluation of data transmission in multiple-channel systems by Zukerman and Rubin (1987) and Rossiter (1987). Models constructed in the packet layer are often solved using matrix-geometric approach, Heffes and Lucantoni, San-qi Li (1987), San-qi Li (1988), or alternatively, they specify a solution in transform domain, Ide (1988). From the analysis of system properties it follows that the methods, which are presented in those papers, can be treated as approximations only. The systematic approach requires that all processes introduced above be described by the model. An insight into validity of renewal process approximation can be obtained from the analysis of the paper by Ramaswami and Latouche (1988).

There are several problems involved in modelling of statistical multiplexing by means of the renewal process. First of all, such approach treats the considered system as a black box, which means that the renewal process parameters do not reflect the physical system parameters. Consequently, the model accuracy depends on the parameter estimation technique and cannot be assessed a priori. Moreover, the techniques proposed by Sriram and Whitt, Heffes and Lucantoni, and Ide are applied to determine the mean delay in systems with infinite capacity, whereas from the practical point of view the blocking probability and delay in systems with finite buffers are of importance.

The analysis of traffic flow in the burst layer is based on the assumption that the number of bursts in progress determines the state of the packet arrival process. The system behavior is modelled by means of the two-dimensional process (U_2, X). The number of packet arrivals is determined knowing U_2 and the size of physical or logical frame. Continuous time version of such description was proposed by Anick, Mitra, and Sondhi (1982). Recently, their model was used to evaluate performance of packet video traffic, Maglaris et.al. (1988), and packet loss in voice multiplexers. Its extension to data/voice systems was described by Mitra (1988). The Anick, Mitra, and Sondhi model was generalize by Kosten (1984) and Stern (1989) so as to describe the multi-service traffic. A discrete time model of the type considered was presented by Descloux (1986).

The analysis of system behaviour in the burst layer is similar, to some extend, to the analysis of Time Assignment Speech Interpolation (TASI) multiplexers, burst switching systems and systems operated according to the movable boundary scheme. TASI performance was investigated by Weinstein (1978). Quality of voice and data transmission in burst switching was considered by Descloux (1986) and by O'Reilly (1986) and O'Reilly and Ghani (1987). O'Reilly uses fluid approximation introduced by Gaver and Lechoczky (1982). Performance of voice transmission in burst and packet switching was compared by San-qi Li (1987). In that last paper, as well as in some previous ones, a tendency is observed to apply Markov models.

DISCRETE TIME MODEL

Consider a two dimensional Markov chain (TDMC) (U, X). In the remainder of the paper we concentrate on system modelling in the burst layer. Therefore, U denotes the number of bursts in progress. TDMC transitions occur at epochs $\kappa \Delta t$. The number of active channels U changes at random independently of the packet queue length. Instead, the number of packets arriving to the buffer in time period Δt is determined by the number of active channels and equals αU.

Denote by L the number of packets which can be transmitted during the time interval Δt: $L = \Delta t / \Delta t_0$, where Δt_0 is the packet transmission time. L may be treated as a size of a virtual frame.

If the number αU of packet arrivals in a given frame is bigger than L, then the $\alpha U - L$ packets must wait in a queue. The queue is unloaded if in a given time interval αU is smaller than L, in which case the queue is decreased by $L - \alpha U$.

Assume that the time scales are chosen so that the observation interval Δt is equal to the packetization delay. The number of packets arriving to the system in one frame L is then equal to U, $\alpha = 1$. In what follows we consider that situation to avoid notational complexity involved in the analysis of the general case.

Denote by $\pi_\kappa(i, x)$ the probability that the TDMC constructed in the burst layer is in the state (i, x), where $i = u$ is a number of bursts in progress and x is a number of packets in a buffer, and let r_{ij} denote a probability that the number of active channels undergoes a transition from i to j during the time Δt. Disregarding the boundary conditions we have:

$$\pi_{\kappa+1}(j, x) = \sum_{i=0}^{K} r_{ij} \pi_\kappa(i, x + L - i) \tag{1}$$

where K is a number of channels active in the call layer. Introduce the matrix notation. Then, for limiting probabilities

$$\pi = \pi \mathbf{R}$$

where π is a row vector of state probabilities and $\mathbf{R}$ is a stochastic matrix of transitions probabilities with ordering corresponding to that of π. Transition probabilities r_{ij} can be obtained by summing probabilities of all possible transitions from i to j that occur in one epoch as a result of different combinations of numbers of arrivals and departures. Descriptions of that type were considered by Viterbi (1986) and Descloux (1988).

Viterbi was able to derive a closed form expression on the mean delay in systems with $L = 1$. That result can be applied to channels in which packets in a burst are not interspersed with silence intervals. The case $L > 1$ is more involved, however, due to the boundary conditions, which change the summation limits in (1) and the form of those equations for states close to boundaries.

The expression $x + L - i$ in Eq. (1) cannot be smaller than 0 or greater than N:

$$0 \le x + L - i \le N \tag{2}$$

where N is a buffer size. Consequently, the following must hold

$$i \le i_{max} = \min\{x + L; \ K\} \tag{3}$$

where K is a number of input channels and

$$i \ge i_{min} = \max\{0; \ x + L - N\} \tag{4}$$

Taking that into account in Eq.(1) we get for $1 \le x \le N - 1$, and $0 \le j \le K$.

$$\pi_{\kappa+1}(j, x) = \sum_{i=i_{min}}^{i=i_{max}} r_{ij} \pi_\kappa(i, x + L - i) \tag{5}$$

According to the description of system operation, the considered Markov chain makes transitions from state (i, x) to states $(j, x - L + i)$, $j = 0 \ldots K$, with the exception of some states close to boundaries. Consider first the left boundary $x = 0$. Define the set A of system states:

$$A = \{(i(x), x) : x = 0, \ldots L - 1, \ i(x) = 0, \ldots L - 1 - x\} \tag{6}$$

Note, that for $(i, x) \in A$ the system moves to states $(j, 0)$ rather than $(j, x - L + i)$ because $x - L + i < 0$. Therefore, Eqs.(5) for states $(j, 0)$, $j = 0, \ldots K$, are as follows:

$$\pi_{\kappa+1}(j, 0) = \sum_{i=i_{min}}^{i=i_{max}} r_{ij} \pi_\kappa(i, L - i) + \sum_{(i,x) \in A} r_{ij} \pi_\kappa(i, x) \tag{7}$$

In a similar way, let us define the set B:

$$B = \{(i(x), x) : x = N + L + 1 - K, \ldots N, \ i = L + 1 + N - x, \ldots K\}, \tag{8}$$

which refers to the right boundary $x = N$. From the states in B the system makes transitions to states (j, N), $j = 0, \ldots K$. This is due to the fact that for $(i, x) \in B$ we have $x - L + i > N$. On the right boundary $x = N$ Eqs. (5) have the form

$$\pi_{\kappa+1}(j, N) = \sum_{i=i_{min}}^{i=i_{max}} r_{ij} \pi(i, N + L - i) + \sum_{(i,x) \in B} r_{ij} \pi_\kappa(i, x). \tag{9}$$

We conclude that the considered Markov chain is described by Eqs. (5) for $x \in [1; N - 1]$, Eqs. (7) for $x = 0$, and Eqs. (9) for $x = N$, where in all cases $j = 0, \ldots K$.

Due to the boundary conditions, the matrix $\mathbf{R}$ has a complicated structure and the closed form solution can be obtained only with a difficulty. Therefore, the model must be solved numerically.

We are interested in a steady state solution

$$\pi(i, x) = \lim_{\kappa \to \infty} \pi(i, x)$$

Substituting $\pi(i, x)$ for $\pi_\kappa(i, x)$ and $\pi_{\kappa+1}(i, x)$ in (5), (7), and (9) gives a set of linear equations in π. We can perform such operation because the considered Markov chain is regular and has a unique steady state solution.

The considered problem has been reduced to solution of linear equations. The task is simple except that the problem dimensionality is high. To illustrate that consider the first stage of multiplexing process, where 130 narrowband voice channels are put onto 1.5 Mbit/s transmission channel. To accomodate fluctuations of the packet arrival rate the system is provided with a buffer having up to 1000 places. Thus $K = 130$, $N = 1000$, and we have 130 000 unknowns.

Several methods can be used to solve Eqs. (5), (7), and (9). We obtained the solution by means of subsequent iterations. That is, starting with the admissible initial solution $\pi_0(i, x)$, Eqs. (5), (7), and (9) were used to compute $\pi_{\kappa+1}(i, x)$ from $\pi_\kappa(i, x)$. The number of multiplications and additions needed to obtain $\pi_\kappa(i, x)$ in that case was K. The problem was programmed using few lines of code. This method saves the computer memory and, what is even more important, avoids division. Its only drawback is slow convergence for systems with high channel utilizations. However, it is preferred to other approaches because computer memory seems to be a critical resource at present, especially for design of systems with high capacity channels (50 Mbit/s and more).

Using the TDMC model we can determine time congestion and packet blocking. Denote by N the buffer size. The time congestion s_2 is defined as a proportion of time that the system spends at the right boundary above L:

$$s_2 = \sum_{i=L+1}^{K} \pi(i, N) \tag{10}$$

The packets start to be blocked when the system enters the region $B = \{(i, x) : i = L + 1 \ldots K; \ x = N - i + L + 1 \ldots N\}$ of the state space close to the right boundary. Therefore, the blocking probability b_2 is defined as follows:

$$b_2 = \sum_{i=L+1}^{K} \sum_{x=N-i+L+1}^{N} (i - L - m + j)\pi(i, x) \tag{11}$$

Some numerical results are shown in Table 1. Several minutes of CPU time (PYRAMID) were needed to compute one set of data. To reduce a computation time the state space was truncated.

Denote by $\bar{u}$ the mean number of bursts in progress. It can be shown that u has the binomial distribution $B(K, p, q)$. For large $\bar{u}$ the binomial distribution can be approximated by the normal distribution $G(\bar{u}, \sigma^2)$ with the mean $\bar{u}$ and variance $\sigma^2 = \bar{u}$. The state space was truncated so that only transitions between states $i \in (i_{min}; i_{max})$ were considered, where $i_{min} = max(\bar{u} - 3\sigma; 0)$ and $i_{max} = min(\bar{u} + 3\sigma; K)$. For that purpose the transition probabilities were modified according to the formulae:

$$r_{i_{min}j} = \sum_{k=0}^{i_{min}} r_{kj}, \qquad r_{ji_{min}} = \sum_{k=0}^{i_{min}} r_{jk}$$

$$r_{ji_{max}} = \sum_{k=i_{max}}^{K} r_{jk}, \qquad r_{i_{max}j} = \sum_{k=i_{max}}^{K} r_{kj}$$

for $j = i_{min}, \ldots i_{max}$.

During computations we did not encountered underflow problems, which could be expected because the total number of states was often greater than one hundred thousands. In such cases some state and transition probabilities were very small. Truncation of the state space was helpful in removing from computations corresponding parameters. At the end of computations the sum of state probabilities was usually equal to 1 with accuracy around 10^{-10}.

TABLE 1. Delay and blocking in statistical multiplexing (K=125, N=500)

p_2	q_2	L	Delay	Blocking
.36	.0167	6	5.4652	.8E-10
.1818	.0175	12	9.2856	.4E-5
.1212	.0184	18	11.4112	.72E-4
.0909	.0194	24	12.2354	.27E-3
.0727	.0205	30	11.9912	.49E-3
.0606	.0217	36	11.2399	.64E-3
Voice				
.0454	.0246	48	8.8176	.57E-3
.0303	.0338	72	3.4363	.56E-4
.0227	.0537	96	.4663	.4842E-8
.0182	.1311	120	.0004	.2232E-12

CONTINUOUS TIME MODEL

The considered problem can also be solved in continuous time. The relatively simple analytical solution, requiring minor computations, can be obtained for systems with infinite buffers. Instead, the finite buffer case is more involved numerically.

Infinite Buffer

As in the previous section, denote by $\pi_t(i, x)$ the probability that i bursts are in progress and that the buffer content does not exceed x. In the considered case i is an integer, and t and x are reals: $i \in [0, K]$ and $x \in [0, \infty)$. In this section we consider systems with infinite buffers. The system capacity is C.

The model formulation is based on the fact that the rate of change of the buffer content when there are i bursts in progress equals $i - C$. The process $U(t)$ is modelled as a birth and death process. This means that at most one of the following transitions to the state (i, x) may occur in the small time interval Δt:

- $(i - 1, x - (i - 1 - C)\Delta t) \rightarrow (i, x)$ (burst arrival) with probability $q[K - (i - 1)]\Delta t$, and
- $(i + 1, x - (i + 1 - C)\Delta t) \rightarrow (i, x)$ (burst departure) with probability $p(i + 1)\Delta t$.

where Δt is measured in units of packetization delay. Note that C corresponds to L in the previous section.

The probability of the event that the number of burst in progress i remains unchanged, $(i, x - (i - C)\Delta t) \rightarrow (i, x)$, equals $1 - q(K - i)\Delta t - pi\Delta t$. Other events have probabilities $o(\Delta t^2)$.

Time behavior of the system is described by the following equations:

$$\pi_{t+\Delta t}(i, x) = (K - i + 1)q\Delta t\pi_t(i - 1, x - (i - 1 - C)\Delta t)$$
$$+(i + 1)p\Delta t\pi_t(i + 1, x - (i + 1 - C)\Delta t))$$
$$+\{1 - [(K - i)q + ip]\Delta t\}\pi_t(i, x - (i - C)\Delta t) + o(\Delta t^2)$$

$i = 0, \ldots K$, where $\Delta x = (i - C)\Delta t$. Dividing both sides of each equation by Δt and taking the limit as $\Delta t \to 0$ yields

$$\frac{\delta \pi_t(i, x)}{\delta t} + (i - C)\frac{\delta \pi_t(i, x)}{\delta x}$$
$$= (K - i + 1)q\pi_t(i - 1, x) - \{(K - 1)q + ip\}\pi_t(i, x) + (i + 1)p\pi_t(i + 1, x)$$

We are interested in the steady-state solution $\delta \pi_t(i, x)/\delta t = 0$ that in stable systems is reached as $t \to \infty$. Denote

$$\pi_\infty(i, x) = P_i(x) \tag{12}$$

We have

$$(i - C)\frac{dP_i(x)}{dx} = (K - i + 1)qP_{i-1}(x) - \{(K - i)q + ip\}P_i(x) + (i + 1)pP_{i+1}(x) \tag{13}$$

for $i = 0 \ldots K$. The boundary conditions for Eqs.(13) are $x \geq 0$ and $P_i(x) = 0$ for $i \notin [0, K]$. In the matrix notation

$$\mathbf{A}\frac{d}{dx}\mathbf{P}(x) = \mathbf{B}\mathbf{P}(x), \tag{14}$$

where $\mathbf{P}(x)$ is a K-dimensional column vector. $\mathbf{A}$ is the diagonal matrix with entries $a_{ii} = i - C$. The matrix $\mathbf{B}$ has the three-diagonal structure,

$$\begin{pmatrix} -Kq & p & & & & \\ Kq & -\{(K-1)q + p\} & 2p & & & \\ & (K-1)q & -\{(K-2)q + 2p\} & 3p & & \\ & & \ddots & & & \\ & & 2q & -\{q + (K-1)p\} & Kp \\ & & & q & -Kp \end{pmatrix}$$

If C is an integer, then the left hand side of $(C + 1)$st equation (13) is equal to zero and the variable $P_C(x)$ can be expressed as a linear combination of $P_{C-1}(x)$ and $P_{C+1}(x)$. This reduces by one the problem dimension.

Denote by ξ the eigenvalue of (14) and by $\mathbf{y}$ the corresponding right eigenvector:

$$\mathbf{A}^{-1}\mathbf{B}\mathbf{y} = \xi\mathbf{y} \tag{15}$$

Solution to Eq.(14) can be written in the form:

$$\mathbf{P}(x) = \sum_{i=0}^{K} d_i \mathbf{y}_i e^{\xi_i x} \tag{16}$$

where coefficients d_i depend on eigenvectors and boundary conditions.

Define the generating function of $\mathbf{y}$:

$$Y(v) = \sum_{i=0}^{K} y_i v^i$$

Multiplying Eqs.(1) by v^i and summing them over i gives

$$\frac{Y'}{Y} = \frac{Kqv - Kq + \xi C}{qv^2 + (p + \xi - q)v - p} \tag{17}$$

Anick, Mitra, and Sondhi (1982) obtained eigenvalues by solving Eq. (17). Moreover, they gave the explicit formulas, which allow one to determine eigenvectors and coefficients d_i.

The considered system has positive and negative eigenvalues. Therefore, having the eigenvalues in explicit form is important because the presence of positive eigenvalues would cause instabilities during numerical computations. The method described by Anick, Mitra, and Sondhi gives the solution which contains negative eigenvalues only:

$$\mathbf{P}(x) = \mathbf{P}(\infty) + \sum_{i=0}^{K - \lfloor C \rfloor - 1} d_i \mathbf{y}_i e^{\xi_i x} \tag{18}$$

Finite Buffer

An approach to the analysis of finite buffer systems in continuous time was described by Tucker (1988). In that paper, the probability G_i of the queue being held at its upper limit $x = N$ is defined as follows:

$$G_i = P_i(\infty) - P_i(N-) \tag{19}$$

where $P_i(N-) = \lim_{x \to N} P_i(x)$. Consider the general solution of the form:

$$\mathbf{P}(x) = \sum_{i=0}^{K} d_i \mathbf{y}_i e^{\xi_i x} \tag{20}$$

Because only the finite values of x are considered there is no need to get rid of positive eigenvalues. Positive and negative eigenvalues of the matrix $\mathbf{A}^{-1}\mathbf{B}$ can be found numerically. The same refers to eigenvectors.

To find d_i we use the boundary conditions. Note that if $i > L$ then the queue is always increasing so its length cannot be zero: $P_i(0) = 0$, $i = \lfloor L \rfloor + 1, \ldots K$. Instead, for $i < L$ the queue is always decreasing which yields $P_i(N-) = 0$, $i = 0, \ldots \lfloor L \rfloor - 1$. Using these boundary conditions in (20) gives the following sets of equations:

$$\sum_{k=0}^{K} d_k \{\mathbf{y}_k\}_i = 0 \qquad L < i \leq K \tag{21}$$

and

$$\sum_{k=0}^{K} \exp(\xi_k N) d_k \{\mathbf{y}_k\}_i = 0 \qquad 0 \le i < L \tag{22}$$

where $\{\mathbf{y}_k\}_i$ denotes the $i-th$ component of $k-th$ eigenvector. Equations (21-22) are solved numerically.

Tucker notes that for large N the exponential function in (22) overflows for positive eigenvalues. One way around this problem is to multiply Eqs.(21-22) by $\exp(\xi_k N)$ converting the potential overflow into a potential underflow. It is advised that for large values of N the states i of probability less than 10^{-5} be excluded from the model.

MIXED TRAFFIC TYPES - DISCRETE TIME ANALYSIS

Assume now that we have J traffic classes with K_j identical sources in each class. The number of packets arriving to the multiplexer from the j-class source in the time interval Δt is α_j.

As previously, denote by L the frame size: $L = \Delta t / \Delta t_0$. Observe that during the time interval Δt the buffer content is increased by

$$\sum_{j=1}^{J} \alpha_j v_j - L,$$

where v_j is a number of bursts in progress in class j. The probability of transition from state $(v_1, \ldots v_J, x - \sum_{j=1}^{J} \alpha_j v_j + L)$ to state $(u_1, \ldots u_J, x)$ is $\prod_{j=1}^{J} r_{v_j u_j}$. We assume that arrivals from different sources are independent. Under that assumption the transition probabilities $r_{v_j u_j}$ can be calculated separately for each traffic class.

The state equations for the considered $J + 1$ dimensional Markov chain are as follows:

$$\pi_{\kappa+1}(u_1, \ldots u_J, x) = \sum_{v_1=0}^{K_1} \cdots \sum_{v_J=0}^{K_J} \prod_{j=1}^{J} r_{v_j u_j} \pi_\kappa(v_1, \ldots v_J, x - \sum_{j=1}^{J} \alpha_j v_j + L) \tag{23}$$

These equations are modified for states close to boundaries $x = 0$ and $x = N$ in a similar way as Eqs. (1). To obtain boundary conditions observe that the term in brackets in the right hand side of Eq.(23) defines the buffer content x' at the time instant κ. We have $0 \le x' < N$ or

$$0 \le x - \sum_{j=1}^{J} \alpha_j v_j + L \tag{24}$$

and

$$x - \sum_{j=1}^{J} \alpha_j v_j + L \le N \tag{25}$$

New summation limits at the left boundary are obtained by noting that the condition (23) is fulfilled when $v_j \le \hat{v}_j$ where

$$\hat{v}_j(x, v_1, \ldots v_{j-1}) = \min\left(\frac{L + x - \sum_{n=1}^{j-1} \alpha_n v_n}{\alpha_j}, \quad K_j\right) \tag{26}$$

Eq. (25) is fulfilled when $v_j > \check{v}_j$ where

$$\check{v}_j(x, v_1 \ldots v_{j-1}) = \max\left(0, \quad \frac{L + x - N - \sum_{m=1}^{j-1} \alpha_m v_m - \sum_{n=j+1}^{J} \alpha_n K_n}{\alpha_j}\right) \tag{27}$$

Using (26) and (27) as summation limits in (23) gives the following state transition equations for states $(v_1, \ldots v_J, x)$, $x = 1, \ldots N - 1$:

$$\pi_{\kappa+1}(u_1, \ldots u_J, x) = \sum_{v_1 = \check{v}_1}^{\hat{v}_J} \cdots \sum_{v_J = \check{v}_J}^{\hat{v}_J} \prod_{j=1}^{J} r_{v_j u_j} \pi_\kappa\left(v_1, \ldots v_J, x - \sum_{j=1}^{J} \alpha_j v_j + L\right) \tag{28}$$

Consider next the set A of states $(v_1, \ldots v_J, x)$ such that

$$x' = x + \sum_{j=1}^{J} \alpha_j v_j - L < 0 \tag{29}$$

We have

$$A = \{(v_1, \ldots v_J, x): \quad x = 0, \ldots L - 1; \quad v_j = 0, \ldots v_j^A, \ j = 1, \ldots J\} \tag{30}$$

where

$$v_j^A(x, v_1, \ldots v_{j-1}) = \frac{L - x - 1 - \sum_{n=1}^{j-1} \alpha_n v_n}{a_j} \tag{31}$$

Assume that the system enters the state $(v_1, \ldots v_J, x) \in A$. Then, it goes to state $(u_1, \ldots u_J, 0)$ rather than $(u_1, \ldots u_J, x')$, where x' is the state defined by Eq. (29). Therefore the equations for states $(u_1, \ldots u_J, 0)$ have the form

$$\pi_{\kappa+1}(\mathbf{u}, 0) = \sum_{v_1 = \check{v}_1}^{\hat{v}_j} \cdots \sum_{v_J = \check{v}_J}^{\hat{v}_J} \prod_{j=1}^{J} r_{v_j u_j} \pi_\kappa\left(\mathbf{v}, -\sum_{j=1}^{J} \alpha_j v_j + L\right) + \sum_{(\mathbf{v}, x) \in A} \prod_{j=1}^{J} r_{v_j u_j} \pi_\kappa(\mathbf{v}, x) \tag{32}$$

where $\mathbf{v} = (v_1, \ldots v_J)$ and $\mathbf{u} = (u_1, \ldots u_J)$.

For states close to the right boundary $x = N$ we define the set B such that for $(\mathbf{v}, x) \in B$:

$$x' = x + \sum_{j=1}^{J} \alpha_j v_j - L > N \tag{33}$$

We have

$$B = \{(\mathbf{v}, x): \ x = N + 1 + L - \sum_{j=1}^{J} \alpha_j K_j, \dots N; \ \ v_j = v_j^B, \dots K_j, \ j = 1, \dots J\} \quad (34)$$

where

$$v_j^B(x, v_1, \dots v_{j-1}) = \max\left(0; \ \frac{N + 1 + L - x - \sum_{m=1}^{j-1} \alpha_m v_m - \sum_{n=j+1}^{J} \alpha_n K_n}{\alpha_j}\right) \quad (35)$$

After the system enters the state $(\mathbf{v}, x) \in B$ it goes to state $(\mathbf{v}, N)$ rather than $(\mathbf{v}, x')$, where x' is defined by Eq.(32). It requires that Eq. (28) for states $(\mathbf{v}, N)$ be modified as follows:

$$\pi_{\kappa+1}(\mathbf{u}, N) = \sum_{v_1=\hat{v}_1}^{\hat{v}_j} \dots \sum_{v_J=\hat{v}_J}^{\hat{v}_J} \prod_{j=1}^{J} r_{v_j u_j} \pi_\kappa(\mathbf{v}, N - \sum_{j=1}^{J} \alpha_j v_j + L) + \sum_{(\mathbf{v},x)\in B} \prod_{j=1}^{J} r_{v_j u_j} \pi_\kappa(\mathbf{v}, x) \quad (36)$$

We conclude that the transitions of the considered multidimensional Markov chain are described by Eqs. (28), (32), and (36).

Again, the model is in the form suitable for writing the computer program. In order to reduce the memory requirements, the state space can be reduced in the similar way as it has been done in Section 2.

MIXED TRAFFIC TYPES - CONTINUOUS TIME MODEL

Statistical multiplexing of mixed traffic types was considered by Kosten (1984). Let us briefly present his approach.

The jth traffic class, $j = 1, \dots J$, is characterized by the following parameters:

- q_j = burst arrival rate
- p_j = burst departure rate
- γ_j = information generation rate

As previously, u_j denotes the number of j sources on. The channel capacity will be denoted by C.

Assuming that there are J traffic classes and that in small time interval only one transition is permitted in each traffic class (from u_j to $u_j + 1$ or $u_j - 1$) Eqs. (13) become:

$$(\sum_{j=1}^{J} \gamma_j u_j - L)\frac{dP_\mathbf{u}(x)}{dx} = \sum_{j=1}^{J} \{q_j(K_j - u_j + 1)P_{u_1 \dots u_j - 1 \dots u_J}(x)$$
$$- [q_j(K_j - u_j) + p_j u_j]P_\mathbf{u}(x) + p_j(u_j + 1)P_{u_1 \dots u_j + 1 \dots u_J}(x)\}$$

for $\mathbf{u} \in \mathbf{U}$. Note that we have $M = (K_1 + 1) \times \dots \times (K_J + 1)$ variables $P_{u_1 \dots u_J}(x)$ and the corresponding number of differential equations. Assume the lexicographical order in the set $\mathbf{U}$. Using the matrix notation gives

$$\mathbf{A}\frac{d\mathbf{P}(x)}{dx} = \mathbf{B}\mathbf{P}(x) \quad (37)$$

Again, finding eigenvalues is crucial for construction of the solution. Denote by ξ the eigenvalue of the matrix $\mathbf{A}^{-1}\mathbf{B}$ and by $\mathbf{y}$ the corresponding right eigenvector. Elements of $\mathbf{y}$ will be denoted $y_{u_1 \ldots u_J}$. Using the multidimensional generating function $Y(v_1, \ldots v_J)$ of $\mathbf{y}$ Eqs. (37) transform to

$$\sum_{j=1}^{J}\{[\xi\gamma_j v_j - (1-v_j)(q_j v_j + p_j)]\frac{\partial Y}{\partial v_j} + (1-v_j)q_j K_j Y\} = C\xi Y \tag{38}$$

Assume that the generating function $Y(v_1, \ldots v_J)$ is separable:

$$Y(v_1, \ldots v_J) = \prod_{j=1}^{J} Y_j(v_j) \tag{39}$$

Using that in Eq.(38) and dividing by Y gives

$$\sum_{j=1}^{J}\{[\xi\gamma_j v_j - (1-v_j)(q_j v_j + p_j)]\frac{1}{Y_j}\frac{dY_j}{dv_j} + (1-v_j)q_j K_j\} = C\xi \tag{40}$$

The decomposition of the problem is based on the observation that Eq.(40) is fulfilled if

$$[\xi\gamma_j v_j - (1-v_j)(q_j v_j + p_j)]\frac{1}{Y_j}\frac{dY_j}{dv_j} + (1-v_j)q_j K_j = \xi c_j \tag{41}$$

for $j = 1, \ldots J$ and

$$\sum_{j=1}^{J} c_j = C \tag{42}$$

Introducing notation

$$\xi_j = \xi\gamma_j \qquad C_j = \frac{c_j}{\gamma_j} \tag{43}$$

equations (41-42) convert to

$$\frac{Y'}{Y} = \frac{K_j q_j v_j - K_j q_j + \xi_j C_j}{q_j v_j^2 + (p_j + \xi_j - q_j)v_j - p_j}, \qquad j = 1, \ldots J \tag{44}$$

$$\sum_{j=1}^{J} \gamma_j C_j = C \tag{45}$$

Note, that Eqs. (44) have exactly the same form as Eqs. (17).

Using the approach developed for one traffic type the capacity C of the system with one traffic type can be expressed as a function of ξ and the system parameters: $C = \Gamma(\xi, K, q, p, k)$. The corresponding equations for the multimode system are

$$C_j = \Gamma(\xi_j, K_j, q_j, p_j, k_j), \quad j = 1, \ldots J$$

or, using (43)

$$C_j = \Gamma(\gamma_j \xi, K_j, q_j, p_j, k_j), \quad j = 1, \ldots J \tag{46}$$

Substituting (46) in (45) gives:

$$C = \sum_{j=1}^{J} \Gamma(\gamma_j \xi, K_j, q_j, p_j, k_j) \tag{47}$$

Knowing capacity C the eigenvalue ξ ccorresponding to the index $k_1 \ldots k_J$ can be found by a numerical inversion of Eq. (47).

Kosten shown that the function Γ has the asymptote at $C = \sum_{j=1}^{J} \gamma_j (K_j - k_j)$. If $\sum_{j=1}^{J} \gamma_j (K_j - k_j)$ is bigger than C then the corresponding eigenvalue $\xi_{k_1 \ldots k_J}$ is negative. It is to be noted that quantities C_i do not have direct physical interpretation. They may take negative values.

Knowing the eigenvalue $\xi_{k_1 \ldots k_J}$ the corresponding right eigenvector $y_{k_1 \ldots k_J}$ can be constructed using Eq.(39) and the procedure developed by Anick, Mitra, and Sondhi for the monomode system.

The general solution to the considered problem can be written as

$$\mathbf{P}(x) = \mathbf{Y} e^{\xi x} \mathbf{a} \tag{48}$$

The matrix $\mathbf{Y} = (y_1, \ldots y_M)$ is composed of the right eigenvectors y_i, $M = \prod_{j=1}^{J}(K_j + 1)$, and $e^{\xi x}$ is the diagonal matrix with elements $e^{\xi_i x}$. $\mathbf{a}$ is the vertical vector of coefficients.

Let us introduce the following order in the set of eigenvalues:

$$\xi_{m+1} \leq \ldots \leq \xi_M < 0 = \xi_1 < \xi_2 \leq \ldots \leq \xi_m \tag{49}$$

That is, there are $M - m$ negative (stable) eigenvalues, one eigenvalue at zero, and $m - 1$ positive (unstable) eigenvalues. We partition $\mathbf{Y}$ and $\mathbf{a}$ as follows

$$\mathbf{Y} = (\mathbf{Y}_u, \mathbf{Y}_s)$$

and

$$\mathbf{a} = \begin{pmatrix} \mathbf{a}_u \\ \mathbf{a}_s \end{pmatrix}$$

where T denotes the vector transposition. The matrix $\mathbf{Y}_u$ and vector $\mathbf{a}_u$ correspond to eigenvalues with indices $i = 1, \ldots m$. $\mathbf{Y}_u$ has the dimension $M \times m$.

We require that the solution $\mathbf{P}$ be bounded. Therefore, the coefficients at the unstable modes $e^{\xi_i x}$ must vanish. Consequently,

$$\mathbf{a}_u = (1, 0, \ldots 0)^T \tag{50}$$

where $dim \ \mathbf{a}_u = m$ and T denotes the vector transposition. The first element corresponds to $\xi_1 = 0$.

For $x = 0$ Eq. (48) gives

$$P(0) = \mathbf{p} = \mathbf{Ya} \tag{51}$$

The vector $\mathbf{p}$, which defines probabilities of the empty buffer is partitioned as follows

$$\mathbf{p} = \begin{pmatrix} \mathbf{p}_o \\ \mathbf{p}_u \end{pmatrix}$$

where $\mathbf{p}_o$ corresponds to overload states such that

$$\sum_{j=1}^{J} \gamma_j u_j > C \tag{52}$$

An important result stated by Kosten (1984) and Stern (1989) is that the number of positive eigenvalues is equal to the number of underload states. That is, $dim \ \mathbf{p}_u = m$ and $dim \ \mathbf{p}_o = M - m$. When the system is in the overload state defined by Eq. (52) the buffer is not empty. Thus

$$\mathbf{p}_o = \mathbf{0} \tag{53}$$

Let us write Eq. (51) as follows

$$\begin{pmatrix} \mathbf{p}_o \\ \mathbf{p}_u \end{pmatrix} = \begin{pmatrix} \mathbf{Y}_u & \mathbf{Y}_s \end{pmatrix} \begin{pmatrix} \mathbf{a}_u \\ \mathbf{a}_s \end{pmatrix} \tag{54}$$

or

$$\begin{pmatrix} 0 \\ 0 \\ \vdots \\ 0 \\ \mathbf{p}_u \end{pmatrix} = \begin{pmatrix} \mathbf{Y}_u & \mathbf{Y}_s \end{pmatrix} \begin{pmatrix} 1 \\ 0 \\ \vdots \\ 0 \\ \mathbf{a}_s \end{pmatrix} \tag{55}$$

which gives

$$
\begin{aligned}
0 &= y_{1_1} &+& \quad \mathbf{Y}_{s_1}\mathbf{a}_s \\
0 &= y_{1_2} &+& \quad \mathbf{Y}_{s_2}\mathbf{a}_s \\
&\vdots \\
0 &= y_{1_{M-m}} &+& \quad \mathbf{Y}_{s_{M-m}}\mathbf{a}_s \\
p_{u_1} &= y_{1_{M-m+1}} &+& \quad \mathbf{Y}_{s_{M-m+1}}\mathbf{a}_s \\
&\vdots \\
p_{u_m} &= y_{1_M} &+& \quad \mathbf{Y}_{s_M}\mathbf{a}_s
\end{aligned}
\tag{56}
$$

where y_{1_i}, $i = 1, \ldots M$ are elements of the eigenvector $\mathbf{y}_1$ corresponding to the eigenvalue $\xi_1 = 0$, and $\mathbf{Y}_{s_k}$, $k = 1, \ldots M$ are rows of the matrix $\mathbf{Y}_s$. Note that y_{1_i} and $\mathbf{Y}_{s_i}$ are known. Therefore, the first $M - m$ equations in (56) can be used to determine the vector $\mathbf{a}_s$ of $M - m$ unknown coefficients. Element y_{1_i} defines the probability that i bursts are in progress and the buffer content does not exceed infinity.

The problem has been reduced to the solution of $M - m$ linear equations. As such it requires of the order of $(M - m)^3$ operations. However, there are several pitfalls in obtaining the solution. To write equations (56) the subset of states $(k_1, \ldots K_J)$ corresponding to negative eigenvalues must be found. This is a difficult task taking into account that the total number of states is $M = \prod_{j=1}^{J}(K_j + 1)$, with M usually much larger than 100 000. To find the eigenvalues, M numerical inversions of (46) are needed. This is also difficult, especially for large J. Knowing the eigenvalues the eigenvector need to be determined from (39). Each eigenvector has M elements.

In view of these difficulties, Kosten advises that for large M the time dependent forward equations, similar to those developed in Section 3, be solved by numerical integration. Note that this method corresponds to the procedure given in Section 4, which, additionally, gives the discretization steps for x and t, needed for numerical integration.

REFERENCES

1. Anick, D., Mitra, D., and Sondhi, M.M., Stochastic theory of data handling system with multiple sources, *Bell Syst. Tech. J.*, vol. 61, no. 8, pp. 1871-1894, 1982.

2. Daigle, J., and Langford, J., Models for analysis of packet voice communication systems, *IEEE J. SAC*, vol. 4, no. 6, pp. 847-855, 1986.

3. Descloux, A., Models for switching networks with integrated voice and data traffic, *Proc. ITC'11, Montreal*, 1986.

4. Descloux, A., Contention probabilities in packet switching networks with strung input processes, *Proc. ITC'12, Torino*, 1988.

5. Gaver, D.P., and Lehoczky, J.P., Channels that cooperatively service a data stream and voice messages, *IEEE Trans. Commun.*, vol. 30, no. 5, pp. 1153-1162, 1982.

6. Heffes, H., and Lucantoni, D., A Markov modulated characterization of packetized voice and data traffic and related statistical multiplex performance, *IEEE J. SAC*, vol. 4, no. 6, pp. 856-868, 1986.

7. Holtzman, J.M., Examining individual traffic streams in superposition, *Proc. ITC Specialists Seminar, Lake Como*, 1987.

8. Ide, I., Superposition of interrupted Poisson processes and its application to packetized voice multiplexers, *Proc. ITC'12, Torino*, 1988.

9. Kosten, L., Stochastic theory of data handling systems with groups of multiple sources, *Proc. Performance of Computer-Communication Systems Seminar, Zurich*, 1984.

10. San-qi Li, A new performance measurements for voice transmission in burst and packet switching, *IEEE Trans. Commun.*, vol. 35, no. 10, pp. 1083-1094, 1987.

11. San-qi Li, Overload control in a finite message storage buffer, *Proc. INFOCOM'88, Ottawa*, 1988.

12. Maglaris, B., Anastassiou, D., Sen, P., Karlsson, G., and Robbins, J.D., Performance models of statistical multiplexing in packet video communications, *IEEE Trans. Commun.*, vol. 36, no. 7, pp. 834-843, 1988.

13. Mitra, D., Stochastic theory of a fluid model of producers and consumers coupled by a buffer, *Adv. Appl. Prob.*, vol. 20, pp. 646-676, 1988.

14. O'Reilly, P., Performance analysis of data in burst switching, *IEEE Trans. Commun.*, vol. 34, no. 12, pp.1259-1263, 1986.

15. O'Reilly, P., and Ghani, S., Data performance in voice switching when the voice silence periods have hyperexponential distributions, *IEEE Trans. Commun.*, vol. 35, no. 10, pp. 1109-1112, 1987.

16. Ramaswami, V., and Latouche, G., Modelling packet arrivals from asynchronous input lines, *Proc. ITC'12 Torino*, 1988.

17. Rossiter, M. H., A switched Poisson model for data traffic, *Australian Telecommunications Research Journal*, vol. 21, no. 1, pp. 53-57, 1987.

18. Sriram, K., and Ward, W., Characterizing superposition arrival processes in packet multiplexers for voice and data, *IEEE J. SAC.*, vol. 4, no. 6, pp. 833-846, 1986.

19. Stern, T. E., Analysis of separable Markov-modulated queueing processes, Technical Report, Columbia University, New York 1989.

20. Tucker, R. C. F., Accurate method for analysis of a packet speech multiplexer with limited delay, *IEEE Trans. Commun.*, vol. 36, no. 4, pp. 479-483, 1988.

21. Viterbi, A. M., Approximate analysis of time synchronous packet networks, *IEEE J.SAC.*, vol. 4, no. 6, pp. 879-890, 1986.

22. Weinstein, C. J., Fractional speech loss and talker activity model for TASI and for packet-switched speech, *IEEE Trans. Commun.*, vol. 26, no. 8, pp. 1253-1257, 1978.

23. Zukerman, M., and Rubin, I., Queueing performance of a multi-channel system under bursty traffic conditions and state-dependentservice rates, *Australian Telecommunication Research Journal*, vol. 21, no. 2, pp. 3-16, 1987.

On the Application of Some Computer-Intensive Statistical Methods in the Earth Sciences

N. I. FISHER
CSIRO Division of Mathematics and Statistics
P.O. Box 218
Lindfield NSW 2070, Australia

INTRODUCTION AND SUMMARY

The diverse statistical problems arising from the Earth Sciences offer great opportunities to the increasing band of developers of computer-intensive statistical methods (CISMs) to demonstrate that CISMs can solve the problems more traditional methods can't. Typical difficulties presented to the statistician are

- too little data
- too much data
- high-dimensional data
- temporally- or spatially-correlated data
- data not well-modelled by 'standard' models

... apart, of course, from the usual tiresome problems of sampling difficulties and lots of noise. Many of the CISMs are so free of assumptions and hence so general in their application that they are basically 'data-analytic tools' which are of greatest use in the exploratory phase of analysis; nevertheless, they also have the potential to allow formal inference under various modest assumptions. Typical of these CISMs are those based on Classification Trees and on 'Projection Pursuit'. Other equally important CISMs are rather more demanding in terms of their adherence to model assumptions but, in return, provide valid statistical inference for previously intractable situations. Prominent among these are so-called Bootstrap methods, and other Monte Carlo methods.

This paper is concerned with describing the ideas behind some of the important CISMs, and giving a few applications to problems arising in Mineral Exploration. Details about mathematical aspects can be found by consulting the referenced articles. No applications are given to problems from the Mining area; however a few remarks about these problems and associated CISMs are made in the final section of the paper. More immediately, the next three sections treat Projection Pursuit-based and similar CISMs, Classification Tree-based methods, and some key aspects of Bootstrap methodology. A few supplementary comments are given in the last section.

EXPLORATORY METHODS FOR HIGH-DIMENSIONAL DATA

This catch-all title encompasses a host of activities, from estimating a multivariate probability density function to "bump-hunting" to fitting multivariate splines. Characteristic of the methods is that they are designed to cope with multivariate data in many dimensions, where the 'curse of dimensionality', as it was styled by Bellman (1961), renders traditional methods effectively either useless or meaningless.

Geochemical exploration is a rich source for data sets of this type: a multi-element analysis of a rock specimen can easily yield measurements of 25 - 30 elements (with measurement scales ranging from parts per billion to large percentages). Typical questions of interest relate to the discrimination of samples from different regions and to prediction of the presence of target minerals, on the basis of the geochemical signatures of the samples to hand.

The method of Projection Pursuit (PP) was introduced by Friedman & Tukey (1974) as a way of finding "...low (one-, two-, or three-) dimensional projections that provide the most revealing views of the full-dimensional data. With these views the human gift for pattern recognition can be applied to help discover effects that may not have been anticipated in advance" (Friedman, 1987).

It was intended to mimic the actions of the user of the PRIM-9 graphical display system (Fishkeller, Friedman & Tukey, 1975), in which the user, aided by graphical input devices, could change the displayed projection of a multivariate point cloud in a continuous fashion. The original application was to a data set from Particle Physics, in which points were clustered around the ends of three skew "rods" joined pairwise; PRIM-9 was used to study the projections obtained when one rod was viewed end-on, so the original focus was on views exhibiting local concentration. More generally, however, one can seek views or projections which are "interesting" in any sense one chooses to define: for example, a one-dimensional projection resulting in a single cluster may well be regarded as supremely uninteresting, and the "index" used to select an interesting projection would concentrate on multimodal projections. In the last several years, projection pursuit ideas have been applied in a number of different settings including regression (PPR - see Friedman & Stuetzle, 1981) and density estimation (PPDE - see Friedman, Stuetzle & Schroeder, 1984). We shall use the former of these to explicate the PP paradigm (*cf.* Friedman & Stuetzle, 1982 §5).

Suppose we wish to model a measured response Y in terms of some (large) vector of explanatory variables $\mathbf{X} = (X_1, ..., X_k)'$ as

$$Y = f(X_1, ..., X_k) + \text{error} \equiv f(\mathbf{X}) + \text{error},$$

based on data $\{(y_i, \mathbf{x}_i), i = 1, ..., n\}$. Choose an initial model, such as

$$Y = f_0 + \text{error}, \quad \hat{f}(\mathbf{x}) \equiv f_0.$$

Then iterate the following procedure –

(i) Find a projection that shows a deviation of the data from the model, that is, find a direction α_j for which the current residuals $\{r_i = y_i - \hat{f}(\mathbf{x}_i)\}$ show a dependence on $\alpha_j' \mathbf{x}_i$

(ii) Approximate this dependence by a smooth function $s_j(\mathbf{x})$, and update the model: $\hat{f}(\mathbf{x}) \to \hat{f}(\mathbf{x}) + s_j(\alpha_j' \mathbf{x})$

– until residuals are obtained for which no projection yields "significant" structure. We end up with a model of the form

$$y = f_0 + \sum_{i=1}^{m} s_j(\alpha_j' \mathbf{x})$$

Important components of this algorithm are, of course, the ability to scan projections rapidly, and to calculate rapid "smoothers" $s_j(.)$; the interested reader is referred to Friedman & Stuetzle (1981, 1982). Note that there is no requirement that the directions $\alpha_1, \alpha_2, \ldots$ be orthogonal. It can be shown that any smooth function of n variables can be represented in the above form, for m sufficiently large. PPR has been implemented in a FORTRAN program called SMART, available from Friedman.

The key elements of PP-based methods are thus: choice of initial model, search for projections showing departure from the model, and updating of the model, until it agrees with the data on all projections. The methods work for high-dimensional data because of the skilful use of computationally efficient methods for local smoothing and for scanning directions.

Space does not permit us to give an application here; however, numerous examples can be found in Friedman (1987), Friedman & Stuetzle (1981, 1982), Friedman, Stuetzle & Schroeder (1984), Breiman & Friedman (1985), and Efron (1988).

Another technique for regression modelling of high-dimensional multivariate data, motivated by the ideas in the next section rather than by PP, is that of Multivariate Adaptive Regression Splines (MARS see Friedman, 1988). The recursive partitioning regression model (a companion technique to the classification tree method discussed in §3), fits a model of the form

$$\hat{f}(x_1, \ldots, x_k) = f_j(x_1, \ldots, x_k) \text{ if } (x_1, \ldots, x_k)\epsilon R_j, j = 1, \ldots, m$$

where $R_1, \ldots, R_m$ constitute a partition of the region of interest. In its simplest form, the f_j's are constant. The MARS method generalizes this approach to one in which the model takes the form of an expansion in product spline basis functions, with the number of basis functions, the degree of each product, and the knot locations all being determined automatically by the data. Unlike the recursive partitioning model, it yields a continuous model with continuous derivatives.

[Further reading: Huber, 1985; Jones & Sibson, 1987; Efron, 1988.]

CLASSIFICATION TREE METHODS

A common problem in Geology is that of classifying a rock specimen as one of, say, k categories based on measuring a number of variables for that specimen. The classification is effected by some rule derived from a training set - that is, similar measurements made on other specimens whose true categories are known. There is a large literature, both statistical and otherwise, devoted to ways of devising appropriate rules, as can be seen by perusing a few books on Multivariate Analysis. Some of the methods are heavily dependent upon assumptions (e.g. that the measurement vectors have multivariate normal distributions, possibly with common covariance matrix); others make very few assumptions (e.g. they depend on the data only through the rank-order values across specimens of each variable in the measurement vector). For the most part, the methods make decisions about categories based on linear or more complicated functions of possibly all the variables measured. However, in much geochemical work, basing decisions on rules of this type is manifestly unsatisfactory, for at least two reasons:

(i) the functions may have no clear interpretation

(ii) some of the variables may be categorical (either ordered or unordered) and hence totally unsuited to incorporation into a linear or nonlinear combination of variable values.

A technique which offers a different approach to determining rules for classifying specimens can be based on classification trees, or so-called "recursive partitioning". This area has received relatively little attention in the statistical literature, but is being very actively pursued in other places, particularly in studies on Artificial Intelligence under the guise of Machine Learning or Expert System development. Yet, some of the earliest references to tree-based methods can be found in statistical journals (e.g. Morgan & Sonquist, 1963). So, what is the basis of classification trees that allows them to avoid the difficulties described above?

Suppose that observations are available on n specimens, with the typical observation Y comprising measurements on p linear variables $X_1, \ldots, X_p$ and q categorical variables $C_1, \ldots, C_q$. The classification of Y into one of k mutually distinct categories $A_1, \ldots, A_k$ is assumed known.

Let $x_{11}, \ldots, x_{1n}$ be the set of measurements made on the variable X_1. Order these values to obtain

$$x_{11}^* \leq x_{12}^* \leq \ldots x_{1n}^*$$

say, and find that interval $(x_{1j}^*, x_{1,j+1}^*)$ which splits the observations $Y_1, \ldots, Y_n$ into two groups, or nodes, in such a way that the within-node homogeneity or purity is maximised (in some sense). Repeat this for each of the other linear variables $X_2, \ldots, X_p$. Perform similar searches for the categorical variables which are ordinal; nominal categorical variables can be assigned nodes based on each value of the given variable, or alternatively, to obtain a binary split, partitions of the set of possible values into two subsets can be checked (possibly, a very computationally expensive business). Then choose, from among the total of $p + q$ competing initial splits, the one which maximises the within-node homogeneity.

This process can now be repeated on the new nodes, until we obtain a final set of nodes which are 100% pure (i.e., all observations in that node belong to a single category). We then have a classification tree.

Of course, we have begged many questions in this terse description. What does "within-node homogeneity" mean, and consequently, how is "goodness-of-split" to be measured? What if the resulting tree is huge? How can we prune it back to something reasonable (i.e. a moderate-sized tree whose terminal nodes are still reasonably pure)? And how good is the predictive power of the tree (i.e. when used on a new datum, what will be the probability of correct classification)? It is not our purpose here to pursue these important issues, which are, of course, vital in implementing the general method. The interested reader should consult the book by Breiman et al. (1984), which explores them in detail, and which explains the operation of an associated software package CART; see also Segal (1988). Rather, we make a few observations on why this (computer-intensive) approach is of great potential application to the Earth sciences.

The first point to note is that this approach focuses on the sort of criteria used intuitively by geoscientists. Thus, in the geochemical setting, it is customary to think in terms of high or low levels of particular elements or minerals, or ratios of these. This is precisely the type of criterion used here (with ratios accommodated by the simple expedient of supplementing the set of linear predictor variables with those ratios thought to be of possible use). (Note that the program CART also enables the user to look at splits based on searching through arbitrary linear combinations of the X's, if that is desired). The second point to note is that categorical predictor variables (above/below water table, weathered/unweathered terrain, etc.) can be used on an equal basis. Thirdly, there is no requirement for a major distributional assumption such as multivariate normality, or equal covariance matrices, assumptions which are quite untenable in many applications since the data can manifest very long-tailed

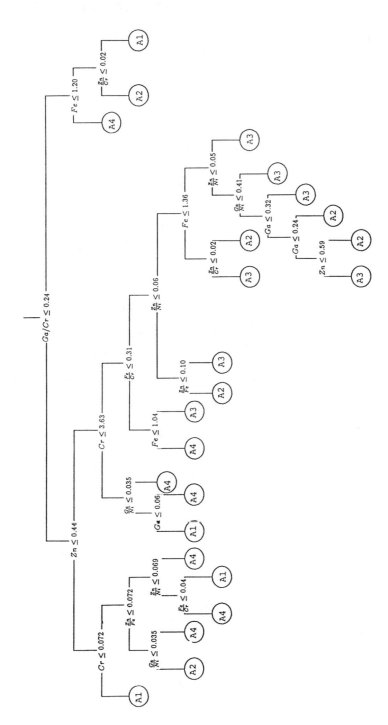

FIGURE 1.   Classification tree for geochemical data.   Cases go left if inequality is true, otherwise right. Ultimate classifications are shown circled.

distributions: note, indeed, that when the method is used in its simplest form, with no ratios or linear combinations included, the resulting tree depends only on the ranks of individual measurements $x_{i1}, \ldots, x_{in}$ amongst themselves, rather than on the measurements x_{ij} themselves. Other points will become evident in the following example, a partial analysis of some geochemical data.

It is not possible, for proprietary reasons, to give a full description of the data set, however the following information should suffice to indicate how the method can work. The training set consists of 4 different groups A_1, A_2, A_3, A_4, with A_1 and A_2 being quite similar, and A_3 and A_4 likewise. Primary interest centres on finding ways of discriminating all four, although it would also be useful to develop a procedure for separating $\{A_1, A_2\}$ from $\{A_3, A_4\}$ since the former corresponds to a non-ore-bearing environment and the latter to an ore-bearing environment. Samples of sizes 87, 133, 177 and 87 are available on A_1, A_2, A_3 and A_4 respectively, with each observation consisting of measurements on the elements Cr, Fe, Ga, Ni and Zn (abstracted from a set of 20); ratios of each element concentration to all the others were also used, giving a total of 15 possible predictor variables.

Using CART, we can obtain the classification tree shown in Figure 1. The way to use the tree to classify a new measurement should be clear from the display. Because the tree has been pruned back from the maximum tree grown, some, if not all, the terminal nodes will not be pure, containing representatives of two or more classes. Such nodes are then classified according to the most populous class present.

The immediate question is: how good is the tree? We might ask initially, how well it classifies its own training set. The proportions of correct and incorrect classifications are shown in Table 1. However, this is a bit like being asked to referee one's own paper (or, in another context, decide one's own salary). A rather better appreciation can be formed by using part of the data to grow the tree, and then estimating misclassification probabilities from the rest of the data. If this is done for 10 random splits of the entire data set into subsets in the ratio 90:10 and then averaging the

TABLE 1. Learning sample classificiation probability matrix for classification tree in Figure 1

TRUE CLASS

		1	2	3	4
Predicted	1	0.87	0.14	0.05	0.03
Class	2	0.09	0.81	0.06	0.02
	3	0.01	0.04	0.83	0.11
	4	0.02	0.02	0.07	0.83

TABLE 2. Cross validation classification probability matrix for classification tree in Figure 1

TRUE CLASS

		1	2	3	4
Predicted	1	0.82	0.13	0.04	0.02
Class	2	0.08	0.64	0.09	0.11
	3	0.00	0.17	0.74	0.10
	4	0.10	0.06	0.13	0.76

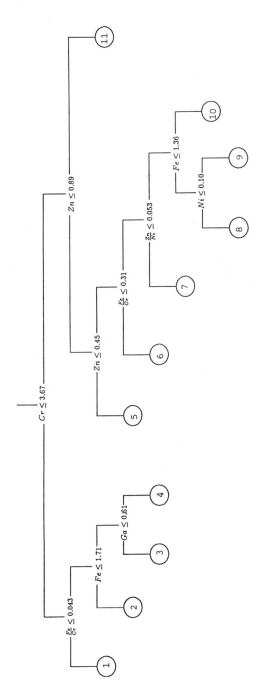

FIGURE 2.   Classification tree for geochemical data, with class probabilities being calculated.   Circles indicate terminal nodes.   (As in Figure 1, cases go left if inequality is satisfied).   Corresponding class probabilities for terminal nodes are given in Table 3.

resulting misclassification matrices, we get the cross-validated misclassification results in Table 2. Note that the quality of our tree, as judged by the cross-validated misclassification matrix, is not as good as the quality suggested by using the results in Table 1, pointing up the overly optimistic picture painted by using the first method.

In the context of these data, we may not be interested so much in correct or incorrect classification as in estimating the probability that a given specimen falls into a particular class. Since our objective is now different, different rules are appropriate for deciding on "goodness-of-split" in the construction of a tree, and a different tree results. One possible tree of this type is shown in Figure 2, with the associated class probabilities given in Table 3.

[Further reading: Hunt, Marin & Stone, 1966; Quinlan 1988.]

BOOTSTRAP METHODS

The Bootstrap was introduced by Efron (1979a) as a means of assessing the variability of estimates in situations where conventional methods are not valid. Much traditional statistical analysis assumes either that a parametric model is appropriate to the data set, or that the sample size is sufficiently large that certain Normal distribution approximations (based on the Central Limit Theorem) are acceptable. However, there are a variety of situations in which neither of these assumptions obtains, and we still wish to draw a valid formal inference from our data. For this purpose, Efron devised the Bootstrap, probably the single most important contribution to Statistics in the last 10 years.

The discussion here is intended solely to explain the basic idea behind the Bootstrap. Although this idea is incredibly simple, the application of the Bootstrap in practical problems involves a number of considerations if it is to be used to best advantage. A few comments will be made about such matters in the sequel.

Consider the following standard problem: given a sample of independent measurements $X_1, \ldots, X_n$ from a population with distribution function $F(x)$, calculate an estimate $\hat{\mu}$ of the population mean $\mu = E(X) = \int x dF(x)$, and an assessment of the variability of $\hat{\mu}$. Asking about the variability of $\hat{\mu}$ is, of course, asking about the sampling distribution of $\hat{\mu}$, i.e., its distribution resulting from infinitely many random samples of size n drawn independently from $F(x)$, and this distribution almost always depends on the form of F itself. If this form is known (up to the value of some parameters), we can usually calculate the distribution $G(\hat{\mu}|F)$ say, of $\hat{\mu}$, from which such things as the variance of $\hat{\mu}$, or a 95% confidence interval for μ, can be computed. Even if F is unknown, the distribution may be well-approximated by the Normal (or Gaussian) distribution if $n > 25$ or 30. However, if n is small and F unknown, what is to be done?

The essence of the Bootstrap approach is to estimate F from the data to hand (e.g. by $\hat{F}(x) = (1/n) \sum I[X_i \leq x]$) where $I(A)$ is the indicator function of the set A), and then to approximate the true sampling distribution $G(\hat{\mu}|F)$ by $\hat{G}(\hat{\mu}) \equiv G(\hat{\mu}|\hat{F})$; the estimated variability of $\hat{\mu}$ is then gauged using $\hat{G}$. We use a simple numerical example to clarify the details.

Stage 0 Suppose we observe a random sample

$$\mathcal{X} = \{0.7 \; 5.5 \; 11.2 \; 11.3 \; 13.9 \; 14.2 \; 14.5 \; 15.2 \; 18.4 \; 19.7\},$$

of size 10, with $\hat{\mu} = \bar{X} \equiv 12.46$ being the mean of all ten values in $\mathcal{X}$. There are four stages to the bootstrap method.

TABLE 3. Class probabilities for terminal nodes for classification tree in Figure 2.
(Small errors in class probabilities due to rounding).

Node	# of Cases	Proportion of Cases	Class	Cases	Class probability
1	46	0.0950	1	3	0.07
			2	3	0.07
			3	7	0.15
			4	33	0.72
2	71	0.1467	1	54	0.76
			2	3	0.04
			3	2	0.03
			4	12	0.17
3	44	0.0909	1	17	0.39
			2	18	0.41
			3	7	0.16
			4	2	0.05
4	16	0.0331	1	2	0.13
			2	0	0.00
			3	2	0.13
			4	12	0.75
5	35	0.0723	1	6	0.17
			2	18	0.51
			3	0	0.00
			4	11	0.31
6	65	0.1343	1	1	0.02
			2	1	0.02
			3	55	0.85
			4	8	0.12
7	30	0.0620	1	0	0.00
			2	19	0.63
			3	10	0.33
			4	1	0.03
8	11	0.0227	1	0	0.00
			2	0	0.00
			3	9	0.82
			4	2	0.18
9	14	0.0289	1	0	0.00
			2	14	1.00
			3	0	0.00
			4	0	0.00
10	104	0.2149	1	0	0.00
			2	18	0.17
			3	82	0.79
			4	4	0.04
11	48	0.0992	1	4	0.08
			2	39	0.81
			3	3	0.06
			4	2	0.04

Stage 1 Resampling

Using a pseudo-random number generator, draw a random sample of 10 values, with replacement from $\mathcal{X}$. Thus, we might obtain the *bootstrap resample*,

$\mathcal{X}^* = \{0.7\ 18.4\ 0.7\ 11.3\ 15.2\ 19.7\ 19.7\ 0.7\ 5.5\ 0.7\}$.

Note that some of the original sample values appear more than once, and others not at all. We should stress that $\mathcal{X}^*$ is of the same size as $\mathcal{X}$.

Stage 2 Calculation of bootstrap estimate

The mean of all ten values in $\mathcal{X}^*$ is $\hat{\mu}_1^* = \bar{X}_1^* = 9.26$.

Stage 3 Repetition

Repeat Stages 1 and 2 a large number of times to obtain a total of B bootstrap estimates $\hat{\mu}_1^*, \ldots, \hat{\mu}_B^*$. (Commonly, for real-valued parameter estimation, the value $B = 200$ is used.)

Stage 4 Confidence interval

Sort the bootstrap estimates into increasing order to obtain

$\hat{\mu}_{(1)}^* \leq \hat{\mu}_{(2)}^* \leq \cdots \leq \hat{\mu}_{(200)}^*$.

For example, we might get

7.12, 7.28, 8.83, 9.90, 9.13, ..., 15.56, 15.61, 15.63, 15.66, 15.87, 16.01.

This set of numbers constitutes the bootstrap distribution $\hat{G}(\hat{\mu})$.

For a $100(1 - \alpha)\%$ confidence interval, let $L = [B\alpha]$, the integer part of $B\alpha$, $U = B - L + 1$; the desired bootstrap confidence interval is $(\hat{\mu}_{(L)}^*, \hat{\mu}_{(U)}^*)$. So, for $\alpha = 0.05$ and $B = 200$, we get $L = 5, U = 196$, and the 95% confidence interval $(9.13, 15.61)$.

[Note that, if we were in the powerful (but unlikely) position of knowing the true underlying distribution, and chose to draw our resamples $\mathcal{X}^*$ from F instead of from the estimate $\hat{F}$ at Stages 2 and 3, we would then be simulating the distribution G exactly. On occasion, we can get close to this form of statistical serendipity – in situations where it is reasonable to make an assumption about the parametric form of F, up to a few unspecified parameters. Then, we estimate these unknown parameters, substitute for them in F to get $\hat{F}_0$ say, and resample from $\hat{F}_0$ at Stages 2 & 3. Readers will immediately recognise a striking resemblance to Monte Carlo estimation: the two methods aren't quite the same when the "fully optioned-up" version of the Bootstrap is being used, although the results are indistinguishable in large samples.]

See Hall & Titterington (1989) and Fisher & Hall (1989a) for further details.

What we have described is the "naive" bootstrap applied to one of the simplest of estimation problems. In fact, a small modification to the procedure gives a greatly improved result in terms of length of confidence interval, and coverage accuracy (how close it is to a $100(1 - \alpha)\%$ interval). The modification requires us to use a "pivotal" (or at least, asymptotically pivotal) quantity when constructing the bootstrap estimates. A pivotal is a quantity whose (asymptotic) distribution is free of unknown parameters. Since we know from the Central Limit Theorem that

$\sqrt{n}(\bar{X} - \mu)/\sigma$ is asympotically Normal $N(0,1)$-distributed (for $\sigma^2 = \int(x - \mu)^2 dF(x)$ finite), $(\bar{X} - \mu)/\sigma$ is a pivotal quantity. The modified bootstrap procedure for calculating a $100(1 - \alpha)\%$ confidence interval for μ, now written out for a general sample, goes as follows:

Stage 0* For the data set $\mathcal{X} = \{X_1, \ldots, X_n\}$ calculate $\bar{X}$ and s^2 (the estimate of σ^2)

Stages 1* and 2* Generate a resample $\mathcal{X}^* = \{X_1^*, \ldots, X_n^*\}$ and calculate the values $\bar{X}^*, s^{*2}$, and then the pivotal or studentized value $\bar{Y}^* = (\bar{X}^* - \bar{X})/s^*$.

Stage 3* Repeat Stage 2* to obtain a total of B values $\bar{Y}_1^*, \ldots, \bar{Y}_B^*$.

Stage 4* Sort the $\bar{Y}^*$-values into increasing order to get $\bar{Y}_{(1)}^* \leq \ldots \leq \bar{Y}_{(B)}^*$ say. With $L = [B\alpha], U = B - L + 1, (\bar{Y}_{(L)}^*, \bar{Y}_{(U)}^*)$ is an interval containing $100(1 - \alpha)\%$ of the values; then

$$(\bar{X} - s\bar{Y}_{(U)}^*, \bar{X} - s\bar{Y}_{(L)}^*) \equiv (\hat{\mu}_{(L)}, \hat{\mu}_{(U)})$$

say is a $100(1 - \alpha)\%$ bootstrap confidence interval (strictly, a percentile-t bootstrap confidence interval) for the parameter μ, with the sort of improvements over the naive bootstrap noted above (see e.g. Hall, 1988).

The significance of this is not so much in its application to the mean μ, but in its application to calculating confidence regions for parameters which are expressible as functions of (possibly multivariate) means. Thus, if the parameter of interest is a function of μ, $\sigma(\mu)$ say, we can get percentile-t bootstrap confidence interval for σ directly as

$$(\theta(\hat{\mu}_{(L)}), \theta(\hat{\mu}_U)).$$

Parameters of this form arise naturally in the Earth Sciences (and elsewhere) with Directional Data, which are data measured as (unit) directions in two or three dimensions, such as directions of palaeocurrent flow, normals to bedding planes, or directions of magnetisation of rock specimens. The mean direction of a random unit vector $\mathbf{X}$ is defined to be

$$E(\mathbf{X})/\|E(\mathbf{X})\|$$

and is estimated from a random sample $\mathbf{X}_1, \ldots, \mathbf{X}_n$ by the direction of the vector resultant, that is, by

$\sum \mathbf{X}_i / \|\sum \mathbf{X}_i\|$ a trivial function of the average $\sum \mathbf{X}_i / n$. The pivoting principle described above can be used (suitably modified for multivariate means) to calculate "efficient" confidence regions for the unknown mean direction (see Fisher & Hall, 1989a).

Another situation in which bootstrap methods seem to offer the only reasonable line of attack on a problem is in the context of the so-called Fold Test in Palaeomagnetism. It is a common occurrence for palaeomagnetic samples to be collected from rocks which have undergone some degree of folding. In the simplest case, one can think of the folded surface as having the two or more fold limbs, as shown in the sketch in Figure 3. If the rocks were magnetised prior to folding, samples from sites on different limbs would tend to have different mean directions (Fig.3a); conversely, relatively comparable mean directions indicate that folding pre-dated magnetisation (Figure 3b). Other knowledge of the Earth's magnetic history then permits

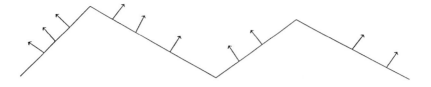

(a)   Magnetisation pre-dated folding

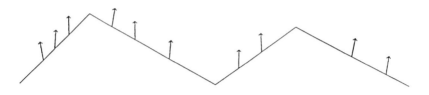

(b)   Magnetisation post-dated folding

FIGURE 3.   Schematic explanation of pre- and post- folding magnetisation of rocks.

assessment of the age of the rock.

There are a number of formulations of this problem, depending on how much is assumed. In its most general setting, one assumes only that the cluster of directions will be more concentrated under one model than under the other (rather than assuming that the correct model will have them aligned more or less exactly, allowing for statistical fluctuations). In this case, a form of bootstrap test seems to be the only thing possible, even with large samples of data available at each site (Fisher & Hall, 1989c). Another possibility is that, say, after correcting for folding, the mean directions are comparable. With "large" amounts of data (i.e. at least 25 - 30 measurements) available at the sampling site on each limb, a fairly straightforward test is possible. Typically, however, only a few measurements are made, and there is the added complication that data from different sites tend to be dispersed differently about their mean directions, which invalidates the usual tests of the hypothesis of equal mean directions. A bootstrap test can be based on the large-sample procedure, and combined with the assumption of a parametric model for each individual sample to give a test which is valid for sample sizes as small as 3 at each site! The reason to be so confident of its validity is that the bootstrap methods can actually be applied to the bootstrap test itself, to check on its actual rejection rate (as against its nominal rate of, say, 5%), were the hypothesis of equal means true; this is known as the *iterated* bootstrap method.

TABLE 4.

Summary statistics for 9 sites from Table 1 of Lackie (1988). The mean directions are in coordinates (Declination, Inclination); the values $\hat{\kappa}_i$ are robust estimates of the concentration parameter for the best-fitting Fisher distribution. (Adapted from Table 1, Fisher and Hall, 1989a).

Site	# of specimens	mean direction (D_i, I_i)	$\hat{\kappa}_i$
1	4	357.1, 77.7	76.6
2	3	106.2, 84.2	54.2
3	3	209.6, 82.2	33.3
4	3	168.8, 75.6	27.2
5	3	249.7, 79.5	1025.8
6	3	304.5, 70.3	12.2
7	3	313.3, 78.1	701.8
8	3	114.0, 61.0	15.2
9	3	354.9, 84.5	32.0

To see this in action, consider the following example, described in more detail in Fisher & Hall (1989b). The data are summarised in Table 4, and plotted in an equal-area projection in Figure 4. The plotted mean directions are based on extremely small samples, and come from distributions with widely differing concentrations. A bootstrap test of the hypothesis of equal mean directions leads to a P-value of 0.002, based on $B = 2000$ bootstrap re-samples. However, an iterated or double bootstrap procedure indicates that this P-value is too low (the test tends to reject the hypothesis too often when it is true), and our P-value should be re-calibrated to 0.006 using the graph in Figure 5.

[Further reading: Efron 1979b, 1981, 1988; Efron & Gong 1982; Efron & Tibshirani 1986.]

CONCLUDING REMARKS

There are a couple of supplementary points worth noting. The first is that, whilst these methods have wide applicability in the Earth Sciences, this applicability is much more potential than realised: the basic principles of good data analysis, in particular careful exploratory data analysis, which are now well-established in other scientific areas are still largely unknown in the geosciences. Most analysis of multivariate data is still done by pouring the data into a big black (or these days, creamed-coloured) box containing a large number of classical multivariate tools (principal components, principal coordinates, linear discriminant analysis, factor analysis,...) and selecting from the pages of beautiful laser-printer output the results (chosen without regard to method) most in accord with preconceptions (i.e., to quote Lewis, 1984, commenting on a paper read to the Royal Statistical Society, "...bugger the data!"). Yet these classical methods are heavily model-dependent – on models which are frequently not justified in the earth sciences – and require moderate delicacy when used in cases where the model may be credible. The computer-intensive methods offer a compelling alternative line of analysis, particularly at the exploratory phase. Secondly, we have made no reference to the use of CISMs in Mining. In fact, Monte Carlo methods have been used for a number of years in problems of ore reserve estimation (so-called Geostatistics), where a technique known as Conditional Simulation is employed to assess quantities like the amount of recoverable reserves resulting from selection of a particular cut-off grade. The

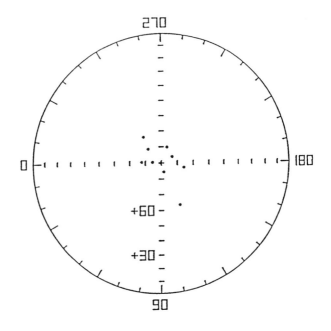

FIGURE 4. Equal-area plot of 9 site mean directions, pre-folding.

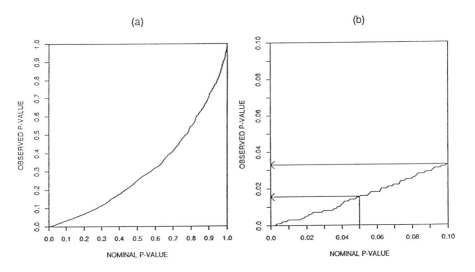

FIGURE 5. Plot of nominal and actual P-values to calibrate bootstrap test of the hypothesis that the mean directions in Figure 4 are drawn from populations with common mean direction. (a) Complete plot. (b) Blow-up of left-hand section of plot, showing which actual significance probabilities correspond to nominal values of 0.05 and 0.10.

essential ingredient of this and many other mining applications is that a (possibly spatial) stochastic process has to be simulated, which leads us rapidly to areas of research barely touched on, to date.

REFERENCES

1. Bellman, R.E., *Adaptive Control Processes*. Princeton, NJ: Princeton University Press, 1961.

2. Breiman, L. and Friedman, J.H., Estimating optimal transformations for multiple regression (with Discussion). *J. Am. Statist. Assoc.* vol 80, pp 580 - 619, 1985.

3. Breiman, L., Friedman, J.H., Olshen, R.A. and Stone, C.J., *Classification and Regression Trees*. California: Wadsworth, 1984.

4. Efron, B., Bootstrap methods: another look at the jackknife, *Ann. Statist.* vol 7, pp 1 - 26, 1979a.

5. Efron, B., Computers and the theory of Statistics: thinking the unthinkable. *SIAM Review* vol 21, pp 460 - 480, 1979b.

6. Efron, B., The Jackknife, the Bootstrap and Other Resampling Plans. *SIAM CBMS-NSF* Monograph #38, 1981.

7. Efron, B., Computer-intensive methods in statistical regression. *SIAM Review* vol 30, pp 421 - 449, 1988.

8. Efron, B., and Gong, G., A leisurely look at the bootstrap, the jackknife and cross-validation. *Am. Statistician* vol 37, pp 36 - 48, 1983.

9. Efron, B., and Tibshirani, R., Bootstrap methods for standard errors, confidence intervals, and other measures of statistical accuracy. *Statistical Science* vol 1, pp 54 - 77, 1986.

10. Fisher, N.I., and Hall, P., Bootstrap confidence regions for directional data. *J. Am. Statist. Assoc.* To appear, 1989a.

11. Fisher, N.I. and Hall, P., New statistical methods for directional data I: Bootstram comparison of mean directions and the Fold test in Palaeomagnetism. *Geophys. J. R. astr. Soc.* To appear, 1989b.

12. Fisher, N.I., and Hall, P., A general statistical test for the effect of folding. Submitted for publication, 1989c.

13. Fishkeller, M.A., Friedman, J.H., and Tukey, J.W., PRIM-9: An Interactive Multidimensional Data Display and Analysis System. Report SLAC-PUB-1408, Stanford, CA: Stanford Linear Accelerator Center, 1975.

14. Friedman, J.H., Exploratory projection pursuit. *J. Am. Statist. Assoc.* vol 82, pp 249 - 266, 1987.

15. Friedman, J.H., Multivariate adaptive regression splines. Technical Report #102, Dept. of Statistics, Stanford University, 1988.

16. Friedman, J.H., and Stuetzle, W., Projection pursuit regression. *J. Am. Statist. Assoc.* vol 76, pp 817 - 823, 1981.

17. Friedman, J.H., and Stuetzle, W., Projection pursuit methods for data analysis. Pp. 123 - 147 in Modern Data Analysis, R.L.Launer and A.F.Siegel, Editors, New York: Academic Press, 1982.

18. Friedman, J.H., Stuetzle, W., and Schroeder, A., Projection pursuit density estimation. *J. Am. Statist. Assoc.* vol 79, pp 599 - 608, 1984.

19. Friedman, J.H., and Tukey, J.W., A projection pursuit algorithm for exploratory data analysis. *IEEE Trans. Computers*, Ser. C, vol 23, pp 881 - 889, 1974.

20. Hall, P., Theoretical comparison of bootstrap confidence intervals (with Discussion). *Ann. Statist.* vol 16, pp 927 - 985, 1988.

21. Hall, P., and Titterington, D.M., The effect of simulation order on level accuracy and power of Monte Carlo tests. *J. R. Statist. Soc.* B 51. To appear, 1989.

22. Huber, P.J., Projection Pursuit (with Discussion). *Ann. Statist.* vol 13, pp 435 - 525, 1985.

23. Hunt, E., Marin, J., and Stone, P., *Experiments in Induction.* New York: Academic Press, 1966.

24. Jones, M.C., and Sibson, R., What is Projection Pursuit? (with Discussion). *J. R. Statist. Soc.* A vol 150, pp 1 - 37, 1987.

25. Lackie, M.A., The palaeomagnetic and magnetic fabric of the Late Permian Dundee Rhyodacite, New England, 157-165, *New England Oregon-Tectonics and Metallogenesis*, UNE Armidale, NSW, 14-18 November 1988, Editor J.D. Kleeman, Department of Geology and Geophysics, UNE, Armidale Australia 2351.

26. Lewis, T., Discussion of P. J. Green "Iteratively reweighted least squares for maximum likelihood estimation, and some robust and resistant alternatives", *J. R. Statist. Soc.* B vol 46, pp 149 - 192, 1984.

27. Morgan, J.N., and Sonquist, J.A., Problems in the analysis of survey data, and a proposal. *J. Am. Statist. Assoc.* vol 58, pp 415 - 434, 1963.

28. Quinlan, J.R., Induction, knowledge and expert systems. Pp 253 - 271 in Artificial Intelligence Developments and Applications, J.S. Gero and R. Stanton, Editors.Amsterdam: North Holland, 1988.

29. Segal, M.R., Regression trees for censored data. *Biometrics* vol 44, pp 35 - 47, 1988.

The Impact of Cray Supercomputers in Science and Engineering Simulations

C. MARINO
Industry Science and Technology Department
Cray Research, Inc.
Minneapolis, Minnesota, USA

INTRODUCTION

Since introducing the first commercially successful vector processor in 1976, Cray Research's mission has been to develop, market and support the most powerful large-scale scientific computer systems. Today these machines are used to simulate diverse physical phenomena in research laboratories and production engineering environments throughout the world. At the end of 1988 over 235 Cray computer systems had been installed at over 165 customer sites. Cray Research will continue to meet its mission and enjoy wide customer acceptance by offering systems which provide one or two orders of magnitude better performance than commonly available computers.

Performance is the most important distinguishing factor between true supercomputers and other large scale systems, but it is often difficult to quantify. In some cases the peak computational rate measured in millions of floating-point operations per second (MFLOPS) is sufficient for scientific problems. However, the peak rate is only achievable under certain conditions which vary among architectures. Several standard benchmark suites (Livermore loops, NASA kernels, LINPACK (Dongarra (1988)) are commonly used to measure Fortran performance, but they do not take into account factors such as memory size or organization which may limit problem size, or I/O bandwidth which may be more important than CPU power for large problems. They do show that true supercomputers provide between 5 to 30 times the speed of other large scale systems, such as minisupercomputers for many scientific and engineering applications.

What performance really means is the quality of the scientific insight gained from running a simulation experiment in the mathematical laboratory. Real world constraints such as deadlines, accuracy requirements, financial constraints and others, dictate what is feasible and what is intractable. Furthermore, application software reliability and the ability of distributing application functionability across hetereogeneous systems are major components of total system performance.

Increasingly, computational scientists and engineers demand not only high performance from their computer systems, but also transparent solution environments. As a result, scientific computing environments are being shaped as much by the availability of versatile and efficient software and networking tools as by the availability of high-performance computers. Network communications must provide high-speed channels that

49

connect mainframes to each other and to high-performance graphics
workstations. Furthermore, visualization and animation will become
more important as more powerful systems generate larger volumes of data
per unit of time.

CRAY PRODUCTS

This paper discusses high-performance computing with examples drawn
from Cray Research's experience and plans. The company's hardware
design efforts proceed down two distinct paths, currently represented
in the CRAY Y-MP and the CRAY-2 computer systems.

Cray currently offers three series of supercomputers. The CRAY-2
series, the CRAY Y-MP series of computer systems and the Cray X-MP
EA/se systems. These series of computer systems are built on the
foundation established by the field-proven CRAY-1 and CRAY X-MP
generations of computer systems. Table 1 lists some characteristics of
the current product line which spans about an order of magnitude in
terms of performance and capacity. Memory size is given in megawords
(1 MW = $2^{20} \approx 10^6$ words). Each word contains 64 bits (8 bytes) of
data and 8 error correction bits. Clock period is given in nanoseconds
(1 ns = 10^{-9} second). Peak speed, for all Cray systems, is based on one
floating-point add and one floating-point multiply result being
delivered by each processor per cloak period. Individual model names
are formed by appending the number of processors and the number of
megawords of memory to the base name. For example, a CRAY Y-MP8/464
system has four processors and 64 MW of memory, and can be upgraded to
eight processors.

CRAY X-MP EA/se

Each CPU in a CRAY X-MP EA/se system can run in either X-mode or Y-mode
under software control. X-mode uses the instruction set of the earlier
CRAY X-MP models which uses 24-bit addressing so X-mode programs are
limited to 16 MW of memory. Y-mode uses the upward-compatible CRAY Y-
MP instruction set which provides access to all of memory by using 32-
bit addressing. Other features include flexible hardware chaining of
vector instructions, four parallel memory ports per processor and
shared registers for efficient multiprocessing.

TABLE 1. Latest Cray Product Lines

Model	Number of CPUs	Memory size (MW)	Clock period (ns)	Peak perform (MFLOPS) 1 CPU System	
CRAY X-MP EA/se	1	4, 8, 16	8.5	235	235
CRAY Y-MP	1, 2, 4, 8	32, 64, 128	6.0	333	2667
CRAY-2	4	256, 512	4.1	488	1951
CRAY-2S	2, 4	64, 128	4.1	488	1951

The CRAY X-MP EA/se models are specially packaged to serve both first-time supercomputer users and dedicated departmental projects with large-scale computational requirements. They require less site preparation and electrical power, but they don't support SSD's. The Cray X-MP EA/se is shown in Figure 1.

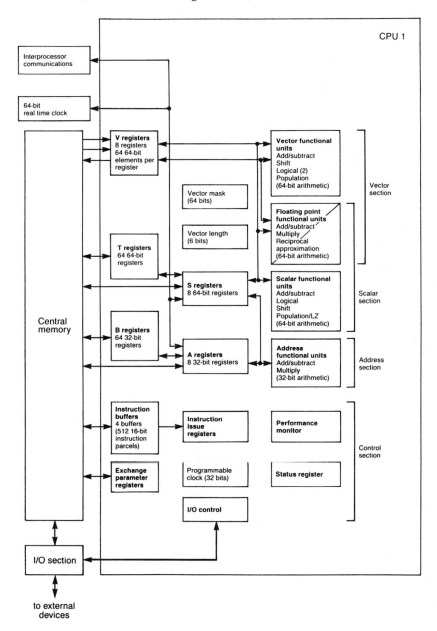

FIGURE 1.   CRAY X-MP EA/se system configuration

51

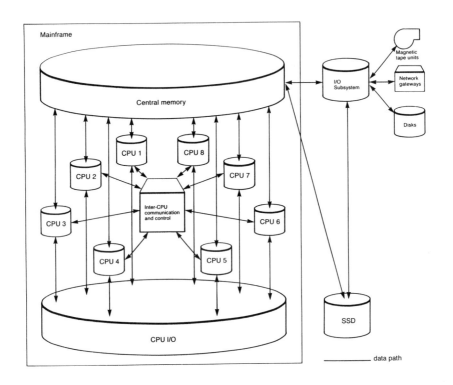

FIGURE 2.   CRAY Y-MP/8 system configuration

CRAY Y-MP SERIES

The top-of the-line CRAY Y-MP system, with eight processors, provides
about 30 times the performance of a CRAY-1 computer. Central memory is
arranged in 256 interleaved banks and has a bank cycle time of 30 ns.
Both its bipolar memory and 2500-gate array logic chips employ ECL
technology. Single-processor models are field upgradable to dual-
processor systems and memory can be field upgraded to the full 128 MW.
A fully configured Cray Y-MP 8 is shown in Figure 2.

The speed advantage of large-scale integrated circuitry is exploited in
the design of the CRAY Y-MP system. The system uses 2500-macrocell
array chips, a leap in integration level of two orders of magnitude
compared to the 16-gate arrays used in its predecessor, the CRAY X-MP
system. The higher level of logic-circuit integration allowed CRAY
Y-MP system designers to package the equivalent processing power of an
entire CRAY 1/S module onto just two chips, and an entire CRAY Y-MP
processor onto a single module. Improvements in silicon bipolar
technology during the past two to three years have resulted in
significantly decreased basic gate delay times, further improving
hardware performance. These developments not only improve performance,
but also improve reliability by reducing the total number of components
needed. Figure 3 and 4 illustrate schematics of a CRAY Y-MP 2 and CRAY
Y-MP 8, respectively.

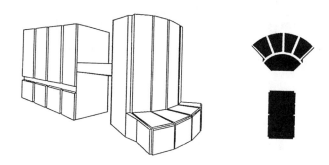

FIGURE 3.   CRAY Y-MP 2

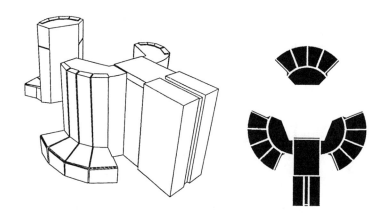

FIGURE 4.   CRAY Y-MP 8

CRAY-2 SERIES

The CRAY-2 system offers the fastest clock speed and largest memory of
any commercially available computer. Its extremely dense modules are
40% integrated circuits by volume and incorporate three-dimensional
interconnects. The computer logic, memory and DC power supplies are
immersed in an inert fluorocarbon liquid inside a 45 inch tall cabinet.
This efficient cooling method reduces and stablizes the operating
temperature of the chips and increases system reliability. The huge
common memory enables users to solve problems that would otherwise be
impractical and enhances interactive response by allowing hundreds of
jobs to reside in memory at once. The CRAY-2S models offer less
capacity, but use faster static MOS memory components. The faster chip
access time and the elimination of the need for refreshing memory
improves per-processor throughout by 10 to 40 percent over CRAY-2
systems which use dynamic memory parts. The Cray 2 is shown in Figure
5.

The CRAY-2 series of computer systems, for example, incorporates novel
packaging strategies at the system level that contribute significantly

to performance. The design of these systems includes circuit interconnections in three dimensions, directly linking neighboring circuit boards within a module. The cooling requirement for these systems is met through the use of liquid immersion technology, in which an inert fluoro-carbon liquid circulates through a system directly contacting the cicuitry.

FUTURE DIRECTIONS

The Cray supercomputers that will be available in the early 1990's will continue the trends of faster clock speed, larger and faster memory and greater parallelism. Silicon circuits will be packed more densely and Gallium Arsenide (G_aA_s) will be used for the first time in a commercial supercomputer. The improved performance of the next generations of Cray computers will allow its users to run more economical simulations or explore even more complex problems while protecting their considerable investment in software.

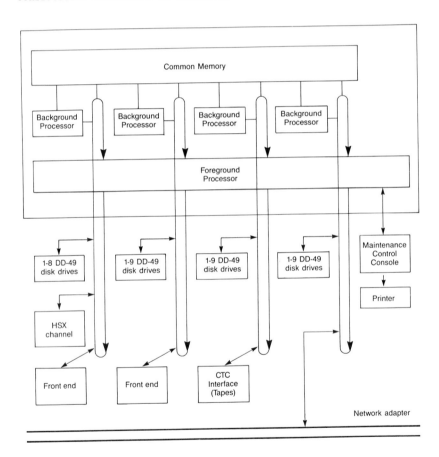

FIGURE 5.   CRAY 2 system configuration

HARDWARE AND SOFTWARE COMPONENTS

The following paragraphs outline the hardware and software components common to the CRAY-2 and CRAY Y-MP series of supercomputers. All models are "well balanced" and "general purpose" which means they have fast, flexible CPUs, large easily accessible memories, sufficient memory and I/O bandwidth, a standard operating system, compilers and libraries that take advantage of the hardware and a wide selection of application software to deliver the power to the user.

CPU

Each Cray CPU contains eight 64 element vector registers, eight 64-bit scalar registers and eight 32-bit address registers. These registers hold operands for the functional units which perform floating-point, integer, logical and other operations. The vector functional units are segmented, so that after a brief startup period, one result is delivered per clock period. Since the functional units are independent, several operations such as a memory reference, floating-point add and multiply and address calculation can execute in parallel. Fast backup registers are provided for holding intermediate results. Floating-point numbers (64-bit quantities) consist of a signed magnitude binary coefficient and biased exponent. Characters are represented internally in ASCII and stored 8 per word. Semaphore registers are provided for inter-processor communications in multiple CPU systems.

MEMORY

The large real memory (up to 4 Gbytes) in a Cray system is accessible by all processors. Since there is no caching or paging, memory latency does not increase with problem size. Vector memory references can proceed contiguously, with an arbitrary constant stride, or randomly using gather/scatter instructions. Common memory is arranged in interleaved banks that can be accessed independently and in parallel during each clock period. All memory references are protected by single-bit error correction, double-bit error detection (SECDED) logic.

I/O AND CONNECTIVITY

Cray systems are equiped with a dedicated I/O processor to handle communication and data handling tasks. High-density disk drives which can sustain transfer rates of 9.6 Mbytes/sec at the user job level can be striped together with resultant I/O performance proportional to the number of drives used. High performance on-line tapes allow large datasets to be handled easily.

Several 6-Mbyte/sec channels are available for connections to other computers or networks up to 1 km away using a fiber optic link. Front end interfaces (FEIs) which compensate for differences in channel widths, word size, logic level and control protocols are available for a wide variety of computers including DEC, IBM and CDC systems. An FEI-3 can connect to a VME device like a SUN, IRIS or Apollo workstation to allow easy access to Ethernet. Cray systems can connect directly to a Network Systems Corporation (NSC) HYPERchannel local area network. The Cray Research HSX-1 high-speed channel provides point-to-point communications (at up to 100 Mbytes/sec) to very fast devices like graphics frame buffers to other Cray systems. An SSD can be connected to a Cray Y-MP series computer with one or two 1000-

Mbyte/sec channels to essentially eliminate most I/O wait time.

OPERATING SYSTEM AND COMPILERS

The UNICOS operating system delivers the full power of the hardware in both interactive and batch environments. UNICOS, which is based primarily on the AT&T UNIX System V operating system and in part on the Fourth Berkeley Software Distribution (BSD), has been substantially enhanced for a supercomputing environment. Besides the features associated with Unix and TCP/IP connectivity like sockets, Network File system (NFS) File System (NFS) and X-Windows, UNICOS adds features like asynchronous I/O, multi-level security and user multitasking. Standard software includes optimizing and vectorizing Fortran 77, C and Pascal compilers, optimized scientific, I/O, multitasking and mathematical libraries, debugging and performance monitoring tools and a Cray assembler (CAL). In addition, Ada, LISP and SIMSCRIPT 11.5 compilers are available to support many existing simulation programs. CFT77 produces highly optimized (scalar and vector) object code from standard Fortran and provides automatic multitasking (Autotasking). The higher levels of parallelism that future systems will offer will make software efficiency, particularly compiler efficiency, a key to achieving the highest possible levels of computer performance.

PARALLEL ARCHITECTURES

Increasing parallelism is perhaps the most challenging hardware trend in large-scale systems.

Multiprocessor Cray supercomputers have exploited parallelism by emphasizing the power of individual processors primarily and the application of parallel processing secondarily. For the past decade Cray Research has been doubling the number of processors about every three years, with the current high end occupied by the eight-processor CRAY Y-MP system.

Achieving high levels of parallelism in general applications remains a software challenge, although recent software and algorithm advances hold promise for more general uses of these systems.

Parallel processing in particular must be addressed effectively by software. Cray Research has offered three generations of multitasking software for Fortran programs. The first generation required that users insert subroutine calls into their codes that implement a basic set of multitasking functions. This type of multitasking, which we now call macrotasking, offers significant speedup for some applications. Typically, it is used to divide a job into relatively large segments. However, because it requires users to modify their programs, it has not become widely used.

The next generation of Cray multitasking software featured a capability we call microtasking. This feature requires users to insert compiler directives into the text of their programs as comment statements, which are translated by a preprocessor into Fortran code. This feature is more user friendly than macrotasking and has been used to exploit parallelism at the loop level with little synchronization overheard. Because the original Fortran source code is not modified and performance is good, microtasking has been fairly well received by the user community.

The third generation of parallel processing software features what we call autotasking. This capability is easy to see and can provide additional performance improvements similar to those that users enjoy with vectorizing compilers. Autotasking detects opportunities for parallel processing automatically and generates code to execute parallel regions on multiple processors. It offers all of the advantages of microtasking, with the added advantage of automation. Of course, with any of these methods, the improvements seen with parallel processing will depend on the proportion of code that can be processed in parallel.

APPLICATIONS SOFTWARE

Cray Research, through its Industry, Science & Technology Department works closely with many third-party software vendors and customers to convert, optimize and maintain hundreds of application programs for Cray computers (Cray (1989)). Software exists for simulating a wide variety of physical phenomena from molecular motion lasting picoseconds to the formation of galaxies over billions of years. Today, application code development is expanding into areas such as manufacturing process design and control, communication systems, intelligent systems, geophysical analysis, robotics, network computing, and management information systems. Lisp is used in a wide range of symbolic processing applications, including artificial intelligence, natural language processing for speech recognition and other applications, image processing, automatic theorem proving, and rule-based expert systems that emulate human experts in field such as geological exploration and aerospace and automotive design. In the longer term, scientific and engineering computing is expected to combine numerical and symbolic processing. Eventually, facilities will have to be provided for knowledge-based programs to drive numeric programs, and for the results of the numeric programs to be fed back into the symbolic programs. To support ongoing technological development, Cray Research promotes and sponsors technical symposiums, conferences and workshops.

NETWORK ENVIRONMENTS AND VISUALIZATION

To meet demands for ease of use and greater network speed, vendors software must develop to enhance and network connectivity as well as to improve system performance. Scientific networks are typically made up of supercomputers, high-performance workstations, graphics devices, and the local area networks that link them together. Networking software must provide users with a total and transparent solution environment with high levels of interactivity and sophisticated visualization capabilities. The achievement of this goal requires software that can partition applications out to various types of hardware, with each section of an application being paired with the appropriate device.

The need to integrate large heterogeneous networks efficiently mandates that vendors adhere to industry standards in software design, including standards that cover protocols, visualization software, compilers, and networking software. Large scale networking environments also must support advanced scientific visualization capabilities. Visualization must be considered a critical technology for computational research particularly in light of the enormous data throughput that future multiprocessor systems will provide. Fast versatile, and interactive visualization may become the primary means by which researchers will interpret the output from future large-scale systems.

An ideal scientific visualization program would handle incoming data from a simulation or model of any sort. The creation of such a general purpose visualization package should be possible, at least in theory, because virtually any simulation involves some number of independent variable generating some number of dependent variables. The challenge of scientific visualization, therefore, lies in devising a presentation strategy that reveals clearly the relationships among variables while being adaptable to a wide range of applications.

SUPERCOMPUTER CLASS APPLICATIONS

It may not seem important to be able to run a simple analysis in one second instead of ten, unless it is a Monte Carlo simulation that requires thousands of repetitions to yield statistically meaningful results. It may not be important to run an analysis in a minute instead of half hour, unless it is one of dozens of iterations in a design optimization run. These are some examples of situations where an engineer faced with a simulation task will look for the fastest computer on his network that will run his software. The following examples highlight simulation areas where even the most powerful computers available are barely adequate and anything slower is insufficient.

FLUID DYNAMICS

The simulation of fluid flows using Computational Fluid Dynamics (CFD) is an important supercomputer application. Only recently has it been possible to solve the full set of Navier-Stokes equations, including time dependent terms, compressibility and viscosity in reasonable detail for a full aircraft. Enough confidence has been gained from correlating CFD results to wind tunnel data that CFD is now being used for analyzing behaviour in flow conditions that cannot be duplication in existing wind tunnels. Important flow regimes that are now being modelled on Cray supercomputers include: hypersonic flight through the rarefied gas at the edge of the atmosphere, the flow within the core of a turbine engine, the combusting flow through a piston engine cylinder, and the creeping flow and solidification of molten plastic or metal in molds.

The NASA Ames Research Center has long been a pioneer in CFD technology development including the use of supercomputers. The latest advance in analytical capability at the center is the Numerical Aerodynamic Simulation (NAS) program. The NAS program is centered around a CRAY Y-MP8/832 system, including a 256 MW SSD, and a CRAY-2S/4-128 system networked to thousands of researchers throughout the country. A centralized facility like this is essential for many of the complex analysis problems being addressed by NASA and contractor engineers such as the design of the National Aeorspaceplane and advanced jet engines (NASA (1988)). An important outcome of the NAS program has been the development of software, distributed over a HYPERchannel network using standard TCP/IP protocols, to provide researchers with interactive three-dimensional graphical visualizations on an IRIS workstation of flow fields simulated on a supercomputer.

An example of a supercomputer class CFD simulation problem, is the calculation of retreating blade stall on a helicopter rotor as reported by (Narramore, (1988)) of Bell Helicopter TEXTRON, Inc. (BHTI). Several CFD methods have been developed to solve for the flow around helicopter blades with various degrees of sophistication. Potential

methods in particular are fairly quick to run and provide useful
results for the advancing blade side where viscous effects are small.
However, to properly predict the flow field around the retreating blade
of a rotor in forward flight, three-dimensional, unsteady,
compressible, separated, nonlinear, and viscous phenomena must be
modelled. Only very recently has algorithm development and
computational power progressed to the point where this analysis can be
attempted.

Table 2 describes the particulars of this simulation. Although this
analysis is quite complicated it is only a first step toward the goal
of developing computer programs that can simulate the viscous flow
field around the total rotorcraft including blades, wakes, bodies and
wings. The use of a supercomputer during this process enables an
engineer to get useful results and gain valuable experience while
applying a new tool that has not yet been optimized. The best way to
evaluate the large amount of generated data is to use a graphics
workstation to display various portions of the flow field from
different vantage points. Several animation sequences of scalar
quantities such as pressure coefficient contour lines and vector
quantities such as velocity were recorded on video tape for a
presentation at the American Helicopter Society technical forum
meeting.

STRUCTURAL ANALYSIS

A recent example that highlights the need for the resources that only a
supercomputer can provide is the redesign and structural integrity
verification of the space shuttle Solid Rocket Motor (SRM)
(Christensen, (1988)). In the two and a half years following the
Challenger spaceship accident, engineers at Morton Thiokol and NASA
were required to analyze the entire SRM structure, redesign the joints
between segments and make pretest predictions before each of the many
recertification tests. The tight schedule and close scrutiny by the
press and public meant that very detailed numerical models had to be
built rapidly to provide accurate and timely results.

TABLE 2. CFD simulation example

Problem description	Simulate the three-dimensional retreating blade stall characteristics on a helicopter rotor.
Numerical algorithm	Hybrid time-marching, Reynolds-averaged, unsteady, three-dimensional Navier-Stokes solver.
Code	GIT3DNS (Georgia Institute of Technology and BHTI)
Computational grid	Sheared parabolic grid which moves with the blade. 121 X 19 X 45 mesh, 103455 nodes, 2.4 MW of memory
Problem domain	Single blade sweeping from 225° to 315° azimuth. Flight direction is 180° azimuth.
Solution	17,800 iterations, CPU time per iteration: 2.67 seconds. Total CRAY-2S CPU time: 47,500 sec. (13.2 hours)
Output	(8 words) X (103455 nodes) X (151 output steps) = 1 Gbyte

When the original SRM design was completed in the early 1970's, the software technology and computational resources of the time limited the numerical analysis to 2-D axisymmetric models. The subsequent availability of supercomputers running off-the-shelf modern finite element software has allowed engineers to build a family of global, component and detail models incorporating 3-D solid elements, substructures, contact elements, plasticity, viscoelasticity and large displacements.

Typical of the analyses performed was the simulation of the transient load imparted on a 7 degree portion of the cylindrical joint. The finite element model consisting of 31,421 degrees of freedom was subjected to pressure and axial loads in 19 discrete steps. The analysis, which was run three separate times, required 2.5 Gbytes of disk space and 181 total iterations to converge. Each run took under 15 hours of CRAY Y-MP CPU time. It was estimated that the same analysis would have taken nearly a week on an IBM 3090 or about a month on a VAX 8700.

Another area of finite element analysis where supercomputers have proven cost effective is crash simulation. An accurate crash simulation program must take into account large strains and displacements, non-linear material effects, and rapidly changing surface contacts. Explicit time integration is usually used since the memory requirement is small and no system of equations has to be solved (Chedmail, (1986)). Therefore, time increments must be on the order of a microsecond to assure numerical stability. Crash simulations typically take on the order of ten hours of CPU time to analyze an impact that occurs in one tenth of a second. A mini-supercomputer might be able to perform this analysis in several days, but that may be unacceptable if there are many engineers waiting for the results.

A good example is the work done by Ford Motor Company in England (Stanger (1988)) to simulate a side impact text procedure being proposed by the European Experimental Vehicles Committee (EEVC) (Bloch (1988)). The test consists of a moving barrier supported by a rigid trolley impacting the side of a stationary car. The barrier is made of composite foam blocks which are designed to have a stiffness similar to the front of an average European passenger car. The barrier and its trolley have a combined mass of 950 kg and are moving at 50 km/hr at impact. The key goals of the simulation were to assess the accuracy and stability of the numerical results, the cost and duration of the analysis and the availability and usefulness of the results.

A description of the simulation is contained in Table 3. A physical test was conducted to validate the simulation in terms of visual comparison of structural failure and specific velocity and displacement profiles. The computer model matched the overall behaviour of the test vehicle very well although some of the spring potentiometers used to measure deformations in the physical test failed under the very high accelerations experienced during the impact. The finite element model can therefore give more reliable and detailed information than is possible from an instrumented test vehicle. Unlike the physical test which is very expensive to duplicate and is not fully repeatable, the computer simulation can be easily reused to study effects of design changes.

TABLE 3. Crash simulation example

Problem description	Simulate a 50 km/hr side impact crash between a proposed EEVC test barrier and a small 3-door hatchback car.
Numerical algorithm	Explicit, Lagangian finite element code with non-linear materials and complex contact conditions.
Code	RADIOSS (Mecalog SA, Paris, France)
Mesh size	7000 elements
Problem domain	100 ms of simulated time, 100,000 time steps.
Solution	8.5 hours of CRAY Y-MP CPU time.
Post-processing	600 PATRAN images showing contours of plastic strain

A video tape showing the quality of the correlation between the physical test specimen and the computer simulation has been prepared and will be shown in conjunction with this paper. An additional couple hours of Cray CPU time was used to generate the hundreds of images needed for this animation. In the future, crash simulation codes will probably use finer grids to reduce the sensitivity to modelling errors and incorporate mechanisms to handle tearing of welded joints, will include modelling of the air bag, and dynamics of the occupants.

WEATHER FORECASTING

Computational weather forecasting requires the fastest computers available, along with numerical methods that fully exploit the hardware. Modern forecasting systems operate on global computational grids consisting of hundreds of thousands of points to realistically represent the growth of new weather systems several days in advance. The numerical algorithms used are ideal candidates for multiprocessing because they are inherently parallel, the solution is time critical and the executable code requires a lot of memory. The European Center for Medium-Range Weather Forecasting (ECMWF) (Simmons (1986)) has been using an operational multitasked forecast model that runs at 335 MFLOPS for several years on its four processor CRAY X-MP system. The system comprises a repeated cycle of 6-hour forecasts followed by analysis of incoming observations and interpolation of results to provide a starting point for the next 6-hour forecast. Once a day a 10 day forecast is generated during a 2.75 hour run that uses 3.5 MW of memory and 15 MW of SSD scratch space. Accurate weather forecasts are economically valuable to the agricultural and transportation industries in particular. The Air Force Global Weather Center estimates its Cray supercomputer helped it improve world-wide forecasting by 10 to 20 percent. That allows the U.S. Military Airlift Command to save 1-3% of its billion dollar a year fuel bill.

RESERVOIR SIMULATION

Reservoir simulation is a numerical tool used by petroleum engineers to select the management strategy that maximizes the oil and gas recovery from existing reservoirs (Shiles (1988)). The mathematical model is based on second order, non-linear partial differential equations which describe the flow of hydrocarbon and other fluids through porous rock. A finite difference method is usually used to solve the equations on a 3-D orthogonal grid. The large number of possible fluids, the complexity of the chemical and thermodynamic processes and the need for tens of thousands of grid blocks for an adequately detailed reservoir description requires enormous computing power for field-size simulations.

Since the results of such simulations are used to make decisions involving the allocation of expensive assets to enhance recovery by a few percent, there can be no compromise on accuracy. Many man-years of effort have been put into optimizing these simulators for supercomputer execution. The potential pay off is impressive. Arco Oil and Gas Company estimates it was able to increase production of its Prudhoe Bay oil field by 2% due to simulations run on its Cray supercomputer. This translates into 160 million barrels of oil worth billions of dollars.

FUTURE TRENDS

The need to perform real-time processing will influence future supercomputer architectures. Networks will evolve to process batch and interactive processes as well as incoming data from physical sensors in real time and to apply the results to the control of other devices in the network. Nuclear reactor accident mitigation is an area where such capabilities now look promising. Operators will have the option of running faster-than-real-time simulations of reactor transients and playing out multiple scenarios or receiving guidance from expert systems before responding to a transient. If model results can be compared to sensor data on the fly then operators can respond with a high degree of certainty that they have correctly diagnosed the source and likely outcome of an accident and can respond accordingly. Such a capability requires parallel processing efficiency, adaptable interfaces for data input, and a capability for combined numeric-symbolic processing.

System software and applications in future architectures will play as important a role as hardware. We can expect increasing synergism among them, and resulting system performance that far outstrips that of today's systems. Large-scale simulation and modelling already constitute a third scientific methodology which, along with experimental and theoretical approaches, opens new doors to research and produce development. As future systems provide expanded capabilities in performance and ease of use, research scientists and engineers will be among the first to reap the benefits.

ACKNOWLEDGEMENTS

Special thanks to Mr. Douglas Petesch, Engineering Applications Specialist of the Industry, Science and Technology Department of Cray Research, who provided me with valuable material for this paper and assisted in collecting the results for some of the examples.

REFERENCES

1. Dongarra, J.J., Performance of Various Computers Using
   Standard Linear Equations Software in a Fortran
   Environment, Technical Memorandum No. 23, Mathematics and
   Computer Science Division, Argonne National Laboratory,
   Argonne, Illinois, September 30, 1988.

2. *Directory of Applications Software For Cray Supercomputers*,
   Cray Research, Inc., January 1989.

3. NASA's 'Super' Computer Model, *NASA Tech Briefs*, vol.12,
   pp. 12-14, 1988.

4. Narramore, J.C., Sankar, L.N., and Vermeland, R., An Evaluation
   of a Navier-Stokes Code for Calculations of Retreating Blade
   Stall on a Helicopter Rotor, American Helicopter Society 44th
   Annual Forum, Washington, D.C., June 1988.

5. Christensen, N.G., Supercomputing Gives a Boost to the Shuttle
   Solid Rocket Motor Redesign, *Forum Int. Science and Engineering
   Symposium on Cray Supercomputers*, Minneapolis, Minnesota,
   pp. 109-126, 1988.

6. Chedmail, J.F., DeBois, P. et al., Numerical Techniques,
   Experimental Validation and Industrial Applications of Structural
   Impact and Crashworthiness Analysis for the Automotive Industries,
   *Int. Conf. on Supercomputer Applications in the Automotive
   Industry, Zurich, Switzerland*, ed. C. Marino, Computational
   Mechanics Publications, Southampton, Great Britain, pp. 127-145,
   1986.

7. Stanger, J. et al., A Numerical Simulation of the proposed
   European Side Impact Test Procedure, *Second Int. Conf. on
   Supercomputer Applications in the Automotive Industry*, Seville,
   Spain, ed. C. Marino, pp. 9-20, 1988.

8. Bloch, J.A. and Cesari, D., Validation of a Side-Impact Test
   Procedure Using a Mobile Deformable Barrier, *International Journal
   of Vehicle Design*, vol.9, pp. 383-399, 1988.

9. Simmons, A.J., Numerical Prediction: Some Results from Operational
   Forecasting at ECMWF, *Anomalous Atmospheric Flows and Blocking*,
   Advances in Geophysics, vol.29, pp. 305-338, 1986.

10. Shiles, G., Petroleum Reservoir Modeling: Good to the Last Drop,
    Cray Channels, pp. 16-18, 1988.

Boundary Integral Equation Calculations for Flow in Fractured and Heterogeneous Media

D. E. MEDINA and J. A. LIGGETT
Hollister Hall
School of Civil and Environmental Engineering
Cornell University
Ithaca, New York, USA

INTRODUCTION

The boundary integral equation method (BIEM) has now emerged as a standard method for the calculation of many problems. One of its most common uses is in problems of potential flow — and especially free surface problems — but the total includes a wide variety of applications. This paper addresses an application in potential flow, specifically, flow in porous media which is governed by Darcy's law

$$\underline{v} = - K \nabla\Phi \tag{1}$$

in which $\Phi = p/\rho g + h$, the head in the fluid, $\underline{v}$ is the discharge, p is pressure, ρ is fluid density, g is gravity, h is elevation, and K is the hydraulic conductivity (which will be considered a scalar herein but could be a tensor with little change in what follows). Since conservation of mass demands that $\underline{v}$ be divergence free, eqn. (1) leads immediately to Laplace's equation

$$\nabla^2\Phi = 0 \tag{2}$$

Thus, this paper treats potential flow with complex geometry and with singularities in two and three dimensions.

REVIEW OF THE BOUNDARY INTEGRAL EQUATION METHOD

The basis of most boundary integral techniques is the divergence theorem or Green's first identity. The relationship between a volume integral and a surface integral is

$$\int_V \nabla \cdot \underline{v} \, dV = \int_{\partial V} \underline{v} \cdot \underline{n} \, dA \tag{3}$$

in which V is a volume defined by its boundary ∂V, $\underline{v}$ is any vector which is differentiable in V and on its boundary, and

$\underline{n}$ is the unit outward normal to the boundary ∂V. Thus, eqn. (3) translates a volume integral into a surface integral. Green's second identity is easily derived from the first. The vector $\underline{v}$ is replaced by a scalar function times the divergence of another scalar

$$\underline{v} = U\nabla W \qquad (4)$$

in which both U and W are at least twice differentiable in the volume (area) and on its boundary. Then (3) becomes

$$\int_V (U\nabla^2 W + \nabla U \cdot \nabla W)\ dV = \int_{\partial V} U\nabla W \cdot \underline{n}\ dA \qquad (5)$$

Suppose that (5) had been written for the reverse combination of U and W as that expressed by (6). Then it would have been

$$\int_V (W\nabla^2 U + \nabla W \cdot \nabla U)\ dV = \int_{\partial V} W\nabla U \cdot \underline{n}\ dA \qquad (6)$$

Subtracting (6) from (5) gives

$$\int_V (U\nabla^2 W - W\nabla^2 U)\ dV = \int_{\partial V} (U\nabla W - W\nabla U) \cdot \underline{n}\ dA \qquad (7)$$

The scalar product of the divergences cancels in the subtraction since the order of the product does not matter. Of course, eqn. (7) applies equally well to areas and lines

$$\int_A (U\nabla^2 W - W\nabla^2 U)\ dA = \int_{\partial A} (U\nabla W - W\nabla U) \cdot \underline{n}\ ds \qquad (8)$$

Eqns. (7) and (8) are Green's second identity.

Green's third identity is a special case of the second. The arbitrary function W is chosen such that it is a solution of Laplace's equation, $\nabla^2 W = 0$, everywhere in the solution region and on its boundary except at a singular point. Then the first term of the right side of (7) is obviously zero. Our actual choice of W is

$$W = 1/r \qquad r = (x^2 + y^2 + z^2)^{1/2} \qquad (9)$$

where r is the distance between a "base point" (observation point) and a "target point" (field point). The base point, p, can be anywhere in or out of the solution region, but the target point, Q, is on the surface of the region (figure 1). The fact that W=1/r is a solution to Laplace's equation is shown by writing that equation in spherical coordinates

$$\nabla^2 q = \frac{1}{r^2}\frac{\partial}{\partial r}\left[r^2\frac{\partial q}{\partial r}\right] + \frac{1}{r^2 \sin\theta}\frac{\partial}{\partial\theta}\left[\sin\theta\frac{\partial q}{\partial\theta}\right] + \frac{1}{r^2 \sin^2\theta}\frac{\partial^2 q}{\partial\phi^2} \qquad (10)$$

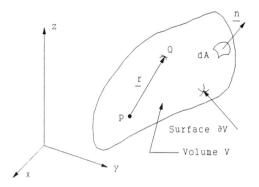

FIGURE 1. The solution domain with observation point (base point) p and target point Q.

in which the spherical coordinates are (r,θ,ϕ) and q is any sufficiently smooth scalar. Since W is a function only of r, only the first term on the right of (10) applies and that is zero. However, $1/r$ is not smooth everywhere; it goes to infinity as r goes to zero. Thus, W is not really a solution to Laplace's equation everywhere since there is a singular point.

The function $1/r$ is called a free space Green's function for Laplace's equation in three dimensions. A free space Green's function is characterized by being a solution of the governing equation, but it does not satisfy the boundary conditions of a problem. There are also Green's functions which satisfy both the governing differential equation for a problem and at least a part of the boundary conditions. These can be used in connection with some specific problems. Suppose that the base point (where r=0) is chosen inside the region and eqn. (7) is applied by choosing a surface for the line integration that excludes the singular point where r=0. That surface includes a sphere, σ, of radius r_0 surrounding the base point (figure 2); thus, the integral on the right side of (7) is to be taken over the area surrounding the volume, ∂V, plus the area of the surface of the sphere, $\partial\sigma$. The sphere can be made as small as desired and, since U is everywhere smooth, U can be considered a constant within that sphere as the radius, r_0, tends toward zero. Further, we specify that U is a solution to Laplace's equation, without singularities, in the entire region. Then (7) can be applied to the small sphere

$$\int_\sigma U\nabla^2(1/r)\ dV = \int_{\partial\sigma} \left[U\nabla(1/r) - \frac{1}{r}\nabla U\right]\cdot\underline{n}\ dA \tag{11}$$

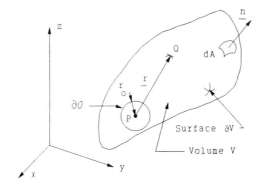

FIGURE 2. The observation point p isolated by a small sphere.

The unit normal, $\underline{n}$, is directed out of the small sphere. The second term in the integral on the right is zero since the area of the sphere goes to zero as r_o^2 whereas the term is divided only by r. The first term in the right hand integral is

$$U\triangledown(1/r) = -\frac{U}{r^2} \tag{12}$$

The area of the sphere is $4\pi r_o^2$, and as r_o tends to zero

$$\int_\sigma U\triangledown^2(1/r) \ dV = -4\pi U_o \tag{13}$$

in which U_o is the value of U in the singular point. Eqn. (7) becomes

$$-4\pi U_o = \int_{\partial V} (U\triangledown(1/r) - \frac{1}{r}\triangledown U)\cdot\underline{n} \ dA \tag{14}$$

The derivation of (14) assumed that the base point was completely inside the volume in order that the excluding sphere can be taken completely surrounding the base point. It can be generalized as

$$-\alpha U_o = \int_{\partial V} [U\triangledown(1/r) - \frac{1}{r}\triangledown U]\cdot\underline{n} \ dA \tag{15}$$

If the base point is outside the volume it need not be excluded by a small sphere and $\alpha=0$; if the base point is on a smooth part of the boundary only one-half the sphere is inside and $\alpha=2\pi$. In general the base point could be at an angle on the boundary (figure 3), in which case α would be the fraction of the sphere that is inside the volume times 4π. That is,

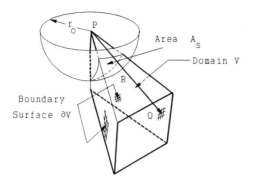

FIGURE 3. The solid angle at point P as a fraction of the surrounding sphere.

$$\alpha = \lim_{r_o \ 0} \frac{A_s}{r_o^2} \qquad (16)$$

where A_s is the part of the sphere inside the volume. α is the solid angle that the boundary makes at the base point. In summary

$$\alpha = \begin{cases} 0 \text{ if the singular point is outside the volume} \\ 4\pi \text{ if the singular point is inside the volume} \\ 2\pi \text{ if the singular point is on a smooth boundary} \\ \text{the solid angle if the singular point is at a kink} \end{cases}$$

Green's third identity relates the value at a point to a surface integral with the provision that the functional relationship .represented by that value satisfies Laplace's equation.

Green's third identity can be derived in two dimensions by the same technique and using ln R (the free space Green's function for Laplace's equation in two dimensions) in place of $1/r$. The sphere which excluded the singular point in three dimensions becomes a circle in two dimensions. Eqn. (15) becomes

$$\alpha U_o = \int_{\partial A} [U\nabla(\ln R) - (\ln R)\, \nabla U] \cdot \underline{n} \ ds \qquad (17)$$

where R is the distance between the base point and the target point in two dimensions and

$$
\alpha = \left\{
\begin{array}{l}
0 \text{ if the singular point is outside the area} \\
2\pi \text{ if the singular point is inside the area} \\
\pi \text{ if the singular point is on a smooth boundary} \\
\text{the boundary angle if the singular point is at a kink}
\end{array}
\right.
$$

Green's third identity is equivalent to eqn. (2).

Although eqns. (15) and (17) are equivalent to (2), Laplace's equation is not valid if the conductivity of the solution region is not constant. In that case it is common to divide the region into subregions, called zones, in which the conductivity can be considered constant. In porous media problems the conductivity is often approximately constant in a subregion or stratum. The conditions along the interzonal interface that links zones i and j are

$$
\Phi_i = \Phi_j \quad \text{and} \quad K_i \frac{\partial \Phi}{\partial n_i} + K_j \frac{\partial \Phi}{\partial n_j} = 0 \quad \text{on } \partial A_{ij} \text{ or } \partial V_{ij} \quad (18)
$$

in which ∂A_{ij} or ∂V_{ij} is the interface between zone i and zone j and n_i is the outward normal from zone i. On such interfaces there are four unknowns initially, Φ and $\partial \Phi / \partial n$ with respect to each zone.

In the remainder of the paper these equations are applied to the problem of flow in zoned porous media with special emphasis on the flow in fractured media. There are two algorithms: in the first the flow is considered to pass entirely through the fractures with the surrounding rock being too impermeable to participate in a material way; in the second that approximation is relaxed and the entire three dimensional problem is solved.

TWO DIMENSIONAL ZONAL FLOW

In this section we consider the flow through fractured porous media. Each fracture represents a two dimensional zone — the flow is considered uniform across the narrow gap of the fracture — but the fractures themselves may be arranged in a three dimensional space with arbitrary orientation. In such a case an interzonal boundary is frequently the intersection of more than two zones and the conditions of eqn. (18) become

$$
\Phi_i = \Phi_j = \Phi_k = \Phi_\ell = \cdots
$$

$$
K_i \frac{\partial \Phi}{\partial n_i} + K_j \frac{\partial \Phi}{\partial n_j} + K_k \frac{\partial \Phi}{\partial n_k} + K_\ell \frac{\partial \Phi}{\partial n_\ell} + \cdots = 0
\qquad (19)
$$

where $i, j, k, \ell, \ldots$ are all the zones that intersect the line. The solution to this problem comes from (15) with standard boundary conditions on the exterior boundaries and (19) applied to internal boundaries. Linear, conforming elements are used to discretize the equations and the boundary conditions.

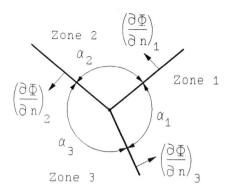

FIGURE 4. Three zones in a plane.

Singularities at Zonal Intersections

Consider a two dimensional problem where three zones, all in the same plane, intersect at a point as in figure 4. The point of intersection represents a singularity in the flow and the character of the singularity can be analyzed (Liggett and Liu, 1983). Along the interzonal boundaries, conditions (18) apply. At points a small distance from the intersection as shown in the figure there is a total of four unknown quantities: Φ (taken as equivalent at all points), and $(\partial \Phi / \partial n)_i$, $i=1,2,3$ after conditions (18) have been used. Eqn. (15) can be applied to each of the three zones with the point of intersection as an observation point, giving three equations. Unfortunately, the system is still one equation short of being determinate. Apparently, this situation is the result of a non-unique solution. The singularity at the intersection causes the solution region to be multiply connected and additional solutions can be obtained by adding arbitrary circulation in the intersection point. There are two apparent remedies. The first is to set the circulation equal to zero in a "path equation"

$$\int_C \underline{v} \cdot \underline{t} \ ds = 0 \tag{20}$$

in which C is a circle about the singular point and $\underline{t}$ is the unit tangent vector. Eqn. (20) is approximated

$$K_1 \left\{ \left[\frac{\partial \Phi}{\partial n} \right]_1 + \left[\frac{\partial \Phi}{\partial n} \right]_2 \right\} \alpha_1 + K_2 \left\{ \left[\frac{\partial \Phi}{\partial n} \right]_2 + \left[\frac{\partial \Phi}{\partial n} \right]_3 \right\} \alpha_2$$

$$+ K_3 \left\{ \left[\frac{\partial \Phi}{\partial n} \right]_3 + \left[\frac{\partial \Phi}{\partial n} \right]_1 \right\} \alpha_1 = 0 \tag{21}$$

which is used to close the system.

A second approach is to avoid the singularity — as all corner problems can be avoided — by using non-conforming elements (figure 5). The nodes are not placed on the end of the element and thus there is no node at the singularity. The number of equations and the number of unknowns are equal. In general non-conforming elements have the advantage of easier programming, but the potential is not continuous at the ends of the element; the elements are C^{-1} continuous. Also there are more nodes (twice as many for the two dimensional problem when using linear interpolation functions) compared with conforming elements and a high price is paid in efficiency with no gain — perhaps even a loss — in accuracy. In the present case there is the compromise of making the element that intersects the singularity semi-conforming while using conforming elements elsewhere.

Three Dimensional Configurations

When the zones can be arranged in a general three dimensional configuration, continuity of flow is ensured by eqn. (19) and the non-uniqueness of the solution is made worse. Consider the example of figure 6a where five zones intersect at a point. The normal derivatives in each zone are indicated by the arrows in figure 6b which have been placed a short distance from the point of intersection. The subscripts on $\partial\Phi/\partial n$ indicate the zone to which the normal derivative belongs and the superscripts are used to differentiate two or more normal derivatives from the same zone. Before application of (19) there are 10 unknown normal derivatives. Eqn. (19) applies along each of the four dividing boundaries and (15) can be written in each of the five zones, which leaves a deficiency of two equations. However, there can be two independent path equations: (a) using zones 1, 2, and 3; and (b) using zones 4, 5, and 3. Additional path equations would be a combination of the above. In the case that all the zones lie in the same plane the sum of the angles in the path equation is 2π; in the three dimensional configuration that is not the case.

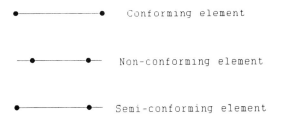

FIGURE 5.   Conforming and non-conforming elements.

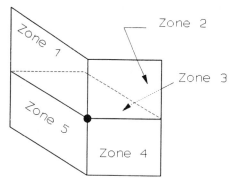
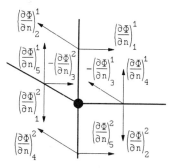

FIGURE 6a. A five zone inter-section with two path equations.

FIGURE 6b. Detail of the corner of figure 3a showing the normal derivatives.

Since there is an infinite number of ways zones can intersect in three dimensions, it is difficult to write a computer program with general path equations. There are two approximate remedies: the use of non-conforming or semi-conforming elements or the introduction of an approximate path equation. We have preferred the approximate path equation in which derivatives are set equal for a zone on different sides of the singularity and these values are used to represent a path equation. Although there is no justification for that procedure, tests of various cases have shown that it has little effect on the solution at a location more than one element away from the singularity. Thus, if each of the boundary segments of a zone is divided into three or more elements, the treatment of the singularity makes little difference in the result except in the immediate vicinity of the singularity. Consider the example of figure 7 which was based on a

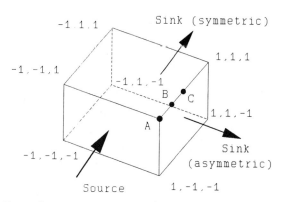

FIGURE 7. Flow around the sides of a cube.

TABLE I. Results for symmetric and asymmetric flow.

	Symmetric Flow					
	Potential			Normal Derivative		
Point	A	B	C	A	B	C
Generic path eqn.	1.000	0.954	0.893	0.023	0.000	0.000
Exact path eqn.	1.000	0.953	0.893	0.000	0.000	0.000

	Asymmetric Flow					
	Potential			Normal Derivative		
Point	A	B	C	A	B	C
Generic path eqn.	1.000	0.970	0.936	0.023	0.025	0.025
Exact path eqn.	1.000	0.969	0.936	0.027	0.025	0.026

Hele-Shaw model of Mathers and Hembroff (no date). The model consisted of nested cubes in which there was a small space between the outer cube and the inner cube so the Hele-Shaw apparatus consisted of six planes. Fluid entered the model through a source in the center of the front face and a sink which was either in the center of the rear face (symmetric flow) or the center of a side face (asymmetric flow). Simulations were made using both the proper path equation and a "generic" equation at the corners. Using just four elements per side the results are as follows in Table I.

Since the corner of the cube should be a streamline in the symmetric flow example, the zero normal derivative using the path equation is exactly correct. Using the program with generic corners, the node at point A gives an incorrect value for the normal derivative, but all other internal point agree with the exact solution to three significant digits. With this sort of accuracy, we have considered that it is not generally worthwhile to provide an exact path equation for general calculation.

THREE DIMENSIONAL ZONAL FLOW

The assumption that all the flow passes through the fractures provides a quick, economical solution to many problems. However, that approximation may not be valid and it is the three dimensional solution, where the flow can pass through the surrounding media as well as the fractures, that is often needed. Although the conductivity in the fractures may be several orders of magnitude greater than in the surrounding rock, the available area for flow in the rock may be several orders of magnitude greater than in the fractures. Thus, it is not clear that the quantity of fluid that passes through the rock is small compared with the quantity that passes through the fractures. Also, the pore volume in the fracture is almost certainly very small compared to the pore volume in the rock. The larger storage capacity of the rock has a

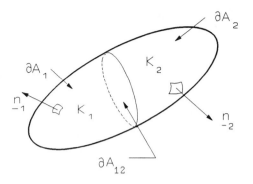

FIGURE 8. Notation for the two zone problem.

major effect on the transport and storage of pollutant

Integral Formulation

We have taken the point of view that the entire problem is a
large zoned medium and the best approach is a general three
dimensional, zoned program. In the case of three dimensional
problems it is especially important that the number of variables be reduced to the minimum. Wei and Liggett (1982) used
the interfacial conditions (18) to eliminate the potential on
the interfacial boundaries in favor of the normal derivative.
Zheng (1983) used a similar technique to eliminate the normal
derivative in favor of the potential. The latter method is
preferred to avoid the interzonal singularities. Consider a
solid that is divided into two zones (figure 8). For a point
on the interface between the zones eqn. (17) is written for
zone 1 and multiplied by K_1; eqn. (17) is then written for
zone 2 and multiplied by K_2. After adding these two equations and applying (18) the result is

$$
-(\alpha_1 K_1 + \alpha_2 K_2)\Phi = \int_{\partial A_1} K_1 \left[\Phi(Q) \frac{\partial}{\partial n_1} \frac{1}{r} - \frac{1}{r} \frac{\partial \Phi(Q)}{\partial n_1} \right] ds
$$

$$
+ \int_{\partial A_2} K_2 \left[\Phi(Q) \frac{\partial}{\partial n_2} \frac{1}{r} - \frac{1}{r} \frac{\partial \Phi(Q)}{\partial n_2} \right] ds \qquad (22)
$$

$$
+ \int_{\partial A_{12}} \left[K_1 \frac{\partial}{\partial n_1} \frac{1}{r} + K_2 \frac{\partial}{\partial n_2} \frac{1}{r} \right] \Phi \, ds
$$

Note that the interface normals have opposite signs, $\partial/\partial n_1 =
-\partial/\partial n_2$. For a region divided into N zones by M interfaces

$$- \Phi(p) \sum_{n=1}^{N} \alpha_n K_n = \int_{\partial A_n + \partial A_d} K \left[\Phi(Q) \frac{\partial}{\partial n} \frac{1}{r} - \frac{1}{r} \frac{\partial \Phi(Q)}{\partial n} \right] ds \tag{23}$$

$$+ \sum_{i=1}^{M} \int_{\partial A_{ij}} (K_i - K_j) \Phi(Q) \frac{\partial}{\partial n} \frac{1}{r} ds$$

where ∂A_d is the Dirichlet boundary and ∂A_n is the Neumann boundary and the K in the first integral must take on the value in the particular zone, the boundary of which is being integrated.

Discretization and Integrals

The most flexible discretization of the boundary is the use of triangular elements. We have chosen plane triangular elements with linear interpolation functions for both the potential and its normal derivative. One reason for this choice is that it is important to express the integrals in closed form using elementary functions. In the case of fractured media, and for many other applications, some of the zones have a large aspect ratio (ratio of length to width) — perhaps to 10^5. In the case of finite differences or finite elements, the element size must generally be based on the smaller dimension (e.g., the width of a fracture instead of the length of a fracture), a situation that would create a large number of elements, leading to excessive computation times and unwanted detail. Since a finite difference or finite element type of approximation for the derivatives is not included in the boundary integral equation method, the element size can be chosen without regard to the dimensions provided that the integration is done accurately. Numerical quadrature can produce accurate results in the case that the integrand is either regular or singular, but there is a difficult case when the integrand is "nearly singular". In that instance neither regular quadrature nor singular quadrature produces satisfactory results. In fracture flow this case occurs when integrating with the observation point on one side of the fracture with the target element directly across the fracture. There are also other examples not associated with fracture flow: when the element spacing becomes very small or when an interior solution is sought at a point very near a boundary.

In the case of linear elements the interpolations are

$$\Phi = a_1 \xi + b_1 \varsigma + c_1 \qquad \frac{\partial \Phi}{\partial n} = a_2 \xi + b_2 \varsigma + c_2 \tag{24}$$

in which the a,b,c are constants where ξ, ς, η is a local coordinate system with the origin in the observation point, the ξ-ς plane parallel to the target element, and the η coordinate normal to the target element. Denoting the nodal values of the potential and the normal derivative as $\{\phi\}$ and $\{\partial \Phi/\partial n\}$

$$\{\phi\} = \begin{Bmatrix} \Phi_1 \\ \Phi_2 \\ \Phi_3 \end{Bmatrix} \qquad\qquad \left\{\frac{\partial\phi}{\partial n}\right\} = \begin{Bmatrix} (\partial\Phi/\partial n)_1 \\ (\partial\Phi/\partial n)_2 \\ (\partial\Phi/\partial n)_3 \end{Bmatrix}$$

The constants are given by

$$a_1 = \{V_\varsigma\}^T\{\phi\} \qquad\qquad a_2 = \{V_\varsigma\}^T\left\{\frac{\partial\phi}{\partial n}\right\}$$

$$b_1 = \{V_\xi\}^T\{\phi\} \qquad\qquad b_2 = \{V_\xi\}^T\left\{\frac{\partial\phi}{\partial n}\right\} \qquad (25)$$

$$c_1 = \{V_{\xi\varsigma}\}^T\{\phi\} \qquad\qquad c_2 = \{V_{\xi\varsigma}\}^T\left\{\frac{\partial\phi}{\partial n}\right\}$$

in which

$$\{V_\varsigma\}^T = \frac{1}{D}\left\lfloor \varsigma_2-\varsigma_3 \quad \varsigma_3-\varsigma_1 \quad \varsigma_1-\varsigma_2 \right\rfloor$$

$$\{V_\xi\}^T = \frac{1}{D}\left\lfloor \xi_3-\xi_2 \quad \xi_1-\xi_3 \quad \xi_2-\xi_1 \right\rfloor$$

$$(26)$$

$$\{V_{\xi\varsigma}\}^T = \frac{1}{D}\left\lfloor \xi_2\varsigma_3-\xi_3\varsigma_1 \quad \xi_3\varsigma_1-\xi_1\varsigma_3 \quad \xi_1\varsigma_2-\xi_2\varsigma_1 \right\rfloor$$

$$D = \xi_1(\varsigma_2-\varsigma_3) - \varsigma_1(\xi_2-\xi_3) + \xi_2\varsigma_3 - \xi_3\varsigma_2$$

The integrals of (23) are defined in general as

$$I_{ij} = -\int_{A_j} \left[\Phi\,\frac{\eta_i}{r^3} + \frac{1}{r}\,\frac{\partial\Phi}{\partial n}\right]\,dA = -\left\{k_2^e\right\}^T\{\phi\} - \left\{k_1^e\right\}^T\left\{\frac{\partial\Phi}{\partial n}\right\}$$

$$(27)$$

$$\left\{k_2^e\right\} = \{V_\varsigma\}I_3^\xi + \{V_\xi\}I_3^\varsigma + \{V_{\xi\varsigma}\}I_3$$

$$\left\{k_1^e\right\} = \{V_\varsigma\}I_1^\xi + \{V_\xi\}I_1^\varsigma + \{V_{\xi\varsigma}\}I_1$$

The separate integrals are

$$I_1 = \int \frac{1}{r}\,dA \qquad I_1^\xi = \int \frac{\xi}{r}\,dA \qquad I_1^\varsigma = \int \frac{\varsigma}{r}\,dA$$

$$(28)$$

$$I_3 = \int \frac{1}{r^3}\,dA \qquad I_3^\xi = \int \frac{\xi}{r^3}\,dA \qquad I_3^\varsigma = \int \frac{\varsigma}{r^3}\,dA$$

The above integrals are sufficient to find the boundary solutions or the potential at a boundary point or the potential at an interior point. The velocity at interior points comes from a derivative of the integral (10). To illustrate, consider the integral of (17) in the ξ direction

$$- \alpha \frac{\partial \Phi}{\partial \xi} = \int \left[- \frac{3\Phi\eta}{r^5} - \frac{\partial \Phi}{\partial n} \frac{\xi}{r^3} \right] dA \qquad (29)$$

Note that Φ is not differentiated under the integral sign since that value refers to the target element whereas the velocity is needed at the observation point. Again assuming linear elements, the following integrals are contained in (16)

$$I_3^{\xi\xi} = \int \frac{\xi^2}{r^3} dA \qquad I_3^{\zeta\zeta} = \int \frac{\zeta^2}{r^3} dA \qquad I_3^{\xi\zeta} = \int \frac{\xi\zeta}{r^3} dA$$

$$\qquad (30)$$

$$I_5 = \int \frac{1}{r^5} dA \qquad I_5^{\xi} = \int \frac{\xi}{r^5} dA \qquad I_5^{\zeta} = \int \frac{\zeta}{r^5} dA$$

$$I_5^{\xi\xi} = \int \frac{\xi^2}{r^5} dA \qquad I_5^{\zeta\zeta} = \int \frac{\zeta^2}{r^5} dA \qquad I_5^{\xi\zeta} = \int \frac{\xi\zeta}{r^5} dA$$

Integration

All 15 of the above integrals are expressible in terms of elementary functions — logarithms and inverse tangents — although the technique for doing so is not immediately obvious in all cases. The area integrals I_1, I_3, and I_5 are converted to line integrals around the target triangle. First, the divergence theorem is used to write

$$\int_{\Delta} \nabla^2 V \, dA = \int_{\partial\Delta} \nabla V \cdot \underline{\nu} \, ds \qquad (31)$$

in which Δ represents the triangle with $\partial\Delta$ its boundary and $\underline{\nu}$ is the unit normal to the side being integrated (figure 9). In the case of I_1 let

$$\nabla^2 V_1 = \frac{1}{r} = \frac{1}{\left[R^2 + \eta^2 \right]^{1/2}} \qquad R = \left[\xi^2 + \zeta^2 \right]^{1/2} \qquad (32)$$

Then V is a solution to Laplace's equation. In cylindrical coordinates with only the radial terms retained

$$\frac{1}{R} \frac{\partial}{\partial R} \left[R \frac{\partial V_1}{\partial R} \right] = \frac{1}{\left[R^2 + \eta^2 \right]^{1/2}} \qquad V_1 = r + \eta \ln(r-\eta) - \eta \ln R \qquad (33)$$

Using (25) in (26)

$$I_1 = \int_{\partial\Delta} \frac{r^2}{r^2 - \eta^2} \frac{\partial r}{\partial\nu} \, ds + \eta\theta_0 \qquad (34)$$

in which θ_0 is a two dimensional angle which accounts for the singularity in the case that the normal projection of the observation point falls into or on the boundary of the target triangle ($\theta_0 = 2\pi$ if the projection falls inside the triangle, $\theta_0 = \pi$ if the projection falls on a side of the triangle, and θ_0 is the vertex angle in the case the projection falls on a vertex). Taking a sub-local (p,q) coordinate system in which the p axis is parallel to a side of the triangle (figure 9), the integral of (34) is

$$\int_{p_2^{(2)}}^{p^{(1)}} \frac{qr}{r^2 - \eta^2} \, dp = \left[q \ln(p+r) - \frac{\eta}{2} \tan^{-1} \frac{2pqr\eta}{p^2\eta^2 - q^2 r^2} \right]_{p^{(2)}}^{p^{(1)}} \tag{35}$$

Summing over the three sides of the triangle produces the final result

$$I_1 = \sum_{i=1}^{3} \left\{ \frac{\eta}{2} \tan^{-1} \frac{2p_i q_i \eta \, (p_i^2 + q_i^2 + \eta^2)^{1/2}}{p_i^2 \eta^2 - q_i^2 (p_i^2 + q_i^2 + \eta^2)} \right. \tag{36}$$

$$\left. - q_i \ln\left[p_i + (p_i^2 + q_i^2 + \eta^2)^{1/2} \right] \right\}_{p_i^{(1)}}^{p_i^{(2)}} + \eta \theta_0$$

I_3 and I_5 are done in a similar manner.

All integrals with ξ or ς to the first power in the numerator or with $\xi\varsigma$ can be done with straightforward area integration over the triangle. Those with ξ^2 or ς^2 are divided into parts by addition and subtraction. For example

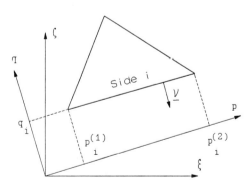

FIGURE 9. The target triangle with $\xi - \varsigma$ and p-q coordinate systems.

$$I_3^{\xi\xi} = \int \frac{\xi^2 + \eta^2}{r^3} \, dA - \int \frac{\eta^2}{r^3} \, dA \tag{37}$$

$$I_5^{\xi\xi} = \int \frac{\xi^2 + \eta^2 - \frac{2}{3} r^2}{r^5} \, dA - \int \frac{\eta^2}{r^5} \, dA + \frac{2}{3} \int \frac{1}{r^3} \, dA \tag{38}$$

The first integral in each of these expressions can be done by straightforward area integration; the other integrals have been done as described above (η is a constant in the integration). A listing of all 15 integrals can be found in Medina and Liggett (1989).

A ZONAL ANOMALY

The direct boundary integral equation method loses accuracy, sometimes disastrously, in those zoned problems in which the conductivity ratio for adjacent zones is high. The limit of the ratio which produces inaccuracy is strongly dependent on the geometry of the problem and the discretization.

Examples of Accuracy Loss

Consider the two dimensional, three zone problem of figure 10 in which the conductivities of zones 1 and 3 are equal but zone 2 may have a much lower conductivity. Figure 11 shows the solution at point B, which is in zone 2 at coordinates x=0.75, y=0.5, as a function of the conductivity ratio, $\mu = K_1/K_2$. Consider curve 1, which is done with the two dimensional counterpart of eqn. (10) using 8 elements on each boundary segment. The solution in the point is valid until μ reaches about 10^3. At that value it departs badly from the true solution which is given by curve 5. Using the Galerkin method (in place of collocation) for satisfying the integral

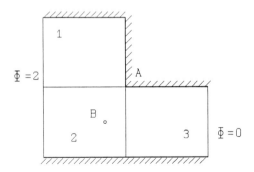

FIGURE 10. A three zoned problem with a strong singularity.

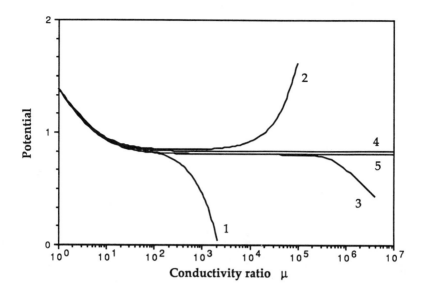

FIGURE 11. Solution in point B of figure 6 for various formulations of the boundary integral equation method and various point spacing.

equation helps the solution; using a finer discretization also improves the accuracy (curve 2, figure 11). The combination of the Galerkin method with fine discretization maintains a good solution to a high conductivity ratio (curve 3), but becomes an expensive calculation. However, if μ becomes large enough, the solution eventually becomes very bad in the region of low conductivity. This behavior does not occur in the regions of high conductivity where the solution remains valid.

To find the reason for the bad solutions consider (10) written for two dimensions and specifically for an interior point

$$2\pi K_2 \Phi(B) = \int_{\partial A} K\left[\Phi(Q) \frac{\partial}{\partial n} \ln R - \ln R \frac{\partial \Phi(Q)}{\partial n}\right] ds \qquad (39)$$

$$+ \int_{\partial A_{12}+\partial A_{23}} (K_2 - K_1) \Phi(Q) \frac{\partial}{\partial n_2} \ln R \, ds$$

where ∂A is now the total exterior boundary. If K_2 is very small, there is no effective integration over the exterior boundary of the low conductivity region; the point B resembles a point outside of the solution region more than a point inside the region. The solution that results comes from the subtraction of large numbers, which requires high accuracy to avoid losing all of the significant digits.

In the case of figure 10 a strong singularity forms at point A. The presence of this singularity aggravates the inaccuracy in the basic boundary element solution and degrades the numbers which make up the solution in point B. Use of finer discretization improves the accuracy; use of double precision arithmetic does not help since it is the numerical approximation and not the arithmetic accuracy that causes the problem. This latter point can be illustrated by the example of figure 12 where the center zone is the one with low conductivity. In that case the boundary integral equation method gives the "exact" solution because the behavior of the potential and the normal derivative are no worse than linear on any part of the boundary. Any deterioration of accuracy is the result of the finite number of digits in the computer and in this case the use of double precision does delay the onset of the inaccurate solution in the low conductivity region. Using single precision the solution is essential exact up to $\mu \approx 10^5$ whereas double precision extends that range to $\mu \approx 10^{13}$.

It is interesting that the indirect boundary element method does not have this difficulty (Medina, et al, 1989; see also curve 4, figure 11). In the indirect method a distribution of sources, σ, along the boundary is found from the equations

$$\Phi(p) = \int_{\partial A} \sigma \ln R \, dA \qquad (40)$$

$$\frac{\partial \Phi(p)}{\partial n_p} = \int_{\partial A} \sigma \frac{\partial}{\partial n_p} \ln R \, dA - \pi \sigma \qquad (41)$$

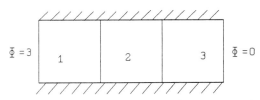

FIGURE 12. A three zone problem without a singularity.

in which the derivative in (41) is taken along the normal at the observation point. Once the source distribution, σ, is found, (40) can be used to find the potential in any point, inside or outside of the region. Since (40) does not contain the conductivity, the accuracy is not a function of low conductivity regions. Unfortunately, the indirect boundary element method has other disadvantages — primarily that it is difficult to find accurate solutions on and near corners and boundaries, and it is somewhat less efficient — that prevent it from becoming the formulation of choice.

An Accurate Method

The fact that the accuracy of the solution in high conductivity regions is not affected by the presence of the low conductivity regions suggests an algorithm for the direct boundary integral equation method. It is simply a multi-step procedure in which (1) the entire problem is solved in the usual manner and (2) the solution is repeated after eliminating the zones of highest conductivity and using the solution on the interzonal boundaries of those zones as boundary conditions for the zones of lower conductivity. If there is range of conductivities that differ by several orders of magnitude, the second step may have to be repeated several times, each time eliminating the remaining zones of highest conductivity.

This algorithm has been tested in both two and three dimensions and provides a satisfactory solution. Curve 5 of figure 11 was obtained using that method. In that case the accuracy was verified using extremely fine discretization (128 elements per boundary segment). Although the algorithm requires multiple solutions, the successive problems become smaller as the zones of higher conductivity are discarded. It appears that the efficiency of the method is competitive with the single solution of the indirect method.

CONCLUSIONS AND SUMMARY

The boundary integral equation method provides several techniques for the calculation of flow through zoned and fractured media. For the three dimensional arrangement of fractures a fast, pseudo three dimensional method is available if the flow through the solid medium can be neglected. If the flow through the solid medium is important, a full three dimensional calculation is necessary. In the methods described herein the fractured medium is treated no differently than zoned media, but it is necessary that the integrals be expressed exactly, as opposed to the use numerical quadrature, to maintain accuracy without using tiny elements in and near zones of large aspect ratio.

In three dimensional problems the boundaries, both exterior and interior, must be discretized with two dimensional elements. Triangular elements seem to be the best choice since they can fit the boundaries well and can represent curved surfaces. We have found that a mesh generator is a practical necessity for the generation process in order to avoid errors in the geometric description of the solution region and the element mesh. An advantage of the boundary integral equation method is that all programs can be run on microcomputers, although in the case of the three dimensional calculations it is easy to exhaust the memory.

ACKNOWLEDGEMENTS

This work was supported in part by the National Science Foundation (Grant Number ECE-8610119), by Battelle Office of Waste Technology Development, and by Xerox Corporation.

REFERENCES

1. Liggett, J.A., and Lui, P.L-F., *The Boundary Integral Equation Method for Porous Media Flow*, Allen and Unwin, London, 1983.

2. Mathers, W.G., and L.R. Hembroff, Analysis of flow through fractured rock by the Hele-Shaw method, draft manuscript, no date.

3. Medina, D.E., and J.A. Liggett, Exact integrals for three dimensional boundary element potential flow, Submitted for publication, 1989.

4. Medina, D.E., Lean, M.H., and Liggett, J.A., Boundary elements in zoned media: Direct and indirect methods, Submitted for publication, 1989.

5. Wei, L-Y., and Liggett, J.A., Zoned boundary elements – an economical calculation, *Proceedings of the International Conference on Finite Element Methods, Beijing China*, Science Press, pp. 816-820, 1982.

6. Zheng, J.D., Location of free surface for zoned problems – A new economical boundary element method, in Boundary Elements, eds. C.A. Brebbia, T. Futagami, and M. Tanaka, *Proceedings of the Fifth International Conference on Boundary Elements, Hiroshima, Japan*, Springer-Verlag, pp. 85-94, 1983.

The Conjugate Gradient Method on Vector and Parallel Supercomputers

G. MEURANT
CEA, Centre d'Etudes de Limeil-Valenton
BP 27, 94195
Villeneuve St. Georges Cedex, France

INTRODUCTION.

Solving linear systems of equations is one of the most computing intensive tasks that concern numerical analysts. Many mathematical models of physical processes (for instance fluid mechanics, nuclear and plasma physics, electromagnetism, oil reservoir simulation) lead to complicated systems of linear or non linear partial differential equations (p.d.e.) in 2 or 3–dimensional domains. Then, these systems are discretized (i. e. reduced to finite dimension) using finite difference or finite element methods. Large and generally sparse linear systems arise from these discretization processes. The problem we are interested in is solving these linear systems on modern supercomputers.

During the last fifteen or twenty years, there has been a rapid development of fast computers dedicated to large scale scientific computation. All the most efficient machines available today, use some form of vector or parallel computing or even both. The maximum theoretical peak speeds of these computers range from 100 Mflops (Millions of floating point operations per second) to 1–5 Gflops. However, on practical applications the speed can be much lower. It is therefore interesting to develop algorithms well suited to these architectures to be able to get the most out of these "super–machines".

Very large (thousands to hundred of thousands of unknowns) and sparse (a few non–zero terms in each row) linear systems are commonly solved using iterative techniques. The most popular and efficient method for symmetric positive definite systems is the Conjugate Gradient (CG) method used in connection with a preconditioner. In this paper, we review some of the techniques which have been recently developed to efficiently use CG on vector and parallel computers.

The outline of the paper is as follows. Section 2 recalls CG algorithm and the problems we are faced with on supercomputers. Section 3 deals with the choice of preconditioners on vector machines and describes some numerical experiments on the Cray X–MP, Y–MP, the ETA 10–E, Alliant FX/80 and Convex C–210 computers. Problems specific to parallel computations are explained in Section 4, where the use of Domain Decomposition techniques is advocated. Numerical experiments on a Cray Y–MP/832 using the 8 processors in parallel and Autotasking show the usefulness of these techniques. The conclusion gives some directions for the future research in this area.

THE CONJUGATE GRADIENT METHOD.

The Conjugate Gradient method was developed (in the 50's) by Hestenes & Stiefel (1952) as a direct method. Unfortunately, it did not compare very favorably with Gaussian elimination. However, Reid (1971) showed that the method can be efficiently used as an iterative technique for systems arising from p.d.e.'s discretization as, in some cases, a good approximate solution is obtained after a few steps. But, unfortunately, one can find some problems for which the rate of convergence is too slow for the method to be of any practical use. The CG rate of convergence primarily depends on the condition number of the matrix of the given system (i.e. the ratio of the largest to the smallest eigenvalues) and also on the distribution of the eigenvalues. For bad distributions, the convergence can be painfully slow.

The idea was then proposed to modify the problem to get a matrix with better spectral properties and to use CG on a preconditioned system. If we would like to solve

$$Ax = b,$$

we premultiply this equation by a symmetric positive definite matrix M^{-1} to get,

$$M^{-1}Ax = M^{-1}b.$$

The matrix M is called the preconditioner and is heuristically chosen to have a better eigenvalue distribution, Concus, Golub & O'Leary (1976).

The preconditioned CG algorithm is described by the following formulas:
given an initial guess x^0, compute $r^0 = b - Ax^0$; then for $k = 0, 1, \ldots$ until convergence do

$$Mz^k = r^k$$

$$\beta_k = \frac{(r^k, z^k)}{(r^{k-1}, z^{k-1})}, \quad \beta_0 = 0,$$

$$p^k = z^k + \beta_k p^{k-1},$$

$$\alpha_k = \frac{(r^k, z^k)}{(Ap^k, p^k)},$$

$$x^{k+1} = x^k + \alpha_k p^k,$$

$$r^{k+1} = r^k - \alpha_k Ap^k.$$

We can already see that most of the operations in this algorithm are defined in terms of vectors; so, CG is likely to be an interesting method for vector computers.

Let us now define the problem we are going to use as a test bed. Suppose, for the sake of simplicity, that we are interested in solving the following class of continuous problems :

$$-\frac{\partial}{\partial x}\left(a(x,y)\frac{\partial u}{\partial x}\right) - \frac{\partial}{\partial y}\left(b(x,y)\frac{\partial u}{\partial y}\right) + cu = f,$$

in $\Omega \subset R^2, \quad u|_{\partial\Omega} = 0 \quad$ or $\quad \frac{\partial u}{\partial n}|_{\partial\Omega} = 0,$

Ω being a rectangle.

With the standard 5 point finite differences scheme on a regular mesh and row–wise ordering of the mesh points, we obtain a block tridiagonal linear system :

$$A\,x\ =\ b,$$

with

$$A = \begin{pmatrix} D_1 & A_2^T & & & \\ A_2 & D_2 & A_3^T & & \\ & \ddots & \ddots & \ddots & \\ & & A_{n-1} & D_{n-1} & A_n^T \\ & & & A_n & D_n \end{pmatrix}.$$

With usual hypotheses on the coefficients of the equation, matrices D_i are point tridiagonal strictly diagonally dominant, and matrices A_i are diagonal. It is well known that A is a positive definite symmetric M-matrix.

As an example and as we are only interested into the computers performances, we will solve the simple "model problem"

$$-\Delta u = f \quad \text{in } \Omega =]0,1[\times]0,1[$$

$u|_{\partial \Omega} = 0$

The right hand side f is computed such that the solution of the discrete problem is the vector of values at mesh points of the function $x(1 - x)y(1 - y)exp(xy)$. The starting guess x^0 is 0 and the stopping criterion is based on reduction of the norm of the relative residual.

Matrices of such a regular sparsity pattern, are usually stored by diagonals, each diagonal being a vector; then the multiplication Ap^k in CG is done by diagonals with vector operations. The basic operations we have to do for each CG iteration are :

- 3 saxpys or saxpy–like operations
- 2 dot products
- 1 matrix vector product
- 1 solution of $Mz^k = r^k$.

One must not forget also, the computation of the stopping criteria and the test for convergence. Basic CG (without preconditioning) is well suited for vector computation, as there exist fast implementations of the 3 first operations on all today vector computers. As an example, we consider in Table 1 the asymptotic Mflops rates for these operations on one processor of a Cray X–MP/416, Cray Y–MP/832, the ETA 10–E, the Alliant FX/80 (8 processors) and the Convex C–210.

Eight processors were considered for the Alliant because it is the only way to get good performances and parallelization was automatic.

The main problems come with the solution of $Mz^k = r^k$, as vectorization depends on the chosen method. As we will see in the next section, there exist different methods that are more or less vectorizable.

The previous results show that we can get very good Mflops rates with CG on vector computers. However, the user is only interested in the computing time he has to spend for the solution of his problem. Unfortunately, as the rate of convergence of basic CG is related to the eigenvalue distribution, it is very slow for examples like the model problem where the eigenvalues are regularly distributed. To reach the given criterion with $\varepsilon = 10^{-6}$ for a problem of order 36100, we need 279 iterations, which means 1.14 seconds on one processor of the Cray X–MP.

TABLE 1. Results for elementary operations

operation	Cray X–MP	Cray Y–MP	ETA 10–E	C 210	FX/80
saxpy	185	219	378	16	20
dot product	209	276	187	14	35
matrix vector	169	236	187	20	24

In the next section, we will see how to choose the preconditioning matrix for having both a good rate of convergence and an acceptable computational speed.

VECTOR PRECONDITIONERS.

Some good preconditioners have been proposed during the last years, the most popular one being the Incomplete Choleski (IC) decomposition, Meijerink & Van der Vorst (1977). However, it has been shown that for structured problems like the ones we are looking at, the block Incomplete Choleski decomposition (INV) and its variations (INV2), Concus, Golub & Meurant (1985), Concus & Meurant (1986), are more efficient. Unfortunately, these preconditioners contain either solves for triangular (IC), tridiagonal (INV) or pentadiagonal (INV2) systems. These operations are not directly defined in vector mode. Some special methods (like the cyclic reduction algorithm) can be used to solve this problem.

Another approach is to modify the preconditioner in order to regain the vector speed. The problem is to know if, modifying the preconditioner, we still get a good convergence rate. This approach has been taken by Van der Vorst (1982) and Meurant (1984). Van der Vorst (1982, 1986a, 1986b) proposed to use truncated Neumann series to approximately solve the bidiagonal systems that arise in forward and backward solves for IC. We will denote this method by VDV. In Meurant (1984), we propose to use a banded approximation to the inverses of the tridiagonal matrices that appear in the block forward and backward solves in the INV method. We denote this modified method by INVV.

TABLE 2. Percentages of basic operations

operation	no precond	IC	INV	INV2	VDV	INVV
saxpy	32.	3.3	2.5	1.8	11.9	10.4
dot product	19.9	2.1	1.6	1.2	7.8	6.8
matrix vector	47.9	4.9	3.7	2.8	18.	15.8
stop test	0.	0.9	0.7	0.5	3.2	3.1
precond	0.	88.6	91.4	93.5	58.6	63.8

TABLE 3. Results for the Cray X–MP

precond	no. of it.	time (s)	Mflops
DIAG	279	1.14	177
IC	91	3.43	35
INV	40	1.17	40
INV2	30	1.68	26
VDV	87	0.8	144
INVV	43	0.45	169

TABLE 4. Results for the Cray Y–MP

precond	no. of it.	time (s)	Mflops
DIAG	279	0.87	229
IC	91	2.7	45
INV	40	0.87	54
INV2	30	1.06	41
VDV	87	0.62	188
INVV	43	0.36	222

These methods are efficient as they allow to get vector speed as well as to have a good rate of convergence for most problems. For a problem of order 36100 (i.e. a 190 × 190 mesh), Table 2 shows the percentages of basic operations for the different preconditioners with a stopping test based on the l–2 norm of the residual on a Cray X–MP.

We see that in the fancy scalar preconditioners like VDV or INV, we spend most of the time in the preconditioning part. This illustrates their poor performance on vector computers.

Table 3 gives the number of iterations, the computing time and the Mflops rates achieved during the iterations with these methods on a Cray X–MP with the CFT77 2.0 compiler, using a stopping criterion based on the generalized residual z^k. DIAG is a diagonal preconditioner.

Table 4 gives the same information for the Cray Y–MP (a 6.4 ns clock model) with the CFT77 3.0 compiler.

There are some minor discrepancies between the times and Mflops results because the Mflops results have been rounded to integers.

These results show that it is possible to have similar numbers of iterations for efficient preconditioners with a high Mflops rate as for more scalar classical methods. As we noted before, most of the computing time is spent in solving $Mz^k = r^k$, so it does not pay to try to improve too much the speed of the other parts : saxpys, sdots, etc. . .

We remark that on this particular problem, CG without preconditioning can be competitive with more elaborate approaches. This is even more noticeable on computers which are faster on saxpy operations as the ETA (linked triads) or some Japanese machines. This explains why some people were claiming that CG with no preconditioning is the good way to solve problems on vector supercomputers. However, there are more difficult problems where the difference in the number of iterations is larger and already on that easy problem the block method is faster.

It is interesting to compare the speed of different supercomputers on CG algorithms with different preconditioners. This is done on Figure 1 where we give the Mflops rates for some methods and supercomputers : Cray X–MP, Cray Y–MP, ETA 10–E, Convex C 210 and Alliant FX/80 (8 processors). Except for the Alliant, all the results are for one processor and a problem size of 10 000. The z^k norm was used in these computations. The preconditioners are ordered from left to right from the most vectorizable to the most scalar one. When a performance is 0, it means that the measurements were not available.

We can remark that the relative merits of the considered supercomputers are different on different methods; for instance, the ETA machine was performing well on vectorized preconditioners, but poorly on more scalar methods. This confirms the fact, noted previously (Meurant (1986)) that the "best algorithm" is a machine dependant concept.

It seems difficult to get better speeds and better computational times with the "standard" version of CG. One way to improve the Mflops rates will be to reduce memory traffic doing "several iterations at a time". However, we have seen that most of the time is spent is the preconditioning part of the algorithm and there is little hope for improvement on that side.

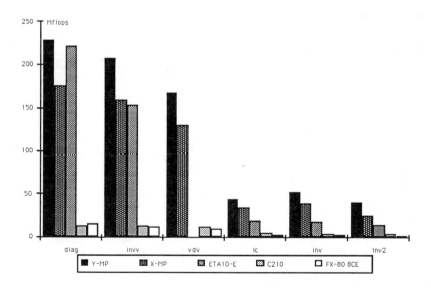

FIGURE 1. Mflops rates for different computers and preconditioners

Another way to vectorize the computation is to use different orderings of the mesh points (i.e. the unknowns), see Duff & Meurant (1988). This method is also used, as we will see in the next section, to introduce parallelism in CG. Unfortunately, the most "vectorizable orderings" usually exhibit a slow rate of convergence.

PARALLEL PRECONDITIONERS.

There are two things we have to be concerned about using CG on parallel computers. The first one is how to remove the data dependancies that prevent the efficient use of basic CG on parallel machines. The second one is to derive preconditioners which are well suited for parallel computations.

Regarding the first issue, it is obvious that if CG is well suited to vector computation (i.e. SIMD parallel computations, which means that the same operation is done on different sets of data), it is not at all the same for MIMD (different operations on possibly different data) parallel computations. CG is a very sequential algorithm; almost every step as to be completed before the next step can begin. Although some kind of pipelining can be introduced, the only operations that can be done in parallel are the computations of x^k and r^k and possibly of Ap^k.

However, some of the data dependancies can be removed, noticing that (in exact arithmetic) we have $(Mz^i, z^j) = 0, i \neq j$. Then, it follows $(r^{k+1}, z^{k+1}) = \alpha_k^2 (M^{-1}Ap^k, Ap^k) - (r^k, z^k)$.

The modified algorithm is :

$$x^0 \text{ given}, r^0 = b - Ax^0, Mz^0 = r^0, p^0 = z^0.$$

For each k until convergence,

$$Mv^k = Ap^k,$$

$$(v^k, Ap^k), \qquad (p^k, Ap^k), \qquad (r^k, z^k)$$

$$\alpha_k = \frac{(r^k, z^k)}{(p^k, Ap^k)}$$

$$s_{k+1} = \alpha_k^2 (v^k, Ap^k) - (r^k, z^k)$$

$$\beta_{k+1} = \frac{s_{k+1}}{(r^k, z^k)}$$

$$x^{k+1} = x^k + \alpha_k p^k$$

$$r^{k+1} = r^k - \alpha_k Ap^k$$

$$z^{k+1} = z^k - \alpha_k v^k$$

$$p^{k+1} = (z^k - \alpha_k v^k) + \beta_{k+1} p^k$$

In this algorithm we use a predictor–corrector for the dot product (r^{k+1}, z^{k+1}). First, we predict a value to be able to compute β_{k+1}; then, we correct the value after having computed the new vectors, to improve the stability of the algorithm.

With this new form of the CG algorithm, we can compute in parallel the 3 dot products and after a small scalar section, all the needed vectors. Another way to improve the degree of parallelism will be to use the three–term recurrence form of CG, see Concus, Golub & O'Leary (1976). However, as we have seen before, for preconditioned CG, the most important part is the solve of $Mz^k = r^k$. So, we must develop preconditioners well adapted to parallel computation. Of course, this is not a problem, for instance we can choose a diagonal preconditioner; the difficult task is to have both parallelism and efficiency.

As the standard Incomplete Cholesky decompositions are not parallel because of the recursions, one way to introduce parallelism is to use different orderings of the unknowns. In Duff & Meurant (1988), different orderings were tried, not necessarily for the sake of parallel computations, but to understand the effect of the numbering scheme of the unknowns on the rate of convergence. However, some of the orderings we tried can lead to some amount of parallelism. For a complete description of these ordering schemes, we refer to Duff & Meurant (1988); they were :
• the well known red–black ordering, which lead to a large degree of parallelism
• the zebra ordering, which is a line red–black; the degree of parallelism is limited by the number of rows in the mesh,
• the one–way dissection ordering, the domain being divided in four regions, which makes it suitable for four processor computers,
• the nested dissection ordering,
• a four color ordering,
• a new ordering, due to H. Van Der Vorst, which consists of numbering the mesh points from the corners to the inside of the domain; the natural parallelism is for four processors.

For all these orderings, we use an Incomplete Choleski decomposition with the same sparsity pattern as the permuted original matrix. Therefore, as the number of operations per iteration is the same for all the orderings, the number of iterations is a good measure of the efficiency.

The numerical results in Table 5 were obtained for a 30 × 30 mesh, the norm for the stopping criterion was $\| \cdot \|_\infty$ and $\varepsilon = 10^{-6}$, the starting guess was a random vector.

From these results, we see that, unfortunately the most parallel orderings like red–black give the worst results. However, orderings with a limited amount of parallelism, like zebra (line red–black) give a larger number of iterations but they are not too far from the standard results; so, it does pay to use these orderings on parallel machines with a small number of processors. With a very large degree of parallelism, no preconditioner or red–black orderings may be more efficient.

A more systematic way to introduce parallelism in the preconditioner is to use Domain Decomposition ideas. The domain Ω is divided into subdomains, for instance strips, see Figure 2; within each subdomain Ω_i, we will be able to use whatever preconditioner we want. The main problem is to define the preconditioner we use on the interfaces between subdomains.

Many solutions have recently been devised to this problem. There has been many different algorithms developed in the last years both for finite difference and finite element methods. A few of these are given in the following, Björstad & Widlund (1986), Bramble, Pasciak & Schatz (1986), Chan & Resasco (1985), Dryja, Proskurowski and Widlund (1987), Golub & Mayers (1984), P.L. Lions (1988), Meurant (1988).

TABLE 5. Results for different orderings

ordering	no. of iterations
row	23
red–black	38
zebra	28
one–way diss	24
nested diss	25
4 colors	33
van der vorst	20

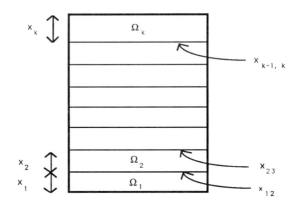

FIGURE 2. Domain Decomposition into stripes

A first conference dedicated to DD methods was organized in Paris in 1987 (Glowinski, Golub, Meurant, Periaux (1988)) and a second one has been held in Los Angeles in 1988 (Chan, Glowinski, Periaux, Widlund (1988)). The two methods we describe below have been proposed in Meurant (1987, 1988a, 1988b).

The first one is the following : we divide the domain into stripes $\Omega_i, i = 1, \ldots, k$; we renumber the unknowns in such a way that the components of x related to the subdomains appear first and then the ones for the interfaces, using the row–wise ordering within each subdomain (whose unknowns are denoted x_l) and interface (whose unknowns are denoted $x_{l,l+1}$). Suppose for the sake of simplicity what we take two subdomains.

After reordering the system to be solved is:

$$Ax = \begin{pmatrix} B_1 & 0 & C \\ 0 & B_2 & E \\ C^T & E^T & B_{1,2} \end{pmatrix} \begin{pmatrix} x_1 \\ x_2 \\ x_{1,2} \end{pmatrix} = \begin{pmatrix} b_1 \\ b_2 \\ b_{1,2} \end{pmatrix}.$$

Each B_i is related to a subdomain Ω_i,

$$B_i = \begin{pmatrix} D_i^1 & (A_i^2)^T & & & \\ A_i^2 & D_i^2 & (A_i^3)^T & & \\ & \ddots & \ddots & \ddots & \\ & & A_i^{m_i-1} & D_i^{m_i-1} & (A_i^{m_i})^T \\ & & & A_i^{m_i} & D_i^{m_i} \end{pmatrix}, \quad i = 1, 2.$$

D_i^j and $B_{i,j}$ are tridiagonal matrices and m_i is the number of mesh lines in Ω_i.

The matrices C and E have a special structure, which we will take advantage of;

$$C = \begin{pmatrix} 0 \\ \vdots \\ 0 \\ C^{m_1} \end{pmatrix}, \qquad E = \begin{pmatrix} E^1 \\ 0 \\ \vdots \\ 0 \end{pmatrix}.$$

This structure comes from the fact that, for the 5 point scheme, an interface is only related to the line above (C^T) and to the line below (E^T). C^{m_1} and E^1 are diagonal matrices.

As a preconditioner we take

$$M = L \begin{pmatrix} M_1^{-1} & 0 & 0 \\ 0 & M_2^{-1} & 0 \\ 0 & 0 & M_{12}^{-1} \end{pmatrix} L^T, \quad \text{with} \quad L = \begin{pmatrix} M_1 & 0 & 0 \\ 0 & M_2 & 0 \\ C^T & E^T & M_{12} \end{pmatrix}.$$

Then

$$M = \begin{pmatrix} M_1 & 0 & C \\ 0 & M_2 & E \\ C^T & E^T & M_{12}^* \end{pmatrix},$$

with,

$$M_{12}^* = M_{12} + C^T M_1^{-1} C + E^T M_2^{-1} E.$$

In order that M be a "good" conjugate gradient preconditioner, we should have $M^{-1}A$ as close as possible to the identity matrix in the sense that the eigenvalues should be close to 1 or the eigenvalues of $M^{-1}A$ occur in clusters;

It is natural to take M_i as an approximation to B_i and M_{12} as an approximation to $B_{12} - C^T M_1^{-1} C - E^T M_2^{-1} E$.

Some methods (Bramble, Pasciak & Schatz (1986), Golub & Mayers (1984)) have used fast solvers for the subdomains ($M_i = B_i$), when the problem allows such a use i.e. regular domains and separable problems. Then, it is more natural to iterate only for the interface unknowns.

Here, we are interested in finding preconditioners for cases where the subproblems can be as complex as the original one, even with strongly discontinuous coefficients. We do not wish to be concerned with the location of discontinuities. So, a good choice for M_i is a block preconditioner INVV, see Concus, Golub & Meurant (1985).

In this particular case we can take for M_1 a block LU approximation :

$$M_1 = (\Delta_1 + L_1) \, \Delta_1^{-1} \, (\Delta_1 + L_1^T),$$

L_1 is the block lower triangular part of B_1, Δ_1 is a block diagonal matrix whose blocks are given by,

$$\begin{cases} \Delta_1^1 = D_1^1, \\ \Delta_1^j = D_1^j - A_1^j \, trid((\Delta_1^{j-1})^{-1}) \, (A_1^j)^T, \quad j = 2, \ldots, m_1. \end{cases}$$

We can do the same for M_2 but, because we need to compute $M_2^{-1} E$ and E has a special sparse block structure, it is more interesting to take for M_2 a block UL approximation :

$$M_2 = (\Sigma_2 + L_2^T) \, \Sigma_2^{-1} \, (\Sigma_2 + L_2),$$

L_2 is the block lower triangular part of B_2, Σ_2 is a block diagonal matrix,

$$\begin{cases} \Sigma_2^{m_2} = D_2^{m_2}, \\ \Sigma_2^j = D_2^j - (A_2^{j+1})^T \, trid((\Sigma_2^{j+1})^{-1}) \, A_2^{j+1}, \quad j = m_2 - 1, \ldots, 1. \end{cases}$$

Then, we define M_{12} as an approximation to the Schur complement,

$$M_{12} = B_{1,2} - (C^{m_1})^T \, trid((\Delta_1^{m_1})^{-1}) \, C^{m_1} - (E^1)^T \, trid((\Sigma_2^1)^{-1}) \, E^1.$$

With suitable hypothesis and the same techniques as in Concus, Golub & Meurant (1985) , it is possible to show that the preconditioner M just constructed is positive definite.

The preconditioner M is used with the conjugate gradient method. It is fairly easy to generalize the method to more than 2 subdomains; we denote it by INVDD. At each iteration of the Conjugate Gradient we have to solve a linear system with matrix M; we see that we can solve for the subdomains in parallel. There is a potential problem for the interface solve when there are more than two subdomains, but this can be solved allowing a parallel solve for the interfaces; see Meurant (1988a, 1988b) for details.

In the following, we present a method called the "incomplete twisted factorization", which can be seen as a variation of INVDD, see Meurant (1987, 1988a). It is easier to explain this factorization with an example. Suppose we have 4 subdomains. Then, as a preconditioner we take

$$M = \Theta \; \Theta^T,$$

where Θ has the following structure

$$\begin{pmatrix} M_1 & N_2^T & & \\ & M_2 & & \\ & N_3 & M_3 & N_4^T \\ & & & M_4 \end{pmatrix},$$

where M_1, M_3 are block lower bidiagonal, M_2, M_4 are block upper bidiagonal, N_2^T, N_4^T have only a non zero block in the lower left corner, N_3 has only a non zero block in the upper right corner. This means that Θ has alternatively a lower block diagonal, an upper one, a lower one and finally an upper one as indicated in Figure 3.

The non diagonal blocks are equal to the corresponding ones in A. The diagonal blocks can be determined begining by the top, the bottom and the "middle" elements.

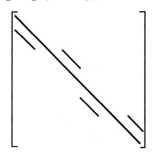

FIGURE 3. Structure of the twisted factorization factor

The formulas are similar to the ones for the Domain Decomposition preconditioner; details can be found in Meurant (1988a).

The following figures present the numerical results we obtained on a Cray Y–MP/832 using the eight processors in parallel. The Cray Autotasking feature was used, which means that most of the loop level parallelism was automatically discovered by the compiler. The only manual intervention was putting four directives to tell the compiler that loops containing calls to subroutines can be safely parallelized.

We use a random starting vector, the z^k norm with $\varepsilon = 10^{-6}$. We can see that, even with small problems, we get very good performance with this algorithm; the Autotasking overhead and the memory contention seem small in this example.

Figure 4 shows the number of iterations for the two methods as a function of the number of subdomains, that is the number of parallel processors we can naturally use in the preconditioning part of CG. Figure 5 gives the Mflops rate we can achieve with INVDD as a function of the problem size (the number of discretization points in one direction) for eight processors; the same information is given for INVkP on figure 6. Figure 7 shows the speed–up defined as the ratio of execution time on one processor to the time for eight processors.

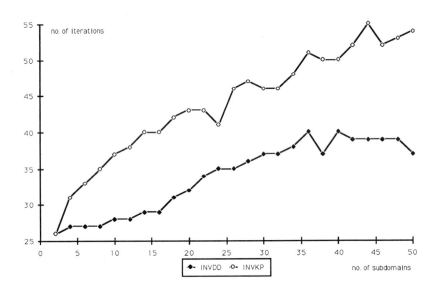

FIGURE 4. Number of iterations

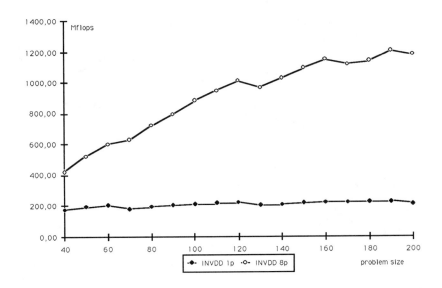

FIGURE 5. Mflops for INVDD

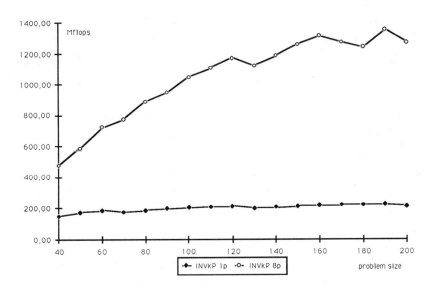

FIGURE 6. Mflops for INVkP

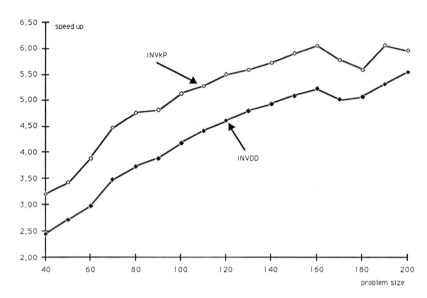

FIGURE 7. Speed–up for INVDD and INVkP

CONCLUSIONS.

We have shown some techniques for the efficient use of preconditioned CG on parallel supercomputers. The main conclusion is that, even on vector computers, some elaborate preconditioners can be much faster than diagonal preconditioners.

On parallel supercomputers, two promising approaches are the Incomplete Choleski decomposition using some reordering schemes for the unknowns and Domain Decomposition techniques that allow to break down the problem into pieces that are given to each processor. Although these methods imply some increase in the number of iterations, the benefit in the computational speed we can reach is, by far, more important. So it is clearly beneficial to employ these techniques for the practical use of CG on parallel vector computers. For massively parallel computers, finding a good preconditioner is still an open problem, but some Domain Decomposition techniques seem promising.

REFERENCES

1. P. Björstad & O.B. Widlund, Iterative methods for the solution of elliptic problems on regions partitioned into substructures. *SIAM J. on Numer. Anal.* v 23, n 6, pp.1097–1120, 1986.

2. J.H. Bramble, J.E. Pasciak & A.H. Schatz, The construction of preconditioners for elliptic problems by substructuring. I. *Math. of Comp.* v 47, n 175, pp.103–104, 1986.

3. T. Chan & D. Resasco, A domain decomposition fast Poisson solver on a rectangle. Yale Univ. report YALEU/DCS/RR 409, 1985.

4. T. Chan, R. Glowinski, J. Periaux & O. Widlund, *Proceedings of the second international symposium on domain decomposition methods for partial differential equations.* SIAM, 1988.

5. P. Concus, G.H. Golub & G. Meurant, Block preconditioning for the conjugate gradient method. *SIAM J. Sci. Stat. Comp.*, v 6, pp.220–252, 1985.

6. P. Concus & G. Meurant, On computing INV block preconditionings for the conjugate gradient method. *BIT* v 26, pp.493–504,1986.

7. P. Concus, G.H. Golub & D.P. O'Leary, A generalized conjugate gradient method for the numerical solution of elliptic partial differential equations. in *Sparse matrix computations*, J.R. Bunch & D. Rose Eds, Academic Press, 1976.

8. M. Dryja, W. Proskurowski & O. Widlund, Numerical experiments and implementation of domain decomposition method with cross points for the model problem. in *Advances in computer methods for PDEs-VI*, IMACS pp.23–27,1987.

9. I.S. Duff & G. Meurant, The effect of ordering on preconditioned conjugate gradients. Harwell Laboratory CSS division, submitted to BIT, 1988.

10. R. Glowinski, G.H. Golub, G. Meurant & J. Periaux, *Proceedings of the first international symposium on domain decomposition methods for partial differential equations.* SIAM, 1988.

11. G.H. Golub & D. Mayers, The use of preconditioning over irregular regions. In *Computing methods in applied science and engineering VI*, R. Glowinski & J.L. Lions Eds, North–Holland, 1984.

12. M.R. Hestenes & E. Stiefel, Methods of conjugate gradients for solving linear systems. *J. Res. Nat. Bur. Stand.* v 49, pp.409-436,1952.

13. P.L. Lions, On the Schwarz alternating method I. in ref. 10. pp.1–42,1988.

14 J.A. Meijerink & H.A. Van Der Vorst, An iterative solution method for linear systems of which the coefficient matrix is a symmetric M–matrix. *Math. Comp.* v 31, pp.148–162,1977.

15. G. Meurant, The block preconditioned conjugate gradient method on vector computers. *BIT* v 24, pp.623–633,1984.

16. G. Meurant, The conjugate gradient method on supercomputers. *Supercomputer* v 13, pp.9–17,1986.

17. G. Meurant, Multitasking the conjugate gradient method on the CRAY X–MP/48. *Parallel Computing* v 5, pp.267–280,1987.

18. G. Meurant, Domain decomposition vs block preconditioning. In ref. 10.

19. G. Meurant, Conjugate gradient preconditioners for parallel computers In *Proceedings of the third SIAM conference on parallel processing for scientific computing,* Los Angeles 1987, SIAM, 1988.

20. G. Meurant, Domain Decomposition Preconditioners for the Conjugate Gradient Method. to appear in *Calcolo,* 1988.

21. J.K. Reid, On the method of conjugate gradients for the solution of large sparse systems of linear equations. In *Large sparse sets of linear equations,* J.K. Reid ed., Academic Press, 1971.

22. H.A. Van Der Vorst, A vectorizable variant of some ICCG methods. *SIAM J. Sci. Stat. Comput.* v 3, pp. 86–92, 1982.

23. H.A. Van Der Vorst, The performance of Fortran implementations of preconditioned conjugate gradients on vector computers. *Parallel Computing* v 3 pp.49–58, 1986.

24. H.A. Van Der Vorst, (M)ICCG for 2D problems on vector supercomputers. Report A–17, Data Processing Center, Kyoto University, Japan, 1986 .

Nonlinear Integer Programming with Applications in Manufacturing and Process Engineering

B. A. MURTAGH
Department of Industrial Engineering
The University of New South Wales
Kensington, New South Wales 2033, Australia

INTRODUCTION

Nonlinear integer programming has many application areas in industry and commerce. Although algorithms for integer linear programming are well established there remains a considerable amount of experimentation in addressing the nonlinear case.

Balas and Mazzola (1984) proposed a linearization approach involving the replacement of general nonlinear functions of binary (0,1) variables appearing in inequality constraints with a family of equivalent linear inequalities. Duran and Grossmann (1986) proposed an outer approximation algorithm, involving an integer linear programming master problem. and an inner nonlinear programming subproblem. The subproblem is solved with the integer variables held fixed, and the master problem is formed by linearizing the functions at the solution of the subproblem. Several other approaches have been described in the literature, and a good review of earlier work is given by Cooper (1981).

This paper covers two aspects of nonlinear integer programming. Firstly it describes the development of a direct search method for attaining an integer-feasible solution from an initially relaxed (continuous) solution. Also it describes applications to specific examples in manufacturing and process engineering.

The next section describes the structure of the problem being addressed, and the third section describes the branch-and-bound strategy for searching partially relaxed solutions. This strategy may be a useful adjunct to the direct search method described in the following section. The subsequent two sections describe computational experience in two application areas.

STRUCTURE OF THE PROBLEM

It is assumed that the problem can be described in the following form, and that a bounded feasible solution exists:

$$\text{minimize} \qquad f^\circ(\underline{x}) + \underline{c}^T \underline{x}$$

$$\text{subject to} \qquad \underline{f}(\underline{x}) + A_1 \underline{x} = \underline{b}_1 \qquad\qquad (1)$$

$$A_2(\underline{x}) + A_3 \underline{x} = \underline{b}_2$$

$$\underline{\ell} \leq \underline{x} \leq \underline{u}$$

$$x_j \text{ integer}, \quad j \in J$$

There are n variables and m constraints, m < n.

Some (assumed small) proportion of the variables $\underline{x}$ are assumed to be nonlinear in either the objective function and/or the constraints, and some (also assumed small) proportion of the variables are required to be integer-valued.

The same structure without the integer requirements forms the basis of the MINOS large-scale nonlinear programming code (Murtagh and Saunders (1982, 1987)). This involves a sequence of "major iterations", in which the first-order Taylors series approximation terms replace the nonlinear constraint functions to form a set of linear constraints, and the higher order terms are adjoined to the objective function with Lagrange multiplier estimates. The set of linear constraints (excluding bounds) are then written in the form:

$$A\underline{x} = \lfloor B:S:N \rfloor \begin{bmatrix} \underline{x}_B \\ \underline{x}_S \\ \underline{x}_N \end{bmatrix} = \underline{b} \qquad\qquad (2)$$

B is $m \times m$ and non-singular, $\underline{x}_N$ are "non-basic" variables which are held at one or other of their bounds. $\underline{x}_B$ and $\underline{x}_S$ are referred to as "basic" and "superbasic" variables respectively, and in order to maintain feasibility during the next step $\Delta\underline{x}$ they must satisfy the equation

$$B\Delta\underline{x}_B + S\Delta\underline{x}_S = 0 \qquad\qquad (3)$$

Thus the superbasics are seen as the driving force, since the step $\Delta\underline{x}_S$ determines the whole step $\Delta\underline{x}$. The key to the success of the algorithm in MINOS is the assumption that the dimension of $\underline{x}_S$ remains small. This can be assured if the proportion of nonlinear variables is small, but it turns out to be true in many instances in practice even when all the variables are nonlinear.

Similar assumptions will be made about the structure of nonlinear integer programs. It will be assumed that the proportion of integer variables in the problem is small. The concept of superbasic variables is used in presenting a direct search approach in the section after next.

THE BRANCH-AND-BOUND METHOD

Nearly all commercial linear programming codes use the branch-and-bound procedure for solving integer linear programming problems. It is

reasonable to extend the approach to nonlinear functions and this can
be done in a straightforward manner. This section will give a brief
description of the branch-and-bound method as it can be used in
conjunction with the direct search approach described in the next
section.

The problem (1) is solved as a continuous nonlinear program, ignoring
the integrality requirements. Suppose the solution x_j, $j \in J$ is not
all integer. We set

$$x_j = \lfloor x_j \rfloor + f_j, \qquad 0 \leq f_j < 1 \tag{4}$$

where $\lfloor x_j \rfloor$ is the integer component of x_j.

The approach is to generate two new subproblems, with additional bounds,
respectively

$$\ell_j \leq x_j \leq \lfloor x_j \rfloor \tag{5}$$

and

$$\lfloor x_j \rfloor + 1 \leq x_j \leq u_j \tag{6}$$

for a particular variable $j \in J$. This process of splitting the problem
is called "branching". One of these new subproblems is now stored in
a master list of problems remaining to be solved, and the other solved
as a continuous problem. The process of branching and solving a sequence
of continuous problems is repeated for different integer variables,
$j \in J$, and different integers $\lfloor x_j \rfloor$. The logical structure of the method
is often represented as a tree. Each node of the tree represents a
subproblem solution. The branching of each node will terminate if one
of the following three criteria are satisfied:

1. The subproblem has no feasible solution.

2. The solution of the subproblem is no better than the current best
    known integer feasible solution.

3. The solution is integer feasible (to within a pre-defined level
    of tolerance).

The "bounds" on the best possible integer solution are provided by
the current best known integer feasible solution (which provides an
upper bound), and the best of the remaining partially integer solutions
on the master list of problems to be solved (which provides a lower
bound). It is usual to terminate the branch-and-bound procedure when
the difference between these two bounds is within some pre-defined
relative tolerance.

The rate of convergence of the procedure is sensitive to the choice
of variable, $j \in J$, on which to branch, and also the choice of node
to backtrack to once the branching from a particular node is discon-
tinued.

For nonlinear integer programs it must be implicitly assumed that the
problem is locally convex, at least in the neighbourhood of the original
continuous solution which contains integer feasible solutions. Otherwise, the bounds discussed above are inadequate. It would not be valid

to terminate the branching process under criterion 2 above, and it also would not be valid to terminate the procedure when the difference between the two bounds is sufficiently small.

A SEARCH PROCEDURE USING SUPERBASIC VARIABLES

This section will describe recent research in the development of a search procedure using the concept of superbasic variables. The work is a development of ideas initially presented by Mawengkang and Murtagh (1986), where in particular the quadratic assignment problem was discussed.

The approach assumes that the continuous problem is solved, and seeks an integer-feasible solution in the close neighbourhood of the continuous solution. The general philosophy is to leave non-basic integer variables at their bound (and therefore integer valued) and conduct a search in the restricted space of basics, superbasics, and nonbasic continuous variables, $j \notin J$.

Suppose at the continuous solution an integer variable is basic at a non-integer value

$$(\underline{x}_B)_{i'} = \lfloor x_{Bi'} \rfloor + f_{i'}, \quad 0 < f_{i'} < 1 \tag{7}$$

Suppose also that a chosen non-basic non-integer variable $(x_N)_{j*}$ is being released from its lower bound. Four possible outcomes may occur.

1. A basic variable $i \neq i'$ hits its lower bound first

2. A basic variable $i \neq i'$ hits its upper bound first

3. The basic variable $(x_B)_{i'}$ becomes an integer

4. The non-basic $(x_N)_{j*}$ hits its upper bound first

Corresponding to these possibilities we should compute:

$$\theta_1 = \min_{i \neq i' \mid \alpha_{ij*} > 0} \left\{ \frac{(\underline{x}_B)_i - \ell_i}{\alpha_{ij*}} \right\} \tag{8}$$

$$\theta_2 = \min_{i \neq i' \mid \alpha_{ij*} < 0} \left\{ \frac{u_i - (\underline{x}_B)_i}{-\alpha_{ij*}} \right\} \tag{9}$$

$$\theta_3 = \frac{1 - f_{i'}}{-\alpha_{i'j*}}, \quad \text{if } \alpha_{i'j*} < 0$$

$$= \frac{f_{i'}}{\alpha_{i'j*}}, \quad \text{if } \alpha_{i'j*} > 0 \tag{10}$$

$$\theta_4 = u_{j*} - \ell_{j*} \tag{11}$$

where

$$\alpha_{ij} = (B^{-1}\underline{a}_j)_i, \text{ and } \underline{a}_j \text{ is the column of A corresponding to the non-basic } (x_N)_j.$$

Therefore we have

$$\theta^* = \min\{\theta_1, \theta_2, \theta_3, \theta_4\} \tag{12}$$

If $\theta^* = \theta_1$ the basic variable $(x_B)_i$ becomes non-basic at ℓ_i and $(x_N)_{j*}$ replaces it in B. $(x_B)_{i'}$ stays basic with a new value (non-integer). If $\theta^* = \theta_2 (x_B)_i$ becomes non-basic at U_i and $(x_N)_{j*}$ replaces it in B as above. If $\theta^* = \theta_3$ then $(x_B)_{i'}$ is made superbasic at an integer value and $(x_N)_{j*}$ replaces it in B. If $\theta^* = \theta_4$ then $(x_N)_{j*}$ remains non-basic, but now at its upper bound, and $(\underline{x}_B)_{i'}$ stays basic with a new value (non-integer).

Similar ratios can be calculated for the case of $(x_N)_{j*}$ being released from its upper bound. Eventually all the integer variables will become non-basic or superbasic.

The superbasics can be varied at will, subject to preserving the feasibility of the basic variables. Thus a search through the neighbourhood system, as defined by Scarf (1986), will verify the (local) optimality of the integer-feasible solution obtained.

It will be evident that there is ample scope for numerical experimentation in implementing the approach. The assumption that the proportion of integer variables is small becomes a key issue in ensuring that the interchange operations can take place; fortunately many practical problems have this characteristic. (Note also that it is assumed there is a full set of slack variables present.)

Experimentation conducted thus far suggests the preferred choice of i' is given by

$$\min\{f_{i'}, 1-f_{i'}\} \leq \min\{f_i, (1-f_i)\} \tag{13}$$

among the integer variables. Also, in choosing the non-basic (continuous) j^* experimentation suggests the preferred criterion is

$$\min_j \left\{ \left| \frac{d_j}{\alpha_{i'j}} \right| \right\} \quad j=1\ldots n-m-s$$

where d_j is the jth component of the reduced gradient vector. The reasoning behind this criterion is that it measures the deterioration of the objective function value per unit change in the basic variable $(x_B)_{i'}$.

The terms $\alpha_{i'j}$ are calculated by firstly producing a vector $\underline{z}^T = \underline{e}_{i'}^T B^{-1}$, and then calculating the inner product $\alpha_{i'j} = \underline{z}^T \underline{a}_j$. Once a particular j* is chosen, the full vector $\underline{\alpha}_{j*} = B^{-1}a_{j*}$ is calculated for the ratio tests, equations (8)-(10).

Since the procedure determines a local optimal solution in the neighbourhood of the original continuous solution, there may be some merit in seeking the assurance of a branch-and-bound procedure for fathoming all possible integer solutions. There would be little cost in this, as the solution obtained by the above procedure should provide a tight bound which will serve to curtail the branching process very rapidly under criterion 2 in the previous section.

APPLICATION TO FLEXIBLE MANUFACTURING SYSTEMS

A flexible manufacturing system is an automated multi-product manufacturing system consisting of numerically controlled machines linked with automated tool-change handling devices. Machine operations, part movements and tool interchanges are all computer controlled.

The loading problem is to assign the operations of selected jobs to the appropriate machines, subject to technological and capacity constraints. A model described by Shanker and Tzen (1985), consists of the following decision variables:

x_{ij}^k = $\begin{cases} 1 \text{ if operation k of job i is assigned on machine j} \\ 0 \text{ otherwise} \end{cases}$

x_i = $\begin{cases} 1 \text{ if job i is selected} \\ 0 \text{ otherwise} \end{cases}$

$0_j, U_j$ = overload and underload, respectively, on machine j.

for j = 1...N, the number of machines
          i = 1...M, the number of jobs
          k = 1...K_i, the number of operations in job i

Data required to define the constraints is as follows:

p_{ij}^k = processing time of operation k of job i on machine j

s_{ij}^k = number of tool slots requried for operation k of job i on machine j

t_j = tool slot capacity of machine j

L = length of scheduling period

B_i^k = set of machines available for operation k of job i

$Z_{i_1,i_2,..i_p}^{k_{i_1},..k_{i_p}}$ = number of slot duplications when operation k_{i_1} of job i_1, operation k_{i_2} of job i_2,...operation k_{i_p} of job i_p, are processed on the same machine.

Constraints in the model can now be defined: Tool slot capacity:
the simple form would be:

$$\sum_{i=1}^{M1} \sum_{k=1}^{K_i} S_{ij}^k \, x_{ij}^k \leq t_j, \qquad j=1\ldots N.$$

However, it is unnecessary to assign any tool more than once to the
same machine; different operations may require the same tools and only
one tool can be used by a machine at a time. By using tool slot
duplications the constraint can be expressed in the form:

$$\sum_{i=1}^{M} \sum_{k=1}^{K_i} S_{ij}^k x_{ij}^k - \sum_{i_1=1}^{M-1} \sum_{i_2=i_1+1}^{M} \sum_{k_{i_1}=1}^{K_{i_1}} \sum_{k_{i_2}=1}^{K_{i_2}} z_{i_1,i_2}^{ki_1,ki_2} x_{i_1}^{ki_1} x_{i_2}^{ki_2} + \ldots$$

$$\ldots (-1)^{p+1} \sum_{i_1=1}^{M-p+1} \sum_{i_2=i_1+1}^{M-p+2} \ldots \sum_{i_p=i_{p-1}+1}^{M} \sum_{k_{i_1}=1}^{K_{i_1}} \sum_{k_{i_2}=1}^{K_{i_2}} \ldots \sum_{k_{i_p}=1}^{K_{i_p}} z_{i_1,i_2,\ldots ip}^{ki_1,ki_2,\ldots kip} x_{i_1}^{ki_1} x_{i_2}^{ki_2} \ldots x_{i_p}^{kip}$$

$$\leq t_j \qquad j=1\ldots N \tag{14}$$

If it is desired to keep the routing of each job unique in order to
balance the system workload, a constraint could be expressed in the
form

$$\sum_{j \in B_i} x_{ij}^k \leq 1 \qquad i=1\ldots M, \; k=1\ldots K_i \tag{15}$$

Also, as a job cannot be split, its operation assignment must equal
the total operation required

$$\sum_{k=1}^{K_i} \sum_{j=1}^{N} x_{ij}^k = x_i K_i, \quad i=1\ldots M \tag{16}$$

Finally, a time constraint appears in the form

$$\sum_{i=1}^{M} \sum_{k=1}^{K_i} p_{ij}^k x_{ij}^k + U_j - O_j = L, \qquad j=1\ldots N \tag{17}$$

An objective function can be expressed in terms of balancing the
workload over the N machines, and minimising the number of late jobs.
The function can be expressed as

$$\text{minimize } F = \sum_{j=1}^{N} w_{1j} O_j + \sum_{j=1}^{N} w_{2j} U_j - \sum_{i=1}^{M} \frac{w_{3i} x_i}{\max\{W, R_i - 2L\}}$$

where

w_{1j} = weighting applied to overload on machine j

w_{2j} = weighting applied to underload on machine j

w_{3i} = weighting applied to selection of job i. (If it has a
remaining time, R_i, of less than two periods at the time
of scheduling, W which is a paramater with suitably small
positive value, is invoked.)

A specific example cited by Shanker and Tzen (1985) consists of four
machines with eight jobs to be loaded. This gives rise to a problem
with 33 continuous variables, 40 binary variables, and 25 constraints.

At the continuous solution 30 of the binary variables were already integer values, and the procedure described in the previous section applied successfully to the other 10, converging in 5.4 records CPU time on a VAX 11/780.

APPLICATIONS IN PROCESS ENGINEERING

Many of the applications of nonlinear integer programming in process engineering are in the area of heat exchanger network analysis and process flowsheet synthesis. In both these applications a "superstructure" allowing all possible configurations is postulated, and the integer programming problem is one of determining which parts of the configuration should be included and which parts excluded.

For this type of problem the integer variables appear linearly in the objective function and constraints, even though there are nonlinearities involving the continuous variables. Duran and Grossmann (1986) and Kocis and Grossmann (1986) were able to exploit this structure applying their outer approximation algorithm to these problems.

One example cited by Duran and Grossmann (1986) is one of simultaneously determining the optimal structure and operating parameters for a process to satisfy a given design specification. There are 8 binary variables, each associated with a process unit and denoting its potential existence in the optimal configuration. There are also 9 bounded parameters such as flowrates of materials, and 23 inequalilty constraints. The procedure described in Section 3 converged to the optimal solution in 16 seconds on a VAX 11/780, compared to 26 seconds for the outer approximation algorithm. This example is described in detail in the next section.

Another example, cited by Kocis and Grossmann (1986) consists of a flowsheet synthesis problem in which it is required to select the best configuration among several alternations in order to produce produce C from chemicals A and B. The problem containts 4 binary variabies, 128 continuous variables, 111 equality constraints and 14 inequality constraints. The procedure converged to the optimal solution in 120 seconds on a VAX 11//80.

NUMERICAL EXAMPLE

This is an example of a synthesis problem faced routinely in process engineering design at the venture analysis phase of construction of a grass-roots complex. A superstructure representing all possible configurations of process units is defined, and the integer programming problem is to choose which units should exist. The decision variables are defined as follows:

y   is a binary variable which has a value of 1 if the unit appears in the optimal configuration.

x   is a continuous variable which represents a process design parameter (flowrate of materials)

The objective is to minimize annualized investment and operating costs in producing a specified range of products. The superstructure of the synthesis problem is shown in Fig. 1.

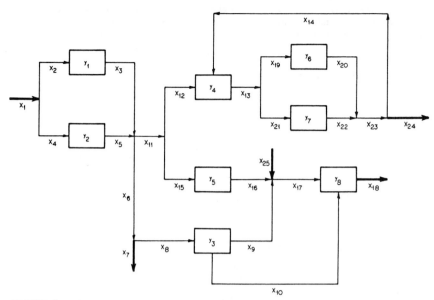

FIGURE 1. Superstructure of Process Synthesis Problem

The formulation is as follows

minimize
$$z = 5y_1 + 8y_1 + 6y_2 + 10y_4 + 6y_5 + 7y_6 + 4y_7 + 5y_8 - 10x_3 - 15x_5 + 15x_{10}$$
$$+ 80x_{17} + 25x_{19} + 35x_{21} - 40x_9 + 15x_{14} - 35x_{25} + \exp(x_3)$$
$$+ \exp(x_5/1.2) - 65\ln(x_{10} + x_{17} + 1) - 90\ln(x_{19} + 1)$$
$$- 80\ln(x_{21} + 1) + 120$$

subject to

$$-1.5\ln(x_{19} + 1) - \ln(x_{21} + 1) - x \leq 0$$
$$-\ln(x_{10} + x_{17} + 1) \leq 0$$
$$-x_3 - x_5 + x_{10} + 2x_{17} + 0.8x_{19} + 0.8x_{21} - 0.5x_9 - x_{14} - 2x_{25} \leq 0$$
$$-x_3 - x_5 + 2x_{17} + 0.8x_{19} + 0.8x_{21} - 2x_9 - x_{14} - 2x_{25} \leq 0$$
$$-2x_{17} - 0.8x_{19} - 0.8x_{21} + 2x_9 + x_{14} + 2x_{25} \leq 0$$
$$-0.8x_{19} - 0.8x_{21} + x_{14} \leq 0$$
$$-x_{17} + x_9 + x_{25} \leq 0$$
$$-0.4x_{19} - 0.4x_{21} + 1.5x_{14} \leq 0$$
$$0.16x_{19} + 0.16x_{21} - 1.2x_{14} \leq 0$$
$$x_{10} - 0.8x_{17} \leq 0$$
$$-x_{10} + 0.4x_{17} \leq 0$$
$$\exp(x_3) - Uy_1 \leq 1$$
$$\exp(x_5/1.2) - Uy_2 \leq 1$$
$$x_9 - Uy_3 \leq 0$$
$$0.8x_{19} + 0.8x_{21} - Uy_4 \leq 0$$
$$2x_{17} - 2x_9 - 2x_{25} - Uy_5 \leq 0$$
$$x_{19} - Uy_6 \leq 0$$
$$x_{21} - Uy_7 \leq 0$$
$$x_{10} + x_{17} - Uy_8 \leq 0$$
$$y_1 + y_2 = 1, \quad y_4 + y_5 \leq 1$$
$$-y_4 + y_6 + y_7 = 0$$
$$y_3 - y_8 \leq 0$$

$$y \in \{0,1\}^8, \quad a \leq x \leq b, \quad x = (x_j : j = 3,5,10,17,19,21,9,14,25) \quad \in R^9$$
$$a^T = (0,0,0,0,0,0,0,0,0), \quad b^T = (2,2,1,2,2,2,2,1,3), \quad U = 10$$

111

Only one binary variable, y_5, was nonbasic at the continuous solution; y_7 was superbasic and all the rest basic. The non-integer binary variables were integerized using the approach described. Both the optimal continuous and integer values of the variables are shown in Table 1.

TABLE 1. Results for Process Synthesis Problem

Variable	Solution	Solution
x_3	1.90293	0.0
x_5	2.0	2.0
x_9	0.65940	0.0
x_{10}	0.52752	0.46784
x_{14}	0.41111	0.26667
x_{17}	0.65940	0.58480
x_{19}	2.0	2.0
x_{21}	1.08333	0.0
x_{25}	0.0	0.58480
y_1	0.57055	0
y_2	0.42945	1
y_3	0.06594	0
y_4	0.30833	1
y_5	0.0	0
y_6	0.2	1
y_7	0.10833	0
y_8	0.11869	1
ν	15.08219	68.00974

CONCLUSIONS

Numerical experience thus far suggests the approach described in this paper shows considerable promise. Further experimentation is warranted, and current research is directed toward implementing a post-solution branch and bound procedure. Applications are not restricted to manufacturing and process engineering, but practical problems in these fields have the desirable property of having proportionally few integer variables.

REFERENCES

1. Balas, E. and Mazzola, J.B., "Nonlinear 0-1 Programming: I. Linearization Techniques, II. Dominance Relations and Algorithms *Mathematical Programming* Vol. 30, pp1-45, 1984.

2. Cooper, M.W., A Survey of Methods for Pure Nonlinear Integer Programming, *Management Science* Vol. 27, pp353-361, 1981.

3. Duran, M.A. and Grossmann, I.E., An Outer-Approximation Algorithm for a Class of Mixed-Integer Nonlinear Programs. *Mathematical Programming* Vol. 36, pp307-339, 1986.

4. Kocis, G.R. and Grossman, I.E., A Relaxation Strategy for the Structural Optimization of Process Synthesis, Annual AIChE Meeting, Miami, Nov. 1986.

5. Mawengkang, H. and Murtagh, B.A., Solving Nonlinear Integer Programs with Large-Scale Optimization Software, *Annals of Operations Research* Vol. 5 pp.425-437, 1986.

6. Murtagh, B.A. and Saunders, M.A., A Projected Lagrangian Algorithm and its Implementation for Sparse Nonlinear Constraints, *Mathematical Programming Study* Vol. 16, pp84-117, 1982.

7. Murtagh, B.A. and Saunders, M.A., MINOS 5.1 User's Guide, Report SOL 83-20R, Stanford University. Dec. 1983, Revised Jan. 1987.

8. Scarf, H.E., Neighbourhood Systems for Production Sets with Indivisibilities. *Econometrica*, Vol. 54, pp.507-532, 1986.

9. Shanker, K. and Tzen, Y.J., A Loading and Dispatching Problem in a Random Flexible Manufacturing System. *International Journal of Production Research* Vol. 23, pp.579-595, 1985.

Computational Methods in Petroleum Reservoir Simulation

M. D. STEVENSON, M. KAGAN, and W. V. PINCZEWSKI
Centre for Petroleum Engineering
University of New South Wales
Kensington, NSW 2033, Australia

INTRODUCTION

Petroleum fluids (oil and gas) are found in naturally occurring underground geological structures or reservoirs. The reservoir rock is porous with highly tortuous but continuous paths for fluids to flow. The pore space is occupied by hydrocarbons and formation water and the production characteristics of the reservoir are determined by a complex interaction of fluid flow and mass transfer processes. Up to three phases may flow simultaneously and mass transfer may take place between the phases. Viscous, gravity and capillary forces, phase compressibilities, and reservoir heterogeneity and geometry influence the fluid flow process. Mass transfer processes are affected by phase behaviour, phase compositions, and the nature of injected solvents and chemicals.

In order to provide realistic predictions of reservoir behaviour for different development schemes and production strategies, it is necessary to describe the important physical and chemical processes taking place in the reservoir and to correctly account for the complex interaction between the processes. This can only be done with mathematical modeling (reservoir simulation). The construction of a reservoir model requires the following steps: a description of the major physical and chemical mechanisms in the reservoir; a mathematical formulation of the conservation equations - resulting in coupled systems of highly non-linear partial differential equations; a discretized numerical model to approximate these equations; and finally, a computer code capable of efficiently solving the systems of algebraic equations generated by the numerical model.

The objective of the present paper is to present an overview of the basic physical phenomena of multi-phase flow in hydrocarbon reservoirs and to discuss some of the more important problems which arise in building reservoir simulation models. A major area of present concern is the application of high-order numerical methods to reduce the level of non-physical or numerical dispersion introduced by low-order methods currently used. Results are presented of computations which demonstrate the increase in the computational work necessary to implement these methods. For the reader interested in further detail, excellent comprehensive reviews of reservoir simulation may be found in Peaceman (1977), Aziz and Settari(1979), Ewing (1983) and Wheeler(1988).

MATHEMATICAL FORMULATION

Consider the case of three-phase immiscible flow of oil, gas and water in a porous media with no mass transfer between the phases. Since there are only three phases present,

$$S_o + S_g + S_w = 1 \tag{1}$$

where, S_o, S_g and S_w are the phase saturations (the fraction of pore space occupied by a particular phase), and the subscripts o, g and w refer to the water, oil and gas phases respectively.

Darcy's Law relates phase velocities to phase pressures,

$$v_i = -\frac{K k_{ri}}{\mu_i}(\nabla P_i - \rho_i g \nabla z) \qquad (i = o, w, g) \tag{2}$$

where v_i is the phase velocity, K is the intrinsic rock permeability, k_{ri} is the relative permeability for the phase, μ_i is the phase viscosity, P_i is the phase pressure, and $\rho_i g \nabla z$ is the gravity term.

The relative permeabilities k_{ri} are assumed to be monotone functions of phase saturation,

$$k_{rw} = k_{rw}(S_w)$$

$$k_{rg} = k_{rg}(S_g) \tag{3}$$

$$k_{ro} = k_{ro}(S_w, S_g)$$

k_{rw} and k_{rg} are empirical functions and must be determined experimentally for the water-oil and gas-oil systems respectively. The oil phase relative permeability k_{ro} is interpolated from the two-phase data using the Aziz and Settari (1979) modification of the model proposed by Stone (1973), written as,

$$k_{ro}(S_w, S_g) = \frac{(1 - S_w - S_g) k_{row} k_{row}}{(1 - S_w)(1 - S_g)} \tag{4}$$

The individual phase pressures in Eq.2 are related by the capillary pressure which is defined as,

$$P_{cgo}(S_g) = P_g - P_o \tag{5}$$

$$P_{cow}(S_w) = P_o - P_w$$

The capillary pressures P_{cgo} and P_{cow} are also assumed to be monotone functions of gas and water saturations respectively, and must be determined experimentally. Typical two-phase relative permeability and capillary pressure functions are shown in Figure 1 and Figure 2.

The mass conservation equations for each phase, with Eq.2 written for phase velocity, are

$$\frac{\partial(\phi \rho_i S_i)}{\partial t} = \nabla \cdot \frac{\rho_i K k_{ri}}{\mu_i}(\nabla P_i - \rho_i g \nabla z) + q_i \qquad (i = o, w, g) \tag{6}$$

where, q_i are the phase mass injection or production rates (source or sink terms) which are used to model wells. The system of Eq.6 are coupled by constraints (Eqs.1,3 and 5). The phase densities are also functions of phase pressures and are obtained from an equation of state. For a slightly compressible liquid it is convenient to write,

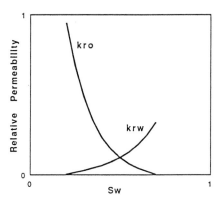

FIGURE 1: Typical relative permeability

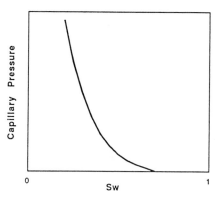

FIGURE 2: Typical capillary pressure

$$c_i = \frac{1}{\rho_i} \frac{\partial \rho_i}{\partial P_i}\bigg|_T \qquad (i = o, w) \tag{7}$$

where c_i is the phase compressibility at constant temperature, T.

Separating the variables in Eq.7 and denoting ρ_{ib} as the density of phase i at pressure P_{ib}, we obtain,

$$\rho_i = \rho_{ib} e^{c_i(P_i - P_{ib})} \qquad (i = o, w) \tag{8}$$

If the compressibility, c_i, is small (as it usually is for the oil and water phases), a further simplification is possible. A truncated Taylor series expansion for the exponential term in Eq.8 yields,

$$\rho_i = \rho_{ib} \left\{ 1 + c_i \left(P_i - P_{ib} \right) \right\} \qquad (i = o, w) \tag{9}$$

It is important to recognise the nature of the system of conservation equations (Eq.6). Consider the case of single-phase flow ($k_r = 1$, $S = 1$) of a slightly compressible liquid (ρ given by Eq.9) with no gravitational effects and constant ϕ, K and μ. It is easily shown that Eq.6, with the phase subscript suppressed for clarity, may be written as,

$$o\phi \frac{\partial P}{\partial t} = \frac{K}{\mu} \left(\nabla^2 \Gamma + c (\nabla P)^2 \right) + \frac{q}{\rho} \tag{10}$$

Since c is usually very small, $c(\nabla P)^2 \ll \nabla^2 P$ and the Eq.10 reduces to,

$$c\phi \frac{\partial P}{\partial t} = \frac{K}{\mu} \nabla^2 P + \frac{q}{\rho} \tag{11}$$

Eq.11 is clearly parabolic. For steady-state incompressible flow the equation is elliptic.

Since Eq.6 resembles Eq.11 we may suppose that this system is also parabolic with diffusion-like properties. Unfortunately, this is usually not the case. To demonstrate this consider the classical Buckley-Leverett problem of one-dimensional, two-phase, incompressible flow of oil and water with no gravity effects, no source or sink terms, and constant K, ϕ and μ. The system (Eq.6) becomes,

117

$$\phi \frac{\partial S_o}{\partial t} = \frac{\partial}{\partial x} \left(\lambda_o \frac{\partial P_o}{\partial x} \right) \tag{12}$$

$$\phi \frac{\partial S_w}{\partial t} = \frac{\partial}{\partial x} \left(\lambda_w \frac{\partial P_w}{\partial x} \right) \tag{13}$$

where the phase mobility, λ_i, is defined as,

$$\lambda_i = \frac{K k_{ri}}{\mu_i} \qquad (i = o, w) \tag{14}$$

For this case, we have the auxiliary relations,

$$S_o + S_w = 1 \tag{15}$$

and

$$P_{cow} = P_o - P_w \tag{16}$$

Adding Eqs.12 and 13, with $\partial S_o / \partial t = -\partial S_w / \partial t$ from Eq.15, gives,

$$\frac{\partial}{\partial x} \left(\lambda_o \frac{\partial P_o}{\partial x} \right) + \frac{\partial}{\partial x} \left(\lambda_w \frac{\partial P_w}{\partial x} \right) = 0 \tag{17}$$

Substituting Eq.16 gives,

$$\frac{\partial}{\partial x} \left(\lambda_T \frac{\partial P_o}{\partial x} \right) - \frac{\partial}{\partial x} \left(\lambda_w \frac{\partial P_{cow}}{\partial x} \right) = 0 \tag{18}$$

where the total mobility is defined as, $\lambda_T = \lambda_o + \lambda_w$. Eq.18 is usually referred to as the *Pressure* equation. Following Spillette, Hillestad and Stone (1973) we define the water fractional flow, f_w, as

$$f_w(S_w) = \frac{\lambda_w}{\lambda_o + \lambda_w} = \frac{\lambda_w}{\lambda_T} \tag{19}$$

Utilising Eq.14, Darcy's law may now be written as,

$$v_o = -\lambda_o \frac{\partial P_o}{\partial x} \quad \text{and} \quad v_w = -\lambda_w \frac{\partial P_w}{\partial x} \tag{20}$$

which together with Eq.16, may be used to expand the RHS of Eq.13 as,

$$\frac{\partial}{\partial x} \left(\lambda_w \frac{\partial P_w}{\partial x} \right) = -\frac{\partial}{\partial x} (f_w v_o) - \frac{\partial}{\partial x} (f_w v_w) - \frac{\partial}{\partial x} \left(f_w \lambda_o \frac{\partial P_{cow}}{\partial x} \right) \tag{21}$$

Defining v_T, the total velocity as, $v_T = v_o + v_w$, we may write for Eq.21,

$$\frac{\partial}{\partial x} \left(\lambda_w \frac{\partial P_w}{\partial x} \right) = -\frac{\partial}{\partial x} (f_w v_T) - \frac{\partial}{\partial x} \left(f_w \lambda_o \frac{\partial P_{cow}}{\partial x} \right) \tag{22}$$

Substitution of Eq.22 into Eq.13, with the observation that for incompressible flow v_T is a constant, yields

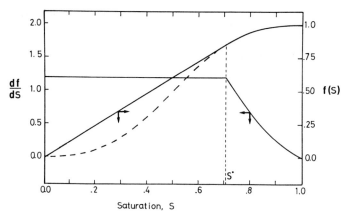

FIGURE 3: Fractional flow functions for two-phase problem.

$$\phi\left(\frac{\partial S_w}{\partial t}\right) + v_T\left(\frac{\partial f_w}{\partial x}\right) + \frac{\partial}{\partial x}\left(f_w\lambda_o\frac{\partial P_{cow}}{\partial x}\right) = 0 \tag{23}$$

Eq.23 is known as the total velocity form of the *Saturation* equation. We note that Eq.18 – the *Pressure* equation, and Eq.23 – the *Saturation* equation, are an alternative form for the system conservations Eqs.12 and 13. Although the pressure Eq.18 is elliptic , the saturation Eq.23 is a non-linear convection-diffusion equation. For cases where capillary effects are small, ie. physical dispersion is low, or when the convective term dominates, eg. flow near wells, the equation assumes many of the properties of a first-order hyperbolic equation.

For the classical Buckley-Leverett problem, with zero capillary pressure, Eq.23 reduces to,

$$\frac{\partial S_w}{\partial t} + \frac{\partial f_w}{\partial x} = 0 \tag{24}$$

where we may set $v_T/\phi = 1$ without loss of generality. The nonlinear results from the dependence of phase relative permeability on saturation. For relative permeabilities which are quadratic in each of the phase saturations this leads to the following expressions for water fractional flow and its first derivative as functions of water saturation, Taggart and Pinczewski (1987).

$$f_w = \frac{S_w^2}{S_w^2 + \frac{\mu_o}{\mu_w}(1 - S_w)^2} \tag{25}$$

and

$$\frac{df_w}{dS_w} = \frac{2\frac{\mu_o}{\mu_w}S_w(1 - S_w)}{\left[S_w^2 + \frac{\mu_o}{\mu_w}(1 - S_w)^2\right]^2} \tag{26}$$

The form of these functions for the case $\mu_o/\mu_w = 1$ is shown in Figure 3. Eqs.24-26 may have multiple weak solutions which satisfy the differential equations only in an integral sense. These solutions satisfy the relevant Rankin-Hugoniot jump condition and the specified initial and boundary conditions. The physically relevant solution must be found as the the limit of travelling wave solutions for the corresponding convection-diffusion problem as the diffusion coefficient tends to zero.

119

As will be shown later, the hyperbolic form of the saturation equation is a source of major difficulty in reservoir simulation. Moreover, Bell et al (1986), have shown that the corresponding problem for three-phase, incompressible flow in porous media in the absence of diffusive fluxes may be ill-posed. The system of conservation equations for this case may be written as,

$$\frac{\partial S}{\partial t} + \frac{\partial f}{\partial x} = 0 \tag{27}$$

where S and f are now the vector quantities:

$$S = [S_w, S_g]^T \quad \text{and} \quad f = [f_w, f_g]^T \tag{28}$$

The quasi-linear form of Eq.27 is written as,

$$\frac{\partial S}{\partial t} + \left[\frac{\partial f}{\partial S}\right] \frac{\partial S}{\partial x} = 0 \tag{29}$$

where the term in square brackets is a 2 × 2 matrix,

$$\left[\frac{\partial f}{\partial S}\right] = \frac{1}{\lambda_T^2} \left[\begin{array}{cc} ((\lambda_g + \lambda_o)\lambda_{ww}' - \lambda_w \lambda_{ow}') & -\lambda_w \left(\lambda_{gg}' + \lambda_{og}'\right) \\ -\lambda_g \left(\lambda_{ww}' + \lambda_{ow}'\right) & (\lambda_w + \lambda_o)\lambda_{gg}' - \lambda_g \lambda_{og}' \end{array} \right] \tag{30}$$

where $\lambda_{i,j}'$ denotes the derivative of λ_i with respect to S_j ie.,

$$\lambda_{i,j}' = \frac{\partial \lambda_i}{\partial S_j} \tag{31}$$

Bell et al (1986) have shown that if $\partial f/\partial S$ has real and distinct eigenvalues for all allowable saturations then the system (Eq.29) is hyperbolic and can be analysed by standard techniques for hyperbolic equations. However, the form of the empirical relative permeability curves usually used in reservoir simulation does not guarantee that $\partial f/\partial S$ always has real eigenvalues. In regions of the saturation space where the eigenvalues are complex the system is elliptic and unstable ie., the problem is ill-posed when diffusive forces are ignored. When capillary or diffusive pressure is included the instability is removed and the system is linearly well-posed.

Since compositional simulators include the effects of mass-transfer between phases, mass is no-longer conserved within each phase as it is for the immiscible flow problems discussed above. Here, the conservation equations are written for individual components which may be distributed in all three phases. If the phases are denoted o, g, w for the oil, gas and water phases respectively, and if C_{io}, C_{ig} and C_{iw} are the mass fractions of component i in the respective phases, then the total mass flux for the ith component may be written as,

$$C_{ig}\rho_g \tilde{V}_g + C_{io}\rho_o \tilde{V}_o + C_{iw}\rho_w \tilde{V}_w \quad (i = 1, N_c) \tag{32}$$

The mass of component i per unit volume is,

$$\phi \left(C_{ig}\rho_g S_g + C_{io}\rho_o S_o + C_{iw}\rho_w S_w\right) \quad (i = 1, N_c)$$

and the conservation equation for the ith component is,

$$\frac{\partial}{\partial t}\left[\phi\left(C_{ig}\rho_g S_g + C_{io}\rho_o S_o + C_{iw}\rho_w S_w\right)\right] = q_i$$
$$+\nabla \cdot \left[C_{ig}\rho_g \tilde{V}_g + C_{io}\rho_o \tilde{V}_o + C_{iw}\rho_w \tilde{V}_w\right] \qquad (i = 1, N_c) \tag{33}$$

where, N_c is the total number of components in the system.

The distribution of components between phases in a multi-phase system is a complex function of phase pressure, temperature and composition and must be obtained from thermodynamic models such as the Peng-Robinson equation of state, (Peng and Robinson (1977)). Although this makes compositional simulators considerably more complex in their detail than immiscible simulators, the mathematical structure of the problem is similar.

FINITE-DIFFERENCE DISCRETIZATION

Finite-difference discretization techniques, because of their relative simplicity and computational efficiency, are the preferred method for approximating the system of conservation laws for multiphase, multi-dimensional flow in reservoir simulation. We demonstrate this technique by considering immiscible two-dimensional, two-phase (oil and water), slightly compressible flow with no mass-transfer between the phases. The system of conservation equations describing this flow is,

$$\frac{\partial\left(\phi\rho_i S_i\right)}{\partial t} = \nabla \cdot \frac{\rho_i K k_{ri}}{\mu_i}\left(\nabla P_i - \rho_i g \nabla z\right) + q_i \quad (i = o, w) \tag{34}$$

As shown previously, the above equation contains only two independent unknowns − choose one of the phase saturation, S_w, and one of the phase pressures, P_o.

Time-stepping Schemes

From the viewpoint of the reservoir engineer time discretization schemes must be stable, robust and computationally efficient. Since forward-differencing (explicit) methods are only stable for time steps constrained by

$$\Delta t \leq k(\Delta x)^2$$

where, Δt and Δx are the temporal and spatial mesh sizes respectively and k is some constant (Aziz and Settari(1979)), we use only backward-differencing (implicit) methods.

Spatial Discretization

The application of implicit time-stepping schemes to the system Eq.34 requires the simultaneous solution of very large systems of non-linear algebraic equations at each time step. The computational workload is very high since the solution involves the application of some form of linearization method which usually involves an iterative linear solution technique at each linearization step.

An alternative approach which greatly reduces computational time and storage is to de-couple the system of conservation laws (Eq.34) or the equivalent set of pressure (Eq.18) and saturation (Eq. 23) equations and solve sequentially (Peaceman (1977).) The pressure

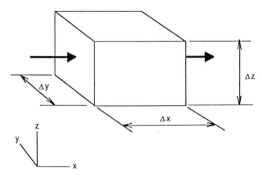

FIGURE 4: Reservoir grid block.

equation is solved implicitly for pressure, lagging all saturation dependencies to the old time level. Having obtained the pressure solution at the new time level, the saturation equation is solved explicitly for saturation. This procedure is referred to as the IMPES method (IMplicit Pressure, Explicit Saturation). Although the method is fast because it solves for only one unknown per grid block, it is subject to a severe time step restriction for problems having difficult nonlinearities. It's use in reservoir simulation is therefore limited to problems of moderate degree of difficulty. For more difficult problems it is necessary to carry out a fully implicit simultaneous solution.

It is possible to write finite-difference approximations for the system Eq.34 by expanding the unknown functions in a Taylor series and developing finite-difference quotients from a truncation of the resulting series. An alternative approach, the preferred method in reservoir simulation, is to write the difference equations in a manner analogous to that used to develop the differential equations themselves. ie. an Eulerian control-volume formulation. For a discretized element of reservoir volume – or block (see Figure 4) – Eq.34 may be approximated by,

$$\frac{\Delta m_{i,K}}{\Delta t} = \sum_F f_{i,K,F} + q_{i,K} \qquad (i = o, w) \tag{35}$$

where, $m_{i,K}$ is the mass rate of phase i in block K which may be expressed as,

$$m_{i,K} = V_{b,K} \phi_K S_{i,K} \rho_{i,K} \qquad (i = o, w) \tag{36}$$

where, $V_{b,K}$ is the block volume, ϕ_K is the block average porosity, $S_{i,K}$ is the block average saturation, $\rho_{i,K}$ is the block average density evaluated at the block average pressure $P_{i,K}$ and $f_{i,K,F}$ is the total mass rate of phase i across face F of the block K. For a rectangular block (cartesian grid) the faces are designated N, S, E and W for the north, south, east and west directions respectively (see Figure 5). The convective flux term is defined as,

$$f_{i,K,F} = A_F (\rho v)_{i,F} \tag{37}$$

where A_F is the area of block face F, and the mass flux $(\rho v)_{i,F}$ is defined at the centre of the block face F. The density in Eq. 37 is evaluated at some interpolated block face pressure.

In Eqs.34 and 35 wells are modelled as point sources and sinks. This is sufficient if we specify only phase injection or production rates. Peaceman (1983) has demonstrated that the block pressure, for a block containing a well, is approximately equal to the actual well radial flowing pressure at a small distance from the wellbore. If it is necessary to specify

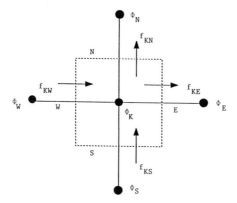

FIGURE 5: Low-order (5-point) scheme.

well bottom hole pressures rather than mass injection or production rates it is necessary to introduce a well model which relates the specified bottom hole pressure to the block pressure. This is usually done by simply assuming steady state-radial flow, for which,

$$P_i - P_{well} = \frac{q_i \mu_i}{2\pi K k_{r,i} h \rho_i} \ln\left(\frac{r_e}{r_w}\right) \tag{38}$$

where P_i is the block pressure, P_{well} is the specified well pressure, h is the block thickness, r_e is the effective well block radius and r_w is the wellbore radius. Eq.38 is used to eliminate the source/sink term q_i from Eq.35 when it is not specified directly. In practice it is common to specify only one individual phase injection or production rate, or a total rate for all phases. In such cases capillary pressure is neglected and Eq.38 is written for each of the phases present. These equations are then be used to evaluate the required individual phase rates.

Eq.35 is written for each block in the reservoir model. For a reservoir model discretized into N blocks, there will be a total of $2N$ such equations.The large number of blocks normally used in reservoir simulation leads to very large sets of equations – tens of thousands or more. These systems of non-linear algebraic equations are solved using Newton's method and for this, it is convenient to write Eq.35 in residual form as,

$$R_{i,K} = \frac{\Delta m_{i,k}}{\Delta t} - \sum_F f_{i,K,F} - q_{i,K} \qquad (K = 1, N \text{ and } i = o, w) \tag{39}$$

The functional form of the mass flux $f_{i,K,F}$ has not yet been specified. However, it must be a function of the two unknowns, S_w andP_o, at block K and its neighbours.

Newton's method

We may write Eq.39 in vector form as,

$$R_K = \frac{\Delta m_K}{\Delta t} - \sum_F f_{K,F} - q_K \tag{40}$$

where, R_K, m_K, $f_{K,F}$ and q_K are vectors given by

$$R_K = \begin{bmatrix} R_{o,K} \\ R_{w,K} \end{bmatrix} \quad m_K = \begin{bmatrix} m_{o,K} \\ m_{w,K} \end{bmatrix} \quad f_{K,F} = \begin{bmatrix} f_{o,K,F} \\ f_{w,K,F} \end{bmatrix} \quad q_K = \begin{bmatrix} q_{o,K} \\ q_{w,K} \end{bmatrix} \tag{41}$$

It is convenient to group the residuals and independent unknowns for each block into vectors, R and X respectively,

$$R(X) = \begin{bmatrix} R_1 \\ \vdots \\ R_K \\ \vdots \\ R_N \end{bmatrix} \quad X = \begin{bmatrix} X_1 \\ \vdots \\ X_K \\ \vdots \\ X_N \end{bmatrix} \quad X_K = \begin{bmatrix} P_{o,K} \\ S_{w,K} \end{bmatrix} \tag{42}$$

R and X are partitioned vectors with subvectors as components. Each subvector contains NEQ elements where NEQ is the number of independent unknowns or equations per block (two for the present case). We note that we seek the solution vector, X, such that the residual vector is zero, ie,

$$R(X) = 0 \tag{43}$$

Newton's method solves Eq.43 for X iteratively with the iteration update for iteration, n, written as,

$$X_{n+1} = X_n - J^{-1}(X_n)R(X_n) \tag{44}$$

$J(X_n)$ is the Jacobian matrix evaluated at iteration n. The Jacobian matrix is a square $N \times N$ matrix whose coefficients are square NEQ $\times$ NEQ sub-matrices. The elements of the Jacobian are given by,

$$J_{K,M} = \frac{\partial R_K}{\partial X_M} \tag{45}$$

where, R_K and X_M are given by Eq.42. To evaluate the elements of the Jacobian we must specify the functional form of the convective flux $f_{i,K,F}$. The actual form depends on the difference scheme used. For the standard low-order, five-point schemes normally used in reservoir simulation (Peaceman (1977) and Aziz and Settari (1979)),

$$f_{i,K,F} = A_F \rho_{i,F} T_{i,K,F} (\Phi_{i,K} - \Phi_{i,F}) \quad (F = N, S, E, W) \tag{46}$$

$T_{i,K,F}$ is the transmissibility for flow across face F and is simply the Darcy law flow coefficient evaluated at the face conditions,

$$T_{i,K,F} = \frac{\lambda_{i,F}}{\Delta x} \tag{47}$$

$\Phi_{i,K}$ and $\Phi_{i,F}$ are the block phase potentials defined in Figure 5, where,

$$\nabla \Phi_{i,F} = \nabla P_{i,K} - \rho_{i,K} g \nabla z \tag{48}$$

Eqs.45 and 46 are used to evaluate the elements of the Jacobian. If the blocks are ordered, ie. numbered in a systematic manner, then the Jacobian will have a particular structure.

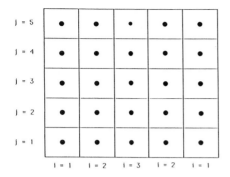

FIGURE 6: Natural ordering of grid blocks.

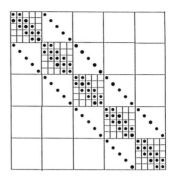

FIGURE 7: Two-dimensional low-order Jacobian matrix.

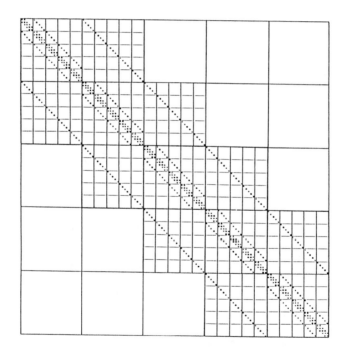

FIGURE 8: Three-dimensional low-order Jacobian matrix.

125

The natural ordering scheme shown in Figure 6 for the two-dimensional problem considered results in a Jacobian which is a sparse $N \times N$ partitioned matrix having elements 2×2 sub-matrices. The structure of the Jacobian for a 5×5, 2-D uniform grid is shown in Figure 7. The structure of the Jacobian for a three-dimensional $5 \times 5 \times 5$ regular grid is shown in Figure 8. It is clear that the Jacobian has a block tridiagonal form. The gaps in the diagonals arise from the application of Neumann boundary conditions at the edges of the grid. Although the structure of the Jacobian is symmetric, the matrix itself is usually asymmetric.

The structure of the Jacobian may also be modified by the introduction of wells. To see how this comes about consider a well opened to flow from two non-neighbouring blocks (say blocks 7 and 17 shown in Figure 6) with the well constrained to a specified total oil rate, $q_{o,T}$. The well model, Eq.38, is written for each block as,

$$P_{o,7} - P_{well} = q_{o,7} \left[\frac{\mu_o}{2\pi K k_{ro} h \rho_o} \ln\left(\frac{r_e}{r_w}\right) \right]_7 \tag{49}$$

$$P_{o,17} - P_{well} = q_{o,17} \left[\frac{\mu_o}{2\pi K k_{ro} h \rho_o} \ln\left(\frac{r_e}{r_w}\right) \right]_{17} \tag{50}$$

$$q_{o,T} = q_{o,7} + q_{o,17} \tag{51}$$

From these equations we easily obtain,

$$q_{o,17} = \frac{P_{o,17} - P_{o,7} + q_{o,T} \left[\frac{\mu_o}{2\pi K k_{ro} h \rho_o} \ln\left(\frac{r_e}{r_w}\right) \right]_7}{\left[\frac{\mu_o}{2\pi K k_{ro} h \rho_o} \ln\left(\frac{r_e}{r_w}\right) \right]_7 + \left[\frac{\mu_o}{2\pi K k_{ro} h \rho_o} \ln\left(\frac{r_e}{r_w}\right) \right]_{17}} \tag{52}$$

a similar expression can be obtained for $q_{o,7}$. Clearly, substitution for the source/sink terms, used to model wells, can result in additional coupling between equations for blocks intersected by a well.

Limitations of Low-order Methods

A number of important reservoir engineering problems have solutions characterised by low levels of physical dispersion (Taggart and Pinczewski (1985, 1987)). Examples are high velocity immiscible flows in the vicinity of wells, improved oil recovery process where small slugs of fluid are injected into the reservoir, thermal recovery processes, and miscible displacements where compositional effects are important. For these conditions convective terms dominate the conservation equations (Eq.29) which assume a wave-like or hyperbolic nature.

Central differencing of the first derivative transport term, $\partial f / \partial x$ has $\mathcal{O}\left((\Delta x)^2\right)$ accuracy, if f is sufficiently smooth, but causes stability problems (Taggart and Pinczewski (1987)). The stability problem is overcome by the use of low-order, single-point upwind differencing. This amounts to using an $\mathcal{O}\left((\Delta x)^2\right)$ accurate approximation to the term,

$$\frac{\partial f}{\partial x} - \frac{v_x \Delta x}{2} \frac{\partial^2 f}{\partial x^2} \tag{53}$$

in the place of $\partial f / \partial x$, and results in the introduction of numerical dispersion or non-physical diffusion of size $v_x \Delta x / 2$. This stabilizes the finite-difference scheme by making

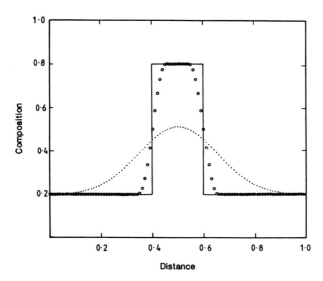

FIGURE 9: Pulse convection for low-order and high-order methods.

the resulting matrix more diagonally dominant by moving one of the off-diagonals of the centred scheme onto the diagonal. The result of applying single-point, upwind differencing to the standard problem of pure convective transport of a square wave on a periodic grid (Taggart and Pinczewski (1987)) is shown in Figure 9. The computed solution is clearly excessively diffused and is therefore unacceptable.

Extending the application of low-order differencing to multiple dimensions introduces further problems. For two-phase flow in two-dimensions in the absence of physical dispersion the application of upwind differencing to the transport terms in the x and y directions, results in an artificial dispersion term of the form,

$$-\frac{v_x \Delta x}{2} \frac{\partial^2 f}{\partial x^2} - \frac{v_y \Delta y}{2} \frac{\partial^2 f}{\partial y^2} \tag{54}$$

This term has obvious directional properties and results in solutions which are sensitive to the orientation of the computational grid relative to the principal flow direction. The grid-orientation effect is illustrated by computing solutions for a repeated five-spot pattern flood (Taggart and Pinczewski (1987)). Figure 10 shows that considerations of symmetry allow the use of either a parallel (main flow direction parallel to the grid) or a diagonal (main flow direction along the diagonal) grid to solve this problem. The results shown in Figure 11, for oil recovery at water breakthrough, clearly

A number of alternative solution techniques have been proposed to overcome the shortcomings of low-order differencing. These include the method of characteristics, modified random choice methods (Concous and Proskurowski (1979)), and various flux-updating schemes. Although the method of characteristics is capable of producing exact solutions for certain simple problems, its use as a general simulation tool is limited by the complexity of the computer codes required to apply the method to general three-dimensional multi-phase flow problems. Complexity also eliminates the random choice and flux-updating methods, all of which are based on the method of characteristics. Although variational methods (finite elements eg. Mercer and Faust (1976)), spectral

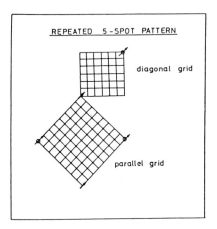

FIGURE 10: Diagonal and parallel grids.

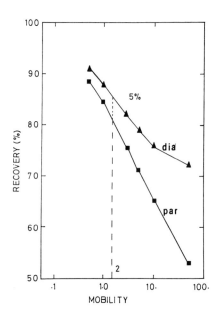

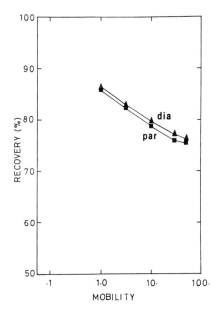

FIGURE 11: Recovery at breakthrough for five-spot flood for parallel and diagonal grids with low-order differencing.

FIGURE 12: Recovery at breakthrough for five-spot flood for parallel and diagonal grids with high-order differencing.

methods, and related weighted residual schemes, such as collocation (Allen and Pinder (1982)) have been shown to produce solutions that are more accurate and less prone to grid orientation sensitivity than those obtained with simple finite-difference schemes, the practical implementation of these methods to reservoir engineering problems is again limited by their complexity and computational expense. The inherent simplicity and generality of finite-difference techniques make them attractive and the majority of commercial simulators use this form of approximation (Peaceman (1977) and, Aziz and Settari (1979)).

HIGH-ORDER METHODS

Nine-point finite-difference schemes (Yanosik and McCracken (1976)), based on linear combinations of the conventionally differenced flow equations for parallel and diagonal grids have been shown to reduce grid orientation sensitivity considerably without incurring the computational expense of variational methods. However, these methods are known to produce physically unrealistic *bullet-like* displacement fronts (Ko and Au (1979)) and offer no significant improvements in truncation error over the simpler five-point schemes.

Taggart and Pinczewski (1985-1987) have demonstrated that third-order finite difference approximations greatly reduce grid-orientation sensitivity in multi-dimensional displacements. Because of the greater accuracy of the higher-order approximations and the resulting smaller truncation error, they allow the introduction of only a small (by comparison with single-point, upwind differencing) amount of rotationally invarient dispersion which dominates the rotationally varient truncation errors. Figure 9 and Figure 12 show the results of applying third-order accurate differencing for the previously discussed one-dimensional convection and the two-dimensional, five-spot test problems. The level of non-physical dispersion for the one-dimensional convection problem and the sensitivity of the solution to grid-orientation for the two-dimensional problem are both greatly reduced.

The practical implementation of high-order difference schemes is relatively straightforward. The block convection rates in Eq.39, $f_{i,K,F}$ are simply evaluated more accurately. This requires higher-order estimates for saturation dependent transmissibilities and phase potential gradients at block faces and these introduce additional diagonals in the Jacobian matrix. The Jacobians resulting from high-order approximation techniques are considerably less sparse than those obtained with low-order approximation methods and are generally ill-conditioned (Wong (1986)). The result is a considerable increase in the computational time and storage requirements both to evaluate the elements of the Jacobian and to obtain the inverse.

To compare the computer workload for high-order and low-order finite-difference methods we have conducted numerical experiments using the previously discussed five-point scheme and a high-order scheme similar to that described by Taggart and Pinczewski (1987) for their uniformly third-order accurate scheme. Here, we employ their 13-point spatial template to evaluate the saturation dependent mobilities and a nine-point template (Yanosik and McCracken (1976)) for pressure (see Figure 13). Inter-block mass fluxes are evaluated with,

$$f_{i,K,F} = A_F \rho_{i,F} T_{i,K,F}^H \nabla \Phi_{i,K,F}^H \quad (i = o, w) \tag{55}$$

where, $T_{i,K,F}^H$ and $\nabla \Phi_{i,K,F}^H$ are the high-order approximations for transmissibility and potential gradient respectively. As before,

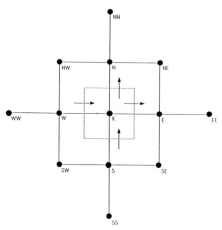

FIGURE 13: High-order (13-point) scheme.

$$T_{i,K,F}^{H} = \frac{\lambda_{i,F}^{H}}{\Delta x} \tag{56}$$

where,

$$\begin{aligned}
\lambda_E^H &= \frac{1}{12}[-\lambda_W + 7\lambda_K + 7\lambda_E - \lambda_{EE}] + \frac{1}{16}\frac{|v_x|}{v}[-\lambda_W + 5\lambda_K \\
&\quad -5\lambda_E + \lambda_{EE} - \lambda_N - \lambda_S + \lambda_{NE} + \lambda_{SE}]
\end{aligned} \tag{57}$$

Eq.57 is written for the E-face. The subscript i has been omitted for clarity. Similar expressions may also be written for the other three faces. The nine-point template for pressure, is written as,

$$\Delta\Phi_E^H = \frac{1}{12}[\Phi_{NE} - \Phi_N + 10(\Phi_E - \Phi_K) + \Phi_{SE} - \Phi_S] \tag{58}$$

Similar expressions may also be written for the other block faces.

The solution procedure is similar to that already described for the low-order methods. However, the Jacobian resulting from the high-order scheme is considerably less sparse. Whereas, the lower-order Jacobian for the two-dimensional, two-phase problem has 5 diagonals each element being a 2×2 submatrix, the high-order Jacobian has 13 diagonals with each element being a 2×2 submatrix. For three-dimensions the low-order Jacobian has 7 diagonals, whilst the high-order Jacobian has 33 diagonals. Figure 14 and Figure 15 show the Jacobian matrices which result from applying the high-order method to two- and three-dimensional problems respectively.

COMPARISON OF WORKLOAD FOR HIGH- AND LOW- ORDER SCHEMES

The system of nonlinear algebraic equations resulting from the discretization technique employed is solved at each time-step using Newton's method. Each Newton's iteration requires the calculation of a Jacobian and the solution of the resulting sparse linear system of equations,

$$Ax = b \tag{59}$$

Because practical reservoir simulation problems are large – involving many tens of thousands of blocks – the solution of the linear system is usually carried out using an

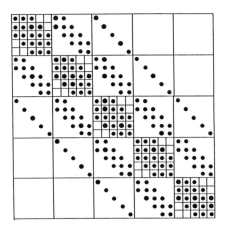

FIGURE 14: Two-dimensional high-order Jacobian matrix.

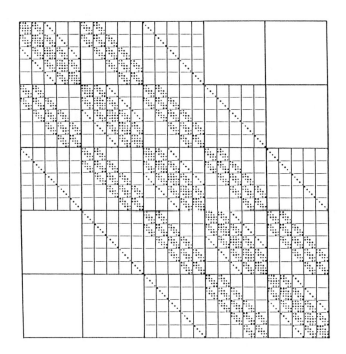

FIGURE 15: Three-dimensional high-order Jacobian matrix.

131

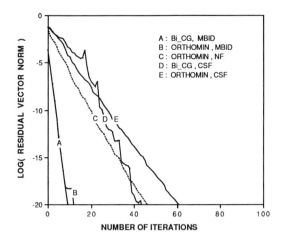

FIGURE 16: Low-order residual norm vrs number of iterations.

iterative procedure. Storage and computational time are primary considerations in selecting the actual procedure. The matrix A is usually sparse, banded, asymmetric and may have some complex eigenvalues. The asymmetry grows with the increasing dominance of convective over diffusive terms.

The most successful solvers all utilize some form of matrix preconditioning accelerated by a conjugate gradient like method. Without preconditioning convergence rates are very slow. We have tested a number of such methods. The preconditioners tested are Nested Factorization (NF) developed by Appleyard and Chershire (1983) – a highly recursive factorization which exploits the nested block tridiagonal structure of A, Column Sum Factorization (CSF) – a simplified varient of NF which is easily extended to systems resulting from the application of high-order discretization methods, pointwise LU incomplete decomposition(ILU) (Langtangen (1989)), and a modified block incomplete decomposition (MBID) proposed by Efrat and Tismenetsky(1986). MBID is easily vectorised and was developed for implementation on parallel processor machines. The other preconditioners are inherently recursive and therefore more difficult to perform in parallel. The preconditioners were used with ORTHOMIN (Vinsome (1976)) and the Bi-conjugate Gradient Method (Fletcher (1976)) accelerators.

All the tests were carried out on a single processor machine running in scalar mode. The test problem was a two-dimensional, two-phase displacement on a five-spot using a diagonal 20×20 grid (800 unknowns). The problem was initialized with a severe permeability variation which results in a highly ill-conditioned Jacobian matrix and makes the problem very difficult to solve. The Jacobian was generated at a time approaching breakthrough and a time step of 0.1 days. Figure 16 and Figure 17 show the norm of the residual vector, $\|r\|$, plotted against the number of iterations for the combinations of preconditioners and accelerators investigated.

Figure 16 shows that for the low-order Jacobian all the methods are successful in reducing the norm of the residual. As expected the better the preconditioner, the faster the rate of convergence. The best of the preconditioners tested was MBID. This preconditioner produced similar rates of convergence both with the ORTHOMIN accelerator and the Bi-conjugate Gradient Method. The overall performance of all of the methods was similar with the total CPU time for for a simulation through to breakthrough within 30% for all

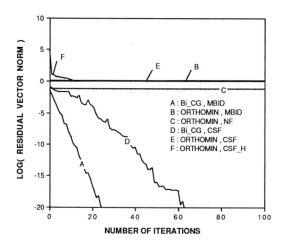

FIGURE 17: High-order residual norm vrs number of iterations.

the methods tested. The best method was Bi-conjugate Gradient-MBID. Although this method required the lowest number of iterations, the work load per iteration was highest.

Only two preconditioners were tested for the high-order Jacobian – MBID and CSF. It is difficult to apply the the other preconditioners to high-order systems. The greater bandwidth of the high-order Jacobian results in a further degradation of the condition number of the matrix and makes the problem considerably more difficult to solve then the corresponding low-order problem. ORTHOMIN failed to converge for both the MBID and CSF preconditioners. The Bi-conjugate Gradient Method converged with both preconditioners. The rate of convergence was fastest with MBID.

ACKNOWLEDGEMENTS

This work was supported in part by grants from the State of New South Wales Department of Energy and the Commonwealth of Australia Department of Resources and Energy under the National Energy Research Development and Demonstration Program.

REFERENCES

1.Allen, M.B. and Pinder, G.F., The Convergence of Upstream Collocation in The Buckley-Leverett Problem, SPE 10978 presented at the 57th Annual Fall Technical Conference, New Orleans, Sept. 26, 1982.

2.Appleyard, J.R. and Cheshire, I.M., Nested Factorization, SPE 12264 presented at the Reservoir Simulation Symposium, San Francisco, CA, Nov. 15, pp. 111-120, 1983.

3.Aziz, K., and Settari, A.,*Petroleum Reservoir Simulation*, Applied Science Publishers, London, 1979.

4.Bell, J.B., Trangenstein, J.A. and Stubin, G.R., Conservation Laws of Mixed Type Describing Three Phase Flow in Porous Media, *S.I.A.M. J.Appl. Math.*, vol. 46, pp. 1000-1017, 1986.

5.Concous, P. and Proskurowski, W., Numerical Solution for a Non-linear Hyperbolic Equation by the Random Choice Method, *J.Comp.Phys.*, vol. 30, pp. 153-166, 1979.

6.Ewing, R.E. (Ed), *The Mathematics of Reservoir Simulation, SIAM*, Philadelphia, 1983.

7.Fletcher, R., Conjugate Gradient Methods for Indefinite Systems, *Lectures Notes in Mathematics No.506*, Springer-Verlag, Berlin, 1976.

8.Ko, S.C.M. and Au, A.D.K., A Weighted Nine-point Finite-Difference Scheme for Eliminating the Grid Orientation Effect in Numerical Reservoir Simulation, SPE 8248 presented at the 54th Annual Fall Technical Conference and Exhibition of SPE, Las Vegas, Sept. 23, 1979.

9.Langtangen, H.P., Conjugate Gradient Methods and ILU Preconditioning of Non-Symmetric Matrix Systems with Arbitrary Sparsity Patterns, *Inter.J.Numer.Methods in Fluids*, vol.9, pp 213-233, 1989.

10.Mercer, J.W. and Faust, C.R., The Application of Finite Element Techniques to Immiscible Flow in Porous Media, Proceedings of the First International Conference on Finite Elements in Water Resources held at Princeton University, July 1976.

11.Peaceman, D.W.,*Fundamentals of Numerical Reservoir Simulation*, Elsevier, New York, 1977.

12.Peaceman, D.W., Interpretation of Well-Block Pressures in Numerical Reservoir Simulation with Non-square Grid Blocks and Anisotropic Permeability, *Soc.Pet.Eng.J.*, vol. 23, pp. 531-543, 1983.

13.Peng, D.Y. and Robinson, D.B., A Rigourous Method for Predicting the Critical Properties of Multi-component Systems from an Equation of State, *AIChE J.*, vol. 23, pp. 137-144, 1977.

14.Spillette, A.G., Hillestad, J.G. and Stone, H.L., A High-Stability Sequential Solution Approach to Reservoir Simulation, paper SPE 4542, presented at 48th Annual Meeting of SPE, Las Vegas, Nev., 1973.

15.Stone, H.L., Estimation of Three Phase Relative Permeability and Residual Oil Data, *J.Can.Petrol.Tech.*, vol.12, pp. 53-61, 1973.

16.Taggart, I.J. and Pinczewski, W.V., Simulation of Enhanced Recovery by Higher Order Difference Techniques, *CTAC-1985* pp. 579-597, 1985.

17.Taggart, I.J. and Pinczewski, W.V., Simulation of Compositional Flooding Using Third Order Techniques, SPE 16702 presented at 62nd Annual Technical Conference and Exhibition of SPE, Dallas, TX, Sept. 27, 1987.

18.Taggart, I.J. and Pinczewski, W.V., The Use of Higher-Order Differencing Techniques in Reservoir Simulation, *Soc.Pet.Eng.Res.Eng.*, vol. 2, pp. 360-372, 1987

19.Vinsome, P.K.W., Orthomin, an Iterative Method for Solving Sparse Sets of Simultaneous Linear Equations, SPE 5729 presented at the SPE/AIME Fourth Symposium of Numerical Simulation of Reservoir Performance, Los Angeles, CA, Feb. 19, 1976.

20.Wheeler, M.F. (Ed), *Numerical Simulation in Oil Recovery*, Springer Verlag, New York, 1988.

21.Wong, Y.S., Preconditioned Conjugate Gradient Methods Applied to Certain Symmetric Linear Systems, *Intern.J.Comp.Math.*, vol.19, pp. 177-200, 1986.

22.Yanosik, J.L. and McCracken, T.A., A Nine Point, Finite-Difference Reservoir Simulator for Realistic Prediction of Adverse Mobility Ratio Displacements, SPE 5734 presented at the Fourth Symposium of Numerical Simulation of Reservoir Performance, Los Angeles, CA, Feb. 19, 1976.

ALGORITHMS AND SUPERCOMPUTING

Case Studies in Parallel Programming

D. ABRAMSON
Division of Information Technology
CSIRO
c/o Department of Communication and Electrical Engineering
Royal Melbourne Institute of Technology
P.O. Box 2476V
Melbourne 3001, Australia

INTRODUCTION

This paper contains case studies of two computationally expensive programs, and discusses the techniques used for producing parallel versions of the code. The first program uses simulated annealing for constructing school timetables. Simulated annealing is a combinatorial optimisation technique based on statistical mechanics. It is used for finding minimal cost solutions to problems which are too large for exhaustive search techniques. It models a set of cooling atoms which are governed by inter-particle forces. The atoms are cooled slowly in order to produce a configuration with a very low system energy. The problem with slow cooling is that it requires a large amount of processor time. An extremely elegant technique for improving the execution speed is to use parallel processors. It is possible to divide the atoms into partitions and then cool them separately. Synchronisation is required only when atoms interact with others in different partitions. This type of parallelisation can show good speedup, and can reduce many hours of processor time to minutes of elapsed time. The paper will present some results gathered on an Encore MULTIMAX shared memory machine.

The second application is a printed circuit router program. Computing the paths of the tracks in printed circuit boards is an extremely expensive process. It involves finding a path for each wire which does not cross any previously created tracks. The paths are found using a maze router algorithm, which advances a wavefront into unfilled cells. A parallel decomposition of the problem consists of routing more than one wire at a time. Because of the relationships between tracks, it is initially difficult to see a good parallel algorithm. However, wires tend to be clumped together thus exhibiting a large degree of locality. Our parallel algorithm involves using a divide and conquer technique for partitioning the circuit board. A recursive program divides the board into two partitions, and passes the relevant wires to each partition. Wires which are not totally contained in the two sub partitions are held until the sub partitions have been routed. The recursion continues until either a fixed maximum depth is reached, or there are no wires left. Routing is then performed on the way back up the recursive tree. The scheme cannot show the same linear speedup found in the simulated annealing code, but does benefit from a small number of processors. Again, a number of test runs on an Encore MULTIMAX are presented.

The paper contrasts two different partitioning strategies, and illustrates their strengths and weaknesses. The programming paradyme that has been used is the *shared memory* model, in which many processes share access to structures in one global memory pool. The paper comments on the use of message passing and dataflow machines as alternatives for the shared memory model.More details on both of these applications can be found in other papers, namely Abramson (1988) and Abramson and Freidin (1989) .

TIMETABLE PROBLEM and SIMULATED ANNEALING

The problem of creating a valid timetable involves scheduling classes, teachers and rooms in such a way that no *teacher, class or room* is used more than once per period. For example, if a class must meet twice a week, then it must be placed in two different periods to avoid a clash. The timetable is to be distributed across a fixed number of periods per week. A class consists of a number of students. We will assume that students have already been grouped into classes. In each period a class is taught a *subject*. It is possible for a subject to appear more than once in a period. A particular combination of a teacher, a subject, a room and a class is called an *element*. An element may be required more than once per week. The combination of an *element* and a frequency is called a *requirement*. Thus, the timetabling problem can be phrased as scheduling a number of requirements such that a requirement, teacher, class or room does not appear more than once per period.

It is possible to define an *objective* or *cost function* for evaluating a given timetable. This function calculates the number of clashes in any given timetable. An acceptable timetable has a cost of 0. The cost of any period is of the sum of three components: a class cost, a teacher cost and a room cost.

The class cost is the number of times each of the classes in the period appears in that period, less one if it is greater than zero. Thus, if a class appears no times or once in a period then the cost of that class is zero. If it appears many times the the class cost for that class is the number of times less one. The class cost for a period is the sum of all class costs. The same computation applies for teachers and rooms. The cost of the total timetable is the sum of the period costs.

SIMULATED ANNEALING

Simulated annealing (SA) is a Monte-Carlo technique which can be used to find solutions to optimisation problems. A good review of the theory and practice can be found in Van Laarhoven and Aarts (1987) . The technique simulates the cooling of a collection of hot vibrating atoms. When the atoms are at a high temperature they are free to move around, and tend to move with random displacements. However, as the mass cools the inter-particle bonds force the atoms together. When the mass is cool, no movement is possible, and the configuration is frozen. If the mass is cooled quickly then the final system energy may not be minimal. However, if it is cooled slowly, then the final energy may be the lowest possible. At any given temperature a new configuration of atoms is accepted if the system energy is lowered. However, if the energy is higher, then the configuration is accepted only if the probability of such an increase is lower than that expected at the given temperature. This probability is given by $P(\Delta E) = e^{-\Delta E/KT}$, where K is Boltzmann's constant.

By modelling optimisation problems as a set of randomly vibrating atoms, it is possible to find optimal, or sub-optimal, solutions. Many optimisation problems can be considered as a number of *objects* which need to be scheduled such that an objective function is minimised. The vibrating atoms are replaced by the objects, and the value of the objective function replaces the system energy. An initial schedule is created by randomly scheduling the objects, and an initial cost (c_0) and temperature (T_0) are computed. Subsequent permutations are created by randomly choosing two objects, interchanging them, and computing a change in cost (Δc). If $\Delta c \leq 0$ then the change is accepted. However, if $\Delta c > 0$ then the probability of that change is calculated,

$$P(\Delta c) = e^{-\Delta c/T}.$$

If the probability is greater than a randomly selected value in the range $(0,1)$ then the change is accepted. After a number of successful permutations the temperature is decreased by a cooling rate, R, such that $T_n = T_{n-1} * R$.

One of the advantages of simulated annealing over algorithms which always seek a better solution (hill climbing algorithms) is that simulated annealing is less likely to get caught in local minima, because the cost can increase as well as decrease.

APPLYING SA TO THE TIMETABLING PROBLEM

The application of simulated annealing to the timetabling problem is relatively straight forward. The atoms are replaced by elements. The system energy is replaced by the timetable cost. An initial allocation is made in which elements are placed in a randomly chosen period. The initial cost and an initial temperature are computed. The cost is used to reflect the quality of the timetable, just as the system energy reflects the quality of a substance being annealed. The temperature is used to control the probability of an increase in cost and relates to the temperature of a physical substance. At each iteration a period is chosen at random, called the **from** period, and an element randomly selected from that period. Another period is chosen at random, called the **to** period. The change in cost is calculated from two components:

1) The cost of removing the element from the **from** period
2) The cost of inserting the element in the **to** period.

The change in cost is the difference of these two components. The element is moved if the change in cost is accepted, either because it lowers the system cost, or the increase is allowed at the current temperature. Unlike the classic simulated annealing technique which would actually swap two elements, an element is removed from one period and placed into another. This allows the number of elements in one period to increase or decrease, and for all periods to a contain different numbers of elements. If two elements were swapped then it would not be possible to change the number of elements per period.

The cost of removing an element consists of a class cost, a teacher cost and a room cost. Likewise, the cost of inserting an element consists of a class cost, a teacher cost and a room cost. If after removing an element from a period the number of occurrences of that class is > 0, then the class cost saving is 1. Similarly, if there are one or more occurrences of the teacher after that teacher has been removed then the teacher saving is 1. This technique also applies for rooms. The cost of inserting an element can be calculated using the same basic technique. In this way it is possible to determine the change in cost incrementally without recalculating the cost of the entire timetable. This attribute is particularly useful when the parallel version of the algorithm is implemented.

A PARALLEL ALGORITHM

Simulated annealing, while very effective at solving the timetabling problem, can be extremely slow (for example, a data set for a real school took 14 hours of processor time on a SUN 3/60). The elapsed time taken to run the simulated annealing algorithm described can be improved by using a parallel algorithm rather than a serial one. In the serial algorithm, each permutation of the elements is performed sequentially, and the new configuration either accepted or rejected. A new configuration is not generated until the previous one is performed. However, it is possible to perform multiple permutations concurrently, providing each permutation is independent of the other permutations.

A parallel algorithm can be implemented by assigning multiple *processes* to the task of permuting the timetable. The timetable must be held in a *shared* memory area accessible to all processes. Each process independently chooses an element to move (from a **from** period), and a **to** period. In order to prevent other processes from choosing the same element and **to** period, they must *lock* the element. It is not actually desirable to lock the entire **to** period, as this would severely limit the number of concurrent swaps which were possible. Instead, they only need lock the *teacher, class* and *room* in the **to** period. Similarly, the *teacher, class* and *room* must be locked in the **from** period. These items must be locked so that no other process can effect the cost computation of a given process. The incremental cost computation technique allows a process to calculate the change in cost without recomputing the cost of the entire timetable. Once these items have been locked a process can determine the change in cost independently from all other potential swaps. If a process chooses an element, teacher, class or room which is already locked, then it must abandon the choice and try another.

The maximum number of concurrent processes depends on the size of the timetable. If there are too many processes for a given number of elements, then the number of processes abandoned swaps will be too high. Every time a choice is abandoned the effective speedup is decreased.

The *locks* described above can be implemented by simple read-modify-write variables in shared memory. A process can read a lock, and write a special marker value into the lock with an indivisible cycle. If the process reads the special lock value then it knows that the lock is already current and can abandon the choice. Such read-modify-write variables are not uncommon for multiprocessor machines. True semaphores are not required because the process does not wish to suspend when a lock is already claimed. Deadlock is not possible because a process *backs-off* any transaction which it cannot complete.

Implementing the parallel algorithm on a shared memory multiprocessor is relatively simple. The timetable must be held in shared memory, together with the lock variables. Once the timetable has been initialised the master process can *fork* and spawn as many child processes as necessary. Each child process permutes the timetable until the system is frozen, or the timetable has been solved. Each process can maintain its own temperature, or access a shared temperature variable. Similarly, each child process may share a common random number generator or maintain its own. If they use separate random number generators then each must use a different initial seed to avoid the same pseudo random sequences.

EXPERIMENTAL RESULTS FOR PARALLEL ALGORITHM

The parallel algorithm described in the previous section has been implemented on a conventional shared memory multiprocessor, a 10 processor Encore MultiMax. Some test data was presented to the parallel program, and the effective speedup was measured. The results are shown in Table 1. The execution time is shown for the purely serial code on the Encore, and then the parallel code using 1, 2, 4, 6 and 8 processors. It was not possible to use all 10 processors because of external demands on the machine. The Peak speedup is defined as the time for the single process run divided by the smallest multiprocess time. The peak speedup for the small problems is low because there are not sufficient resources to keep the processors busy. In general, the larger the problem, the greater the speedup. It can be seen from these examples that whilst not ideal, the speedup in many cases is significant.

When evaluating a parallel algorithm, it is important not just to consider speedup, but also absolute speed. The efficiency of the parallel algorithm can be expressed as the time taken to solve the problem using the parallel code with one processor, divided by the time taken using a serial version of the program with one processor. A number of experiments have indicated that the parallel code varies from equal, to at worst two times slower than the serial version. Thus, in this worst case two processors are required before the parallel code overcomes the cost of the locking and synchronisation code. The serial time is shown in Table 1. In most of these

TABLE 1 - Results of Parallel Execution

Test Data	Number Elements	Serial Time	1	2	4	6	8	Speedup
				Time in Secs for Number of Processors				
1	100	43	41	20	16	13	13	3.2
2	150	79	97	52	29	27	22	4.3
3	200	139	180	87	54	38	33	5.4
4	250	211	255	142	85	71	72	4.4
5	300	390	729	409	218	157	137	5.3
6	400	402	529	288	159	103	78	6.7
7	600	807	842	528	256	165	150	5.6
8	600	774	906	473	265	194	135	6.7
9	2252	5700	6900	3840	2100	1440	1020	6.8

examples the parallel version was only slightly slower than the serial code. Large timetable data sets could easily be expected to use up to 32 MIMD processors, providing a significant speedup. Further, it is possible to omit much of the locking code, which would reduce the cost of the parallel solution substantially by lowering the overheads. This technique is currently being investigated.

ROUTING USING LEE'S ALGORITHM

Routing a printed circuit board involves finding paths for the wires that connect integrated circuits. The wires are not allowed to cross on any given layer, however, more than one layer may be used to connect all wires. There are many different routing algorithms, and a good summary can be found in IEEE (1983). The oldest and most general technique uses a wavefront, which is advanced from the source to the destination. First described by Lee in Lee (1961), this scheme finds the shortest path between any two points, and is based on a fixed grid of cells. Cells either contains a value indicating that the board position is either occupied by a hole, circuit track, or part of an advancing wave. The cell corresponding to the source location is initialised with a low score value, e.g. 1, and is then added to a current wave front list. Then all of the cells surrounding the current wave front are set to the score 2, and are added to the wave front list. The initial cell is subsequently removed from the wave front list. This process continues until the wave either meets the target destination, or entirely fills the board. If the destination is not touched, then there is no path from the source to the destination. If the wave meets the target, then a path is traced by tracing a path of decending cell value from the destination until the source is reached. This basic technique is enhanced to cater for practical considerations in the routing of printed circuit boards and integrated circuits.

A PARALLEL ROUTER

The basic parallel algorithm attempts to detect wires that are *unrelated* and route them concurrently. For example, wires that are at different ends of a board are likely to be completely independent, and thus could be routed at the same time. These wires are grouped by recursively dividing the board into regions, and passing the wires completely contained in a region to the router separately. Wires which are contained in more than one region are held until all those at lower levels have either been successfull routed, or could not be routed because of congestion.

At each stage, the router calls itself, and splits the region it has been passed into two sections. It created three wire lists; two for those wires completely contained in each region, and one for those wires which cannot be partitioned. This process continues until either there are no wires left to route, or some preset maximum depth has been reached. After the routine has called itself recursively, it proceeds to route the wires it had held onto, and also attempts to route those which could not be connected by the recursive call. Thus, each call to the router accepts a wire list, and returns a wire list containing the wires it could not connect.

This process continues until it returns to the root of the call tree. If all of the wires have been connected then the router terminated, however, if there are wires left then it starts the process over again with two new layers. It is worth noting that this recursive procedure effectively sorts and routes the wires by length, because the shortest wires are passed down to the lower call levels, and are thus routed first.

In itself, this recursive algorithm is sequential. However, by replacing the recursive call with a recursive *parallel* procedure call, each section can be routed concurrently. There is no need for any communication between the processes after the procedure call because they are all operating in separate areas of the board. If the algorithm is being executed on a shared memory multiprocessor, the board should be placed in shared memory to avoid copying it between calls. The reference parameters used in the call must also be placed in shared memory so that the results can be returned to the calling processs. If the machine is a distributed message passing machine, then the wire lists and board contents must be transmitted to the processor executing the called code.

EXPERIMENTAL RESULTS

The parallel router was exercised on a circuit board which measured 5 inches by 8 inches and contained 559 wires between 52 integrated circuits. Whilst this is quite a small circuit, it was large enough to demonstrate the effectiveness of the router without using enormous amounts of processor time. The program was run using 1, 2, 4 and 8 processors. During each run two traces were maintained. The first logged the number of active processes against time. The second logged the number of wires routed against time, and shows the progress of the program. These traces are plotted for each of the test runs and are shown in Figures 1 and 2, although they are not all included in this paper due to lack of space. The activity traces show how much concurrency is being extracted at any point in time. They all exhibit the same basic form; they start by using the maximum number of processors, but diminsh as the run proceeds. The reason for this is that the divide and conquer algorithm proceeds to the bottom of the recursion tree where it routes as many wires conncurrently as possible. Consequently, it consumes as many processes as possible. However, the long wires are held until the shorter ones have been routed. The progress traces also all have the same form. The shorter wires are routed first and are connected fairly quickly. The more processors available, the quicker these wires are connected. The longer wires take more time to connect for a few reasons. First, the basic maze routing algorithm execution time is proportional to the square of the wire length. Second, the board is already congested by the shorter wires, and thus the longer wires have to route around them. Third, there are fewer processors which can be applied to the longer wires because they span more than one board region. Because of these reasons the curves all flatten out, and the time to route the entire board is dictated by the time to route the long wires.

The results show that increasing the number of processors from 1 to 8 has only decreased the execution time from 180 minutes to 100 minutes. Because of these problems, an additional level of concurrency was added to the program. In the modified scheme, the wave front for each wire is advanced by more than one process, thus speeding the routing of individual wires. This technique is not nearly as efficient as wiring separate wires concurrently because using more than one process to advance the wave front requires substantial interprocess communication and synchonisation, whereas the divide and conquer technique does not require any interprocess communication until all wires at a level are routed. However, the parallel wave front approach has the ability to speed the routing of the long wires, which are responsible for controlling the time to connect the entire board. Using this scheme, the execution time can be further reduced to 50 minutes. Whilst not optimal this can constitute a significant speedup. Work is continuing on increasing the effciency of the algorithms used in the router.

CONCLUSION

Whilst it is too early to report, this scheme seems to require no more cycles to achieve the same cost solution, and is much faster because the locking overhead is not present. With this approach, the parallel code could easily be as fast as the serial version.
This paper has demonstrated how concurrency can be applied to two different computationally expensive programs. Both use different programming techniques to divide the problem. The simulated annealing code is best solved using a number of different worker processes, each of which manipulate and relax a central shared structure. Workers are created once when the program starts. They resolve their resource contention using simple memory locks, without the need for complex semaphore or monitor software. Deadlock is prevented by using a backoff technique. The simulated annealing code exhibits quite good speedup, especially for large problems. One inefficiency in the algorithm is that it requires shared resources to be locked so that two processes do not try and move the same tuple. This locking can make a single process program as much as two times slower than the serial code, although most of the examples shown in this paper are within 30% of the serial speed. Some work has been done on the effect of removing the locks altogether, and allowing the algorithm to proceed with inaccurate values for the costs.

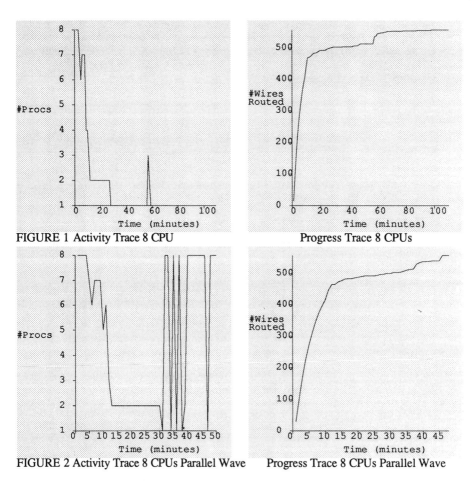

FIGURE 1 Activity Trace 8 CPU   Progress Trace 8 CPUs

FIGURE 2 Activity Trace 8 CPUs Parallel Wave   Progress Trace 8 CPUs Parallel Wave

The router code uses a divide and conquer algorithm rather than a fixed number of workers. Each time the board is divided a new process is created, and each time a procedure call is made a process is destroyed. This fork-join approach has been largely discouraged because the process creation and destructions costs are usually quite large. However, in this problem, the cost of routing a number of wires in an area is so high, that the process creation and destruction costs are not relevant. The algorithm performs quite well on problems with a very high degree of wire locality, because there are very few wires kept at higher levels of the call tree. However, as the number of non-local wires becomes significant, the algorithm performs very badly. This performance degradation is not because of the implementation, but because the wires held are longer than those passed on, and thus take much longer to route (The complexity increases roughly as the square of the distance between the wires). To make matters worse, they are being routed by fewer processes; thus the algorithm slows down from two factors. To improve the performance of this program we have added a second level of concurrency, which takes over when there are not enough recursive workers to occupy the multiprocessor. In the second scheme the wave front is advanced by a number of worker processes, in much the same manner as they are used in the simulated annealing code.

Both of these programs have been implemented on shared memory machine. It is almost impossible to map the simulated annealing efficiently onto a message passing machine because, whilst the timetable itelf can be split across many machines, information about the numbers of teachers, classes and rooms in each period must be available to all processors. Further, this

143

information is quite volatile, and could not be copied around efficiently. The printed circuit router could be mapped onto a message passing machine, but the board structure would need to be transmitted between processors each time a split and join occured. For large circuits, this could involve quite a large amount of data. Another disadvantage is that the processor which runs the process at the root of the recursive tree requires sufficient memory to hold the entire board.

Key sections of these programs have also been written in a functional language called SISAL (McGraw (1985)) for execution on a prototype dataflow machine as described in Abramson and Egan (1988). Dataflow machines allow easy extraction of concurrency, and SISAL allows easy expression of concurrency without program *side-effects*. The simulated annealing code has also been written in a low level dataflow language (DL1) and run on the prototype machine. The results of the simulated annealing code written in DL1 were similar to the shared memory results. Many problems have been encountered in expressing the problem in the SISAL, mostly because the *single assignment* property of the language makes it very difficult to perform parallel updates on shared structures without excessive data movement.The low level dataflow language did not present any major obstacles other than its lack of expressive power as a general purpose programming language. The concurrency was extracted easily, indicating that the dataflow multiprocessor was an appropriate execution environment for this class of problem.

The router was coded in SISAL without difficulty, and allowed easy expression of the concurrency. However, because of the nature of the problem, the SISAL implementation required large amount of data to be copied and rebuilt. Research is continuing on the best way to specify such problems with minimal data movement, without restricting the concurrency.

ACKNOWLEDGEMENTS

The Parallel Systems Architecture Project is a joint project between the Commonwealth Scientific and Industrial Scientific Organisation (CSIRO) and the Royal Melbourne Institute of Technology (RMIT).The programs used for producing the experimental results were written by Tony Starr, Junko Freidin, Ha Nguyen and Mark Gazzola. The members of the Parallel Systems Architecture Project, in particular Dr. G. Egan, assisted with various parts of this work. Thanks must also go the the RMIT Department of Computer Science for providing access to their Encore multiprocessor.

REFERENCES

1. Abramson, Using Simulated Annealing to Solve School Timetables: Serial and Parallel Algorithms, RMIT Technical Report TR-112--069R, 1988, submitted for publication.

2. Abramson and Freidin, A Parallel Router for Printed Circuit Boards, RMIT Technical Report, 1989, in preparation.

3. Abramson and Egan, An Overview of the RMIT/CSIRO Parallel Systems Architecture Project, *The Australian Computer Journal,* Vol 20, No 3, pp 113 - 121, 1989.

4. IEEE,*Transactions on Computer-Aided Design of Integrated Circuits and Systems*, October 1983, Volume CAD-2, Number 4, 1983, ISSN 0278-0070.

5. Lee, An Algorithm for Path Connection and its Applications, *IRE Transactions on Electronic Computers*, Vol EC-10, No 3, pp 346 - 352, September 1961.

6. McGraw, SISAL: Streams and Iteration in a Single Assignment Language, Language Reference Manual, Lawrence Livermore National Laboratories Technical Report, M-146, March 1985.

7. Van Laarhoven and Aarts, *Simulated Annealing: Theory and Applications*, D. Reidel Publishing Company, Kluwer Academic Publishers Group, 1987.

Interactive Access to Subroutine Libraries: An Interlink for SENAC

K. A. BROUGHAN
University of Waikato
Hamilton, New Zealand

INTRODUCTION

In this paper questions concerning the interaction between symbolic and numeric software systems are discussed. Particular emphasis is given to the software system SENAC which contains an interactive interface ("interlink") to the NAG Library of numeric subroutines. The relationship between numeric and symbolic algorithms, the development of problem solving environments for scientific computation, and general questions relating to the interlink concept are also considered.

Natural scientific computer languages need to interface to Fortran because the fortran subroutine libraries represent a body of such a valuable heavily used, well tested, robust and efficient code that they would be too expensive to replace.

Languages for the symbolic manipulation of algebraic and analytic expressions have been developing during the past two decades. Here the attempt to provide natural mathematical and scientific languages has been more obvious than is the case with the subroutine libraries.

The symbolic languages provide the types of facility that are needed at the first level of development for natural scientific languages. For example they may be used interactively, permit the use of mixed symbolic and numeric expressions, have automatic formatting of output, and allow reassignment of output and input, Calmet and van Hulzen (1983). These are some of the features that users of symbolic languages enjoy. After appropriate interfacing has taken place a symbolic language can become the host for both symbolic and comprehensive numeric computation. The input syntax and fundamental data types remain those of the host language whose facilites are greatly expanded through the functions and subroutines from the interfaced numeric subroutine library.

In addition to being able to call numeric routines, there are advantages in having a symbolic language directly available. Problem solving activity usually includes a symbolic phase wherein the problem is posed in a mathematical form and then discretised. This is then followed by a solution phase wherein numeric algorithms are employed. The symbolic phase is greatly assisted through the use of a symbolic system, especially in the case of practical problems where the number of variables involved is large.

The numerical subroutine library becomes a great deal easier to use. The time spent in preparing code to call a subroutine is significantly reduced. A user may readily call a routine several times with different parameter settings. Several related routines could be run and their output compared, since this is so easy to do. The more sophisticated routines now become more accessible because of the reduction in programming time.

SENAC

SENAC is an interactive symbolic manipulation language for numeric and algebraic computation. The user is able to describe problems in a mathematical form and obtain answers using numeric routines from the NAG Library with none of the programming overhead required when using a language such as Fortran.

The software has been designed for people who use mathematical methods to solve problems. This includes scientists, engineers, statisticians, applied mathematicians, economists, university research students in educational institutions, research organisations, companies and government departments. An individual or organisation that currently uses a numeric subroutine library would be a potential user of SENAC. Individuals with high performance workstations and currently looking at a new generation interactive mathematical problem solvers would also be potential users, The Gibbs Group (1984), Wilson (1984).

FEATURES

The combination of a symbolic manipulation language with a high-level tailored interface to a comprehensive subroutine library make this system unique. The fact that similar interfaces may be written to other libraries or individual user programs means that this software breaks new ground for the scientific computation field. Further interfaces and tools for automated interface development by users are under development.

SENAC includes the following features:
- An easy to use natural user interface with extensive error checking.
- Interactive input of symbolic and numeric expressions.
- Two dimensional symbolic expression output formatting.
- Matrix, list and sparse matrix expression data types.
- User defined functions and programs for symbolic and numeric transformation.
- Differentiation of expressions.
- Translation of user defined functions into Fortran for SENAC calculation or external use.

These features are included in the first release. The features scheduled for inclusion in the second release includes: limits, taylor series, laurent series and fourier series; polynomials, rational function and rational expression data types; factorisation of polynomials with integer coefficients; symbolic integration of expressions including the Risch algorithm.

SENAC will contain an implementation of the full Risch algorithm for the symbolic integration of rational function integrands which contain nested exponentials and logarithms. SENAC is able to solve the following integration problems. None of these examples can be completed successfully by the Risch algorithm of the symbolic language MACSYMA.

1. $\int \left(1 - \dfrac{1}{x \log(x)^2}\right) \exp\left(\dfrac{1}{\log(x)} + x\right) dx$

$= \exp\left(\dfrac{1}{\log(x)} + x\right)$

2. $\int \dfrac{\exp(\exp(x+1)/x)}{\exp(x^2)} \left(\dfrac{\exp(x+1)}{x} - \dfrac{\exp(x+1)}{x^2} - 2x\right) dx$

$= \dfrac{\exp(\exp(x+1)/x)}{\exp(x^2)}$

3. $\int -\dfrac{(\exp(x)+1)\log(1/\log(x)+1)}{(\exp(x)+x)^2} - \dfrac{1}{x(\exp(x)+x)(1/\log(x)+1)\log(x)^2} dx$

$= \dfrac{\log(1/\log(x)+1)}{(\exp(x)+x)}$

4. $\int \dfrac{1}{x\log(\exp(x)+1)} - \dfrac{\exp(x)\log(x)}{(\exp(x)+1)\log(\exp(x)+1)^2} dx$

$= \dfrac{\log(x)}{\log(\exp(x)+1)}$

5. $\int \dfrac{(x+1)\exp(x/\exp(1/x))}{\exp(2/x)} dx$

$= \dfrac{(x - \exp(1/x))\exp(x/\exp(1/x))}{\exp(1/x)}$

6. $\int \dfrac{\exp(1/\log(x))}{x\log(x)^3} dx$

$= \dfrac{(\log(x)-1)}{\log(x)} \exp(1/\log(x))$

SENAC will contain partially probabilistic algorithms for factorising large integers into prime factors. Examples:

TABLE 1. Factorization Examples

Number	Factors	Min:Sec
12403928231	(1831)(6774401)	0:3
52199328221	(2333)(25339)(883)	0:3
4385264716677487	(8707)(1472789)(601)(569)	0:7
499823456233	(73709)(6781037)	0:7
1046648839426789	(4889)(594989)(3301)(109)	0:15
72208953972803	(1151)(1307)(448597)(107)	0:26

Number	Factors	Min:Sec
344176664219723	(1765891)(194902553)	0:48
27813602910515039	(129719)(611953)(350377)	0:59
2004583724863781	(1497557)(52127)(25679)	1:6
4830863254898621	(48912491)(98765431)	1:18
168826792187441653	(636190001)(265371653)	2:10
4652171006191243	(82471201)(56409643)	2:54
27793221391729	(417523)(524149)(127)	3:43
112725377331627187	(529510939)(212885833)	5:47
32018664494925586909	(4278255361)(7484047069)	6:0
14514000917331368813	(4999465853)(2903110321)	12:36
1919240390842624114697	(54410972897)(35273039401)	15:13
2550696547957308071911	(77158673929)(33057806959)	15:22
613211391952174304227	(311347)(1969543281137041)	17:10
134354439210659932541377	(487824887233)(275415303169)	31:59
54373131685714568603971	(328006342451)(165768537521)	38:40
33533386858766252958079231	(37243)(701788271390597999)(1283)	125:52
4518912801236515512194249	(30197)(3750587875813)(39899809)	144:9

NUMLINK

In addition to symbolic language features the software includes a comprehensive interface Numlink from SENAC to the NAG Library of numeric subroutines.

The advantage of using a Library routine is its efficiency, robustness and correctness. The disadvantage is the time required to solve problems. Finding the routine, understanding the documentation, writing code including subprograms for function and derivative evaluation and formatting input and output are all time consuming and give rise to errors. It can take a week or more to do all that is necessary to obtain answers from many library routines.

Numlink retains the advantages and all but eliminates the coding problem by using interface routines to automatically perform all of the steps in the coding exercise which could possibly be automated. The time to get results is reduced from days to minutes without loss in functionality. The scope and power of the NAG Library become more readily available to users.

Input and output variables are classified into two types: those relating to the problem to be solved such as the name of an input function, endpoint of a function domain and so on, and those relating to algorithm tuning or analysis which the user may not need to change frequently. These latter variables are placed in the environment and need never be looked at unless the user wishes. Subprograms required to be supplied by the user are written automatically and incorporated in SENAC in their compiled form. There is a significant reduction in the complexity of the library call, which then reflects the essence of the problem in a natural manner.

AN EXAMPLE

The user wishes to find the minimum value of a function defined on a subrectangle of four dimensional euclidean space using a numerical subroutine. The function definition is as follows:

$$f(x, y, z, w) := (x + 10y)^2 + 5(z - w)^2 + (y - 2z)^4 + 10(x - w)^2$$

The domain is the cartesian product of the real intervals

$$[1, 3], [-2, 0], [1, 3], [0, 4]$$

and the iteration is to begin at the point with coordinates

$$[2, -1, 2, 2]$$

Suppose the user selects the routine e04lbf from the NAG Library to perform the minimization. There is a daunting amount of work involved writing code which can call this routine. Initialization of variables, writing subroutines for function values and derivatives, formatting of the output all require detailed work. This should be unnecessary for a problem with such a simple description.

rather than
....

many many lines of FORTRAN code
.....

E04LBF(N,FUNCT,HESS,MONIT,IPRINT,MAXCAL,ETA,TOL,
1 STEPMX,IBOUND,BL,BU,X,HESL,LH,HESD,ISTATE,F,G,IW,
1 LIW,W,LW,IFAIL)
....

more more lines of FORTRAN code
end

in the SENAC environment, the call could be made in the following manner:
f(x, y, z, w) := (x+10*y)**2 +5*(z-w)**2 +(y-2*z)**4 + 10*(x-w)**2;
rectangle: [[1, 3], [-2, 0], [1, 3], [0, 4]];
initial: [2, -1, 2, 2];
e04lbf(f , rectangle, initial) ? and produce the output
[2.4338, [1.0, -0.0852, 0.4093, 1.0]]

where the first element of the list represents the minimun value of f and the last the point at which the minimum occures.

Other subroutine parameters such as the tolerance and maximum number of algorithm iterations are available in the environment, and may be inspected and changed should the user so wish.

Input and output data are always valid SENAC data types and so may be manipulated interactively - in particular the output from one call becomes the input for another with no special formatting.

All the features of the NAG Library become accessible to the user through SENAC - these features include fourier transforms, special functions, sorting, linear equations, eigenvalue equations, linear algebra operations, ordinary differential equations, integral equations, partial differential equations, quadrature, curve and surface fitting, polynomial zeros, roots of equations, non-linear optimisation, linear programming for small problems, basic statistics, non-parametric statistics, analysis of variance, correlation, regression and time series analysis.

DOCUMENTATION

The user documentation consists of a user manual for SENAC which includes a typeset example for each SENAC function, a user manual for the Numlink interface to the NAG Library, again including an example program for each interfaced function. The manuals are available on-line as part of the system.

HARDWARE AND SOFTWARE REQUIREMENTS

SENAC has been written in Franz LISP. It is currently available for VAX/VMS and Sun-3 processors. Although Franz LISP has been implemented on a wide range of machines a port of the system to a version or versions of common lisp is to commence in the very near future. This should enable the system to migrate to machine families such as the Sun-4, DECstation, pVAX and Cray.

SIMILAR SYSTEMS

The author is not familiar with any mathematical software that comes close to SENAC in terms of its tailored interface to the NAG Library and its realisation of the interlink concept. Because the NAG Library continues to develop at the hands of high level experts in scientific computation and numerical analysis, the on-going adaptation of SENAC to emerging prallel architectures and improvements in algorithms design is ensured. Systems such as Macsyma, Reduce, Mathematica, Maple all have more comprehensive symbolic manipulation features - but these offer some but little relief to a person who needs to use a subroutine library. Some systems and languages offer facilities for interaction with other languages and translation from a symbolic language to a numeric language. The work and expertise required for users to write their own interfaces for each problem being solved is such that normally these interfaces would not be written.

STATUS OF THE PROGRAM

The current features and implementations are listed above. The interface is now at the level of Mark 13 of the NAG Library which was released late in 1988. The printed manual has been written and typeset. Release notes and the on-line documentation are nearing completion. An upgrade of the Sun-3 version to SunOS 4.0 is being planned for the near future.

DEVELOPMENT OF INTERFACES

The work of constructing interfaces may be reduced considerably through using an "interface compiler". Such a program takes a description of a Fortran subroutine and information supplied by the user. It would output a Lisp (object) program. The Lisp program would enable the Fortran subroutine to be called interactively from a host symbolic language. Using an interface compiler, interfaces for a users own subroutines as well as Libraries could be constructed readily. A partial compiler has been written by the Mathematical Software Project at the University of Waikato and used extensively in the production of Numlink.

REFERENCES

1. J. Calmet and J. A. van Hulzen, in *Computer Algebra and Symbolic and Algebraic Computation*, ed B. Buchberger et al, pp. 221-243, Springer-Verlag, 1983.

2. The Gibbs Group, *Gibbs - a programming environment and workstation for scientists*, Cornell University Center for Theory and Simulation in Science and Engineering, CTSSE 84-7, 1984.

3. K. G. Wilson, Planning for the future of U.S. scientific and engineering computing, *Comm.ACM*, vol. 27, pp. 279-280, 1984.

Adaptive Grids and Super-Computing

**G. F. CAREY, E. BARRAGY, S. BOVA, R. MCLAY, A. PARDHANANI, and
K. C. WANG**
University of Texas at Austin
Austin, Texas 78712, USA

INTRODUCTION

The need to compute solutions to complex large scale scientific and engineering problems continues to spur developments in both supercomputing hardware and in algorithms. Vector processors are now routinely applied and parallel or parallel-vector systems increasingly available (e.g., see Carey (1989)). Even with these advances in hardware, efficient methods and fast algorithms are needed if the true gains are to be achieved. Here we briefly examine some issues related to adaptive grids with vector and parallel processing, drawing on several related investigations and numerical experiments by the authors.

DISCUSSION

Grid generation has become a major "bottleneck" in design analysis of large systems. Various techniques ranging from simple algabraic grid interpolants to methods that require PDE solution are regularly applied (e.g., Thompson *et al*, (1985)). The PDE methods are more effective but may require computational effort comparable to that for the problem in question. Block methods are widely used and seem suited to parallel supercomputing because they naturally lend themselves to a "divide and conquer" approach. That is, the domain can be decomposed to a number of blocks on which grids can be generated and optimized concurrently. If the parallel system has coarse granularity (such as the CRAY or Alliant systems) then the number of blocks can be chosen equal to the number of processors (or a multiple thereof).

The grid generation algorithm then proceeds as follows:

1. Partition domain to a number of blocks.

2. Define the grid point coordinate data on the block boundaries.

3. Concurrently generate the interior block grids by either algebraic, interpolating or PDE solution strategies.

For good accuracy and high efficiency the grid should be clustered in regions where the solution gradients are large. (See, e.g., Shephard and Gallagher (1980).) This can be viewed as an optimization problem in which the grid points are to be redistributed to minimize an objective function. Other properties such as smoothness of the grid and grid orthogonality may also be included in the objective function so that these properties of the grid are preserved. For example, in Pardhanani and Carey (1988), we extend the ideas of Brackhill and Salzman (1982) and use a Polak-Ribiere gradient scheme in an optimization algorithm for 2D and 3D grid optimization. The objective function has the form

$$F(v) = E(v) + \alpha_1 S(v) + \alpha_2 O(v) + P_1(v) + P_2(v) \tag{1}$$

where parameters α_1, α_2 weight the smoothness functional $S(v)$ and orthogonality functional $O(v)$; $E(v)$ is the error functional and $P_1(v)$, $P_2(v)$ are exterior and interior penalty functionals to prevent grid tangling near the boundaries. By sacrificing storage on the CRAY2 (uniprocessor calculations) we are able to enhance vectorization of the grid optimization algorithm for 3D grid redistribution.

The grid optimization procedure is computationally intensive but clearly lends itself to parallel computation. The previous block generation scheme, for instance, can be extended to include optimization within blocks as a concurrent computation. The algorithm can be most simply modified by adding the additional steps:

4. Solve the analysis problem (in parallel) on the current grid.

5. Optimize the interior grids concurrently within blocks.

Provided the regions where clustering is important are remote from the block interfaces, this scheme works well. If the approach is further extended as a Schwarz-type method with overlapping domains then global optimization is possible with data transfer between overlapping subdomains.

The grid in Figure 1 was generated on overlapping blocks with single grid layer overlap and interior grid points redistributed within the blocks. The problem corresponds to shocked compressible flow past a cylinder and the grid clustering near the shocks on the cylinder upper and lower surfaces is evident in the figure. The problem was solved using an Euler equation difference solution scheme on 2, 4, 6 and 8 grid blocks using the Alliant FX-8 vector-parallel computer. Speedup for multiprocessor solution was close to linear (8 blocks of equal size with 8 processors gave a speedup of approximatelly 7.7).

Instead of redistributing the grid one can introduce grid refinement procedures. Here the grid is adaptively refined from an initial coarse grid using local cell refinement in those areas where solution error estimates indicate the grid is inadequate. Various data structures and adaptive refinement schemes have been introduced (e.g., see Rheinboldt and Mesztenyi (1980)). Our approach is an extension of the procedure developed by Bank and Sherman (1980). Quadrilateral elements and quadrilateral bricks are subdivided by the refinement algorithm to quartets and octets, respectively, of subelements. This defines a quad-tree or octree data structure that can be traversed efficiently by means of a pointer system linking different levels in the tree. To maintain a well-graded mesh, we require that no two adjacent elements in the mesh be more than one level apart in the tree. At the interface

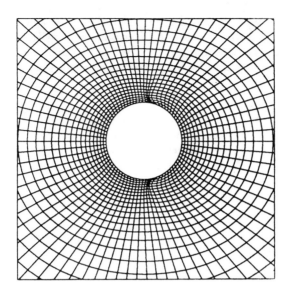

FIGURE 1. Redistributed grid for shocked flow problem using parallel block method.

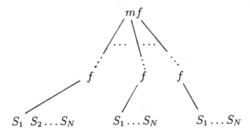

FIGURE 2. Quad-tree data structure for quadrilateral refinement.

between adjacent elements at different levels constraints are enforced to maintain
global continuity of the solution correctly. A sketch of the quad-tree for quadrilateral refinement to quartets is shown in Figure 2. This scheme has been implemented
for both bilinear and biquadratic elements and applied to a variety of applications
in, for instance, semiconductor device modelling and viscous fluid flow.

In the algorithm the solution on the current grid is post-processed to determine
where local refinement is needed, the grid is refined and the previous solution interpolated to the new grid points. A few iterative sweeps provide the solution on
the refined grid. Multigrid strategies can also be exploited in this setting. Hence
the solution and grid evolve together during this iterative process.

The iterative solver employed in our calculations is of conjugate gradient type
and can be vectorized over the elements in the current mesh. We have been developing element-by-element iterative methods for vector and parallel processing (Carey

155

et al, (1988)) and have applied them in conjunction with the adaptive refinement procedure. In particular, the biconjugate gradient method (BCG) is well-suited to nonsymmetric systems such as those encountered in transport processes involving convection. The basic algorithm is: for iterate $n + 1$ update solution $\boldsymbol{u}^n$ as

$$\boldsymbol{u}^{n+1} = \boldsymbol{u}^n + \beta_n \boldsymbol{p}^n \tag{2}$$

where $\boldsymbol{p}^n$ is the step direction and the step length β_n requires computation of the matrix-vector product $\boldsymbol{A}\boldsymbol{p}^n$ where $\boldsymbol{A}$ is the system matrix for the linear sparse system $\boldsymbol{A}\boldsymbol{u} = \boldsymbol{b}$. Here the matrix-vector products are the computationally intensive part of the calculation. They may be recast in terms of element matrix-vector products,

$$
\begin{aligned}
\boldsymbol{A}\boldsymbol{p} &= \left(\sum_e \hat{\boldsymbol{A}}_e \right) \boldsymbol{p} \\
&= \left(\sum_e \hat{\boldsymbol{A}}_e \hat{\boldsymbol{p}}_e \right), \quad \hat{\boldsymbol{p}}_e = X_e(\boldsymbol{p}) \\
&= \sum_e \widehat{\boldsymbol{A}_e \boldsymbol{p}_e}
\end{aligned}
\tag{3}
$$

where $\hat{\boldsymbol{A}}_e$, $\hat{\boldsymbol{p}}_e$ are element contributions expanded to system size and X_e is an extraction operator for element e. The matrix vector product reduces to global accumulation of element matrix-vector products $\boldsymbol{A}_e \boldsymbol{p}_e$. These in turn can be vectorized over the elements. That is, instead of assembling $\boldsymbol{A}$, we construct a 3-D array A_{eij}, for e ranging over the elements and i, j over the local element nodes. Then the loop on e becomes the innermost loop in the matrix-vector product calculation. Vector lengths are then of the order of the number of elements. Parallelization for coarse-granularity architectures is still feasible since the outer loops can be parallelized over the local node indices. Vectorization of the EBE scheme has been investigated on the CRAY and ETA uniprocessors and significant processor speeds achieved. For example, on the CRAY XMP24 system at the University of Texas High Performance Computing Center, we obtain uniprocessor performance at 150 Megaflops for the matrix vector products.

The adaptive refinement scheme has been applied to Newtonian and non-Newtonian viscous flow applications. In Figure 3 the grids after 2 and 3 refinement steps are shown for a die swell problem using a power-law fluid with power law index $n = 0.8$. The initial axisymmetric domain is rectangular with a uniform 10×5 grid on the half-section. The free surface location on the upper half of the right side is adjusted during flow solution. There is a strong singularity where the free surface emerges from the cylinder and the adaptive refinement procedure grades the mesh strongly in this region. This results in a robust scheme that determines die swell ratio more accurately and efficiently than previous methods.

The element-by-element scheme can be parallized over the elements by reversing the previous loop order so that the element loop is outermost. This approach is suited best to parallel scalar calculations if the elements are low order (due to the short vector lengths). Vectorization becomes efficent when high-order elements (spectral or p schemes) are considered.

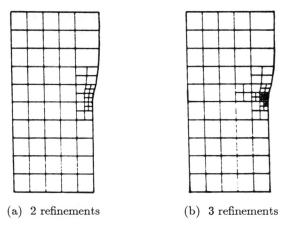

(a) 2 refinements   (b) 3 refinements

FIGURE 3. Parallel speedup for EBE scheme on Alliant.

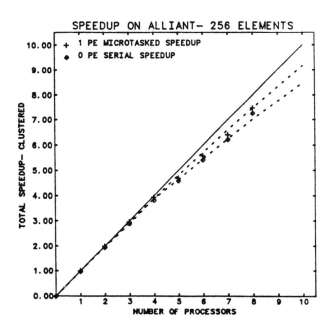

FIGURE 4. Meshes for axisymmetric die-swell problem using
refinement-solution procedure.

Parallel scalar calculations on the Alliant FX-8 computer using uniform grids
produce speedup close to linear (Figure 4). If the approach is applied with our adaptive refinement procedure the performance is still good. For example, matrix-vector
product computations for the Alliant with one processor and scalar calculations
takes .06 seconds and with 8 processors .009 seconds (speedup approximately 6.5)
in computations of a nonlinear semiconductor problem with adaptive refinement to
a final grid containing 700 highly graded elements (Sharma and Carey (1989)). The

vector-parallel scheme for matrix-vector product calculations with bilinear elements required .011 second and .0026 second for 1 and 8 processors, respectively.

The parallel element-by-element scheme has also been applied directly to grid generation and solution for the full potential equation (Carey and Barragy (1988)). Here the grid is generated by computing the approximate solution to Laplaces equation (corresponding to a conformal map PDE generator). A variational finite element method is used to generate the symmetric positive sparse system for the grid and parallel solution by element-by-element conjugate gradient iteration conducted. As before parallel scalar speedup close to optimal linear speedup is achieved.

CONCLUDING REMARKS

Special algorithms involving adaptive grid redistribution or adaptive refinement can be combined to significant advantage with iterative schemes on vector and vector parallel processors. Domain decomposition and nested iteration strategies are straightforward and particularly useful.

Acknowledgments. This research has been supported in part by the National Science Foundation and Department of Energy.

REFERENCES

1. Bank, R. E., and A. H. Sherman, A Refinement Algorithm and Dynamic Data Structure for Finite Element Meshes, Report TR-166, CNA, University of Texas at Austin, 1980.

2. Barragy, E., and G. F. Carey, A Parallel Element-by-Element Solution Scheme, *Int. J. Num. Meth. Eng.*, 26, 2367–2382, 1988.

3. Brackbill, J. V., and J. S. Salzman, Adaptive Zoning for Singular Problems in Two Dimensions, *J. Comp. Phys.*, 46, 342–368, 1982.

4. Carey, G. F., E. Barragy, R. McLay, and M. Sharma, Element-by-Element Vector and Parallel Computations, *Comm. Appl. Num. Meth.*, 4, 299–307, 1988.

5. Carey, G. F., M. Sharma, and K. C. Wang, A Class of Data Structures for 2D and 3D Adaptive Mesh Refinement, *Int. J. Num. Meth. Eng.*, 26, 2607–2622, 1989.

6. Pardhanani, A., and G. F. Carey, Optimization of Computational Grids, *Num. Meth. PDE's*, 4, 95–117, 1988.

7. Rheinboldt, W. C., and C. K. Mesztenyi, On a Data Structure for Adaptive Finite Element Mesh Refinement, *ACM Trans. Math. Soft.*, 6, 2, 166–187, 1980.

8. Sharma, M. and G. F. Carey, Semiconductor Device Simulation Using Adaptive Refinement and Flux Upwinding, *IEEE Trans. CAD of Integrated Circuits and Systems*, 1989 (in press).

9. Shepard, M., and R. Gallagher, Finite Element Grid Optimization, ASME PVP-38, 1980.

10. Thompson, J. F., Z. V. A. Warsi and C. N. Mastin, Numerical Grid Generation, North Holland, 1985.

Creeping Flow in the Earth's Mantle: Iterative Finite Difference Methods with Vector Processing

G. F. DAVIES
Research School of Earth Sciences
Australian National University
G.P.O. Box 4
Canberra ACT 2601, Australia

INTRODUCTION

The underlying cause of continental drift and plate tectonics is thermal convection in the earth's silicate mantle, which extends to a depth of about 2900 km, nearly half way to the center of the earth. This convection is sustained by heat generated from radioactive decay in the mantle and seems to be driven mainly by the negative buoyancy of the cool thermal boundary layer at the earth's surface.

Two distinctive features of this convection are the insignificance of inertial forces and the strong temperature dependence of the rheology of the mantle. With flow velocities of about 50 mm/year and a viscosity of the order of 10^{21} Pa s, the Reynolds number is about 10^{-23}, and the term creeping flow applies. At prevailing temperatures (about 1400 C), the viscosity drops by about a factor of 10 with every 100 C increase in temperature. The rheology is linear or weakly non-linear within the mantle, but near the surface it is strongly non-linear, culminating in plastic or brittle behavior in the upper 50 km or so.

These features yield a distinctive form of convection (Fig. 1) in which the top boundary layer is stiff because of its low temperature, but broken into about 10 large "plates" that are in relative motion (hence the term "plate tectonics"). This cool, stiff boundary layer, known as the lithosphere, is about 100 km thick. The mechanical behavior of the plates largely determines the near-surface location of upwellings and downwellings and hence the large-scale structure of the flow: upwelling where plates separate at "spreading centers" and downwelling where plates converge at "subduction zones". Furthermore the spreading is observed to be symmetric (equal amounts of new sea floor attach to each plate), but the subduction is asymmetric (only one plate descends). A result is that plates grow and shrink and are odd shapes and sizes, and the flow is thus time-dependent.

From the point of view of numerical modelling, mantle convection presents several challenges. The top boundary layer is relatively thin (Rayleigh number is about 10^7), but it contains very large viscosity gradients: at least a factor of 1000 increase in viscosity is required to achieve reasonable stiffness, and this occurs in less than 3% of the layer

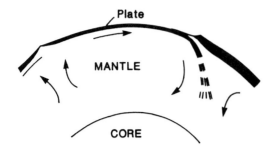

FIGURE 1. Sketch of convection in the earth's silicate mantle. The
cool surface boundary layer (black) is strong, but broken into plates
that are part of, and largely drive, convection in the underlying mantle.

thickness. The faults separating the plates are effectively of zero
thickness, so that the horizontal velocity within plates is spatially
constant but changes discontinuously at plate margins. The plates are
thus difficult to incorporate into models, but important observational
constraints, such as seafloor topography and small spatial variations in
the gravity field, depend directly on the flow structure and hence on
the presence of plates.

Two approaches to including plates have been successful in recent years.
A temperature-dependent viscosity can reproduce the stiffness of the
lithosphere, but this alone merely yields a stiff, static lithosphere.
One way to mobilise the plates is to artificially reduce the viscosity at
designated plate margins. Another way is to impose the plate velocities
as a boundary condition on the top surface: this gives a better
simulation of the plates, but special devices are then needed to get
realistic stresses and topography from the model. Essentially, the
structure of the flow, temperature and viscosity are established with the
surface velocities imposed, and then an instantaneous flow field is
recalculated with a free-slip surface, from which stresses and topography
are obtained. Details and a quantitative justification of these methods
have been given by Davies (1988, 1989).

Timestepping involves solution of an advective-diffusive heat equation,
and at each timestep the flow equation must be solved. Timestepping the
heat equation is relatively routine and is done here using an
alternating-direction implicit (ADI) method with upwind differencing.

The difficult part of the calculation is solving the flow equation, which
is fourth-order with rapidly varying coefficients. For the case of
constant viscosity (or purely depth-dependent viscosity), efficient,
direct fourier transform methods can be used. Iterative methods are
dictated both by the size of the finite difference grids needed and the
fact that the solution from the previous timestep is available as an
estimate of the solution for the new timestep. The method currently used
is an ADI method combined with conjugate gradient (CG) acceleration (or
equivalently, a CG method with an ADI preconditioner).

The discussion here will describe some features of the method used to
date, as well as briefly looking at the promise of multigrid methods.
The ADI/CG method has been run on a Fujitsu supercomputer with over 90%

vectorization, including the vectorised solution of pentadiagonal matrices.

ITERATIVE ADI SOLUTION OF THE FLOW EQUATION

The flow equation is derived from the Navier-Stokes equation with inertial terms neglected. In two dimensions it can be written in terms of a streamfunction, ψ:

$$[\eta(\psi_{xx} - \psi_{yy})]_{xx} - [\eta(\psi_{xx} - \psi_{yy})]_{yy} + 4[\eta\psi_{xy}]_{xy} = T_x \tag{1}$$

where η is the viscosity, T is temperature and x and y are cartesian coordinates.

For constant viscosity equation (1) reduces to the biharmonic equation, for which direct solutions can be obtained efficiently using Fourier transforms in one direction. For moderate viscosity variations (up to a factor of about 30 on a 64x64 grid), the terms involving viscosity gradients can be collected on the right hand side, with a biharmonic operator on the left, and the equation solved by simple iteration. For more extreme problems, this method fails to converge adequately.

ADI Iterative Method

An iterative ADI method proposed originally for the biharmonic equation by Conte and Dames (1958) has been extended to equation (1) (Houston and DeBremaecker, 1974) and found to have relatively good convergence properties (Davies, 1988). The two-stage equations are:

$$\psi^{n+1} = \psi^n + r[T_x - (\eta\psi^{n+1}|_{yy})_{yy} + (\eta\psi^n|_{xx})_{yy}$$
$$+ (\eta\psi^n|_{yy})_{xx} - (\eta\psi^n|_{xx})_{xx} - 4(\eta\psi^n|_{xy})_{xy}] \tag{2a}$$
$$\psi^{n+2} = \psi^{n+1} - r[(\eta\psi^{n+2}|_{xx})_{xx} - (\eta\psi^n|_{xx})_{xx}] \tag{2b}$$

where n denotes the iteration number and the derivative notation denotes finite difference approximations. Equation (2a) is implicit in y and (2b) is implicit in x. This is not an obvious way to split equation (1), and Conte and Dames did not reveal their rationale, but trials have revealed that it gives more rapid convergence than more obvious splittings. However, errors accumulate more rapidly, so double precision is necessary for all but the smallest grids.

The factor r is analogous to a time step, and its value may be varied with both iteration number and position. Here it was written as $r = \omega/c_{ij}$, where c_{ij} is the coefficient of ψ^{n+1}_{ij} in the right-hand side, and ω is independent of position. The value of ω is important, and the value which optimises convergence is best found empirically.

Different values of ω cause different wavelengths of the solution to converge fastest. An empirical example of convergence rate as a function of ω and of solution wavelength is illustrated in Fig. 2. A horizontal velocity was imposed on the top surface of the sketched box that varied sinusoidally with x. For each wavelength, the number of iterations for convergence to 1 part in 10^{-5} was found for a series of values of ω. It is evident that a single value of ω causes only a relatively narrow range of wavelengths to converge rapidly. Thus a better procedure is to cycle

163

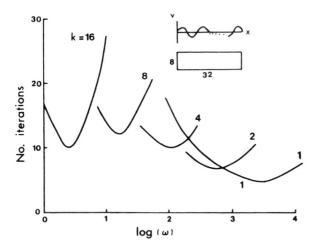

FIGURE 2. Convergence rate as a function of wavenumber, k, and iteration parameter, ω, for the ADI solution of equation (1).

through a range of values of ω in successive iterations. A series of about 5 values would be desirable in the example shown. A geometric progression of values from near 1 to a maximum approximated by $0.1m^3$ has been found to be satisfactory (where m is the number of grid intervals in the longest direction).

We note for the next section that equations (2) can be written in terms of corrections in a compact matrix form:

$$(I + V)\underline{\varepsilon}^{n+1} = \underline{b}^n - A\underline{\psi}^n \tag{3a}$$

$$(I + H)\underline{\delta}^{n+2} = \underline{\varepsilon}^{n+1} \tag{3b}$$

where A, V and H are matrices, $\underline{\varepsilon}^{n+1} = \underline{\psi}^{n+1} - \underline{\psi}^n$, $\underline{\delta}^{n+2} = \underline{\psi}^{n+2} - \underline{\psi}^n$ and $\underline{b} = r\underline{T}_x$. V and H are pentadiagonal matrices: an efficient vector algorithm for solving these is described below. The two stages can be written in a single matrix operation as

$$M\underline{\delta}^{n+2} = (I + V)(I + H)\underline{\delta}^{n+2} = \underline{b}^n - A\underline{\psi}^n \tag{3c}$$

Conjugate Gradient Acceleration

Even with these refinements, the ADI method is still unsatisfactory with large grids (64x256). It was found that the conjugate gradient method can be used as an accelerator to speed convergence by a factor of 6 or more. The difference form of equation (1) can be written as a matrix equation

$$A\underline{x} = \underline{b} \tag{4}$$

where x is a vector of all the unknown values on the grid. The conjugate gradient method (Reid, 1971) is based on minimising the associated quatratic form

164

$$f(\underline{x}) = \underline{x}^T A \underline{x}/2 - \underline{b}^T \underline{x} \tag{5}$$

which also minimises the residue

$$\underline{r}^n = \underline{b} - A\underline{x}^n \tag{6}$$

where x^n is the nth approximation to x. The vector that minimises (5) can be written as the sum of a series of orthogonal vectors which can be defined by a Gram-Schmidt procedure. Iterative conjugate gradient methods simply use the first vectors in this series, defined as orthogonal to $\underline{r}^n$. However, this method alone still tends to be slow to converge for large problems.

It has been found that the conjugate gradient method can be combined with other iterative methods with results superior to either used separately (Kershaw, 1978; Khosla and Rubin, 1981). Suppose M is a matrix which approximates A and is "easy" to solve. An example is the matrix expression (3c) of equations (2). Then M^{-1} can be used to "precondition" A, in effect minimising the residue

$$\underline{e}^{n+1} = \underline{x}^{n+1} - \underline{x}^n = M^{-1}\underline{b} - M^{-1}A\underline{x}^n \tag{7}$$

This is accomplished by solving

$$M\underline{e}^{n+1} = -\underline{r}^n \tag{8}$$

Since (8) is equivalent to solving equations of the form (2), the ADI method is integrated into the conjugate gradient method.

This method is more robust than the first iterative method mentioned above. Solutions have been obtained on a 64x256 grid with a zero-stress top boundary and viscosity variations of a factor of 1000. With a velocity boundary condition and a 128x256 grid, viscosity variations of at least 3000 are possible, although about 100 iterations may be necessary.

On the Australian National University's Fujitsu VP100 supercomputer, one iteration on a 64x256 grid takes about 0.23 s. This is about 300 times the speed on a Vax 11/780. With the vectorized pentadiagonal matrix solver described below, vectorization is about 92% (this is a measure of the ratio of the time used on the pipelined "vector processor" to the total vector plus scalar processor time).

"Parallel" Solution of Pentadiagonal Matrices

At the heart of the ADI solution of the flow equation is the solution of a pentadiagonal matrix for every row and column of the grid, so an efficient algorithm is essential. The traditional method of solving band matrices is based on an LU decomposition, which is equivalent to Gaussian elimination. However this method is not suitable for vector or parallel computation because it involves recursive operations between adjacent matrix elements. An alternative "cyclic reduction" method exists for tridiagonal matrices. In this method, every second row is eliminated, yielding a new tridiagonal matrix with only half the elements, and the process is repeated until a single row remains. Solution of this is trivial, and a series of back-substitutions yields the complete solution. This method can be generalized to pentadiagonal matrices, although the algebra becomes tedious with larger bandwidths.

The cyclic reduction method is suitable for vector or parallel computers. However, it is not as efficient as it might be because the vector length is halved with each recursion, and for other than 2^n-1 equations a special solution step for the last equations is required. Hockney and Jessop (1981) proposed a variation on this for tridiagonal matrices called "parallel cyclic reduction". In this method the vector length is the same for each recursion and there is no restriction on the number of equations. The solution is obtained with $\log_2(n)$ recursions and no backsubstitutions. The total number of operations is in fact greater than for "serial" cyclic reduction, but for vector or parallel machines the total speed can be greater for long vector lengths.

I have succeeded in generalizing this method to pentadiagonal matrices. Write the i-th equation in the form

$$a_i x_{i-2} + b_i x_{i-1} + c_i x_i + d_i x_{i+1} + e_i x_{i+2} = f_i, \quad 1 \le i \le n \tag{9}$$

For tridiagonal matrices ($a_i = e_i = 0$), Hockney and Jessop proposed extending this set for $i < 1$ and $i > n$ using $b_i = 1$ and $a_i = c_i = f_i = 0$. For pentadiagonal matrices, the required extensions are:

$$i \le 0: \quad a_i = d_i = e_i = f_i = 0, \quad b_i = c_i = 1$$

$$a_1 = 0, \quad b_1 = 1, \qquad a_2 = 1 \tag{10}$$

$$e_{n-1} = 1, \qquad d_n = 1, \quad e_n = 0$$

$$i \ge n+1: \quad a_i = b_i = e_i = f_i = 0, \quad c_i = d_i = 1$$

After this the algorithm is the same as for the tridiagonal case. The cyclic reduction operation is applied to every row of the matrix in each recursion, and the solution is obtained in $\underline{f}$ after $\log_2(n)$ recursions. Full details will be given elsewhere (Davies, 1989, in preparation).

SOME RESULTS

An example of the results from the method described here is shown in Fig. 3. This calculation was done on a 64x256 finite difference grid with a temperature dependent viscosity ranging from 1 (dimensionless) in the hottest fluid to 1000 in the coldest fluid. The fluid is uniformly heated throughout (simulating radioactive heating), the sides and bottom are insulating and the top is held at a temperature of 0. All sides are free slip. Because the heat input is prescribed, rather than the temperature difference across the layer, a Rayleigh number is defined as

$$R_Q = \frac{g\alpha Q D^5}{K\kappa\nu} \tag{11}$$

where g is gravity, α is thermal expansion, Q is the heat input per unit volume, D is the fluid depth, K is conductivity, κ is thermal diffusivity and ν is kinematic viscosity. This Rayleigh number is the Nusselt number times the more usual Rayleigh number. The value used for Fig. 3 was 10^7, still about a factor of 200 lower than for the mantle.

The viscosity has been reduced in narrow zones near the top at $x = 0$ and $x = 3$, and the intervening material given a little "push" to the right.

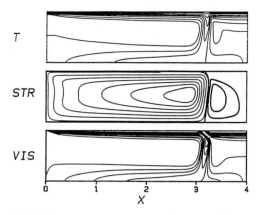

FIGURE 3. Example of a convection model including a simulated plate.
The panels show isotherms (T), streamlines (STR) and isoviscous lines
(VIS). The viscosity is temperature dependent except where it is reduced
near x = 1 and x = 3 to simulate faults between plates. The left plate
extends from x = 1 to x = 3 and moves to the right, descending under the
stationary right plate.

These low viscosity "faults" serve to free the otherwise stiff upper
boundary layer, and the Figure shows a near-steady state where there is a
"plate" on the left moving to the right and subducting at x = 3 under the
right plate, which is stationary. Note that the stiffness of the moving
plate keeps the velocity nearly constant in both its horizontal and
descending segments.

This is an example of a fully dynamic simulation of convection with
plates. It successfully reproduces the observed forms of heat flux out
of the sea floor, sea floor topography and the gravity perturbations due
to thermal density variations in the interior (Davies, 1989).

DISCUSSION

The method described here has been found to be reasonably robust and
efficient for a fairly difficult problem. However, some required
circumstances involving larger grids, larger viscosity variations or
faults simulated by internal boundaries so degrade convergence that the
method is inadequate. It seems that when large values of the iteration
parameter ω are used to converge the long-wavelength components, the
short-wavelength components, which are amplified by the large local
viscosity gradients, can diverge. Using only small values of ω prevents
this divergence, but then the long-wavelengths converge only extremely
slowly. An attempt to avoid the problem by filtering the corrections was
not successful.

The most promising candidate for a better method is a multigrid method
(Brandt, 1977). The essence of this method is to calculate a series of
corrections on successively coarser grids, and the coarse-grid solutions
are then interpolated back to successively finer grids. On each grid, a
very simple point or line iteration is used to "smooth" the wavelength
components near that grid size before moving to the next grid. In this
way the various wavelength components of the solution remain relatively

167

decoupled and the succession of grid sizes efficiently transmits
information across the entire grid.

Some very attractive properties are claimed for this method (Denning,
1987). The amount of work per iteration scales as $n^2\log(n)$ for an $n{\times}n$
grid, and the number of iterations is independent of n. In addition, on
a highly parallel computer with hypercube architecture, switching between
grids can be done very efficiently. It seems that not only does the
multigrid have nearly optimal scaling properties, but that it is
optimally suited to some emerging parallel computer architectures. It is
also claimed to be very robust and versatile (Brandt, 1977).

REFERENCES

1. Brandt, A., Multi-level Adaptive Solutions to Boundary-value
   Problems, *Math. Comput.*, vol.31, pp. 333-390, 1977.

2. Conte, S.D. and Dames, R., *Math. Tables Aids Computation*,
   vol.12, pp. 198-205, 1958.

3. Davies, G.F., Role of the Lithosphere in Mantle Convention,
   J. Geophys. Res., vol.93, pp. 10451-10466, 1988.

4. Davies, G.F., Mantle Convection with a Dynamic Plate:
   Topography, Heat Flow and Gravity Anomalies, *Geophys. J.*,
   forthcoming, 1989.

5. Denning, P.J., Multigrids and Hypercubes, *Amer. Scientist*,
   vol.75, pp. 234-238, 1987.

6. Hockney, R.W., and Jessop, C.R., *Parallel Computers*, Adam
   Hilger, Bristol, 1981.

7. Houston, M.H. and DeBremaecker, J.C., ADI Solution of Free
   Convection in a Variable Viscosity Fluid, *J. Comput. Phys.*,
   vol.16, pp. 221-239, 1974.

8. Kershaw, D.S., The Incomplete Cholesky-conjugate Gradient
   Method for the Iterative Solution of Systems of Linear
   Equations, *J. Comput. Phys.*, vol.26, pp. 43-65, 1978.

9. Khosla, P.K. and Rubin, S.G., A Conjugate Gradient
   Iterative Method, *Computers and Fluids*, vol.9, pp. 109-121,
   1981.

10. Reid, J.K., *On the method of Conjugate Gradients for the
    Solution of Large Sparse Systems of Linear Equations*, in
    Large Sparse Systems of Linear Equations, edited by J.K.
    Reid, Academic Press, London, pp. 231-253, 1971.

Supercomputing and Parameter Estimation in NMR Spectroscopy

M. H. KAHN
Computer Sciences Laboratory
Research School of Physical Sciences
Australian National University
G.P.O. Box 4
Canberra ACT 2601, Australia

INTRODUCTION

In one-dimensional nuclear magnetic resonance (NMR) spectroscopy the time domain signal, referred to as the free induction decay (FID) is a record of the magnetisation of the precessing spins in the molecule at discrete times after a known sequence of radio frequency excitation pulses. This magnetisation is measured in quadrature and is thus a complex random variable which can be theoretically modelled as a sum of damped complex exponentials plus complex white noise. That is,

$$y(n) = \sum_{k=1}^{K} \mu_k exp\{i\phi_k\} exp\{(-b_k + i2\Pi f_k)\Delta t n\} + e(n) \tag{1}$$

for $n = 1, 2, \ldots, N$. In this model, μ_k is the amplitude of the kth resonance, ϕ_k its phase, b_k the damping factor, f_k the frequency of the resonance, Δt is the time increment between observations of the FID and $e(n)$ is complex white noise.

More information about the structure of the spin system of a molecule can be obtained from two-dimensional NMR spectroscopy, where another dimension is achieved by varying the length of time between the radio frequency excitation pulses prior to measuring the FID. The model used is then

$$y(m,n) = \sum_{k=1}^{K}\sum_{\ell=1}^{L} \mu_{k\ell} exp\{i\phi_{k\ell}\} exp\{(-b_k + i2\Pi f_k)\Delta t_1 m\} exp\{(-c_\ell + i2\Pi g_\ell)\Delta t_2 n\}$$

$$+e(m,n) \tag{2}$$

for $m = 1, \ldots, M, n = 1, \ldots, N$.

In practice it is often difficult to keep the chemical solution stable enough to collect long series of data. A typical 2D NMR signal has 512 rows of 1024 complex points where the rows represent the first dimension. For such a data set the number of rows is

169

too truncated to cover the decay in that dimension. These experiments are expensive to run and the chemist may have only limited time available to collect the data. It is therefore imperative that any method for estimating parameters makes efficient use of the restricted data set and leads to reliable parameter estimates. The derivation of confidence intervals for these estimates could also help in the acceptance or rejection of hypothesised chemical structures.

To date any attempt at developing improved estimation procedures has been hampered by excessive computational time on conventional computers. Many of the techniques in use involve making incorrect assumptions about the data such as stationarity to arrive at a tractable computational problem. The advent of supercomputers such as the Fujitsu VP-100 at ANU enables a reappraisal of existing techniques and the development of improved estimation procedures based on sound statistical theory. These procedures must be tested on a variety of typical data sets which will have up to several hundred resonances. Even very thorough reviews such as Stephenson (1988) use examples with only a few peaks. Methods which perform well on such data may not be satisfactory when applied to an FID with many peaks. The comparison of a broad range of techniques applied to realistic data requires supercomputer speed and memory.

As the estimation of the parameters of the 2D model in (2) can be reduced to a series of 1D problems, the discussion will now centre on the 1D model as in (1).

TRADITIONAL METHODS OF FID MODELLING

The conventional procedure is to calculate the Fourier transform of the FID to obtain a spectrum from which estimates of the parameters are made by visual inspection. Techniques such as zero-filling and apodisation or windowing are used to increase the resolution of the spectrum but have not proved satisfactory at separating very close peaks, especially for severely truncated data sets. Noise in the FID can produce spurious peaks in the spectrum as can initial experimental conditions. These problems have led to investigations of methods of estimating the parameters in (1) directly from the FID.

This is a significant computational exercise as, for even moderately complex molecules, K can be of the order of several hundred.

Theoretically, the parameters can be estimated by minimising the sum of squares

$$\phi(\mathbf{r}, \phi, \mathbf{b}, \mathbf{f}) = \sum_{n=1}^{N} (y(n) - \mu(n))^2 \tag{3}$$

where $\mu(n) = \sum_{k=1}^{K} \mu_k exp\{i\phi_k\} exp\{(-b_k + i2\Pi f_k)\Delta tn\}$ and $\mathbf{r}$, ϕ, $\mathbf{b}$, $\mathbf{f}$ represent the vectors of r_k, ϕ_k, b_k and f_k respectively. This is a nonlinear optimisation problem which is known to be difficult to solve even for very small K. For large K other methods must be investigated even if only to provide initial estimates for the minimisation.

Most published techniques are based on Prony's method for fitting a sum of exponentials. The model is written as

$$\mu(n) = \sum_{k=1}^{K} c_k z_k^n \tag{4}$$

for $n = 1, \ldots, N$, where c_k and z_k are complex and $y(n) = \mu(n) + e(n)$.

It can be shown that the $\mu(n)$ satisfy a difference equation

$\mu(n) = -\sum_{k=1}^{K} b_k \mu(n - k)$

and the polynomial

$B(z) = b_0 z^K + b_1 z^{K-1} + \ldots + b_K$

with $b_0 = 1$ has complex roots $z_1, \ldots, z_k$ as given by Equation (4). The coefficients $b_0, b_1, \ldots, b_K$ are referred to as the Prony parameters.

Prony's method involves three steps as follows:

1. Solve for the b_k as the least squares solution to an overdetermined system of linear equations;

2. Find the zeros of the resulting polynomial $B(z)$ which gives estimates of the frequencies and damping factors;

3. Use these estimates to solve another least squares problem to estimate the c_k and thus the amplitudes and phases.

Authors such as Kumaresan and Tufts (1982) and Barkhuijsen et al (1987) have proposed different numerical methods for solving the least squares problem in step 1. The motives for the different choices range from minimising the effect of experimental noise to minimising computer time and memory. With the availability of increased computational power the latter has become less of a consideration and more emphasis can be put on choosing a technique which explicitly copes with the noise and uses numerical techniques with improved accuracy.

A STATISTICALLY RIGOROUS VARIATION TO PRONY'S METHOD

Prony's method is statistically suboptimal because the first least squares problem is an approximation to a true least squares approach for maximizing the likelihood of the estimates. Osborne and Smyth (1987) also show that Prony's method is not statistically consistent. This means that parameter estimates based on data points in a fixed time interval will not converge to their true values as the number of data points increases.

Osborne and Smyth show that minimising the sum of squares in (3) as a function of the 4K parameters is equivalent to minimising the objective function $\psi(\mathbf{b})$ with respect to the vector of Prony parameters $\mathbf{b}$ where

$$\psi(\mathbf{b}) = \mathbf{b}^T \mathbf{Y}^T (\mathbf{B}^T \mathbf{B})^{-1} \mathbf{Y} \mathbf{b} \tag{5}$$

where $\mathbf{B}^T$ denotes the transpose of $\mathbf{B}$. The matrix $\mathbf{Y}$ is an $(n-k) \times (k+1)$ matrix of real data and the matrix $\mathbf{B}$ is an $n \times (n-k)$ Toeplitz matrix with $K+1$ bands taking the real values $b_0, b_1, \ldots, b_k$. This separation of the linear and nonlinear parameters in the model is due to Golub and Pereyra (1973).

The same optimisation problem is posed by Bresler and Macovski (1986) for complex data such as the NMR FID and the objective function is then

$$\psi(\mathbf{b}) = \mathbf{b}^H \mathbf{Y}^H (\mathbf{B}^H \mathbf{B})^{-1} \mathbf{Y} \mathbf{b} \tag{6}$$

where $\mathbf{B}^H$ denotes the Hermitian transpose of $\mathbf{B}$. As $\psi(\mathbf{b})$ is a nonlinear function of the $\mathbf{b}$, an iterative method must be used. It is of interest to note that replacing $\mathbf{B}^H \mathbf{B}$ by the identity matrix gives the first step of the traditional Prony's method.

For the $(k+1)$th iteration, Bresler and Macovski treat $\mathbf{B}$ as a constant matrix calculated in terms of the estimates of $\mathbf{b}$ at the kth iteration. This method also is not consistent; the expectation of the objective function has a stochastic component which overwhelms the deterministic component as N increases. The argument is based on the consistency proof of Osborne and Smyth.

Working with real data, Osborne and Smyth set the derivative of the constrained objective function to zero and then solve an iterative nonlinear eigenvalue problem. This gives consistent estimates of the Prony parameters and thus the frequency and damping. The constraint is $\mathbf{b}^T \mathbf{b} = 1$.

The situation is more complicated when complex data is used. The objective function $\psi(\mathbf{b})$ is not a differentiable function of a complex vector $\mathbf{b}$ and the problem must be posed explicitly in terms of the real and imaginary data sequences. These are, for $y_n = y_{1n} + i y_{2n}$,

$$y_{1n} = \sum_{k=1}^{K} \mu_k exp\{-b_k \Delta tn\} \cos(2\Pi f_k \Delta tn + \phi_k) + e_{1n} = \mu_{1n} + e_{1n},$$

$$y_{2n} = \sum_{k=1}^{K} \mu_k exp\{-b_k \Delta tn\} \sin(2\Pi f_k \Delta tn + \phi_k) + e_{2n} = \mu_{2n} + e_{2n}.$$

As the μ_{1n} and μ_{2n} satisfy the same difference equations, the two data sets should estimate the same Prony parameters. However there are now $2K$ frequencies to be estimated, the f_k and their complex conjugates. In each individual set of data it is impossible to differentiate between true f_k from the complex model (1) and their corresponding conjugates. It can be shown that by combining the estimates of all parameters from both the real and the imaginary data, the conjugate terms cancel in theory and approximately cancel in practice.

For large values of K this now becomes an involved computational task including iterative eigenvalue problems to solve for $2K$ eigenvalues, finding zeros of polynomials of order $2K$ and solving the overdetermined system of linear equations for the linear parameters. Osborne and Smyth's method has been shown to perform well on small problems but it has yet to be applied to problems with many parameters. Clearly this is fertile ground for the use of supercomputers to enable detailed study of this procedure in a realistic time frame.

STATE SPACE FORMALISM

The second step in a Prony procedure is to find the zeros of a large order complex polynomial. In practice this becomes complicated if the roots cluster near the unit circle. To avoid this problem Barkhuijsen et al (1987) have proposed using a state space formalism to develop a method to solve for the frequency and damping directly. Their approach is based on the work of Kung et al (1983) and relies on the factorisation of a data matrix into the product of two full rank matrices called the observability and control matrices. Any factorisation of the data matrix into a product of two full rank matrices can be reduced to this theoretical factorisation. The singular value decomposition of the data matrix provides a suitable factorisation.

For the model in (4), where $z_k = exp(-b_k + i2\Pi f_k)\Delta t$, the data matrix $\mathbf{X}$ can be expressed as

$$
\mathbf{X} =
\begin{bmatrix}
y_1 & y_2 & \cdots & y_M \\
y_2 & y_3 & \cdots & y_{M+1} \\
\cdot & & & \cdot \\
\cdot & & & \\
y_L & y_{L+1} & \cdots & y_N
\end{bmatrix}
=
\begin{bmatrix}
\mathbf{e}^T \\
\mathbf{e}^T\mathbf{Z} \\
\cdot \\
\cdot \\
\cdot \\
\mathbf{e}^T\mathbf{Z}^L
\end{bmatrix}
\begin{bmatrix}
\mathbf{c}, \mathbf{Z}\mathbf{c}, \ \ldots \ , \mathbf{Z}^M\mathbf{c}
\end{bmatrix}
= \mathbf{X}_\ell\mathbf{X}_r,
$$

where $L = N - M + 1$, $\mathbf{e}^T$ is a row vector with elements equal to one, $\mathbf{Z}$ is a $K \times K$ diagonal matrix whose diagonal elements are equal to z_k, $k = 1, \ldots , K$, and $\mathbf{c}$ is a column vector whose entries are the complex amplitudes $c_1, \ldots , c_K$.

The matrix $\mathbf{X}_\ell$ is equated to the matrix of left singular vectors in the truncated singular value decomposition of $\mathbf{X}$,

$$\mathbf{X} = \mathbf{U}_K\mathbf{\Lambda}_K\mathbf{V}_K^H,$$

where $\mathbf{\Lambda}_K$ is the truncated $K \times K$ matrix of singular values, $\mathbf{U}_K$ is the left singular vector matrix and $\mathbf{V}_K$ the matrix of right singular vectors. The Vandermonde structure of $\mathbf{X}_\ell$, where each row is equal to the previous one multiplied by $\mathbf{Z}$, is utilised to solve for the matrix $\mathbf{Z}$, thus giving the frequencies and dampings.

On simulated data sets this procedure has performed extremely well at separating very close peaks. With real FID's the presence of experimental artifacts and lack of knowledge of the value of K makes the estimation problem more difficult. However, by setting the value of K larger than expected, some of the experimental artifacts appear as resonances with anomalous dampings and amplitudes. Noise peaks can be identified by a combination of small amplitude and large damping or even very small positive damping which would theoretically be zero.

To perform efficiently, this technique requires the data matrix to have dimensions such that $\mathbf{XX}^H$ is a square matrix of size $L \times L$ with L greater than K. If L is too close to K, the matrix $\mathbf{XX}^H$ becomes of less than full rank in computational terms although theoretically still of full rank. This leads to inaccurate parameter estimates. So L must be chosen to be quite large. For example, for K equal to 252, a value of L of about 500 seems to be suitable. Finding the singular value decomposition of a matrix of this size can make prohibitive demands for both time and memory on a conventional computer. However on a current generation supercomputer using NAG (1983) algorithms it becomes quite practical.

A major disadvantage of this method is that there is no formal modelling of the noise. This means no statistical inferences can be made about the parameter estimates. If these estimates are used as initial starting values in a minimisation of the sum of squares in (3), the resulting parameter estimates can be displayed with confidence intervals. Surprisingly, with all the simulated data sets processed in the current project to date, use of either a conjugate-gradient method or a Levenberg-Marquadt algorithm leads to insignificant changes in the parameter estimates. It is difficult to know whether this is due to the accuracy of the state space estimates or the inadequacies of the minimisation technique.

ESTIMATING AMPLITUDES AND PHASES

Some remarks need to be made on the final step in the estimation procedure, that of estimating amplitudes and phases. For this, the frequency and damping estimates are substituted into (4) leading to N equations in the K complex unknowns expressed as $\mathbf{y} = \mathbf{Zc} + \mathbf{e}$ where $\mathbf{y} = (y_1, \dots, y_N)^T$, $\mathbf{c} = (c_1, \dots, c_K)^T$ and

$$
\mathbf{Z} = \begin{pmatrix} z_1 & z_2 & \cdots & z_K \\ & \cdot & & \\ & \cdot & & \\ & \cdot & & \\ z_1^N & z_2^N & \cdots & z_K^N \end{pmatrix}.
$$

Care needs to be taken with finding the least squares solutions of these parameters. The damping in the values of the data and the values of z_K^N as $n \to \infty$ needs to be considered and a weighted least squares problem is obtained by premultiplying both sides of the equation by a suitable scaling matrix. A singular value decomposition of $\mathbf{Z}$ is used and this is obtained by a series of Householder decompositions rather than forming the matrix $\mathbf{Z}^H\mathbf{Z}$ directly as this can lead to problems with numerical accuracy.

DISCUSSION

All the calculations discussed above can be carried out using standard numerical packages such as NAG (1983) and SSLII (1988) modified for vector processors. The speed with which calculations are completed indicates the subroutines used are highly vectorizable. Attention to a few small details in coding the main program gives vectorization rates of up to 99% so the calculation is eminently suited to a vector processor.

Further advantages in using a supercomputer can be seen in estimating the parameters in the 2D model. Equation (2) can be expressed as

$$y(m,n) = \sum_{\ell=1}^{L} A_\ell exp\{-(c_\ell + i2\Pi g_\ell)\Delta t_2 n\} + e(m,n)$$

where

$$A_\ell = \sum_{k=1}^{K} \mu_{k\ell} exp\{i\phi_{k\ell}\} exp\{(-b_k + i2\Pi f_k)\Delta t_1 n\}.$$

This indicates a two-stage estimation procedure in which frequency and damping estimates in one dimension are used to estimate the A_ℓ which then leads to L further estimation problems to solve for the remaining parameters. If each of these estimation problems is performed using the state space modelling technique, the CPU time on a VAX 11/750 is estimated to be $L + 1$ days, each 1D problem requiring approximately 24 hours processing. Recall that L may be several hundred. On the ANU VP-100 supercomputer, however, the processing time for each 1D problem is 3 minutes. Thus the estimation of the 2D model can realistically be carried out.

Estimation of parameters of a model represented by a large number of complex exponentials has applications much broader than NMR spectroscopy. For example, in seismology, models of damped exponentials are used to simultaneously estimate the speed of propogation of seismic surface waves and their decay. In areas where the expense or difficulty of collecting data leads to a severly truncated record of the decay of the signal, the use of a supercomputer to ensure reliable parameter estimation can not only compensate for deficiencies in the data but also indicate to analysts whether the inferences they wish to make are supportable by the data.

ACKNOWLEDGEMENT

The calculations upon which this work is based were carried out using the Fujitsu VP-100 of the Australian National University Supercomputer Facility.

REFERENCES

1. Barkhuijsen H., de Beer R., Bovee W.M.M.J. and Van Ormondt D., Retrieval of Frequencies, Amplitudes, Damping Factors, and Phases from Time Domain Signals Using a Linear Least-Squares Procedure, *Journal of Magnetic Resonance*, vol 61, pp 465-481, 1985.

2. Barkhuijsen H., de Beer R. and Van Ormondt D., Improved Algorithm for Noniterative Time-Domain Model Fitting to Exponentially Damped Magnetic Resonance Signals, *Journal of Magnetic Resonance*, vol 73, pp 553-557, 1987.

3. Bresler Y. and Macovski A., Exact Maximum Likelihood Parameter Estimation of Superimposed Exponential Signals in Noise, *IEEE Transactions on Acoustics, Speech, and Signal Processing*, vol ASSP-34, no 5, pp 1081-1089, 1986.

4. Golub G.H. and Pereyra V., The Differentiation of Pseudo-Inverses and Non-Linear Least Squares Problems whose Variables Separate, *SIAM Journal of Numerical Analysis*, vol 10, no 2, pp 413-432, 1973.

5. Kumaresan R. and Tufts D.W., Estimating the Parameters of Exponentially Damped Sinusoids and Pole-Zero Modelling in Noise, *IEEE Transactions on Acoustics, Speech, and Signal Processing*, vol ASSP-30, no 6, pp 833-840, 1982.

6. Kung S.Y., Arun K.S. and Bhaskar Rao D.V., State Space and Singular-Value Decomposition-Based Approximation Methods for the Harmonic Retrieval Problem, *Journal of the Optical Society of America*, vol 73, pp 1799-1811, 1983.

7. Numerical Algorithms Group, NAG FORTRAN Library Manual, Numerical Algorithms Group, Oxford, 1983.

8. Osborne M.R. and Smyth G.K., A Modified Prony Algorithm II: Exponential Function Fitting, *Technical Report No 30*, Program in Statistics and Applied Probability, University of California, Santa Barbara, June 1987.

9. Scientific Subroutine Library, FACOM FORTRAN SSLII User's Guide, Fujitsu Australia Ltd, Sydney, 1988.

10. Stephenson D.S., Linear Prediction and Maximum Entropy Methods in NMR Spectroscopy, *Progress in NMR Spectroscopy*, vol 20, pp 515-626, 1988.

Three-Dimensional Unstructured Meshes

A. N. STOKES
CSIRO Division of Mathematics and Statistics
Private Bag 10
Clayton, Victoria, 3168, Australia

INTRODUCTION

Stokes (1988) reported progress on an algorithm for generating an unstructured triangular mesh, with density specified by a metric function. There has been further progress, and experience with its use, which is reported here. Unstructured meshes have also received increasing attention elsewhere (e.g. Mavripilis 1987), with the development of better algorithms for solving the resulting linear systems, analogous to corresponding methods for structured meshes (Thompson, Warsi *et al* 1985).

Considerable effort has gone into developing a corresponding algorithm to produce a tetrahedral mesh in three dimensions. Like the two-dimensional version, it proceeds by generating somehow a crude initial mesh, and then using a series of transformations to subdivide it, with the aim of producing regular tetrahedra scaled according to an arbitrary continuous mesh density function. These transformations are chosen to reduce the length of the longest line segment at each stage.

The transformations required are much more complicated than in two dimensions, where just two operations form the basis of the strategy. Further, many of them tend to upset the ratio of elements to faces, and if corrective action is not taken, a satisfactory mesh cannot be obtained. To choose, check and implement the transformations, it has been necessary to formalise an algebraic scheme for describing them.

The required subdivision scheme has now been satisfactorily implemented, together with a moving front numbering scheme for reducing bandwidth. Still lacking is a scheme for reducing an arbitrary polyhedron with triangular faces to a formally correct initial tetrahedral mesh.

There are some other ways of generating tetrahedral meshes described in the literature. Some (Yerry and Shephard, 1984), (Cavendish, Field et al 1985) are more geometrical, while a more flexible method (Nguyen, 1982) requires the user to supply interior nodes.

MESH SUBDIVISION

As with the two-dimensional mesh, the subdivision is driven by the need to eliminate long line segments. The length function used can be varied to achieve a mesh of concentration designed to be optimal for some purpose. Once a line to be removed is identified, some rearrangement must be implemented to do the removal. This may involve creating a new node, and in this way the mesh grows.

In regular meshes the ratio number elements : number faces : number lines : number nodes is approximately constant, any variation being usually caused at the boundaries. For example, an infinite cubic array of nodes, with each cube divided into six tetrahedra, has the ratio 6:12:7:1. Many other regular meshes have the same or similar ratios, and it seems that any substantial deviation from that ideal leads to an unsatisfactory mesh.

In two dimensions, the corresponding ratio is 2:3:1, and it so happens that the simplest subdivision manoeuvre, which is to remove a line, create a node at the centre of the resulting quadrilateral, and connect the node to the four corners, results in the gain of exactly 2 elements, 3 lines and 1 node. In three dimensions, corresponding subdivisions have varying effects, corresponding to the more complicated shapes which may surround a line. For example, if a line adjoining three tetrahedra is similarly removed and a node inserted and connected, the result is the creation of 3 elements, 6 faces, 4 lines and 1 node. Clearly, doing this often would produce an excess of nodes.

In general, line splitting does tend to create too many nodes; to remain in balance, lines being split would have to adjoin on average 6 elements, while the average in a 6:12:7:1 mesh is 36/7, or about 5. So it is necessary to include some other operations which can restore the balance by creating more elements, faces and nodes. To define such operations, which must retain the topological integrity of the mesh, it is useful to create some appropriate nomenclature. This is done in the next section.

THE ALGEBRA OF ORIENTED SIMPLEX MESHES

Setting out the algebra associated with elements more formally not only helps clarify the exposition, but is an aid to creating a computer program. An n-dimensional oriented simplex mesh consists of a number of n-tuples of distinct integers:

$$(p_1, ...p_n)$$

called *elements*. The integers are referred to as *nodes*. For each element, an even permutation is an equivalence relation.

Faces and joining

There are n subsets of the n-1 distinct nodes of an element, and these are called *cafes* (as an odd permutation of *faces*). Each cafe has two equivalence classes under the operation of even permutation, and one is called the *face* of the element. The one chosen is that which, when the missing element is added at the end, forms a valid permutation of the element.

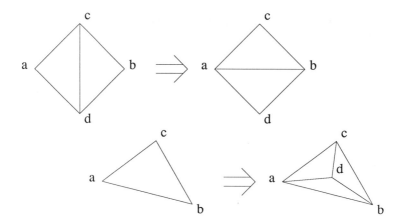

FIGURE 1. Triangle rearrangements I and II.

Where two elements have a common cafe, they are said to be *joined*. They are also called *connected*, but connectedness is extended as a transitive relation. An *oriented mesh* may now be defined as *a set of connected elements in which each face occurs only once*.

Here brevity has overtaken clarity. Some cafes will occur only once; those faces are the *boundary* of the mesh. Other cafes will occur in an even and odd permutation; these are *interior*, and join their two elements. But no cafe can occur three times, for then at least two would be related by an even permutation.

Although I do not attempt to define a formal topology here, the term topology will be loosely used to refer to the properties of the mesh which are independent of coordinates.

Rearrangements

Each step in subdividing a mesh consists of selecting a connected subset of elements, and arranging the nodes in a different combination, preserving the outward faces. New nodes may be created. For small subsets, this can be done in only a few distinct ways.

Examples - Two dimensions

I. Rearrangement - the elements (abc) and (bad) have the four boundary faces (bc), (ca), (ad), and (db), with the common cafe (ab). If reconstituted as the elements (dca) and (cdb), then the boundary faces are the same, but the common cafe is now (cd). A notable feature of this rearrangement is that it preserves the number of faces, elements and nodes.

II. Node creation - the element (abc) can be replaced by the three elements (abd), (bcd), (cad), where d is new. These have the shared cafes (ad), (bd), and (cd), and the same boundary.

A practical scheme of mesh subdivision must use both these manoeuvres, for rearrangement creates no new nodes, and creation removes no lines. They are also sufficient, for any mesh P whose nodes are a superset of mesh Q can be created using them. There is also

III. Node removal - the submesh consists of all the elements containing the interior node (i.e. the node is part of no boundary face), and the rearrangement consists of remeshing the polygon formed from the boundary.

Examples - Three dimensions

In 3-d, node creation and removal are much the same. But rearrangement is more complicated. The joined elements (abcd) and (bace) cannot be rearranged to form any other pair, but can only form three: (abed), (bced) and (caed). A submesh of three elements will contain five or six nodes, and can only be rearranged to form submeshes of two or four elements respectively.

The simplest submesh which can be rearranged to a similar mesh is that with four elements and six nodes. This is equivalent to subdividing an octahedron, and that can be done about any of the three diameters.

In general a subregion can be characterised by the number of outer faces, lines and nodes that it has, because these may not be changed in rearrangement. A table showing these arrangements of these important cases is given below.

Splitting a line, with production of a new node, involves rearranging the entire space occupied by tetrahedra adjoining the line. This produces two equivalent adjacent submeshes. If the line was ab, and the adjacent tetrahedra *(abcd), (abde),...*, then with the new node z the tetrahedra are *(azcd), (azde),...* and *(zbcd), (zbde),....*

TABLE 1. Possible rearrangements of the simplest sub-meshes.

Number of boundary			Arrangements		
nodes	lines	faces	1	2	3
5	9	6	(abcd) (bace)	(abed) (bced) (caed)	
6	12	8	(abcd) (abde) (abef) (abfc)	(ceda) (ceaf) (cefb) (cebd)	(dfea) (dfac) (dfcb) (dfbe)

MORE ON TWO-DIMENSIONAL MESHES - THEIR USE AS A BASE FOR A THREE-DIMENSIONAL MESH

The two-dimensional algorithm can be used to subdivided surfaces in a three-dimensional space. This is moderately useful in itself, but becomes more useful as a tool in building up a representation of a three-dimensional object.

Treating surfaces in three dimensions

The algorithm for planar subdivision is essentially topological, with geometric checks and geometric priorities. It therefore works quite well whether the nodes are given two or three (or more) co-ordinates. There are two fairly minor difficulties:

a) Placement of new nodes. In a plane, this is fairly easy; somewhere in the middle of the vacant quadrilateral (in step II) will do. But in three-dimensions, the corresponding linear formula leads to nodes placed on plane surfaces in the shape of the original elements. To improve this extra information is needed, but supplying and using it is not easy.

b) Checking for geometrical faults. The next section develops the use of the area as a check for various difficulties. The idea is that the consistent topological ordering of the element nodes should produce positive areas, in a plane. On a surface, there is no such simple check.

Generating an initial mesh in two and three dimensions

In two dimensions, a triangular mesh can be produced within an arbitrary polygon by just eliminating triangular regions adjacent to nodes where there is an acute angle, with precautions against the intrusion of other nodes. This does not generalise well to three dimensions, partly because the number of checks required increases greatly, but also there seem to be possibilities of insoluble deadlock that do not arise in two dimensions.

There is an alternative which, in two dimensions, is very fast when it works. A topologically adequate mesh is created without attention to geometry. That means that the nodes may appear ordered clockwise in some elements. For example, in Figure 2, if the outer element is *adc*, then the inner one must be *bcd*. This problem disappears after the rearrangement shown.

Of course, the rearrangement affects neighboring elements, and may itself create problems. It is possible to devise a strategy, based on the elimination of negative area (the area of clockwise triangles) which often produces a satisfactory mesh very quickly. But occasionally the strategy will lead to a loop with no termination, and there seems to be no certain way of avoiding this.

This currently imperfect method generalises well to three dimensions, converging very fast when it succeeds.

FIGURE 2. Using rearrangement I to repair a defective mesh fragment.

Extending a two-dimensional mesh into three dimensions

Existing mesh generators in three dimensions generally work by slicing the space in some way. The two-dimensional generator described here also has a slice mode of operation, in which a layer of prisms can be built on a triangular mesh base, reproducing on the upper surface a mesh topologically identically to the lower one, with corresponding nodes. The coordinates of these nodes can be assigned by some mapping function; the obvious one is just a shift of the z-value, if the space to be meshed is cylindrical (in the sense that all horizontal sections are similar).

Simplex meshes have been used throughout this work because of the ease of subdividing them, and so for further progress it is desirable to divide the prisms into tetrahedra. This might be thought a trivial problem, but there is a difficulty. Each face of a prism is a rectangle, and can be divided into triangles by either of two diagonals. The way this is done may be regarded as a direction applied to the bottom mesh line segment; a segment could be said to progress towards the node to which the diagonal of the adjacent rectangle connects (see Figure 3). Then a prism whose base triangle is directed in a loop cannot be divided into tetrahedra (in a way consistent with those diagonals). To see this, consider in the figure below the tetrahedron lying above the triangle (abc). Its apex must be one of d,e or f, and whichever it is must be connected by line segments to all the points a,b and c. Only one connection is vertical, so two must be diagonals.

It is therefore necessary to be able to assign directions to sides in a triangular mesh in such a way that no loops are formed. Fortunately the moving front numbering scheme can provide this assignment of directions. Essentially, one assumes that the front itself is directed, and the sides retain this direction after the front has passed through.

Use of this extensibility of two-dimensional meshes has proved to be the best way of generating initial meshes in three dimensions. Indeed, for suitable, approximately cylindrical, shapes it can generate satisfactory final meshes. For unsuitable shapes, further rearrangement is needed.

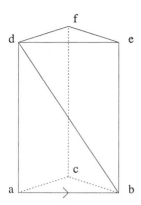
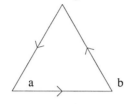
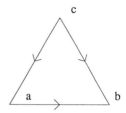

FIGURE 3. The orienta-   A prism with faces divided   A prism with faces divided
tion of the diagonal lines as-   like this cannot be divided   like this can be divided into
signs a direction to the base   into tetrahedra.   tetrahedra.
lines.

PRACTICAL THREE-DIMENSIONAL MESH GENERATION

The only really satisfactory way found so far of generating an initial tetrahedral mesh
is by building up from a two-dimensional base, as described above. A length metric
is defined, and rearrangements of the kind described in Sec 3 applied until either a
prescribed number of nodes has been created or a prescribed maximum length reached.

At each step in the subdivision the longest line is found (from an ordered structure)
and one of the following actions taken:
a) Find the longest line in the mesh. If it is on the boundary, form a new node, and
end this step.
b) If the longest line is interior, and there are four adjacent tetrahedra, try internal
rearrangements. If any one gives a shorter line, accept it, and end this step. Otherwise
proceed.
c) If the interior line adjoins three tetrahedra, perform a rearrangement to eliminate
it.
d) If the interior line adjoins four or more tetrahedra, split it.

The following table shows the effect of these actions:

TABLE 2. Effect of rearrangement and line splitting stratagems.

Action	Nodes	Change in number of Lines	Faces	Elements
a) with n elements adjoining	+1	+(n+2)	+(2n+1)	+n
b)	0	0	0	0
c)	0	-1	-1	-1
d) with n elements adjoining	+1	+(n+1)	+2n	+n

Example

A unit cube was built on a unit square, and divided (using ordinary length as the criterion) until 100 nodes had been created. There were then 638 line segments, 1103 faces and 564 elements. The average length of the line segments was 0.26, with a standard error (used as a measure of variation) of 0.058. The longest remaining line had length 0.324.

The ratio of nodes:lines:faces:elements is fairly similar to that for the regular mesh with cubic lattice nodes; there is a slight relative preponderance of nodes and lines, but that is probably the effect of the boundaries. It would be possible to design a subdivision strategy which explicitly checked this ratio, but so far that has not been necessary.

A run was also done using an algorithm which does line splitting only, without the various rearrangements. The ratios then were 100 : 479 : 694 : 314, the average length was 0.32, the standard deviation 0.08, and the longest line 0.5.

REFERENCES

1. Cavendish, J. C., Field, D. A. and Frey, W. H. An approach to automatic three-dimensional finite element mesh generation. *Int. J. Numer. Meth. Eng.,* Vol 21., pp 329-348, 1985.

2. Mavrapilis, D. J., Multigrid solution of the two-dimensional Euler equations on unstructured triangular meshes. *AIAA Journal* , Vol. 26 pp. 824-831, 1987.

3. Nguyen, N. Ph., Automatic mesh generation with tetrahedral elements. *Int. J. Numer. Meth. Eng.,* Vol 18, pp 273-280, 1982.

4. Stokes, A.N., Generation of a triangular mesh for a general two-dimensional region, *Computational Techniques and Applications:CTAC-87,* North-Holland, Amsterdam, 1988.

5. Thompson, Joe F., Warsi, Z. U. A. and Mastin, C. W., *Numerical grid generation - foundation and applications,* North-Holland, New York 1985.

6. Yerry, M. A. and Shephard, M. S. Automatic Three-Dimensional mesh-Generation by the modified-octree technique, *Int. J. Numer. Meth. Eng.,* Vol. 20, pp 1965-1990, 1984.

A Complete Three-Dimensional Euler Code for Calculating Flow Fields in Centrifugal Compressor Diffusers

I. TEIPEL and A. WIEDERMANN
Institute for Mechanics
University of Hannover
Appelstrasse 11
D-3000 Hannover 1, FRG

INTRODUCTION

Theoretical investigations of flow fields in vaned diffusers of high-loaded centrifugal compressor units have remained a complicated problem because several flow phenomena have to be taken into account. Transonic effects play an important part due to high rotational speeds. A strong interaction of the flow of the impeller and that in the vaned diffuser causes supersonic regimes and shock waves which are located in the vaneless and semivaned space in front of the blades of the diffuser (Stein and Rautenberg 1985). Jet and wake regions are formed at the impeller exit and consequently the flow pattern at the diffuser entrance is characterized by strong fluctuations. In addition energy gradients from hub to shroud are produced in the impeller due to secondary flow motion (Krain 1984). Therefore the flow angles vary considerably between the side walls, and the flow material with low energy content is concentrated near the upper wall (shroud side) of the diffuser. The unsteady effects prove to be damped out by mixing processes within a short distance in the diffuser as being shown by Stein and Rautenberg (1985). The assumption of time-averaged flow variables in all directions is therefore sufficient, and an unsteady analysis can be avoided. On the other hand the variation of the total pressure in axial direction has to be taken into account and in general a three-dimensional approach has to be chosen for calculating inviscid flow fields even if the diffuser is equipped with cylindrical blades. In order to obtain a better adjustment of the channel geometry to the varying flow angles Jansen and Rautenberg (1982) designed a three-dimensionally twisted blade passage. Teipel and Wiedermann (1987) applied a quasi-three-dimensional approach for calculating flow fields in the twisted diffuser channel based on the well-known theory given by Wu (1952). Here in this paper flow fields are calculated with a new complete three-dimensional Euler code and the results are compared with experimental data.

DESCRIPTION OF THE METHOD

Three-dimensional flow field analysis requires the availability of computers with large storage capacities and fast processors. In the past much effort has been made for approximating three-dimensional flow fields using simpler methods. A very common approach, known as through-flow analysis, had been developed by Marsh (1968) and proved to be well suited for solving flow fields in the meridional plane of multi-stage turbomachines. Instead of the three components of the momentum equation mass-averaged relations àre solved by assuming axial symmetry. A better representation of the complete three dimensional flow field was achieved by Krimerman and Adler (1978) who were among the first to perform an iterative coupling of the results in several S_1- and S_2- surfaces as being suggested by Wu. Recently Teipel and Wiedermann (1987) applied this theory for solving flow fields in centrifugal compressor diffusers. To meet the requirements of transonic flows a time marching scheme has been chosen.

If an explicit scheme is applied a restrictive stability criterion, known as CFL-condition has to be obeyed, and consequently the time step size is fixed. However, the increasing number of fast computers and vector processors has reduced the problem concerning high CPU-times. Especially with vector computers Euler codes prove to be very efficient. As the rapid developement in computer technology provides additional storage capacities of more than one order of magnitude with reasonable costs, Euler codes are now often used for calculating flow fields around complete configurations in aircraft research (Eberle and Misegades 1986) as well as in propulsion applications (Celestina et.al. 1986).

At last it should be mentioned that in addition to Euler codes recently attempts have been made to solve the time-dependent parabolized Navier-Stokes equation for viscous compressible flow in turbomachines (Dawes 1988). The results offer insight into very important details of the real fluid flow because separation domains and cross-flows are obtained. Especially in cases of strong adverse pressure gradients as they occur in radial diffusers the application of a Navier-Stokes code would give a more realistic view than an Euler code. However, a mesh with a large number of grid nodes of about 10^5 and more is necessary to obtain results with sufficient accuracy (Nozaki et.al. 1987). That means large memory requirements and very high CPU-times even if a vector computer is available. Therefore in these investigations an Euler code has been preferred, because it is the authors' purpose to present an economical tool for calculating pressure distributions in centrifugal diffuser channels.

THE GOVERNING EQUATIONS

As mentioned above transonic regimes including shock waves are inherent with flow fields in diffusers of high-loaded centrifugal compressor units. With this condition in mind a conservative formulation of the equations is chosen which can be written:

$$\frac{\partial \vec{U}}{\partial t} + \frac{\partial \vec{F}}{\partial r} + \frac{\partial \vec{G}}{\partial \theta} + \frac{\partial \vec{H}}{\partial z} + \vec{J} = \vec{0} \tag{1}$$

with the vectors

$$\vec{U} = \begin{bmatrix} \varrho \\ \varrho u_r \\ \varrho u_\theta \\ \varrho u_z \end{bmatrix} \quad \vec{F} = \begin{bmatrix} \varrho u_r \\ \varrho u_r^2 + p \\ \varrho u_r u_\theta \\ \varrho u_r u_z \end{bmatrix} \quad \vec{G} = \begin{bmatrix} \varrho u_\theta / r \\ \varrho u_r u_\theta / r \\ (\varrho u_\theta^2 + p)/r \\ \varrho u_z u_\theta / r \end{bmatrix} \quad \vec{H} = \begin{bmatrix} \varrho u_z \\ \varrho u_r u_z \\ \varrho u_\theta u_z \\ \varrho u_z^2 + p \end{bmatrix} \quad \vec{J} = \begin{bmatrix} 0 \\ \varrho(u_r^2 u_\theta^2)/r \\ 2\varrho u_r u_\theta / r \\ 0 \end{bmatrix} \tag{2}$$

Here u_r, u_Θ and u_z mean velocity components in a cylindrical coordinate system, p and ϱ denote pressure and density. Assuming isoenergetic conditions an algebraic solution for calculating the total enthalpy h_{tot} can be found if only the steady state solution is of interest

$$h_{tot} = \frac{\gamma}{\gamma-1} \frac{p}{\varrho} + \frac{\vec{u} \cdot \vec{u}}{2}$$ (γ : ratio of specific heats)   (3)

A boundary fitted coordinate system is introduced containing $55 \times 23 \times 9$ grid nodes. A three-dimensional plott of the mesh configuration is shown in Fig.1. The mesh can be refined at the side walls and the blades by applying algebraic functions to meet strong gradients. A mapping into a regular computational domain will be performed. The transformed governing equations are integrated by an explicit MacCormack scheme.

The flow variables at the entrance and the exit of the computational domain are calculated by applying three dimensional characteristic theory. At the solid boundaries different procedures have been used along the blades and the side walls. In both cases the wall pressure is obtained as a solution of the normal component of the momentum equation. At the blades the velocity components are calculated with the solid wall condition and characteristic relations. At the side walls this procedure fails, and, therefore, the wall condition is enforced by applying a reflective principle as being suggested by Holmes and Tong (1984).

A note should be given about the boundary conditions at the impeller exit. Here the complicated flow pattern has to be considered very carefully. As only the steady-state solution is of interest the time-dependent jet-wake structure of the impeller flow has been modelled with time-averaged variables. The axial gradient of the total pressure p_t from hub to tip is simulated by prescribing a function for p_t which can be derived from experiment. Jansen and Rautenberg (1982) showed that in spite of entropy gradients the total enthalpy can be assumed as uniform across the whole diffuser entrance regime. Consequently the algebraic energy equation, eq.3 remains valid in all considered cases, although the flow is strongly rotational within the whole computational domain.

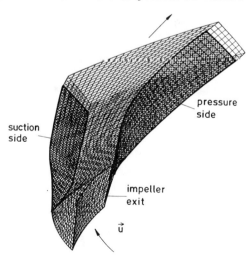

pressure side

suction side

impeller exit

$\vec{u}$

FIGURE 1.
Three-dimensional graph of
the computational
mesh configuration

DISCUSSION OF THE RESULTS

This section is devoted to the discussion of the results obtained by the present method. The validation of the new code will be proved by comparing theoretical flow fields with measurements which were carried out at the Institute of Turbomachinery at the University of Hannover (Jansen and Rautenberg 1982, Stein and Rautenberg 1985). To demonstrate the improvement in three-dimensional flow analysis achieved with the present Euler code the results will be compared with those obtained with the earlier quasi-three-dimensional approach of the authors.

To verify the range of application of the code an operating point was chosen where the rotational speed of the impeller is 18,000 rpm. In this case the Mach number at the impeller exit amounts about 1.25 in the average, and a complicated transonic flow pattern including shocks will be expected. Measured flow fields are shown in Fig.2 obtained at the shroud and hub side. The graphs contain domains of constant pressure which are related to the maximum total pressure at the impeller exit. The considered operating point is located near the surge line as can be reckognized by the strong pressure rise in the vaneless and semivaned region whereas there is only a small pressure recovery observed in the vaned diffuser further downstream. The same pressure distribution had been also referred to for comparison purposes by Teipel and Wiedermann (1987).

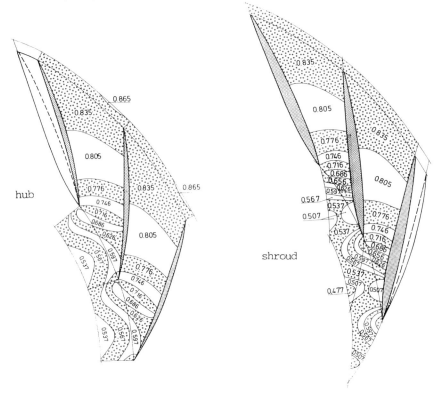

FIGURE 2. Regions of constant pressure p/p_t at the shroud and hub; measurements by Jansen et al (1982); n_{red} = 18,000 rpm

Calculated flow fields are shown in Fig. 3. In spite of the neglect of viscous forces the results correspond well with experiments. Especially the curved shape of the isobars in the vaneless region is in close agreement with measurements. As indicated by experiments strong pressure gradients can be found in front of the vaned diffuser.

A direct comparison of the flow fields obtained by the present code with the results of the earlier quasi-three-dimensional method clearly shows that now one is able to get a considerably better representation of the flow phenomena. If the pressure field at the hub side is obtained by the earlier scheme the isobars accumulate strongly in front of the blades (Fig.4) while there is no shock observed by experiments, see Figs. 2 and 3. In contrast with those results the shape of the isobars is now predicted in a closer agreement with the measured data. At the hub side as well as at the shroud side an improved prediction can be seen with the new code.

The pressure fields obtained at S_2- stream sheets are shown in Fig.5 as a projection into the meridional plane. Due to the prescribed axial entropy profile at the impeller exit there is a pressure rise from the lower to the upper wall. The leading edge of the blade is located at $\lambda = r/r_{IE} = 1.15$ (r_{IE}: impeller exit radius) and is indicated by an accumulation of isobars.

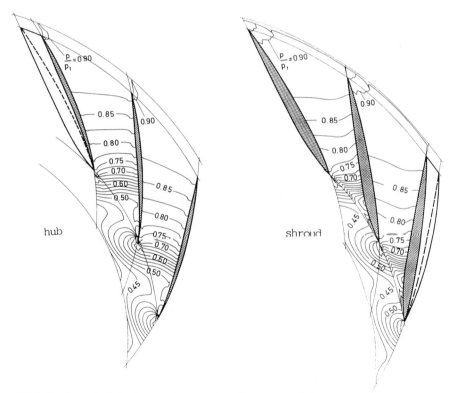

FIGURE 3. Computed isobars p/p_t at the shroud and hub; complete three-dimensional analysis; n_{red} = 18,000 rpm.

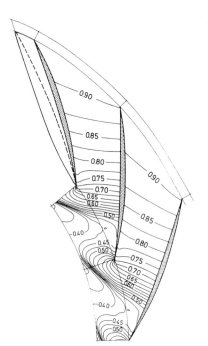

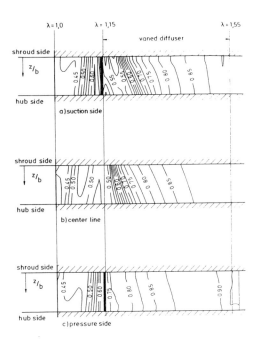

FIGURE 4.
Computed isobars p/p_t,
obtained with a quasi-three-
dimensional approach
(Teipel and Wiedermann 1987)

FIGURE 5. Isobars p/p_t in the
meridional plane, n_{red} = 18,000 rpm

b: diffuser width

PROFILES OF THE FLOW ANGLE IN FRONT OF THE VANED DIFFUSER

With the three-dimensional code flow profiles can be calculated at different radial locations and, thus, the developement of the three-dimensional structure of the flow in front of the diffuser can be studied.
As an example the distribution of the absolute flow angle will be considered. At a rotational speed of 18,000 rpm the measured and predicted
flow angle profiles at λ = 1.10 are shown in Fig.6. Because of secondary flow within the impeller fluid with low energy content is displaced towards the shroud side. This fact is indicated by low figures of
the flow angle in this regime. On the whole the measured profile provides a varying flow angle in axial as well as in circumferential direction.

In Fig.6 the calculated flow angles are shown at the impeller exit
(λ = 1.0), at λ = 1.1 and in the entrance plane of the vaned diffuser
(λ = 1.15). Caused by the diffuser geometry and the axial variation of
the total pressure at the impeller exit an increase of the flow angles
from the shroud to the hub can be observed which becomes more pronounced towards the entrance into the vaned part of the diffuser. Due to
numerical requirements the observed total pressure profiles cannot be
simulated completely because one is not able to model the low figures
nearby the upper wall.

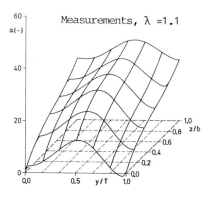

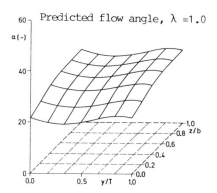

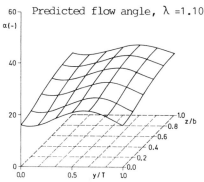

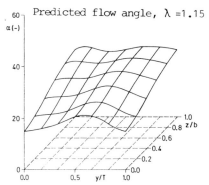

FIGURE 6. Measured flow angle, λ =1.1 and predicted flow angle profiles at different locations; n_{red} = 18,000 rpm

EFFICIENCY OF THE CODE

The new Euler code is easier to handle with than the earlier quasi-three-dimensional scheme because the latter consists of several indivi-dual programs. These are required to perform data transfer between the different stream sheets and to calculate derivatives of the function for the stream surfaces, which may be twisted in a complicated manner. In order to achieve a reliable representation of the three-dimensional flow field a couple of iterations has to be completed in case of the quasi-three-dimensional code. Because of large CPU-times only one ite-ration cycle had been done. If one introduces three S_1- and S_2- stream sheets the CP-time was about half an hour at a CYBER 990. Within the same amount of computational time the new Euler code produces complete three-dimensional flow fields so that in summary it has proved to be more efficient. With a parallel processor the CPU-time can be reduced significantly of at least one order of magnitude. If a reduced time is formed by relating the required computational time to the number of ti-me steps and grid nodes, one obtains about $9.4 \cdot 10^{-5}$ sec at a CYBER 990 and $6 \cdot 10^{-6}$ sec at a CRAY-XMP. Thus, only few minutes are needed for calculating three-dimensional flow fields with more than 10,000 grid nodes at a vector computer.

CONCLUDING REMARKS

A new Euler code for calculating complete three-dimensional flow fields
in vaned diffusers has been described. For dealing with transonic flows
a time marching scheme has been applied. A comparison of the new code
with a quasi-three-dimensional approach shows that the present code is
simpler to handle with and more efficient for calculating flow fields
with compareable accuracy. It is very fast at a vector computer. However, the most important shortcoming of the present code is the neglect
of viscous effects. Nevertheless, if the rotational flow at the impeller exit is prescribed by a corresponding total pressure profile, this
method proves to be an efficient tool for providing valuable details
concerning flows in high-loaded centrifugal compressor diffusers.

REFERENCES

1. Celestina, M., Mulac, R.A., and Adamczyk, J.J., A Numerical Simula-
    tion of the Inviscid Flow Through a Counter-Rotating Propeller,
    ASME-Paper No. 86-GT-138, 1986.

2. Dawes, W.N., Development of a 3D Navier Stokes Solver for Applica-
    tion to all Types of Turbomachinery, ASME-Paper 88-GT-70, 1988.

3. Eberle, A., and Misegades, K., Euler Solution for a Complete Fighter
    Aircraft at Subsonic and Supersonic Speeds, AGARD CP 412, 1986.

4. Holmes, D.G., and Tong, S.S., A Three-Dimensional Euler Solver for
    Turbomachinery Blade Rows, ASME-Paper 84-GT-79, 1984.

5. Jansen, M., and Rautenberg, M., Design and Investigations of a Three
    Dimensional Twisted Diffuser for a Centrifugal Compressor, ASME-Pa-
    per 82-GT-102, 1982.

6. Krain, H., Experimental Observation of the Flow in Impellers and
    Diffusers, VKI-LS 1984-9, 1984.

7. Krimerman, Y., and Adler, D., The Complete Three-Dimensional Cacula-
    tion of the Compressible Flowfiels in Turbo Impellers, *Journal of
    Mechanical Engineering Science*, vol. 20, pp. 277 - 285, 1978.

8. Marsh, H., A Digital Computer Program for the Through-Flow Fluid Me-
    chanics in an Arbitrary Turbomachine Using a Matrix Method, ARC R&M
    No. 3709, 1968.

9. Nozaki, O., Nakahashi, K., and Tamura, A., Numerical Analysis of
    Three-Dimensional Cascade Flow by Solving Navier-Stokes Equations,
    Proc. of the 1987 Tokyo Int. Gas Turbine Congress, Paper No. 87-
    TOKYO-IGTC-43, 1987.

10. Stein, W., and Rautenberg, M., Flow-Measurements in Two Cambered
    Vaned Diffusers with Different Passage Widths, ASME-Paper 85-GT-46,
    1985.

11. Teipel, I., and Wiedermann, A., Three-Dimensional Flowfield Calcula-
    tion of High-Loaded Centrifugal Compressor Diffusers, *Journal of
    Turbomachinery*, vol. 109, pp. 20 - 26, 1987.

12. Wu, Ch.-H., A General Theory of Three-Dimensional Flow in Subsonic
    and Supersonic Turbomachines of Radial-, Axial- and Mixed Flow Ty-
    pes, NACA TN D 2604, 1952.

Fast Triangulation of Planar Domains

J. C. TIPPER
Department of Geology
Australian National University
G.P.O. Box 4
Canberra ACT 2601, Australia

INTRODUCTION

Numerical methods for modelling and analysing spatially distributed data commonly require that the domain in which the data lie be first triangulated. The development of efficient, general-purpose triangulation algorithms has thus assumed a considerable importance: Cavendish (1974), Cavendish et al. (1985) and Watson & Philip (1984) refer to many of those that have been proposed. The triangulation problem is simply stated: given a set of points in the plane lying within a closed domain defined by some or all of those points, subdivide that domain entirely into non-overlapping triangles, the vertices of which are the points of the original set, supplemented, as necessary, by extra points added as the triangulation is being made. It should be noted that the problem of triangulating a domain is not the same as that of just triangulating a set of points in the plane, but that the ability to solve that latter problem is prerequisite.

Although straightforward in concept, the problem of triangulating a domain poses some serious questions. What, for instance, is the optimal triangulation of the original data points? Is this triangulation ever adequate in itself, or must extra points always be added? Which criteria are appropriate to judge the number and positioning of these extra points? In this paper these questions are addressed and a new triangulation algorithm described. This algorithm (1) is applicable to any planar domain, (2) gives triangulations that are well suited both to interpolation and to finite-element and finite-difference modelling, (3) is controlled directly by the configuration of the original data points (and hence is fully automatic), and (4) is simple to implement and fast to run.

SELECTING A BASIS FOR TRIANGULATION

Triangulating a set of points in the plane — the prerequisite for triangulating a domain — is a process that can be carried out in many different ways. The various algorithms that implement these are of greatly varying computational complexity, and computational load must of course be taken into account in selecting which one to use. Of greater importance, however, are the properties of the triangulations that the different algorithms produce. For the present purpose the triangulation of any given data set must (1) preserve the neighbourhood relationships implicit in the original data point configuration, (2) introduce a minimum of distortion when used subsequently for interpolation or modelling, and (3) be efficient to compute.

For any set P = {P_1, P_2 ..., P_N} of N points in the plane, there is just one triangulation that fulfils these requirements — the Delaunay triangulation (Miles, 1970; Sibson, 1978), the geometric dual of the Voronoi diagram (Fig. 1; Preparata & Shamos, 1985). The Voronoi diagram is the set V = {V_1, V_2 ..., V_N} of convex polygons of which the element V_i bounds that region of the plane within which all points are closer to P_i than to any other element of P. The Delaunay triangulation of P is formed by connecting all point pairs, P_i and P_j, for which the corresponding Voronoi polygons, V_i and V_j, are adjacent. As each Voronoi polygon provides the most natural definition of the spatial neighbourhood of its associated data point, such point pairs as P_i and P_j are often termed 'natural' neighbours. The Delaunay triangulation thus encapsulates completely the natural neighbourhood relationships for P.

A unique property of the Delaunay triangulation is that it is locally equiangular (Sibson, 1978): Delaunay triangles, on average, are as close as possible to equilateral. This is critically important both for interpolation and for modelling, as the use there of triangles that are far from equilateral is widely recognised to lead to considerable distortion (McCullagh, 1981; Watson & Philip, 1984). Sibson (1981) has shown how the Delaunay triangulation is used in 'natural neighbour interpolation'.

The computation of the Delaunay triangulation is a process for which the asymptotically optimal running time is $\Theta(N \log N)$. Algorithms which achieve this have been described by Lee & Schachter (1980) and Preparata & Shamos (1985). If, however, the points are pre-sorted in either the x- or y-coordinate direction, an algorithm developed by Tipper (forthcoming) can be used. This algorithm, running in almost linear time, is inherently iterative, not recursive, and provides a rapid, efficient way of computing the Delaunay triangulation, especially for large sets of points.

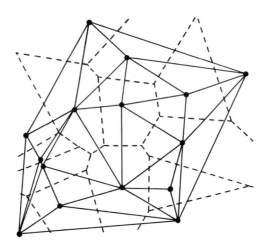

FIGURE 1.   A data set of 15 points.   Delaunay triangulation shown by solid lines; Voronoi diagram shown by broken lines.

The domain triangulation algorithm introduced here has two parts. The first obtains a triangulation of the domain using just the original data points. This initial triangulation is then modified in the second part of the algorithm by adding extra points to the data set, internally or on the domain boundary as necessary, until an acceptable final triangulation is achieved. The Delaunay triangulation basis is maintained throughout.

The initial triangulation

For simply connected, convex domains, the Delaunay triangulation of the original data points is a suitable initial triangulation of the domain: any of the existing Delaunay algorithms can be used. If, however, as is generally the case, the domain has re-entrant boundaries or is multiply connected, this is not so. The initial triangulation must instead be obtained by (1) segmenting the domain into simply connected, convex sub-domains, (2) constructing the Delaunay triangulations for the data points in each of these independently, and (3) patching these together to give the Delaunay triangulation of the entire set of points <u>conditioned on the domain boundaries</u>. An illustration of this sequence is given in Fig. 2. (Note that, in patching together the Delaunay triangulations from abutting sub-domains, a swapping process is used, starting at the shared boundary and propagating as far into each sub-domain as is necessary. This process, originally due to Lawson (1977), is an important component of the iterative Delaunay algorithm of Tipper (forthcoming). For that reason it is that Delaunay algorithm which is always used here.)

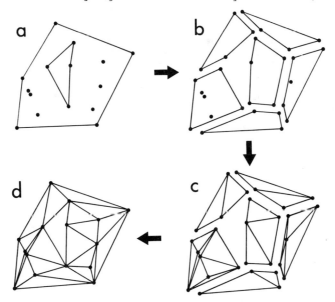

FIGURE 2. Obtaining the initial triangulation for multiply connected domains with non-convex boundaries. (a) Original data set of 15 points. Domain boundaries outlined. (b) Segmentation into six simply connected, convex sub-domains. For clarity these have been moved apart. (c) Delaunay triangulations of the separate sub-domains. (d) Delaunay triangulation for the entire set of points, conditioned on the given boundaries, obtained by patching together the separate sub-domain triangulations. Compare with Fig. 1

Deficiencies in the initial triangulation

The initial triangulation for any data set, being Delaunay, must apparently satisfy the requirements for a domain triangulation that were stated earlier. In most instances, however, it will nevertheless have two very serious deficiencies. The first is that, despite its overall local equiangularity, the initial triangulation will almost certainly contain many triangles that are far from equilateral, especially at the domain boundaries. This seems to be the case for the great majority of natural data sets. The second deficiency is that the initial triangulation will commonly contain obtuse-angled triangles. These will render it unusable for certain finite-difference techniques (Macneal, 1953; see also Heinrich, 1987).

To remove these deficiencies in the initial triangulation, extra points must be added to the original data set until a final triangulation results that has neither obtuse-angled nor markedly inequilateral triangles. The number of added points should be kept as low as possible, and the final density of points should reflect the complexity of the original data set. In the more complex areas of the domain there should be relatively more points; in the simpler ones, fewer. The Delaunay basis must, of course, be retained.

Adding extra points

The present algorithm uses three strategies in modifying the initial triangulation: addition of new points, either within the domain (AI) or on its boundaries (AB); removal of previously added points, either within the domain (RI) or on its boundaries (RB); repositioning of previously added points (M).

The addition of new internal points (AI) is done by adding them at the circumcentres of all existing obtuse-angled triangles (provided that those circumcentres do not fall outside the domain). The addition of new boundary points (AB) is done by bisecting existing boundary segments that subtend obtuse-angles internally (Fig. 3). In either variant of the addition strategy all the new points are added at once and the Delaunay triangulation then updated.

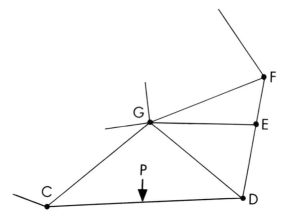

FIGURE 3. Changing the boundary point configurations. In strategy AB, a new point is added at P (CP = PD) because angle CGD is obtuse. In strategy RB, point E is removed because angle DGF is acute and E is not one of the original data points. If it were an original data point it would be retained.

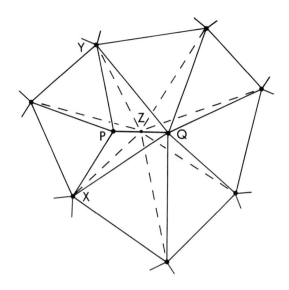

FIGURE 4. Removing internal points (strategy RI). XPYQ is a
'boomerang' of two abutting obtuse-angled triangles, XPQ and YPQ. The
ratio XY/PQ measures the thinness of the boomerang. Points P and Q,
neither of which is an original data point, are replaced by point Z (PZ =
ZQ), and the triangulation is modified to that shown by the broken lines.
It is then tested to ensure it is Delaunay.

The removal of previously added points on the boundary (RB) is done
simply by deleting all such points whose immediate boundary neighbours
subtend acute-angles internally (Fig. 3) : all such points are removed at
once, and the Delaunay triangulation updated. Strategy RB is then
applied again and again until no further point removal is possible. In
strategy RI, previously added internal points are removed if they define
'boomerangs' (Fig. 4), pairs of adjacent obtuse-angled triangles. The
'boomerangs' are eliminated in turn, not simultaneously, the relatively
thinnest first, until none remain: the Delaunay triangulation is updated
after each point has been removed.

The repositioning of points (M) is done simply by moving each one to the
centroid of its associated Voronoi polygon: it is applied
to all points except those in the original data set. The point furthest
from its Voronoi polygon centroid is moved first, and the Delaunay
triangulation updated. The point that is then furthest from its centroid
is next moved, and so on until no more movement is possible. Strategy M
is carried out every time points have been added or removed, i.e. after
strategies AI, AB, RI, or RB have been completed.

Modifying the initial triangulation is an iterative process. Which of
the various strategies is most appropriate at any stage (AI-M, AB-M,
RI-M, or RB-M) is determined by (1) the current Delaunay triangulation
(specifically, the number of obtuse-angled triangles in it (a) on the
domain boundaries, and (b) within the domain), (2) by the past iteration
history, and (3) by the requirement to minimise the number of added
points. A log is maintained of the progress of the algorithm in reducing
the number of obtuse-angled triangles, and this is used both to select
the strategy to be implemented in the next iteration and to detect
potential looping. The final triangulation is produced as soon as the
first configuration is achieved that has no obtuse-angled triangles.

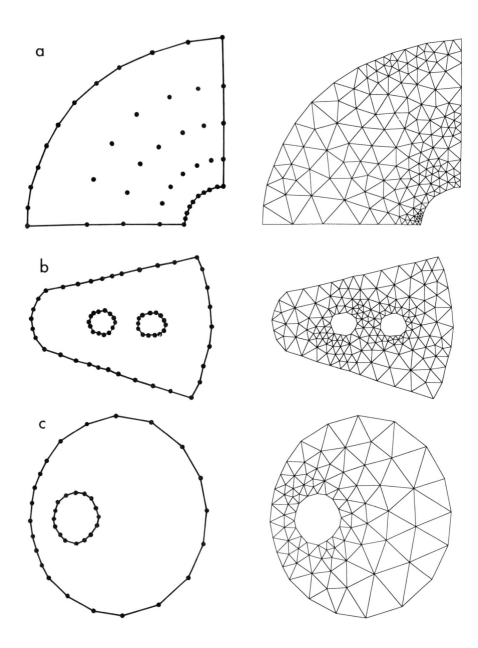

FIGURE 5. Test data sets, showing original point and domain boundary
configurations, and final triangulations. (a), (b), (c), (d) are similar
to those used by Cavendish (1974). (a) 43 original data points; 178
points in final triangulation; 100 secs cpu-time (VAX 8700 operating
under VMS); 94 iterations from initial to final triangulation. (b) 62
points; 181 points; 80 secs; 28 iterations. (c) 38 points; 91 points; 30
secs; 25 iterations. (d) 10 points; 85 points; 75 secs; 270 iterations.
(e) 74 points; 235 points; 185 secs; 141 iterations.

d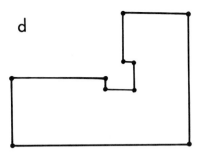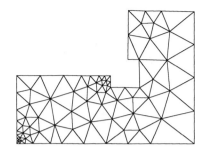

e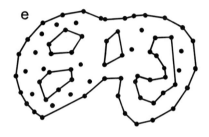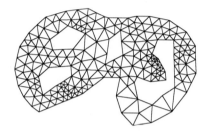

FIGURE 5.   (Continued).

DISCUSSION: THE ALGORITHM IN PRACTICE

The algorithm has been implemented in a FORTRAN program and applied to a large number of test data sets.   Five of these are reproduced here (Fig. 5).   The first four sets are similar to those originally used by Cavendish (1974) to test his own domain triangulation algorithm, and provide a variety of point configurations.   (Note that the sets used here are not identical to those that Cavendish used, but have been deliberately made less regular and less symmetric, in order to introduce a greater degree of spatial variation in the complexity of the point patterns.)   The fifth data set is the most complex, a multiply connected domain with both convex and re-entrant boundaries, with both internal and boundary data points, and with substantial spatial variation in its complexity.

The triangulations of these data sets that the new algorithm produces (Fig. 5) are very acceptable by almost any criteria.   The algorithm's run-time performance is, however, rather difficult to assess.   There seems no good correlation between the size of the original data set and either the number of iterations needed to achieve the final triangulation, or the actual cpu-time used.   It is the detailed spatial configuration of the original data set, not its size, that seems to determine both the final triangulation and how difficult that triangulation is to compute.   That is not unreasonable, of course, but a corollary is that the run-time performance of the algorithm is not in practice predictable even for data sets that are superficially similar. Fortunately the algorithm is sufficiently fast on realistic data sets (Fig. 5) that this lack of performance predictability is relatively unimportant.

REFERENCES

1. Cavendish, J.C., Automatic triangulation of arbitrary planar
    domains for the finite element method, *International Journal for
    Numerical Methods in Engineering,* vol. 8, pp. 679-696, 1974.

2. Cavendish, J.C., Field, D.A. and Frey, W.H., An approach to
    automatic three-dimensional finite element mesh generation,
    International Journal for Numerical Methods in Engineering, vol.
    21, pp. 329-347, 1985.

3. Heinrich, B., *Finite Difference Methods on Irregular Networks,*
    pp. 1-206, Birkhäuser Verlag, Basel, 1987.

4. Lawson, C.L., Software for C^1 surface interpolation, in
    Mathematical Software III, ed. J.R. Rice, pp. 161-194, Academic
    Press, New York, 1977.

5. Lee, D.T. and Schachter, B.J., Two algorithms for constructing a
    Delaunay triangulation, *International Journal of Computer and
    Information Sciences,* vol. 9, pp. 219-242, 1980.

6. Macneal, R.H., An asymmetrical finite difference network, *Quarterly
    of Applied Mathematics,* vol. 11, pp. 295-310, 1953.

7. McCullagh, M.J., Creation of smooth contours over irregularly
    distributed data using local surface patches, *Geographical
    Analysis,* vol. 13, pp.51-63, 1981.

8. Miles, R.E., On the homogeneous planar Poisson point process,
    Mathematical Biosciences, vol. 6, pp. 85-127, 1970.

9. Preparata, F.P. and Shamos, M.I., *Computational Geometry,*
    pp. 1-390, Springer-Verlag, New York, 1985.

10. Sibson, R. Locally equiangular triangulations, *The Computer
    Journal,* vol. 21, pp. 243-245, 1978.

11. Sibson, R. A brief description of natural neighbour interpolation,
    in *Interpreting Multivariate Data,* ed. Barnett, V, pp. 21-36,
    Wiley, New York, 1981.

12. Tipper, J.C. A straightforward iterative algorithm for the planar
    Voronoi diagram, *Information Processing Letters,* forthcoming.

13. Watson, D.F. and Philip, G.M. Systematic triangulations, *Computer
    Vision, Graphics, and Image Processing,* vol. 26, pp. 217-223, 1984.

On the Optimal Implementation of LU Decomposition Methods for Solving Linear Equations

NAI-KUAN TSAO
Wayne State University
Detroit, Michigan 48202, USA

INTRODUCTION

Given a system of linear equations $Ax = b$, where A is an n by n non-singular matrix, b and x are n-vectors, the LU decomposition method first find a uni-diagonal lower triangular matrix L and an upper triangular matrix U such that $A = LU$ and then find x as $x = U^{-1}y$, where $y = L^{-1}b$. The individual elements of L and U are given as

$$l_{ij} = (a_{ij} - \sum_{k=1}^{j-1} l_{ik}u_{kj})/u_{jj}, \quad i \geq j+1, \tag{1a}$$

$$u_{ji} = a_{ji} - \sum_{k=1}^{j-1} l_{jk}u_{ki}, \quad j \geq i. \tag{1b}$$

From the above equations we see that any l_{ij} is basically a_{ij} modified by an inner product of size $j-1$ followed by a division. Similarly, any u_{ji} is a_{ji} modified by an inner product of size $j-1$. It is also obvious that the j-th column of L is a function of the previous $j-1$ columns of L as well as the first $j-1$ components of the j-th column of U, and the j-th row of U is a function of the previous $j-1$ rows of U and the first $j-1$ components of the j-th row of L.

The actual implementation of (1a) and (1b) requires three loops using indices i, j, and k. If we denote by ijk-form (see Dongarra(1984) or Ortega(1988)) an implementation with the outer-most loop indexed by i and inner-most loop indexed by k, then we have many different implementations. Let either i or j be the inner-most index of the loops, then we have four different implementations as follows:

ikj-**form:**
for $i = 1$ to n do
  for $k = 1$ to $i - 1$ do
    $l_{ik} = a_{ik}/a_{kk}$
    for $j = k + 1$ to n do
      $a_{ij} = a_{ij} - l_{ik}a_{kj}$
jki-**form:**
for $j = 1$ to n do
  for $k = 1$ to $j - 1$ do
    for $i = j + 1$ to n do
      $a_{ij} = a_{ij} - l_{ik}a_{kj}$
    for $i = j + 1$ to n do
      $l_{ij} = a_{ij}/a_{jj}$

kij-form:
for $k = 1$ to $n - 1$ do
   for $i = k + 1$ to n do
     $l_{ik} = a_{ik}/a_{kk}$
     for $j = k + 1$ to n do
       $a_{ij} = a_{ij} - l_{ik}a_{kj}$

kji-form:
for $k = 1$ to $n - 1$ do
   for $i = k + 1$ to n do
     $l_{ik} = a_{ik}/a_{kk}$
   for $j = k + 1$ to n do
     for $i = k + 1$ to n do
       $a_{ij} = a_{ij} - l_{ik}a_{kj}$

Note the final U is stored in place in A. Furthermore, there is a common feature shared by the above forms: that each a_{ij} is updated after each iteration of the computation $a_{ij} = a_{ij} - l_{ik}a_{kj}$. Assuming that each floating point arithmetic operation incurs a round-off operation, then it is easily shown that the same L and U are obtained using anyone of the above four different forms. Thus the numerical behavior of the above forms can be represented by the kji-form.

On the otherhand, if k is used as the inner-most loop index, then the computation of (1a) and (1b) involves essentially the evalution of inner-products. In such cases different orders of evaluation of the inner-product will give different results for L and U.

In this paper we discuss the accuracy issue in comparing the kji-form with the various forms using inner-product in the implementation of the LU decomposition stage. Some preliminary results are first given, followed by the error complexity results of the various algorithms for the LU decomposition and the numerical experiment and concluding remarks.

SOME PRELIMINARY RESULTS

Given a normalized floating-point system with a t-digit base β mantissa, the following equations can be assumed to facilitate the error analysis of general arithmetic expressions using only $+, -, *,$ or $/$ operations (see Wilkinson(1963)):

$$fl(x\#y) = (x\#y)\Delta, \quad \# \in \{+, -, *, /\} \tag{2}$$

where

$$|\Delta| \leq 1 + u, \ u \leq \begin{cases} \beta^{1-t}/2 & \text{for rounded operations} \\ \beta^{1-t} & \text{for chopped operations} \end{cases}$$

and x and y are given machine floating-point numbers and $fl(.)$ is used to denote the computed floating-point result of the given argument. We shall call Δ the unit Δ-factor.

In general one can apply (2) repeatedly to a sequence of arithmetic steps, and the computed result z can be expressed in rational form as

$$z = \frac{z_p}{z_q}, \ z_p = \sum_{i=1}^{\lambda(z_p)} z_{pi}\Delta^{\sigma(z_{pi})}, \ z_q = \sum_{j=1}^{\lambda(z_q)} z_{qj}\Delta^{\sigma(z_{qj})} \tag{3}$$

where each z_{pi} or z_{qj} is an exact product of error-free data, and Δ^k stands for the product of k possibly different Δ-factors. We should emphasize that all common factors between the numerator and denominator should have been factored out before z can be expressed in its final rational form of (3). We shall also call such an exact product of error-free data a basic term, or simply a term. Then $\lambda(z_p)$ or $\lambda(z_q)$ is the total number of such terms whose sum constitutes z_p or z_q, respectively, and $\sigma(z_{pi})$ or $\sigma(z_{qj})$ gives the possible number of round-off occurrences associated with z_{pi} or z_{qj} during the computational process. We define the following two measures:

maximum error complexity:

$$\sigma(z_p) \equiv \max_{1 \le i \le \lambda(z_p)} \sigma(z_{pi}), \quad \sigma(z_q) \equiv \max_{1 \le j \le \lambda(z_q)} \sigma(z_{qj}). \tag{4}$$

cumulative error complexity:

$$s(z_p) \equiv \sum_{i=1}^{\lambda(z_p)} \sigma(z_{pi}), \quad s(z_q) \equiv \sum_{j=1}^{\lambda(z_q)} \sigma(z_{qj}). \tag{5}$$

Different algorithms used to compute the same z can then be compared using the above error complexity measures and the comparative numbers of basic terms created by the algorithms.

For convenience we will use $\bar{z}_q$ and $\bar{z}_p$ to represent the 3-tuples

$$\{\lambda(z_q), \sigma(z_q), s(z_q)\}, \quad \{\lambda(z_p), \sigma(z_p), s(z_p)\},$$

respectively, so that the computed z of (3) is fully characterized by $\bar{z} \equiv \frac{\bar{z}_p}{\bar{z}_q}$. The unit Δ-factor is then defined as

$$\bar{\Delta} \equiv \{1, 1, 1\}. \tag{6a}$$

In division-free computations any computed z will have only the numerator part z_p. The following lemma is useful in dealing with such intermediate computed results:

Lemma 1. Given x and y with their associated $\bar{x}_p$ and $\bar{y}_p$,

(i) if $z = xy$, then

$$\bar{z}_p \equiv \bar{x}_p \bar{y}_p = \bar{y}_p \bar{x}_p = \{\lambda(x_p)\lambda(y_p), \quad \sigma(x_p) + \sigma(y_p), \quad s(x_p)\lambda(y_p) + \lambda(x_p)s(y_p)\},$$

(ii) if $z = x \pm y$, then

$$\bar{z}_p \equiv \bar{x}_p + \bar{y}_p = \bar{y}_p + \bar{x}_p = \{\lambda(x_p) + \lambda(y_p), \quad \max(\sigma(x_p), \sigma(y_p)), \quad s(x_p) + s(y_p)\}.$$

Proof. The results can be obtained easily by expressing x and y as

$$x = \sum_{i=1}^{\lambda(x_p)} x_{pi} \Delta^{\sigma(x_{pi})}, \quad y = \sum_{j=1}^{\lambda(y_p)} y_{pj} \Delta^{\sigma(y_{pj})}$$

and applying (4), (5) and the definition of $\lambda(z_p)$ to find $\bar{z}_p$. **Q.E.D.**

For Δ^i one can apply Lemma 1 and obtain easily that

$$\overline{\Delta^i} = \bar{\Delta}^i = \{1, i, i\}. \tag{6b}$$

For general floating-point computations, we have the following lemma:

Lemma 2. Given x and y with their associated $\bar{x} = \frac{\bar{x}_p}{\bar{x}_q}$, $\bar{y} = \frac{\bar{y}_p}{\bar{y}_q}$,

(i) if $z = fl(x \pm y)$ and there is no common factors between x_q and y_q, then

$$\bar{z} = \frac{\bar{z}_p}{\bar{z}_q} = \frac{\bar{x}_q \bar{y}_p \bar{\Delta} + \bar{y}_q \bar{x}_p \bar{\Delta}}{\bar{x}_q \bar{y}_q};$$

(ii) if $z = fl(x * y)$ and there is no common factors between x_p and y_q or between y_p and x_q, then

203

$$\bar{z} = \frac{\bar{z}_p}{\bar{z}_q} = \frac{\bar{x}_p \bar{y}_p \bar{\Delta}}{\bar{x}_q \bar{y}_q};$$

(iii) if $z = fl(x/y)$ and there is no common factors between x_p and y_p or between x_q and y_q, then

$$\bar{z} = \frac{\bar{z}_p}{\bar{z}_q} = \frac{\bar{x}_p \bar{y}_q \bar{\Delta}}{\bar{y}_p \bar{x}_q}.$$

Proof. First we apply (2) to each case and obtain

$$z = fl(x \# y) = (x \# y)\Delta, \quad \# \in \{+, -, *, /\}.$$

The results can then be obtained easily by using Lemma 1. **Q.E.D.**

Often one needs to add up an extended sum given by

$$z^* = \sum_{i=1}^{k} x_i, \quad \bar{z}^* = \frac{\bar{z}_p^*}{\bar{z}_q^*} = \frac{1}{\bar{z}_q^*} \sum_{i=1}^{k} \bar{z}_{pi}^* \tag{7a}$$

where

$$2^{k-2}\lambda(z_{p1}^*) \le 2^{k-2}\lambda(z_{p2}^*) \le 2^{k-3}\lambda(z_{p3}^*) \le \ \dots \ \le 2^0\lambda(z_{pk}^*), \tag{7b}$$

$$\sigma(z_{p1}^*) + k - 1 \le \sigma(z_{p2}^*) + k - 1 \le \sigma(z_{p3}^*) + k - 2 \le \ \dots \ \le \sigma(z_{pk}^*) + 1. \tag{7c}$$

If it is desired to find $z = fl(\sum_{i=1}^{k} x_i)$, then certainly we have $\bar{z} = \frac{\bar{z}_p}{\bar{z}_q}$ and $\bar{z}_p$ can be found as $\bar{y}$ of

$$y = fl(\sum_{i=1}^{k} z_{pi}^*) \tag{7d}$$

where the z_{pi}^*'s satisfy (7b) and (7c) and the order of summation is to be specified. One would like to select an order to minimize the error complexities of y. Let us denote by left-to-right the order of summation as follows:

$$\sum_{i=1}^{k} z_{pi}^* = ((\dots ((z_{p1}^* + z_{p2}^*) + z_{p3}^*) + \dots) + z_{pk}^*).$$

We have the following theorem:

Theorem 1. Given (7a) through (7c) and it is desired to find y of (7d), then the optimal order for the evaluation of y to minimize $\sigma(y)$ and $s(y)$ is left-to-right and

$$\lambda(y) = \sum_{i=1}^{k} \lambda(z_{pi}^*), \quad \sigma(y)_{min} = \sigma(z_{pk}^*) + 1,$$

$$s(y)_{min} = \sum_{i=1}^{k} s(z_{pi}^*) - \lambda(z_{p1}^*) + \sum_{i=1}^{k}(k - i + 1)\lambda(z_{pi}^*).$$

Proof. The result for $\lambda(y)$ is obvious. For error complexities we prove by induction on k. For $k = 2$ the theorem is trivial. Let us assume the theorem is true for $k \le r$. For $k = r + 1$, then the final addition is done as

$$y = fl(y_\mu + y_\nu), \quad \mu + \nu = r + 1$$

where

$$y_\mu = fl(\sum_{j=1}^{\mu} z_{p\zeta_j}^*), \quad y_\nu = fl(\sum_{j=1}^{\nu} z_{p\xi_j}^*),$$

$$\zeta_1 < \zeta_2 < \ \dots \ < \zeta_\mu, \quad \xi_1 < \xi_2 < \ \dots \ < \xi_\nu,$$

$$\{\zeta_1, \zeta_2, \ \dots \ , \zeta_\mu, \xi_1, \xi_2, \ \dots \ , \xi_\nu\} = \{1, 2, \ \dots \ , r + 1\}.$$

By Lemmas 1 and 2, we have

$$\sigma(y) = \max(\sigma(y_\mu) + 1, \ \sigma(y_\nu) + 1), \quad s(y) = s(y_\mu) + s(y_\nu) + \lambda(y).$$

There are three cases:

(i) $\nu = 1$, $y_\nu = z^*_{p,r+1}$.

In this case we have from the induction assumption that

$$\sigma(y_\mu)_{min} = \sigma(z^*_{pr}) + 1, \quad s(y_\mu)_{min} = \sum_{i=1}^{r} s(z^*_{pi}) - \lambda(z^*_{p1}) + \sum_{i=1}^{r}(r - i + 1)\lambda(z^*_{pi})$$

where y_μ is evaluated in the order of left-to-right. Thus

$$\sigma(y)_{min(i)} = \max(\sigma(z^*_{pr}) + 2 , \ \sigma(z^*_{p,r+1}) + 1) = \sigma(z^*_{p,r+1}) + 1,$$

$$s(y)_{min(i)} = \sum_{i=1}^{r+1} s(z^*_{pi}) - \lambda(z^*_{p1}) + \sum_{i=1}^{r+1}(r + 1 - i + 1)\lambda(z^*_{pi})$$

and the theorem is true.

(ii) $\nu = 1$, $y_\nu = z^*_{pi_1}$, $i_1 \neq r + 1$.

In this case we have

$$\sigma(y_\mu)_{min} = \sigma(z^*_{p,r+1}) + 1, \quad s(y_\mu)_{min} = \sum_{j=1}^{r} s(z^*_{p\zeta_j}) - \lambda(z^*_{p\zeta_1}) + \sum_{j=1}^{r}(r - j + 1)\lambda(z^*_{p\zeta_j}).$$

Hence

$$\sigma(y)_{min(ii)} = \sigma(z^*_{p,r+1}) + 2 > \sigma(y)_{min(i)},$$

$$s(y)_{min(ii)} = \sum_{j=1}^{r+1} s(z^*_{pj}) - \lambda(z^*_{p\zeta_1}) + \sum_{j=1}^{r}(r - j + 1)\lambda(z^*_{p\zeta_j}) + \lambda(y).$$

Now

$$s(y)_{min(ii)} - s(y)_{min(i)} = \sum_{j=i_1+1}^{r+1} \lambda(z^*_{pj}) - (r + 1 - i_1)\lambda(z^*_{pi_1}) \ > \ 0.$$

Hence the theorem is also true.

(iii) $\mu > 1$, $\nu > 1$.

We have

$$\sigma(y_\mu)_{min} = \sigma(z^*_{p\zeta_\mu}) + 1, \quad \sigma(y_\nu)_{min} = \sigma(z^*_{p\xi_\nu}) + 1,$$

$$s(y_\mu)_{min} = \sum_{j=1}^{\mu} s(z^*_{p\zeta_j}) - \lambda(z^*_{p\zeta_1}) + \sum_{j=1}^{\mu}(\mu - j + 1)\lambda(z^*_{p\zeta_j}),$$

$$s(y_\nu)_{min} = \sum_{j=1}^{\nu} s(z^*_{p\xi_j}) - \lambda(z^*_{p\xi_1}) + \sum_{j=1}^{\nu}(\nu - j + 1)\lambda(z^*_{p\xi_j}).$$

Hence

$$\sigma(y)_{min(iii)} = 2 + \max(\sigma(z^*_{p\zeta_\mu}) , \ \sigma(z^*_{p\xi_\nu})) = 2 + \sigma(z^*_{p,r+1}) > \sigma(z_p)_{min(i)},$$

$$s(y)_{min(iii)} = \sum_{j=1}^{r+1} s(z^*_{pj}) + \sum_{j=1}^{\mu}(\mu - j + 1)\lambda(z^*_{p\zeta_j})$$
$$+ \sum_{j=1}^{\nu}(\nu - j + 1)\lambda(z^*_{p\xi_j}) - \lambda(z^*_{p\zeta_1}) - \lambda(z^*_{p\xi_1}) + \lambda(y).$$

Now

$$s(y)_{min(iii)} - s(y)_{min(i)} = s_0 - \sum_{j=1}^{r}(r - j + 1)\lambda(z^*_{pj}) + \lambda(z^*_{p1})$$

where

$$s_0 = \sum_{j=1}^{\mu}(\mu - j + 1)\lambda(z^*_{p\zeta_j}) + \sum_{j=1}^{\nu}(\nu - j + 1)\lambda(z^*_{p\xi_j}) - \lambda(z^*_{p\zeta_1}) - \lambda(z^*_{p\xi_1})$$

$$> \sum_{j=1}^{r+1} \lambda(z_{pj}^*).$$

By assumption we have

$$j\lambda(z_{p,r-j+2}^*) - j\lambda(z_{p,r-j+1}^*) \geq j\lambda(z_{p,r-j+1}^*).$$

Summing up the above for $j = 1$ to r and telescoping, we have

$$\sum_{j=2}^{r+1} \lambda(z_{pj}^*) - r\lambda(z_{p1}^*) \geq \sum_{j=1}^{r}(r - j + 1)\lambda(z_{pj}^*)$$

or

$$\sum_{j=1}^{r+1} \lambda(z_{pj}^*) \geq (r + 1)\lambda(z_{p1}^*) + \sum_{j=1}^{r}(r - j + 1)\lambda(z_{pj}^*).$$

Hence

$$s(y)_{min(iii)} - s(y)_{min(i)} \geq \sum_{j=1}^{r+1} \lambda(z_{pj}^*) - \sum_{j=1}^{r}(r - j + 1)\lambda(z_{pj}^*) + \lambda(z_{p1}^*) > 0$$

and the theorem is true in all cases. **Q.E.D.**

ERROR COMPLEXITY ANALYSIS

For simplicity we assume that A and b are error-free with $\bar{a}_{ij} = \bar{b}_i = \bar{1} = c_1 = c_{1\cdot}$. Then we have the following theorem (see Tsao(1989)):

Theorem 2. The computed L and U using the kji-form are such that

$$\bar{l}_{ik} = \frac{c_k}{c_{k\cdot}}\bar{\Delta}, \quad \bar{u}_{kk} = \frac{c_{k\cdot}}{c^*(1,k-1)}, \quad \bar{u}_{kj} = \frac{c_k}{c^*(1,k-1)}, \quad c^*(1, k - 1) = \prod_{r=1}^{k-1} c_{r\cdot}, \quad j > k,$$
$$c_{j+1} = c_{j+1\cdot} = c_j c_{j\cdot}\bar{\Delta} + c_j c_j \bar{\Delta}^3.$$

From the above theorem we can also easily deduce that

$$\lambda(c_{j+1}) = \lambda(c_{j+1\cdot}) = 2\lambda^2(c_j), \tag{8a}$$

$$\sigma(c_{j+1}) = \sigma(c_{j+1\cdot}) = 3 + 2\sigma(c_j). \tag{8b}$$

If the elements of L and U are evaluated using the inner-product forms of (1a) and (1b), respectively, then error-wise they are equivalent to the following implementation:

$$l_{ij} = fl((\sum_{k=1}^{j} x_k)/u_{jj}), \quad x_1 = a_{ij}, \quad x_k = fl(-l_{i,k-1}u_{k-1,j}), \quad 2 \leq k \leq j, \tag{9a}$$

$$u_{ji} = fl(\sum_{k=1}^{j} x_k), \quad x_1 = a_{ji}, \quad x_k = fl(-l_{j,k-1}u_{k-1,i}), \quad 2 \leq k \leq j. \tag{9b}$$

We have the following theorem:

Theorem 3. The optimal order for the evaluation of the inner-products in (9a) and (9b) is left-to-right and the computed L and U satisfy Theorem 2.

Proof. We prove by induction on j. For $j = 1$ the theorem is trivial. Assume the theorem is true for $j \leq r$. For $j = r + 1$, we consider (9b) first and define

$$z^* = \sum_{i=1}^{r+1} x_i, \quad x_1 = a_{r+1,i}, \quad x_k = -l_{r+1,k-1}u_{k-1,i}\Delta, \quad 2 \leq k \leq r + 1.$$

Hence

$$\bar{z}^* = \sum_{k=1}^{r+1} \bar{x}_k = \frac{1}{\bar{z}_q^*}\sum_{k=1}^{r+1} \bar{z}_{pk}^*, \quad \bar{z}_q^* = \prod_{\mu=1}^{r} c_{\mu\cdot}, \quad \bar{z}_{p1}^* = c_1 \prod_{\mu=1}^{r} c_{\mu\cdot},$$

$$\bar{z}_{pk}^* = c_{k-1}^2 \bar{\Delta}^2 \prod_{\mu=k}^{r} c_{\mu\cdot}, \quad 2 \leq k \leq r + 1.$$

By (8a) and (8b) it is easily shown that

$$\lambda(z_{pk}^*) = 2\lambda(z_{p,k-1}^*), \quad \sigma(z_{pk}^*) - \sigma(z_{p,k-1}^*) = 3 > 1$$

which satisfy (7b) and (7c). Therefore by Theorem 1 the optimal order for the evaluation of inner-products in (9b) is left-to-right. Let

$$z = fl(\textstyle\sum_{k=1}^{r+1} x_k), \quad y = fl(\textstyle\sum_{k=1}^{r+1} z_{pk}^*).$$

By repeated applications of (2) to y and by induction we have

$$\bar{y} = \bar{z}_{p1}^* \bar{\Delta}^r + \textstyle\sum_{k=2}^{r+1} \bar{z}_{pk}^* \bar{\Delta}^{r-k+2} = c_{r+1}.$$

Hence

$$\bar{u}_{r+1,i} = \bar{z} = \frac{z_p}{z_q} = \frac{\bar{y}}{\bar{z}_q} = \frac{c_{r+1}}{\bar{z}_q}$$

for any $i \neq r + 1$. For $i = r + 1$ we simply replace c_{r+1} by c_{r+1}^*. This proves the theorem for (9b). The proof for (9a) is similar and is skipped. **Q.E.D.**

NUMERICAL EXPERIMENT AND CONCLUDING REMARKS

For the numerical experiment a set of 1000 matrices of order 8 are generated. The coefficients of the matrix A are random numbers in the range from 0 to 1. The computed LU decompositions are obtained in single precision using a Sun/3 system. The unit round-off is 2^{-24}. The exact LU decompositions are obtained in double precision with a unit round-off less than 2^{-48}. The number of significant digits for each l_{ij} is calculated as

$$-log_{10}(|l_{ij}^{exact} - l_{ij}^{computed}|/|l_{ij}^{exact}|).$$

A similar formula can be devised for each u_{ij}. The average number of significant digits is simply the cumulative number of significant digits divided by the number of components accumulated. Let e_{ij} be the average number of significant digits of the (i, j)-th component of the computed L and U using optimal order in the evaluation of (9a) and (9b). Similarly, let f_{ij} be the same except now the order of evaluation of (9a) and (9b) is right-to-left, the reverse of left-to-right. Then Table 1 gives the values of e_{ij} and Table 2 gives the same for f_{ij}.

TABLE 1. $e_{ij}, 1 \leq i, j \leq 8$

	$j = 1$	$j = 2$	$j = 3$	$j = 4$	$j = 5$	$j = 6$	$j = 7$	$j = 8$
$i = 1$	∞	∞	∞	∞	∞	∞	∞	∞
$i = 2$	7.815	7.456	7.453	7.480	7.481	7.476	7.486	7.495
$i = 3$	7.820	7.195	7.065	7.054	7.047	7.073	7.015	7.006
$i = 4$	7.789	7.203	6.902	6.689	6.721	6.752	6.701	6.711
$i = 5$	7.815	7.207	6.897	6.530	6.439	6.440	6.389	6.459
$i = 6$	7.836	7.191	6.893	6.519	6.332	6.161	6.237	6.214
$i = 7$	7.824	7.207	6.902	6.506	6.352	6.101	5.999	6.029
$i = 8$	7.804	7.209	6.896	6.557	6.322	6.088	5.958	5.867

TABLE 2. $f_{ij}, 1 \leq i, j \leq 8$

	$j=1$	$j=2$	$j=3$	$j=4$	$j=5$	$j=6$	$j=7$	$j=8$
$i=1$	∞	∞	∞	∞	∞	∞	∞	∞
$i=2$	7.815	7.456	7.453	7.480	7.481	7.476	7.486	7.495
$i=3$	7.820	7.195	7.038	7.057	7.030	7.060	7.028	6.999
$i=4$	7.789	7.203	6.878	6.668	6.685	6.699	6.717	6.705
$i=5$	7.815	7.207	6.875	6.501	6.399	6.399	6.373	6.428
$i=6$	7.836	7.191	6.860	6.498	6.286	6.152	6.174	6.179
$i=7$	7.824	7.207	6.863	6.502	6.298	6.080	5.973	5.991
$i=8$	7.804	7.209	6.866	6.527	6.267	6.062	5.887	5.820

We see from the above two tables that the first two rows and columns of Table 1 and Table 2 are identical. This is not suprising as these elements are evaluated in exactly the same sequence regardless of the chosen order of inner-product evaluation. However, this is not so for the remaining elements: in general $e_{ij} \geq f_{ij}$ with the exceptions of e_{37} and e_{47}. This confirms the validity of Theorem 3 for this set of examples using the prescribed orders of inner-product evaluation. Experiments with higher order matrices give similar conclusions.

ACKNOWLEDGEMENT

This work was supported in part by NASA while the author was visiting ICOMP, NASA Lewis Research Center, Cleveland, Ohio.

REFERENCES

1. Dongarra, J., Gustavson, F., and Karp, A., Implementing Linear Algebra Algorithms for Dense Matrices on a Vector Pipeline Machine, *SIAM Review*, vol. 26, pp. 91-112, 1984.

2. Ortega, J. M., *Introduction to Parallel and Vector Solution of Linear Systems*, Plenum Press, New York, 1988.

3. Tsao, N. K., On the Equivalence of Gaussian Elimination and Gauss-Jordan Reduction in Solving Linear Equations, NASA Technical Memorandum 101466, 1989.

4. Wilkinson, J. H., *Rounding Errors in Algebraic Processes*, Prentice-Hall, Englewood Cliffs, N.J., 1963.

BOUNDARY INTEGRALS AND ELEMENTS

Numerical Studies of Some Integral Equations Arising in Compressible Flow

J. A. BELWARD
Department of Mathematics
The University of Queensland
St. Lucia, Q. 4067, Australia

INTRODUCTION

This paper presents a numerical study of a pair of integral equations which arise in a model problem in two dimensional laminar compressible viscous flow. Each is a Fredholm equation of the first kind on a finite interval with a difference kernel with a logarithmic singularity. In the underlying physical problem a finite width strip is inclined at a small angle to a uniform stream of viscous compressible fluid. The Navier-Stokes equations are linearised by a method used by Oseen for incompressible flows. Integral representations of the velocity field may be combined with the non-slip conditions to give a vector equation which uncouples for small angles of incidence to give two scalar equations. All flows are dependent on two non-dimensional parameters, the Reynolds number R and Mach number M.

The equations were set up by Homentcovschi (1981) who gave a comprehensive asymptotic analysis and numerical solutions for small Reynolds numbers. The numerical results were obtained by a series expansion method. If a quadrature rule (or an orthogonal expansion) is used instead, the restriction to small Reynolds numbers may be relaxed and the dependence of the solutions on the Mach number is more easily investigated. The current study presents such a method, solutions are obtained over a wide range of parameter values.

Earlier Miyagi and Nishioka (1980) used a quadrature rule to obtain numerical results in a study of trailing edge flows. A quadrature rule was used to solve the Oseen integral equation for incompressible fluids; the solution was approximated by a piecewise constant function, except that the first and last values were weighted by $1/\sqrt{(1 \pm x)}$. This method of including the singular behaviour of the solution converges at a rate of $O(1/\sqrt{N})$, where N is the number of intervals used. Miyagi and Nishioka (1980) did not report on the accuracy of the method in their studies, however it gave good accuracy for low values of N ($N = 16$ in many cases) on the problem reported here.

The current study demonstrates that despite the poor
theoretical rate of convergence, the numerical scheme
described above may be used to obtain solutions of the
integral equations over the full range of parameter values,
recapturing the asymptotic results of Homentcovschi while
obtaining results in the intermediate ranges. While the
reliability of the asymptotic methods used by Homentcovschi
is well established the expansions are non-uniform and
without error bounds. Thus the numerical results obtained
here also provide valuable information on the range of
accuracy of the asymptotic analysis.

THE INTEGRAL EQUATIONS

The equations for the lift and drag, Homentcovschi (1981), are

$$\int_{-1}^{1} f_D(t) \, k_D(x-t) \, dt \;=\; -1 \; ; \quad \int_{-1}^{1} f_L(t) \, k_L(x-t) \, dt \;=\; -\vartheta \qquad (1, \ 2)$$

The functions f_D and f_L are the drag and lift per unit
length exerted by the fluid on an elemental segment. The
kernels are

$$k(u) \;=\; \frac{\sigma}{2\pi} \, e^{\sigma u} \; (K_0(\sigma|u|) \; \pm \; sgn(u) \, K_1(\sigma|u|)) \; + \; \frac{\alpha}{2\pi} e^{-\alpha \gamma u} \left\{ K_0(\alpha|u|) \mp \right.$$

$$(3)$$

$$sgn(u) K_1(\alpha|u|) \; - \; \left(\int_1^{\gamma} \frac{e^{\alpha u t}}{(t^2 - 1)^{1/2}} dt \; \pm \; \int_1^{\gamma} \frac{e^{\alpha u t} \, t}{(t^2 - 1)^{1/2}} dt \right) H(\gamma - 1) \right\}$$

where $k_D(x-t)$ corresponds to taking the upper signs and
$k_L(x-t)$ the lower signs in (2.3) . H is the unit step
function, K_ν are modified Bessel functions, ϑ is the
small angle of incidence The Reynolds number R is defined
as $R = U\ell\rho\mu^{-1}$, U, ℓ, μ, and ρ being the free stream
velocity at infinity, strip width, density and viscosity. R,
α, σ, γ and M are related by

$$\sigma \;=\; 4\alpha/3 \;=\; R/4, \quad \gamma \;=\; 2/M^2 - 1. \qquad (4, \ 5)$$

$M = U/a$ where a is the velocity of sound in uniform flow.

As $\gamma \to \infty$ we recapture viscous incompressible flow and as
$R \to \infty$ we obtain inviscid flows. In the limit as $\gamma \to \infty$
the integrals in (3) become the modified Bessel Functions
K_0 and K_1. Each kernel has only one singularity, a
logarithmic singularity at $u = 0$. For further details the
reader is referred to Homentcovschi (1981).

The integral equations may be classified as being of Oseen
type. From the arguments given in Belward (1984) a

uniqueness theorem for the underlying physical problem will imply an existence theorem for the integral equations (1) and (2).

The net force on the strip is the primary focus of interest in applications; its components are the integrated skin friction and local lift, viz.

$$c_D = \int_{-1}^{1} f_D(t) \, dt \quad , \qquad c_L = \int_{-1}^{1} f_L(t) \, dt \quad . \tag{8, 9}$$

ASYMPTOTICS

Pointwise in u, we have the following asymptotic approximations to the kernels. For large γ ,

$$k(u) = \frac{\sigma}{2\pi} e^{\sigma u} (K_0(\sigma|u|) \pm sgn(u) K_1(\sigma|u|)) \mp \frac{1}{2\pi u}(1\pm\frac{1}{\gamma}) + O(\frac{1}{\gamma^2}) \tag{10}$$

(upper signs correspond to $k_D(u)$ and lower signs to $k_L(u)$).
For large Reynolds numbers in subsonic flow

$$k_D(u) = \left(\frac{\sigma}{2\pi u}\right)^{1/2} H(u) - \frac{1}{(1-M^2)^{1/2} 2\pi u} + O(\sigma^{-1/2}) \tag{11}$$

$$k_L(u) = \frac{(1-M^2)^{1/2}}{2\pi u} - \frac{H(u)}{(2u)^{3/2} \pi^{1/2} \sigma^{1/2}} + O(1/\sigma) \tag{12}$$

and in supersonic flow

$$k_D(u) = \left(\frac{\sigma}{2\pi u}\right)^{1/2} H(u) + \frac{1}{2(M^2-1)^{1/2}} \delta(u) + O(\sigma^{-1/2}) \tag{13}$$

$$k_L(u) = \frac{(M^2-1)^{1/2}}{2} \delta(u) - \frac{H(u)}{(2u)^{3/2} \pi^{1/2} \sigma^{1/2}} + O(1/\sigma) \tag{14}$$

$\delta(u)$ is the Dirac delta function.

From these expansions Homentcovschi (1981) deduced the following expansions for the total drag and lift. For high Reynolds number in subsonic flow

$$c_D = \frac{2}{(\pi\sigma)^{1/2}} + \frac{3.664}{\pi^2 \sqrt{(1-M^2)} \, \sigma} \quad ; \quad c_L = \frac{\pi}{\sqrt{(1-M^2)}} + \frac{7.72 \, \sigma^{-1/2}}{(2\pi)^{3/2} (1-M^2)} \tag{15, 16}$$

while in supersonic flow

$$c_D = \frac{2}{(\pi\sigma)^{1/2}} - \frac{1}{2\sqrt{(M^2-1)} \, \sigma} \quad ; \quad c_L = \frac{2}{\sqrt{(M^2-1)}} - \frac{2\sigma^{-1/2}}{\sqrt{\pi} \, (M^2-1)} \tag{17, 18}$$

213

These expansions are non-uniform in M and R and no error bounds are placed on them; the computed values of c_D and c_L given later confirm their validity, give guidance on their accuracy, and provide results over all parameter ranges.

KERNEL EVALUATION

The aim is to develop a numerical scheme effective over as large a range of parameter values as is reasonably possible; kernel evaluation therefore needs careful consideration.

For the Bessel functions, Chebyshev expansions given by Clenshaw (1962) were used ; these give 16 digit accuracy over all real arguments. The functions $e^x K_s(x)$ $(s = 0,1)$ are $O(1/\sqrt{x})$ as $x \to \infty$;these must be evaluated as a single term otherwise separate calculation of the exponential and Bessel functions for large arguments causes under- and overflow problems.The fifth and sixth terms in (3) contribute in the subsonic regime only. For moderate values of α and γ they only require a quadrature rule which takes account of the $(t^2-1)^{1/2}$ term . When α and γ become larger a product rule can be used to take account of the exponential term; see Belward (1986) where a Filon type rule is given.

For positive values of u the multiplier $e^{-\alpha\gamma u}$ should be combined with the integrands of the fifth and sixth terms; these terms are then $O(\alpha\gamma u)$ and $O(1)$ for large arguments. For negative values of u both the third and fifth terms (and the fourth and sixth) may become large leading to loss of accuracy through catastrophic cancellation. By monitoring the magnitudes of these terms the need to prevent this problem may be detected and countered as follows. The third and fifth (and the fourth and sixth) terms may be combined using an integral representation of the Bessel functions to give terms of the form

$$\int_{\gamma}^{\infty} \frac{e^{\alpha u t}\, t^s}{(t^2 - 1)^{1/2}}\, dt, s=0,1$$

This integral may be evaluated adaptively by calculating its value on (γ, A), estimating the remainder on (A, ∞) and adjusting the value of A if necessary. The computed values gave good agreement with their asymptotic expressions given in Homentcovschi (1981).

Of the two equations, the equation for the lift is is the more difficult to solve. The drag kernel is always positive, but the kernel for the lift has sign changes hence the possibility of nontrivial solutions of the homogeneous equation is raised.

THE NUMERICAL SCHEME

The basic method consists of approximating f by a piecewise

214

constant function on a set of equally spaced intervals except
on the first and last where the constant is weighted by
$1/\sqrt{(1+x)}$ and $1/\sqrt{(1-x)}$. Such singularities are to be
expected in the solution of Oseen type integral equations.
Although this method will ultimately suffer from a slow
convergence rate, it provided useful results for Miyagi and
Nishioka (1980) in a study of Oseen's approximation in
trailing edge flows.

If the equation is denoted by

$$Kf = g \tag{19}$$

then this approximation scheme defines an operator K given
explicitly by

$$K_N f = f_1 \int_{-1}^{t_1} k(x,t)/\sqrt{(1+t)}\,dt + \sum_{j=2}^{N-1} f_j \int_{t_{j-1}}^{t_j} k(x,t)\,dt + f_N \int_{t_{N-1}}^{1} k(x,t)/\sqrt{(1-t)}\,dt$$

A linear system for $f_1, f_2, \ldots f_N$ may be obtained by
collocating at the N points $(t_{i-1}+ t_i)/2,$ $i=1,2,\ldots N.$
If f has inverse square root singularities at 1 and -1 it
is a simple matter to show that

$$Kf - K_N f = O(N^{-1/2})$$

when $K(x,t)$ is a smooth function of t; the log singularity
degrades this to $N^{-1/2}\log N,$ which in practice is rarely
distinguishable from $N^{-1/2}$. This error dependence is due to
the fact that as the subdivision is refined the contribution
from the singularities becomes vanishingly small. For low
and moderate values of N a more significant contribution is
made to the singularities and this probably accounts for the
acceptable accuracy for low values of N. In addition the
ill-posed nature of first kind equations means that highly
accurate methods are not likely to be more successful since
any approximation scheme must ultimately diverge and optimal
errors are reached much sooner by high order methods. The
ill-posed nature does not manifest itself because of the weak
singularity of the kernel and we are most concerned with the
total lift and drag, integrals of f_D and f_L.

Since the kernel is a difference kernel and the quadrature
points are equally spaced the coefficient matrix of the
linear system is a Toeplitz matrix apart from the first and
last columns. It may be computed with $4N-3$ separate
calculations. The integrals corresponding to the first and
last columns and the main diagonal have logarithmic and/or
algebraic singularities. After these were taken into account,
simple or composite 20 point Gaussian rules were used.

RESULTS

The graph plots and tables in this section summarise the main
features of the results for the computation of the lift and

TABLE 1. Values of the drag at low Reynolds number. C
denotes current method, *M* denotes Homentcovschi's results

R	*M* = 0			*M* =.8		*M* = 1.33		*M* =2.0	
	C	H	Formula (6.1)	C	H	C	H	C	H
0.25	11.0	-	11.0	7.57	7.65	6.62	6.76	7.05	7.18
0.5	6.44	6.40	6.46	4.61	5.48	3.93	3.92	4.23	4.16
0.75	4.77	4.68	4.81	3.51	3.52	2.93	2.92	3.18	3.16
1.0	3.87	3.75	3.93	2.92	2.92	2.40	2.38	2.61	2.55
1.5	2.91	2.84	3.00	2.29	2.28	1.83	1.78	2.01	1.96
2.0	2.39	2.35	2.51	1.94	1.98	1.52	1.49	1.68	1.61

drag. They were calculated for a continuous range of Reynolds
number from zero to 10^3 and a range of Mach numbers from zero
to 5 . There would be no difficulty in continuing for larger
Mach numbers, however in the case of the lift equation the
present scheme probably needs to be modified to provide
results for larger values of the Reynolds number.

Tables 1 and 2 compare the values of c_D and c_L at low
Reynolds numbers. The formulae

$$c_D = \frac{\pi}{\sigma (1 - \ln (\sigma/4) - \Gamma)} \quad ; \quad c_L = \frac{\pi}{\sigma (1 + \ln (\sigma/4) + \Gamma)} \qquad (20,21)$$

(where Γ is Eulers constant) given by Homentcovschi (1981)
are also compared at $M = 0$. Homentcovschi's results were
read from graphs in his paper; where blank entries appear no
results were available.

Table 3 compares results for high Reynolds numbers with the
asymptotic formulae given in section 3. For Reynolds numbers
of 5×10^3 values of the lift were not computed since the
solutions appeared unstable for values of *N* up to 64.

TABLE 2. Values of the lift at low Reynolds numbers, C
denotes the current method, H denotes Homentcovschi's results.

R	*M* = 0			*M* = .8		*M* = .33		*M* = 2.0	
	C	H	Formula (6.2)	C	H	C	H	C	H
0.25	18.9		19.5	11.7	11.2	9.29	8.59	8.53	7.45
0.5	12.6	12.3	13.3	8.02	8.02	5.93	5.84	5.32	5.10
0.75	10.1	9.45	11.3	6.71	6.88	4.69	4.76	4.12	4.07
1.0	8.84	8.02	10.5	6.06	6.30	4.03	4.01	3.48	3.44
1.5	7.42	-	10.6	5.45	-	3.33	3.29	2.79	2.64
2.0	6.65	-	12.5	5.21	-	2.97	2.86	2.41	2.18

TABLE 3. Comparison of asymptotic results with the numerical
scheme at high Reyolds number. C denotes computed values, A
denotes the asymptotic fomulae (15)-(18).

$R = 10^3$	C_D		C_L		$R = 5 \times 10^3$	C_D	
M	A	C	A	C	M	A	C
0.80	.07879	.07340			0.80	.03340	.03242
1.35	.06915	.07066	2.127	2.158	1.35	.03147	.03213
1.83	.07006	.07153	1.279	1.296	1.83	.03165	.03231

Figures 1 and 2 show the dependence of the lift and drag
on the Mach number, for different Reynolds numbers. The
graphs are piecewise linear plots. Near $M = 1$ the results
were examined carefully, stepping in small increments from
above and below unity. It appears from these results that
the model equations need to be carefully examined in the
transonic regime. In order to obtain some experimental
information about the convergence rate of the method, three
quantities were monitored. These were the net force (either
c_D or c_L), f_1, the singularity strength at the leading edge
and $f(0)$ the value of the solution at the centre of the
strip. Assuming an error dependence given by (22) α was
estimated from (23).

$$e_N \cong N^{-\alpha}, \qquad 2^{-\alpha} = \frac{q_{4N} - q_{2N}}{q_{2N} - q_N} \qquad (22,23)$$

q_i being one of the three quantities mentioned above.
Applying this to c_D and c_L, with $N = 8$, 12, and 16, the mean
of 17 sets of results was 1.98 with a standard deviation

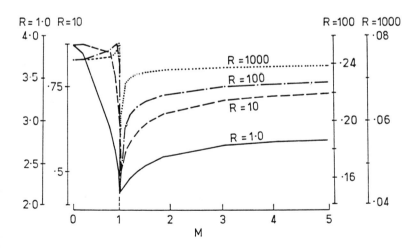

FIGURE 1. Drag vs. Mach number

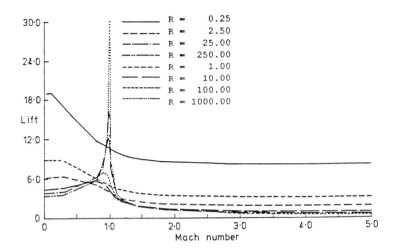

FIGURE 2. Lift vs Mach number.

of *.16,* and greatest and least observed values of *2.26* and *1.71.* Applying the same method to the computed values of *f(0),* the mean was observed to be *1.89,* with a standard deviation of *.12* and greatest and least observed values of *2.03* and *1.64.* If this evidence is accepted as an indication of quadratic convergence then it shows that the local nature of the approximation by piecewise constants is dominant here, and enough of the end point singularities are included to mask the theoretical *O(1/√N)* convergence rate.

No conclusive convergence rates were observed for the singularity strengths; in the supersonic case, the values of α were widely scattered around unity while in the subsonic case there was no distinct pattern.

REFERENCES

1. Homentcovschi,D. Oseen flow of a compressible fluid past a flat plate. *Quart. Appl. Math.* vol. 39, pp. 221-237, 1981.

2. Miyagi, T., & Nishioka M. Oseen velocity distributions in the wake of a flat plate. *J. Fluid Mech.,* vol. 97, pp.145-155, 1980.

3. Belward,J.A. Existence of solutions of integral equations ofOseen type. *Bull. Austral. Math. Soc.,* vol. 29, pp. 57-66 1984.

4. Clenshaw, C.W. Mathematical Tables, Vol. 5, Chebyshev series for mathematical functions. Issued by H.M.S.O., London, 1962.

5. Belward, J.A. An exponential version of Filon's rule. *J. Comp. and Appl. Math.,* vol. 14, pp. 461-466, 1986.

The Boundary Integral Method Applied to Underwater Explosion Bubble Dynamics

J. P. BEST
Materials Research Laboratory, DSTO
Melbourne, Australia

A. KUCERA
Department of Mathematics
La Trobe University
Bundoora, Victoria 3083, Australia

INTRODUCTION

When an underwater explosion occurs we observe two principal damage causing mechanisms. The immediate response is the propagation of a shock wave into the water with a speed $c \sim 1500ms^{-1}$. On a larger time scale we observe the oscillations of a bubble containing the remnants of detonation. During these oscillations a weight of water is displaced which is of the same order as the displacement of a typical target (destroyer). As a consequence, we find that in some circumstances, the potential for damage due to the bubble motion may be comparable with, or exceed that due to the primary shock wave.

An extensive experimental and theoretical effort was directed towards the study of underwater explosions during World War II. The principal workers were Herring in the U.S. and Taylor in the U.K., with Herring (1941) being the first to give the equations describing the upwards motion of a bubble of gas due to buoyancy. A principal assumption in this work was that the bubble remained spherical throughout its oscillatory motion. In the next section of this paper, we present the elements of this simple theory. Subsequent experimental studies by Taylor & Davies (1950) and Bryant (1950) established that this approximation does not even hold approximately as the bubble contracts towards its first minimum. During this phase of motion the elements of a re-entrant jet threading the bubble from the rear are observed. Because of the large deformations involved, perturbation methods are of limited utility.

In recent times the Boundary Integral Method (B.I.M.) has been utilised extensively in the study of bubble dynamics (see for instance Blake et. al. (1986), Guerri et. al. (1981), Wilkerson (1989)). We may reformulate the problem of the underwater explosion via Green's theorem leading to an expression for the potential as an integral over the surface of the bubble; the starting point for the boundary integral formulation. In this application a cubic spline representation of the bubble surface is implemented leading to much improved accuracy over previous applications to cavitation bubble dynamics which made use of linear and quadratic surface elements.

We then present the results of calculations describing the rise of an underwater explosion bubble under buoyancy forces. The results of these calculations throw considerable light on our understanding of the explosion bubble phenomenon. For moderate explosions, a re-entrant jet penetrates the upper surface of the bubble after the first oscillation providing

219

direct evidence in support of the postulate of Snay (see Holt (1977)) that for deep explosions the bubble rises with an attached vortex ring for much of its motion. For smaller explosions we do not observe a jet until after the first minimum, during which time a thin jet grows as the bubble re-expands. These results are qualitatively similar to some of the experimental results of Lauterborn (1980).

THE ELEMENTARY MODEL

The earliest attempts at modelling the rise of underwater explosion bubbles due to buoyancy proceed via the assumption of a spherical bubble oscillating in an inviscid, incompressible and irrotational fluid ('Triple I' fluid). The bubble is characterised by its radius $R^*(t^*)$ and the flow field is described by a velocity potential, ϕ, which is a solution of Laplace's equation, that is,

$$\nabla^2 \phi = 0. \tag{1}$$

The contents of the bubble are assumed ideal and the thermodynamic processes assumed adiabatic. We thus introduce the adiabatic coefficient, γ, and may write the pressure, p, within the bubble as

$$p = p_0 (R_0^* / R^*)^{3\gamma}, \tag{2}$$

where p_0 is the initial pressure within the bubble and R_0^* is its initial radius. In order to determine a system of equations describing the bubble motion we require an expression for the velocity potential of the flow induced by the bubble's oscillatory motion. The appropriate expression is

$$\phi = -\frac{R^{*2} \dot{R}^*}{|\mathbf{r}|} - \frac{1}{2} \frac{R^{*3} U^* \cos\theta}{|\mathbf{r}|^2}, \tag{3}$$

where U^* is the velocity of the bubble centroid, $\mathbf{r}$ is the position vector of some point in the flow field and θ is the angle between $\mathbf{r}$ and the vertical. The source-like term of (3) arises from the changing volume of the bubble whereas the dipole-like term is due to the translational motion of the bubble centroid (Lighthill (1986)). Although we are considering the motion of a bubble in an infinite fluid we will introduce a reference plane (e.g. the ocean surface) from which we will measure the depth of the bubble, h^*. We will further suppose that the far-field pressure in this plane is equal to the atmospheric pressure, p_a. The appropriate geometry is shown in Figure 1(a). In accord with this convention we may write the Bernoulli equation as

$$\frac{\partial \phi}{\partial t^*} + \frac{1}{2} |\nabla \phi|^2 + p/\rho + gz = p_a/\rho, \tag{4}$$

where ρ is the density of the fluid, g is the gravitational acceleration and the reference plane is at $z = 0$. Since pressure is continuous across the boundary of the bubble we may evaluate the Bernoulli equation at the surface of the bubble with the help of (2), whence we obtain

$$\frac{\partial \phi}{\partial t^*} + \frac{1}{2} |\nabla \phi|^2 + p_0/\rho (R_0^*/R^*)^{3\gamma} + gz = p_a/\rho. \tag{5}$$

If we now integrate this expression over the surface of the bubble we obtain the equation of motion

$$R^* \ddot{R}^* + \frac{3}{2} \dot{R}^{*2} - \frac{1}{4} U^{*2} = p_0/\rho (R_0^*/R^*)^{3\gamma} - (gh^* + p_a/\rho), \tag{6}$$

where the potential of (3) has been used. This equation may also be derived via considerations of energy conservation.

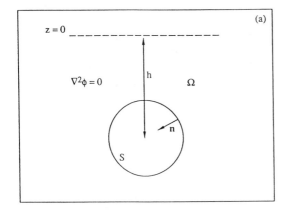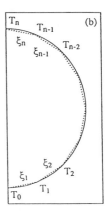

FIGURE 1. (a) Geometry used for explosion bubble motion in an infinite fluid. (b) The discretisation of the bubble surface for the boundary integral algorithm.

We establish another equation of motion by considering momentum conservation within the formalism of the Kelvin Impulse. In the absence of boundaries the time rate of change of the Kelvin Impulse is equal to the buoyancy force. With the z-component of the Kelvin Impulse given as (Lighthill (1986))

$$I = \frac{2}{3}\pi\rho R^{*3}U^*,$$ (7)

this condition yields

$$\frac{d}{dt^*}(R^{*3}U^*) = 2gR^{*3},$$ (8)

an equation named after Herring (1941), who first derived it in the study of explosion bubbles. The description is completed with the observation that

$$U^* = -\frac{dh^*}{dt^*} \quad \text{and} \quad R^*(0) = R_0^*, \quad h^*(0) = h_0^*, \quad \dot{R}^*(0) = 0, \quad U^*(0) = 0.$$ (9)

In proceeding to numerical solutions of these equations we introduce the non-dimensional variables and constants

$$t = \frac{t^*}{R_0^*}\sqrt{\frac{p_a}{\rho}}, \ U = U^*\sqrt{\frac{\rho}{p_a}}, \ R = R^*/R_0^*, \ h = h^*/R_0^*, \ \alpha = p_0/p_a \quad \text{and} \quad \delta = gρR_0^*/p_a,$$ (10)

in which notation the system of equations become

$$R\ddot{R} + \frac{3}{2}\dot{R}^2 - \frac{1}{4}U^2 = \alpha\left(\frac{1}{R}\right)^{3\gamma} - (1+\delta h), \quad \frac{d}{dt}(R^3U) = 2\delta R^3, \quad U = -\frac{dh}{dt},$$ (11)

where the dot now denotes differentiation with respect to non-dimensional time, and the initial conditions for the system are

$$R(0) = 1, \quad \dot{R}(0) = 0, \quad U(0) = 0, \quad h(0) = h_0,$$ (12)

with h_0 somewhat arbitrary.

NUMERICAL ANALYSIS

In this section we explain, in some detail, how the Boundary Integral Method (B.I.M.) is used to determine the motion of the underwater explosion bubble in the presence of the buoyancy forces, in the restricted case of axisymmetric flow.

In order to apply the B.I.M. we recast the problem in terms of its boundary integral formulation, namely

$$c(p)\phi = \int_S \left(\frac{\partial \phi}{\partial n} G - \phi \frac{\partial G}{\partial n} \right) dS, \quad \text{with} \quad c(p) = \begin{cases} 4\pi & \text{if} \quad p \in \Omega, \\ 2\pi & \text{if} \quad p \in S, \end{cases} \tag{13}$$

where ϕ is the potential, $\frac{\partial \phi}{\partial n}$ is the normal velocity with n the normal exterior to the domain, Ω, of solution and G is the Green's function. S is the surface of the explosion bubble. (See Figure 1(a).)

The next step is to divide the bubble surface S into a collection of subintervals and represent the true surface with cubic splines. On each segment element the splines are parameterized with respect to the segment lengths and the cylindrical co-ordinates of the bubble surface are given by

$$\left. \begin{array}{l} r(\xi) = a_{rj} + b_{rj}(T_j - \xi) + c_{rj}(T_j - \xi)^2 + d_{rj}(T_j - \xi)^3 \\ z(\xi) = a_{zj} + b_{zj}(T_j - \xi) + c_{zj}(T_j - \xi)^2 + d_{zj}(T_j - \xi)^3 \end{array} \right\}, \quad \text{with} \ T_{j-1} \leq \xi \leq T_j, \tag{14}$$

where the subscript j is used to denote the line segment (i.e. $j = 1, \ldots, n$), the T_j's are the progressive arc lengths (see Figure 1(b)), while the constants $a_{rj}, b_{rj}, c_{rj}, d_{rj}$ and $a_{zj}, b_{zj}, c_{zj}, d_{zj}$ are the spline coefficients. The progressive arc lengths are calculated by performing the following segment length integrations

$$\xi_j = \int_{T_{j-1}}^{T_j} \sqrt{\left(\frac{dr}{d\xi} \right)^2 + \left(\frac{dz}{d\xi} \right)^2} \, d\xi, \quad T_0 = 0, \quad \text{and} \quad T_j = T_{j-1} + \xi_j, \tag{15}$$

numerically and repeating the process until the scheme converges. Similarly, the potential, ϕ, and its normal derivative, $\psi(= \partial \phi / \partial n)$, are represented using cubic splines and linear interpolations respectively;

$$\left. \begin{array}{l} \phi(\xi) = a_{\phi j} + b_{\phi j}(T_j - \xi) + c_{\phi j}(T_j - \xi)^2 + d_{\phi j}(T_j - \xi)^3 \\ \psi(\xi) = \psi_{j-1}[(T_j - \xi)/\xi_j] + \psi_j[(\xi - T_{j-1})/\xi_j] \end{array} \right\}, \quad \text{with} \ T_{j-1} \leq \xi \leq T_j. \tag{16}$$

Choosing the collocation points to be the node points (r_i, z_i) of the subintervals, the discretized form of Eq. (13) yields the following set of $(n + 1)$ equations in the $(n + 1)$ unknowns, ψ_i,

$$2\pi \phi_i + \sum_{j=1}^{n} A_{i,j} = \sum_{j=1}^{n} (B_{i,j} \psi_{j-1} + C_{i,j} \psi_j) \quad \text{for} \quad i = 0, \ldots, n, \tag{17}$$

where the coefficients $A_{i,j}, B_{i,j}$ and $C_{i,j}$ are given by

$$\left. \begin{array}{l} A_{i,j} = \int_{T_{j-1}}^{T_j} \phi(\xi) \left\{ \int_0^{2\pi} \frac{\partial}{\partial n} \left(\frac{1}{|p_i - q(\xi, \theta)|} \right) d\theta \right\} d\xi, \\ B_{i,j} = \int_{T_{j-1}}^{T_j} \left(\frac{T_j - \xi}{\xi_j} \right) \left\{ \int_0^{2\pi} \left(\frac{1}{|p_i - q(\xi, \theta)|} \right) d\theta \right\} d\xi, \\ C_{i,j} = \int_{T_{j-1}}^{T_j} \left(\frac{\xi - T_{j-1}}{\xi_j} \right) \left\{ \int_0^{2\pi} \left(\frac{1}{|p_i - q(\xi, \theta)|} \right) d\theta \right\} d\xi. \end{array} \right. \tag{18}$$

The inner integrals in (18) are evaluated analytically, resulting in complete elliptic integrals of the first and second kind, whereas the outer integrals are evaluated numerically using Gaussian and Gauss-Legendre quadrature rules.

With the initial shape and potential distribution on the surface of the bubble known, the B.I.M. involves solving Eq. (17) for the normal velocity $\frac{\partial\phi}{\partial n}$ (the ψ_i's), which together with the tangential velocity, $\frac{\partial\phi}{\partial\xi}$ (obtained by differentiating (16)), allows us to calculate the velocity $\mathbf{u}$ of a fluid particle on the surface of the bubble,

$$\left.\begin{aligned} u(\xi) &= \frac{\partial\phi}{\partial\xi}\frac{dr}{d\xi} - \frac{\partial\phi}{\partial n}\frac{dz}{d\xi} \\ v(\xi) &= \frac{\partial\phi}{\partial\xi}\frac{dz}{d\xi} + \frac{\partial\phi}{\partial n}\frac{dr}{d\xi} \end{aligned}\right\}, \text{ with } T_{j-1} \leq \xi \leq T_j, \tag{19}$$

where u and v are the radial and vertical components of the velocity respectively. We then update the surface and the potential on the surface using the Euler scheme,

$$\left.\begin{aligned} r_i(t+\Delta t) &= r_i(t) + u_i\Delta t + 0(\Delta t)^2, \\ z_i(t+\Delta t) &= z_i(t) + v_i\Delta t + 0(\Delta t)^2, \\ \phi_i(t+\Delta t) &= \phi_i(t) + \left(\frac{\partial\phi}{\partial t} + |\mathbf{u}|^2\right)_i \Delta t + 0(\Delta t)^2, \end{aligned}\right\} \text{ for } i = 0,\ldots,n, \tag{20}$$

where the value of $\frac{\partial\phi}{\partial t}$ is obtained by using Bernoulli's equation evaluated at the surface of the bubble (Eq. (5)). To ensure the accuracy of the updating procedure, Δt is chosen to satisfy the following condition

$$\Delta t = \frac{\Delta\phi/\left(\frac{v_0}{v}\right)^\gamma}{\max_i\left[1 + \frac{1}{2}|\mathbf{u}|^2\right]}, \tag{21}$$

with $\Delta\phi$ some constant. The time step, Δt, as chosen has the following desirable features. For large fluid velocities at the surface of the bubble smaller time increments are required in order to adequately represent this fast motion. Similarly, for large compressions ($v \approx v_0$) smaller time increments are necessary to capture the motion of the bubble surface, including the phenomena of rebound which is characterised by very high accelerations.

COMPUTATIONAL RESULTS

The motion of bubbles generated by explosions of two different strengths has been determined via the B.I.M., and in this section we present the results of these computations and make comparisons with the predictions of the elementary model. In the following computations the bubble is initially spherical with a radius of 1.0 and the depth of the detonation taken as 200.0 (non-dimensional quantities). The initial radial velocity is zero. The buoyancy parameter, δ, was chosen to be 0.049. (Thus the calculations describe the motion of an explosion bubble of initial radius 0.5m at an initial depth of 100m below the ocean surface.) The adiabatic co-efficient, γ, was chosen to be 1.4. For specific explosives we can obtain an appropriate value of γ from empirical data. (For instance TNT is described by $\gamma = 1.25$.)

Figure 2 shows the growth and collapse of an explosion bubble for which $\alpha = 540$ ($p_0 = 540$ atm). During the expansion phase the bubble is observed to remain approximately spherical. The illustration of the collapse phase (Figure 2(b)), however, establishes the invalidity of the particular assumption of the simple model that the bubble remains spherical throughout its motion. We observe that as the bubble collapses the underside becomes flattened and subsequently a high speed liquid jet develops and ultimately penetrates the upper surface of the bubble. Beyond this time the current model cannot proceed. Of particular interest is the speed of the collapse phase with the jet forming and completely penetrating the bubble in about 2% of the bubble lifetime. (Lifetime here referring to the period between initiation and the time when the jet completely penetrates the bubble.) This result provides direct evidence in support of the postulate of Snay (see Holt (1977)) that after the first minimum the bubble generated by a deep explosion will rise with an attached vortex ring. It is clear that the penetration of the upper surface by the re-entrant jet will generate a multiconnected toroidal bubble with an associated circulation. To model this as a ring vortex is most likely an oversimplification as during the initial phases of the motion any vorticity generated by the impact will be localised over the region in the neighbourhood of the impact site.

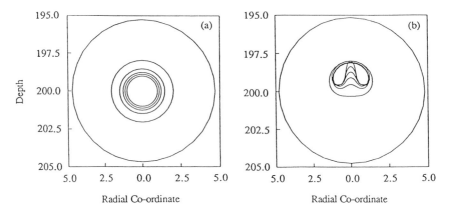

FIGURE 2. Bubble profiles for the growth and collapse of an explosion bubble characterised by $\alpha = 540$. (a) The growth phase. Successive profiles are at times 0.0000 (innermost), 0.0238, 0.0366, 0.0592, 0.1236, 1.1517 (outermost). (b) The collapse phase. Successive profiles at times 1.3903 (outermost), 2.8692, 2.8937, 2.8946, 2.9131, 2.9215, 2.9299 (innermost). Note the short time over which the high speed jet grows.

The presence of a compressible gas inside the bubble raises the question of whether the high pressures generated during the contracted phases of the bubble motion have the capacity to arrest the jet formation process. In order to initiate an investigation of this question we present the results of the computation of the motion of a bubble characterised by $\alpha = 108$ ($p_0 = 108$ atm). The bubble profiles are illustrated in Figure 3. The initial expansion phase is not shown because the bubble remains approximately spherical during this time, in a similar manner to the expansion illustrated in Figure 2(a). Figure 3(a) shows the collapse phase of this bubble. As the bubble collapses the underside becomes slightly flattened. Up until the time when the bubble rebounds, the approximation that the bubble is spherical is not unreasonable. The rebound phase, as illustrated in Figure 3(b), demonstrates an intriguing result. As the bubble re-expands, we observe that the asymmetry in the flow field induced by the ambient pressure gradient continues to drive a thin liquid jet through the bubble. Although jet formation has been resisted as the bubble is near its minimum, the reducing pressure as the bubble re-expands allows the jet to grow during this phase of the motion. The continued growth of the jet is in qualitative agreement with some experimental studies made by Lauterborn (1980) in which laser generated bubbles in the neighbourhood of boundaries are observed to be penetrated by a liquid jet which continues to grow as the bubble rebounds.

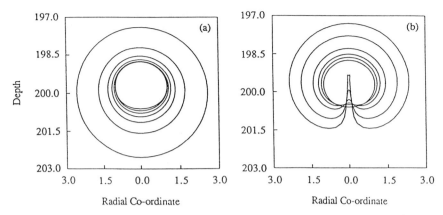

FIGURE 3. Bubble profiles for the collapse and rebound of an explosion bubble characterised by $\alpha = 108$. (a) The collapse phase. Collapse occurs after the approximately spherical growth phase. Successive profiles are at non-dimensional times 1.0757 (outermost), 1.5577, 1.6532, 1.6931, 1.7178, 1.7370, 1.7548 (innermost). (b) The rebound phase. Successive profiles are at times 1.7555 (innermost), 1.7946, 1.8220, 1.8685, 1.9940, 2.1886 (outermost). Observe the growth of the thin jet as the bubble re-expands.

These limited results conclusively establish that the assumption of the simplistic theory that the bubble remains spherical throughout its motion is erroneous. It is pertinent to inquire as to what effect this assumption has upon the gross characteristics of the bubble motion, in particular its migration upwards. Figure 4 shows a comparison of the upwards migration of the explosion bubbles of Figures 2 and 3 with that upwards motion predicted by the elementary model of spherical bubble dynamics. For much of the motion the two results are indistinguishable. Examination of the bubble profiles of Figures 2 and 3 establishes that it is during these phases that the bubble has retained its spherical shape. It is only when the shape departs significantly from sphericity that the results diverge, with the elementary theory considerably overstating the upwards motion of the bubble. As the jet forms the fluid momentum becomes concentrated in the jet rather than the upwards motion of the bubble, resulting in a reduced upwards displacement of the bubble centroid.

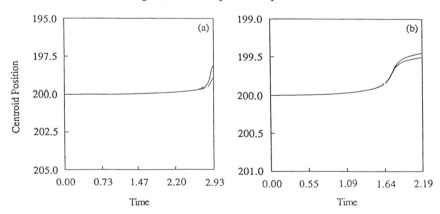

FIGURE 4. Comparison of centroid position predicted by the spherical model and that predicted by the B.I.M. In each case the upper curve is that predicted by the elementary spherical model. (a) The explosion of Figure 2. (b) The explosion of Figure 3.

The war-time experiments of Taylor & Davies (1950) established that the elementary model overstated the upwards rise of the explosion bubble. It is clear that the asymmetry of the bubble shape is a primary factor in this regard.

CONCLUSIONS

In this paper we have presented the results of simulations of the motion of bubbles produced by underwater explosions. This study differs from previous ones in that the motion has been examined beyond the first oscillation. We have established the plausibility of the proposition that bubbles generated by deep explosions rise for much of their lifetime with an attached vortex ring. Furthermore, we have demonstrated theoretically the growth of a re-entrant jet during the re-expansion of a bubble, and shown that the non-spherical nature of the bubble is a most significant factor in the over estimates of upwards motion made by elementary theory.

ACKNOWLEDGEMENTS

We would like to thank Professor J.R. Blake for his encouragement and advice throughout the course of this study. Partial support for this research was provided by the Australian Research Grants Scheme. This work was carried out by J.P.B. in the Department of Mathematics, The University of Wollongong.

REFERENCES

1. Blake, J.R., Taib, B.B. and Doherty, G., Transient cavities near boundaries. Part 1. Rigid Boundary, *J. Fluid Mech.* Vol. 170, pp.479-497, 1986.

2. Bryant, A.R., Photographic measurements of the size, shape and movement of the bubble produced by 1– oz charges of polar ammon gelignite detonated underwater at a depth of 3 feet, in *Underwater Explosion Research* Vol. II, Office of Naval Research, Department of the Navy, Washington, D.C., 1950.

3. Guerri, L., Lucca, G. and Prosperetti, A., A numerical method for the dynamics of non-spherical cavitation bubbles, in *Proc. 2nd Int. Colloq. on drops and bubbles*, California, pp. 175-181, 1981.

4. Herring, C., Theory of the pulsations of the gas bubble produced by an underwater explosion, NDRC Division 6 Report C4-sr20, 1941.

5. Holt, M., Underwater explosions, *Ann. Rev. Fluid Mech.*, Vol. 9, pp.187-214, 1977.

6. Lauterborn, W., in *Cavitation and Inhomogeneities in Underwater Acoustics*, W. Lauterborn, ed. New York: Springer-Verlag, 1980.

7. Lighthill, J., *An Informal Introduction to Theoretical Fluid Mechanics*, O.U.P., Oxford, 1986.

8. Taylor, G.I. and Davies, R.M., The motion and shape of the hollow produced by an explosion in a liquid, in *Underwater Explosion Research* Vol. II, Office of Naval Research, Department of the Navy, Washington, D.C., 1950.

9. Wilkerson, S.A., Boundary integral technique for explosion bubble collapse analysis, ASME Conference on Energy-Sources Technology Conference and Exhibition, 22-25 January, 1989.

Water Coning in Oil Reservoirs

S. K. LUCAS
Department of Mechanical Engineering
University of Sydney
Sydney, NSW 2006, Australia

J. R. BLAKE
Department of Mathematics
University of Wollongong
Wollongong, NSW 2500, Australia

A. KUCERA
Department of Mathematics
LaTrobe University
Bundoora, Victoria, 3083, Australia

INTRODUCTION

Muskat(1949) describes water coning as 'that observed in many oil wells, usually when producing at high rates, in which water gradually, and frequently suddenly, displaces part or all of oil production and comes to the surface in place of the oil'. Oil is usually trapped in interstices of a porous rock like sandstone or limestone, with an impermeable layer of rock above to trap it. Water, being more dense, is often found as a layer below, as shown in Figure 1(a). The rock above will cause a natural pressure that forces the oil to the surface when a well is drilled. This is known as 'primary oil recovery' (Muskat(1949), Bear(1972)).

When pumping of oil occurs, a suction pressure is generated, which tends to cause the oil-water interface to rise towards the well, while gravity forces due to density differences counterbalance this. If equilibrium is reached, a stable oil-water interface will develop. If well forces are greater, the interface will reach the well, producing water from the well, which is clearly undesirable. We define the critical pumping rate as such that any greater rate would cause water breakthrough. Our objective is to find the shape of this steady state oil-water interface for a fixed pumping rate. This raising of the interface, and possibly breaking through, is known as water coning in oil reservoirs.

Darcy's law is that most commonly used for fluid movement in porous media, and it was Muskat(1949) who first suggested explicitly applying Darcy's law for single-phase flows to each phase in a multiphase flow. Oil recovery studies have successfully used this since (see Hinch(1985)), as well as the assumption of a sharp interface between fluids. It must also be noted that Darcy's law, and all models based on it, are macroscopic in scope only; we deal with geometries far larger than that of the single pore.

We should realise that models for stable fluid flow are often not strictly stable. In an oil reservoir, as oil is removed, the internal geometry and pressures will alter, altering the stability criterion. Additionally, we note that a constant critical pumping rate may not maximise the well's production. We may wish to alter pumping rates from significantly above, to significantly below the critical rate (Ewing(1983)). This paper will, however, consider only the steady state problem.

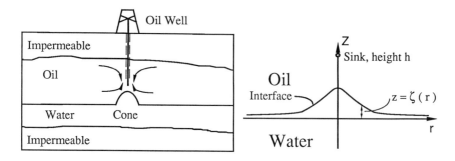

FIGURE 1. (a) Water coning and (b) Geometry of the model

EQUATIONS

We will consider an homogeneous medium of constant permeability k occupying all space, as in Figure 1(b). We assume the fluids are immiscible and that, apart from on the interface, all pore spaces are filled with one or other of the fluids. We can then apply Darcy's law independently to each region.

Assume the upper fluid has density ρ_1 and viscosity μ_1 and the lower fluid has density $\rho_2(> \rho_1)$ and viscosity μ_2. The fluids are separated by an interface $z = \zeta(r,t)$. Darcy's law in both fluids gives

$$\mathbf{u}^{(i)} = -\frac{k}{\mu_i}\nabla \hat{p}^{(i)} \quad \text{for} \quad i = 1, 2, \tag{1}$$

where $\hat{p}^{(i)} = p^{(i)} + \rho_i g z$ is the modified pressure, $\mathbf{u}^{(i)}$ is the velocity, μ_i the viscosity and ρ_i is the density. In the upper fluid (oil) $i = 1$, while $i = 2$ in the lower fluid (water). Taking (1) with the incompressibility condition gives Laplace's equation for the modified pressures, with boundary conditions $\mathbf{u}^{(i)}$ and $\hat{p}^{(i)}$ tending to zero at infinity, the dynamic boundary condition that

$$p^{(1)} = p^{(2)} \quad \text{on} \quad z = \zeta(r,t), \tag{2}$$

and the material boundary condition for particles on the interface, which in cylindrical polar co-ordinates is

$$\frac{\partial \zeta}{\partial t} + u\frac{\partial \zeta}{\partial r} - v = 0 \quad \text{on} \quad z = \zeta(r,t), \tag{3}$$

where u is the radial velocity, and v is the axial velocity.

For a sink of volume flow rate m, modelled by a point sink and using the convention that $m \geq 0$, we have the expression for p^* the suction pressure in an infinite porous medium at height h in the upper fluid as

$$p^* = \frac{-m\mu_1}{4\pi k \sqrt{(z-h)^2 + r^2}}. \tag{4}$$

228

Since we are considering a steady state solution, we require no initial shape, and so our essential unknown, the interface ζ, can be considered independent of time. In addition $\hat{p}^{(2)} = 0$ since the steady state solution requires no flow in the lower region. Thus, dropping the superscript for pressure in the upper fluid, we need to solve for pressure $\hat{p}$ given by $\hat{p} = p^* + p'$, where p' is the pressure variation from that due to a sink in an infinite fluid (p^*).

If we take the dynamic boundary condition and use the definition of modified pressure, we can scale lengths with respect to h and pressure with respect to $m\mu_1/kh$ to get the dimensionless form of the dynamic boundary condition as

$$\tilde{\zeta} + F\tilde{p} = 0 \quad \text{on} \quad z = \tilde{\zeta}(r), \tag{5}$$

where $\tilde{\zeta}$, $\tilde{p}$ are dimensionless height and pressure respectively, and $F = m\mu_1/kh^2(\rho_2 - \rho_1)g$ is the only dimensionless parameter, which represents the balance between the suction force of the sink and the gravitational restoring force of the denser fluid. Finally, we need the dimensionless suction pressure, which from (4) becomes, with z, r both dimensionless parameters

$$p^* = \frac{-1}{4\pi\sqrt{(z-1)^2 + r^2}}. \tag{6}$$

BOUNDARY INTEGRAL METHOD

We have, for a smooth function ϕ that satisfies Laplace's equation in a domain Ω with a smooth surface S, Green's integral formula (Blake, Taib and Doherty(1986), Blake and Kucera(1988)) as

$$c\phi(p) + \int_S \phi(q)\frac{\partial G}{\partial n}\,dS = \int_S \frac{\partial\phi(q)}{\partial n}G\,dS, \tag{7}$$

where $p \in \Omega + S$, $q \in S$, with $\partial/\partial n$ being the normal derivative outward from S,

$$c = \begin{cases} 1, & p \in \Omega \\ \frac{1}{2}, & p \in S \end{cases} \quad \text{and} \quad G = \frac{1}{4\pi|p-q|}. \tag{8}$$

We can rewrite (7) for our problem as

$$cp(a) = p^* + \int_S \left(\frac{\partial p(b)}{\partial n}G - p(b)\frac{\partial G}{\partial n}\right)\,dS, \tag{9}$$

where $a \in \Omega + S$ and $b \in S$. Since we are considering a steady state problem, there is no motion normal to the interface, therefore $\partial p/\partial n = 0$. If we multiply (9) throughout by $(-F)$ and then apply the dynamic boundary condition (5), we obtain the integral representation for the interface shape as

$$c\zeta = -Fp^* - \int_S \zeta\frac{\partial G}{\partial n}\,dS, \tag{10}$$

where S is the surface $z - \zeta(r) = 0$ for $r : 0 \to \infty$. The part of S at infinity can be ignored due to boundary conditions. Our domain Ω is here the upper fluid (oil), so the outward normal will be down into the lower fluid. Now we are interested in points on the boundary $z = \zeta(r)$, which we actually want to find, so we take $c = 1/2$. After extensive manipulation, including analytic integration through the θ angle, we have the formulation

$$\frac{1}{2}\zeta(r) = -Fp^* - \int_0^\infty \lambda \zeta(\lambda) K(r, \lambda)\, d\lambda, \tag{11}$$

where

$$K(r, \lambda) = \frac{1}{\pi \left[(\lambda + r)^2 + (\zeta(\lambda) - \zeta(r))^2\right]^{3/2}} \left[\frac{E(k)}{1 - k^2} \left\{ \frac{2r\zeta'}{k^2} - (\lambda - r)\zeta' + (\zeta(\lambda) - \zeta(r)) \right\} \right.$$

$$\left. - \frac{2r\zeta' K(k)}{k^2} \right], \tag{12}$$

$$k^2 = \frac{4r\lambda}{(\lambda + r)^2 + (\zeta(\lambda) - \zeta(r))^2}, \qquad p^* = \frac{-1}{4\pi\sqrt{(\zeta - 1)^2 + r^2}}, \tag{13}$$

$\zeta' \equiv \partial \zeta / \partial \lambda$, and K, E are complete elliptic integrals of the 1st and 2nd kinds. The problem is now specified as a nonlinear Fredholm integral equation of the second kind for the unknown interface $z = \zeta(r)$, in which the suction pressure p^* is specified. Equation (11) can then be solved by a fixed-point iteration technique, starting with the 1st order small parameter approximation $\zeta^0(r)$ (Blake and Kucera(1988)) as an initial approximation where $\zeta^0(r) = 1/2\pi\sqrt{r^2 + 1}$. Our method is to choose N equally spaced points on the interval $[0, r_{max}]$, and find $\zeta^1(r)$ as the left hand side of (11) based on these points, substituting $\zeta^0(r)$ in the right hand side. This procedure is repeated, producing a sequence of fuctions which will converge pointwise to $\zeta(r)$.

Equation (11) was solved using the QUADPACK numerical integration package (Piessens et al(1983)). Investigation of the kernel function K from (12) shows a singularity when $\lambda = r$. QUADPACK is designed to be able to deal with singularities, as long as they are at the end points of the integral's interval.Thus, the integral was split into the parts $0 \leq \lambda \leq r$ and $r \leq \lambda < \infty$, and calculated using the routines DQAGS and DQAGI respectively. QUADPACK had no difficulty with out required accurracy. $\zeta^n(r)$ was interpolated using a clamped cubic spline from 0 to r_{max} with end conditions being $\zeta'^n(0) = 0$ (since it is only at the critical rate that a cusp occurs) and $\zeta'^n(r_{max}) = \zeta'^0(r_{max})$ where $\zeta'^0(r)$ is taken from the 1st order small parameter expansion above. Further, we take for $r > r_{max}$, $\zeta^n(r) \equiv \zeta^0(r)$.

QUADPACK requires a function for the integrand, not just a set of points. Thus we used a cubic spline interpolation to produce a smooth differentiable piecewise polynomial S. A cubic spline gives enough undetermined constants so that the curve is not only continuously differentiable on the interval, but has a continuous second derivative as well. The much simpler linear interpolation was tried between points, but was found to be unacceptable. The discontinuities in the derivative of S caused QUADPACK to be much slower, and convergence was much slower. In addition, accurate answers required a point density far higher than what was considerd acceptable. Table 1 gives an indication of the quick convergence of the cubic spline method, not only in iterations required for convergence, but also in the required number of points in the spline.

TABLE 1. Convergence for increasing point density, range 5

	F = 1.0				F = 2.0			
Points	25	50	100	Points	25	50	100	200
Iterations	4	4	4	Iterations	9	11	11	11
r = 0.0	0.1863	0.1863	0.1863	r = 0.0	0.5660	0.5637	0.5637	0.5637
1.0	0.1162	0.1162	0.1162	1.0	0.2376	0.2376	0.2376	0.2376
2.0	0.0712	0.0712	0.0712	2.0	0.1419	0.1419	0.1419	0.1419
3.0	0.0502	0.0502	0.0502	3.0	0.0998	0.0998	0.0998	0.0998

There is one problem with this procedure. When $r = 0$, the integral becomes split into the two parts: $0 \to 0$ and $0 \to \infty$ with the singularity at 0. $0/0$ errors were found in the first interval. We get around this problem by simplifying the integration at $r = 0$, giving the expression

$$c\zeta(0) = -Fp^* - \int_0^\infty \lambda\zeta(\lambda) \left\{ \frac{\zeta(\lambda) - \zeta(0) - \lambda\frac{\partial\zeta}{\partial\lambda}}{2[\lambda^2 + (\zeta(\lambda) - \zeta(0))^2]^{3/2}} \right\} d\lambda. \qquad (14)$$

This is singular at 0 only, and was successful when used.

RESULTS

(i) Point Sink

We reproduce the results of Blake and Kucera(1988) in Figure 2, showing the interface shape and slope for $F = 0.2, 0.4, \ldots, 2.0$. The slope graph is useful in recognising cusping. The graphs were produced with a 100pt spline in the range $0 \to 5$ ($r_{max} = 5$), and then reproduced on the negative axis for aesthetic reasons. Comparisons with the small parameter solution showed that $r_{max} = 5$ lead to a smooth transition, and so was an acceptable value. We also see that, if there was a non-unique solution to (11), the one we have found is what we want

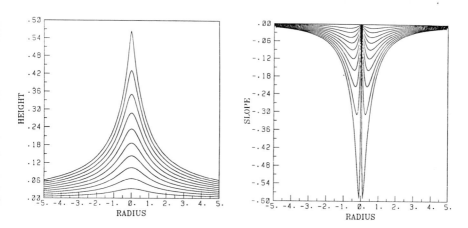

FIGURE 2. (a) Interface shape and (b) Slope for $F = 0.2, 0.4, \ldots, 2.0$, 100pt spline, range $0 \to 5$

TABLE 2. Critical Suction, $\zeta(0)$, and iterations to convergence for various F

N	F_c	$\zeta(0)$	Iterations
25	2.031	0.6512	50
50	2.043	0.6424	23
100	2.048	0.6541	18
200	2.050	0.6626	24

because it matches the small parameter solution. Table 1 indicates that the accuracy of 100 point splines is more than acceptable, and shows the speed of convergence.

Of most interest here is the critical pumping rate F_c, the value of F such that we must have $F \leq F_c$ for a stable interface. We have enforced a zero derivative at $r = 0$ on the assumption that the interface will stay smooth up to critical F. Our value of F_c has been found by a simple testing process, to find the maximum F that leads to a converged cone shape. When $F > F_c$, our model would produce an iteration with ζ near $0 > 1.0$ our dimensionless point sink height, indicating fluid breakthrough. Further iterations oscillated wildly beyond heights of ± 1. Table 2 shows the results of F_c for increasing N, and shows that that too is converging. Also, the slope graph of Figure 2(b) can be of use in finding the shape of the interface as we approach the critical pumping rate F_c. We can see how the magnitude of the maximum of the slope increases and is moving inwards as F increases. This shows that the interface is approaching a cusp, agreeing with the hodograph method results of Bear and Dagan(1964).

(ii) Multiple Sinks/Sources

Equation (11) shows that the integrand does not depend on F. In fact, it is only the portion $-Fp^*$ where sink strength is specified. Thus, it is quite simple to extend to multiple sinks and/or sources at various heights with various sink strengths. Scaling previously was done to put the single sink at height 1; here scaling to put one of the sinks/sources at 1, and scaling

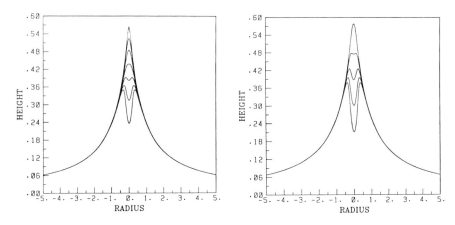

FIGURE 3. (a) Interface shape for $F = 2.2$ at $h = 1.0$ and $F = -0.2$ at $h = 1.0, 0.9, \ldots, 0.4$ and (b) for $F = 2.5$ at $h = 1.0$ and $F = -0.3$ at $h = 0.8, 0.7, \ldots, 0.4$, 100pt spline over range $0 \to 5$

232

others to this is a simple solution. Thus, for the multiple sinks/sources case, (11) becomes

$$\frac{1}{2}\zeta(r) = -\left[\sum_{i=1}^{n} \frac{-F_i}{4\pi\sqrt{(\zeta - h_i)^2 + r^2}}\right] - \int_0^\infty \lambda\zeta(\lambda)K(r,\lambda)\,d\lambda, \qquad (15)$$

where there are n sinks at height h_i, and each sink has strength F_i (a negative F_i means a source). The summation takes into account all sinks and sources, which are all placed at $r = 0$ to retain the axisymmetric nature of the problem.

Our main aim here was to place a small source below the larger sink in an attempt to suppress the water cone directly under the well. This has been illustrated in Figure 3. In Figure 3(a), we can see that the sum of F's is 2.0, which is just under the critical F for a single sink, and so having both sink and source at $h = 1.0$ is possible. In an effort to extract oil at a greater rate than for F_c for a single sink we may consider Figure 3(b). Those curves shown have the lower source at heights between 0.4 and 0.8, since other ranges did not lead to a stable interface shape. When the source went down further, the model breaks down. As the source goes down between the humps produced by this configuration, its push outwards will cause the interface near this low source to become many valued in r, which is beyond the capabilities of our current program.

(iii) Line Sink

In reality, an oil well is more of a line of suction, rather than a single point. Thus, we extend our model to a vertical line with a total suction F over its length. Our line sink is the limiting case of having n sinks of suction F/n equally spaced along a finite line segment, and then taking the limit as $n \to \infty$. We use the endpoints as $(0,0,h_1)$ and $(0,0,h_2)$ $(h_2 > h_1)$. Using the parametric representation of this line as $z = h_1 + t(h_2 - h_1)$ as $t : 0 \to 1$, then we have our basic equation as

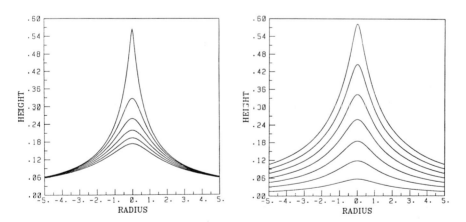

FIGURE 4. (a) Interface shape for $F = 2.0$ over line sink from 1 to $1+n$, $n = 0.0, 0.5, \ldots, 2.5$ and (b) $F = 0.5, 1.0, \ldots, 3.5$ over line from 1 to 2, 100pt spline, range $0 \to 5$

$$c\zeta(r) = -F \int_0^1 \frac{-1}{4\pi\sqrt{\zeta - (h_1 + t(h_2 - h_1)))^2 + r^2}} \, dt - \int_0^\infty \lambda\zeta(\lambda)K(r,\lambda) \, d\lambda. \tag{16}$$

For scaling considerations, taking $h_1 = 1$ and finding h_2 in an appropriate manner is a simple solution.

We have shown the results for two major sets of line sinks. Figure 4(a) has a line sink of constant $F(= 2.0)$, as the length of the sink is increased. As expected, the interface height goes down as a larger proportion of suction occurs further from the interface. The case of a line of length zero corresponds to that of a point sink. Figure 4(b) shows the results for a line sink for varying F, and constant length (here $h_1 = 1.0$ and $h_2 = 2.0$). Naturally, higher values of F are possible, and we can also see that interface curves are much smoother.

CONCLUSION

We have developed a boundary integral method to predict the steady state height of the interface between oil and water in an oil reservoir under the influence of an oil well pumping out oil to the surface. The models of a point sink, multiple point sinks/sources and a line sink have been implemented.

We must emphasise that this is far from a complete model for water coning in oil reservoirs. We have dealt with a steady state problem, and so ignored the more interesting complexities of the time dependent problem. We have also stayed with an axisymmetric problem; a general 3D implementation would be a natural extension. In addition, some of our basic assumptions are simplifications. Of importance is assuming a constant permeability k. At the depth of oil bearing rock, the vertical permeability is typically 1/10 to 1/100 of the horizontal permeability. While this can be resolved by a stretching of the vertical axis, we have the problem that oil bearing rock is rarely homogeneous, and is finite in size. Detailed knowledge of any particular section of rock would be required for a more exact solution. Since this is rarely known, we must consider the above to be more a qualitative than quantitative analysis of the problem of water coning in oil reservoirs.

REFERENCES

1. Bear, J., *Dynamics of Fluids in Porous Media*, McGraw-Hill, N.Y., 1972.

2. Bear, J., Dagan, G., Some exact solutions of interface problems by means of the hodograph method, *J.Geophys.Res.*, vol. 69, pp. 1563-1572, 1964.

3. Blake, J.R., Kucera, A., Coning in Oil Reservoirs, *Math. Scientist*, vol. 13, pp. 36-47, 1988.

4. Blake, J.R., Taib, B., Doherty, G., Transient cavities near boundaries part 1. Rigid boundary, *J.Fluid Mech*, vol. 170, pp. 479-497 1986.

5. Ewing, R.E., *The Mathematics of Reservoir Simulation*, SIAM, Philadelphia, 1983.

6. Hinch, E.J., The recovery of oil from underground reservoirs, *PCH*, vol. 6, pp. 601-622, 1985.

7. Muskat, M. *Physical Principles of Oil Production*, McGraw-Hill, N.Y., 1949.

8. Piessens, R., De Doncker-Kapenga, E., Uberhuber, C.W., Kahauer, D.K., *QUADPACK, A Subroutine Package for Automatic Integration*, Springer-Verlag, Berlin, 1983.

Application of a Boundary Integral Method to the Magnetostatics of Thin Sheets in Three-Dimensions

J. R. MOONEY
CSIRO, Advanced Computing Support Group
G.P.O. Box 664
Canberra, 2601, Australia

INTRODUCTION

The general magnetostatic problem is to determine the perturbation (i.e. induced field) to a given excitation field caused by the presence of magnetically permeable bodies in three dimensions. The particular problem addressed in this paper is to make an accurate determination of the induced field where the permeable bodies consist of thin sheets of material (shell objects) of high permeability with uniform dimensionless thickness h comparable in magnitude to the reciprocal relative permeability μ_r. It is also preferable if such bodies can be modelled numerically by a single sheet of surface finite elements (e.g. triangles) with an implied thickness parameter, rather than modelling the exact shell surface. For this application the main requirement is to be able to evaluate the induced field at points exterior to the magnetic bodies.

The difficulty that arises with shell objects of high permeability is illustrated by the case of the spherical shell. For a spherical shell of radius a and dimensionless thickness h (thickness/radius) elementary analysis shows that the induced dipole moment $\mathbf{p}$ of the shell (see Eq. (6)) is given by

$$\mathbf{p} = 4\pi a^3 \frac{3h}{3h + 4.5\mu_r}\mathbf{H}_0, \tag{1}$$

when placed in a uniform magnetic field of intensity $\mathbf{H}_0$.

When h, μ_r are both small and of comparable magnitude (say 0.01 to 0.001) the strength of the induced field (measured by $\mathbf{p}$) is highly sensitive to the ratio h/μ_r. For fixed h, as the permeability is increased (μ_r decreased) $\mathbf{p}$ is increased; in the limit of infinite permeability ($\mu_r = 0$) $\mathbf{p}$ attains a maximum value (in this case equal to that of the solid sphere of infinite permeability). For fixed μ_r, as the thickness is reduced $\mathbf{p}$ is decreased, approaching zero as h approaches zero, corresponding to the absence of a magnetic object.

A similar analysis applies to the case of a very oblate spheroid which approximates a thin circular plate, with the excitation field parallel to the plate. For this object for small h and small μ_r

$$\mathbf{p} \sim \pi a^3 \frac{h}{\mu_r + 0.1875\pi h} \mathbf{H}_0, \tag{2}$$

where a is the radius of the disc, h is dimensionless mean thickness (spheroid volume is $\pi a^3 h$) and $\mathbf{H}_0$ is a uniform excitation field in the plane of the plate. [The response of the plate to an excitation field orthogonal to the plate is very small, in fact $O(h)$].

This paper proposes a suitable numerical method for determining the induced field for a single shell object in a general excitation field. This shell boundary method is based on established boundary integral methods for electro- and magnetostatic problems: these methods have advantages of accuracy and efficiency (Lindholm (1980)). The shell method will be shown to correctly model the dependence of the field on the ratio h/μ_r. The shell boundary method has been incorporated into a package that solves the general magnetostatic problem for an arbitrary set of magnetic objects, both shell and non-shell.

SHELL BOUNDARY METHOD

Firstly, the boundary integral method is briefly sketched as applied to an ordinary non-shell object. See Lindholm (1980) and Stratton (1941). The object occupies a volume V, bounded by a surface S. The magnetic permeability of the object is μ_1, the external permeability is μ_0 and the reciprocal relative permeability is $\mu_r = \mu_0/\mu_1$. $\mu_r > 0$ and usually $\mu_r < 1$. For a steel object $\mu_r \sim 0.001$.

$\mathbf{H}_0(\mathbf{r})$ is the given excitation magnetic field intensity vector, $\mathbf{r}$ is position vector. $\mathbf{H}(\mathbf{r})$ is the total field vector, and $\tilde{\mathbf{H}}(\mathbf{r})$ is the field contributed by induction in the magnetic object, i.e. the perturbation field. The problem is to find $\mathbf{H}(\mathbf{r})$, or equivalently $\tilde{\mathbf{H}}(\mathbf{r})$, for a given excitation field $\mathbf{H}_0(\mathbf{r})$. Typically $\mathbf{H}_0(\mathbf{r})$ will contain a uniform part (e.g. the Earth's field) and contributions from sources such as known dipoles or current coils.

Throughout the paper the units of distance, magnetic field intensity, magnetostatic potential and dipole moment are the metre, the ampere/metre, the ampere and the ampere(metre)^2 respectively.

For simplicity we consider the case where $\mathbf{H}_0(\mathbf{r})$ and $\tilde{\mathbf{H}}(\mathbf{r})$ are conservative everywhere and hence expressible as the negative gradients of magnetostatic potentials $\phi_0(\mathbf{r})$, $\tilde{\phi}(\mathbf{r})$ everywhere. The total magnetostatic potential $\phi(\mathbf{r})$ is the sum of $\phi_0(\mathbf{r})$ and $\tilde{\phi}(\mathbf{r})$. In principle the magnetostatic problem can be solved entirely in terms of ϕ. By elementary magnetostatic theory ϕ is harmonic inside V and outside V, is continuous everywhere and satisfies the following boundary conditions. At infinity $\nabla\phi$ is asymptotic to $-\mathbf{H}_0$ and at the surface of the body

$$\left(\frac{\partial\phi}{\partial n}\right)_- = \mu_r \left(\frac{\partial\phi}{\partial n}\right)_+, \quad \text{on } S, \tag{3}$$

where '$\partial/\partial n$' denotes differentiation along the outward normal to S. The '+' sign denotes the outward side of S, '−' the inner side.

We define the kernel functions (scalars)

236

$$p(\mathbf{r}, \mathbf{r}_s) = \frac{1}{4\pi|\mathbf{r} - \mathbf{r}_s|}, \qquad q(\mathbf{r}, \mathbf{r}_s) = \frac{(\mathbf{r} - \mathbf{r}_s) \cdot \mathbf{n}(\mathbf{r}_s)}{4\pi|\mathbf{r} - \mathbf{r}_s|^3}.$$

Here $\mathbf{n}(\mathbf{r}_s)$ denotes the unit normal vector to surface S at point $\mathbf{r}_s$. We further define the surface function

$$\sigma(\mathbf{r}_s) = \left(\frac{\partial\phi}{\partial n}\right)_- - \left(\frac{\partial\phi}{\partial n}\right)_+, \qquad \mathbf{r}_s \in S.$$

σ is simply the jump discontinuity in normal component of magnetic field intensity at the surface S and has the unit ampere/metre.

Application of Green's theorem to ϕ together with the boundary conditions leads to the following pair of boundary integral equations.

$$\phi(\mathbf{r}) = \phi_0(\mathbf{r}) - \frac{(1 - \mu_r)}{\mu_r} \int_S q(\mathbf{r}, \mathbf{r}_s)[\phi(\mathbf{r}_s) - \phi(\mathbf{r})] \, dA, \qquad \mathbf{r} \in S. \tag{4}$$

$$\tilde{\phi}(\mathbf{r}) = \int_S p(\mathbf{r}, \mathbf{r}_s)\sigma(\mathbf{r}_s) \, dA, \qquad \text{all } \mathbf{r}. \tag{5}$$

The integrals subscripted by 'S' here and below denote integrals taken over the whole surface S, $\mathbf{r}_s$ is position vector on S and dA is the element of surface area. These equations are derived by considering a point just inside S and just outside S and obtaining a pair of limiting equations for a point on S. The second equation above is derived for $\mathbf{r} \in S$ but holds for all $\mathbf{r}$.

The basis of the numerical boundary integral method for non-shell objects is to solve Eq. (4) for $\phi(\mathbf{r})$ over S and then to solve Eq. (5) for $\sigma(\mathbf{r}_s)$ over S. Equation (5) shows that the perturbation potential induced on the object is equivalent to that produced by a layer of magnetic monopoles of density σ over S. Once σ has been determined $\tilde{\phi}$, and hence ϕ, can be determined at any required point; $\tilde{\mathbf{H}}(\mathbf{r})$ can be found similarly.

The total induced dipole moment of the object is the vector

$$\mathbf{p} = \int_S \sigma(\mathbf{r}_s)\mathbf{r}_s \, dA. \tag{6}$$

The induced far field is asymptotic to that of a magnetic dipole of moment $\mathbf{p}$. There is of course no net charge in the magnetic field.

We now develop the theory further for the case where the body is a shell object of uniform dimensionless thickness h, where $h \sim \mu_r \ll 1$. The shell may be open or closed: if open, the edge is a closed space curve C. The method is inspired by the known stable behaviour of the potential equation when the reference point $\mathbf{r}$ is close to the surface S (Lindholm (1980)) and the continuous nature of ϕ itself. We assume the potential is constant through the thickness of the shell. Surface S is taken to be one

237

side of the object and the other side is taken to be a second surface S', derived from S by uniform translation along the unit normal a distance hL over the whole surface, where L is the maximum diameter of the object. If the object has an edge the total surface is the union of S, S' and an edge strip of width hL centred on C. We apply boundary integral equation (4) using the total surface, but for now ignoring an edge strip: thus for $\mathbf{r} \in S$

$$
\phi(\mathbf{r}) = \phi_0(\mathbf{r}) + \frac{(1 - \mu_r)}{\mu_r} \int_S q(\mathbf{r}, \mathbf{r}_s)[\phi(\mathbf{r}_s) - \phi(\mathbf{r})] \, dA
$$
$$
- \frac{(1 - \mu_r)}{\mu_r} \int_{S'} q(\mathbf{r}, \mathbf{r}_s')[\phi(\mathbf{r}_s) - \phi(\mathbf{r})] \, dA', \tag{7}
$$

where $\mathbf{r}_s' = \mathbf{r}_s + hL\mathbf{n}(\mathbf{r}_s)$ over S. dA' is the element of area on S'.

We now treat h as a small parameter and expand terms in Eq. (7) and retain only dominant terms. Thus

$$
q(\mathbf{r}, \mathbf{r}_s') = q(\mathbf{r}, \mathbf{r}_s) + q_1(\mathbf{r}, \mathbf{r}_s)h + O(h^2), \quad \mathbf{r} \neq \mathbf{r}_s, \tag{8}
$$

where

$$
q_1(\mathbf{r}, \mathbf{r}_s) = -\frac{1}{4\pi} \frac{L}{|\mathbf{r} - \mathbf{r}_s|^3} + \frac{3}{4\pi} \frac{L[(\mathbf{r} - \mathbf{r}_s) \cdot \mathbf{n}(\mathbf{r}_s)]^2}{|\mathbf{r} - \mathbf{r}_s|^5}.
$$

Also by elementary differential geometry

$$
dA' = [1 + K(\mathbf{r}_s)Lh + O(h^2)]dA, \tag{9}
$$

where $K(\mathbf{r}_s)$ is the sum of principal curvatures (twice the mean curvature) of S at $\mathbf{r}_s$. If the surface is convex away from $\mathbf{n}(\mathbf{r}_s)$ then $K(\mathbf{r}_s)$ is taken as positive. Defining the shell kernel

$$
q_{sh}(\mathbf{r}, \mathbf{r}_s) = -q_1(\mathbf{r}, \mathbf{r}_s) - \frac{K(\mathbf{r}_s)L(\mathbf{r} - \mathbf{r}_s) \cdot \mathbf{n}(\mathbf{r}_s)}{4\pi|\mathbf{r} - \mathbf{r}_s|^3},
$$

the limiting equation thus obtained for potential from Eq. (7) for small h is

$$
\phi(\mathbf{r}) = \phi_0(\mathbf{r}) + \frac{(1 - \mu_r)}{\mu_r} h \int_S q_{sh}(\mathbf{r}, \mathbf{r}_s)[\phi(\mathbf{r}_s) - \phi(\mathbf{r})] \, dA, \quad \mathbf{r} \in S. \tag{10}
$$

If the magnetic shell has an edge C then to the right hand side of Eq. (10) we add the term

$$
\phi_C(\mathbf{r}) = -\frac{(1 - \mu_r)}{\mu_r} \frac{hL}{4\pi} \int_C \frac{(\mathbf{r} - \mathbf{r}_s) \cdot \hat{\mathbf{n}}(\mathbf{r}_s)}{|\mathbf{r} - \mathbf{r}_s|^3}[\phi(\mathbf{r}_s) - \phi(\mathbf{r})] \, ds, \tag{11}
$$

where $\hat{n}(\mathbf{r}_s)$ is the outward normal to C in the shell plane, and ds is the element of arc length.

As it stands Eq. (10) is not acceptable because the kernel $q_{sh}(\mathbf{r}, \mathbf{r}_s)$ is singular for $\mathbf{r}_s$ near $\mathbf{r} \in S$. The difficulty arises because expansion (8) breaks down for $\mathbf{r}_s$ near $\mathbf{r}$. However Eq. (10) is corrected by replacing kernel $q_{sh}(\mathbf{r}, \mathbf{r}_s)$ by the modified kernel $\tilde{q}_{sh}(\mathbf{r}, \mathbf{r}_s)$, defined by replacing all negative powers of $|\mathbf{r} - \mathbf{r}_s|$ by matching powers of $[|\mathbf{r} - \mathbf{r}_s|^2 + h^2 L^2]^{1/2}$. This follows by considering the detailed geometry of the shell for $\mathbf{r}_s$ near $\mathbf{r} \in S$. The final shell boundary equation for ϕ on S is thus obtained:

$$\phi(\mathbf{r}) = \phi_0(\mathbf{r}) + (1 - \mu_r)\frac{h}{\mu_r} \int_S \tilde{q}_{sh}(\mathbf{r}, \mathbf{r}_s)[\phi(\mathbf{r}_s) - \phi(\mathbf{r})]\, dA + \phi_C(\mathbf{r}), \quad \mathbf{r} \in S. \tag{12}$$

Terms retained in Eq. (12) are $O(h/\mu_r)$: terms neglected are $O(h^2/\mu_r)$ arising from the expansion of $q(\mathbf{r}, \mathbf{r}'_s)$ in Eq. (8) and the relationship of dA' to dA, Eq. (9) . However terms $O(h)$ have also been neglected in Eq. (7) arising from the variation of ϕ through the thickness of the shell. Such terms are $O(h/\mu_r N_-)$, that is $O(hN_+)$ by virtue of boundary condition (3); defining the norms

$$N_\pm = \left[\int_S \left(\frac{\partial \phi}{\partial n}\right)^2_\pm dA/ \text{ area } S \right]^{1/2}.$$

Now $h \sim \mu_r \ll 1$ and the above terms are essentially $O(h)$ and are small relative to the terms retained in Eq. (12).

The integrals in Eqs. (11), (12) are continuous as functions of $\mathbf{r}$ and are in a form suitable for numerical discretization and solution. Once a numerical representation of ϕ over S has been found, a numerical representation of σ over S can be found by solving a discrete analogue of Eq. (5). Here we regard σ as a single surface distribution of magnetic monopole: this is sufficient for determination of the field at points at medium distance from S. Once σ has been determined, the perturbation field at exterior points can readily be calculated, as discussed above.

The magnetostatic problem for a non-shell object can be formulated and solved satisfactorily using just the monopole function σ. However it appears to be very difficult to do this for a shell object. The difficulty can be appreciated by considering the spherical shell: a formulation based on σ alone loses essential information about the distribution of monopole density between inner and outer surfaces. This distribution is sensitive to the ratio h/μ_r. Thus for shell objects it appears to be essential to use a method based on the potential.

MAGNETOSTATIC PACKAGE

The shell method described above for handling magnetic shell objects has been incorporated into a computer package. In this package shell objects are represented by single sheets of triangular elements. A thickness parameter h is required for each object. The boundary integral equations for ϕ and σ are discretized using constant triangular elements for both functions in turn and using the centroids of the triangles as a set of collocation points. The potential equation also requires the mean curvature and

TABLE 1. x-components (p_x) of dipole moments for simple shell objects.

object	μ_r	h	p_x (numerical)	p_x (analytic)
spherical shell	0.0100	0.0100	4.9104	5.0265
spherical shell	0.0100	0.0050	3.0619	3.1416
circular plate	0.0010	0.0010	1.8951	1.9770
circular plate	0.0005	0.0010	2.7777	2.8847

this is estimated numerically by examining the directional variation of the normals of neighbouring triangles. Allowance is made for sharp edges.

The package has been designed to calculate the induced field of a general set of magnetic objects of different permeabilities, both shell and non-shell objects. Objects of infinite permeability are allowed, as a special case. The excitation field can be a general combination of fields arising from given magnetic dipoles, current coils and a uniform field. Non-shell objects are handled similarly to shell objects but potential and monopole distributions are found based on boundary integral equations (4), (5), again using constant triangular elements. The method has been extended to cope with the non-conservative fields arising from coils.

The magnetostatic package has been developed for the Department of Defence in a collaborative project between CSIRO and Paxus ComNet (Almond & Mooney (1989)). The package has been designed and coded to run on the CSIRO Cyber 205 vector computer. The package is oriented specifically towards the determination of the magnetic field round naval ships.

The package solves the multi-object problem iteratively, i.e. the field for one object is determined in the temporarily given fields of all the other objects and the process is continued cyclically to convergence. The solution for the potential and monopole distributions of each object is carried out using full matrix and linear equation solving routines. The coding of the package is designed to take full advantage of the vector architecture of the Cyber 205 computer.

EXAMPLES

We present examples of calculations of induced fields of simple shell objects for which analytic results are known: the spherical shell and the circular plate. Here we just examine the dipole moment induced by a uniform excitation field of unit strength along the positive x-axis, in the plane of the plate.

Results obtained using the magnetostatic package are presented in Table 1. The spherical shell is of unit radius and is discretized by 1280 quasi-uniform triangles. The plate is also of unit radius and discretized by 1024 triangles. The dipole moments calculated from the analytic formulae (1), (2) are also given. The table lists just the x-components of the dipole moments (p_x).

The results for the spherical shell are correct to within 2.3% ; the results for the plate are correct to within 4% . In the latter case some of the discrepancy is due to the real geometric difference between the plate, i.e. a truncated cylinder with parallel faces, and the approximating very oblate spheroid of equal volume and equal cross-section area. In all cases spurious numerical y- and z-components of the dipole moments are below the 0.1% level.

It is clear from the results that the shell boundary method based on the potential integral equation (12) is numerically stable and correctly models the sensitivity of the induced field to the ratio of shell thickness to reciprocal relative permeability.

It is interesting to note that in the case of the plate if the edge potential term ϕ_C is omitted from Eq. (12) then the resultant dipole moment is reduced by 13% (case $h = \mu_r = 0.001$) — an unacceptable error. The edge potential must be included: of course in general this term is $O(h/\mu_r)$, just as for the main potential terms.

ACKNOWLEDGEMENT

I wish to thank Dr F. R. de Hoog for some very useful discussions.

REFERENCES

1. Almond, R. and Mooney, J. R., A program to calculate the magnetostatic field in the vicinity of magnetically permeable objects, Users' guide, Paxus ComNet, March 1989.

2. Lindholm, D. A., Notes on Boundary Integral Equations for Three-Dimensional Magnetostatics, *IEEE Trans. Magn.*, vol. MAG-16, no. 6, pp. 1409–1413, 1980.

3. Stratton, J. A., *Electromagnetic Theory*, pp. 165–172, McGraw-Hill, New York, 1941.

A Boundary Element Method for Free Surface Flow Problems

R. ZHENG and R. I. TANNER
Department of Mechanical Engineering
The University of Sydney
Sydney, NSW 2006, Australia

INTRODUCTION

The Boundary Element Method (BEM) has received increasing attention in the last decade. Its mathematical treatments and its applications in potential and elasticity problems can be found in several textbooks (*e.g.*, Brebbia *et al.*, 1984). The basis of the BEM is flexible enough to allow the formulation to be applied also to Newtonian and Non-Newtonian fluid mechanics (Bush and Tanner, 1983; Tanner, 1988). The purpose here is to describe a BEM for treating creeping viscous free surface flows with surface tension.

In free surface flow problems the governing partial differential equations are defined on domains which are not known in advance due to the presence of free boundaries, and there are three boundary conditions on the free surface to be satisfied:
(i) zero normal velocity;
(ii) zero tangential traction;
(iii)balance of normal traction and surface tension force.
Generally numerical techniques must be used because of the non-linear nature of the equations. One may apply two of the above conditions directly in any particular numerical scheme and use the other one as the criterion for the location of the free surface. In most previous attempts a "kinematic scheme" was employed (Tanner1973, Nickell *et al.*1974; Reddy and Tanner1978): The free surface location is found by first imposing the traction boundary conditions on an assumed free surface and computing the corresponding flow field; the kinematic boundary condition together with the current approximation to the flow field is then used to update the free boundary, and the cycle is repeated until convergence is reached. Instead, some authors (*e.g.*, Orr and Scriven1978; Silliman and Scriven1980; Ruschak1980; Satio and Scriven1981) employed a "normal traction scheme". That is, the normal traction condition is used as the criterion for updating the free surface. The most common numerical technique for the free surface flow problems has been the Finite Element Method (FEM) in which one finds the complete solution field whether or not this is of interest. For a creeping Newtonian flow, the domain-type solution scheme, although successful, seems to be wasteful since the update procedure in fact requires knowledge of the boundary solution only. The BEM is therefore a particularly economical technique for this class of problem. We have used the BEM for a numerical simulation of calendering problems (with an exit free surface), using the " kinematic scheme " but ignoring surface tension effects (Zheng and Tanner1988).

The present paper describes an implementation of an isoparametric boundary element method for 2-D flow problems. A normal traction free surface update scheme based on the idea suggested by Orr and Scriven (1978) is employed. Ingham and Kelmanson (1984) also used the traction conditions as a criterion for updating the free surface in a biharmonic boundary integral equation (BBIE) method (in which the equations are formulated in terms of stream function, velocity, vorticity and vorticity gradient), but their method differs from ours since it requires an analytic form of the free surface shape. We shall use the direct integral formulation, which is formulated in terms of boundary velocities and tractions, and the free surface update scheme will be implemented employing the Broyden iterative method. No analytic surface shape is assumed. We demonstrate the method by solving a problem of a plane extrusion of a Newtonian fluid with surface tension.

GOVERNING EQUATIONS AND NUMERICAL SCHEME

Governing Equations

In this work the isothermal, steady state, creeping flows of incompressible Newtonian fluids is considered. The governing equations and the constitutive equation are, respectively,

$$\nabla \cdot \sigma = 0, \tag{1}$$
$$\nabla \cdot \mathbf{u} = 0, \tag{2}$$
$$\sigma = -p\,\mathbf{I} + 2\,\eta\,\mathbf{D}, \tag{3}$$

where σ is the total stress tensor and $\mathbf{u}$ is the velocity vector, $\mathbf{D}$ is the rate-of-strain tensor, η is the viscosity, p is the hydrostatic pressure which arises due to the incompressibility constraint and $\mathbf{I}$ is the unit tensor.

The reduction of eq.(1-3) to a set of boundary integral equations is standard (Brebbia *et al.*1980) and the procedure yields

$$C_{ij}(\mathbf{x})u_j(\mathbf{x}) = \int_\Gamma u_{ij}^*(\mathbf{x},\mathbf{y})t_j(\mathbf{y})d\Gamma(\mathbf{y}) - \int_\Gamma t_{ij}^*(\mathbf{x},\mathbf{y})u_j(\mathbf{y})d\Gamma(\mathbf{y}), \tag{4}$$

where Γ is the boundary of the solution domain Ω, $\mathbf{x},\mathbf{y} \in \Omega$, $u_j(\mathbf{y})$ is the j-component of velocity at $\mathbf{y}$, $t_j(\mathbf{y})$ is the j-component of boundary traction at $\mathbf{y}$, $u_{ij}^*(\mathbf{x},\mathbf{y})$ is the i-component of velocity field at $\mathbf{x}$ due to a "stokeslet" in j-direction at $\mathbf{y}$ and $t_{ij}^*(\mathbf{x},\mathbf{y})$ is its associated traction. The plane-flow solutions for u_{ij}^* and t_{ij}^* are well-known (Brebbia *et al.*1984). $C_{ij}(\mathbf{x})$ depends on local geometry: $C_{ij}(\mathbf{x}) = \delta_{ij}$ if $\mathbf{x} \in \Omega$, and $C_{ij}(\mathbf{x}) = \frac{1}{2}\delta_{ij}$ if $\mathbf{x} \in \Gamma$ and Γ is a smooth surface.

Discretization Scheme

Isoparametric elements are used to discretise the boundary integral equation (4). Along the boundary, each element has three node points, and the coordinates at any point in the element are approximated as

$$x_i(\xi) = N^\alpha(\xi)x_i^\alpha, i = 1, 2, \alpha = 1, 2, 3, \tag{5}$$

Note that Einstein's summation convention applies to repeated index α. In eq. (5), N^α is the quadratic shape function of node α . The same shape functions are used for the variables $\mathbf{u}$ and $\mathbf{t}$:

$$u_i(\mathbf{x}(\xi)) = N^\alpha(\xi)u_i^\alpha, \tag{6}$$
$$t_i(\mathbf{x}(\xi)) = N^\alpha(\xi)t_i^\alpha. \tag{7}$$

Using these elements, a system of algebraic equations results which can be written in matrix form as

$$Gt = Hu, \tag{8}$$

where t, u are the global nodal traction vector and velocity vector respectively and G, H are coefficient matrices. The elements of G are contributions from the integral $\int N^\alpha u_{ij}^* d\Gamma$ over each boundary element. Similarly the elements of H involve the integral $\int N^\alpha t_{ij}^* d\Gamma$, and C_{ij} is absorbed into H. With the isoparametric elements, a problem appears when a node is situated at a corner which can have multi-valued traction depending on the side under consideration. We treat this problem in the manner described previously by Tran-Cong and Phan-Thien (1988) for a 3-D BEM scheme.

Free Surface Update Scheme

When the motion is steady, at a free surface the boundary conditions are

$$u \cdot n = 0 \tag{9}$$
$$t \cdot s = 0 \tag{10}$$
$$t \cdot n = \kappa(Ca)^{-1} \tag{11}$$

where n and s are the unit normal and the unit tangent on the free boundary, respectively. Suppose the free surface shape is defined as $y = h(x)$, in terms of which

$$n = \begin{pmatrix} -(\partial h/\partial x)[1 + (\partial h/\partial x)^2]^{-1/2} \\ [1 + (\partial h/\partial x)^2]^{-1/2} \end{pmatrix}, \tag{12}$$

$$s = \begin{pmatrix} [1 + (\partial h/\partial x)^2]^{-1/2} \\ (\partial h/\partial x)[1 + (\partial h/\partial x)^2]^{-1/2} \end{pmatrix}, \tag{13}$$

and the curvature is given by

$$\kappa = (\partial^2 h/\partial x^2)[1 + (\partial h/\partial x)^2]^{-3/2} \tag{14}$$

The capillary number which appears in (11) is defined as $Ca \equiv \eta U/\sigma$, where η is the fluid viscosity, U is the characteristic velocity and σ is the surface tension coefficient. In the calculation a cubic spline interpolation is used to estimate the curvature at each node along the free surface. The main attractive feature of the cubic spline interpolation is that it has a continuous second derivative on each node.

It can be seen from eq.(9-11) that, if there is no surface tension (i.e. $Ca \rightarrow \infty$ and hence $t \cdot n = 0$), the traction components satisfying (10) and (11) are known to be zero everywhere on the free surface. However, the velocity field on the free surface is not known in advance. It is therefore convenient in this case to take zero tractions to be given in eq.(8), and leave the free surface velocity field to be computed iteratively; a new free surface can be easily sought by drawing a streamline, the slope of which is parallel to the velocity field. On the other hand, if surface tension is involved, neither velocity nor traction fields on the free surface are known a priori, because they depend on surface orientation, and the normal traction condition also depends on second derivatives of surface position. In this case we incorporate eq.(9) and (10) as additional equations in the system of equations so that the normal velocity and tangential traction boundary conditions are automatically satisfied, while leaving the normal traction and the tangential velocity to be computed iteratively, i.e., we find new free surface locations by solving eq.(11) which, after discretization, reduces to a non-linear set of algebraic equations.

Gartling et al.(1973) and Engelman et al. (1981) surveyed various commonly used iterative techniques for the non-linear system of equations resulting from the finite-element discretization of fluid mechanics problems. They show that the choice of solution scheme is partly problem-dependent, but the Newton-Raphson method nearly always performs well. In the case where the function is so complicated that it is either impractible or impossible to obtain explicit expressions for the Jacobian matrix, there are several modifications of Newton's method such as the Broyden method can be used to save labour in evaluating the Jacobian matrix. In the present work we use the Broyden's method, which is explained briefly here. Consider a set of m simultaneous non-linear algebraic equations in m unknowns

$$\mathbf{F}(\mathbf{h}) = \mathbf{0}. \tag{15}$$

In the normal traction scheme the left hand side of the equations is the residual of the normal traction boundary condition. The solution vector $\mathbf{h}$ contains the unknown y-coordinates of the free surface collocation points. Suppose $\mathbf{h}_k$ is the kth approximation to the solution of (15), and $\mathbf{F}_k$ is written for $\mathbf{F}(\mathbf{h}_k)$, then Newton's method is defined by

$$\mathbf{h}_{k+1} = \mathbf{h}_k - \mathbf{J}_k^{-1}\mathbf{F}_k, \tag{16}$$

where $\mathbf{J}_k$ is the Jacobian matrix, $\partial F_j/\partial h_i$, evaluated at $\mathbf{h}_k$. The algorithm of Broyden's method is summarized as follows:

(i) Assume an initial estimate solution $\mathbf{h}_0$ and obtain an initial inverse Jacobian matrix $\mathbf{J}_0^{-1}$.

(ii) Compute $\mathbf{F}_k = \mathbf{F}(\mathbf{h}_k)$.

(iii) Compute $\mathbf{d}_k = -\mathbf{J}_k^{-1}\mathbf{F}_k$.

(iv) Compute $\mathbf{h}_{k+1} = \mathbf{h}_k + \mathbf{d}_k$.

(v) Compute $\mathbf{F}_{k+1}$. Test $\mathbf{F}_{k+1}$ for convergence, if not, then

(vi) Compute $\mathbf{f}_k = \mathbf{F}_{k+1} - \mathbf{F}_k$.

(vii) Compute

$$\mathbf{J}_{k+1}^{-1} = \mathbf{J}_k^{-1} + \frac{(\mathbf{d}_k - \mathbf{J}_k^{-1}\mathbf{f}_k)\mathbf{d}_k^T}{\mathbf{d}_k^T\mathbf{J}_k^{-1}\mathbf{f}_k}\mathbf{J}_k^{-1}. \tag{17}$$

(viii) Replace $\mathbf{h}_k$ by $\mathbf{h}_{k+1}$, $\mathbf{J}_k^{-1}$ by $\mathbf{J}_{k+1}^{-1}$, and repeat from step (ii).

The approximation to the inverse Jacobian $\mathbf{J}_0^{-1}$ is obtained in the way suggested by Broyden (1969). That is, we begin with $\mathbf{J}_0^{-1} = \mathbf{I}$ (the mth order unit matrix), and perform the first m iterations described above with step (iii) replaced by $\mathbf{d}_1 = \delta\mathbf{e}_1$, $\mathbf{d}_2 = \delta\mathbf{e}_2$, ..., $\mathbf{d}_m = \delta\mathbf{e}_m$, where $\mathbf{e}_1$, $\mathbf{e}_2$, ..., $\mathbf{e}_m$ are the first, second, ..., mth columns of the unit matrix respectively and δ is an arbitrary non-zero quantity which is taken to be 0.001 in the algorithm. We obtain $\mathbf{J}_{m+1}^{-1}$, which is then used as the initial matrix $\mathbf{J}_0^{-1}$.

RESULTS AND DISCUSSION

As a test of the accuracy of the isoparametric BEM program we firstly applied the program to a parallel plate Poiseuille flow and a lubrication flow in a thrust bearing. In the first case

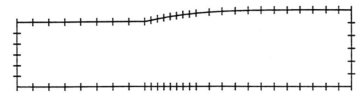

FIGURE 1. Boundary element discretization.

the results were essentially "exact". In the second case the BEM results were in excellent agreement with the corresponding analytical results from lubrication theory (Tanner1988) which are themselves approximate, of course.

We now turn to a free surface flow problem — the flow from a slit die. Because of the symmetry about the centreline of the slit, only the top half of the field is considered. The numerical convergence with mesh refinement was checked by employing discretizations consisting of 46, 60 and 70 boundary nodes. The discretization with 60 nodes was found to be sufficient to get satisfactory convergence. The results to be presented below were obtained using 70 nodes as shown in Fig. 1. The boundary conditions are (apart from those at the free surface): at the inlet the velocity profile is a fully developed Poiseuille flow with mean velocity U; at the wall the fluid satisfies the no-slip condition; at the cross-section far downstream of the free sheet it is supposed that the flow is parallel to the symmetry axis and there is no axial traction. The velocity component normal to the centreline is set to zero, and zero axial traction is imposed on the centreline. All the computations were performed on a VAX-11/780 computer. Convergence of the process is determined in terms of the norm of the residual of the normal traction conditions (if the normal traction scheme is used) and the maximum difference between two successive approximations of the position (*i.e.*, the y-coordinate) of the free boundary. When the former falls below 10^{-8} and the latter falls below 10^{-5}, iterations are interrupted. To start the iterative procedure, a guess of the boundary position is required. The initial jet profile we used is a flat sheet with the same width as the slit.

The convergence behaviour (Fig. 2) depends on the value of the capillary number, Ca. When $Ca \to 0$, the normal traction scheme tends toward second-order convergence. When $Ca \to \infty$, it tends toward first-order convergence. The convergence behaviour of the kinematic scheme (with the Broyden method) for $Ca \to \infty$ showed that, in this case where surface tension is negligible, the kinematic scheme converges faster than the normal traction scheme. Of course, the speed of convergence also depends on the initial guess of the free surface position.

Fig. 3 shows the final free surface profiles at $Ca \to \infty$ obtained from the normal traction scheme and the kinematic scheme respectively. The results are in excellent agreement, though they are not exactly identical. It has been noticed that most finite element solutions predict a slightly greater swelling ratio than the boundary element solutions. The result we obtained from our BEM program with "constant elements" predicted 17.7% swelling (with 58 elements), while the present results obtained using isoparametric elements are 18.6%, 18.4% and 18.4% corresponding to 46, 60 and 70 boundary nodes respectively, which can be compared with $19.0 \pm 0.2\%$ reported by Tanner (1988).

When surface tension is taken into account, we use the normal traction scheme only. The swelling ratio and the free surface shape at various Ca numbers are shown in Figs. 4 and 3 respectively. It is clear that the surface tension reduces the swell and tends to straighten the meniscus profile. In Table 1 we present our BEM results of swelling ratios and the correspond-

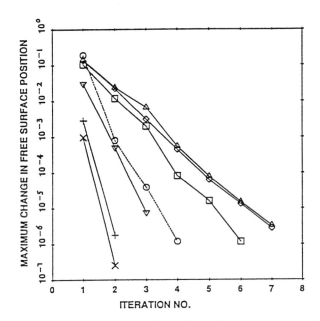

FIGURE 2.   Convergence behaviour. *Normal traction scheme*: ×: $Ca = 0.001$; +: $Ca = 0.01$; ∇: $Ca = 0.1$; □: $Ca = 1.0$; ◇: $Ca = 10.0$; △: $Ca = \infty$. *Kinematic scheme*: ○: $Ca \to \infty$.

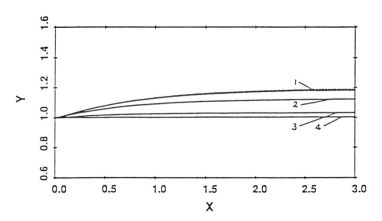

FIGURE 3.   Free surface profiles for different Ca values: The broken line is the result using the "kinematic scheme" for $Ca = \infty$ and the solid lines are the results using the "normal traction scheme",for Ca: (1) ∞; (2) 1.0; (3) 0.1; (4) 0.01.

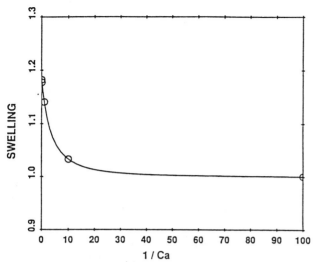

FIGURE 4.   The effect of surface tension on the swelling ratio.

TABLE 1.   Comparison of swelling ratios obtained from the BEM, FEM and BBIE methods.

Ca	present results	FEM results (Ruschak)	BBIE results (Ingham *et al.*)
∞	1.1841	– – –	– – –
1000	1.1822	1.182	1.1781
10	1.1726	1.172	1.1654
1	1.1123	1.100	1.1044
0.1	1.0237	1.017	1.0202
0.01	1.0039	1.002	1.0031
0.001	1.0004	– – –	1.0003

Table header spanning: Swelling ratios

ing results obtained from the FEM (Ruschak1980) and BBIE (Ingham and Kelmanson1984) respectively. The results from these independent numerical methods are seen to be in good ageement.

CONCLUSION

In this work the isoparametric elements are successfully implemented in a two dimensional BEM program. A scheme employing the normal traction condition to calculate free surfaces by the Broyden iteration is described. For Newtonian fluid, the prediction of the extrudate swell and the free surface shape has been demonstrated. A wide range of surface tensions is included in the calculations. It is shown that the scheme can be efficient, especially for high surface tension. Unlike the usual domain-type solution methods such as FEM and FDM, there is no need for the present method to evaluate the field variables in the interior of the solution domain at each iteration.

ACKNOWLEDGEMENT

This research was supported by an Australian Research Council (ARC) Program Grant which are gratefully acknowledged.

249

REFERENCES

1. Brebbia, C. A., Telles, J.C.F. and Wrobel, L.C., *The Boundary Element Method in Engineering*, 2nd ed., Pentech Press, Great Britain, 1980.

2. Broyden, C. G., A New Method of Solving Nonlinear Simultaneous Equations, *Computer Journal*, vol.12, pp. 94-99, 1969.

3. Bush, M. B. and Tanner, R. I., Numerical Solution of Viscous Flows Using Integral Equation Methods, *Int. J. Num. Meth. Fluids*, vol.3, pp. 71-92, 1983.

4. Engelman, M. S., Quasi-Newton Methods in Fluid Mechanics, *Int. J. Numer. Meth. Engng.*, vol.17, pp. 707-718, 1981.

5. Gartling, D. K., Nickell, R. E. and Tanner, R. I., A Finite Element Convergence Study for Accelerating Flow Problems, *Int. J. Numer. Meth. Engng.*, vol.11, pp. 1155-1174, 1973.

6. Ingham D. B. and Kelmanson, M. A., *Boundary Integral Equation Analyses of Singular, Potential, and Biharmonic Problems*, Lecture Notes in Engineering, vol.7, Springer-Verlag, New York, 1984.

7. Nickell, R. E., Tanner, R. I. and Caswell, B., The Solution of Viscous Incompressible Jet and Free Surface Flows Using the Finite Element Method, *J. Fluid Mech.*, vol.65, pp. 189-206, 1974.

8. Orr, F. M. and Scriven, L. E.,Rimming Flow: Numerical Simulation of Steady, Viscous, Free-surface Flows with Surface Tension, *J. Fluid Mech.*, vol.84, pp. 145-165, 1978.

9. Reddy, K. R., and Tanner, R. I., Finite Element Solution of Viscous Jet Flows with Surface Tension, *Computers and Fluids*, vol.6, pp. 83-91, 1978.

10. Ruschak, K. J., A Method for Incorporating Free Boundaries with Surface Tension in Finite Element Fluid Flow Simulators, *Int. J. Num. Meth. Engng.*, vol.15, pp. 639-648, 1980.

11. Saito, H. and Scriven, L. E., Study of Coating Flows by the Finite Element Method, *J. Comput.Phys.*, vol.42, pp. 53-76, 1981.

12. Silliman, W. J. and Scriven, L. E., Separating Flow near a Static Contact Line: Slip at a Wall and Shape of a Free surface, *J. Comput. Phys.*, vol.34, pp. 287-313, 1980.

13. Tanner, R. I., *Engineering Rheology*, Oxford University Press, Revised Edition, 1988.

14. Tanner, R. I., Die-swell Reconsidered: Some Numerical Solutions Using a Finite Element Program, *Applied Polymer Symposium*, vol.20, pp. 201-208, 1973.

15. Tran-Cong, T. and Phan-Thien, N., Three Dimensional Study of Extrusion Processes by Boundary Element Method. Part 1: An Implementation of High Order Elements and Some Newtonian Results, *Rheologica Acta*, vol.27, pp. 21-30, 1988.

16. Zheng, R. and Tanner, R. I., A Numerical Analysis of Calendering, *J. Non-Newt. Fluid Mech.*, vol.28, p. 149-170, 1988.

COMBINED METHODS AND COMPARISONS

A Combined Spectral-Finite Element Method for Solving Unsteady Navier-Stokes Equations

GUO BEN-YU and CAO WEI-MING
Shanghai University of Science and Technology
Shanghai, PRC

INTRODUCTION

When we study the boundary layer stability, the unsteady separation of viscous fluid flow and other related problems, we have to solve Navier-Stokes equations with semi-periodic boundary conditions. Many authors developed numerical methods to solve such problems. For instance, Murdock (1977), Ingham (1984), Biringen (1984), Milinazzo and Saffman (1985), and Guo Ben-yu (1988) used combined spectral-finite difference method. Recently, Canuto et al. (1984) proposed a combined spectral-finite element method. In this paper, we construct a combined Fourier spectral-finite element scheme for solving two-dimensional, semi-periodic, unsteady Navier-Stokes equations. Surely such problems can also be solved by a spectral-difference method, but it is difficult to extend it to three-dimensional problems with non-rectangular domains. On the contrary, the method in this paper can be generalized easily to more complicated problems.

THE SCHEME

Let

$$I_x = \{x/0 < x < 1\} \ , \quad I_y = \{y/0 < y < 2\pi\} \ , \quad Q = I_x \times I_y \ .$$

Denote by $U = (U^{(1)}, U^{(2)})$, P, f and $\nu > 0$ the velocity, the ratio of pressure over density, the body force, and the kinetic viscosity, respectively. We consider the two-dimensional unsteady Navier-Stokes equations as follows:

$$\begin{cases} \dfrac{\partial U}{\partial t} + (U \bullet \nabla) U + \nabla P - \nu \nabla^2 U = f \ , & \text{in } Q \times (0,T] \ , \\[2mm] \nabla \bullet U = 0 \ , & \text{in } Q \times [0,T] \ , \\[2mm] U\big|_{t=0} = U_o \ , & \text{in } Q \ . \end{cases} \tag{1}$$

Assume that U, P and f have the period 2π for the variable y, and that $U(0,y,t) = U(1,y,t) = 0$.

In addition, we assume $P\epsilon\tilde{L}^2(Q)$, with

$$\tilde{L}^2(Q) = \{u\epsilon L^2(Q) \ / \ \textstyle\iint_Q u \ dx \ dy = 0\} \ .$$

For $\mu \geq 0$, we denote by $H^\mu(Q)$, $\|\bullet\|_\mu$ and $|\bullet|_\mu$ the classical Sobolev space, its norm and semi-norm respectively. In particular, we define $L^2(Q) = H^0(Q)$ with the norm $\|\bullet\|$ and the inner product $(\bullet,\bullet)$. Furthermore, define

$$C_p^\infty(Q) = \{u\epsilon C^\infty(\bar{Q}) \ / \ u \text{ has the period } 2\pi \text{ for the variable y}\} \ ,$$

$$C_{o,p}^\infty(Q) = \{u\epsilon C_p^\infty(Q) \ / \ u(0,y) = u(1,y) = 0, \text{ for all } y\epsilon I_y\} \ ,$$

and denote by $H_p^\mu(Q)$ and $H_{o,p}^\mu(Q)$ the closure of $C_p^\infty(Q)$ and $C_{o,p}^\infty(Q)$ in $H^\mu(Q)$ respectively.

The generalized solution of (1) is the pair $(U(t),P(t))\epsilon[H_{o,p}^1(Q)]^2 \times \tilde{L}^2(Q)$ satisfying

$$\begin{cases} \dfrac{\partial}{\partial t}(U(t),v) + ((U(t)\bullet\nabla)U(t),v) - b(v,P(t)) + \nu a(U(t),v) \\ \qquad = (f(t),v), \ v\epsilon[H_{o,p}^1(Q)]^2 \\ b(U(t),w) = 0 , \quad w \ \epsilon \ \tilde{L}^2(Q) , \\ U(0) = U_o , \end{cases} \qquad (2)$$

where

$$a(u,v) = \textstyle\iint_Q (\nabla u)(\nabla v) \ dx \ dy \ ,$$

$$b(u,v) = \textstyle\iint_Q (\nabla\bullet u) \ v \ dx \ dy \ .$$

We introduce a trilinear form $J(\bullet,\bullet,\bullet) : \ [(H^1(Q))^2]^3 \rightarrow \mathfrak{R}^1$ as follows:

$$J(u,w,v) = \tfrac{1}{2}\{((w\bullet\nabla)u,v) - ((w\bullet\nabla)v,u)\} \ .$$

Clearly,

$$J(u,w,v) + J(v,w,u) = 0 , \qquad (3)$$

and if $\nabla\bullet w = 0$, then

$$J(u,w,v) = ((w\bullet\nabla)u,v) \ .$$

Thus (2) is equivalent to

$$\begin{cases} \dfrac{\partial}{\partial t}(U(t),v) + J(U(t), \ U(t),v) - b(v,P(t)) + \nu a(U(t),v) \\ \qquad = (f(t),v), \ v\epsilon[H_{o,p}^1(Q)]^2 \\ b(U(t),w) = 0 , \quad w\epsilon\tilde{L}^2(Q) , \\ U(0) = U_0 \end{cases} \qquad (4)$$

For numerical solution of (4), we shall approximate the first equation directly. While to tackle the incompressible condition, we adopt the idea of artificial compression. That is to approximate the second equation of (4) by the following equation:

$$\beta \frac{d}{dt}(P(t),w) + b(U(t),w) = 0 , \qquad w \in \tilde{L}^2(Q) ,$$

where $\beta > 0$ is a small parameter.

Now we construct the scheme. Firstly, we divide I_x into M_h subintervals, with the nodes $0 = x_0 < x_1 < \ldots < x_{M_h} = 1$. Let

$$I_{x,\ell} = (x_{\ell-1}, x_\ell), \quad h_\ell = x_\ell - x_{\ell-1}, \quad h = \max_{1 \le \ell \le M_h} h_\ell, \quad h' = \min_{1 \le \ell \le M_h} h_\ell .$$

We assume furthermore that there exists a positive constant C_0 independent of the divisions of I_x such that $h/h' \le C_0$. For any integer $k \ge 0$, we denote by $\mathcal{P}_k$ the set of all the polynomials defined on $\mathfrak{R}^1$ of degree $\le \bar{k}$. We define the finite element subspaces in the non-periodic direction as follows:

$$\tilde{S}_h^k = \{u \;/\; u|_{I_{x,\ell}} \in \mathcal{P}_k, \; 1 \le \ell \le M_h\} , \qquad S_h^k = \tilde{S}_h^k \cap H_0^1(Q) .$$

Next, suppose N be a positive integer. We define the subspaces for Fourier spectral approximations as follows:

$$S_N = \text{Span } \{e^{ijy} \;/\; |j| \le N\} .$$

We define the following finite dimensional subspace as the trial function space for the velocity

$$X = \{S_h^{k+1} \boxtimes S_N\} \times \{S_h^{k+2} \boxtimes S_N\} .$$

The trial function space for the pressure P is

$$Y = \{\tilde{S}_h^k \boxtimes S_N\} \cap \tilde{L}^2(Q) .$$

Let P_N be the orthogonal projection operator from $L^2(I_y)$ onto S_N, and Π_h^k be the piecewise Lagrange's interpolation of order k from $C(\bar{I}_x)$ onto S_h^k, i.e. for any $u \in C(\bar{I}_x)$, $\Pi_h^k u \in S_h^k$ satisfies

$$(\Pi_h^k u)(x_{\ell-1} + \tfrac{m}{k} h_\ell) = u(x_{\ell-1} + \tfrac{m}{k} h_\ell) , \qquad 0 \le m \le k , \; 1 \le \ell \le M_h .$$

Let τ be the spacing in time t, and $S_\tau = \{t = \ell\tau \;/\; 0 \le \ell \le [\tfrac{T}{\tau}]\}$. Define

$$u_t(t) = \tfrac{1}{\tau}(u(t + \tau) - u(t)) .$$

The combined Fourier spectral-finite element scheme for (4) is to find the pair $(u(t), p(t)) \in X \times Y$ for all $t \in S_\tau$, such that

$$\begin{cases} (u_t(t), v) + J(u(t) + \delta\tau u_t(t), u(t), v) - b(v, p(t) + \theta\tau\, p_t(t)) \\ \qquad + \nu a(u(t) + \sigma\tau\, u_t(t), v) = (P_{N^o}\, \Pi_h^{k+1} f(t), v), \qquad v\epsilon X, \\ \beta(p_t(t), w) + b(u(t) + \theta\tau\, u_t(t), w) = 0, \qquad w\epsilon y, \\ u(0) = P_{o}\, \Pi_{h_N}^{k+1} U_o, \qquad p(0) = 0, \end{cases} \qquad (5)$$

where δ, $\sigma \geq 0$ and $\theta > \frac{1}{2}$ are parameters.

Remark 1: Let P be the L^2-projection operator from $\tilde{L}^2(Q)$ onto Y. Then it follows from the second equation of (5) that

$$p(t + \tau) = p(t) - \frac{\tau}{\beta}\,[\theta P(\nabla\bullet(t + \tau)) + (1 - \theta)\, P(\nabla\bullet(t))]. \qquad (6)$$

By substituting (6) into the first equation of (5), we can get a linear equation for $u(t + \tau)$. As soon as it is solved, we get $p(t + \tau)$ by (6) directly. In this way, we can solve the velocity and the pressure separately. This is indeed one of the advantages of the artificial compression treatment.

THE CONVERGENCE THEOREM

Suppose B be a Banach space, and $I \subset \mathfrak{R}^1$ be an interval. We define

$$L^2(I;B) = \{u \,/\, u\colon I \to B, \; \|u\|_{L^2(I;B)} = (\int_I \|u(t)\|_B^2 dt)^{\frac{1}{2}} < \infty\}.$$

Similarly, we define the spaces $C(I;B)$ and $H^\mu(I;B)$, et al..

For convenience, we recall the definition of the non-isotropic Sobolev spaces as follows: for $r \geq 0$ and $s \geq 0$,

$$H^{r,s}(Q) = L^2(I_y, H^r(I_x)) \cap H^s(I_y, L^2(I_x)),$$

equipped with the norm

$$\|u\|_{H^{r,s}(Q)} = (\|u\|^2_{L^2(I_y, H^r(I_x))} + \|u\|^2_{H^s(I_y, L^2(I_x))})^{\frac{1}{2}}.$$

For $r \geq 1$ and $s \geq 1$, we define

$$M^{r,s}(Q) = H^{r,s}(Q) \cap H^1(I_y, H^{r-1}(I_x)) \cap H^{s-1}(I_y, H^1(I_x)),$$

with the norm

$$\|u\|_{M^{r,s}(Q)} = (\|u\|^2_{H^{r,s}(Q)} + \|u\|^2_{H^1(I_y, H^{r-1}(I_x))} + \|u\|^2_{H^{s-1}(I_y, H^1(I_x))})^{\frac{1}{2}}.$$

Besides, we denote by $H_p^{r,s}(Q)$ the closure of $C_p^\infty(Q)$ in $H^{r,s}(Q)$, $H_{o,p}^{r,s}(Q)$ and $M^{r,s}(Q)$ the closures of $C_{o,p}^\infty(Q)$ in $H^{r,s}(Q)$ and $M^{r,s}(Q)$ respectively.

For describing the errors of the numerical solutions, we introduce the following notation:

$$E(u,v,t) = \| u(t) \|^2 + \beta \| v(t) \|^2 + \nu\tau(\sigma + \theta) \, | \, u(t)|_1^2$$

$$+ \, \nu\tau \sum_{t' \leq t-\tau} |u(t')|_1^2 \ .$$

Theorem 1: Let (U,P) be the solution of (4), and (u,p) the solution of (5). Assume that $U \in C(0,T;M_{o,p}^{r,s}(Q)) \cap H^1(0,T;H^1(Q)) \cap H_p^{r-1,s-1}(Q)) \cap H^2(0,T;$ $L^2(Q))$, $P \in C(0,T;H_p^{r-1,s-1}(Q)) \cap H^1(0,T;L^2(Q))$, $f \in C(0,T;H_p^{r-1,s-1}(Q))$ with $r \geq 1$, $s \geq 1$ and $\bar{r} = \min (r,k+2)$. Suppose $\beta = 0(\tau^2)$, $h = 0(N^{-\mu})$ and $\tau = 0(N^{-\lambda})$ with $\mu \geq 1$, $\lambda \geq 0$. Let h, N^{-1}, and τ be sufficiently small. If one of the following two conditions is satisfied,

(i) $\sigma > \frac{1}{2}$, $\theta > \dfrac{\sigma}{2\sigma - 1}$, and

$$\lambda > \max \, (\frac{1 + \mu}{3} , 1 + 3\mu - 2\mu\bar{r} , 3 + \mu - 2s) \ ; \tag{7}$$

(ii) $\sigma \leq \frac{1}{2}$ and $\tau(C_1 h^{-2} + N^2) < \dfrac{2\theta - 1}{\nu(\sigma + \theta(1 - 2\sigma))}$;

where C_1 is a positive constant depending only on C_0, then there exists a positive constant C_2 depending only on U, P, f and ν, such that

$$E(U - u, P - p, t) \leq C_2(\tau^2 + h^{2(\bar{r}-1)} + N^{2(1-s)}) \ .$$

Remark 2: If $\delta = \theta > \frac{1}{2}$, then condition (7) is not necessary.

NUMERICAL RESULTS

In this section, we examine the numerical performances of scheme (5). The velocity U is determined by

$$U^{(1)}(x,y,t) = - 0.1 \cos y \exp (x + \sin y + 0.1t) \ ,$$

$$U^{(2)}(x,y,t) = 0.1 \exp (x + \sin y + 0.1t) \ ,$$

while the pressure P is taken to be zero constantly. Besides, we assume that the kinetic viscosity $\nu = 10^{-2}$, and choose the body force f such that U, P and f satisfy (4) exactly.

We consider only the case of $k = 0$ in scheme (5), i.e. the finite element subspaces in x direction for $u^{(1)}$, $u^{(2)}$ and p are piecewise linear, piecewise quadratic and piecewise constant, respectively. For comparison, we also solve (4) by finite element method, in which the domain $\bar{Q}$ is divided into $M_h(2N + 1)$ congruent small rectangles, each with the length $h_x = 1/M_h$ and the width $h_y = 2\pi/(2N + 1)$. We take the trial spaces for $u^{(1)}$, $u^{(2)}$ and p to be piecewise linear, piecewise biquadratic and piecewise constant separately. The finite element scheme is constructed similarly to (5) by artificial compression treatment. Besides, for saving the time of computation, we approximate the nonlinear terms explicitly (i.e. $\delta = 0$), in both the spectral-finite element scheme and the finite element scheme, but we always approximate the linear terms implicitly (i.e. $\theta = \sigma = 1$).

For describing the accuracy, we define the discrete L^2-normed relative errors for $u^{(1)}$ and $u^{(2)}$ as follows:

$$E_1(u^{(\ell)}, t) = \left\{ \frac{\sum\limits_{(x,y)\in Q^{(\ell)}} \left| U^{(\ell)}(x,y,t) - u^{(\ell)}(x,y,t) \right|^2}{\sum\limits_{(x,y)\in Q^{(\ell)}} \left| U^{(\ell)}(x,y,t) \right|^2} \right\}^{\frac{1}{2}}, \quad \ell = 1, 2,$$

where

$$Q^{(1)} = \{(x,y) \ / \ x = j_1 h_x, \ y = j_2 h_y, \ 1 \le j_1 \le M_h - 1, \ 1 \le j_2 \le 2N + 1\}.$$

$$Q^{(2)} = \{(x,y) \ / \ x = j_1 h_x/2, \ y = j_2 h_y/2, \ 1 \le j_1 \le 2M_h - 1,$$

$$1 \le j_2 \le 2(2N + 1)\} .$$

The numerical results show that

(1) For the same mesh sizes M_h and N, the spectral-finite element scheme gives more accurate results than the finite element scheme does (see Table 1). However, the computations required by each are almost the same.

(2) We observe from Theorem 1 that if $\beta = O(\tau^2)$, then the artificial compression term $\beta \frac{\partial p}{\partial t}$ does not affect the order of the error $E_1(u^{(\ell)}, t)$. However, in practice, the value of β determines the structure of the linear equations derived from scheme (5). An excessive small β would result in a large condition number of the corresponding coefficient matrix (in fact, the matrix tends to be singular, as β tends to zero), and a greater requirement of computation. Consequently, we shall take into account of the matrix structure apart from the approximation order, when we select an appropriate β. In the example of the section, the results for $\beta = 10^{-2}$ and 10^{-3} are better than for $\beta = 10^{-4}$ (see Table 2).

TABLE 1. $\tau = 10^{-2}$, $\beta = 10^{-3}$, T = 5.0

		N = 4		N = 8	
		$E_1(u^{(1)}, t)$	$E_1(u^{(2)}, t)$	$E_1(u^{(1)}, t)$	$E_1(u^{(2)}, t)$
SPECTRAL-F.E.M.	$M_h = 5$	0.21379E-2	0.98124E-3	0.76151E-3	0.82940E-3
	$M_h = 10$	0.19619E-2	0.60285E-3	0.26212E-3	0.35686E-3
F.E.M.	$M_h = 5$	0.18724E-1	0.10882E-1	0.38954E-2	0.24330E-2
	$M_h = 10$	0.16047E-1	0.90782E-2	0.38383E-2	0.22202E-2

TABLE 2.   Scheme (5), $\tau = 10^{-2}$, $\beta = 10^{-3}$, $t = 5.0$

	N = 4		N = 8	
	$E_1(u^{(1)},t)$	$E_1(u^{(2)},t)$	$E_1(u^{(1)},t)$	$E_1(u^{(2)},t)$
$\beta = 10^{-2}$	0.21396E-2	0.97437E-3	0.75778E-3	0.82725E-3
$\beta = 10^{-3}$	0.21379E-2	0.98124E-3	0.76151E-3	0.82940E-3
$\beta = 10^{-4}$	0.20537E-2	0.11868E-2	0.91537E-3	0.11504E-2

ACKNOWLEDGEMENT

The authors gratefully acknowledge the support of K.C. Wong, Education Foundation, Hong Kong.

REFERENCES

1.   Biringen, S., Final stages of transition to turbulence in plane channel flow, J. Fluid Mech., Vol. 148, pp. 413-442, 1984.

2.   Canuto, C., Maday, Y., and Quarteroni, A., Combined finite element and spectral approximation of the Navier-Stokes equations, Numer. Math., Vol. 44, No. 2, pp. 201-217, 1984.

3.   Guo Ben-yu, A spectral-difference method for solving two-dimensional vorticity equations, J. Comput. Math., Vol. 6, No. 3, pp. 238-257, 1988.

4.   Ingham, D.B., Unsteady separation, J. Comput. Phys., Vol. 53, No. 1, pp. 90-99, 1984.

5.   Milinazzo, F.A., and Saffman, P.G., Finite-amplitude steady waves in plane viscous shear flows, J. Fluid Mech., Vol. 160, pp. 281-295, 1985.

6.   Murdock, J.W., A numerical study of nonlinear effects on boundary-layer stability, AIAA Journal, Vol. 15, No. 8, pp. 1167-1173, 1977.

A Galerkin Method for Wind Induced Current in a Two-layered Stratified Sea

K. T. JUNG and B. J. NOYE
Applied Mathematics Department
The University of Adelaide
G.P.O. Box 498
Adelaide 5001, South Australia

INTRODUCTION

This paper is a sequel to Jung and Noye (1988). It continues the development of a spatially three-dimensional layered model with an arbitrary variation of vertical eddy viscosity within each layer. Specifically, a two-layered hydrodynamic model is formulated. An essential feature of the method is that horizontal components of current are expanded through the vertical in terms of the fourth order B-spline functions together with time and horizontally varying coefficients. A system of equations is then derived using the Galerkin method from which the coefficients of the expansion are determined by using a standard type of finite difference method for the horizontal integration, along with appropriate time stepping methods. Once the coefficients are computed, a continuous current profile can be obtained through the depth. The discontinuity in the depth variation of density requires a definition of a weighted scalar product.

Variational approaches have been used previously in other disciplines to handle strong discontinuities in the coefficient of the second-order diffusivity operator. By using linear basis functions Javandel and Witherspoon (1969) have developed a variational model for fluid flow in anisotropic multilayered aquifers, and Desai and Johnson (1973) solved a one-dimensional consolidation equation with piecewise varying coefficients. Thermal interactions between the soil and the atmosphere were investigated by Garder and Raymond (1974) in a similar manner. Use of B-spline functions of order four $(n_o = 4)$ is advantageous over the above mentioned lower-order finite element methods because the B-spline functions have continuous derivatives up to n_o-2.

The method is applied to the computation of wind drift current in a two-layered open sea. The influence of the two-layered structure of the vertical eddy viscosity on the current profiles and the time variation of the surface current is examined.

GOVERNING EQUATIONS FOR A TWO-LAYER FLUID

The linearised σ-transformed equations of continuity and momentum governing the motion of a two-layer stratified sea of constant undisturbed depth h on a rotating frame may be written as follows (Jung and Noye, 1988):

$$\frac{\partial \zeta_1}{\partial t} - \frac{\partial \zeta_2}{\partial t} + h \left(\frac{\partial}{\partial x} \int_0^{\xi_1} u_1 d\sigma + \frac{\partial}{\partial y} \int_0^{\xi_1} v_1 d\sigma \right) = 0, \tag{1}$$

$$\frac{\partial \zeta_2}{\partial t} + h \left(\frac{\partial}{\partial x} \int_{\xi_1}^1 u_2 d\sigma + \frac{\partial}{\partial y} \int_{\xi_1}^1 v_2 d\sigma \right) = 0, \tag{2}$$

K. T. Jung has a post-graduate Scholarship from the University of Adelaide.

$$\frac{\partial u_1}{\partial t} - \gamma v_1 = -g\frac{\partial \zeta_1}{\partial x} + \frac{\partial}{h^2 \partial \sigma}\left(N_1\frac{\partial u_1}{\partial \sigma}\right), \tag{3}$$

$$\frac{\partial v_1}{\partial t} + \gamma u_1 = -g\frac{\partial \zeta_1}{\partial y} + \frac{\partial}{h^2 \partial \sigma}\left(N_1\frac{\partial v_1}{\partial \sigma}\right), \tag{4}$$

$$\frac{\partial u_2}{\partial t} - \gamma v_2 = -g\frac{\rho_1}{\rho_2}\frac{\partial \zeta_1}{\partial x} - g\left(\frac{\rho_2-\rho_1}{\rho_2}\right)\frac{\partial \zeta_2}{\partial x} + \frac{\partial}{h^2 \partial \sigma}\left(N_2\frac{\partial u_2}{\partial \sigma}\right), \tag{5}$$

$$\frac{\partial v_2}{\partial t} + \gamma u_2 = -g\frac{\rho_1}{\rho_2}\frac{\partial \zeta_1}{\partial y} - g\left(\frac{\rho_2-\rho_1}{\rho_2}\right)\frac{\partial \zeta_2}{\partial y} + \frac{\partial}{h^2 \partial \sigma}\left(N_2\frac{\partial v_2}{\partial \sigma}\right), \tag{6}$$

in which t denotes time; x, y, z are the Cartesian spatial coordinates, with z the depth below the undisturbed surface which contains the x and y axes; $\sigma = -z/h$; $\xi_1 = h_1/h$ where h_1 is the depth of the upper layer; ζ_1 and ζ_2 are the free surface elevation and the interfacial displacement, respectively, Also u_1, u_2 and v_1, v_2 are horizontal components of currents in the x, y directions, respectively; ρ_1, ρ_2 are water densities; N_1, N_2 are vertical eddy viscosity coefficients, with the subscripts 1,2 denoting values for the upper and lower layers, respectively; g is the acceleration due to gravity; and γ is the Coriolis parameter given by $\gamma = 2\omega_e \sin\phi_e$, where ω_e is the circular frequency of the earth's rotation and ϕ_e is the latitude which is considered positive in the Northern Hemisphere and negative in the Southern Hemisphere.

In order to solve the system of eqs. (1) to (6) for $\zeta_1, \zeta_2, u_1, u_2$ and v_1, v_2, boundary conditions have to be specified at the sea surface and at the sea bottom, together with suitable interfacial conditions. In the σ-transformed coordinate system these are given as follows:

at the undisturbed level of the sea surface

$$-\frac{\rho_1}{h}\left(N_1\frac{\partial u_1}{\partial \sigma}\right)_0 = \tau_{sx}, \qquad -\frac{\rho_1}{h}\left(N_1\frac{\partial v_1}{\partial \sigma}\right)_0 = \tau_{sy}; \tag{7}$$

at the undisturbed level of the interface

$$u_1(\xi_1) = u_2(\xi_1), \qquad v_1(\xi_1) = v_2(\xi_1), \tag{8}$$

$$-\frac{\rho_1}{h}\left(N_1\frac{\partial u_1}{\partial \sigma}\right)_{\xi_1} = -\frac{\rho_2}{h}\left(N_2\frac{\partial u_2}{\partial \sigma}\right)_{\xi_1}, \qquad -\frac{\rho_1}{h}\left(N_1\frac{\partial v_1}{\partial \sigma}\right)_{\xi_1} = -\frac{\rho_2}{h}\left(N_2\frac{\partial v_2}{\partial \sigma}\right)_{\xi_1}; \tag{9}$$

and at the sea bed a slip condition

$$-\frac{\rho_2}{h}\left(N_2\frac{\partial u_2}{\partial \sigma}\right)_1 = \tau_{bx}, \qquad -\frac{\rho_2}{h}\left(N_2\frac{\partial v_2}{\partial \sigma}\right)_1 = \tau_{by} \tag{10}$$

may be defined. Each of the derivatives is evaluated at the level of the σ coordinate denoted by the suffix. The stress terms τ_{sx}, τ_{sy} represent the x and y components of the surface wind stress and τ_{bx}, τ_{by} represent the corresponding components of the bottom stress.

GALERKIN SOLUTIONS USING B-SPLINES IN THE VERTICAL

Fig. 1 illustrates the basis sets of the fourth order B-splines defined through the vertical water column. The intervals $[0, 1]$ are subdivided into units of length $\Delta \nu_r = (\nu_{r+1} - \nu_r)$, by the partition

$$0 = \nu_0 < \cdots < \nu_r < \cdots < \nu_m = 1. \tag{11}$$

The knots, denoted by ν_r may have arbitrary separation. The r th B-spline of order four has non-zero positive values over the knot interval $\nu_{r-4} < \sigma < \nu_r$, with its values and derivatives vanishing at ν_{r-4} and ν_r. Two sets of three supporting knots are defined to complete the basis set outside of the domain, that is ν_{-3} to ν_{-1}, and ν_1 to ν_3. The knot positions are constrained to coincide with the undisturbed interface levels to avoid any undesirable oscillations in numerical results (Axelsson and Barker, 1984).

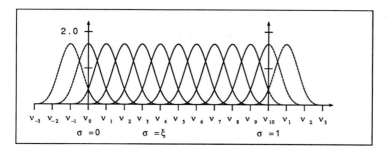

FIGURE 1. Configuration of the fourth-order B-spline functions for $m = 10$.

Galerkin solutions are sought with respect to the scalar product

$$\langle \phi, \psi \rangle = \int_0^{\xi_1} \phi_1 \cdot \psi_1 \, d\sigma + \frac{\rho_2}{\rho_1} \int_{\xi_1}^1 \phi_2 \cdot \psi_2 \, d\sigma , \tag{12}$$

where ϕ_j, ψ_j is a pair of square integrable functions defined over the jth layer.

For simplicity, it is assumed that the vertical variation of eddy viscosity has a fixed structure over the domain, that is,

$$N = N_j(x, y, \sigma, t) = \alpha(x, y, t)\mu_j(\sigma) \qquad \text{for} \quad \xi_{j-1} \leq \sigma \leq \xi_j,$$

where $j = 1, 2$.

$$\tag{13}$$

The two components of horizontal current, U and V, at a depth σ, are expanded using m basis functions of B-splines, so that

$$U(x, y, \sigma, t) = \sum_{r=1}^m A_r(x, y, t) \, M_r(\sigma) , \tag{14}$$

$$V(x, y, \sigma, t) = \sum_{r=1}^m B_r(x, y, t) \, M_r(\sigma) , \tag{15}$$

where the coefficients A_r and B_r are to be determined.

Taking the scalar product (4) of (5) with M_k gives

$$\frac{\partial \langle U_k \rangle}{\partial t} - \gamma \langle V_k \rangle = - \sum_{\ell=1}^2 g\left(\frac{\rho_\ell - \rho_{\ell-1}}{\rho_1} \right) \frac{\partial \zeta_{\ell-1}}{\partial x} \, a_{\ell-1,k} + I_k, \tag{16}$$

in which

$$\langle U_k \rangle = \langle U, M_{i,k} \rangle = \sum_{j=1}^2 \left(\frac{\rho_j}{\rho_1} \int_{\xi_{j-1}}^{\xi_j} U_j \, M_k \, d\sigma \right) , \tag{17}$$

$$\langle V_k \rangle = \langle V, M_k \rangle = \sum_{j=1}^2 \left(\frac{\rho_j}{\rho_1} \int_{\xi_{j-1}}^{\xi_j} V_j \, M_k \, d\sigma \right) , \tag{18}$$

$$a_{\ell,k} = \sum_{j=\ell+1}^2 \int_{\xi_{j-1}}^{\xi_j} M_k \, d\sigma = \int_{\xi_\ell}^1 M_k \, d\sigma, \tag{19}$$

$$I_k = \frac{\alpha}{H^2} \sum_{j=1}^2 \left\{ \frac{\rho_j}{\rho_1} \int_{\xi_{j-1}}^{\xi_j} \frac{\partial}{\partial \sigma} \left(\mu_j \frac{\partial U_j}{\partial \sigma} \right) M_k \, d\sigma \right\}, \qquad k = 1, \cdots, m. \tag{20}$$

For convenience, $\rho_0 = 0$ is assumed in (16).

Integrating (20) by parts, incorporating the sea surface and bottom boundary conditions (7) and (10), and the interfacial condition (9), and substituting (13) and (14) into the resulting equations, gives

$$\sum_{r=1}^{m} \frac{\partial A_r}{\partial t} \int_0^1 M_r M_k \, d\sigma \; - \; \gamma \sum_{r=1}^{m} B_r \int_0^1 M_r M_k \, d\sigma$$

$$= -\sum_{\ell=1}^{2} g\left(\frac{\rho_\ell - \rho_{\ell-1}}{\rho_1}\right) \frac{\partial \zeta_{\ell-1}}{\partial x} a_{\ell-1,k} - \frac{\tau_{bx}}{\rho_2 H} M_k(1) + \frac{\tau_{sx}}{\rho_1 H} M_k(0) \qquad (21)$$

$$- \frac{\alpha}{H^2} \sum_{r=1}^{m} A_r \sum_{\ell=1}^{2} \left(\frac{\rho_\ell}{\rho_1} \int_{\xi_{\ell-1}}^{\xi_\ell} \mu_\ell \frac{dM_r}{d\sigma} \frac{dM_k}{d\sigma} d\sigma\right), \qquad k = 1, \cdots, m.$$

Similarly,

$$\sum_{r=1}^{m} \frac{\partial B_r}{\partial t} \int_0^1 M_r M_k \, d\sigma \; + \; \gamma \sum_{r=1}^{m} A_r \int_0^1 M_r M_k \, d\sigma$$

$$= -\sum_{\ell=1}^{2} g\left(\frac{\rho_\ell - \rho_{\ell-1}}{\rho_1}\right) \frac{\partial \zeta_{\ell-1}}{\partial y} a_{\ell-1,k} - \frac{\tau_{by}}{\rho_2 H} M_k(1) + \frac{\tau_{sy}}{\rho_1 H} M_k(0) \qquad (22)$$

$$- \frac{\alpha}{H^2} \sum_{r=1}^{m} B_r \sum_{\ell=1}^{2} \left(\frac{\rho_\ell}{\rho_1} \int_{\xi_{\ell-1}}^{\xi_\ell} \mu_\ell \frac{dM_r}{d\sigma} \frac{dM_k}{d\sigma} d\sigma\right), \qquad k = 1, \cdots, m,$$

where

$$\tau_{bx} = \rho_2 k_b \sum_{r=1}^{m} A_{i,r} M_{i,r}(1), \qquad \tau_{by} = \rho_2 k_b \sum_{r=1}^{m} B_{i,r} M_{i,r}(1). \qquad (23)$$

Substituting (14) and (15) into the equation of continuity (9), and rearranging the set of resultant equations, yields,

$$\frac{\partial \zeta_j}{\partial t} + \sum_{r=1}^{m} H\left(\frac{\partial A_r}{\partial x} + \frac{\partial B_r}{\partial y}\right) a_{j,r} = 0, \qquad (24)$$

where $j = 0, 1$. Once the coefficients of B-spline expansions are computed from (21) to (24), along with the appropriate initial conditions and lateral boundary conditions, the two components of horizontal current at a depth σ are obtained from (13) and (14).

WIND DRIFT CURRENTS IN A TWO-LAYERED OPEN SEA

We apply the Galerkin models to wind drift currents in two-layered horizontally unbounded seas induced by a step-function wind stress. The two-layered structure and the vertical eddy viscosity are assumed to be independent of the horizontal coordinates. Both surface and interfacial displacement are suppressed. The thickness of each layer is kept constant with respect to time.

Fig. 2 shows the variation of density and eddy viscosity considered in this study. Although the models described in Section 3 allow for an arbitrary variation of eddy viscosity within each layer, in the interest of demonstrating the accuracy of the models the depth variation of eddy viscosity is prescribed in a piecewise-linear manner as sketched in Figure 2b. The Coriolis parameter is taken as $\gamma = -0.9178 \times 10^{-4} \, s^{-1}$, representative of Bass Strait, Australia, at latitude 39° S.

Integration on the staggered grid system leads to the introduction of periodic boundary conditions for all t. The size of the time steps Δt can be chosen in a flexible manner since

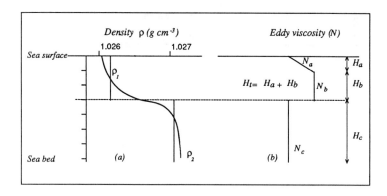

FIGURE 2. A schematic variation of (a) density and (b) eddy viscosity through the vertical.

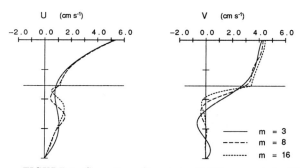

FIGURE 3. Current profiles at $t = 21/4T_c$ computed with an increasing number of B-spline functions. Computed using a step-function wind stress ($\tau_{sx} = 1$ dyne cm^{-2}) with $N_a = N_b = 300$, $N_c = 10$ $cm^2 s^{-1}$ $H_t = 25$, $H_c = 40m$.

the propagation of the fast-moving surface waves is suppressed. The Courant-Friedrich-Lewy condition is no longer a restriction. However, to ensure high accuracy of solutions a time step of 9 seconds has been chosen.

It is evident from Fig. 3 that as the number of B-splines are increased, the representation of the high shear region, in the proximity of the interface, is improved for each component of current. For a given number of B-spline functions, a high concentration of knots across the interface is important in determining current profiles accurately. If an insufficient number of B-splines are used with a uniform distribution of knots, regions of high shear are obviously smoothed out.

It is also seen from Fig. 4 that the current profiles in stratified conditions are characterised by the presence of high shear in the proximity of the surface layer. The continuity require-ments of the shear stresses force the velocity to change abruptly at $z = -h\xi_1$, the degree of the velocity change depending upon the ratio $\rho_1\mu_1/(\rho_2\mu_2)$. The shear within the surface layer is sustained predominantly in the U component of the current which is parallel to the direction of the wind stress at the sea surface. Current profiles in the surface layer computed with $H_a = 40m$ is similar to those of homogeneous seas.

The top graph in Fig. 5 shows the sensitivity of the U component of the surface current to

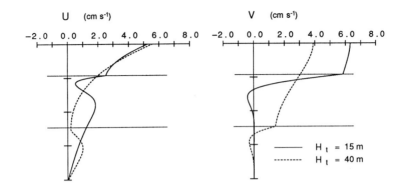

FIGURE 4. Current profiles at $t = 21/4T_c$ computed using a step-function wind stress $(\tau_{sx} = 1\ dyne\ cm^{-2})$ with $H_t = 15$ and $40m$, using $h = 65m$, $N_a = N_b = 300$, $N_c = 10\ cm^2 s^{-1}$, $k_b = 0.2\ cms^{-1}$.

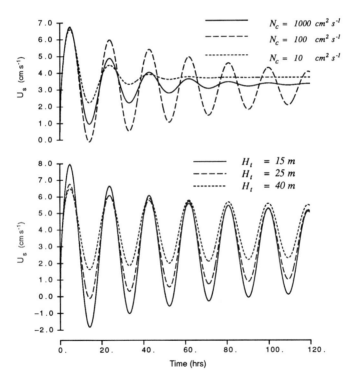

FIGURE 5. Time variations of u_s and v_s induced by a step-function wind stress $(\tau_{sx} = 1\ dyne\ cm^{-2})$. (a) Computed with $k_b = 0.2\ cms^{-1}$, $H_t = 25$, $H_b = 40m$, $N_t = 300$ and $N_c = 10$; 100; $1000\ cm^2s^{-1}$. (b) Computed with $k_b = 0.2\ cms^{-1}$, $N_a = N_b = 300$ and $N_c = 10\ cm^2s^{-1}$. $H_t = 10$; 25; $40m$.

266

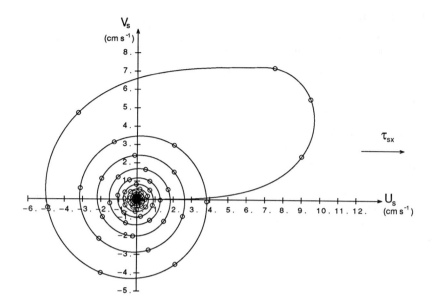

FIGURE 6. Hodograph of u_s and v_s (with o denoting three-hourly values in the one domain system) induced by a constant wind pulse with $\tau_{sx} = 1$ $dyne\ cm^{-2}$ and a duration of $T_c/4$ hours. Computed with $k_b = 0.2$ cms^{-1}, $N_a = 60$, $N_b = 300$ and $N_c = 10$ $cm^2 s^{-1}$. $H_a = 10$, $H_b = 15$, $H_c = 40m$.

changes in N_c. It is apparent that the values of N_c and H_t are important in determining the rate of damping of the inertial oscillations in the surface currents. When N_c is increased the wind momentum can more rapidly penetrate to the bottom layer. This leads to an enhanced initial damping of the surface current. It is evident that wind drift surface currents are characterised by an oscillation of inertial period $T_c = 2\pi/\gamma \simeq 18.2$ hours. The bottom graph in Fig. 5 shows that the rate of damping of the surface current is substantially influenced by the surface mixed layer depth. The initial decay of the surface current will last until the system reaches a steady state. For details on the wind induced inertial oscillation of the one domain system, see Davies (1985).

The time variation of the U and V components of the surface current computed with a surface wall layer are displayed as an hodograph in Fig. 6. The wind stress is maintained for $T_c/4$ hours. Initially, the U surface current intensifies in the direction of the wind stress and then, due to the rotational effects, a V component of surface current is developed. Since the wind's energy is supplied for a finite period, the steady state solution is zero, hence the surface current rapidly changes its direction and the current oscillates inertially around zero after the wind stops. The present approach enables one to introduce a surface wall layer in a discontinuous manner.

CONCLUSIONS

A numerical method for modelling wind induced motion in a two-layered stably stratified sea in which an arbitary vertical variation of vertical eddy viscosity may be used has been developed. The ability of the present method to produce the surface current and current profiles accurately offers significant advantages over the eigenfunction method (Jung and Noye, 1988). Also, there is no evidence of numerical oscillations when a sufficient number of B-spline functions are used with a distribution of knots concentrated at the interface.

267

REFERENCES

1. Axelsson, O., and V.A. Barker, *Finite Element Solution Of Boundary Value Problems: Theory And Computation.* Academic press, 1984.

2. Davies, A.M., Application of a sigma coordinate sea model to the calculation of wind-induced currents. *Cont. Shelf Res.*, vol. 4, pp. 389–423, 1985.

3. Desai, C.S., and Johnson, L.D., Evaluation of some numerical schemes for consolidation. *Int. J. Num. Methods Eng.,* vol. 7, pp. 243–254, 1973.

4. Garder, A., and Raymond, W.H., Galerkin approximations to vertical temperature profiles across the earth's surface. *Monthly Weather Rev.*, vol. 102, pp. 426–432, 1974.

5. Javandel, I., and Witherspoon, P.A., A method of analyzing transient fluid flow in multilayered aquifers. *Water Resour. Res.*, vol. 5, pp. 856–869, 1969.

7. Jung, K., and Noye, J., A spectral method of determining depth variations of currents in a two-layer stratified sea. *Computational Techniques and Applications: CTAC 87*, ed. J. Noye, pp. 333–346, North-Holland, 1988.

A Comparative Study of Analytical and Numerical Solutions for Water Movement in Unsaturated Porous Materials

K. K. WATSON and V. SARDANA
School of Civil Engineering
The University of New South Wales
Kensington, NSW, 2033, Australia

G. C. SANDER, W. L. HOGARTH, and I. F. CUNNING
Division of Australian Environmental Studies
Griffith University
Nathan, Qld, 4111, Australia

J.-Y. PARLANGE
Department of Agricultural Engineering
Cornell University
Ithaca, New York 14853, USA

INTRODUCTION

Exact solutions of the nonlinear transport equation describing single phase water movement in an unsaturated porous material under constant flux infiltration have recently been presented by Sander et al.(1988) and Broadbridge and White (1988). Both papers make use of a diffusivity function of a form first proposed by Fujita (1952) and both have followed the general solution approach detailed by Rogers et al.(1983) although this solution was initially applied to oil and water infiltration. An attractive feature of the solutions is the manner in which realistic soil water characteristics can be described by the specified diffusivity and conductivity functions. At this time there is no exact solution available for constant concentration infiltration although a quasi-analytical solution (Philip, 1969) has been available for many years. However, for comparative purposes the solution of Fujita (1952) for the zero gravity case of constant concentration absorption is available and forms a suitable benchmark solution for a comparative study of the constant concentration boundary condition with its attendant very steep pressure head gradients at early times. Sander et al.(1989) have also developed an exact solution for the nonhysteretic redistribution of water following the development of a prescribed infiltration profile. Detailed studies comparing the exact solutions with numerical solutions using finite difference approximations of the flow equation have been given by Watson et al.(1989a) for the constant flux infiltration and the constant concentration absorption cases whilst the nonhysteretic redistribution case has been described by Watson et al.(1989b).

In this paper the detailed results given by Watson et al. (1989a, 1989b) are summarized in order to give a 'global' view of the accuracy of the numerical procedures and the effect of grid spacing on the accuracy at early times. The next section of this paper summarizes the numerical approach used and this is followed by an outline of the significant equations relating to the three analytical solutions. The paper concludes with the presentation and discussion of the comparative results.

NUMERICAL APPROACH

The movement of water in a rigid, homogeneous one-dimensional vertical system under isothermal conditions may be written (see e.g. Philip, 1969)

$$\partial\theta/\partial t = \partial[K(\partial h/\partial z)]/\partial z + \partial K/\partial z \tag{1}$$

where θ is the volumetric water content, h the soil water pressure head, K the hydraulic conductivity, z the vertical ordinate positive upward and t the time. Equation 1, which follows directly from the combination for an unsaturated flow system of Darcy's law and the equation of continuity, is usually rewritten to eliminate either h or θ. In the former case the resulting equation is

$$\partial\theta/\partial t = \partial[D(\theta)\partial\theta/\partial z]/\partial z + \partial K(\theta)/\partial z \tag{2}$$

where the diffusivity function

$$D(\theta) = K(\theta) \ dh/d\theta \tag{3}$$

The alternative pressure head form may be written

$$C(h)\partial h/\partial t = \partial[K(h)\partial h/\partial z]/\partial z + \partial K(h)/\partial z \tag{4}$$

where the specific water capacity

$$C(h) = d\theta/dh \tag{5}$$

The form of the flow equation given in eq. 2 is that used in analytical and quasi-analytical studies whilst that given in eq. 4 is preferred for numerical simulations because of the modelling flexibility afforded by such a potential form.

Whisler and Watson (1968) described a computer-based numerical solution of eq. 4 which used an implicit finite difference approach in the analysis. The computer program has been continually updated and, in its present form, is able to handle all types of constant and time-dependent boundary conditions, material variations including spatial heterogenity, and intermittency in the application of water. This latter facility is necessary for the simulation of naturally-occurring environmental systems involving time-dependent water inputs at the soil surface.

Two basic parametric relationships describing the hydrologic characteristics of unsaturated porous materials are required as input for the numerical analysis. These are the $h(\theta)$ and $K(\theta)$ relationships. The former relationship exhibits marked hysteresis during the wetting-draining cycle requiring, for systems involving the intermittency of water application, the inclusion of a hysteresis model in the dataset. Although such hysteretic datasets are available in the numerical package for a range of materials, the use of these is not necessary in the present study as the exact redistribution solution presented relates solely to an assumed nonhysteretic material.

EXACT SOLUTIONS

a) Constant Flux Infiltration

Although eq. 2 has been written in terms of z positive upwards in order to be consistent with the sign convention of the numerical approach, most available analytical solutions use the reverse convention with z positive downwards; this practice will be followed in this section. Equation 2 becomes

$$\partial\theta/\partial t = \partial[D(\theta)\partial\theta/\partial z]/\partial z - \partial K(\theta)/\partial z \tag{6}$$

It is convenient to introduce the reduced variables

$$\Theta = (\theta - \theta_i)/(\theta_{sat} - \theta_i) \qquad \text{and} \tag{7}$$

$$K(\Theta) = [K(\theta) - K(\theta_i)]/(\theta_{sat} - \theta_i) \tag{8}$$

where θ_i and θ_{sat} are the initial and saturated water contents respectively. The Fujita diffusivity function may be written

$$D(\Theta) = D_o/(1 - \upsilon\Theta)^2 \tag{9}$$

where D_o and υ are constants determined from the soil properties. From Rogers et al.(1983) the most general $K(\Theta)$ which allows the linearization of the system is

$$K(\Theta) = (K_1 + K_2\Theta + K_3\Theta^2)/(1 - \upsilon\Theta) \tag{10}$$

With $K(0) = 0$ and $K(1) = K_{sat}$ it can be shown that $K_1 = 0$ and $K_2 = K_{sat}(1 - \upsilon) - K_3$

Equation 6 is solved subject to the conditions

$$t = 0, \ z \geq 0, \ \Theta = 0 \tag{11a}$$
$$t > 0, \ z = 0, \ -D(\Theta)(\partial\Theta/\partial z) + K(\Theta) = Q \tag{11b}$$

where Q is the reduced constant surface flux given by

271

$$Q = [q - K(\theta_i)]/(\theta_{sat} - \theta_i) \tag{12}$$

where q is the actual constant surface flux.

Dimensionless variables X and τ may then be defined as

$$X = Qz/D_O \; ; \quad \tau = Q^2 t/D_O \tag{13}$$

Space precludes the detailing of the set of solution equations (see Sander et al. (1988), equations 10a, 10b and 10c). However, these equations give the water content profile $\Theta(X, \tau)$ through an intermediate parameter ξ and the relationships $\Theta(\xi,\tau)$ and $X(\xi,\tau)$.

A range of solutions is available depending on the conductivity function chosen which in turn depends on the values specified for K_2 and K_3. If $K_3 = 0$ and $K_2 = K_{sat}(1 - \upsilon)$ then

$$K(\Theta) = [K_{sat}(1-\upsilon)\Theta]/(1 - \upsilon\Theta) \quad \text{and} \tag{14}$$
$$K(h) = K_{sat} \exp[\alpha(h - h_{WEV})] \tag{15}$$
$$\text{where} \quad \alpha = [K_{sat}(1-\upsilon)]/D_O \tag{16}$$

and h_{WEV} is the water entry value of the porous material.

For comparison purposes the conductivity relationship given by eq. 14 has been used in the numerical simulation. Equation 15 gives the required input data for $K(h)$ while the specific water capacity is obtained from

$$C(h) = d\Theta/dh = [K_{sat}(1-\upsilon)\Theta](1-\upsilon\,\Theta)/D_O \tag{17}$$

This can be integrated readily to give $h(\Theta)$, the remaining data input component for the numerical analysis.

b) Constant Concentration Absorption.

For horizontal absorption the flow equation becomes in terms of reduced variables

$$\partial\Theta/\partial t = \partial[D(\Theta)\partial\Theta/\partial x]/\partial x \tag{18}$$

where x is the spatial variable. The exact solution for this equation was presented by Fujita (1952) using $D(\Theta)$ as given by eq. 9 with $\upsilon \leq 1$. The initial and boundary conditions may be written

$$t = 0, \; z>0, \; \Theta = 0 \; ; \quad t \geq 0, \; z=0, \; \Theta = \Theta_o \tag{19}$$

If $h = 0$ at $x = 0$ then $\theta_o = \theta_{sat}$ and $\Theta_o = 1$. The absence of the gravity term in eq. 18 results in eq. 9 being a sufficient specification of the soil properties. This also gives flexibility in the selection of the $K(\Theta)$ and $h(\Theta)$ relationships for the numerical simulation since the requirement of a particular $D(\Theta)$ value can be met by a variety of $K(\theta) - h(\theta)$ combinations providing $K(\Theta)dh/d\Theta$ for a particular Θ value equals $D(\Theta)$. The solution equations given below are those presented by Sander et al. (1986), namely

$$\upsilon\Theta \;\; = 1 - (1 - \upsilon\Theta_o)/\{1 - \upsilon\Theta_o + 2^{-1}\upsilon s'\pi^{1/2}\exp\ (\upsilon^2 s'^2/4)$$
$$\operatorname{erfc}[2^{-1}\upsilon(s'^2 - 4\upsilon^{-2}\ln\ \omega)^{1/2}]\} \tag{20}$$

$$x(D_o t)^{-1/2} = \upsilon(s'^2 - 4\upsilon^{-2}\ln\ \omega)^{1/2}/(1 - \upsilon\Theta) - \upsilon s'\omega/(1 - \upsilon\Theta_o) \tag{21}$$

where $S' = S\ D_o^{-1/2}$ with S being the Sorptivity and

$$S'\ \operatorname{erfc}\ (\upsilon S'/2).\exp\ (\upsilon^2 s'^2/4)\ =\ 2\Theta_o/\pi^{1/2} \tag{22}$$

In eqs. 20 and 21 ω is an intermediate parameter with the range $0 \leq \omega \leq 1$. By varying ω from 0 to 1 in eqs. 20 and 21 the $\Theta(x,t)$ profiles are obtained.

c) Nonhysteretic Redistribution

The details of this case have been presented by Sander et al. (1989). The flow equation is again given by eq. 6 with the $D(\Theta)$ and $K(\Theta)$ functions being in the form of eqs. 9 & 14.

The initial and boundary conditions are

$$t = 0,\ \Theta = \Theta_i(z) \tag{23a}$$
$$t > 0,\ z = 0,\ -D(\Theta)(\partial\Theta/\partial z) + K(\Theta) = 0 \tag{23b}$$
$$t > 0,\ z = L,\ -D(\Theta)(\partial\Theta/\partial z) + K(\Theta) = 0 \tag{23c}$$

where L is the length of the column and $\Theta_i(z)$ is prescribed in form. The development of the exact solution follows the same general procedure as that used in the constant flux solution with an intermediate parameter again appearing in the solution equations. Sander et al. (1989) also present the equation for the equilibrium water content profile developed at long times following the redistribution process in the finite length column.

In the numerical analysis the soil properties are handled in a similar manner to that described previously. Since the exact solution relates only to nonhysteretic redistribution the inclusion of a domain model of hysteresis in the dataset becomes unnecessary with the wetting and draining sequence at any point being assumed to 'follow' the same $h(\theta)$ relationship. From a numerical perspective the significance of this particular case relates to the scrutiny it provides of the automatic decision-making process built into the 'hysteresis' subroutine whereby the reversal mode at a given profile elevation is initiated.

COMPARISON OF RESULTS

Space limitations allow only a small part of the comprehensive comparisons that have been carried out to be included here. These covered not only the range of times over which the flow process occurred but also the effect of grid spacing on the accuracy achieved. The parameter used for comparative purposes is the actual volumetric water content θ. In the following tables the values of θ have been listed to six decimal places. It is recognized that this is at least two orders of magnitude more accurate than the requirements of normal usage; however, this level of accuracy does allow the precise departures from the exact theoretical values to be established.

a) Constant Flux Infiltration.

The values used in the numerical simulation were as follows: θ_{sat} = 0.35 cm^{3}cm^{-3}, θ_i = 0.06 cm^{3}cm^{-3}, υ = 0.85, K_{sat}/q = 1.25, K_{sat} = 0.1 cm min^{-1} and D_O = 2.75862 cm^2 min^{-1}. This particular D_O value allowed the dimensionless variable X to equal $|z|/10$ thus obviating the necessity for interpolation in obtaining specific comparable values from the listed outputs. The comparisons for constant flux infiltration at selected elevations for t = 0.3625 min (τ = 0.01) and t = 36.2500 min (τ = 1.0) and for two Δz values are given in Table 1. The excellent correspondence is immediately apparent. There is also no 'drift' from the agreement indicated in Table 1 at intermediate depths. For t = 0.3625 min a simulation with the coarser grid spacing of 0.5cm gave θ values of 0.08996 and 0.07152 at the surface and z = -1.0 cm respectively, indicating the need of a fine grid spacing at early times if high-order accuracy is to be achieved.

To check the accuracy of time-to-ponding calculations (i.e. h = 0 at the surface) a case was run with K_{sat}/q = 0.80. At ponding, the wet front was located at approximately z = -150 cm. Accordingly, a profile of 180 cm was used for the simulation with Δz = -0.2 cm. The theoretical time-to-ponding was 212.341 min

TABLE 1. Volumetric Water Content θ (cm^{3}cm^{-3})

| | | t = 0.3625 min | | | t = 36.2500 min |
| | | Numerical | Numerical | | Numerical |
z (cm)	Exact	Δz = -0.01cm	Δz = -0.1cm	Exact	Δz = -0.1cm
0	0.090652	0.090652	0.090640	0.246262	0.246261
-0.2	0.086118	0.086119	0.086107		
-0.4	0.081947	0.081948	0.081937		
-0.6	0.078176	0.078176	0.078168		
-0.8	0.074826	0.074827	0.074822		
-1.0	0.071908	0.071909	0.071909		
-2.0	0.063125		0.063135	0.237000	0.237000
-6.0				0.215433	0.215433
-10.0				0.189694	0.189693
-20.0				0.117019	0.117019
-30.0				0.072257	0.072259
-40.0				0.060357	0.060435

TABLE 2.   Volumetric Water Content θ (cm^3cm^{-3})

z (cm)	Exact	Numerical
0	0.350 000	0.350 000
-1.0	0.347 307	0.347 309
-2.0	0.344 318	0.344 321
-5.0	0.333 025	0.333 028
-10.0	0.300 692	0.300 699
-15.0	0.229 094	0.229 111
-20.0	0.107 894	0.107 914
-21.8	0.080 148	0.080 162

with the numerical result determined by linear interpolation from
closely spaced adjacent time steps being 212.326 min.   As would
be expected the θ data again showed excellent correspondence in
this simulation.

b)   Constant Concentration Absorption

The main problem in simulating a constant concentration case
numerically, particularly where high accuracy is required at
early times, is the steepness of the pressure head gradient
immediately below the surface.   With a fine grid spacing (say
0.01cm), which is necessary for early-time accuracy, and an
initial uniform pressure head in the profile of say -1000 cm it
is apparent that across the first node (for h = 0 at the surface)
the pressure head gradient is 100,000.   A gradient of this
magnitude causes stability difficulties.

For this horizontal flow case the same soil data was used as
previously with Δz = -0.04cm and L = 36cm.   The comparisons are
given in Table 2 for t = 7.0 min

As with the previous case the comparisons are again very good.
Satisfactory simulations were also carried out where the constant
surface water content was prescribed as a value less then θ_{sat}.

c)   Nonhysteretic Redistribution

The data for this case is given in Table 3.   The close
correspondence indicates the satisfactory operation both of the
'automatic' reversal techniques written into the numerical

TABLE 3.   Volumetric Water Content θ(cm^3cm^{-3})

z (cm)	t = 7.380 min Exact	t = 7.380 min Numerical	t = 147.60 min Exact	t = 147.60 min Numerical	Equilibrium Exact	Equilibrium Numerical
0	0.277548	0.277550	0.156344	0.156344	0.099982	0.099970
-10	0.281332	0.281334				
-20	0.271558	0.271559	0.223081	0.223082	0.112425	0.112414
-40	0.142170	0.142168			0.127556	0.127551
-50	0.061306	0.061309				
-60			0.089954	0.089955	0.145443	0.145447
-80			0.058880	0.058881*	0.165895	0.165906
-100					0.188405	0.188419

* These values relate to the z = -76 cm elevation.

program and the lower boundary specification for the zero flux condition. The data used in the simulation were $\Delta z = -0.2$cm, θ_i = 0.057cm^{3}cm^{-3} and L = 100cm. The larger Δz is possible because of the considerably smaller potential gradients that occur during redistribution.

CONCLUSION

The comparisons presented above confirm both the accuracy and flexibility of the numerical package. The correspondence between the exact and numerical solutions for the three cases considered is excellent and emphasizes not only the accuracy of the iterative finite difference approach adopted but also the adequacy of the logic and execution of the reversal procedures used in the nonhysteretic redistribution case. On the basis of the results of this study the numerical package can be used with confidence in modelling more complicated systems for which exact solutions are not available.

REFERENCES

1. Broadbridge, P., and White, I., Constant rate rainfall infiltration: A versatile nonlinear model 1. Analytic solution, *Water Resour. Res.*, vol. 24, no.1, pp.145-154, 1988.

2. Fujita, H., The exact pattern of a concentration-dependent diffusion in a semi-infinite medium, 2, *Textile Res.*, vol.22, pp.823-827, 1952.

3. Philip, J.R., Theory of infiltration, *Adv. Hydrosci.*, vol.5, pp.215-296, 1969.

4. Rogers, C., Stallybrass, M.P., and Clements, D.L., On two phase filtration under gravity and with boundary infiltration: Application of a Bäklund transformation, *Nonlinear Anal. Theory Methods Appl.*, vol.7, pp.785-799, 1983.

5. Sander, G.C., Kühnel, V., Brandyk, T., Dooge, J.C.I., and Kane, J.P.J., Analytical solutions of the soil moisture flow equation, Civ. Eng. Dep., Univ. College Dublin, EEC/NBST Proj. CLI-038-EIR (H) Rep. no.3, 1986.

6. Sander, G.C., Parlange, J.-Y., Kühnel, V., Hogarth, W.L., Lockington, D., and O'Kane, J.P.J., Exact nonlinear solution for constant flux infiltration, *J. Hydrol.* vol.97, pp.341-346, 1988.

7. Sander, G.C., Cunning, I.F., Hogarth, W.L., and Parlange, J.-Y., Exact solution of nonlinear, nonhysteretic redistribution in a vertical soil column of finite length. Submitted for publication, 1989.

8. Watson, K.K., Sardana, V., and Sander G.C., Comparative study of exact and numerical solutions for constant flux infiltration and constant concentration absorption. To be submitted for publication, 1989.

9. Watson, K.K., Sander, G.C., Sardana, V., Hogarth, W.L., and Cunning, I.F., Nonhysteretic redistribution: A comparison of exact and numerical solutions. To be submitted for publication, 1989.

10. Whisler, F.D. and Watson, K.K., One-dimensional gravity drainage of uniform columns of porous materials, *J.Hydrol.*, vol.6, pp.277-296, 1968.

A Comparison of Different Techniques in the Simulation of Elasto-Hydrodynamic Point Contact Problems

Z. R. XU and A. K. TIEU
Department of Mechanical Engineering
The University of Wollongong
Wollongong, 2500, Australia

INTRODUCTION

For the highly stressed contact machine components such as gears, bearings and cams, designers must consider the demand of minimum film thickness to keep the machine runing smoothly and efficiently. In the past few decades, many researches in elastohydrodynamic lubrication problems have been carried out, but because of the immense complexity and difficulty of EHL point contact problems a majority of them were restricted to line contact. Russian scientists Ertel and Grubin were considered the first to solve the EHL line contact problems in 1949. The breakthrough in EHL point contact problem was made by Ranger et al (1975) and Hamrock and Dowson (1976) who developed regression formulae for point contacts. Ranger et al derived doubtful results that the film thickness would be increased with loads. Evans and Snidle (1982) developed inverse method to solve EHL point contact problems under the condition of heavy load. Evans and Snidle (1981) obtained a number of solutions in which the largest difference reached 27% when compared with the solutions from Hamrock and Dowson regression formulae. The purpose of this work is to apply the direct iterative method on two different techniques to get some fundamental solutions in EHL point contact problems and to compare their results.

NOTATION

a	Hertzian contacting radius
D	deformation matrix
E	elastic modulus or composite elastic modulus $2/E=(1-\mu_1^2)/E_1+ (1-\mu_2^2)/E_2$
F	applied normal load
G	dimensionless materials parameter $E\alpha$
h	film thickness, h_m= minimum film thickness; h_c= central point film thickness;
p	hydrodynamic pressure
R	reduced radius, $1/R= 1/R_1+1/R_2$
s	seperation of undeformed spheres
u	mean tangential speed parameter, $u=(u_1+u_2)/2$
U	dimensionless speed parameter, $U=u\,\eta_0/ER$
W	dimensionless load parameter, $W=F/ER^2$; local deflection
α	pressure- viscosity coefficient
η	lubricant viscosity, η_0=viscosity at zero pressure
μ	Poisson's ratio, or dimensionless parameter in eq.(9) $\mu=\rho/\eta$
ρ	lubricant density
δ	local deformation

BASIC EQUATIONS

The elasto-hydrodynamic problem is governed by the Reynolds equation and elastic deformation equation. These two equations are coupled and can not be solved independently, because the solution of the Reynolds equation depends on film thickness which is modified by elastic deformation. On the other hand, the elastic deformation is determined by pressure distribution which in turn is generated by the convergent film gap. In addition, in EHL contact problems the lubricant density and viscosity in the Reynolds equation are functions of pressure distribution.

Elastic deformation. When two elastic spheres contact together, the problem can be treated as an equivalent sphere and plane in contact. In order to evaluate the deformation from the arbitrary pressure in the contact area, a discrete numerical method developed by Hamrock and Dowson (1976) was used in this paper. For point (k,l) the elastic deformation from distributed force over the whole contacting area can be expressed as:

$$\delta_{k,l} = \frac{2}{\pi} \sum_{i=1}^{imax} \sum_{j=1}^{jmax} P_{i,j} \, D_{k,l,i,j} \tag{1}$$

Where D is a four dimensional array, for any element it means that the deformation contribution of point (k,l) from unit force acting on point (i,j) and of course k,l cover the whole contacting area.

Composition of film shape. For any point in the contact area, the film thickness is composed from three components:

$$h(x,y) = s(x,y) + \delta(x,y) + h_0 \tag{2}$$

Viscosity. The relationship between pressure and viscosity can be described by different expressions. For simplicity one parameter exponential relationship was used in this paper:

$$\eta = \eta_0 e^{\alpha p} \tag{3}$$

Density. According to Dowson and Higginson (1971), the lubricant density can be taken to be given the following relationship:

$$\rho = \rho_0 \left(1 + \frac{1+ 0.6p}{1+ 1.7p} \right) \tag{4}$$

Reynolds equation. Generally, the hydrodynamic pressure is governed by the Reynolds equation:

$$\frac{\partial}{\partial x}\left(\frac{\rho h^3}{\eta} \frac{\partial p}{\partial x} \right) + \frac{\partial}{\partial y}\left(\frac{\rho h^3}{\eta} \frac{\partial p}{\partial y} \right) = 12\, u\, \frac{\partial}{\partial x}(\rho h) \tag{5}$$

In elastohydrodynamic point contact problems, the variation of pressure and gradient of pressure distribution are very large, so that the reduced pressure is often used to substitute the pressure p in the Reynolds equation. Usually, two different substitutions are applied: one is reduced pressure q, and another is by Φ substitution.

Reynolds equation with reduced pressure q. The reduced pressure q is defined as:

$$q = \eta_0 \int_0^p \frac{d\omega}{\eta(\omega)} \tag{6}$$

Substitute q into the Reynolds equation (5), the Reynolds equation can be derived:

$$\frac{\partial}{\partial x}\left(\rho h^3 \frac{\partial q}{\partial x}\right) + \frac{\partial}{\partial y}\left(\rho h^3 \frac{\partial q}{\partial y}\right) = 12\, u\, \eta_0\, \frac{\partial}{\partial x}(\rho h) \qquad (7)$$

If we consider q as pressure, this equation represents an isoviscous hydrodynamic problem. When the pressure becomes high, q will approach $1/\alpha$ which is the upper limit. With the substitution q, the unstable problem in the contact area could be improved.

Reynolds equation with Φ substitution. To improve the convergent problem in contacting area, another substitution can be used: $\qquad \Phi = P\, H^{3/2}$
$\qquad\qquad\qquad (8)$

Substitute eq. (8) into eq.(5), the following form of Reynolds equation can be derived:

$$H^{3/2}\left[\frac{\partial}{\partial x}\left(\mu \frac{\partial \phi}{\partial x}\right) + \frac{\partial}{\partial y}\left(\mu \frac{\partial \phi}{\partial y}\right)\right] - \frac{3}{2}\phi\left\{\frac{\partial}{\partial x}\left(\mu H^{1/2}\frac{\partial H}{\partial x}\right) + \frac{\partial}{\partial y}\left(\mu H^{1/2}\frac{\partial H}{\partial y}\right)\right\} = \frac{12\, Ua}{R}\frac{\partial(\rho H)}{\partial x} \qquad (9)$$

FINITE DIFFERENCE APPROXIMATION

This method is often used for the numerical solution of Reynolds equation. In EHL point contact problems, because of its high coupling and high pressure features involved, nonuniform grid is a more suitable choise of grid structure. As a result the stability of numerical iterative procedure could be improved. Fig. 1 shows the finite difference grid structure. With nonuniform grid structure a relative wide space in the inlet area can be used, so that the lubricant starvation problem may be avoided. Along the x direction in the upstream region, the element width is large and reduced gradually, and in the trailing area where the pressure is very high, finer grid is used. After center line, the element width is the same and the minimum width along x direction is a/18. Along the y direction the element width is uniform (a/7). The dimensionless computational domain is: along x direction in the inlet area m=4.0970 and in the outlet area n=1.1099 respectively. Along the y direction the side boundary distance is l=1.4286. With the finite difference approximation, equation (5) can be changed into an algebraic equation and Gauss-Seidel iteration with over-relaxation procedure is applied. Comparing with the ordinary iterative procedure, over-relaxation method is more effective, because for each iteration the new value of pressure distribution is used.

Boundary Conditions and Initial Pressure. Along the periphery the pressure was set to ambient, that is p=0. In cavitation area Reynolds condition was used, if p < 0 then set p=0. Hertzian initial pressure distribution was adopted for iteration, but modified pressure distribution could also be adopted to accelerate the iterative procedure (for example, numerical results of pressure distribution obtained previously from the same load).

Numerical procedure of iteration. The whole numerical method was a complicated iterative procedure which consisted of nested loops. The principle is that assume at first an initial pressure distribution and obtain a corresponding film shape from elastic theory. Then through the hydrodynamic mechanics a new pressure distribution can be obtained. If these two pressures are not convergent, use the new pressure to calculate the new film shape and get

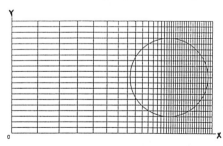

FIGURE 1. Finite difference grid structure

another pressure distribution. Continue this iterative procedure until these two pressure distributions are converged. Before the new film shape is evaluated, under-relaxation of pressure distribution is necessary to sustain a stable film shape.

Iterative procedure of reduced pressure q. The iterative method was similar to that developed by Ranger et al (1975). It included three nested loops: reduced pressure loop, load loop and pressure loop. In the load loop, the difference of calculated load and desired or applied load must satisfy the convergent criterion: $|F^c - F^d| / F^d < 0.0005$ (10)

If not satisfied, adjust the film thickness by modifying h_0 in equation (2) and reenter the hydrodynamic solver. The whole convergence was measured by the sum of absolute pressure residuals divided by the sum of pressures. The relative error adopted in this paper was 0.05.

However different values were used in this work. It was found that there was no substantial difference between 0.05 and 0.01 (the load range was below 30N), especially for film thickness. $\sum |p_2 - p_1| / \sum p_1 < 0.01 - 0.05$ (11)

If not satisfied, under-relaxation for pressure distribution was neccessary and returned to the calculation of elastic deformation. The similar convergence criteria were used in the inner loop.

Iterative procedure of substitution Φ. There were three nested loops in the iterative procedure which was similar to that developed by Hamrock and Dowson (1981). The initial pressure distribution was Hertzian pressure. The difference between this procedure and q solution was that the deformation was calculated in the pressure loop not in the outer loop. The three iteration loops were:

1). The inner loop (Φ loop): a Gauss-Seidel with over-relaxation iteration solution of the Reynolds equation (9). During the iterative procedure, density and viscosity at each nodal point were kept constant. This procedure continued until the convergence criterion was satisfied:

$$\sum_{j=1}^{n-1} \sum_{i=1}^{m-1} \frac{\left| \phi_{i,j}^{n+1} - \phi_{i,j}^{n} \right|}{\phi_{i,j}^{n+1}} < 0.1$$ (12)

Where m, n= grid number along x and y coordinates, Φ superscripts represented the different iterations. If (12) was not satisfied, then the new value of $\Phi_{i,j}$ was used as the initial value for the next iteration.

2). The pressure loop: With the new pressure distribution from the Φ loop the following pressure convergence criterion was used:

$$\sum_{j=1}^{n-1} \sum_{i=1}^{m-1} \frac{\left| P_{i,j}^{n} - P_{i,j}^{n-1} \right|}{P_{i,j}^{n}} < 0.1$$ (13)

If it was not satisfied, then re-calculated the elastic deformation. Before the new pressure distribution was determined from the inner loop, updated density and viscosity must be considered from the under-relaxed pressure distributions.

3). The load loop (outer loop): With the converged pressure distribution from the second loop, the resultant pressure distribution was integrated over the contact area. If the calculated load was close to initial applied load F^d, the overall iterative procedure was finished, otherwise the constant h_0 of equation (2) must be modified and re-enter the hydrodynamic solver. The over-all convergence criterion was the same as expression (10).

TABLE 1.   Numerical solutions of film thickness from two methods

unit: μm

Case No. and dimensionless parameters			Central film thickness	Minimum film thickness	Average film thickness
1	$W=1.3060 \times 10^{-7}$	Φ	0.3465	0.2603	0.3464
	$U=1.7117 \times 10^{-11}$	q	0.3499	0.2610	0.3481
2	$W=2.6120 \times 10^{-7}$	Φ	0.3489	0.2427	0.3406
	$U=1.7117 \times 10^{-11}$	q	0.3486	0.2436	0.3409
3	$W=3.2651 \times 10^{-7}$	Φ	0.6661	0.4915	0.6463
	$U=5.1352 \times 10^{-11}$	q	0.6681	0.4967	0.6465

Case No. and dimensionless parameters			Central film thickness	Minimum film thickness	Average film thickness
4	$W=3.2651 \times 10^{-7}$	Φ	0.6661	0.4915	0.6463
	$U=5.1352 \times 10^{-11}$	q	0.6681	0.4967	0.6465
5	$W=3.2651 \times 10^{-7}$	Φ	0.8871	0.6779	0.8648
	$U=8.5586 \times 10^{-11}$	q	0.8886	0.6757	0.8608

NUMERICAL SOLUTIONS

With the methods illustrated above, the elasto-hydrodynamic lubrication point contact problem could be solved from different methods. For the purpose of comparison the same dimensionless parameters, computational domain and grid structure were used. Although the iterative loop and the formation of Reynolds equation were different, the results shown the same characterestics of point contact problems. Table 1 lists some results obtained from different numerical procedures including average film thickness h_{av} (defined as the average film thickness over the Hertzian contact circle). Results of iterative procedure of Φ parameter were under-estimated comparing to those of reduced pressure q, but the difference was not significant. It shows that two different iterative procedures converge to similar results.

Fig. 2 shows plots of pressure distribution and film thicknesse of case 1. The value of dimensionless pressure distribution contour $-p/p_0$, (p_0=pressure at center point) and the difference between two contour lines is 0.1. For film thickness the contour plots (the value of contour lines is h/h_m and the difference between two contour lines is 0.1) show the similar shapes about the two results: the minimum film thickness is located at one whole crescent shaped region. But the pressure distribution represents some difference especially in outlet area (Fig. 2). In q solution, the spike effect is clear. The reason why the pressure distributions were different was that these two procedures had different convergence factors, expressions (11) and (13), which represented the different stages of the numerical procedures and hence different accuracy of the two methods. The comparison means that the convergence factor $\varepsilon=0.1$ is more difficult to achieve. When the load and speed were higher, the instability of pressure became more serious. The difference between these two techniques may be exaggerated. But even in this condition, the film thicknesses were almost identical.

The numerical procedure of Φ solution developed in this paper is considered more effective and accurate. In the iterative procedure of reduced pressure q more relaxation factors were

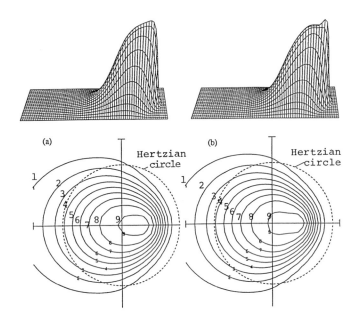

FIGURE 2.1 Three dimensional pressure plots and contours for case 1,Table 1 (value of contour lines is p/p_0), p_0=pressure at coordinate origin. (a) Φ solution (b) q solution

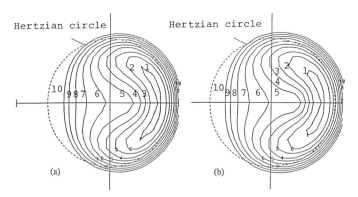

FIGURE 2.2 Contour plots of film thickness for table1 case 1, the value of contour lines is h/h_m (a) Φ solution (b) q solution

used, and during the iteration process, pressure greater than $1/\alpha$ was truncated to this limited value. All these inevitably resulted in errors and with the progress of iteration this accumulative error would be increased.

Comparing with q solution, the iterative procedure of Φ solution is relatively simple. For some cases which were difficult to be solved through q solution, converged solution could be obtained through the Φ solution. Table 2 shows another solutions solved by this method. The dimensional load parameter in Table 2 is 40N. The EHL direct iteration process is time consuming especially when the load is heavier.

TABLE 2. Numerical solution of point contact problem for elliptical contact

unit: μm

Case No. and dimensionless parameters			Central film thickness	Minimum film thickness	Average film thickness
6	$W=5.2242\times10^{-7}$	k=1	0.3475 (0.3312)	0.2158 (0.2805)	0.3347
7	$U=1.7117\times10^{-11}$	k=4	0.5056 (0.4537)	0.4015 (0.3897)	0.5327

(Figures in the parentheses are results from H.D. regression formulae)

Parameters and physical constants used in the numerical calculation:

$R=2.54 \times10^{-2}$ m $\qquad\qquad \mu_1=\mu_2=0.3$

$E_1 = E_2 =1.08 \times10^{11}$ N/m^2 $\qquad \eta_0= 0.516$ Ns/m^2

$\alpha=1.0 \times10^{-8}$ N/m^2 $\qquad\qquad \rho_0=880$ kg/m^2

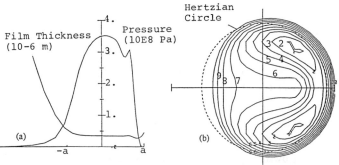

FIGURE 3. (a) Pressure distribution and film thickness along x-axis of contact center line (b) contourline (= h/h$_m$) of dimensionless film thickness (case 6)

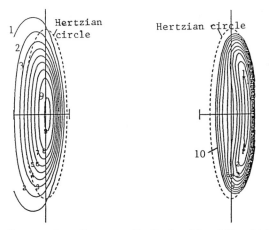

FIGURE 4. Contour plots of pressure distribution (a) and film thickness (b) for elliptical contact of case 7 (F=40N k=4)

Shown in Fig. 3 is the presentation of pressure distribution, film thickness along x-axis and contour of film thickness of case 6. It shows clearly that the film shape is parallel in this case. Fig. 4 shows results of case 7 for elliptial contact and with the increased elliptical parameter, the spike effects reduces and film thickness increases. The position of minimum film thickness also changes into one cresent shaped lobe which is quite different from the situation of circular contact.

CONCLUSIONS

For the comparison of two different methods, the iterative procedure of Φ substitution is considered better. Because of its relative simpler loop and less under-relaxation factors the accumulative error would be reduced. The iterative procedure of Φ solution is considered more effective and accurate. However there is some difference of pressure distributions in spike region between the two techniques.

In this paper, just forward-iterative method was considered to solve the EHL point contact problem. When the load becomes large, this method will break down. Based on numerical smoothing method, some results in medium load range could also be obtained with this method [4]. There is some potential in improving this method further so that some severe load condition could be considered or to accelerate the iterative procedure, for example by introducing inverse or multigrid methods [5,7].

REFERENCES

1. Dowson, D., and Higgingson, G. R., *Elastohydrodynamic Lubrication* , Pergamon Press , 1971.

2. Evans, H. P., and Snidle, R. W. The Isothermal Elastohydrodynamic Lubrication of Spheres, *Journal of Tribology,* Vol. 103., pp. 547-55,1981.

3. Evans, H. P., and Snidle, R. W., The Elastohydrodynamic Lubrication of Point Contacts at Heavy Loads, *Proceedings of the Royal Society,* London, Series A, Vol.382,pp.183-199, 1982.

4. Hamrock, B. J., and Dowson, D., Isothermal Elastohydrodynamic Lubrication of Point Contacts, Part 3-Fully Flooded Results, *Journal of Lubrication Technology,* Vol. 2, pp. 264-276, 1977.

5. Lubrecht, A. A., Film thickness Calculations in Elastohydrodynamicaly Lubricated Circular Contacts, Using a Multigrid Method, *Journal of Tribology,* Vol. 110., pp. 502-507, 1988.

6. Ranger, A. P., Ettles, C. M. M., and Cameron, A., The Solution of the Point Contact Elastohydrodynamic Problem, *Proceedings of the Royal Society,* London, Series A,Vol. 346, pp. 227-244,1975.

7. Zhu, D., and Wen, S. Z. , A Full Numerical Solution for the Thermoelasto-Hydrodynamic Problem in Elliptical Contacts, *Journal of Tribology,* Vol. 106, pp.246-254,1984.

FINITE DIFFERENCE METHODS AND APPLICATIONS

Finite Difference Solutions
of the Navier-Stokes Equations
on Staggered and Nonstaggered Grids

S. W. ARMFIELD
Centre for Water Research
University of Western Australia
Nedlands, Australia, 6009

INTRODUCTION

The most common finite-difference scheme in current use for solving the Navier-Stokes equations is the SIMPLE algorithm, defined on a staggered grid. The grid used is that which offsets the velocities half a mesh width in their respective coordinate directions from the pressure and other scalar variables. The most likely reason for the popularity of the staggered grid scheme is that it is robust, stable and relatively straightforward to code.

Despite these advantages the staggered mesh scheme does have drawbacks. When used with a non-uniform mesh the discretization is more difficult, particularly for the non-linear terms, and an interpolation operator for each of the velocities, and for the scalar variables, must be calculated. This adds considerably to the programming and computation time and is further exacerbated if adaptive, multi-grid or non-orthogonal mesh approaches are used. It is therefore suggested that if a satisfactory scheme on a non-staggered mesh can be developed it will be more appropriate for problems involving non-uniform and non-orthogonal grids, particularly if multi-grid and adaptive gridding techniques are employed.

Over the years there have been many non-staggered mesh schemes appear in the literature, such as, for example, the Chorin (1968) scheme, and more recently, the schemes of Rhies and Chow (1982) and Strikwerda (1984). It appears that the success or otherwise of these methods is determined primarily by two factors, integrability and regularity. A finite difference scheme is said to be regular elliptic if it has the properties of an elliptic operator independent of the gridsize, it is integrable if it satisfies, in discrete form, the integrability constraint for all intermediate solutions. These factors are presented and discussed in detail by Strikwerda (1984), and are covered briefly in the analysis section of this paper.

The important components of the schemes with respect to the use or otherwise of staggered meshes is the differencing of the pressure gradient terms and the divergence equation. The Chorin scheme is a central differencing approach with the pressure gradient term and the divergence equation differenced using the same two point operator. Chorin then constructs a Poisson equation for the pressure directly from the pressure gradient and the divergence equation. In this manner he produces a scheme that will satisfy the integrability constraint. However the resultant discrete Laplace operator for the pressure is a sparse five point scheme, with intervals of $2\Delta h$ between the points, and is therefore not regular elliptic. Such a scheme results in the pressure splitting into two weakly coupled fields for high Reynolds number flow, with a subsequent degradation of the solution.

The Strikwerda scheme uses a regular elliptic differencing. The scheme can be considered similar to that of Chorin, with the addition of regularizing terms to the pressure gradient and the divergence equation, so that each has a different quite complex three point operator. In this way the problem of loss of regularity and solution splitting is no longer encountered. However the construction of a Poisson equation directly from the pressure gradients and the divergence is now extremely complicated. Strikwerda does not attempt to construct such an operator, instead obtaining the pressure in a manner that will not preserve integrability, but adding a correction term to the divergence equation, thereby ensuring the integrability constraint is satisfied.

Rhies and Chow constructed a non-staggered mesh scheme in a manner that essentially mimics the staggered mesh scheme. They do this by producing another velocity field that is equivalent, with equivalent differencing for the pressure terms, to the staggered scheme. To avoid the problem of obtaining all the discrete operators for this field on the staggered grid, which essentially takes the problem directly back to the staggered mesh resulting in no advantage, they are obtained by averaging the operators on the non-staggered mesh. The Poisson equation is then constructed from the divergence of the staggered velocities in exactly the same way as on the staggered mesh. The resultant scheme is regular and satisfies the integrability constraint, but is however rather complex.

Non-staggered schemes have therefore been developed and used, with apparent success. The advantage of using such schemes on non-uniform meshes is such that one can only attribute the continued use of the staggered scheme for such problems either to extreme conservativism, or that the schemes are not satisfactory or have been presented in a manner that makes them not entirely accessible.

Doubtless there are many other possible differencings of the pressure gradient and divergence equation that will give stable, robust, and accurate schemes on non-staggered meshes. The scheme presented in the present paper is one such approach. It is considered to be easier to implement and less complex than either the Rhies and Chow scheme or the Strikwereda scheme.

In the present paper the numerical method and some analysis is briefly presented. A more detailed presentation is given in Armfield (1989). Results are obtained using this scheme, and the conventional staggered mesh SIMPLE scheme, for natural convection in a cavity. Comparison of results show both schemes have the same accuracy. Timing results indicate the non-staggered scheme on a fixed mesh is more efficient than the staggered scheme on the non-uniform grid used for the current results. It is likely that on a uniform grid the non-staggered scheme would be slightly less efficient.

NUMERICAL METHOD

Although results are presented for natural convection flow, for the purpose of presenting the method in a simple manner in this section the fluid is assumed to be isothermal. The inclusion of the temperature equation and buoyancy is quite straightforward and does not directly involve the features of the equations discussed here that relate to the use of non-staggered grids.

Governing Equations

The Navier-Stokes equations are expressed in non-dimensional form in Euclidean coordinates, (x, y) with corresponding velocity components (U, V) as follows,

$$U_t + UU_x + VU_y = -P_x + \frac{1}{Re}(U_{xx} + U_{yy}),$$ (1)

$$V_t + UV_x + VV_y = -P_y + \frac{1}{Re}(V_{xx} + V_{yy}),$$ (2)

$$U_x + V_y = 0,$$ (3)

where subscripts indicate partial differentiation, the Reynolds number $Re = \overline{U}L/\nu$, with $\overline{U}$ being a characteristic velocity and L a characterisic length.

Discretization

The domain, which is assumed to be rectangular, is discretized on a non-staggered grid with all variables store at each node. The index i is used to indicate location in the x direction and j indicates location in the y direction. Values of variables at the nodal point (i,j) are given as, for instance x^i, y^j, U^{ij} and so on. Indexes may occur as either superscript or subscript. When values of dependent variables are required at other than nodal points they are obtained by linear interpolation.

Finite volumes are used to convert differential terms in the governing equations in the following way. All second derivatives are approximated by second order central differences as,

$$U_{xx}(x^i, y^j) = (\frac{U^{i+1} - U^i}{\Delta x^{i+1}} - \frac{U^i - U^{i-1}}{\Delta x^i})^j/(\frac{\Delta x^{i+1} + \Delta x^i}{2}) + 0(\Delta x^2),$$
$$= SDU_j^i,$$

where $\Delta x^2 = \Delta x \times \Delta x$, and $\Delta x^i = x^i - x^{i-1}$.

Derivatives occuring in convective terms are approximated using a QUICK scheme, as follows

$$UU_x(x^i, y^j) = (F^{i+1/2,j} - F^{i-1/2,j})2/(\Delta x^{i+1} + \Delta x^i), \qquad (4)$$

where

$$F^{i+1/2,j} = U^{i+1/2}(U^{i+1/2} + SDU^{(i,j)}(Dx^{i+1})^2/8),$$
$$F^{i-1/2,j} = U^{i-1/2}(U^{i-1/2} + SDU^{(i-1,j)}(Dx^i)^2/8),$$

assuming $U^{i,j}$ is positive.

Integration

The above equations are integrated in the following way. Assuming all variables are known at time step t^n, we obtain a first estimate for $(U,V)^{n+1}$, denoted as $(U,V)^{n+1,m}$ with $m = 0$, from,

$$G1(U^{n+1,m}) = G2(U^n) - (P^{i+1} - P^{i-1})^{j,n+1/2,m}/(\Delta x^{i+1} + \Delta x^i),$$
$$G1(V^{n+1,m}) = G2(V^n) - (P^{j+1} - P^{j-1})^{i,n+1/2,m}/(\Delta y^{j+1} + \Delta y^j),$$

$G1 = (L + 1/\Delta t), G2 = (-L + 1/\Delta t)$ and L is in discrete form a block quinta-diagonal matrix with components obtained from the differencing given above. The best available guess is used for $P^{n+1/2,0}$.

We then obtain a correction for P by solving the following equation for Pc, the pressure correction,

$$\left[g^{i+1/2,j}(\frac{Pc^{i+1} - Pc^i}{\Delta x^{i+1}}) - g^{i-1/2}(\frac{Pc^i - Pc^{i-1}}{\Delta x^i})\right]^j 2/(\Delta x^{i+1} + \Delta x^i) +$$

$$\left[g^{j+1/2,i}(\frac{Pc^{j+1} - Pc^j}{\Delta y^{j+1}}) - g^{j-1/2}(\frac{Pc^j - Pc^{j-1}}{\Delta y^j})\right]^i 2/(\Delta y^{j+1} + \Delta y^j) = B^{i,j}, \qquad (5)$$

where g is the inverse of the diagonal of $G1$.

$P^{n+1/2,m}, U^{n+1,m}, V^{n+1,m}$ are then corrected as

$$U^{i,j,n+1,m+1} = U^{i,j,n+1,m} - g^i(Pc^{i+1} - Pc^{i-1})/(\Delta x^{i+1} + \Delta x^i),$$
$$V^{i,j,n+1,m+1} = V^{i,j,n+1,m} - g^j(Pc^{j+1} - Pc^{j-1})/(\Delta y^{j+1} + \Delta y^j),$$
$$P^{i,j,n+1/2,m+1} = RxPc^{i,j} + P^{i,j,n+1/2,m},$$

with Rx typically between 0.5 and 1.

The forcing term in the pressure correction equation is defined to be,

$$B^{i,j} = ((U^{i+1} - U^{i-1})/(\Delta x(i+1) + \Delta x(i)))^{j,n+1,m} + \tag{6}$$
$$((V^{j+1} - V^{j-1})/(\Delta y(j+1) + \Delta y(j)))^{i,n+1,m} +$$
$$\left[g^{i+1,j}(\frac{P^{i+2} - P^i}{\Delta x^{i+2} + \Delta x^{i+1}}) - g^{i-1}(\frac{P^i - P^{i-2}}{\Delta x^i + \Delta x^{i-1}}) \right]^j /(\Delta x^{i+1} + \Delta x^i) +$$
$$\left[g^{j+1,i}(\frac{P^{j+2} - P^j}{\Delta y^{j+2} + \Delta y^{j+1}}) - g^{j-1}(\frac{P^j - P^{j-2}}{\Delta y^j + \Delta y^{j-1}}) \right]^i /(\Delta y^{j+1} + \Delta y^j)$$
$$- \nabla^2 P,$$

where ∇^2 is the discrete laplacian given in Eqn. (5) above. The above equation set is iterated on m until a preset sup-norm covergence criterion is satisfied. Time step $n + 2$ is then obtained from $n + 1$, and so on.

Boundary Values

Boundary values are set by including extra points outside the boundary with a prescribed value. The value at the boundary is then the average of that point and the nearest point in the domain. For the present simulation on all boundaries the velocities are set to zero. The gradient of the pressure correction term is set to zero, to enable Eqn. (5) to be solved. However the pressure itself, which is required at a two locations outside the boundary, is obtained via a second order extrapolation from within the domain.

ANALYSIS

Regularity

To obtain accurate numerical solutions to partial differential equations it is necessary that the discrete operator must have the same character as the continuous operator it represents. Thus for instance a discretized elliptic operator must itself be elliptic, independent of the mesh. Naturally if the discretization is consistent then the operator will be elliptic in the limit, however actual solutions are obtained on finite meshes, thus ellipticity in the limit is not enough. Brandt and Dinar (1978) give an estimate of the discrete elliptic-ity of an operator, which they define as the h−ellipticity. Bube and Strikwerda (1983) further introduce the concept of regularity and define a discrete operator to be regular elliptic if the only non-zero solutions of the symbol are modulo 2π, where the symbol is the representation in wave number space of the linearized difference operator.

This requires that the solution at any given point in the discretized domain depends contin-uously on that at adjacent points, and is therefore seen to be equivalent to the continuous ellipticity condition. Such schemes are called regular elliptic (Bube and Strikwerda 1983). This is applied to the present problem in the following way.

We write the Navier-Stokes equations in discretized form as,

$$L(u) + Dpx(p) = 0,$$

$$L(v) + Dpy(p) = 0,$$
$$Dvx(u) + Dvy(v) = 0,$$

where L, Dpx, Dpy, Dvx, Dvy are all difference operators. The determinant of the symbol is,

$$\hat{L}(\hat{Dpx}\hat{Dvx} + \hat{Dpy}\hat{Dvy}),$$

where the $\hat{\ }$ denotes the Fourier transform. If we assume that the operator L is differenced in such a way as to not affect the regularity of the system, then clearly the second part of the above equation is the important factor. For the standard staggered mesh scheme we obtain,

$$(\hat{Dpx}\hat{Dvx} + \hat{Dpy}\hat{Dvy}) = 4h^{-2}(sin^2\frac{\theta_x}{2} + sin^2\frac{\theta_y}{2}). \tag{7}$$

As this only has zero's for θ equal to zero modulo 2π the standard staggered mesh differencing is seen to be regular elliptic. The standard central differencing for the non-staggered mesh for the pressure and divergence terms gives,

$$h^{-2}(sin^2\theta_x + sin^2\theta_x), \tag{8}$$

which has zeros for $\theta = 0$ modulo π, and therefore is not regular.

The determinant of the symbol for the method described in the present paper is identical to that of the staggered mesh scheme, given above in equation (8), and hence the present scheme is also regular elliptic. Full details are given in Armfield (1989).

Integrability

The integrability constraint requires that,

$$\int_D Div(u,v)dD = \int_{\delta D} n.(u,v)dD, \tag{9}$$

where D is the domain, δD the boundary and n is the unit normal to the boundary. In the present case this becomes,

$$\int_D Div(u,v)dD = 0. \tag{10}$$

In discrete terms we obtain that,

$$\Sigma_D Div(u,v)\Delta x\Delta y = 0. \tag{11}$$

Now we require that this always be satisfied by the intermediate solution. Chorin ensures this by constructing the Poisson equation directly from the pressure gradient and divergence equation. The velocity field will then be such that the integrability constraint will be satisfied to the order the pressure equation is solved (Armfield 1988). In the Strikwerda scheme where it is not possible to construct the pressure equation in such a way a correction term is added to the divergence equation.

For finite volume methods such as the SIMPLE scheme and that presented in the present paper, the differencing ensures that the solution is automatically compatible with the boundary data, and thus no explicit method of ensuring this is necessary. This is because the discrete equations are constructed directly from the momentum balance on each volume.

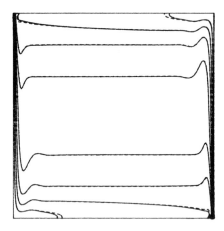

FIGURE 1. Temperature contours. – – – non-staggered, —— staggered.

RESULTS

Results are presented for natural convection in a square cavity ($24 \times 24cms$) at a Rayleigh number of 5×10^7. The top and bottom of the cavity are insulated and the side walls are set to $\overset{+}{-}\, \Delta T$, with all boundaries non-slip. The fluid is initially at rest and isothermal, the heating and cooling is then switched on instantaneously. Results for a comparison between the scheme presented in the present paper and a conventional staggered mesh scheme demonstrate the equivalent accuracy obtained with both methods.

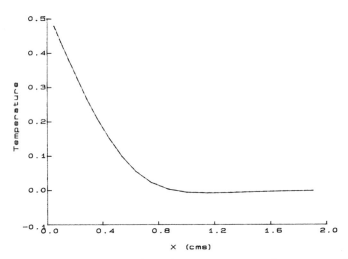

FIGURE 2. Temperature on the wall. – – – non-staggered, —— staggered

292

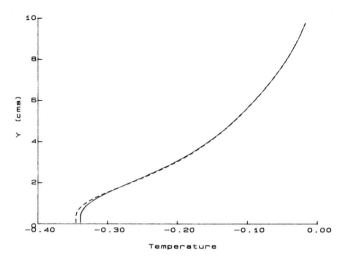

FIGURE 3. Temperature in the intrusion. – – – non-staggered, —— staggered

The staggered mesh scheme has been found to give accurate results for this problem when compared to experimental data, (Patterson and Armfield 1989).

Figure 1 presents the temperature contours for both methods at a time of approximately $t = 540 seconds$, which is attained after 300 time steps. Figure 2 presents temperature profiles in the thermal boundary layer at a location half way up the vertical wall. Figure 3 presents temperature profiles on the floor in the cold intrusion. As can be seen from the contour plots both methods produce close to identical temperature fields. The temperature profiles further demonstrate the similarity of the two solutions, with variations of only around 1%. The fact that similar results are obtained for what is a highly time dependent flow, after 300 time steps, is a good indication that the two methods have equivalent accuracy, both in space and time.

Timing results indicate that the non-staggered scheme uses around 50% of the cpu time of the staggered scheme. The results have been obtained on a long instruction word mini-supercomputer. The increased speed results both from the fact that less operators are calculated, and those remaining are more easily micro-coded. A similar result could be expected on a vector machine, however on a conventional machine the speed up resulting from the more efficient micro-coding, or vectorizing, would not be seen.

CONCLUSIONS

A finite difference scheme for the Navier-Stokes equations on a non-staggered mesh has been developed. The scheme ensures that the regularity and integrability constraints are satisfied.It has been compared to a conventional staggered approach for natural convection in a cavity at high Rayleigh number. This is a difficult flow to simulate due to the range of time, space and velocity scales present, and cannot be successfully simulated using a standard central differencing for the pressure gradient and the divergence, of the type presented in Chorin (1968). It is therefore considered to be a good test of the present scheme. Further testing with alternative flows is required before the scheme can be said to be generally as robust as the staggered approach.

The scheme has been demonstrated to have equivalent accuracy and greater efficiency than a conventional staggered mesh scheme for natural convection in a cavity with a non-uniform mesh. The advantages of using the non-staggered scheme in terms of programming and run time, when non-uniform and/or non-orthogonal meshes are used, are such that the use of such an approach is strongly recommended.

REFERENCES

1. Armfield S., Analysis of SIMPLE type algorithms on staggered and non-staggered grids, Proc.*1st. National Fluid Dynamics Conference*, Cincinnatti Ohio, July 25-28 1988, pp 58-63, AIAA 883528, 1988.

2. Armfield S., Finite difference schemes for the Navier-Stokes equations on staggered and non-staggered grids, submitted to Computers and Fluids, 1989 .

3. Brandt A. and Dinar N., Multi-grid solutions to elliptic flow problems, Proc. *Conference on Numerical solutions of Partial Differential Equations*, Mathematical Research Centre, Madison, WI, October 1978.

4. Bube K. and Strikwerda J., Interior regularity estimates for elliptic systems of difference equations, *SIAM J. Numer. Anal.*, V 20, pp 653-670,1983.

5. Chorin A.J., Numerical solutions of the Navier-Stokes equations, *Math. Comp.*, V 22, pp 745-762,1968

6. Patterson J.C. and Armfield S.W., Transient features of natural convection in a cavity- Experimental and numerical methods and validation, submitted to J. Fluid Mech., 1989.

7. Rhies C.M. and Chow W.L., A numerical study of the turbulent flow past an airfoil with trailing edge separation, AIAA-82-0988, 1982.

8. Strikewerda J.C., Finite difference methods for the Stokes and Navier-Stokes equations, *SIAM J. Sci. Stat. Comput*, V 5, pp 56-68, 1984.

An Explicit Finite Difference Model for Tidal Flows in the Arabian Gulf

M. BASHIR, A. Q. M. KHALIQ, and A. Y. AL-HAWAJ
Department of Mathematics
University of Bahrain
P.O. Box 32038
Bahrain

E. H. TWIZELL
Department of Mathematics and Statistics
Brunel University
Uxbridge, Middlesex, UB8 3PH, UK

INTRODUCTION

In recent years, finite difference models have been successfully applied to simulate water level fluctuations in the Arabian Gulf (see for example, Hansen (1962), Lardner et al. (1982), Murty and El-Sabah (1984) and von Trepka (1968) who used Hansen's scheme. Lardner et al (1982) modified the finite difference scheme of Leendertse (1967) to study various regions in the Arabian Gulf. A detailed discussion and comparison of these schemes is well documented in Provost (1983). Murty and El-Sabah (1984) developed a two dimensional linear numerical model for storm surge in the Arabian Gulf. However, generally satisfactory derivations of the surge profile, particularly in areas of shallow water (Dahran - Bahrain - Qatar) require the use of non-linear equations in which tide surge and their interactions are properly taken into account (Banks (1974)).

In the present paper, a two dimensional explicit finite difference model is developed to reproduce the principal semi-diurnal (M_2) constituent of the tide in the Arabian Gulf. The semi-implicit scheme proposed by Lardner et al (1982) uses a sufficiently large time step to give a good computing economy but damping and phase-error limit the time step to compute the solution. One of the main differences between the schemes of Lardner et al and von Trepka and the present explicit scheme is the way the tidal forcing at the Strait of Hormuz is approximated. The present numerical model differs considerably according to the approximation of the boundary conditions at the open boundary. Another important feature is the simplicity and economy of the scheme. In section 2 the description of the mathematical model is presented. Particular consideration is given to radiation type conditions at the open boundary.

The finite difference scheme is discussed in section 3. Numerical results are described in section 4 and a detailed comparison between the computed and observed values of the amplitudes and phases for the M_2 tide is presented. The model will be tested for storm surge, residual currents, and pollutant dispersion in future papers.

THE MATHEMATICAL MODEL

The basic equations governing the flow (tides, storm surge, wind induced currents etc.) are the Navier-Stokes equation and the equation of continuity (Lardner et al (1982)):

$$\frac{\partial \mathbf{U}}{\partial t} + (\mathbf{U}.\ \nabla\ )\ \mathbf{U} = -\ \frac{1}{\rho}\ \nabla p - \mathbf{g} - 2\ \Omega\ X\ \mathbf{U} +\ \frac{1}{\rho}\nabla.\tau \tag{1}$$

$$\frac{\partial \rho}{\partial t} + \nabla.(\rho\ \mathbf{U}) = 0, \tag{2}$$

where $\mathbf{U}$, t, ρ, p, $\mathbf{g}$, Ω and τ are, respectively, the three dimensional fluid velocity vector, time, density, pressure, the gravitational force per unit mass, the coriolis vector and viscous stress tensor. In the present model the specific mass density is assumed constant and according to this approximation, the above equations are replaced by their averaged versions over the vertical dimension. This leads to a two dimensional system of equations rather than the full three dimensional system above. The depth averaged density ρ and depth average horizontal velocity components u and v are defined as follows:

$$\bar{\rho} = \frac{1}{H} \int_{-h}^{\xi} \rho\, dz\ , \qquad \bar{u} = \frac{1}{H} \int_{-h}^{\xi} u\, dz\ , \qquad \bar{v} = \frac{1}{H} \int_{-h}^{\xi} v\, dz\ ,$$

where $H = h + \xi$ is the total water depth and ξ is the height of the free surface above the reference plane z = 0 (see Fig. 1)

Neglecting the surface shear stresses and atmospheric pressure gradients the governing shallow water equations may then be written as follows (Arnold and Noye (1982) and Lardner et al (1982)):

$$\frac{\partial \xi}{\partial t} + \frac{\partial (H\ u\ )}{\partial x} + \frac{\partial (H\ v\ )}{\partial y} = 0, \tag{3}$$

$$\frac{\partial u}{\partial t} + u\ \frac{\partial u}{\partial x} + v\ \frac{\partial u}{\partial y} - f v + g\ \frac{\partial \xi}{\partial x} - \tau_x = 0, \tag{4}$$

$$\frac{\partial v}{\partial t} + u\ \frac{\partial v}{\partial x} + v\ \frac{\partial v}{\partial y} + f u + g\ \frac{\partial \xi}{\partial y} - \tau_y = 0\ , \tag{5}$$

where τ_x and τ_y are the bottom stresses in the x and y directions given by

$$\tau_x = -D\ u\ \frac{\sqrt{u^2 + v^2}}{H}\ , \qquad \tau_y = -D\ v\ \frac{\sqrt{u^2 + v^2}}{H}$$

and $D = \frac{g}{c^2}$ (c is chezy coefficient) denotes the bottom friction coefficient varying from 2.5×10^{-3} to 2.8×10^{-3} .

Boundary conditions

On the land boundary, the normal velocity component is taken as zero ($\eta.\mathbf{U} = 0$), η being the outward directed normal. On certain open boundaries (river boundary in estuary models) the more appropriate boundary condition is to prescribe the flux of water across the boundary. Across open ocean boundaries, however, it is most appropriate to prescribe the water heights. In the present model at the open boundary (Strait of Hormuz) the radiation type conditions are applied (Beckers and Neves (1985) and Flather (1976)) leading to:

$$\xi(s,t) = \xi_0(s)\ \cos\left(\ \frac{2\pi t}{T} - \Psi(s)\ \right)\ \text{and}\ u = -\ \sqrt{\frac{g}{h}}\ \xi(s,t) \tag{6}$$

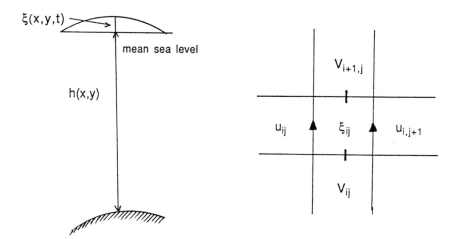

FIGURE 1. Coordinate system

FIGURE 2. Staggered grid

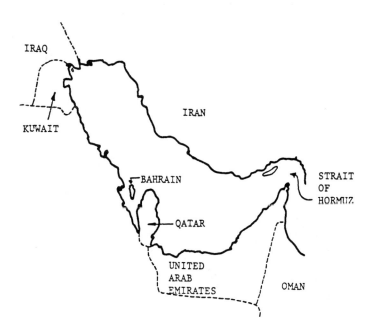

FIGURE 3. Geography of the Gulf

where s denotes the position on the boundary, ξ_0, Ψ and T respectively are the amplitude, the phase and the period of the M_2 Tidal constituent.

The values of ξ_0, h and Ψ are taken from Admiralty charts (1988 a,b) and Flather (1976) noted that the application of the radiation condition in the numerical model may remove the unrealistically large currents and grid scale oscillations in the open boundary, which may be produced by application of conventional open-sea boundary conditions.

Initial conditions

The initial conditions defined throughout the region are as follows:

$$\xi(x,y,t=0) = \xi_0(x,y) \ , \ u(x,y,t=0) = u_0(x,y) \ , \ v(x,y,t=0) = v_0(x,y) \tag{7}$$

The values of ξ_0, u_0, v_0 are generally not known. The computation is started by assuming these values to be zero. Due to the frictional dissipation, the influence of initial conditions becomes negligible after some time and stable conditions are reached after a few cycles of the computation.

THE EXPLICIT FINITE DIFFERENCE SCHEME

The finite difference model is based on staggered grids (Beckers and Neves,1985 and Hansen,1962) given in Fig. 2. The scheme is applied with mesh sizes of Δx and Δy in x and y directions respectively at wet points with time step Δt. The geography of the Gulf is given in Fig. 3. The details of the finite difference scheme are as follows:

$$\frac{\xi_{ij}^{k+1} - \xi_{ij}^{k}}{\Delta t} = - \frac{\left\{ \left(H_{ij}^{k} + H_{i+1,j}^{k} \right) u_{i+1,j}^{k} - \left(H_{ij}^{k} + H_{i-1,j}^{k} \right) u_{ij}^{k} \right\}}{2\Delta x}$$

$$- \frac{\left\{ \left(H_{ij}^{k} + H_{i,j+1}^{k} \right) v_{i,j+1}^{k} - \left(H_{ij}^{k} + H_{i,j-1}^{k} \right) v_{ij}^{k} \right\}}{2\Delta y}$$

$$\frac{u_{ij}^{k+1} - u_{ij}^{k}}{\Delta t} = - \frac{u_{ij}^{k} \left(u_{i+1,j}^{k} - u_{ij}^{k} \right)}{\Delta x} + f v_{ij}^{k} - g \frac{\xi_{ij}^{k+1} - \xi_{i-1,j}^{k+1}}{\Delta x} - \frac{g u_{ij}^{k} \sqrt{ \left(u_{ij}^{k} \right)^2 + \left(v_{ij}^{k} \right)^2}}{c^2 H_{ij}^{k}}$$

$$- \bar{v}_{ij} \begin{cases} \dfrac{\left(3 u_{ij}^{k} - 4 u_{i,j-1}^{k} + u_{i,j-2}^{k} \right)}{2\Delta y} , & \bar{v}_{ij} > 0 \\[2mm] \dfrac{\left(-3 u_{ij}^{k} + 4 u_{i,j+1}^{k} - u_{i,j+2}^{k} \right)}{2\Delta y} , & \bar{v}_{ij} \le 0 \end{cases}$$

where $\bar{v}_{ij} = \frac{1}{4}\left(v^k_{i-1,j} + v^k_{ij} + v^k_{i,j-1} + v^k_{i-1,j+1}\right)$ and $H^k_{ij} = \left(h + \xi^k_{ij}\right)$

$$\frac{v^{k+1}_{ij} - v^k_{ij}}{\Delta t} = -\frac{v^k_{ij}\left(v^k_{i+1,j} - v^k_{ij}\right)}{\Delta y} + fu^{k+1}_{ij} - g\frac{\xi^{k+1}_{ij} - \xi^{k+1}_{i-1,j}}{\Delta y} - g\frac{v^k_{ij}\sqrt{\left(u^k_{ij}\right)^2 + \left(v^k_{ij}\right)^2}}{c^2 H^k_{ij}}$$

$$- \bar{u}_{ij}\begin{cases} \dfrac{\left(3v^k_{ij} - 4v^k_{i-1,j} + v^k_{i-2,j}\right)}{2\Delta x} &, \bar{u}_{ij} > 0 \\[2ex] \dfrac{\left(-3v^k_{ij} + 4v^k_{i+1,j} - v^k_{i+2,j}\right)}{2\Delta x} &, \bar{u}_{ij} \leq 0 \end{cases}$$

where $\bar{u}_{ij} = \frac{1}{4}\left(u^{k+1}_{i,j-1} + u^{k+1}_{ij} + u^{k+1}_{i+1,j} + u^{k+1}_{i+1,j-1}\right)$

The order of consistency of the finite difference scheme is at least one, which is confirmed by numerical experiments. The scheme imposes the usual CFL stability requirement

$$\Delta t < \frac{\Delta x \Delta y}{\sqrt{gH_{max}(\Delta x^2 + \Delta y^2)}}$$

NUMERICAL RESULTS AND DISCUSSION

The proposed finite-difference scheme is applied with a mesh size of 10' (approximately 16.67 km in the x direction and 18 km in the y direction) of latitude and longitude to cover the region with the rectangular grid as shown in Fig. 4. The numerical solution of the shallow water equations (3-5) together with initial and boundary conditions (6-7), deal with the major tidal constituents, (M_2) of the Arabian Gulf. The amplitudes and phases are computed using a time step of six minutes. The computations have been carried out for approximately 100 hrs of real time. Very little change in the computed amplitudes and phases was noticed when the experiment was extended to 125 hrs. The computed and observed values of amplitudes and phases at some stations are given in Table 1. It is noticed that computed amplitudes and phases are in good agreement with the observed values at various stations (see also Figs 5). A detailed study of Table 1 reveals that the numerical model has produced reasonably good results in shallow as well as deep waters. Comparing the values of computed amplitudes at some stations e.g. Bandar Abbas, Bandar Mish'ab and Abu Dhabi, it is noticed that the present model gives better results than those of the Lardner et al (1982,p.438). However, the computed amplitudes are slightly small in the region around the northern tip of Bahrain. Due to the explicit nature of the present model, less CPU time is required as compared to that of Lardner et al (1982). The model is thus computationally economical and efficient.

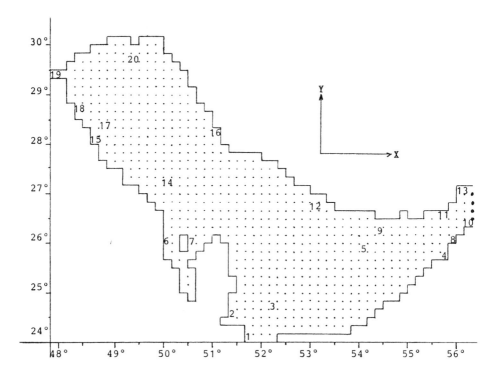

FIGURE 4. Grids for the Gulf.

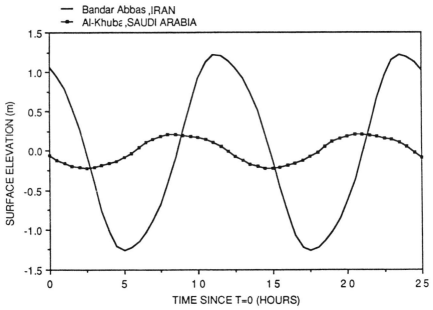

FIGURE 5. Amplitude of M$_2$ Tide

300

TABLE 1. Amplitude (m) and Phase (degree) of the M_2 Tide at the Tidal stations shown in Fig. 4.

Station	Lat. N°		Long E°		Amplitudes		Phases	
					O*	C*	O*	C*
1. Umm al Hatab	24	13	51	52	0.23	0.24	224	225
2. Ras Abu Qumayyis	24	34	51	30	0.31	0.30	185	183
3. Jazirat Arzanah	24	47	52	34	0.05	0.05	217	215
4. Ras Al Khaymah	25	49	55	57	0.54	0.50	326	320
5. Jazirat Sirri	25	54	54	33	0.39	0.34	351	345
6. Al-Khubar	26	17	50	07	0.31	0.28	146	148
7. Umais	25	59	50	53	0.27	0.30	172	180
8. Mina Saqr	25	58	56	03	0.60	0.57	325	320
9. Tunb al Kubra	26	16	55	18	0.54	0.54	329	327
10. Khawr al Quway	26	21	56	22	0.69	0.69	309	308
11. Jazireh-ye-Hangam	26	41	55	54	0.74	0.78	305	312
12. Jazireh-ye-Lavan	26	48	53	23	0.33	0.35	073	076
13. Bandar Abbas	27	10	56	17	1.10	1.20	298	306
14. Jazirat al Jurayd	27	12	49	57	0.41	0.37	124	119
15. Bandar Mish'ab	28	07	48	37	0.25	0.28	003	010
16. Lavar	28	15	51	16	0.50	0.50	169	168
17. Zulaf Oilfield	28	22	49	10	0.10	0.11	340	341
18. Mina Saud	28	44	48	24	0.42	0.40	336	333
19. Shuwaykh	29	21	47	55	0.95	0.98	343	345
20. Khor Musa	29	42	49	24	0.58	0.58	306	307

*O = Observed values are taken from admirality chart (1988a).
*C = Computed values.

REFERENCES

1. Admirality Tide Tables, Vol.2,pp.346-436, The Hydrographer of the Navy, London, 1988a.

2. Admirality Chart No.2858, The Hydrographer of the Navy, London, 1988b.

3. Arnold, R. and Noye, J., in *Numerical Solutions of Partial Differential Equations*, ed. J.Noye, pp.437-453, North -Holland, Amsterdam, 1982.

4. Banks, J.E., A Mathematical Model of a River Shallow Sea System used to Investigate Tide , Surge and their Interaction in the Thames-Southern North Sea Region, Phil.Trans.R.Soc., Vol.A 275, 567-609, 1974.

5. Beckers, P.M. and Neves, R.J., A semi-implicit Tidal Model of the North European Continental Shelf, *Appl.Math Modelling*, Vol.9, pp.395-402, 1985.

6. Flather, R.A., Results from a Storm Surge Prediction Model of the North-West european Continental Shelf for April, November and December, 1973, IOS Rep. No.24, Bidston Observatory, England, pp33, 1976.

7. Hansen W., Hydrodynamical Methods Applied to Oceanographic Problems, *Proc.Symp.Math. Hydrodyn.Phys.Ocean*, Institut fur Meereskunde, Universitat Hamburg, pp.24-34, 1962.

8. Lardner, R.W., Belen, M.S. and Cekirge, H.M., Finite Difference Model for Tidal Flows in the Arabian Gulf, *Compu.Math.Appl.*, Vol.8, No.6, pp.425-444, 1982.

9. Leendertse, J.J., Aspects of Computational Model for Long Period Water Wave Propagation, Rand.Corp.Rep. RM-5294-PR, 1967.

10. Murty, T.S. and El-Sabh, M.I., in Oceanographic Modelling of the Kuwait Action Plan (KAP) Region, UNESCO Reports in Marine Science (Report of a Symposium-Workshop, UPM, Dahran, Saudi Arabia), 1984.

11. Provost, Le.C., MOdels for Tides in the KAP Region, Symposium on Oceanographic Modelling of the Kuwait Action Plan Region, UPM, 1983.

12. von Trepka, L., Investigations of the Tides in the Persian Gulf by Means of a Hydrodynamic Numerical Model, *Proc.Symposium on Mathematical and Hydrodynamic Investigation of Physical Processes in the Sea*, Inst. fur Meer. der Univ. of Hamburg, 10. pp.59-63, 1968.

13. Wilders, P., van Sijin, Th.L., Stelling, G.S. and Fokkema, G.A., A Fully Implicit Splitting Method for Accurate Tidal Computations, *Int. J. Numer.Methods Eng*, Vol.26, pp.2707-2721, 1988.

The Ultimate CFD Scheme with Adaptive Discriminator for High Resolution of Narrow Extrema

B. P. LEONARD
Department of Mechanical Engineering
The University of Akron
Akron, Ohio, USA

H. S. NIKNAFS
Norton Company
Chemical Products Division
Stow, Ohio, USA

INTRODUCTION

Simulation of highly convective flows is one of the most challenging problems in computational mechanics. Even superficially simple one-dimensional pure convection at constant velocity has not been computationally
conquered. This has been referred to as the "ultimate embarrassment"
(Mitchell, 1984). Classical (second-order central) schemes are able to
handle very smooth (low-curvature or long wavelength) profiles fairly
satisfactorily. But a sudden change in gradient (i.e., locally high curvature) excites spurious unphysical oscillations due to the inherent
(third-derivative) dispersion term in the truncation error. Changing to
higher order symmetric schemes produces a (slight) quantitative improvement but not a qualitative improvement in dispersion error. First-order
upwinding can resolve a step monotonically (giving a spreading error-
function), but the results are so artificially diffuse that other profiles are unrecognizable after a short time. Higher odd-order upwind
schemes such as QUICKEST (Leonard, 1979), with dissipative (but not diffusive) leading truncation error, offer much better prospects; but they
are still unable to resolve steps without overshoots and undershoots.
Monotonic high resolution schemes can be constructed for simulating discontinuities. This is achieved by using a subtle blending of a second-
order central base scheme with first-order upwinding (introducing enough
positive artificial diffusion to maintain monotonicity) and first-order
downwinding (with inherent negative artificial diffusion to enhance "compression"). This strategy is the basis of so-called "shock-capturing"
and "TVD" schemes (Yee, 1987). Unfortunately, the best step-resolution
(supercompressive) schemes tend to convert smooth profiles into a series
of steps and plateaus. All such schemes strongly "clip" narrow extrema.

Recently, a universal limiter (UL) has been designed which can be applied
to arbitrarily high order transient interpolation modelling (TIM) of the
advective transport equations (ATE). This ULTIMATE computational-fluid-
dynamic scheme (Leonard, 1988) has several attractive properties. Step
resolution is monotonic and can be made competitive with the best supercompressive shock-capturing schemes by locally increasing the order of
accuracy, using adaptive stencil expansion (which moves with the profile).
Smooth profiles are not corrupted. However, one problem remains: very
narrow extrema are slightly clipped relative to what can be achieved with
an unlimited high order scheme. Near sudden changes in gradient the
limiter needs to be "on"; but near local physical extrema (but not spurious overshoots) it needs to be switched "off". The problem is one of

303

pattern recognition. This paper describes an adaptive discriminator which
can identify local narrow extrema and automatically switch off the universal limiter in such regions.

UNIVERSAL LIMITER

Space does not permit a detailed description of the ULTIMATE scheme; this
can be found elsewhere (Leonard, 1988). A brief description is given here
for convenience. Consider a control-volume face with a convecting velocity, u_f, positive to the right. Assume that the convected scalar, ϕ,
increases monotonically to the right. Label the node value just downstream of the CV face ϕ_D, the adjacent upstream value ϕ_C, and the next
upstream value ϕ_U. Now define normalized variables

$$\tilde{\phi}_f = \frac{\phi_f - \phi_U}{\phi_D - \phi_U} \quad \text{and} \quad \tilde{\phi}_C = \frac{\phi_C - \phi_U}{\phi_D - \phi_U} \tag{1}$$

Figure 1 shows the universal limiter boundaries in the normalized variable diagram--a plot of ϕ_f against ϕ_C. In the monotonic range (given by
$0 \leq \tilde{\phi}_C \leq 1$), note the downwind constraint: $\tilde{\phi}_f \leq 1$; the upwind constraint:
$\tilde{\phi}_f \geq \tilde{\phi}_C$; and the time-step constraint: $\tilde{\phi}_f \leq \tilde{\phi}_C/c$, where c is the Courant
number based on u_f. In the nonmonotonic region, various strategies are
possible. Figure 1 shows the simple case, $\tilde{\phi}_f = \tilde{\phi}_C$.

TEST PROFILES

Figure 2 shows a number of challenging test profiles together with a second-order central simulation (Lax and Wendroff, 1960). The figure shows
a snapshot of uniform convection to the right after the profiles have travelled 45 mesh widths from exact initial conditions. Discontinuity resolution is tested by a simple step, rather than the square wave used by
some authors (Sweby, 1984); smooth behaviour is represented by an isolated
sine-squared wave 20Δx wide; a semi-ellipse 20Δx wide is very challenging

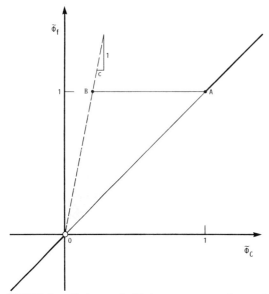

FIGURE 1. Universal limiter constraints shown in the $(\tilde{\phi}_C, \tilde{\phi}_f)$ plane.

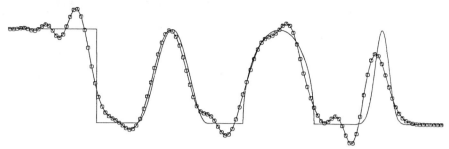

FIGURE 2. Lax-Wendroff simulation of: a step, an isolated sine-squared profile 20Δx wide, a semi-ellipse 20Δx wide, and a narrow Gaussian (σ = 1.9Δx); c = 0.45; 100 time steps.

because of a combination of sudden and gradual changes in gradient; a narrow Gaussian (σ = 1.9Δx) has been chosen to test peak resolution. The Courant number for all cases is c = 0.45, corresponding to 100 time-steps. It is perhaps hardly worth saying that Lax-Wendroff performance is very disappointing for these profiles.

Figure 3 shows first-order upwinding results. These certainly need no comment! Figures 4 and 5 show, respectively, results for two well-known shock-capturing schemes (Roe, 1986): Minmod (a relatively diffuse scheme) and Superbee (a supercompressive scheme). Note the steepening and clipping introduced by Superbee in the semi-ellipse and Gaussian profiles. Figure 6 shows results for ULTIMATE seventh-order upwinding. Note the generally good overall performance--except for a slight clipping of the Gaussian. By contrast, Figure 7 shows results for unlimited seventh-order upwinding. This gives somewhat better resolution of the narrow Gaussian, but undershoots or overshoots are excited near sudden changes in gradient. To get very good resolution of the narrow peak, it seems necessary to go to an unlimited ninth-order upwind scheme, shown in Figure 8; but this gives serious dispersion (as expected) near the discontinuous profiles. The ULTIMATE ninth-order upwind results, seen in Figure 9, again show some clipping of maxima, especially the narrow Gaussian peak.

Adaptive Stencil Expansion

Ideally, some sort of adaptive strategy seems desirable. For example, in smoothly varying (low curvature) regions, an inexpensive (but artificial-diffusion-free) low-phase-error method such as QUICKEST (Leonard, 1988)

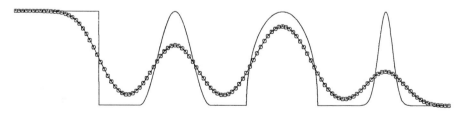

FIGURE 3. Simulation, using first-order upwinding, of the step, sine-squared, semi-ellipse, and narrow Gaussian profiles; c = 0.45; 100 time steps.

is appropriate. By monitoring the (absolute relative) curvature, it is a
simple matter to devise automatic stencil expansion, locally, to ULTIMATE
seventh-order upwinding, say, to give high resolution in regions of strongly varying gradients. This is extremely cost-effective because the more
expensive wide-stencil computations are made only in narrow localized regions of the overall flow domain, by definition. The location of the
wide-stencil calculation automatically moves along with the profile. But
this still leaves the problem of clipping of narrow extrema. For this it
seems desirable to (i) identify physical extrema, (ii) switch off the
limiter near each extremum, and (iii) use a very high order method, locally, depending on the narrowness of the extremum. A method for achieving this is described in the next section.

FIGURE 4. Simulation of the same profiles using the diffusive "Minmod"
TVD scheme; c = 0.45; 100 time steps.

FIGURE 5. Simulation of the same profiles using the supercompressive
"Superbee" TVD scheme; c = 0.45; 100 time steps. Note the artificial
steepening (due to inherent negative artificial diffusivity) and con-
comitant flattening of peaks.

FIGURE 6. Simulation using the ULTIMATE seventh-order upwind scheme.
Step resolution is better than supercompressive TVD schemes but smooth
profiles are not distorted. However, some clipping of peaks occurs,
especially in the case of the narrow (Gaussian) profile.

FIGURE 7. Peak resolution is restored by using an unlimited seventh-order upwind scheme. But now, dispersion occurs near regions of sudden change in gradient.

FIGURE 8. Results of unlimited ninth-order upwinding are similar to seventh-order, with slightly better resolution of the narrow peak.

FIGURE 9. ULTIMATE ninth-order upwinding. Results are similar to those of ULTIMATE seventh-order (Figure 6), with somewhat better step resolution. But clipping still occurs in narrow-peak regions.

CONSTRUCTION OF THE DISCRIMINATOR

The following algorithm attempts to distinguish between artificial numerical peaks, such as those seen in Figures 7 and 8, and true physical peaks (such as those of the exact profiles in the test cases studied). Artificial peaks would be associated with short wave-length numerical oscillations, with rapidly changing value, gradient, and curvature; long wave-length numerical oscillations, such as Lax-Wendroff (Figure 2), are excluded from the basic 3rd/7th-order algorithm. If a local extremum is associated with short wave-length oscillations, the discriminator chooses the limited algorithm, however, if the curvature is of the same sign in adjacent regions, the discriminator relaxes the limiter constraints at the extremum and the immediate upstream and downstream adjacent nodes.

In setting up the convective fluxes (for the right-face of each CV cell), consider an increasing-i DO-loop sweep. For the discriminator, choose a stencil of seven points: ϕ_{i-3}, ϕ_{i-2}, ϕ_{i-1}, ϕ_i, ϕ_{i+1}, ϕ_{i+2}, ϕ_{i+3}. Then compute the differences between each pair of consecutive points:

$$D1 = (\phi_{i-2} - \phi_{i-3}), \ D2 = (\phi_{i-1} - \phi_{i-2}), \ldots D6 = (\phi_{i+3} - \phi_{i+2}) \tag{2}$$

For convenience, assume there is a local maximum; a minimum requires reversal of some of the subsequent inequalities. Limiter constraints are active unless otherwise stated. The algorithm proceeds as follows.
(i) Check if D1 and D2 are both positive and D3 and D4 both negative; if not, go to step (iii); if they are, then:
(ii) check if $|D2| < |D1|$ and $|D3| < |D4|$; if true, switch to an unlimited version at the current i-value and skip the remaining steps; otherwise:
(iii) Check if D2 and D3 are both positive and D4 and D5 both negative; if not, go to step (v); if they are, then:
(iv) Check if $|D3| < |D2|$ and $|D4| < |D5|$; if true, switch to an unlimited version at the current i-value and skip the remaining steps, otherwise:
(v) Check if D3 and D4 are both positive and D5 and D6 both negative; if not, keep limiter constraints active; if they are, then:
(vi) Check if $|D4| < |D3|$ and $|D5| < |D6|$; if true, switch to an unlimited version at the current i-value; if not, proceed with the limiter constraints active.
This discriminator routine is by-passed unless one of ϕ_{i+1}, ϕ_i, or ϕ_{i-1}, is a local extremum. As noted above, limiter relaxation occurs (if at all) in groups of three points, with the extremum in the middle.

Results Using the Discriminator

Figure 10 shows seventh-order upwinding used with the automatic discriminator. Nodes at which the limiter is relaxed are shown by the small arrows. Note how they occur in groups of three with the profiles exhibiting local maxima; limiter constraints automatically remain active near the step discontinuity. Although all profiles are well simulated, seventh order is not quite enough to capture the narrow Gaussian peak. This was also seen in the unlimited case, Figure 7. Using ninth-order upwinding with the discriminator gives excellent results for all profiles--but, of course, such a scheme used globally would be unnecessarily expensive in smooth (low-curature) regions.

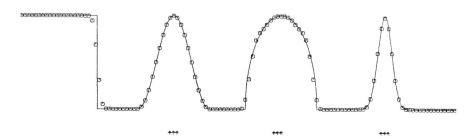

FIGURE 10. Automatic discriminator, relaxing the limiter in peak regions, use with ULTIMATE seventh-order upwinding. Note the generally excellent results for all profiles. Small arrows show nodes where the limiter is automatically relaxed.

Hybrid Strategy

The ultimate scheme, shown in Figure 11, uses an inexpensive ULTIMATE third-order upwind scheme in smooth non-steep regions; i.e., wherever the average (right-face) absolute "curvature"

$$CURVAV = |\phi_{i+2} - \phi_{i+1} - \phi_i + \phi_{i-1}|/2 \tag{3}$$

and "gradient"

$$GRAD = |\phi_{i+1} - \phi_i| \tag{4}$$

are both less than pre-assigned thresholds. If CURVAV exceeds THC1 or the larger THC2, the algorithm switches to ULTIMATE seventh- or ninth-order upwinding, respectively. If GRAD exceeds THG it switches directly to ULTIMATE ninth-order. Near local extrema, the discriminator automatically relaxes the limiter, as before. This is a very cost-effective strategy because the wider-stencil computations are performed only where needed--at a relatively small number of grid-points, by definition. This is even more effective in two and three dimensions. In Figure 11, ninth-order is automatically used in the vicinity of the step, the Gaussian, and high-curvature portions of the other profiles--shown by solid dots. The sine-square peak uses (unlimited) seventh-order, whereas the smoother semi-ellipse maximum region uses (unlimited) QUICKEST. Similar principles can be applied to multidimensional nonlinear systems such as the shallow-water, Euler, and Navier-Stokes equations.

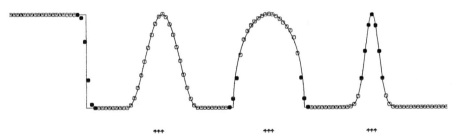

FIGURE 11. Cost-effective hybrid scheme: ULTIMATE QUICKEST in smooth (low-curvature) regions; ULTIMATE seventh-order in moderate-curvature regions; ULTIMATE ninth-order in high-curvature or high-gradient regions (solid dots). Small arrows indicate automatic relaxation of the limiter.

ACKNOWLEDGMENT

Part of the first author's research was supported by the Institute for Computational Mechanics in Propulsion (ICOMP), NASA-Lewis Research Center (under Space Act Agreement C99066G).

REFERENCES

1. Leonard, B.P., A Stable and Accurate Convective Modelling Proce-
    dure Based on Quadratic Upstream Interpolation, *Computer Methods
    in Applied Mechanics and Engineering*, vol. 19, pp 59-98, 1979.

2. Leonard, B.P., Universal Limiter for Transient Interpolation
    Modelling of the Advective Transport Equations: The ULTIMATE
    Conservative Difference Scheme, NASA TM-100916, ICOMP-88-11,
    NASA-Lewis Research Center, September 1988.

3. Mitchell, A.R., Recent Developments in the Finite Element Method,
    in *Computational Techniques and Applications: CTAC-83*, J. Noye
    and C.A.J. Fletcher, eds., pp. 2-14, Elsevier North-Holland, 1984.

4. Roe, P.L., Characteristic-Based Schemes for the Euler Equations,
    in *Annual Reviews of Fluid Mechanics*, vol. 18, M. van Dyke, J.V.
    Wehausen, and J.L. Lumley, eds., pp. 337-365, Annual Reviews Inc.,
    1986.

5. Sweby, P.K., High Resolution Schemes Using Flux Limiters for
    Hyperbolic Conservation Laws, *SIAM Journal of Numerical Analysis*,
    vol. 21, no. 5, pp. 995-1011, 1984.

6. Yee, H.C., Upwind and Symmetric Shock-Capturing Schemes, NASA
    TM-89464, NASA-Ames Research Center, March 1988.

Second and Third Order Two-Level Implicit FDM's for Unsteady One-Dimensional Convection-Diffusion

B. P. LEONARD
Department of Mechanical Engineering
The University of Akron
Akron, Ohio, USA

B. J. NOYE
Department of Applied Mathematics
The University of Adelaide
G.P.O. Box 498
Adelaide 5001, South Australia

INTRODUCTION

Consider unsteady one-dimensional convection and diffusion of a scalar, $\rho(x,t)$,

$$\frac{\partial \rho}{\partial t} + \frac{\partial(u\rho)}{\partial x} = \frac{\partial}{\partial x}\left(D\frac{\partial \rho}{\partial x}\right) + S \tag{1}$$

in which u is velocity, D is a diffusion coefficient and S represents a source term. Equation (1) can be integrated in time over an interval Δt and in space from $-\Delta x/2$ to $+\Delta x/2$, assuming a uniform mesh. This gives

$$\overline{\rho}_i^{n+1} = \overline{\rho}_i^n + (\overline{c\rho})_\ell - (\overline{c\rho})_r + \left[\left(\overline{\alpha\frac{\partial \rho}{\partial x}}\right)_r - \left(\overline{\alpha\frac{\partial \rho}{\partial x}}\right)_\ell\right]\Delta x + \Delta t S^* \tag{2}$$

where the bars represent spatial averaging, the c's are Courant numbers, the α's are diffusion parameters, and S^* is a space-time average. The subscript notation refers to left and right control-volume (CV) faces. Equation (2) is exact; in order to construct a numerical algorithm, approximations are needed for the spatial averages and the CV face values and gradients. At the left face, for example, $(\overline{c\rho})_\ell$ represents a time average of the product, which will be written as the product of time averages, $c_\ell\rho_\ell$, neglecting the correlation term; similarly for the diffusion term. In the following, two related methods are described for estimating ρ_ℓ and $(\partial\rho/\partial x)_\ell$; because of strict CV conservation, the right-face formulas are obtained by increasing i by 1. Consistent formulas are found for $\overline{\rho}_i$. Both methods make use of convective characteristics. Figure 1 shows a sloping characteristic passing through node $(i-1)$ at time-level n. Note the definition of local coordinates (ξ,τ) relative to this characteristic. If $\rho(\xi,\tau)$ were known in the vicinity of the left face, "ℓ", the appropriate time-averages of value and gradient could be calculated. In the first case, a bilinear space-time behaviour is assumed; this results in an implicit method which is second-order accurate in both space and time. In the second case, an additional quadratic spatial term is added, resulting in a third-order implicit method. In each case, the equation being modelled is

$$\overline{\rho}_i^{n+1} = \overline{\rho}_i^n + c_\ell\rho_\ell - c_r\rho_r + \left[\alpha_r\left(\frac{\partial \rho}{\partial x}\right)_r - \alpha_\ell\left(\frac{\partial \rho}{\partial x}\right)_\ell\right]\Delta x + \Delta t S^* \tag{3}$$

TIME AVERAGES

If $\rho = \rho(\xi,\tau)$, the time averaged left-face value is given by, referring to Figure 1,

$$\rho_\ell = \frac{1}{\Delta t}\int_0^{\Delta t} \rho\left(\frac{\Delta x}{2} - u_\ell\tau, \tau\right) d\tau \tag{4}$$

where u_ℓ is the local average convecting velocity near the left face.

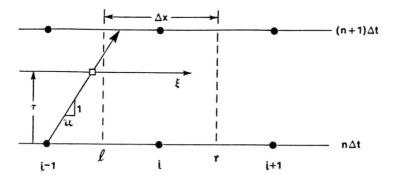

FIGURE 1. Characteristic-based coordinate system (ξ, τ) near left CV face.

The time-averaged left-face gradient is given by

$$\left(\frac{\partial \rho}{\partial x}\right)_\ell = \frac{1}{\Delta t} \int_0^{\Delta t} \frac{\partial \rho}{\partial \xi} \left(\frac{\Delta x}{2} - u_\ell \tau, \tau\right) d\tau \tag{5}$$

Bilinear Space–Time Variation

Assume that, in the vicinity of the left face,

$$\rho = \rho^L(\xi, \tau) = (\rho_{i-1}^n + L_1 \tau) + \left(\frac{\rho_i^n - \rho_{i-1}^n}{\Delta x} + L_2 \tau\right) \xi \tag{6}$$

Note that this gives the correct node values at time-level n $(\tau = 0)$. The coefficients L_1 and L_2 are determined from conditions at time-level $n + 1$:

$$\rho_{i-1}^{n+1} = \rho^L(-u_\ell \Delta t, \Delta t) \quad \text{and} \quad \rho_i^{n+1} = \rho^L(\Delta x - u_\ell \Delta t, \Delta t) \tag{7}$$

After a few lines of algebraic manipulation, the time-averaged left-face value is found to be

$$\rho_\ell^L(i) = \frac{(\rho_i^{n+1} + \rho_{i-1}^{n+1} + \rho_i^n + \rho_{i-1}^n)}{4} + \frac{c_\ell}{6} \left[\left(\rho_i^{n+1} - \rho_{i-1}^{n+1}\right) - \left(\rho_i^n - \rho_{i-1}^n\right)\right] \tag{8}$$

Note that this consists of the arithmetic mean of the node-values surrounding the left face in Figure 1, modified by a space-time "twist" term. The right-face value is obtained by conservation: $\rho_r^L(i) = \rho_\ell^L(i + 1)$.

Similarly, the gradient turns out to be a simple arithmetic mean

$$\left(\frac{\partial \rho^L}{\partial x}\right)_\ell = \frac{1}{2} \left(\frac{\rho_i^{n+1} - \rho_{i-1}^{n+1}}{\Delta x} + \frac{\rho_i^n - \rho_{i-1}^n}{\Delta x}\right) \tag{9}$$

Upstream–Weighted Quadratic Behaviour

In this case assume, for $c_\ell > 0$, that

$$\rho = \rho^Q(\xi, \tau) = (\rho^n_{i-1} + Q_1\tau) + \left(\frac{\rho^n_i - \rho^n_{i-2}}{2\Delta x} + Q_2\tau\right)\xi + \left(\frac{\rho^n_i - 2\rho^n_{i-1} + \rho^n_{i-2}}{2(\Delta x)^2}\right)\xi^2 \tag{10}$$

again agreeing with node values at $\tau = 0$. If $c_\ell < 0$, upstream weighting requires the second-difference to be shifted one node to the right (increase i by 1 in the quadratic term). No time-dependence is assumed in the spatially quadratic term; this retains a tridiagonal structure in the implicit update algorithm. The Q-coefficients are evaluated from

$$\rho^{n+1}_{i-1} = \rho^Q(-u_\ell\Delta t, \Delta t) \quad \text{and} \quad \rho^{n+1}_i = \rho^Q(\Delta x - u_\ell\Delta t, \Delta t) \tag{11}$$

Again, after some algebra, ρ^Q_ℓ can be written

$$\rho^Q_\ell(i) = \rho^L_\ell(i) - \left(\frac{1}{8} - \frac{c^2_\ell}{12}\right)(\rho^n_i - 2\rho^n_{i-1} + \rho^n_{i-2}), \quad c_\ell > 0 \tag{12}$$

$$= \rho^L_\ell(i) - \left(\frac{1}{8} - \frac{c^2_\ell}{12}\right)(\rho^n_{i+1} - 2\rho^n_i + \rho^n_{i-1}), \quad c_\ell < 0 \tag{13}$$

Because the quadratic term in Equation (10) does not contribute to the integral in Equation (5), the formula for the gradient is the same as Equation (9).

SPATIAL AVERAGES

The spatial averages in Equation (3) can be estimated as

$$\overline{\rho}^{n+1}_i = \rho^{n+1}_i + \frac{1}{24}\left(\rho^{n+1}_{i+1} - 2\rho^{n+1}_i + \rho^{n+1}_{i-1}\right) \tag{14}$$

with a similar formula for $\overline{\rho}^n_i$. Since this is consistent with a tridiagonal update, it will be used for both the second-order and third-order algorithms.

UPDATE ALGORITHMS

Second–Order Method

Substitution of Equations (8) and (9) and the corresponding right-face formulas into Equation (3), along with spatial averages corresponding to Equation (14), results in a tridiagonal implicit update formula of the form

$$a^L_{-1}\rho^{n+1}_{i-1} + a^L_0\rho^{n+1}_i + a^L_1\rho^{n+1}_{i+1} = b^L_{-1}\rho^n_{i-1} + b^L_0\rho^n_i + b^L_1\rho^n_{i+1} + \Delta t S^* \tag{15}$$

In this case, the a's and b's are given by

$$a^L_{-1} = \frac{1}{24} - \frac{c_\ell}{4} + \frac{c^2_\ell}{6} - \frac{\alpha_\ell}{4} \tag{16}$$

$$a^L_0 = \frac{11}{12} + \frac{c_r}{4} - \frac{c^2_r}{6} - \frac{c_\ell}{4} - \frac{c^2_\ell}{6} + \frac{\alpha_r}{2} + \frac{\alpha_\ell}{2} \tag{17}$$

$$a^L_1 = \frac{1}{24} + \frac{c_r}{4} + \frac{c^2_r}{6} - \frac{\alpha_r}{4} \tag{18}$$

$$b^L_{-1} = \frac{1}{24} + \frac{c_\ell}{4} + \frac{c^2_\ell}{6} + \frac{\alpha_\ell}{2} \tag{19}$$

$$b^L_0 = \frac{11}{12} - \frac{c_r}{4} - \frac{c^2_r}{6} + \frac{c_\ell}{4} - \frac{c^2_\ell}{6} - \frac{\alpha_r}{2} - \frac{\alpha_\ell}{2} \tag{20}$$

$$b^L_1 = \frac{1}{24} - \frac{c_r}{4} + \frac{c^2_r}{6} + \frac{\alpha_r}{4} \tag{21}$$

Third–Order Method

Taking account of both positive and negative convecting velocities results in a five-point stencil at time-level n. The implicit update can thus be written

$$
\begin{aligned}
a^L_{-1}\rho^{n+1}_{i-1} + a^L_0 \rho^{n+1}_i + a^L_1 \rho^{n+1}_{i+1} &= b^Q_{-2}\rho^n_{i-2} + b^Q_{-1}\rho^n_{i-1} + b^Q_0 \rho^n_i \\
&+ b^Q_1 \rho^n_{i+1} + b^Q_2 \rho^n_{i+2} + \Delta t S^*
\end{aligned} \tag{22}
$$

Note that the $(n+1)$ time-level coefficients are the same as those of the second-order case; but additional terms appear on the right-hand side. The b-coefficients are given by

$$
b^Q_{-2} = -\left(\frac{1}{16} - \frac{c^2_\ell}{24}\right)(c_\ell + |c_\ell|) \tag{23}
$$

$$
b^Q_{-1} = b^L_{-1} + \left(\frac{1}{16} - \frac{c^2_\ell}{24}\right)(c_\ell + 2|c_\ell|) + \left(\frac{1}{16} - \frac{c^2_r}{24}\right)(c_r + |c_r|) \tag{24}
$$

$$
b^Q_0 = b^L_0 - \left(\frac{1}{16} - \frac{c^2_r}{24}\right)(c_r + 3|c_r|) + \left(\frac{1}{16} - \frac{c^2_\ell}{24}\right)(c_\ell - 3|c_\ell|) \tag{25}
$$

$$
b^Q_1 = b^L_1 - \left(\frac{1}{16} - \frac{c^2_r}{24}\right)(c_r - 3|c_r|) - \left(\frac{1}{16} - \frac{c^2_\ell}{24}\right)(c_\ell - |c_\ell|) \tag{26}
$$

$$
b^Q_2 = \left(\frac{1}{16} - \frac{c^2_r}{24}\right)(c_r - |c_r|) \tag{27}
$$

The use of the absolute-values of the Courant numbers automatically takes account of upwinding without using "IF" statements in a computer program.

CONSTANT COEFFICIENTS

For $c_\ell = c_r = c$ and $\alpha_\ell = \alpha_r = \alpha$, Equation (15) becomes, after multiplying through by 24,

$$
\begin{aligned}
(1 - 6c + 4c^2 - 12\alpha)\rho^{n+1}_{i-1} + (22 - 8c^2 + 24\alpha)\rho^{n+1}_i + (1 + 6c + 4c^2 - 12\alpha)\rho^{n+1}_{i+1} \\
= (1 + 6c + 4c^2 + 12\alpha)\rho^n_{i-1} + (22 - 8c^2 - 24\alpha)\rho^n_i + (1 - 6c + 4c^2 + 12\alpha)\rho^n_{i+1} + 24\Delta t S^*
\end{aligned} \tag{28}
$$

Similarly, for constant $c(> 0)$ and α, Equation (22), multiplied by 24, is equivalent to Equation (28) with the addition of the term

$$
-c\left(3 - 2c^2\right)\left(\rho^n_{i-2} - 3\rho^n_{i-1} + 3\rho^n_i - \rho^n_{i+1}\right) \tag{29}
$$

to the right-hand side. Note that for pure convection ($\alpha = 0$) at $c = 1$, Equation (22) becomes (for $S = 0$)

$$
-(1)\rho^{n+1}_{i-1} + (14)\rho^{n+1}_i + (11)\rho^{n+1}_{i+1} = -(1)\rho^n_{i-2} + (14)\rho^n_{i-1} + (11)\rho^n_i + (0)\rho^n_{i+1} \tag{30}
$$

which is consistent with exact point-to-point transfer across one mesh-width: $\rho^n_{i-2} \rightarrow \rho^{n+1}_{i-1}$, $\rho^n_{i-1} \rightarrow \rho^{n+1}_i$, and $\rho^n_i \rightarrow \rho^{n+1}_{i+1}$.

Stability

The von Neumann method has been used to establish the values of (c, α) for which these constant coefficient finite difference equations are stable. This technique was chosen in preference to others, such as the matrix method, since it can be used with consistency to establish convergence of the numerical solutions (see Smith, 1985).

The second order equation (28) is stable if the values of (c, α) lie in the vertically shaded region of Figure 2(a), and the third order equation is stable in the vertically shaded region of Figure 2(b).

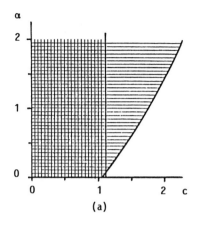
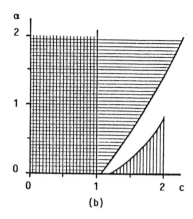

(a)   (b)

FIGURE 2. Regions in which (a) the second-order (b) the third order constant coefficient schemes are stable (vertical shading) and diagonally dominant (horizontal shading)

Diagonal Dominance

Both the second and third order implicit equations may be solved by means of the Thomas (1949) method, if the resulting tridiagonal set of linear algebraic equations has a coefficient matrix which is diagonally dominant. In both cases, this is true if

$$|22 - 8c^2 + 24\alpha| \geq |1 - 6c + 4c^2 - 12\alpha| + |1 + 6c + 4c^2 - 12\alpha| \tag{31}$$

This inequality holds in the horizontally shaded regions of Figures 2(a), (b).

The equations may, therefore, be used to solve (1) in the cross-hatched regions of Figure 2.

Truncation Error

For the constant-coefficient model equations, the truncation error can be evaluated using the Modified-Equivalent-Partial-Differential-Equation (MEPDE) method (Noye and Hayman, 1986). For example, the algorithm given by Equation (28) is equivalent to Equation (1) with additional truncation-error (TE) terms. If the TE terms are considered to be placed on the left-hand side of Equation (1), the leading term corresponding to Equation (28) is a dispersion term (involving a third spatial derivative)

$$TE(2) = u\frac{\Delta x^2}{24}(3 - 2c^2)\frac{\partial^3 \rho}{\partial x^3} + \dots \tag{32}$$

thus confirming the second-order accuracy of this method. Since $TE(2) > 0$ for $0 < c < \sqrt{3/2}$, the numerical wave speed is always too slow when Δx is small (see Noye, 1988) causing numerically propagated peaks to lag behind their true positions.

The leading term in (32) is exactly cancelled by the third-difference term given by the expression (29), thus giving third-order accuracy. The truncation error for the MEPDE corresponding to the third-order method is

$$TE(3) = u\frac{\Delta x^3}{48}\left((1 - c)(3 - 2c^2) - 2\alpha c^{-1}(1 + 2c^2)\right)\frac{\partial^4 \rho}{\partial x^4} + \dots \tag{33}$$

FIGURE 3. Step convection at $P_\Delta = \infty$ (a) Second order (b) Third order

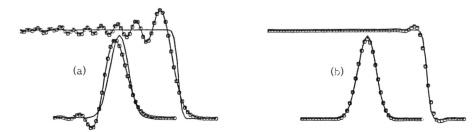

FIGURE 4. Gaussian peak and step $P_\Delta = 100$ (a) Second order (b) Third order

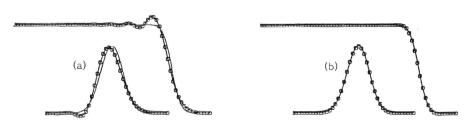

FIGURE 5. Gaussian peak and step $P_\Delta = 20$ (a) Second order (b) Third order

FIGURE 6. Gaussian peak and step $P_\Delta = 5$ (a) Second order (b) Third order

The leading term in (33) represents an error in amplitude of propagated waves, and the choice

$$\alpha = c(1 - c)(3 - 2c^2)/2(1 + 2c^2) \tag{34}$$

minimises this error.

Numerical Experiments

The second and third order methods have been tested by simulating convection-diffusion of a step function, and a Gaussian pulse with variance $\sigma = 3\Delta x$, as initial conditions. In both cases the Courant number chosen was $c = 0.5$, and the simulation was for 130 time-steps; that is, both profiles moved a distance of $65\Delta x$ to the right from their initial position.

Four values of the diffusion number α were used, namely $\alpha = 0, 5 \times 10^{-3}, 2.5 \times 10^{-2}, 10^{-1}$. These correspond to values of the grid Peclet number

$$P_\Delta = c/\alpha \tag{35}$$

given by $P_\Delta = \infty, 100, 20, 5$ respectively. The results are shown in Figures 3-6.

Typically, the second order scheme is more wiggly at smaller Courant numbers; other tests indicate that it remains oscillatory even to $c = 1$ when $\alpha \to 0$. The third order method gives results which are rather insensitive to the value of c used.

CONCLUSION

The fact that the third order scheme involves no dispersion error, such as that which occurs in the second order method, means that wave speed is much better simulated. This is evident in the results of the numerical experiments involving the propagation of a step and a Gaussian pulse; for large values of the grid Peclet number, P_Δ, results from the lower order scheme exhibit large trailing oscillations which reduce in amplitude as P_Δ is decreased, while the oscillations in the higher order scheme are much smaller. When P_Δ is small, such as $P_\Delta = 5$, the results obtained from both methods are similar.

REFERENCES

1. Noye, B.J., and Hayman, K.J., Accurate finite difference methods for solving the advection-diffusion equation, *Computational Techniques and Applications: CTAC-85*, eds. B.J. Noye and R.L. May, pp.137-158, Elsevier Science Publishers B.V. (North-Holland), 1986.

2. Noye, B.J., Finite-difference methods for the one-dimensional transport equation, *Computational Techniques and Applications: CTAC-87*, eds. B.J. Noye and C.J. Fletcher, pp.539-562, Elsevier Science Publishers B.V. (North-Holland), 1988.

3. Smith, G.D., *Numerical Solution of Partial Differential Equations - Finite Difference Methods* (third ed.), Clarendon Press, Oxford, 1985.

4. Thomas, L.H., Elliptic Problems in Linear Difference Equations over a Network, Watson Scientific Computing Laboratory, Columbia University, New York, 1949.

A Model of the Deployment of a Towed Body

R. L. MAY and H. J. CONNELL
Royal Melbourne Institute of Technology
G.P.O. Box 2476 V
Melbourne, Victoria 3001, Australia

INTRODUCTION

This paper describes a model of the dynamic motion of a towed system consisting of a module attached to a thin inextensible cable which is payed out from an aircraft flying horizontally at a constant velocity. Only motion in a vertical plane is considered, and the cable is assumed to be homogeneous, of constant diameter and have no resistance to bending. The aim of the model is to determine the position, velocity and tension of the cable as it is deployed from the aircraft.

A survey of the literature has revealed that while some work has been done on the modelling of airborne systems (see the references of Karlsen (1981)), underwater cables have been subject to more detailed investigation (see for example Sanders (1982) and Ablow & Schechter (1983)). It should be noted that underwater cables are more easily modelled numerically because greater damping is present in the system. Although many of the previous investigations are more general in allowing motion in three dimensions, none consider the payout phase of the system.

MATHEMATICAL MODEL

The motion of the cable is governed by the four partial differential equations

$$\frac{\partial T}{\partial s} = -m\frac{\partial u}{\partial t} + mv\frac{\partial \phi}{\partial t} + f_1, \tag{1}$$

$$\frac{\partial u}{\partial s} = -\frac{mv}{T}\frac{\partial v}{\partial t} - \frac{muv}{T}\frac{\partial \phi}{\partial t} - \frac{vf_2}{T}, \tag{2}$$

$$\frac{\partial v}{\partial s} = \frac{mu}{T}\frac{\partial v}{\partial t} + \left(\frac{mu^2}{T} - 1\right)\frac{\partial \phi}{\partial t} + \frac{uf_2}{T}, \tag{3}$$

$$\frac{\partial \phi}{\partial s} = -\frac{m}{T}\frac{\partial v}{\partial t} - \frac{mu}{T}\frac{\partial \phi}{\partial t} - \frac{f_2}{T}, \tag{4}$$

where

$$f_1 = \frac{\pi}{2}\rho dc_t \left|V\cos\phi - u\right|(V\cos\phi - u) + mg\sin\phi, \tag{5}$$

$$f_2 = \frac{1}{2}\rho dc_n \left|V\sin\phi + v\right|(V\sin\phi + v) - mg\cos\phi, \tag{6}$$

319

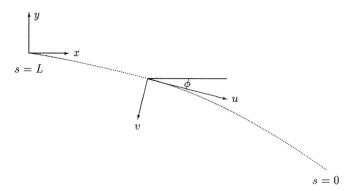

FIGURE 1: Orientation of axes used in the model.

and s is the distance along the cable from the module end, u and v are the tangential and normal components of velocity of the cable relative to the aircraft, ϕ is the angle the cable makes with the horizontal (see Figure 1), T is the cable tension, m is the mass per unit length of the cable, ρ is the density of air, d is the cable diameter, c_t and c_n are the tangential and normal drag coefficients for the cable, V is the speed of the towing aircraft and g is the gravitational acceleration constant. These equations are based on the assumption that the normal and tangential components of drag on the cable are proportional to the square of the normal and tangential velocity components respectively, and are derived from two *momentum equations* and two *compatibility equations*.

There are four boundary conditions. Assuming the module is non-lifting, the boundary conditions at the module end ($s = 0$) are

$$-T + Mg \sin \phi + D_H \cos \phi - D_V \sin \phi - M\frac{\partial u}{\partial t} + Mv\frac{\partial \phi}{\partial t} = 0, \qquad (7)$$

$$Mg \cos \phi - D_H \sin \phi - D_V \cos \phi - M\frac{\partial v}{\partial t} - Mu\frac{\partial \phi}{\partial t} = 0, \qquad (8)$$

where D_V and D_H are the vertical and horizontal components of the drag force on the module and M is the mass of the module. The drag is again assumed to be proportional to the square of the velocity. If the horizontal and vertical drag coefficients of the module are denoted by c_H and c_V and the cross-sectional areas of the module in the planes perpendicular to the vertical and horizontal directions are denoted by A_H and A_V then

$$D_H = \frac{1}{2}\rho c_H A_H \left|V - u \cos \phi + v \sin \phi\right| (V - u \cos \phi + v \sin \phi), \qquad (9)$$

$$D_V = \frac{1}{2}\rho c_V A_V \left|v \cos \phi + u \sin \phi\right| (v \cos \phi + u \sin \phi). \qquad (10)$$

At the aircraft end of the cable ($s = L$) the boundary conditions are

$$u = u_p, \qquad (11)$$
$$v = 0, \qquad (12)$$

where u_p is the known payout velocity of the cable from the aircraft.

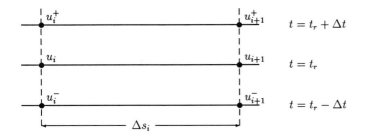

FIGURE 2: Computational grid—variable u shown.

Equations (1) – (12) together with suitable initial conditions define T, u, v and ϕ as functions of s and t. The position of the cable relative to the aircraft at time t (see Figure 1) is found by solving

$$\frac{\partial x}{\partial s} = \cos \phi, \qquad \frac{\partial y}{\partial s} = -\sin \phi, \qquad x(L,t) = y(L,t) = 0. \tag{13}$$

A derivation of the above equations may be found in May & Connell (1988).

NUMERICAL SCHEME

The cable is divided into $n-1$ segments of length Δs_i, $i = 1, 2, \ldots, n-1$ by the nodes located at $s = s_i$, $i = 1, 2, \ldots, n$ where $0 = s_1 < s_2 < s_3 < \ldots < s_n = L$, and

$$\Delta s_i = s_{i+1} - s_i. \tag{14}$$

Approximations to T, u, v and ϕ are calculated at the nodes at times $t_k = t_0 + k\delta t$ for $k = 1, 2, \ldots$ where t_0 is the time at which initial values of these variables are known. At some reference time t_r the values of the variables at node i are denoted by T_i, u_i, v_i and ϕ_i, at time $t_{r-1} = t_r - \Delta t$ by T_i^-, u_i^-, v_i^- and ϕ_i^-, and at time $t_{r+1} = t_r + \Delta t$ by T_i^+, u_i^+, v_i^+ and ϕ_i^+ (see Figure 2).

Equations (1)–(4) are discretized using approximations of the form

$$\frac{\partial T}{\partial s} \approx \alpha \left(\frac{T_{i+1}^+ - T_i^+}{\Delta s_i} \right) + \beta \left(\frac{T_{i+1} - T_i}{\Delta s_i} \right), \tag{15}$$

$$mv\frac{\partial \phi}{\partial t} \approx \frac{m}{2} \left\{ \left(\alpha v_i^+ + \beta v_i \right) \left[\gamma \left(\frac{\phi_i^+ - \phi_i}{\Delta t} \right) + \theta \left(\frac{\phi_i - \phi_i^-}{\Delta t} \right) \right] \right.$$
$$\left. + \left(\alpha v_{i+1}^+ + \beta v_{i+1} \right) \left[\gamma \left(\frac{\phi_{i+1}^+ - \phi_{i+1}}{\Delta t} \right) + \theta \left(\frac{\phi_{i+1} - \phi_{i+1}^-}{\Delta t} \right) \right] \right\}, \tag{16}$$

$$f_1 \approx \frac{1}{2} \left[\left(\alpha f_1^+{}_i + \beta f_{1i} \right) + \left(\alpha f_1^+{}_{i+1} + \beta f_{1i+1} \right) \right], \tag{17}$$

where α, β, γ and θ are weights such that $\alpha + \beta = 1$ and $\gamma + \theta = 1$. These approximations are centred at a point midway between nodes i and $i + 1$ at a time which is dependent on the value of the weights. The boundary conditions at $s = 0$, equations (7) and (8), are differenced in a similar way. At the aircraft end the situation is complicated by the fact that the cable is being payed out. The approximate length of cable payed out in the time interval from $t = t_r$ to $t = t_r + \Delta t$ is calculated using

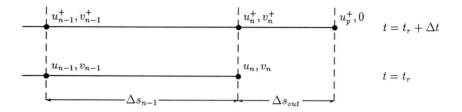

FIGURE 3: Location of node n at time $t = t_r + \Delta t$.

Simpson's rule to give Δs_{out}. This is the distance of node n from the cable end at time $t = t_r + \Delta t$, and as values of the velocity components are known at the end of the cable, linear interpolation may be used to give u_n^+ and v_n^+ (see Figure 3).

These approximations results in a nonlinear system of equations which may be solved for the unknowns T_i^+, u_i^+, v_i^+ and ϕ_i^+, $i = 1, 2, \ldots, n$. These equations were originally solved using an IMSL subroutine, but in order to reduce the storage requirements and execution time a scheme using functional iteration was developed. However both methods fail to converge when the end of the cable moves very rapidly, a situation that is more likely for long cables far from equilibrium.

Once the solution has been found at the "new" time level, a decision must be made as to whether or not an additional segment is needed. A new segment is added if the length of the last segment Δs_{n-1} is larger than some prescribed length Δs_{min} and $\Delta s_{out} > 0$, in which case $\Delta s_n = \Delta s_{out}$ and linear extrapolation used to obtain estimates of T_{n+1}^+ and ϕ_{n+1}^+ at the end of the cable. If $\Delta s_{n-1} < \Delta s_{min}$ or $\Delta s_{out} = 0$ then a new node is not added, but node n is shifted to the cable end. Linear extrapolation is used to calculate values of T_n^+ and ϕ_n^+ at the cable end, and Δs_{n-1} is increased by Δs_{out}. In either case the variables T, u, v and ϕ at the previous time level are also extrapolated to a fictitious node which has the same value of s as the new node. Figure 4 illustrates the way in which the number of elements is increased as the cable is payed out.

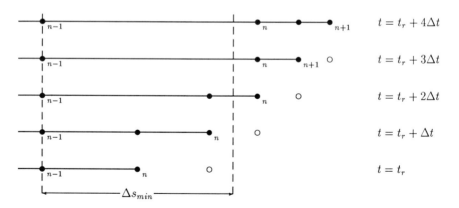

FIGURE 4: The number of nodes is increased at $t = t_r + 3\Delta t$ when $\Delta s_{n-1} \geq \Delta s_{min}$. The fictitious nodes are indicated by the small circles.

The cable is divided into elements which are not necessarily of equal length. Experience has shown that it is desirable to have shorter elements near the module end of the cable so Δs_{min} is multiplied by a factor larger than 1 (typically 1.1) each time a new element is added so that the lengths of the elements increase in an approximately geometric way.

The scheme described allows the numerical solution to be advanced in time once the an approximation to the solution is known at the nodes at two consecutive time steps. Initially the solution is known at one time only, so the first step is carried out with $\gamma = 1$ (and therefore $\theta = 0$). Many numerical experiments were performed to find the best values of the weights α and γ. Although many values of the weights produced a large amount of numerical damping, the numerical scheme was found to converge a solution which was independent of the weights. The values which gave the most rapid convergence were close to $\alpha = \frac{1}{\sqrt{3}}$ and $\gamma = \frac{1}{2} + \frac{1}{\sqrt{3}}$.

RESULTS

In this section some results are presented using the values $V = 150\text{m/sec}$, $m = 3\text{g/m}$, $d = 2\text{mm}$, $c_t = 0.01$, $c_n = 1.2$, $\rho = 1.2256\text{kg/m}^3$, $M = 35\text{kg}$, $c_H = 1.1$, $c_V = 1.4$, $A_H = 0.05\text{m}^2$ and $A_V = 0.6\text{m}^2$.

In the first example, a cable of constant length is released from a horizontal position. To determine the accuracy of the scheme, the three combinations $(11, 0.02)$, $(11, 0.1)$ and $(6, 0.02)$ of the number of nodes n and the time step Δt were used.

A plot of the y-coordinate of the position of the module is shown in Figure 5. This shows that when the module is released it oscillates about its equilibrium position with the small amount of damping in the system causing the amplitude of oscillation to decay slowly with time. The results for the two time steps are plotted and show that while the period of oscillation depends slightly on the time step the amplitude is virtually unchanged. The results for $n = 6$ are not plotted because they can not be

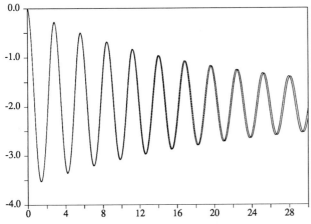

FIGURE 5: The vertical displacement of the module (m) plotted against time (sec) for a 5 metre cable released from rest from a horizontal position. Results for the time steps $\Delta t = 0.02$ and $\Delta t = 0.1$ are shown.

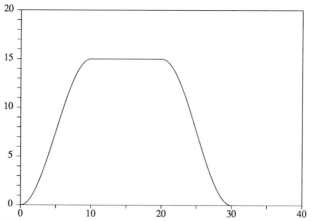

FIGURE 6: Payout velocity (m/sec) plotted against time (sec).

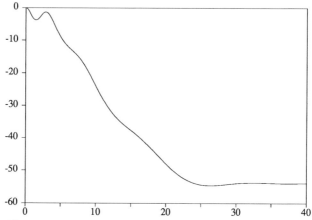

FIGURE 7: Vertical displacement of module (m).

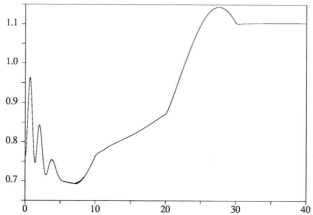

FIGURE 8: Tension (kN) at the aircraft end of the cable.

distinguished from those with $n = 11$, the maximum relative difference in values being less than 0.06%.

In the second example 300 metres of cable are payed out in 30 seconds. The payout velocity is shown in Figure 6. The cable was started from a horizontal position with a length of 5 metres, and in view of the results of the first example, the parameters N=6 and $\Delta t = 0.1$ were tried. However, it was found that a small amount of numerical instability became apparent after about 5 seconds (when the rate of increase of u_p is a maximum), so the time step was reduced to $\Delta t = 0.05$ seconds.

In Figure 7 the y-coordinate of the module is plotted against time. There are a couple of oscillations in the first few seconds before much cable is payed out, but the oscillations rapidly die away as the payout velocity increases. As the cable lengthens the module drops further below the aircraft, and as the payout velocity reduces to zero the module slightly overshoots then slowly oscillates about the equilibrium position for a cable of length 305 metres.

A plot of the tension at the aircraft end of the cable is given in Figure 8, and while the reduction of the time step reduced the numerical instability, some can be still seen in the interval from about 5 seconds to 10 seconds. The tension oscillates in the first 5 seconds, with the mean tension dropping as the payout velocity increases. As the rate of acceleration reduces, the tension increases rapidly until between 10 and 20 seconds, when the payout velocity is constant, the tension increases at a lower but almost constant rate due to the lengthening of the cable. As the cable is slowed the rate of increase in tension again increases until the payout velocity and deceleration both become small when the tension drops off to approximately the equilibrium value at $t = 30$ seconds and remains virtually constant thereafter.

Figure 9 shows the cable shape at 5 second intervals. The positions of the nodes are also shown, and it can be seen that the lengths of successive elements increase by about 10% as the cable lengthens and the number of elements increases from 6 to 39.

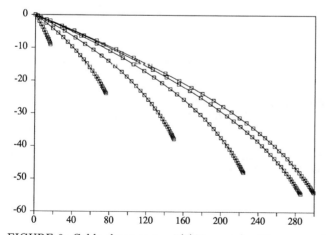

FIGURE 9: Cable shape at $t = 5(5)30$ seconds. The squares indicate the position of the nodes on the cable.

VALIDATION OF MODEL

Validation of the model is extremely difficult because of the lack of suitable experimental data. The only available data is some tension measurements obtained by the Advanced Engineering Laboratory, but unfortunately these were recorded for a cable of more than 5 kilometres length consisting of several sections of increasing diameters (a "stepped cable"). The accuracy of this data is uncertain because the tension gauge had widely spaced calibrations, and the accuracy also depends on the aircraft precisely maintaining a constant altitude and speed. Due to security considerations, details such as the module weight, cable diameter, payout velocity and aircraft speed cannot be given.

Using the cable diameter and mass per unit length of the second section (roughly average values for the entire cable), the model overestimates all but the last measurement because the first section of the cable used in the flight is of a smaller diameter than that used in the model. The first two measurements were made when the diameter of cable being payed out was the same as that used in the model, and it is encouraging that the difference in values is approximately constant.

Clearly the model can not be expected to reproduce the experimental results because they were obtained using a stepped cable whereas the model assumes the cable is of constant diameter and mass per unit length. Nevertheless the model produced results which are consistent with the experimental data, and with some fine-tuning of the drag coefficients would probably accurately reproduce experimental results for a constant diameter cable.

ACKNOWLEDGMENTS

This work was carried out under contract to the Advanced Engineering Laboratory, Defence Research Centre, Salisbury, South Australia. The authors would like to thank J. Barton of the Advanced Engineering Laboratory, for his assistance with this project.

REFERENCES

1. Ablow, C.M. and Schechter S., Numerical Simulation of Undersea Cable Dynamics, *Ocean Engineering*, vol. 10, no. 6, pp. 443–457, 1983.

2. Karlsen, L.K., Large Scale Dynamics of Long Flexible Cables Towed through Air, Technical Note KTH AERO TN 61, Department of Aeronautics, Royal Institute of Technology, Stockholm, Sweden, 1981.

3. May, R.L. and Connell, H.J., Towed Systems Modelling Study, Technical Report No. 10, Department of Mathematics, Royal Melbourne Institute of Technology, 1988.

4. Sanders, J.V., A Three-dimensional Dynamic Analysis of a Towed System, *Ocean Engineering*, vol. 9, No 5, pp. 483–499, 1982

Combined Advection and Diffusion in an Implicit Flux-Corrected Transport Algorithm

R. MORROW
CSIRO Division of Applied Physics
Sydney, Australia, 2070

P. STEINLE
Bureau of Meteorology
Perth, Western Australia, 6004

INTRODUCTION

Gas discharge calculations often involve the transport and diffusion of electrons and diffusion can dominate part of the calculation (Morrow, 1985). The prototype equation we wish to solve efficiently and accurately is

$$\frac{\partial \rho}{\partial t} = -\frac{\partial (w\rho)}{\partial x} + \frac{\partial}{\partial x}\left(D\frac{\partial \rho}{\partial x}\right) \; , \tag{1}$$

where ρ is the particle density, w is the drift velocity and D the diffusion coefficient. For an explicit finite difference solution of this equation the limitation for stable solution of the transport term is that the Courant number $c \leq 1$, where $c = \delta t w/\delta x$, δt is the time step and δx is the mesh size. To overcome this problem we have developed an implicit flux-corrected transport (FCT) scheme (Steinle and Morrow, 1989) which can solve the equation, neglecting diffusion, for $c \gg 1$ and give reasonably accurate non-negative results. However, in some cases (Morrow, 1985) an explicit solution of the equations is limited by the von Neumann condition $s \leq 0.5$ for the solution of the diffusion term, where $s = \delta t D/\delta x^2$. Further, since c depends on $1/\delta x$, s depends on $1/\delta x^2$, and the calculations are performed on a variable mesh there will be regions where diffusion dominates the numerical calculation and regions where it is negligible. Thus we wish to include diffusion in the implicit transport algorithm previously developed so that the one algorithm can be used throughout the calculation.

We introduce second order diffusion to the fourth-order implicit FCT scheme (Steinle and Morrow, 1989) and use this scheme as the high-order solution following Zalesak's method of developing flux-corrected schemes (Zalesak, 1979). We restrict our attention to conditions where diffusion dominates and we find that the fully implicit transport and diffusion scheme provides the required non-negative results for the lower-order solution for the FCT scheme.

NUMERICAL METHODS

We use the implicit fourth-order space- and time-centred scheme developed by Steinle and Morrow (1989) as the high-order solution for an FCT algorithm, and incorporate second order diffusion in the form developed by Morrow and Cram (1985). The equation, written in variable mesh and variable velocity form, is

$$
\delta x_j \overline{\rho}_j + \tfrac{1}{4}[\delta x_{j+1/2} c_{j+1/2}(\overline{\rho}_{j+1} + \overline{\rho}_j) - \delta x_{j-1/2} c_{j-1/2}(\overline{\rho}_j + \overline{\rho}_{j-1})]
$$

$$
+ \delta x_{j+1/2}\frac{(2+c_{j+1/2}^2-6\eta_{j+1/2})}{12}(\overline{\rho}_{j+1} - \overline{\rho}_j) - \delta x_{j-1/2}\frac{(2+c_{j-1/2}^2-6\eta_{j-1/2})}{12}(\overline{\rho}_j - \overline{\rho}_{j-1}) =
$$

$$
\delta x_j \rho_j^n - \tfrac{1}{4}[\delta x_{j+1/2} c_{j+1/2}(\rho_{j+1}^n + \rho_j^n) - \delta x_{j-1/2} c_{j-1/2}(\rho_j^n + \rho_{j-1}^n)]
$$

$$
+ \delta x_{j+1/2}\frac{(2+c_{j+1/2}^2+6\eta_{j+1/2})}{12}(\rho_{j+1}^n - \rho_j^n) - \delta x_{j-1/2}\frac{(2+c_{j-1/2}^2+6\eta_{j-1/2})}{12}(\rho_j^n - \rho_{j-1}^n) \quad , \tag{2}
$$

where ρ_j^n is the density at mesh point j at time level n, $\overline{\rho}_j$ is the high-order solution to the implicit scheme at time level $n+1$, $c_{j+1/2} = \delta t w_{j+1/2}/\delta x_{j+1/2}$, δt is the time step and $w_{j+1/2}$ is the drift velocity defined at the boundary between mesh points j and $j+1$.

We wish to solve this equation on a non-uniform mesh defined as the set $(x_i | i = 1, N)$, so that two interleaved mesh spacings may be specified as $\delta x_{i+1/2} = x_{i+1} - x_i$ and $\delta x_i = \tfrac{1}{2}(\delta x_{1+1/2} + \delta x_{1-1/2})$.

The dimensionless diffusion coefficient is

$$
\eta_{j+1/2} = \delta t D_{j+1/2}/\delta x_{j+1/2}^2 \tag{3}
$$

where $D_{j+1/2} = (D_{j+1} + D_j)/2$.

From Eq.(2) we can derive the high-order flux

$$
\phi_{j+1/2}^H = \delta x_{j+1/2}\left[\frac{(2+c_{j+1/2}^2)}{12}(\overline{\rho}_{j+1} - \overline{\rho}_j - \rho_{j+1}^n + \rho_j^n) + \frac{c_{j+1/2}}{4}(\overline{\rho}_{j+1} + \overline{\rho}_j + \rho_{j+1}^n + \rho_j^n)\right.
$$

$$
\left. - \frac{\eta_{j+1/2}}{2}(\overline{\rho}_{j+1} - \overline{\rho}_j + \rho_{j+1}^n - \rho_j^n)\right] \quad . \tag{4}
$$

The low-order scheme used is the fully implicit scheme

$$
\delta x_j \tilde{\rho}_j + \tfrac{1}{2}[\delta x_{j+1/2} c_{j+1/2}(\tilde{\rho}_{j+1} + \tilde{\rho}_j) - \delta x_{j-1/2} c_{j-1/2}(\tilde{\rho}_j + \tilde{\rho}_{j-1})]
$$

$$
- \delta x_{j+1/2}\eta_{j+1/2}(\tilde{\rho}_{j+1} - \tilde{\rho}_j) + \delta x_{j-1/2}\eta_{j-1/2}(\tilde{\rho}_j - \tilde{\rho}_{j-1}) = \delta x_j \rho_j^n \tag{5}
$$

where $\tilde{\rho}_j$ is the low-order solution at mesh point j and time level $n+1$.

It can be shown that for $s \geq c/2$ this algorithm returns smooth non-negative results (Morrow and Steinle, 1989), as required for the lower-order scheme. (This condition is often fulfilled in gas discharge calculations)

From Eq.(5) we can derive the low-order flux

$$\phi^L_{j+1/2} = \delta x_{j+1/2} \left[\frac{c_{j+1/2}}{2}(\tilde{\rho}_{j+1} + \tilde{\rho}_j) - \eta_{j+1/2}(\tilde{\rho}_{j+1} - \tilde{\rho}_j) \right] \qquad (6)$$

and hence we can derive the antidiffusive flux $\phi_{j+1/2}$ required to transform the lower-order solution into the high-order solution, allowing the high-order solution to be used at all times, except where to do so would introduce unwanted maxima and minima,

$$\phi_{j+1/2} = \phi^H_{j+1/2} - \phi^L_{j+1/2} \ . \qquad (7)$$

After applying either Boris and Book's flux limiter (Boris and Book, 1976), or Zalesak's flux limiter (Zalesak, 1979) to obtain a corrected flux $\overline{\phi}_{j+1/2}$ we can compute a new solution at time level $n+1$

$$\rho^{n+1}_j = \tilde{\rho}_j + \frac{1}{\delta x_j}(\overline{\phi}_{j-1/2} - \overline{\phi}_{j+1/2}) \ . \qquad (8)$$

For $c \leq 1$ we can efficiently solve Eqs.(2) and (5) using the Thomas algorithm (Roache, 1972). Equation(2) is stable for all values of s provided $c \leq 1$, and, unfortunately, is unstable if $s > 0$ and $c > 1$ (Noye, 1986). Equation(5) is unconditionally stable and gives smooth non-negative results for $s \geq c/2$ (Morrow and Steinle, 1989).

Exact Test Solutions. We outline below two exact solutions, one dynamic and the other static, against which the numerical method can be tested.

1) For the exact test solution in the dynamic case we use an initial Gaussian distribution

$$\rho(x, t = 0) = A(t_0)^{-1/2} exp(-\frac{x^2}{4Dt_0}) \ , \qquad (9)$$

where $D = 5 \times 10^5 \ cm^2 s^{-1}$, $t_0 = 1.81 \times 10^{-9} \ s$, and A is a constant which is adjusted to give an initial maximum amplitude of 10. After a time t the distribution becomes (Llewellyn-Jones, 1957)

$$\rho(x,t) = A(t + t_0)^{-1/2} exp(-\frac{x^2}{4D(t + t_0)}) \ . \qquad (10)$$

2) The appropriate boundary condition for electrons being absorbed at an anode is that the density effectively goes to zero at the anode (Braglia and Lowke, 1979). Thus the case of electrons being emitted from a cathode in a uniform electric field, propagating at uniform velocity across a gap, and then being absorbed by the anode is a classic singular perturbation problem (Nayfeh, 1973, and Barton, 1988). The problem has the following exact solution for the steady state (Barton, 1988):

$$\rho(x) = A(1 - exp(wx/D))/(1 - exp(wd/D)) \ , \qquad (11)$$

where A is a constant determined by the flux at the emitting boundary and d is the electrode separation.

RESULTS AND DISCUSSION

The philosophy behind development of the present method has been to maintain the highest possible order of accuracy in a practical method. We particularly wish to avoid numerical diffusion and dispersion, as well as negative density values. Thus we avoid first order methods because of their inherent numerical diffusion. Once we use a higher-order method we run the risk of obtaining oscillatory results even when diffusion dominates over advection. For example if we use the fourth-order method alone, Eq.(2), to transport and diffuse a square-wave of density we obtain the results shown by the dashed curve of Fig.(1). This result was for 10 time steps with $c = 1$ and $s = 10$. Clearly the effects of the sharp change in density due to the square-wave do not diffuse away rapidly and similar results have been obtained for practical gas discharge calculations (Morrow and Steinle, 1989). The effect of using the flux-corrected method is shown as the solid curve in Fig.(1), and clearly the method overcomes the problem, with only a few ripples remaining of the square-wave.

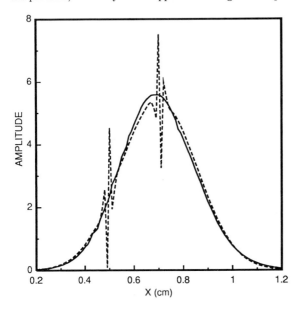

FIGURE 1: Transport and diffusion of a square-wave which was initially of amplitude 10, positioned between x=0.49 and x=0.69; motion is from left to right with $w = 2 \times 10^7$ cm/s, $c = 1$, and $s = 10$. The dashed curve is the solution from high-order method alone; the solid curve is the is the solution using the flux-corrected method.

The high order of accuracy of the method is demonstrated in Fig.(2) where the result of transporting and diffusing an initial Gaussian described by Eq.(9) for 161 time steps, or 80 ns, with $c = 1$ and $s = 5$, is compared with the exact solution given by Eq.(10). The maximum deviation of any point shown from the exact solution is 0.67 %, and there is negligible tendency to "clip" the peak value.

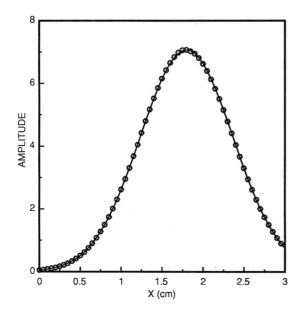

FIGURE 2: Result of transporting and diffusing a Gaussian density profile, $c = 1$ and $s = 5$. The solid curve is the exact solution; the circles represent the flux-corrected solution.

The steady state test of the method is shown in Fig.(3) where the flux-corrected solution after 500 time steps is compared with the exact solution given by Eq.(11) for electrons moving from right to left into an anode, with $c = 1$, $s = 6$, and $w = 2 \times 10^7$ cm/s. The maximum deviation from the exact solution is 0.35 %. Note that far fewer time steps could have been used to achieve this result; however, the result presented shows the extreme stability of the method over many time steps.

In conclusion we can say that the method presented is an extremely accurate and stable method of computing transport and diffusion for conditions where diffusion dominates. The problems of numerical diffusion or oscillatory behaviour due to steep gradients are all over-come by using flux-correction principles. The remaining problem to be addressed is the necessary transition to some other low-order method when $s < c/2$ and the fully implicit method no longer guarantees positive results.

ACKNOWLEDGEMENTS

The authors would like to acknowledge many helpful discussions with Drs. John Noye and Noel Barton. We would also like to thank Vivienne Bowers for her help in preparing this manuscript.

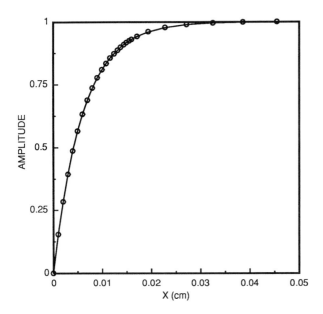

FIGURE 3: Steady state solution for electrons moving from right to left into an anode compared with the exact solution. $c = 1$, $s = 6$, $w = 2 \times 10^7$ cm/s and number of time steps = 500. The solid curve is the exact solution; the circles represent the flux-corrected solution. Note that the mesh expands after the first 20 points.

REFERENCES

1. Barton, N., Private communication, 1988.
2. Boris, J. P. and Book, D. L., in *Methods in Computational Physics*, Ed. John Killeen, pp. 85-130, Academic Press, New York, 1976.
3. Braglia, G. L. and Lowke, J. J., Comparison of Monte Carlo and Boltzmann calculations of electron diffusion to absorbing electrodes, *J. Phys. D.: Appl. Phys.* vol. 12, pp. 1831-1838, 1979.
4. Llewellyn-Jones, F., *Ionization and Breakdown in Gases*, pp. 30-32, Wiley, New York, 1957.
5. Morrow, R., Theory of Negative Corona in Oxygen, *Phys. Rev. A*, vol. 32, pp. 1799-1809, 1985.
6. Morrow, R. and Steinle, P., to be published(1989).
7. Morrow, R. and Cram, L. E., Flux-Corrected Transport and Diffusion on a Non-uniform Mesh, *J. Comput. Phys.*, vol. 57, pp. 129-136, 1985.
8. Nayfeh, A., *Perturbation methods*, p. 110, John Whiley, New York, 1973.
9. Noye, J. *Three-point Two-level Finite Difference Methods for the One-Dimensional Advection Equation*, in Computational Techniques and Applications: CTAC-85, Eds. J. Noye and R. May, pp. 137-158, North Holland, Amsterdam, 1986.
10. Roache, P. J. *Computational Fluid Dynamics*, pp. 345-350, Hermosa, Albuquerque, 1972.
11. Steinle, P. and Morrow, R., An Implicit Flux-Corrected Transport Algorithm, *J. Comput. Phys.*, vol. 80, pp. 61-71, 1989.
12. Zalesak, S. T., Fully Multidimensional Flux-Corrected Transport Algorithms for Fluids, *J. Comput. Phys.*, vol. 31, pp. 335-362, 1979.

Parabolic Multigrid Method
for Incompressible Viscous Flows Using
Group Explicit Relaxation Scheme

S. MURATA and N. SATOFUKA
Department of Mechanical and System Engineering
Kyoto Institute of Technology
Matsugasaki, Sakyo-ku, Kyoto 606, Japan

T. KUSHIYAMA
Department of Mechanical Engineering
Osaka Industrial University
Nakagaito, Daito-shi, Osaka 574, Japan

INTRODUCTION

Recently, in the field of fluid engineering, it has become possible to analyze the practical engineering problems in three dimensions, owing to the installation of the supercomputers. The computational efficiency of the numerical method on the supercomputers depends much on the algorithm, hence it should adapt to the architectural features of the vector and parallel computer. From this point of view, Evans & Youshif(1987) and the authors(1988) have proposed a class of new iterative methods for elliptic partial differential equations, which is called Group Explicit Iterative (GEI) method. Their explicit nature makes them most suitable for the vector and parallel computers. Application of the GEI method to the analysis of steady flows in a square cavity, has indicated that the GEI method is more efficient on the vector computer than other typical iterative methods. On the other hand, as an approach to the efficient analysis of unsteady problem on the supercomputer, the Parabolic Multi-Grid (PMG) method has been provided by Hackbusch(1984) for solving parabolic partial differential equations. This method is characterized by the simultaneous computation of several time levels in one step to the computational process. The computation in each time level can be independently performed by using several processors of the parallel computer. In this paper, a new Parabolic Multi-Grid method has been developed for solving unsteady incompressible viscous flows in a square cavity, by using the GEI method as a relaxation scheme. The computed results, based on the Navier-Stokes equations in vorticity-velocity form, are compared with those by other method or those based on vorticity-streamfunction equations in order to ensure the reliability and the computational efficiency of the present method.

GOVERNING EQUATIONS

The two-dimensional vorticity-velocity Navier-Stokes equations can be written as follows :

$$\zeta_t + (\zeta u)_x + (\zeta v)_y = \frac{1}{Re} (\zeta_{xx} + \zeta_{yy} + f), \tag{1}$$

$$u_{xx} + u_{yy} = -\zeta_y - g, \tag{2}$$

$$v_{xx} + v_{yy} = \zeta_x - h, \tag{3}$$

where t is the time, x,y are Cartesian coordinates, ζ is the vorticity, and u,v are the velocity components in the directions x,y, respectively. Re represents Reynolds number, and f,g,h are the residual functions which are assigned to zero on the finest grid. These equations in vorticity-velocity form have many advantages, e.g. the easy implementation of boundary conditions, the applicability to the three dimensional problem, and so on.

NUMERICAL PROCEDURE

Parabolic Multi-Grid Method

We will explain shortly the concept of the Parabolic Multi-Grid method by taking Eq.(1) as an example. The parabolic equation may be discretized by means of the implicit Euler formula as follows :

$$\frac{\zeta^k - \zeta^{k-1}}{\Delta t} + \Delta_x^{(m)} (\zeta u)^k + \Delta_y^{(m)} (\zeta v)^k = \frac{1}{Re} (\Delta_{xx}^{(m)} \zeta^k + \Delta_{yy}^{(m)} \zeta^k + f^{(m)}) , \tag{4}$$

where Δt is the time step, $\Delta_x, \Delta_y, \Delta_{xx}$ and Δ_{yy} are the difference operators for the derivative terms $\partial/\partial x$, $\partial/\partial y$, $\partial^2/\partial x^2$ and $\partial^2/\partial y^2$, respectively. The superscript k denotes the time level, and the superscript (m) indicates that the functions and operators are defined on the grid level m for $1 \le m \le M$, where M-th grid is the finest grid. Equation (4) is a discrete elliptic problem for the unknown grid function ζ^k and is often solved by the usual Multi-Grid (MG) method, time step by time step. In the PMG method, several equations, corresponding for $1 \le k \le K$, are simultaneously solved in one computational step. Figure 1 shows the computational procedure of the PMG method for the case with K=2. In this figure, δt is the time increment by which the computation proceeds per step. If the parallel

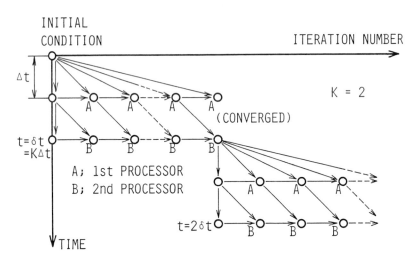

FIGURE 1. Computational procedure of the Parabolic Multi-Grid method.

computer is available, the multi-grid iteration for the time level, $k=1,2$, can be carried out by using the first and second processor, respectively. Equations (2) and (3) may be solved by the usual MG method, subject to the computed result of the vorticity ζ^k at each iteration step. In the present research, Full Approxmation Storage (FAS) mode of the Multi-Grid algorithms, proposed by Brandt(1977), is employed. If Eq.(4) may be represented by

$$(L^k \zeta^k)^{(m)} = \frac{f^{(m)}}{Re} + \frac{(\zeta^{k-1})^{(m)}}{\Delta t} \equiv (F^k)^{(m)},$$ (5)

where L is the elliptic operator, the restriction and prolongation of FAS mode are performed as follow :

$m \leftarrow m-1$ (from coarse to fine grid)

$$\zeta^{(m)} \leftarrow \zeta^{(m)} + P^m_{m-1} \left(\zeta^{(m-1)} - R^{m-1}_m \zeta^{(m)} \right),$$ (6)

$m \leftarrow m+1$ (from fine to coarse grid)

$$\zeta^{(m)} \leftarrow R^m_{m+1} \zeta^{(m+1)},$$ (7)

$$F^{(m)} \leftarrow L^{(m)} \zeta^{(m)} + R^m_{m+1} \left(F^{(m+1)} - L^{(m+1)} \zeta^{(m+1)} \right).$$

In Eqs.(6) and (7), the transfer operators R^m_{m-1}, P^m_{m+1} denote the 9-points restriction and prolongation, respectively. These operations, as shown in Eqs.(6) and (7), are execued according to the following criteria :

$m \leftarrow m-1$ (from coarse to fine grid)
$$(Rs^n)^{(M)} < 1 \times 10^{-6},$$
$$(Rs^n)^{(m)} < 0.25(Rs^n)^{(m+1)} \quad \text{for } 1 \leq m \leq M-1,$$ (8)

$m \leftarrow m+1$ (from fine to coarse grid)
$$Rs^n/Rs^{n-1} > 0.6,$$ (9)

where Rs^n is the L_2-residual of the vorticity at the iteration step n , defined by

$$Rs^n = \sqrt{\sum_{i=1}^N \sum_{j=1}^N (\zeta^n_{i,j} - \zeta^{n-1}_{i,j})^2 / (N-1)^2}.$$ (10)

Group Explicit Relaxation Scheme

The Group Explicit Iterative (GEI) method is employed in the smoothing step of the PMG method. Discretization of the spatial derivative terms by the central difference, reduces Eq.(2) for each grid point (i,j) to a difference equation. The system of four difference equations on the grid points $(i,j),(i+1,j),(i,j+1)$ and $(i+1,j+1)$, are expressed in matrix notation as follows :

$$\begin{pmatrix} a & c & b & 0 \\ c & a & 0 & b \\ b & 0 & a & c \\ 0 & b & c & a \end{pmatrix} \begin{pmatrix} u_{i,j} \\ u_{i+1,j} \\ u_{i,j+1} \\ u_{i+1,j+1} \end{pmatrix}^{n+1} = \begin{pmatrix} d_{i,j} & + u_{i-1,j} & -bu_{i,j-1} \\ d_{i+1,j} & + u_{i+2,j} & -bu_{i+1,j-1} \\ d_{i,j+1} & + u_{i-1,j+1} & -bu_{i,j+2} \\ d_{i+1,j+1} & + u_{i+2,j+1} & -bu_{i+1,j+2} \end{pmatrix}^n,$$ (11)

335

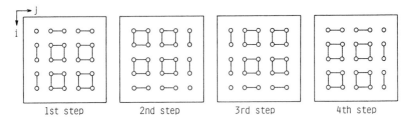

FIGURE 2. Grouping of grid points for the case with N=7.

where

$$a = 1 + \beta^2 + \rho, \quad b = -\beta^2, \quad c = -1,$$
$$d_{i,j} = \Delta x^2 \cdot g_{i,j} -(1+\beta^2-\rho)u_{i,j} +0.5 \cdot \Delta x \cdot \beta(\zeta_{i,j+1} - \zeta_{i,j-1}) \, . \tag{12}$$

In Eq.(12), Δx is the spatial grid size in x-direction, β is the mesh ratio, ρ is the iteration parameter and superscript n denotes the iteration step. The (4×4) matrix on LHS of Eq.(11) can be analytically inversed, hence the solutions at the next step (n+1) are expressed explicitly. The present iterative method has four steps per one iteration because there are four ways of grouping grid points, as shown in Fig.2.

Boundary and Initial Conditions

In this paper, the present method has been applied to the unsteady incompressible viscous flow in a square cavity. The upper wall at y=1 impulsively moves with unit velocity at t=0. Hence, the initial conditions for all variables are given by the stationary state

$$\zeta = 0, \ u = 0, \ v = 0 \qquad \text{at } t = 0, \tag{13}$$

and the boundary conditions for the velocity components u,v are given by

$$u(x,0) = u(0,y) = u(1,y) = 0, \ u(x,1) = 1, \ v = 0 \qquad \text{for } t > 0. \tag{14}$$

The boundary condition for the vorticity is imposed using Wood's condition, which has been reported by Giannattasio & Napolitano(1988) to provide an correct solution. It may be written for the boundary condition at y=0, by using the spacial grid size Δy for y-direction as follows :

$$\zeta(x,0) = -\zeta(x,\Delta y) -\Delta y \cdot g(x,0) +2\{u(x,0)-u(x,\Delta y)\}/\Delta y \, . \tag{15}$$

RESULTS AND DISCUSSIONS

First of all, the results are shown for the case with the total number of time levels in one computational step K=1 in order to check the effect of the total number of grid levels M and the iteration parameter ρ on the computational efficiency. The Reynolds number is 100 and the number of grid points, N, in each directions x,y is 65. Figure 3 shows the correlation between the total number of grid levels, M, and the work units which is needed for the computation up to t=21.5. The work unit is defined in such a way that the computational effort on the finest grid corresponds to 1 work unit. It is found from Fig.3 that the computational efficiency

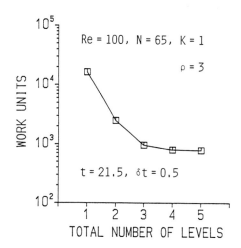

FIGURE 3. Improvement in computational work by the Multi-Grid method.

for the case with M=4,5 is about twenty-one times of that of single grid method, M=1. Convergence history is depicted for the first time step, at t= δt, in Fig.4 that plots the log of the L_2-residuals vs. work units. Though the optimum value of the iteration parameter is shown to be 3, it can not be said that the computational efficiency depends much on the parameter. Figure 5 shows the convergence history at the first three time levels, for the case with M=4, ρ=3 and K=2. The constant convergence rate may be obtained regardless of the mesh size. Next, the time evolution of the vorticity at the center of the cavity, x=0.5 and y=0.5, is shown in Fig.6. The results of two different time steps, δt, are consistent each other, hence it can be said that the unsteady flows in a square cavity

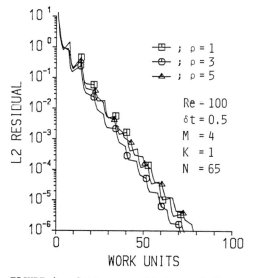

FIGURE 4. Convergence history of the vorticity.

337

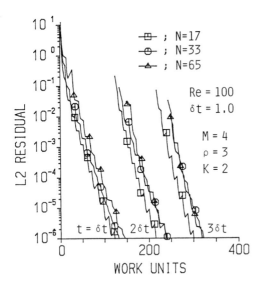

FIGURE 5. Multi-Grid convergence.

are accurately obtained by the present method, with this order of time step. Table 1 shows the comparison of CPU time, which is required for the computation up to t=1, on the vector computer FACOM VP-200. It can be seen from this table that the computational efficiency of the present method for tha case with N=129 and M=3 is about twice of that of the PMG method with Gauss-Geidel relaxation scheme, because the latter relaxation scheme is not yet very efficient on a vector computer due to vector interdependencies between neibouring grid points. In addition, it is obvious that the computational efficiency for the case with K=2 is almost equal to that of the conventional method, K=1, in which Eq.(5) is solved

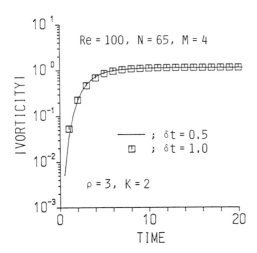

FIGURE 6. Time evolution of ζ at x=y=0.5.

TABLE 1. Comparison of CPU time for the case with N=129 and M=3.

RELAXATION SCHEME	K = 1	K = 2
PRESENT	13.3 (sec)	14.0
GAUSS-SEIDEL	21.4	24.7

time step by time step. If the parallel computer is available, it is expected that the efficiency of the PMG method for the case with K>1 will increase much, because the relaxation of K time levels can be simultaneously carried out by using K processors of the parallel computer. Finally, the results for the case with Re=1000 and N=257, are shown to test the reliability of the computed results by the present method. Figure 7 shows the constant vorticity lines and the velocity vectors in a square cavity in order to check the flow pattern. The vectors indicate only the flow direction and the length has no meaning. One secondary vortex is visible in each bottom corner of the cavity. Figure 8 shows the velocity profiles for u along vertical line and v along horizontal line passing through the geometric center of tha cavity. In this figure, the results by Ghia et al.(1982) are included, which is denoted by square. The results by the present method are in good agreement with their results which are based on the Navier-Stokes equations in vorticity stream-function, completely satisfying the equation of continuity.

CONCLUSIONS

In this research, Parabolic Multi-Grid method with Group Explicit Relaxation scheme has been developed for the unsteady flow in a square cavity. The results may be summarized as follows :

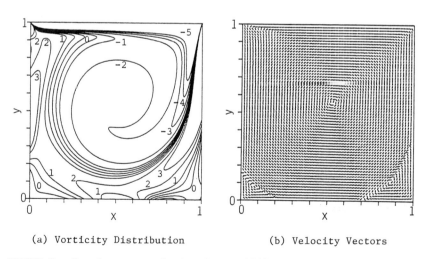

(a) Vorticity Distribution   (b) Velocity Vectors

FIGURE 7.   Steady state solution for Re=1000.

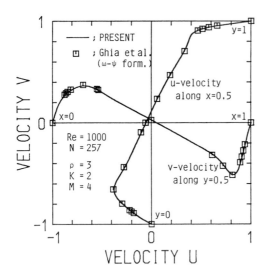

FIGURE 8. Velocity distributions.

(1) The computational efficiency of the present method is higher than that of Multi-Grid method with Gauss-Seidel relaxation scheme, on the vector computer.

(2) The total number of time levels in one computational procedure does not affect so much on the computational work on the scalar compuer. Hence, as the total number of time levels increases, the computational efficiency becomes higher on the parallel computer.

(3) The computed results based on the vorticity-velocity form are in good agreement with those based on the vorticity-steamfunction form.

In this paper, we use the vector computer FACOM VP-200 and the scalar computer M-382 of Kyoto Univercity, made by Fujitsu.

REFERENCES

1. Brandt,A., Multi-Level Adaptive Solutions to Boundary-Value Problems, *Mathematics of Computation*, vol.31, no.138, pp.333-390, April 1977.

2. Evans,D.J. and Yousif,W.S., The Alternating Group Explicit Method (BLAGE) for the Solution of Elliptic Difference Equations, *Intern. J. Comuter Math.*, vol.22, pp.177-185, 1987.

3. Giannattasio,P. and Napolitano,M., Numerical Solutions to the Navier-Stokes Equations in Vorticity-Velocity Form, *Proc. Third Italian Meeting of Computational Mechanics, Palermo*, pp.95-101, 1988.

4. Hackbusch,W, Parabolic Multi-Grid Method, in *Computing Methods in Applied Sciences and Engineering VI*, ed. R.Glowinski and J.L. Lions, pp.189-197, North-Holland, 1984.

5. Murata,S., Satofuka,N. and Kushiyama,T. , Numerical Analysis for Square Cavity Flow Using a New Poisson Solver, in *Computational Fluid Dynamics*, ed. G.de Vahl Davis and C.Fletcher, pp.547-556, North-Holland, 1988.

Implicit Three-Level Finite Difference Methods for the One-Dimensional Constant Coefficient Advection-Diffusion Equation

J. B. NIXON and B. J. NOYE
Applied Mathematics Department
The University of Adelaide
G.P.O. Box 498
Adelaide 5001, South Australia

INTRODUCTION

Two highly stable implicit three-level finite-difference methods for the one-dimensional constant coefficient advection-diffusion equation are described. These are developed using differencing on a (3,1,1) computational stencil. One is unconditionally stable with second-order accuracy, the other is very stable with third-order accuracy. Both involve no numerical diffusion. The two methods are compared, theoretically and in a numerical experiment, with the "leapfrog"/Du Fort-Frankel (1,2,1) explicit method, the only three-level method currently employed to solve this equation. The former are generally found to be more stable and more accurate than the latter.

NOTATION

The one-dimensional advection-diffusion equation, used to describe such processes as the spread of pollutants in a stream, may be scaled and written in the form

$$\frac{\partial \hat{\tau}}{\partial t} + u \frac{\partial \hat{\tau}}{\partial x} - \alpha \frac{\partial^2 \hat{\tau}}{\partial x^2} = 0, \qquad 0 \le x \le 1, \qquad 0 \le t \le T, \tag{1}$$

where $\hat{\tau}(x,t)$ may represent, for example, the pollutant concentration at position x and time t. In the following, u and α are considered to be positive constants representing advection and diffusion parameters, respectively. The solution domain of the problem is covered by a mesh of grid-lines $x = j\Delta x$, $j = 0, \ldots, J$, and $t = n\Delta t$, $n = 0, \ldots, N$, parallel to the time and space co-ordinate axes, respectively and approximations τ_j^n to $\hat{\tau}(j\Delta x, n\Delta t)$ are calculated at the points of intersection of these lines. The constant spatial and temporal grid-spacings are $\Delta x = 1/J$ and $\Delta t = T/N$, respectively. The point $(j\Delta x, n\Delta t)$ is termed the (j,n) grid-point and the derivative $\partial^p \hat{\tau}(j\Delta x, n\Delta t)/\partial x^p$ is denoted $\partial^p \tau/\partial x^p|_j^n$. A finite-difference method (FDM) is termed a "(k,l,m) method" if the finite-difference equation (FDE) for the method involves k, l and m grid-points at the $(n+1)$, n and $(n-1)$ time-levels respectively. A FDM is termed a "p-level method" if the corresponding FDE involves grid-points at the p time-levels $(n+1), \ldots, (n-p+2)$.

J. B. Nixon has a post-graduate scholarship at the University of Adelaide.

EQUIVALENT AND MODIFIED EQUIVALENT PDEs

Consider the FDE

$$\mathcal{L}_\Delta[\tau_j^n] = 0, \tag{2}$$

which is consistent with the given partial differential equation (PDE) (1). Expressing each term of FDE (2) as Taylor series expansions about the (j, n) grid-point yields the equivalent partial differential equation (EPDE)

$$\frac{\partial \tau}{\partial t} + u \frac{\partial \tau}{\partial x} - \alpha \frac{\partial^2 \tau}{\partial x^2} + \sum_{q=2}^{\infty} \sum_{p=0}^{q} C_{p\,q-p} \frac{\partial^q \tau}{\partial x^{q-p} \partial t^p} = 0, \tag{3}$$

which is the PDE actually solved by FDE (2). The leading terms in the truncation error of EPDE (3) with respect to PDE (1) contain derivatives with respect to both space and time and the magnitude of these terms indicate the order of accuracy of FDE (2). However, errors introduced through spatial discretisation may cancel with errors introduced through temporal discretisation. This is accounted for by considering the modified equivalent partial differential equation (MEPDE), found by converting all time derivatives higher than first order in EPDE (3) into derivatives with respect to space (see Warming and Hyett (1974)). An efficient computational procedure for achieving this is described in Noye and Hayman (1986).

All FDEs which are consistent with PDE (1) have MEPDEs of the form

$$\frac{\partial \tau}{\partial t} + u \frac{\partial \tau}{\partial x} - \alpha \frac{\partial^2 \tau}{\partial x^2} + \sum_{p=2}^{\infty} u(\Delta x)^{p-1} \frac{\eta_p(c, s)}{p!} \frac{\partial^p \tau}{\partial x^p} = 0, \tag{4}$$

in which

$$c = u(\Delta t)/(\Delta x) \quad \text{and} \quad s = \alpha(\Delta t)/(\Delta x)^2 \tag{5}$$

are the Courant and diffusion numbers, respectively. The terms in the truncation error of MEPDE (4) with respect to PDE (1) thus involve spatial derivatives only. A FDE (2) approximating PDE (1) is termed "r^{th}-order accurate" if $\eta_p(c, s) = 0$ for $p = 2, \ldots, r$ and $\eta_{r+1}(c, s) \neq 0$ in the corresponding MEPDE (4). If $\eta_2(c, s) \neq 0$, the diffusion coefficient is not α (as required for correct solution of PDE (1)) and the method is termed first-order accurate. Such a FDM is said to introduce "numerical diffusion" into the solution of PDE (1). If the MEPDE (4) corresponding to a FDM is not consistent with PDE (1), the FDM is considered of zero-order.

THE "LEAPFROG"/DU FORT-FRANKEL (1,2,1) METHOD

The "leapfrog"/Du Fort-Frankel (1953) spatially centred (1,2,1) explicit FDE

$$\tau_j^{n+1} = [\{c + 2s\}\tau_{j-1}^n - \{c - 2s\}\tau_{j+1}^n + \{1 - 2s\}\tau_j^{n-1}]/[1 + 2s], \tag{6}$$

has a MEPDE of the general form (4) with

$$\eta_2(c, s) = 2cs, \qquad \eta_3(c, s) = (1 - c^2)(1 - 12s^2), \cdots. \tag{7}$$

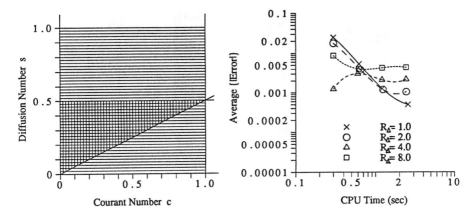

FIGURE 1. Von Neumann stable (horizontal shading) and non-negative (vertical shading) regions in the c-s plane; and $\text{Av}\{|\text{Err}|\}$ graphed against CPU time, for the "leapfrog"/Du Fort-Frankel (1,2,1) zero-order explicit FDE (6).

The zero-order FDE (6) is von Neumann stable (see O'Brien et al. (1950)) in the region

$$0 < c \leq 1, \quad \text{for all} \quad s > 0 \tag{8}$$

(Roache (1972, p61)), shown horizontally shaded in Fig.1. Approximations to PDE (1) with non-negative initial and boundary conditions produced by FDE (6) are guaranteed to be non-negative (see Noye (1988)) in the region defined by Eq.(8) and

$$c/2 \leq s \leq 1/2, \tag{9}$$

shown vertically shaded in Fig.1.

A WEIGHTED (3,1,1) METHOD

If, on a spatially centred (3,1,1) computational stencil, a weighted combination of centred space (CS) and backward space (BS) approximations is used for the first spatial derivative in PDE (1) and the differencing as specified by Richtmyer (1957, p95) is used for the temporal and second spatial derivatives, viz.

$$\left.\frac{\partial \hat{\tau}}{\partial t}\right|_j^n \approx (1+\theta) \times [\text{BT at } (j,n+1)] - \theta \times [\text{BT at } (j,n)], \qquad \left.\frac{\partial^2 \hat{\tau}}{\partial x^2}\right|_j^n \approx [\text{CS about } (j,n+1)],$$

$$\left.\frac{\partial \hat{\tau}}{\partial x}\right|_j^n \approx (1+\phi) \times [\text{CS about } (j,n+1)] - \phi \times [\text{BS at } (n+1)], \tag{10}$$

in which BT denotes backward time approximation, the following FDE is obtained

$$\{c(\phi-1)-2s\}\tau_{j-1}^{n+1} + 2\{1-c\phi+\theta+2s\}\tau_j^{n+1} + \{c(\phi+1)-2s\}\tau_{j+1}^{n+1} = 2\{1+2\theta\}\tau_j^n - 2\theta\tau_j^{n-1}. \tag{11}$$

The MEPDE for this weighted FDE is of the general form (4) with

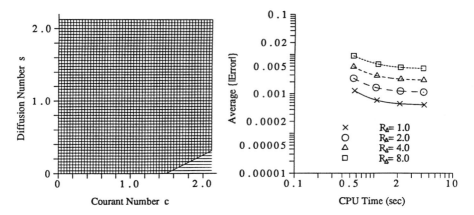

FIGURE 2. Regions of von Neumann stability (horizontal shading) and diagonal dominance (vertical shading) in the c-s plane; and Av{|Err|} graphed against CPU time, for the Richtmyer type (3,1,1) implicit second-order FDE (14).

$$\eta_2(c,s) = \phi + c(2\theta - 1), \qquad \eta_3(c,s) = 1 + 2c^2 - (2\theta - 1)(6s - 3c(2\theta c + \phi)), \cdots. \qquad (12)$$

THE RICHTMYER TYPE (3,1,1) METHOD

To obtain a Richtmyer type (3,1,1) FDM with no numerical diffusion, the weights

$$\theta = 1/2, \qquad \phi = 0, \qquad (13)$$

corresponding to second-order CT and CS differencing respectively, are substituted into FDE (11), yielding the FDE

$$\{2s + c\}\tau_{j-1}^{n+1} - \{3 + 4s\}\tau_j^{n+1} + \{2s - c\}\tau_{j+1}^{n+1} = -4\tau_j^n + \tau_j^{n-1} \qquad (14)$$

which has a MEPDE of the general form (4), with

$$\eta_2(c,s) = 0, \qquad \eta_3(c,s) = 1 + 2c^2, \qquad \eta_4(c,s) = 6c(c^2 - 4s) - 2(s/c), \cdots. \qquad (15)$$

The second-order FDE (14) is unconditionally von Neumann stable, while values produced by this equation are never guaranteed to be non-negative in the first quadrant of the c-s plane. Diagonal dominance (see Noye (1984, p146)) of the coefficient matrix generated by this implicit FDE is assured — in the first quadrant of the c-s plane — when

$$s \geq (2c - 3)/4, \qquad (16)$$

shown as the vertically shaded region of Fig.2. In this region the very efficient Thomas (1949) algorithm may be used to solve FDE (14).

THE OPTIMAL (3,1,1) METHOD

Substituting the following weights into the weighted FDE (11)

344

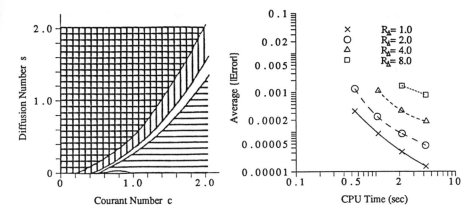

FIGURE 3. Regions of von Neumann stability (horizontal shading) and diagonal dominance (vertical shading) in the c-s plane; and Av{|Err|} graphed against CPU time, for the (3,1,1) optimal third-order FDE (18).

$$\theta = (6s - c^2 + 1)/D, \quad \phi = -2c(2c^2 + 1)/D, \quad \text{where} \quad D = 6(2s - c^2), \tag{17}$$

yields the FDE

$$\{12s^2 - c(c-1)(6s - c(2c-1))\}\tau_{j-1}^{n+1} - \{24s^2 - 6s(2c^2 - 3) + (c^2 - 1)(4c^2 - 1)\}\tau_j^{n+1}$$
$$+\{12s^2 - c(c+1)(6s - c(2c+1))\}\tau_{j+1}^{n+1} = 2\{4c^2 - 12s - 1\}\tau_j^n + \{6s - (c^2 - 1)\}\tau_j^{n-1}. \tag{18}$$

This has a MEPDE of the general form (4) in which

$$\eta_2(c, s) = \eta_3(c, s) = 0,$$
$$\eta_4(c, s) = -c[6s(20s - 6c^2 + 5) + (c^2 - 1)(4c^2 - 1)]/[3(2s - c^2)],$$
$$\eta_5(c, s) = [(c^2 - 1)(36(54c^2 - 1)s^2 - 6c^2(34c^2 - 31)s - c^2(4c^2 - 1)(c^2 + 5))$$
$$+ 1080s^3(4s - 6c^2 + 1)]/[9(2s - c^2)^2], \cdots, \tag{19}$$

resulting in a third-order method. This is the optimal (highest order) FDM for PDE (1) based on the spatially centred (3,1,1) computational stencil. The regions of von Neumann stability and diagonal dominance for FDE (18) — in a portion of the first quadrant of the c-s plane — are shown horizontally and vertically shaded, respectively, in Fig.3. Values produced by this FDE are never guaranteed to be non-negative in the first quadrant of the c-s plane.

A NUMERICAL TEST

Equation (1) has an exact solution

$$\hat{\tau}(x, t) = \frac{1}{\sqrt{4t + 1}} \exp\left\{-\frac{(x - x_0 - ut)^2}{\alpha(4t + 1)}\right\} \tag{20}$$

for boundary conditions described by Eq.(20) with $x = 0, 1$ for $t \geq 0$ and an initial condition described by Eq.(20) with $t = 0$ for $0 \leq x \leq 1$. The initial condition is a Gauss distribution centred at $x = x_0$ with unit pulse height. Given $\alpha > 0$ and $u = 1.0$, a pulse initially centred

345

at $x_0 = -0.5$ reduces in height to $1/\sqrt{5}$ at a position centred at $x = 0.5$ when $t = 1$.

The FDEs (6), (14) and (18) for the "leapfrog"/Du Fort-Frankel (DF) (1,2,1), Richtmyer type (RT) (3,1,1) and optimal (OPT) (3,1,1) methods respectively were used to approximate PDE (1) with exact solution (20) using various values for c and s.

The parameters used for all tests were $u = 1.0$, $T = 1.0$, $x_0 = -0.5$, $\Delta x = 0.01$ ($J = 100$). Tests were carried out for four values of the cell Reynolds number $R_\Delta = c/s$, namely $R_\Delta = 1.0$ ($\alpha = 0.01$), 2.0 (0.005), 4.0 (0.0025) and 8.0 (0.00125). For each value of R_Δ, four values of c were used, namely $c = 0.1, 0.2, 0.4, 0.8$. For the four tests for each value of R_Δ, s was chosen to force $\Delta t = 0.001$ ($N = 1000$), 0.002 (500), 0.004 (250), 0.008 (125) as the value of c was increased. Equation (20) was used to produce values for the first time-level.

The theoretical accuracy (Order), average absolute error (Av{|Err|}) and minimum value of τ (Min{τ}) over all grid-points at the final time-level, and the central processor unit (CPU) time used, are listed in Tab.1 for the parameter values resulting in stable tests of each method. It is noted that errors increase with the product cs for the DF method, and with c^2 for the RT method, respectively. These trends, as well as the variation in errors for the OPT method with changing c and s, can be accounted for by the form of the leading error term in the MEPDEs corresponding to the FDEs of the respective methods. The RT method is less accurate than the DF method for $R_\Delta = 8.0$ due to the large value of $\eta_3(c, s)$ for the former and the small value of $\eta_2(c, s)$ for the latter which, taking into account both the value of the corresponding derivatives and the multiplying coefficients (as given by MEPDE (4)), results in a greater error for the second-order method compared to that for the zero-order method, respectively.

For tests with $R_\Delta \leq 4.0$, all methods produced insignificant negative values irrespective of the appropriate non-negative region, except for the DF method at $(c, s) = (0.8, 0.2)$. Negative values were produced by this test, and all tests with $R_\Delta = 8.0$ of the DF method, as may have been expected (see Fig.1). For $R_\Delta = 8.0$ the RT method also produced negative values as may have been expected, but the OPT method did not. It is expected that for tests with larger R_Δ, the size of both errors and negative values will increase for all FDMs.

The OPT and RT methods require up to twice the amount of CPU time when compared to the DF method for tests with the same parameters. This is due to the effort required to solve the set of linear algebraic equations associated with implicit FDEs. However, for errors of comparable magnitude, the OPT method uses less CPU time than both the RT and DF methods for all R_Δ, since larger values of Δt may be used due to its greater accuracy. This can be seen from the plots of Av{|Err|} against CPU time in Figs.3, 2 and 1, respectively. These figures also show that only when using the RT method with large R_Δ is it more efficient than the DF method, when CPU usage is taken into account.

CONCLUSION

It is evident that successive improvement in results is obtained, for $R_\Delta \leq 4.0$, when the zero-order DF, the second-order RT and the third-order OPT methods are used, in that order. For larger R_Δ the OPT method is the most accurate, although stability considerations

TABLE 1. Numerical Test Results.

R_Δ	c	s	Method	Order	Av{|Err|}	Min{τ}	CPU
1.0	0.1	0.1	DF(1,2,1)	0	4.94×10^{-4}	0.0	2.55 sec
			RT(3,1,1)	2	5.25×10^{-4}	0.0	4.30 sec
			OPT(3,1,1)	3	1.35×10^{-5}	0.0	4.27 sec
	0.2	0.2	DF(1,2,1)	0	1.26×10^{-3}	0.0	1.22 sec
			RT(3,1,1)	2	5.56×10^{-4}	0.0	2.12 sec
			OPT(3,1,1)	3	3.15×10^{-5}	0.0	2.15 sec
	0.4	0.4	DF(1,2,1)	0	5.35×10^{-3}	0.0	0.62 sec
			RT(3,1,1)	2	6.83×10^{-4}	0.0	1.08 sec
			OPT(3,1,1)	3	9.10×10^{-5}	0.0	1.08 sec
	0.8	0.8	DF(1,2,1)	0	2.47×10^{-2}	0.0	0.30 sec
			RT(3,1,1)	2	1.19×10^{-3}	0.0	0.57 sec
			OPT(3,1,1)	3	3.34×10^{-4}	0.0	0.55 sec
2.0	0.1	0.05	DF(1,2,1)	0	1.04×10^{-3}	0.0	2.43 sec
			RT(3,1,1)	2	1.11×10^{-3}	0.0	4.22 sec
			OPT(3,1,1)	3	4.56×10^{-5}	0.0	4.22 sec
	0.2	0.1	DF(1,2,1)	0	1.19×10^{-3}	0.0	1.25 sec
			RT(3,1,1)	2	1.17×10^{-3}	0.0	2.15 sec
			OPT(3,1,1)	3	9.10×10^{-5}	0.0	2.15 sec
	0.4	0.2	DF(1,2,1)	0	3.81×10^{-3}	0.0	0.65 sec
			RT(3,1,1)	2	1.44×10^{-3}	0.0	1.05 sec
			OPT(3,1,1)	3	2.44×10^{-4}	0.0	1.05 sec
	0.8	0.4	DF(1,2,1)	0	1.75×10^{-2}	0.0	0.30 sec
			RT(3,1,1)	2	2.48×10^{-3}	0.0	0.55 sec
			OPT(3,1,1)	3	1.22×10^{-3}	0.0	0.55 sec
4.0	0.1	0.025	DF(1,2,1)	0	2.16×10^{-3}	0.0	2.42 sec
			RT(3,1,1)	2	2.24×10^{-3}	0.0	4.23 sec
			OPT(3,1,1)	3	1.85×10^{-4}	0.0	4.32 sec
	0.2	0.05	DF(1,2,1)	0	2.10×10^{-3}	0.0	1.22 sec
			RT(3,1,1)	2	2.37×10^{-3}	0.0	2.18 sec
			OPT(3,1,1)	3	3.43×10^{-4}	0.0	2.12 sec
	0.4	0.1	DF(1,2,1)	0	2.99×10^{-3}	0.0	0.60 sec
			RT(3,1,1)	2	2.89×10^{-3}	0.0	1.08 sec
			OPT(3,1,1)	3	1.08×10^{-3}	0.0	1.08 sec
	0.8	0.2	DF(1,2,1)	0	1.23×10^{-2}	-1.26×10^{-10}	0.30 sec
			RT(3,1,1)	2	4.89×10^{-3}	0.0	0.55 sec
8.0	0.1	0.0125	DF(1,2,1)	0	4.36×10^{-3}	-6.43×10^{-8}	2.43 sec
			RT(3,1,1)	2	4.48×10^{-3}	-8.20×10^{-8}	4.18 sec
			OPT(3,1,1)	3	8.54×10^{-4}	0.0	4.25 sec
	0.2	0.025	DF(1,2,1)	0	4.25×10^{-3}	-5.48×10^{-8}	1.22 sec
			RT(3,1,1)	2	4.74×10^{-3}	-1.32×10^{-7}	2.17 sec
			OPT(3,1,1)	3	1.46×10^{-3}	0.0	2.15 sec
	0.4	0.05	DF(1,2,1)	0	4.02×10^{-3}	-6.61×10^{-9}	0.62 sec
			RT(3,1,1)	2	5.72×10^{-3}	-2.70×10^{-5}	1.12 sec
	0.8	0.1	DF(1,2,1)	0	8.67×10^{-3}	-2.64×10^{-11}	0.30 sec
			RT(3,1,1)	2	9.31×10^{-3}	-8.28×10^{-3}	0.55 sec

become restrictive, while the RT method becomes less accurate than the DF method. For all tests, the optimal implicit (3,1,1) method produced values with greater accuracy using less CPU time (comparatively) than did the Richtmyer type implicit (3,1,1) method or the "leapfrog"/Du Fort-Frankel explicit (1,2,1) method.

REFERENCES

1. Du Fort, E. C., and Frankel, S. P., Stability Conditions in the Numerical Treatment of Parabolic Differential Equations, *Mathematical Tables and Other Aids to Computation*, vol. 7, pp. 135–152, 1953.

2. Noye, J., Finite Difference Techniques for Partial Differential Equations, in *Computational Techniques for Differential Equations*, ed. J. Noye, pp. 95–354, Elsevier Science Publishers B.V. (North-Holland), Amsterdam, 1984.

3. Noye, J., Finite-Difference Methods for the One-Dimensional Transport Equation, *Computational Techniques and Applications: CTAC-87*, eds. J. Noye and C. Fletcher, pp. 539–562, Elsevier Science Publishers B.V. (North-Holland), Amsterdam, 1988.

4. Noye, J., and Hayman, K., Accurate Finite Difference Methods for Solving the Advection-Diffusion Equation, *Computational Techniques and Applications: CTAC-85*, eds. J. Noye and R. May, pp. 137–157, Elsevier Science Publishers B.V. (North-Holland), Amsterdam, 1986.

5. O'Brien, G. G., Hyman, M. A., and Kaplan, S., A Study of the Numerical Solution of Partial Differential Equations, *Journal of Mathematics and Physics*, vol. 29, pp. 223–251, 1950.

6. Richtmyer, R. D., *Difference Methods for Initial-Value Problems*, vol. 4 of *Interscience Tracts in Pure and Applied Mathematics*, Interscience Publishers, Inc., New York, 1957.

7. Roache, P. J., *Computational Fluid Dynamics*, Hermosa Publishers, Albuquerque, 1972.

8. Thomas, L. H., Elliptic Problems in Linear Difference Equations over a Network, Watson Scientific Computing Laboratory, Tech. Rept., Columbia University, New York, 1949.

9. Warming, R. F., and Hyett, B. J., The Modified Equation Approach to the Stability and Accuracy Analysis of Finite-Difference Methods, *Journal of Computational Physics*, vol. 14, no. 2, pp. 159–179, February 1974.

Some Two-Point Three-Level Explicit FDM's for One-Dimensional Advection

B. J. NOYE
Department of Applied Mathematics
The University of Adelaide
G.P.O. Box 498
Adelaide 5001, South Australia

INTRODUCTION

At the present time fluid motion is generally modelled by numerically solving the governing partial differential equations on high speed digital computers. Finite-difference methods (FDMs), because of their relative simplicity and their long history of successful application, are the most frequently used.

However, in common with other numerical techniques, finite-difference methods of solving the Eulerian equations of hydrodynamics seldom model the advective terms accurately. Errors in the phase and amplitude of waves are usual, especially the former.

The accuracy of several finite-difference methods for solving the equation which governs the one-dimensional advection of a scalar quantity $\hat{\tau}(x,t)$ in a fluid travelling with speed u, namely

$$\frac{\partial \hat{\tau}}{\partial t} + u \frac{\partial \hat{\tau}}{\partial x} = 0, \quad 0 \leq x \leq 1, \ 0 < t \leq T, \tag{1}$$

is investigated in this article. In particular, u is chosen to be a positive constant. With the initial value $\hat{\tau}(x,0)$ prescribed for $0 \leq x \leq 1$, the boundary condition $\hat{\tau}(0,t)$ is defined for $t > 0$; no values are given at $x = 1$.

There are several reasons for studying finite-difference methods for solving this equation. Firstly, an insight is gained into suitable ways of differencing the advective term $u \partial \hat{\tau}/\partial x$ when it occurs in more complicated equations. Secondly, Equation (1) may be used to model advection separately from other physical processes, such as diffusion, over some fraction of the time-step, using the process of time-splitting (see D'Yakonov, 1963, and Yanenko, 1971).

Approximations τ_j^n of the true values $\hat{\tau}(j\Delta x, n\Delta t)$, $j = 1(1)J$, $n = 1(1)N$, where $\Delta x = 1/J$, and $\Delta t = T/N$, the notation $1(1)J$ indicating the set of integers ranging from 1 to J in steps of 1, may be computed from finite-difference equations (FDE's) of the general form

$$\mathcal{L}_\Delta \{\tau_j^n\} = 0. \tag{2}$$

The properties of methods based on these FDE's depend on the Courant number chosen, namely on

$$c = u\Delta t/\Delta x > 0. \tag{3}$$

The analysis used to determine the stability of (2) is that due to von Neumann (see O'Brien et al., 1950). This technique analyses the error propagation equation

$$\mathcal{L}_\Delta \{\xi_j^n\} = 0, \tag{4}$$

in which ξ_j^n is the error in the computed value of τ_j^n. The propagation of each component of the finite Fourier series of the initial error distribution, namely of

$$\xi_j^0 = \exp\{i(mj\Delta x)\}, \quad i = \sqrt{-1}, \quad j = 0(1)J, \tag{5}$$

is assessed, where m is the wave number of the mode being investigated. The solution of (4) with initial condition (5) is of the form

$$\xi_j^n = (G)^n \xi_j^0, \tag{6}$$

where G may be found by substitution of (6) in (4). The stability criterion is, then,

$$|G(c, \beta)| \leq 1 \quad \text{for all} \quad \beta = m\Delta x. \tag{7}$$

The von Neumann procedure was chosen because the results obtained can be used with consistency to ensure convergence of the numerical solution to the exact solution as $\Delta x \to 0$, $\Delta t \to 0$, by invoking Lax's Equivalence Theorem (see Lax and Richtmyer, 1956). This contrasts with the matrix stability method, results from which cannot always be used to establish convergence in this way (see Morton, 1980, Hindmarsh et al., 1984, and Sucec, 1987).

Throughout this work, the modified equivalent partial differential equation (MEPDE) obtained by the method of Noye and Hayman (1986) is used to estimate the accuracy of a FDM. Any finite-difference equation which is consistent with the advection equation, with smooth initial-boundary values, has an MEPDE of the form

$$\frac{\partial \tau}{\partial t} + u\frac{\partial \tau}{\partial x} + \sum_{q=2}^{\infty} \frac{u(\Delta x)^{q-1}}{q!} \eta_q(c) \frac{\partial^q \tau}{\partial x^q} = 0. \tag{8}$$

If $\eta_q(c) = 0$ for $q = 2(1)Q$ and $\eta_{Q+1}(c) \neq 0$, the FDM is said to be of order Q since the truncation error

$$E = \sum_{q=Q+1}^{\infty} \frac{u(\Delta x)^{q-1}}{q!} \eta_q(c) \frac{\partial^q \tau}{\partial x^q}, \tag{9}$$

is $O\{(\Delta x)^Q\}$, $\Delta x << 1$.

It has been shown by Noye (1984a, 1984b) that the coefficients $\eta_q(c)$ of the error in (8) are related to wave response parameters of the FDM. For instance, the amplitude response of an infinitely long sinusoidal wave is the ratio of the amplitude of the wave propagated by the FDM relative to the true amplitude of the wave, after the wave has travelled one wavelength, and is given by

$$\zeta(c, N_\lambda) = \exp\left\{\frac{2\pi^2 \eta_2(c)}{N_\lambda} - \frac{2\pi^4 \eta_4(c)}{3(N_\lambda)^3} + O\{(N_\lambda)^{-5}\}\right\}, \tag{10}$$

in which N_λ is the number of grid spacings in a wavelength. Also the relative wave speed, which is the ratio of the wave speed of the numerical solution relative to the true speed of propagation produced by the advection equation, is given by

$$\mu(c, N_\lambda) = 1 - \frac{2\pi^2 \eta_3(c)}{3(N_\lambda)^2} + \frac{2\pi^4 \eta_5(c)}{15(N_\lambda)^4} + O\{(N_\lambda)^{-6}\}. \tag{11}$$

Ideally, both ζ and μ should be 1 for all values of c and N_λ. The asymptotic forms (10) and (11) may be used to theoretically estimate the accuracy of a given FDM when N_λ is large.

In this article four finite-difference methods for solving (1) are described and compared. The first two of these are second-order accurate and are in common use, namely the two level (2,2) 'box' marching method and the three level (1,2,1) explicit leapfrog method. The notation (p, q) indicates that the computational stencil involves two time levels with p grid points at the $(n+1)$ time level and q points at the n time level; the notation (p, q, r) denotes p, q, r grid points at the $(n + 1)$, n and $(n - 1)$ time levels, respectively. Two new explicit three-level methods are then developed, a compact (1,2,1) scheme of second-order accuracy and a compact (1,2,2) scheme of third-order accuracy. Like the leapfrog method, these are explicit and therefore may be used to advantage on array or vector processors.

To illustrate the relative accuracy of these FDMs investigated, they are used to solve the following simple problem for which the exact solution is known. The initial condition is the infinite train of Gaussian peaks,

$$\hat{\tau}(x, 0) = \exp\{-400(x - 0.5)^2\}, \quad 0 \le x \le 1,$$
$$\hat{\tau}(x + K, 0) = \hat{\tau}(x, 0), \quad K \le x \le K + 1, \quad K = \pm 1, \pm 2, \ldots, \tag{12}$$

for which the exact solution at time T on the infinite spatial domain $-\infty < x < \infty$ is the same train displaced a distance uT in the positive $\dot{x}$-direction. On the interval $0 \le x \le 1$, the numerical solution is obtained using cyclic boundary conditions at $x = 0$ and $x = 1$ and compared with the exact solution. In the tests which follow $u = 0.8$, so that four periods elapse when the value of t is incremented by five.

THE BOX METHOD

The two-point two-level box FDM (see Figure 1(a) and Noye, 1986) is based on the FDE

$$\tau_j^{n+1} = \tau_{j-1}^n + \left(\frac{1-c}{1+c}\right)\left(\tau_j^n - \tau_{j-1}^{n+1}\right), \quad j = 1(1)J. \tag{13}$$

This equation is unconditionally stable in the marching and time-stepping senses.

The MEPDE of (13), obtained using the computer program described in Noye and Hayman (1986), is

$$\frac{\partial \tau}{\partial t} + u\frac{\partial \tau}{\partial x} - \frac{u(\Delta x)^2}{12}(1 - c^2)\frac{\partial^3 \tau}{\partial x^3} + O\{(\Delta x)^4\} = 0, \tag{14}$$

which contains only odd derivatives of x. Clearly (13) is second-order accurate. From this is obtained the amplitude response

$$\zeta(c, N_\lambda) \equiv 1 \tag{15}$$

and the relative wave speed

$$\mu(c, N_\lambda) = 1 + \pi^2(1 - c^2)/[3(N_\lambda)^2] + O\{(N_\lambda)^{-4}\}. \tag{16}$$

The form of (16) indicates that, for large N_λ, the numerical wave is too fast for $0 < c < 1$ and too slow for $c > 1$. As the number of grid spacings in a wavelength decreases, that is, as the length of the component waves in the initial condition become shorter, the wave speed error increases. Therefore, the smaller the component wavelength the faster the numerical speed of propagation if $0 < c < 1$ and N_λ is reasonably large, such as $N_\lambda > 10$. The increased wave speed when $c = 0.4$ is the cause of the dominant peak appearing ahead of the true peak in the Gauss test results of Figure 2(a). It is also the cause of the leading oscillations which have appeared. These are due to the short wavelength components (with small N_λ) travelling faster than the long components (with large N_λ); the effect is emphasised because no damping of wave amplitudes occurs, due to (15). The dominant peak appears attenuated because its component waves are no longer in phase.

THE LEAPFROG THREE-LEVEL METHOD

Discretising the advection equation (1) at the point $(j\Delta x, n\Delta t)$ using centred-difference approximations for both derivatives (see Figure 1(b) and Noye, 1984a) yields the explicit three-level leapfrog equation

$$\tau_j^{n+1} = c\left(\tau_{j-1}^n - \tau_{j+1}^n\right) + \tau_j^{n-1}, \quad j = 1(1)J - 1, \tag{17}$$

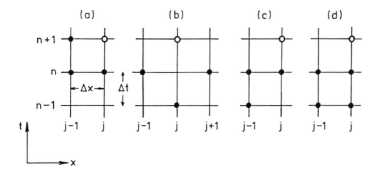

FIGURE 1. The computational stencils for (a) the box method, (b) the leapfrog method, (c) the new (1,2,1) method, (d) the new (1,2,2) method.

which is von Neumann stable for

$$0 < c \le 1. \tag{18}$$

The MEPDE for (17) contains no even derivatives, having the form

$$\frac{\partial \tau}{\partial t} + u \frac{\partial \tau}{\partial x} + \frac{u(\Delta x)^2}{6}(1 - c^2)\frac{\partial^3 \tau}{\partial x^3} + O\{(\Delta x)^4\} = 0. \tag{19}$$

This indicates that (17) is second-order accurate, with amplitude response

$$\zeta(c, N_\lambda) \equiv 1, \tag{20}$$

and relative wave speed

$$\mu(c, N_\lambda) = 1 - 2\pi^2(1 - c^2)/[3(N_\lambda)^2] + O\{(N_\lambda)^{-4}\}. \tag{21}$$

The latter implies that, when N_λ is reasonably large, the numerical wave speed is too slow in the stability range (18). Comparison of (21) with (16) shows that the dominant wave speed error for the leapfrog method is twice the size of that for the box method, but with an opposite sign.

Since (17) is a three-level finite-difference equation, a second-order two-level method is required as a starter to compute the values of τ_j^1 at the first time level. The box method is suitable for this purpose.

A further complication occurs because of the nature of the computational stencil for the leapfrog equation, shown in Figure 1(b); the value of τ_j^{n+1} cannot be found using (17). In fact, to compute all values of τ_J^N in $0 < x \le 1$ requires an extended solution domain with initial values τ_j^0 given for $j = 0(1)J + N$. The extra calculations involved at each time-level to compute τ_j^n for $j = 1(1)J + N - n$, $n = 1(1)N$, greatly increases the computational costs.

Perhaps the only advantage of the leapfrog method over the box method is that values of τ_j^{n+1} may be computed using a purely explicit form in terms of values of τ at previous time levels, not in a marching fashion using previously computed values of τ_{j-1}^{n+1}. Consequently, it may be used to advantage on array or vector processors.

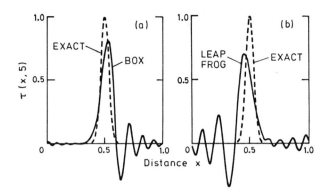

FIGURE 2. Results of the pulse test for (a) the box method, (b) the leapfrog method, with $c = 0.4$.

The result of solving the test problem with the leapfrog method is shown in Figure 2(b). It illustrates the effect of (20) and (21) for $c = 0.4$, namely that the amplitude response is ideal while the wave speed is too slow, becoming slower as N_λ becomes smaller. This causes the numerical peak to lag behind that of the exact solution, and large trailing oscillations to appear. These effects are more pronounced than in the test results for the box method, as anticipated from (21).

THE (1,2,1) TWO-POINT EXPLICIT METHOD

Using the computational stencil shown in Figure 1(c), the advection equation (1) may be discretised at the point $((j - 1/2)\Delta x, n\Delta t)$ using centred-differencing for the space derivative and an average of forward-differencing at the point $(j\Delta x, n\Delta t)$ and backward differencing at the point $((j - 1)\Delta x, n\Delta t)$ for the time derivative. This yields the explicit three-level finite-difference equation

$$\tau_j^{n+1} = \tau_{j-1}^{n-1} + (1 - 2c)(\tau_j^n - \tau_{j-1}^n), \quad j = 1(1)J, \tag{22}$$

which is von Neumann stable for

$$0 < c \leq 1. \tag{23}$$

The MEPDE for (22) is

$$\frac{\partial \tau}{\partial t} + u\frac{\partial \tau}{\partial x} - \frac{u(\Delta x)^2}{12}(1 - c)(1 - 2c)\frac{\partial^3 \tau}{\partial x^3} + O\{(\Delta x)^4\} = 0, \tag{24}$$

which contains no even derivatives and indicates (22) is second-order accurate. The relative wave speed is

$$\mu(c, N_\lambda) = 1 + \pi^2(1 - c)(1 - 2c)/[3(N_\lambda)^2] + O\{(N_\lambda)^{-4}\} \tag{25}$$

and the amplitude response

$$\zeta(c, N_\lambda) \equiv 1. \tag{26}$$

The former indicates that, for large N_λ, the numerical wave speed is too fast for $0 < c < 1/2$, and too slow for $1/2 < c < 1$. Note that, for values of c satisfying (23), the dominant (second-order) error term in (25) is much smaller than the corresponding error terms for both the box and leapfrog methods.

From the computational stencil of (22) shown in Figure 1(c) it is seen that the new (1,2,1) method, like the box method, permits direct evaluation of τ_j^{n+1} on the right-hand boundary $x = 1$. Like the three-level leapfrog method it requires a second-order two-level method as a starter, for example, the box method. Also, as it is purely explicit, it can be used efficiently with array or vector processors.

The result of the application of the pulse test to this method is shown in Figure 3(a). It illustrates the effect of the wave response (25-26) for $c = 0.4$. The fact that the relative wave speed is a little too large, causes the dominant peak in the numerical solution to be only a little ahead of that in the exact solution, and the leading oscillations to be very small indeed.

THE (1,2,2) TWO-POINT EXPLICIT METHOD

The second-order error term in the MEPDE of the (1,2,1) method may be eliminated by combining it with the MEPDE of the box method. The same combination of the corresponding FDEs will then contain no second-order errors.

The (1,2,1) equation (22) may be written

$$\mathcal{L}_\Delta^{121}\{\tau_j^n\} = 0, \tag{27}$$

in which the operator $\mathcal{L}_\Delta^{121}$, in difference form, may be written

$$\mathcal{L}_\Delta^{121}\{\tau_j^n\} = \tau_j^{n+1} + (1 - 2c)\tau_{j-1}^n - (1 - 2c)\tau_j^n - \tau_{j-1}^{n-1}. \tag{28}$$

Alternatively, $\mathcal{L}_\Delta^{121}$ may be written in a differential form based on the modified equivalent equation (24), namely

$$\mathcal{L}_\Delta^{121}\{\tau_j^n\} = 2\Delta t \left[\frac{\partial \tau}{\partial t} + u\frac{\partial \tau}{\partial x} - \frac{u(\Delta x)^2}{12}(1 - c)(1 - 2c)\frac{\partial^3 \tau}{\partial x^3} + \dots\right]_j^n. \tag{29}$$

Similarly, the box equation (13) may be written as

$$\mathcal{L}_\Delta^{BOX}\{\tau_j^n\} = 0, \tag{30}$$

in which the difference form is written at the n and $(n - 1)$ time levels as

$$\mathcal{L}_\Delta^{BOX}\{\tau_j^n\} = (1 - c)\tau_{j-1}^n + (1 + c)\tau_j^n - (1 + c)\tau_{j-1}^{n-1} - (1 - c)\tau_j^{n-1}. \tag{31}$$

Based on the modified equivalent equation (14), it may also be written as

$$\mathcal{L}_\Delta^{BOX}\{\tau_j^n\} = 2\Delta t \left[\frac{\partial \tau}{\partial t} + u\frac{\partial \tau}{\partial x} - \frac{u(\Delta x)^2}{12}(1 - c)(1 + c)\frac{\partial^3 \tau}{\partial x^3} + \dots\right]_j^n. \tag{32}$$

We now consider the (1,2,2) operator

$$\mathcal{L}_\Delta^{122}\{\tau_j^n\} = (1 + c)\mathcal{L}_\Delta^{121}\{\tau_j^n\} - (1 - 2c)\mathcal{L}_\Delta^{BOX}\{\tau_j^n\}. \tag{33}$$

Substitution of (29) and (32) into (33) yields the differential form

$$\mathcal{L}_\Delta^{122}\{\tau_j^n\} = 6c\Delta t \left[\frac{\partial \tau}{\partial t} + u\frac{\partial \tau}{\partial x} + O\{(\Delta x)^3\}\right]_j^n, \tag{34}$$

whereas using (28) and (31) in (33) yields the difference form

$$\mathcal{L}_\Delta^{122}\{\tau_j^n\} = (1 + c)\tau_j^{n+1} + 2c(1 - 2c)\tau_{j-1}^n - 2(1 - 2c)(1 + c)\tau_j^n$$

$$-2c(1 + c)\tau_{j-1}^{n-1} + (1 - 2c)(1 - c)\tau_j^{n-1}. \tag{35}$$

Since substitution of (27) and (30) into (33) gives

$$\mathcal{L}_\Delta^{122}\{\tau_j^n\} = 0, \tag{36}$$

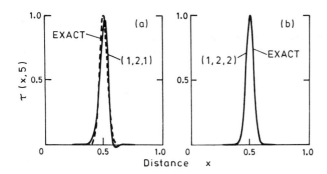

FIGURE 3. Results of the pulse test for (a) the (1,2,1) method, (b) the (1,2,2) method, with $c = 0.4$.

then the form (35) leads to the (1,2,2) explicit finite-difference equation

$$\tau_j^{n+1} = 2(1 - 2c)\tau_j^n + 2c\tau_{j-1}^{n-1} - (1 - 2c)(1 + c)^{-1}\left(2c\tau_{j-1}^n + (1 - c)\tau_j^{n-1}\right), \quad j = 1(1)J. \tag{37}$$

This is von Neumann stable for

$$0 < c \leq 1/2, \tag{38}$$

and has the modified equivalent equation

$$\frac{\partial \tau}{\partial t} + u\frac{\partial \tau}{\partial x} + \frac{u(\Delta x)^3}{72}(1 - c^2)(1 - 2c)\frac{\partial^4 \tau}{\partial x^4}$$

$$+ \frac{u(\Delta x)^4}{540}(1 - c^2)(1 - 2c)(2 - c)\frac{\partial^5 \tau}{\partial x^5} + O\{(\Delta x)^5\} = 0. \tag{39}$$

This verifies the third-order accuracy of (37). It has the amplitude response

$$\zeta(c, N_\lambda) = 1 - 4\pi^4(1 - c^2)(1 - 2c)/[9(N_\lambda)^3] + O\{(N_\lambda)^{-5}\}, \tag{40}$$

and the relative wave speed

$$\mu(c, N_\lambda) = 1 + 4\pi^4(1 - c^2)(1 - 2c)(2 - c)/[135(N_\lambda)^4] + O\{(N_\lambda)^{-6}\}. \tag{41}$$

Clearly, as $N_\lambda \to \infty$ in the stability range $0 < c < 1/2$ both ζ and $\mu \to 1-$; that is, for large values of N_λ both the amplitude and wave speed of simulated waves are too small, the error increasing as N_λ decreases.

Like the (1,2,1) three-level method developed earlier, (37) may be used efficiently on vector or array processors and has a compact computational stencil (see Figure 1(d)) which permits direct evaluation of τ_J^{n+1} at the right-hand boundary. A suitable third-order starting procedure is the two-level (2,3) marching method developed by Noye (1986).

As the wave speed error is now $O\{(N_\lambda)^{-4}\}$ compared with the corresponding error of $O\{(N_\lambda)^{-2}\}$ for the methods described previously, and the slightly displaced small amplitude short waves are numerically damped as indicated by (40), the result obtained at $T = 5$ with the pulse test was practically the same as the exact solution (see Figure 3(b)). At $T = 25$ the numerical simulation produced by the new (1,2,2) FDM is about as accurate as that produced by the (1,2,1) method with $T = 5$. At $T = 50$, the numerical simulation produced by the (1,2,2) method remained better than either the box method or the leapfrog method at $T = 5$.

CONCLUSION

Two new explicit second and third-order two-point (in space) three-level (in time) finite-difference equations have been developed for solving the one-dimensional advection equation. These are spatially more compact than the three-level second-order leapfrog method, and do not require special procedures to compute values at the downstream boundary. As they are purely explicit, they can be used efficiently with array or vector processors, which is not possible with the two-level second-order box method. Furthermore, the second-order compact (1,2,1) method is just as stable as the leapfrog method, and is much more accurate than it or the box method; and the third-order (1,2,2) method, although only stable up to a Courant number of one half, is much more accurate than any of the three second-order methods.

REFERENCES

1. D'Yakonov, E.G., Difference schemes with split operators for multi-dimensional unsteady problems, *U.S.S.R. Computational Mathematics*, vol.4, no.2, pp.92-110, 1963.

2. Hindmarsh, A.C., Gresho, P.M., and Griffiths, D.F., The stability of explicit Euler time-integration for certain finite-difference approximations of the multi-dimensional advection-diffusion equation, *International Journal for Numerical Methods in Fluids*, vol.4, pp.853-897, 1984.

3. Lax, P.D., and Richtmyer, R.D., Survey of the stability of linear finite difference equations, *Communications on Pure and Applied Mathematics*, vol.9, pp.267-293, 1956.

4. Morton, K.W., Stability of finite difference approximations to a diffusion-convection equation, *International Journal of Numerical Methods in Engineering*, vol.15, pp.677-683, 1980.

5. Noye, B.J., Finite difference techniques for partial differential equations, *Computational Techniques for Differential Equations*, ed. B.J. Noye, pp.95-354, North-Holland Mathematics Studies, No.83, 1984a.

6. Noye, B.J., Analysis of explicit finite difference methods used in computational fluid mechanics, *Contributions of Mathematical Analysis to the Numerical Solution of Partial Differential Equations*, ed. A. Miller, pp.106-118, Proceedings of the Centre for Mathematical Analysis, Australian National University, vol.7, 1984b.

7. Noye, B.J., Three-point two-level finite difference methods for the one-dimensional advection equation, *Computational Techniques and Applications: CTAC-85*, eds. B.J. Noye and R.L. May, pp.159-192, North-Holland Publishing Co., 1986.

8. Noye, B.J., and Hayman, K.J., Accurate finite difference methods for solving the advection-diffusion equation, *Computational Techniques and Applications: CTAC-85*, eds. B.J. Noye and R.L. May, pp.137-158, North-Holland Publishing Co., 1986.

9. O'Brien, G.G., Hyman, M.A., and Kaplan, S., A study of the numerical solutions of partial differential equations, *Journal of Mathematics and Physics*, vol.29, pp.223-251, 1950.

10. Sucec, J., Practical stability analysis of finite difference equations by the matrix method, *International Journal of Numerical Methods in Engineering*, vol.24, pp.679-687, 1987.

11. Yanenko, N.N., *The method of fractional steps: the solution of problems of mathematical physics in several variables*, English translation, ed. M. Holt, Springer-Verlag, New York, 1971.

Wave Propagation Characteristics for the Finite Difference Equations of a Three-Dimensional Tidal Model

M. W. STEVENS and B. J. NOYE
Department of Applied Mathematics
University of Adelaide
G.P.O. Box 498
Adelaide 5001, South Australia

INTRODUCTION

The accuracy of a numerical method for solving the governing partial differential equations for tidal flow may be estimated in several ways, the most common being comparison with analytic solutions to the corresponding linearised problem and/or simplification of the sea geometry. One useful alternative is to determine the wave propagation characteristics of the numerical technique. This involves application of the same finite difference technique used in the model to a quasi-linearised form of the governing partial differential equations, and comparison of the resulting numerical solution with the analytic solution to the same problem with an initial condition which is an infinitely long wave train. This procedure also indicates the stability of the finite difference equations.

The latter approach has been used to estimate the accuracy of finite difference solutions of the two dimensional advection-diffusion equations by Leendertse (1967), Abbott (1979) and Noye (1984a), for the linearised depth-averaged tidal equations by Sobey (1970), and for the quasi-linearised depth-averaged tidal equations by Noye (1984b). In the following, this approach is extended to the set of quasi-linearised governing equations for three-dimensional tidal motion.

WAVE PROPAGATION ANALYSIS

The quasi-linearised forms of the governing equations of three-dimensional tidal motion are

$$\frac{\partial \zeta}{\partial t} + h_a \frac{\partial \overline{\mu}}{\partial \chi} + h_a \frac{\partial \overline{\nu}}{\partial \lambda} = 0 , \tag{1}$$

$$\frac{\partial \mu}{\partial t} + \mu_a \frac{\partial \mu}{\partial \chi} + \nu_a \frac{\partial \mu}{\partial \lambda} = -\frac{g}{L^2} \frac{\partial \zeta}{\partial \chi} + \frac{N_\chi}{L^2} \frac{\partial^2 \mu}{\partial \chi^2} + \frac{N_\lambda}{B^2} \frac{\partial^2 \mu}{\partial \lambda^2} + \frac{1}{h_a^2} \frac{\partial}{\partial \eta}(N_\eta \frac{\partial \mu}{\partial \eta}) + \frac{f_a B \nu}{L} , \tag{2}$$

$$\frac{\partial \nu}{\partial t} + \mu_a \frac{\partial \nu}{\partial \chi} + \nu_a \frac{\partial \nu}{\partial \lambda} = -\frac{g}{B^2} \frac{\partial \zeta}{\partial \lambda} + \frac{N_\chi}{L^2} \frac{\partial^2 \nu}{\partial \chi^2} + \frac{N_\lambda}{B^2} \frac{\partial^2 \nu}{\partial \lambda^2} + \frac{1}{h_a^2} \frac{\partial}{\partial \eta}(N_\eta \frac{\partial \nu}{\partial \eta}) - \frac{f_a L \mu}{B} , \tag{3}$$

where μ, ν are the velocity components in the χ, λ directions, ζ is elevation above mean sea level, H is the depth of water, η is the relative depth, so $\eta = 0$ is at the sea surface and the sea floor at $\eta = 1$, g is gravitational acceleration, f is the Coriolis parameter, L, B are distances in the χ, λ directions, t is time,

$$\overline{\mu} = \int_0^1 \mu d\eta , \qquad \overline{\nu} = \int_0^1 \nu d\eta , \tag{4}$$

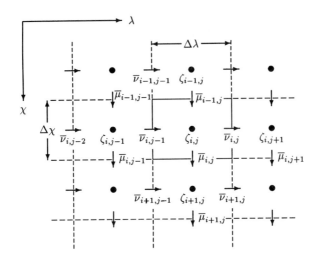

FIGURE 1. Position of variables ζ (grid points $\times$), $\bar{\mu}$ (grid points $\downarrow$) and $\bar{\nu}$ (grid points $\rightarrow$) in the two-dimensional array.

and μ_a, ν_a, h_a and f_a are constants representative of μ, ν, H and f over a tide cycle, in the neighbourhood of (χ, λ, η). The horizontal eddy viscosity coefficients, N_χ and N_λ, are constant and the vertical eddy viscosity coefficient, N_η, is depth dependent. Absence of wind stress on the surface of the water and a "no-slip" condition at the sea floor imply that the boundary conditions are

$$\left.\frac{\partial \mu}{\partial \eta}\right|_{\eta=0} = \left.\frac{\partial \nu}{\partial \eta}\right|_{\eta=0} = 0 \,, \qquad \left.\mu\right|_{\eta=1} = \left.\nu\right|_{\eta=1} = 0 \,. \tag{5}$$

In the absence of horizontal boundaries, for the given wave numbers m_χ, m_λ an analytic solution to Eqns.(1) to (5) can be obtained in the form

$$\zeta(\chi, \lambda, t) = a \exp\{i(\hat{\sigma}t + m_\chi\chi + m_\lambda\lambda)\} \,, \tag{6}$$

$$\mu(\chi, \lambda, \eta, t) = \left(\frac{-g m_\chi m_\lambda a}{L^2 B \hat{\sigma}^2}\right) F_\chi(\eta) \exp\{i(\hat{\sigma}t + m_\chi\chi + m_\lambda\lambda)\} \,, \tag{7}$$

$$\nu(\chi, \lambda, \eta, t) = \left(\frac{-g m_\chi m_\lambda a}{L B^2 \hat{\sigma}^2}\right) F_\lambda(\eta) \exp\{i(\hat{\sigma}t + m_\chi\chi + m_\lambda\lambda)\} \,, \tag{8}$$

where $i = \sqrt{-1}$ and $F_\chi(\eta)$ and $F_\lambda(\eta)$ depend on the choice of the vertical eddy viscosity, N_η, provided that the following relationship is satisfied.

$$\hat{\sigma}^3 = \frac{g h_a m_\chi m_\lambda}{LB} \left(\frac{m_\chi}{L} \int_0^1 F_\chi(\eta) d\eta + \frac{m_\lambda}{B} \int_0^1 F_\lambda(\eta) d\eta\right) \,. \tag{9}$$

The wave propagation analysis requires that this analytic solution to the quasi-linearised equations, has real values for m_χ and m_λ thus making $\hat{\sigma}$ complex. This is not a trivial task but can be solved by using an iterative scheme to find the appropriate value of $\hat{\sigma}$ for given values of m_χ and m_λ, using Eq.(9).

The same method used to discretise the full equations of tidal motion should be used on Eqns.(1) to (4). The discretisation scheme used for illustration is the method described in

Noye and Stevens (1987) with a variable vertical grid, but with uniform grid spacings $\Delta \chi$ and $\Delta \lambda$ in the χ and λ directions and the double subscripted spatial array shown in Figure 1. This gives for Eq.(1)

$$\frac{1}{\Delta t}(\zeta_{i,j}^* - \zeta_{i,j}^n) + \frac{h_a}{\Delta \chi}(\overline{\mu}_{i,j}^n - \overline{\mu}_{i-1,j}^n) + \frac{h_a}{\Delta \lambda}(\overline{\nu}_{i,j}^n - \overline{\nu}_{i,j-1}^n) = 0 , \tag{10}$$

where $\overline{\mu}_{i,j}^n$ and $\overline{\nu}_{i,j}^n$ are calculated from the Simpson approximation to the integrals given by

$$\overline{\mu}_{i,j}^n = \sum_{k=0}^{K} s_k \mu_{i,j,k}^n , \qquad \overline{\nu}_{i,j}^n = \sum_{k=0}^{K} s_k \nu_{i,j,k}^n , \tag{11}$$

where the s_k are the Simpson coefficients for a variable grid.

Then with rearrangement of Eq.(10) and substitution of Eq.(11), $\zeta_{i,j}^*$, which is the first approximation for $\zeta_{i,j}^{n+1}$, may be computed explicitly as follows:

$$\zeta_{i,j}^* = \zeta_{i,j}^n - \frac{h_a \Delta t}{\Delta \chi} \sum_{k=0}^{K} s_k(\mu_{i,j,k}^n - \mu_{i-1,j,k}^n) - \frac{h_a \Delta t}{\Delta \lambda} \sum_{k=0}^{K} s_k(\nu_{i,j,k}^n - \nu_{i,j-1,k}^n) . \tag{12}$$

The corresponding "corrector" equation for ζ based on discretising Eq.(1) using time centred differencing for all terms, is

$$\frac{1}{\Delta t}(\zeta_{i,j}^{n+1} - \zeta_{i,j}^n) + \frac{h_a}{2\Delta \chi} \sum_{k=0}^{K} s_k(\mu_{i,j,k}^{n+1} - \mu_{i-1,j,k}^{n+1} + \mu_{i,j,k}^n - \mu_{i-1,j,k}^n)$$

$$+ \frac{h_a}{2\Delta \lambda} \sum_{k=0}^{K} s_k(\nu_{i,j,k}^{n+1} - \nu_{i,j-1,k}^{n+1} + \nu_{i,j,k}^n - \nu_{i,j-1,k}^n) = 0 . \tag{13}$$

The discretised form for Eq.(2), with Eq.(12) used to replace all the values of ζ^*, is

$$\frac{1}{\Delta t}(\mu_{i,j,k}^{n+1} - \mu_{i,j,k}^n) + \frac{\mu_a}{2\Delta \chi}(\mu_{i+1,j,k}^n - \mu_{i-1,j,k}^n) + \frac{\nu_a}{2\Delta \lambda}(\mu_{i,j+1,k}^n - \mu_{i,j-1,k}^n)$$

$$= +\frac{gh_a \Delta t}{2L^2(\Delta \chi)^2} \sum_{k=0}^{K} s_k(\mu_{i+1,j,k}^n - 2\mu_{i,j,k}^n + \mu_{i-1,j,k}^n)$$

$$+ \frac{gh_a \Delta t}{2L^2 \Delta \chi \Delta \lambda} \sum_{k=0}^{K} s_k(\nu_{i+1,j,k}^n - \nu_{i+1,j-1,k}^n - \nu_{i,j,k}^n + \nu_{i,j-1,k}^n) - \frac{g}{L^2 \Delta \chi}(\zeta_{i+1,j}^n - \zeta_{i,j}^n)$$

$$+ \frac{N_\chi}{L^2(\Delta \chi)^2}(\mu_{i+1,j,k}^n - 2\mu_{i,j,k}^n + \mu_{i-1,j,k}^n) + \frac{N_\lambda}{B^2(\Delta \lambda)^2}(\mu_{i,j+1,k}^n - 2\mu_{i,j,k}^n + \mu_{i,j-1,k}^n)$$

$$+ \left(\begin{array}{l} [N_{k+1} + (2r_k - 1)N_k](\mu_{i,j,k+1}^{n+1} + \mu_{i,j,k+1}^n) \\ -[N_{k+1} - (r_k + 1)(r_k^2 - 3r_k + 1)N_k + r_k^3 N_{k-1}](\mu_{i,j,k}^{n+1} + \mu_{i,j,k}^n) \\ +r_k^2[(2 - r_k)N_k + r_k N_{k-1}](\mu_{i,j,k-1}^{n+1} + \mu_{i,j,k-1}^n) \end{array} \right) /2h_a^2(\Delta \eta_k)^2(r_k + 1)$$

$$+ \frac{f_a B}{4L}(\nu_{i,j-1,k}^{n+1} + \nu_{i,j,k}^n + \nu_{i+1,j-1,k}^n + \nu_{i+1,j,k}^n) , \qquad k = 1 \ldots K-1 , \tag{14}$$

with a similar discrete equation obtained from Eq.(3). These equations, in conjunction with Eq.(13) and the boundary Eqns.(21) and (22) provide $2K + 3$ equations for the $2K + 3$ unknowns $\zeta^\otimes$ and $\mu_k^\otimes$, $\nu_k^\otimes$ for $k = 0 \ldots K$.

We now consider the behaviour of the system of finite difference Eqns.(12) to (13) with the computational element assumed undisturbed by the influence of any boundary. Therefore a solution is sought to the system of Eqns.(10) to (13) given the initial conditions

359

$$\zeta^0(\chi,\lambda) \;=\; \sum_p \zeta_p^* \exp\{i(m_{\chi,p}\chi + m_{\lambda,p}\lambda)\}\,, \qquad -\infty < \chi, \lambda < \infty\,, \tag{15}$$

$$\mu^0(\chi,\lambda,\eta) \;=\; \sum_p \mu_p^* F_{\chi,p}(\eta)\exp\{i(m_{\chi,p}\chi + m_{\lambda,p}\lambda)\}\,, \qquad 0 \le \eta \le 1\,, \tag{16}$$

$$\nu^0(\chi,\lambda,\eta) \;=\; \sum_p \nu_p^* F_{\lambda,p}(\eta)\exp\{i(m_{\chi,p}\chi + m_{\lambda,p}\lambda)\}\,, \qquad 0 \le \eta \le 1\,, \tag{17}$$

where

$m_{\chi,p}$, $m_{\lambda,p}$ are the wave numbers of pth Fourier components in the χ, λ directions,

$F_{\chi,p}(\eta)$, $F_{\lambda,p}(\eta)$ are depth variations of the pth mode, for μ, λ,

$\zeta^n(\chi,\lambda)$, $\mu^n(\chi,\lambda,\eta)$, $\nu^n(\chi,\lambda,\eta)$ are the solutions of Eqns.(11) to (13) at time $n\Delta t$,

ζ_p^*, μ_p^*, ν_p^* are the amplitudes of the Fourier components of ζ^0, μ^0, ν^0.

Since the system of Eqns.(11) to (13) is linear, superposition is valid and only one component of the Fourier series of the initial conditions need be considered. Therefore a solution to the set of finite difference equations is sought, in the form

$$\zeta_{i,j}^n \;=\; \zeta^{\otimes} \exp\{i(\sigma n\Delta t + i\Delta\chi m_\chi + j\Delta\lambda m_\lambda)\}\,, \tag{18}$$

$$\mu_{i,j,k}^n \;=\; \mu_k^{\otimes} \exp\{i(\sigma n\Delta t + i\Delta\chi m_\chi + j\Delta\lambda m_\lambda)\}\,, \qquad k = 0\ldots K\,, \tag{19}$$

$$\nu_{i,j,k}^n \;=\; \nu_k^{\otimes} \exp\{i(\sigma n\Delta t + i\Delta\chi m_\chi + j\Delta\lambda m_\lambda)\}\,, \qquad k = 0\ldots K\,, \tag{20}$$

where $\mu_k^{\otimes}$, $\nu_k^{\otimes}$ are the difference solutions for $\mu^* F_\chi(\eta)$, $\nu^* F_\lambda(\eta)$ at $\eta = \eta_k$.

The "no-slip" bottom boundary equation implies that

$$\mu_K^{\otimes} \;=\; \nu_K^{\otimes} \;=\; 0, \tag{21}$$

and the surface boundary equation with no wind stress implies that

$$a_0\mu_0^{\otimes} + b_0\mu_1^{\otimes} + c_0\mu_2^{\otimes} = 0 \quad \text{and} \quad a_0\nu_0^{\otimes} + b_0\nu_1^{\otimes} + c_0\nu_2^{\otimes} = 0\,, \tag{22}$$

where a_0, b_0 and c_0 are defined by

$$a_0 = \frac{-r_1 - 2}{\Delta\eta_0(r_1 + 1)}\,, \qquad b_0 = \frac{r_1 + 1}{\Delta\eta_0 r_1} \quad \text{and} \quad c_0 = -\frac{1}{\Delta\eta_0 r_1(r_1 + 1)}\,, \tag{23}$$

which gives a second order approximation to the first boundary equation in Eqns.(5).

Assuming that $m_\lambda = \theta_1 m_\chi$ and that $\Delta\lambda = \theta_2\Delta\chi$ then, allowing for the fact that $\mu_{i,j}$ is evaluated at $((i+\tfrac{1}{2})\Delta\chi, j\Delta\lambda)$ and $\nu_{i,j}$ is evaluated at $(i\Delta\chi, (j+\tfrac{1}{2})\Delta\lambda)$, as shown in Figure 1, substitution of Eqns.(18) to (20) into Eqns.(13), (14) and the discretised ν equation, for $k = 1\ldots K-1$ and using the boundary Eqns.(21) and (22), yields the matrix equation

$$\mathbf{A}\underline{v} = \xi\mathbf{B}\underline{v} \tag{24}$$

where $\mathbf{A}$ and $\mathbf{B}$ are partitioned matrices in the form

$$\mathbf{A} = \begin{bmatrix} 1 & -\underline{A}_\chi & -\underline{A}_\lambda \\ -\underline{B}_\chi & \mathbf{A}_\chi & BL^{-1}\mathbf{A_1}(P) \\ -\underline{B}_\lambda & B^{-1}L\mathbf{A_1}(-P) & \mathbf{A}_\lambda \end{bmatrix}, \quad \mathbf{B} = \begin{bmatrix} 1 & \underline{A}_\chi & \underline{A}_\lambda \\ 0 & \mathbf{B_1} & -\mathbf{B_2} \\ 0 & \mathbf{B_2} & \mathbf{B_1} \end{bmatrix},$$

$$\underline{v} = \begin{bmatrix} \zeta^{\otimes} & \mu_0^{\otimes} & \mu_1^{\otimes} & \cdots & \mu_{K-1}^{\otimes} & \nu_0^{\otimes} & \nu_1^{\otimes} & \cdots & \nu_{K-1}^{\otimes} \end{bmatrix}^T,$$

and

$$A_\chi = [A_{\chi,0} \ \ A_{\chi,1} \ \ A_{\chi,2} \ \ \ldots \ \ A_{\chi,K-1}] \ , \qquad A_\lambda = [A_{\lambda,0} \ \ A_{\lambda,1} \ \ A_{\lambda,2} \ \ \ldots \ \ A_{\lambda,K-1}] \ ,$$

$$B_\chi = [\ 0 \ \ B_\chi \ \ B_\chi \ \ \ldots \ \ B_\chi \ ]^T \ , \qquad B_\lambda = [\ 0 \ \ B_\lambda \ \ B_\lambda \ \ \ldots \ \ B_\lambda \ ]^T \ ,$$

$$\mathbf{A}_\varphi = \begin{bmatrix} -a_0 & -b_0 & -c_0 & \ldots & 0 \\ \Gamma_1^- - G_{\varphi,0} & \Gamma_1 - H_{\varphi,1} & \Gamma_1^+ - G_{\varphi,2} & \ldots & -G_{\varphi,K-1} \\ -G_{\varphi,0} & \Gamma_2^- - G_{\varphi,1} & \Gamma_2 - H_{\varphi,2} & \ldots & -G_{\varphi,K-1} \\ \vdots & \vdots & \vdots & \ddots & \vdots \\ -G_{\varphi,0} & -G_{\varphi,1} & -G_{\varphi,2} & \ldots & \Gamma_{K-1} - H_{\varphi,K-1} \end{bmatrix} \ ,$$

in which $\varphi = \chi$ or λ,

$$\mathbf{A}_1(R) = \begin{bmatrix} 0 & 0 & 0 & \ldots & 0 \\ -Q_0 & R - Q_1 & -Q_2 & \ldots & -Q_{K-1} \\ -Q_0 & -Q_1 & R - Q_2 & \ldots & -Q_{K-1} \\ \vdots & \vdots & \vdots & \ddots & \vdots \\ -Q_0 & -Q_1 & -Q_2 & \ldots & R - Q_{K-1} \end{bmatrix} \ ,$$

$$\mathbf{B}_1 = \begin{bmatrix} 0 & 0 & 0 & \ldots & 0 \\ -\Gamma_1^- & 1 - \Gamma_1 & -\Gamma_1^+ & \ldots & 0 \\ 0 & -\Gamma_2^- & 1 - \Gamma_2 & \ldots & 0 \\ \vdots & \vdots & \vdots & \ddots & \vdots \\ 0 & 0 & 0 & \ldots & 1 - \Gamma_{K-1} \end{bmatrix} \ ,$$

$$\mathbf{B}_2 = Diag(\ 0 \ \ P_b \ \ P_b \ \ \ldots \ \ P_b \ )$$

and where

$$\begin{aligned}
\xi &= \exp\{i\sigma\Delta t\} \ , & A_{\chi,k} &= irh_a s_k \sin(\beta_\chi) \ , & A_{\lambda,k} &= \frac{irh_a s_k \sin(\beta_\lambda)}{\theta_2} \ , \\
r &= \frac{\Delta t}{\Delta\chi} \ , & \beta_\chi &= \tfrac{1}{2}m_\chi \Delta\chi \ , & \beta_\lambda &= \tfrac{1}{2}\theta_1\theta_2 m_\chi \Delta\chi \ , \\
S_\eta &= \frac{\overline{N}_a \Delta t}{2(h_a \Delta\eta_k)^2} \ , & B_\chi &= \frac{2iCr_\chi^2}{rh_a}\sin(\beta_\chi) \ , & B_\lambda &= \frac{2iCr_\lambda^2}{rh_a}\sin(\beta_\lambda) \ , \\
Cr_\chi &= \frac{\Delta t\sqrt{gh_a}}{L\Delta\chi} \ , & Cr_\lambda &= \frac{LCr_\chi}{B\theta_2} \ , & \sqrt{gh_a} &= \frac{L\Delta\chi N_L}{\text{Period}} \ ,
\end{aligned}$$

$$H_{\chi,k} = -1 + i\mu'\sin(2\beta_\chi) + i\nu'\sin(2\beta_\lambda) + 4N_\chi'\sin^2(\beta_\chi) + 4N_\lambda'\sin^2(\beta_\lambda) + G_{\chi,k}$$

$$H_{\lambda,k} = -1 + i\mu'\sin(2\beta_\chi) + i\nu'\sin(2\beta_\lambda) + 4N_\chi'\sin^2(\beta_\chi) + 4N_\lambda'\sin^2(\beta_\lambda) + G_{\lambda,k}$$

$$\Gamma_k^- = \frac{S_\eta}{\overline{N}_a(r_k+1)}\left\{r_k^2(2-r_k)N_k + r_k^3 N_{k-1}\right\} \ ,$$

$$\Gamma_k = \frac{S_\eta}{\overline{N}_a(r_k+1)}\left\{-N_{k+1} + (r_k+1)(r_k^2 - 3r_k + 1)N_k - r_k^3 N_{k-1}\right\} \ ,$$

$$\Gamma_k^+ = \frac{S_\eta}{\overline{N}_a(r_k+1)}\left\{N_{k+1} + (2r_k-1)N_k\right\} \ ,$$

$$G_{\chi,k} = 2Cr_\chi^2 \sin^2(\beta_\chi)s_k , \quad P = f'\cos(\beta_\lambda)\exp(i\beta_\lambda)/2 , \quad P_b = f'\cos(\beta_\lambda)\exp(-i\beta_\lambda)/2 ,$$

$$G_{\lambda,k} = 2Cr_\chi^2 \sin^2(\beta_\lambda)s_k , \quad Q_k = 2Cr_\chi Cr_\lambda \sin(\beta_\chi)\sin(\beta_\lambda)s_k ,$$

$$\mu' = \frac{\mu_a \Delta t}{\Delta \chi} , \qquad \nu' = \frac{\nu_a \Delta t}{\theta_2 \Delta \chi} , \qquad f' = f_a \Delta t ,$$

$$N_\chi' = \frac{N_\chi \Delta t}{L^2 \Delta \chi^2} , \qquad N_\lambda' = \frac{N_\lambda \Delta t}{\theta_2^2 B^2 \Delta \chi^2} , \qquad \overline{N_a} = \int_0^1 N(\eta)d\eta .$$

A non-trivial solution of the matrix Eq.(24) exists if and only if ξ is an eigenvalue of this extended eigen-equation The $2K+1$ eigenvalues ξ can be found numerically using a computer package (for example the Nag routine F02GJF).

The parameters β_χ and β_λ may be expressed in terms of N_L, the number of $\Delta \chi$ grid spacings in a wave length of the χ component of the horizontal wave motion, by means of the relations

$$\beta_\chi = \pi/N_L , \qquad\qquad \beta_\lambda = \theta_1\theta_2\pi/N_L , \tag{25}$$

which mean that $m_\chi = 2\pi/(\Delta\chi N_L)$. Therefore the eigenvalues ξ obtained by solving Eq.(24) may be considered to be functions of the non-dimensional parameters N_L, Cr_χ, S_η, θ_1, θ_2, μ', ν', N_χ', N_λ' and f'.

Clearly the finite difference Eqns.(12) to (13) will be stable so long as $|\xi| \leq 1$. In the following this criterion has been used to indicate the range of the values of the non-dimensional parameters for which the difference equations used to solve the quasi-linearised governing equations are stable. However, this does not imply that their solution is a good approximation to the solution of Eqns.(1) to (4). This may be determined, by comparison of the solution of the numerical scheme given by Eqns.(12) to (13) with one component of the initial conditions (15) to (17) to the true solution of Eqns.(1) to (4) given the same initial conditions. The true wave solution is given by Eq.(6) to (8)

Two parameters which quantify differences between the true wave propagating in the χ–λ plane and the corresponding numerical approximate motion, are the ratios of the amplitudes and the speeds of these waves. These can be expressed in terms of the values of $\sigma = i\log(\xi)$ from the numerical wave and $\hat\sigma$ from the true wave (see Stevens and Noye (1989)). The amplitude response per wave period of the numerical method is

$$A_r = \frac{\alpha_N}{\alpha_T} = \exp\left\{2\pi\left(\frac{\text{Im}\{\sigma\}}{\text{Re}\{\sigma\}} - \frac{\text{Im}\{\hat\sigma\}}{\text{Re}\{\hat\sigma\}}\right)\right\} , \quad \text{Re}\{\hat\sigma\} < 0 , \quad \text{Re}\{\sigma\} < 0 . \tag{26}$$

and the relative wave speed is

$$S_r = u_N/u_T = \text{Re}\{\sigma\}/\text{Re}\{\hat\sigma\} . \tag{27}$$

In the expected range of values of θ_1, θ_2, μ', ν', N_χ', N_λ', f', and S_η the values of A_r and S_r can be computed as functions of N_L, the number of grid spacings per wave length, in the χ-direction, of a horizontal plane wave, and the Courant number Cr_χ. Ideally, both A_r and S_r should take the value of one for all values of N_L and Cr_χ, whatever the values of S_η.

For the finite difference method to be stable for a particular value of Cr_χ the $2K+1$ eigenvalues of Eq.(24) must each have modulus less than one for all possible values of N_L. For each different value of N_L and Cr_χ, one of the $2K+1$ eigenvalues will correspond to the true wave solution the others being computational modes. This will be the value of ξ from which σ and therefore A_r and S_r are calculated. Also for this eigenvalue a corresponding eigenvector $\underline{v}$ can be calculated, which can be normalised so that $\zeta^\otimes = a$. This eigenvector can be used to analyse the distribution over depth of the errors introduced by the numerical scheme. By

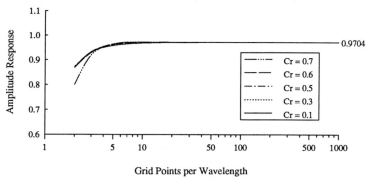

FIGURE 2. Amplitude response for the numerical scheme under consideration

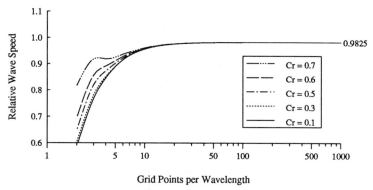

FIGURE 3. Relative wave speed for the numerical scheme under consideration

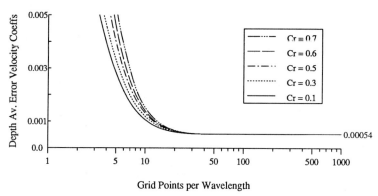

FIGURE 4. Error estimate for the numerically approximated values of $\mu_k^{\otimes}$ and $\nu_k^{\otimes}$.

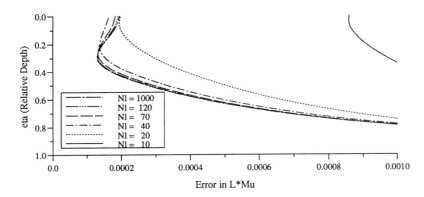

FIGURE 5. Error profiles for $\mu_k^\otimes$ for $Cr = 0.3$.

comparing $\mu_k^\otimes$ and $\nu_k^\otimes$ with the analytic solution of $\mu^* F_\chi(\eta_k)$ and $\mu^* F_\lambda(\eta_k)$ it can be seen by how much the numerical solution differs from the analytic solution and at various depths.

A typical set of graphs obtained by this analysis is given in Figures 2 to 5. Figures 2 and 3 show the amplitude responses and relative wave speeds plotted against N_L for various courant numbers. Figure 4 shows an error estimate for the errors between the $\mu_k^\otimes$, $\nu_k^\otimes$ and their analytic values. Figure 5 is for a specific value of the courant number and show the distribution over depth of the errors in Figure 4 for various values of N_L for the μ velocity component. A graph with similar characteristics is obtained for the ν component.

REFERENCES

1. Abbott, M. B., *Computational Hydraulics*, Pitman, London, 1979.

2. Leendertse, J. J., Aspects of a Computational Model for Long Period Water-Wave Propagation, *Memorandum RM-5294-PR*, 165 pp., Rand Corporation, Santa Monica, Calif., 1967.

3. Noye, B. J., Finite difference techniques for partial differential equations, in *Computational Techniques for Differential Equations*, edited by B.J. Noye, pp. 95-354, North-Holland, Amsterdam, 1984a.

4. Noye, B. J., Wave Propagation Characteristics of a Numerical Model of Tidal Motion, in *Computational Techniques and Applications: CTAC-83*, edited by B.J. Noye and C.J. Fletcher, pp. 360-374, North-Holland, Amsterdam, 1984b.

5. Noye, B. J. and Stevens M. W., A three-dimensional model of tidal propagation using transformations and variable grids, in *Three-Dimensional Coastal Ocean Models*, edited by N.S. Heaps (A. G. U. Monograph Series, Coastal and Estuarine Science, Vol 4), pp. 41-70, 1987.

6. Sobey, R. J., 1970, Finite-difference schemes compared for wave deformation characteristics in mathematical modelling of two-dimensional long-wave propagation, *Tech. Memo. 32*, U.S. Army Corps of Engineers, Coastal Engineering Research Center, Washington, D.C., 1970.

7. Stevens M. W. and Noye, B. J., Wave Propagation Characteristics Of A Three Dimensional Hydrodynamic–Numeric Tidal Model, *Computers and Fluids*, to be published.

Numerical Modeling of Shocks in Metals

M. D. TYNDALL
Materials Research Laboratory
Defense Science and Technology Organisation
P.O. Box 50, Ascot Vale, Victoria, 3032
and Monash University
Clayton, Victoria, 3168, Australia

INTRODUCTION

What happens when two solids collide? For impacts, without large deformations a description of the behaviour in terms of elastic, plastic and shock wave propagation is needed and the main aim of this work is to develop a numerical model which can describe the impacts of solids (involving no large deformations) in one and two dimensions. However, before elastic-plastic properties of impacting solids can be considered a thorough understanding of shocks in solids is required.

In this paper we start by modelling shocks in an inviscid gas. In one dimension the Eulerian inviscid fluid flow equations, for conservation of mass, momentum and energy, are

$$\frac{\partial \mathbf{U}}{\partial t} + \frac{\partial \mathbf{F(U)}}{\partial x} = 0, \qquad \mathbf{U} = \begin{pmatrix} \rho \\ \rho u \\ E \end{pmatrix}, \qquad \mathbf{F(U)} = \begin{pmatrix} \rho u \\ \rho u^2 + p \\ (E+p)u \end{pmatrix}, \tag{1}$$

where ρ is the density, u the particle velocity, E the total energy per unit volume and p the pressure. Heat conduction effects have been ignored in this formulation, but will be discussed later. For an ideal gas we also have the equation of state

$$p = (\gamma - 1)\rho I \tag{2}$$

where I is the internal energy per unit mass and γ the ratio of specific heats. By choosing a different equation of state, which describes the properties of a solid material, similar types of problems can be solved in a solid. For example, we can use the "stiffened-gas" equation of state, if the density varies only slightly from the normal density,

$$p = c_o^2(\rho - \rho_o) + (\gamma - 1)\rho I, \tag{3}$$

where c_o is the sound speed and ρ_o the density of the material, or the Mie-Grüneisen equation of state (Harlow & Amsden, 1971) for larger density variations. Treating a solid as an effectively inviscid fluid is found to be unsatisfactory and the inclusion of viscous and heat conduction terms is required to give an accurate description of shock-wave behaviour.

When trying to model shock waves numerically we encounter many difficulties. Some are due to the steep gradients involved (in theory a shock in an inviscid fluid is a point discontinuity) and others due to the numerical method being used eg. excessive implicit diffusion and non-physical oscillations. When choosing a suitable numerical method to model shock waves, we selected Flux-Corrected Transport (FCT) which gives a monotonic solution and can accurately model any sharp discontinuities.

NUMERICAL SCHEME

If we introduce a uniform grid with $x_i = i\Delta x$ and $t_n = n\Delta t$, where Δx is the grid spacing and Δt the time step, a finite-difference approximation to eq. (1) can be written as

$$U_i^{n+1} = U_i^n - \frac{\Delta t}{\Delta x}\left[T_{i+\frac{1}{2}} - T_{i-\frac{1}{2}}\right], \tag{4}$$

where U_i^n is the value of U at the ith grid point for the nth time step. The $T_{i\pm\frac{1}{2}}$ are called transportive fluxes, and are depend upon both U and F at the nth time step.

The FCT technique, which was first developed by Boris & Book (1973), gives a monotonic solution with a high level of accuracy. To achieve this it constructs a transportive flux which is the weighted average of a flux computed by a low-order *monotonic* scheme and a flux computed by a high-order scheme. The weighting procedure is chosen to ensure that both the monotonicity of the low-order scheme and the accuracy of the high-order scheme are retained to the greatest possible extent. Zalesak (1979) presented a more generalised form of FCT (hereinafter ZFCT) and a weighting procedure that does not require time splitting in multi-dimensions. Compared to the original FCT of Boris & Book (1973) (hereinafter BBFCT), ZFCT has an improved flux-limiting stage, but does not have the low phase errors of BBFCT. Dietachmayer (1987) constructed a scheme that has the low phase errors of BBFCT and the *generality* of ZFCT. This scheme, called "re-ordered ZFCT"(RZFCT), is based on ZFCT but is used in such a way that time levels are not mixed in the anti-diffusive step (see step 5 below).

The procedure is as follows (Dietachmayer, 1987):

(1) Compute the transportive flux $T_{i+\frac{1}{2}}^{L(n)}$ given by some low-order monotonic scheme.

(2) Obtain the updated low-order ('transported and diffused') solution U_i^*, using the low-order monotonic scheme.

$$U_i^* = U_i^n - \frac{\Delta t}{\Delta x}\left[T_{i+\frac{1}{2}}^{L(n)} - T_{i-\frac{1}{2}}^{L(n)}\right]$$

(3) Compute the low-order fluxes $T_{i+\frac{1}{2}}^{L(*)}$ and the high-order fluxes $T_{i+\frac{1}{2}}^{H(*)}$ at ($*$) time level and hence evaluate the 'anti-diffusive' fluxes:

$$A_{i+\frac{1}{2}}^* = T_{i+\frac{1}{2}}^{H(*)} - T_{i+\frac{1}{2}}^{L(*)}$$

(4) Limit the anti-diffusive fluxes

$$A_{i+\frac{1}{2}}^{C(*)} = C_{i+\frac{1}{2}}A_{i+\frac{1}{2}}^* \qquad 0 \le C_{i+\frac{1}{2}} \le 1$$

(5) Implement the antidiffusive step,

$$U_i^{n+1} = U_i^* - \frac{\Delta t}{\Delta x}\left[A_{i+\frac{1}{2}}^{C(*)} - A_{i-\frac{1}{2}}^{C(*)}\right].$$

The critical step in the above is step 4, which is referred to as "flux-correction" or "flux-limiting". If the $C_{i+\frac{1}{2}}$ coefficients in step 4 are zero then the low-order, monotonic solution is obtained, but if they are equal to one (in regions where non-physical oscillations are absent) then the high-order solution is obtained. In regions where non-physical oscillations are present the $C_{i+\frac{1}{2}}$ coefficients are calculated in such a way that at each point sufficient implicit non-linear diffusion is added to damp out these oscillations. Details of this 'flux-limiting' process (in particular how the $C_{i+\frac{1}{2}}$ coefficients are calculated) are described in Zalesak (1979).

No extra phase errors are introduced in RZFCT. Dietachmayer (1987) tested a wide range of schemes, including ZFCT and BBFCT, and found RZFCT to be the most accurate of all the schemes.

SOME ONE-DIMENSIONAL TEST PROBLEMS AND RESULTS

Gases

Two problems were used to test the ability of ZFCT and RZFCT to model shock waves in a gaseous material with the ideal-gas equation of state. Rusanov's method was used to calculate the low-order fluxes and the Lax-Wendroff method was used to calculate the high-order fluxes (for details see Roache (1972) or Sod, 1978).

The shock tube The first problem was the one-dimensional shock tube (see Sod, 1978), where a diaphragm separates two regions of different pressures and densities in a semi-infinite tube, at rest when $t = 0$ s. Once the diaphragm is ruptured ($t > 0$ s), a rarefaction propagates into the high-pressure gas and a shock wave, followed by a contact discontinuity, propagates into the low-pressure gas. Details of the exact solution can be found in Harlow & Amsden (1971) and Tyndall (1988). For the numerical solutions one hundred points were taken each side of the diaphragm (which was positioned at $x = 0$ m with $\Delta x = 0.025$ m. The ratio of specific heats γ in eq. (2) was chosen to be 1.4, and the initial conditions were $\rho = 1.0$ kgm^{-3}, $p = 1.0$ Pa for $x < 0$ m, and $\rho = 0.125$ kgm^{-3}, $p = 0.1$ Pa for $x > 0$ m.

The numerical results to the shock tube problem using ZFCT is shown in Figs. 1a and 1b and this follows the general form of the exact solution. The contact discontinuity has been slightly diffused and the shock front has been well resolved (only two to three grid points wide). This solution also agrees well with the solution given by Sod (1978), using BBFCT but neither of these solutions is monotonic. In the density profile there is an undershoot at the end of the rarefaction ($x = 0$ m) and a slight overshoot in the contact discontinuity ($x = 1.0$ m), with similar behaviour also present in the particle velocity profile. This non-monotonic behaviour is not caused by phase errors or the flux-correction stage but through errors introduced over the crucial first time steps. If instead we integrate numerically starting from the exact solution at some time $t > 0$ s, rather than from $t = 0$ s, these undershoots and overshoots do not appear (Tyndall, 1989).

The shock tube problem was also solved using RZFCT and the results are presented in Figs. 1c and 1d. Comparing these with those described above, RZFCT gives a more accurate solution than ZFCT. In particular the contact discontinuity is less rounded, the shock front is only two points wide and there is only a slight undershoot at the end of the rarefaction. Dietachmayer (1987) tested a range of monotonic advection schemes using the one-dimensional advection equation and also found that the best results were obtained using RZFCT. Thus for the remainder of this paper RZFCT is used exclusively.

Shock hitting a density discontinuity In this problem a one-dimensional shock (of speed $U = 1.67$ ms^{-1}) was propagated into a stationary density discontinuity; the density and pressure profiles before impact are shown in Figs. 2a and 2b. Examining the density at $t = 2.5$ s (Fig. 2c), the shock has propagated into the dense stationary gas and another weaker shock has been reflected from the discontinuity. The shock travelling through the denser gas is two to three grid points wide and the reflected shock is three to four grid points wide and neither shock diffuses as it travels through the gas. The shock velocities after impact are $U = 0.97$ ms^{-1} and $U = -0.58$ ms^{-1}, which are correct to two significant figures, while the other quantities are correct to three significant figures.

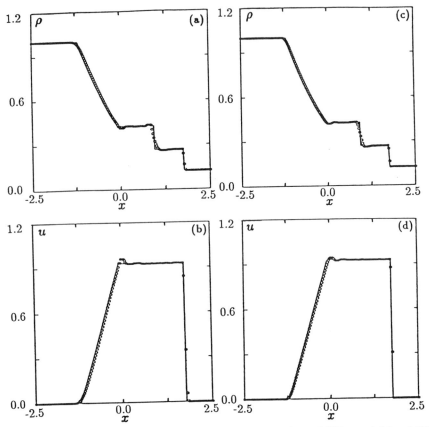

FIGURE 1. The shock tube problem at time $t = 1.0\ s$, with $\Delta x = 0.025\ m$ and $\Delta t = 0.005\ s$. The diaphragm was initially at $x = 0\ m$. The solid line is the exact solution and the circles represent the computed solution. The profiles for (a) density and (b) particle velocity given by ZFCT, and (c) density, and (d) particle velocity given by RZFCT are shown.

Metals

The propagation of a one-dimensional shock and the collision of two shocks were also used to test the ability of ZFCT and RZFCT to model shock waves in a metallic material. Three types of material were used in these simulations, aluminium, copper and stainless steel but since they all behaved similarly, only the results for aluminium are presented here. These problems involved only small density changes, so the "stiffened-gas" equation of state was used where $c_0 = 5380\ ms^{-1}$, $\rho_0 = 2710\ kgm^{-3}$ and $\gamma = 2.7$ for Al. Roe's upwind scheme (see Roe, 1981) was used to calculate the low-order fluxes and the Lax-Wendroff method was used to calculate the high-order fluxes.

The steady shock A one-dimensional weak shock was propagated through aluminium with all variables upstream of the shock and the shock velocity U derived from the Rankine-Hugoniot conditions (Harlow & Amsden, 1971). The initial conditions were $\rho = 2737\ kgm^{-3}$, $p = 790\ MPa$ for $x < 0\ m$ and $\rho = \rho_0$, $p = 0\ MPa$ for $x > 0\ m$. The results of this simulation are shown in Fig. 3 and, although the density ratio is small, the pressure, particle velocity and energy ratios are much larger. Taking this into consideration, the shock front is resolved extremely well, even though it is about eight points wide. The shock propagates through the

368

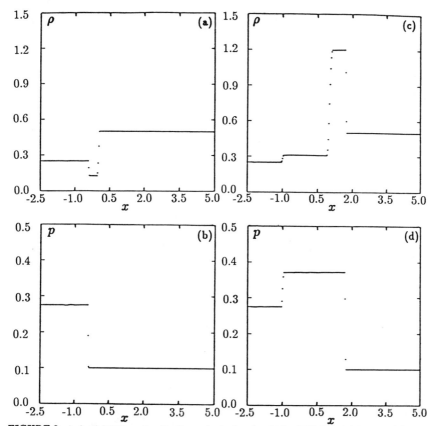

FIGURE 2. A shock hitting a density discontinuity in a gas (using RZFCT), (a) density, (b) pressure at time $t = 0.5$ s and, (c) density, (d) pressure at time $t = 2.5$ s. The density discontinuity was positioned at $x = 0$ m and the shock was initially at $x = -1.25$ m.

aluminium with shock velocity $U = 5450$ ms^{-1}, which is within 0.4% of the exact solution while the density, particle velocity and pressure are all monotonic and involve less that 0.1% error. However the energy is not monotonic, with a small spike at $x = 0$ m and when there is a larger density discontinuity, a corresponding dip can also be seen in the density at this point. In the region of the spike there is a 13.0% error in the energy but elsewhere there is less than 0.5% error. In this problem the spike does not affect the pressure as the energy contribution to the equation of state is much smaller than the other terms.

The non-monotonic behaviour described above is caused by the initial conditions and we can use the steady shock problem to illustrate how this occurs. A general, linearised solution of the fluid flow equations for the density is

$$\rho(x, t) = f(x - (u + c)t) + g(x - (u - c)t) + h(x - ut) \tag{5}$$

where u is particle velocity, c the sound speed in the material and the functions f, g and h are determined by the initial conditions. At $t = 0$ s the shock was at $x = 0$ m and assumed to be a point discontinuity and for a single steady shock we want to excite a shock wave which will travel with wave speed '$u + c$', so that $g = h = 0$ and f is a step function. However when we

369

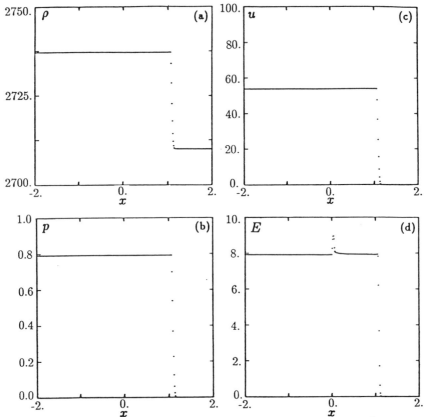

FIGURE 3. The steady shock problem in aluminium at time $t = 2.0 \times 10^{-4}$ s, with $\Delta t = 5.0 \times 10^{-7}$ s and $\Delta x = 0.01$ m. The RZFCT profiles for (a) density ρ (kgm^{-3}), (b) pressure p (GPa), (c) particle velocity u (ms^{-1}) and (d) total energy per unit volume E (MJm^{-3}) are shown. The shock was initially at $x = 0.0$ m.

model the shock numerically there is a slight smoothing of the discontinuity (Fig. 4), causing small amplitude waves which for $t > 0$ s propagate with speeds 'u' and '$u - c$', and the spike in the energy near $x = 0$ m is a small amplitude wave travelling with speed 'u'. Current work involves including viscous and heat conduction terms in the fluid flow equations which may diffuse these small contributions. Alternatively, rather than assuming a shock to be point discontinuity, we could introduce it as a thin region where viscosity and heat conduction are important.

The collision of two shocks The collision of two equal strength shocks in aluminium , with initial shock velocities $U = \pm 5450$ ms^{-1}, is shown in Figs. 5. After the collision the shocks are reflected and travel with shock velocities $U = \pm 5430$ ms^{-1}, which are within 0.4% of the exact shock velocities. Spikes in the energy occur at both the initial shock positions ($x = \pm 1.5$ m) and at the point of collision ($x = 0$ m). The point of collision is where the shocks merge momentarily causing a spike to appear and slight overshoots at the shock fronts. In the region of the spike at $x = 0$ m there is a 6.5% error in the energy and in the region of the overshoots a 1.2% error.

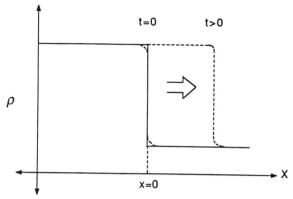

FIGURE 4. Schematic diagram of the steady shock problem. The solid line represents the theoretical shock and the dashed line represents the numerical solution.

CONCLUSIONS

The inviscid fluid flow equations were solved using FCT, for a range of one-dimensional shock wave problems in gases and metals. Heat conduction effects were ignored. FCT was chosen to solve these problems because it gives a *monotonic* solution with a high level of accuracy, especially in regions of steep gradients, such as shock fronts. FCT resolved the shock fronts in gases over no more than four grid points and contact discontinuities over six grid points. In metals, however, FCT spread shock fronts over eight grid points, but, since the pressure, particle velocity and energy ratios were much larger than an equivalent gas problem, this was not unacceptable. In gases, and to a greater extent in metals, the numerical solution exhibited some non-monotonic behaviour, found to be due to effects from the initial conditions. In all problems the initial condition was a point discontinuity, and when this was modelled numerically there was a slight smoothing, causing small amplitude non-monotonic waves to be propagated through the solution. To eliminate these waves an initial condition with a thin viscous region could be used. Apart from the regions of non-monotonic behaviour, the numerical solutions for the problems in gases and solids were within 1.0% of their exact solutions.

Current work includes the addition of viscous and heat conduction terms onto the momentum and energy equations, so that shock waves in solids can be modelled more accurately. Strong shocks will also be modelled using the Mie-Grüneisen equation of state and incorporating elastic-plastic terms in the fluid flow equations.

ACKNOWLEDGEMENTS

I would like to thank Dr Michael Page for his invaluable discussions and help in preparing this paper.

REFERENCES

1. Boris, J.P. and Book, D.L., Flux-Corrected Transport I: SHASTA, A Fluid Transport Algorithm That Works, *J.Comput.Phys*, vol. 11, pp. 38-69, 1973.

2. Dietachmayer, G.S., *On the numerical simulation of small scale intense atmospheric vortices*, PhD thesis, Monash University, 1987.

3. Harlow, F.H. and Amsden, A.A., Fluid Dynamics, Los Alamos Scientific Laboratory, LA-4700, 1971.

4. Roache, P.J., *Computational Fluid Dynamics*, Hermosa Publishers, Albuquerque, 1972.

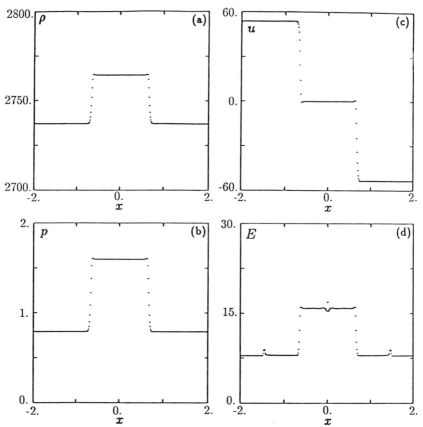

FIGURE 5. The collision of two equal stength shocks in aluminium at time $t = 4.0 \times 10^{-4}$ s. The RZFCT solution is shown for (a) density ρ (kgm^{-3}), (b) pressure p (GPa), (c) particle velocity u (ms^{-1}) and (d) total energy per unit volume E (MJm^{-3}) where $\Delta t = 5.0 \times 10^{-7}$ s and $\Delta x = 0.01$ m. The shocks were initially at $x = \pm 1.5$ m.

5. Roe, P.L., Approximate Riemann Solvers, Parameter Vectors and Difference Schemes, *J.Comput.Phys*, vol. 43, pp. 357-372, 1981.

6. Sod, G.A., A Survey of Several Finite Difference Methods for Systems of Nonlinear Hyperbolic Conservation Laws, *J.Comput.Phys*, vol. 27, pp. 1-31, 1978.

7. Tyndall, M.B., Analysis of an implicit factored scheme, Materials Research Laboratory Report, MRL-R-1117, 1988.

8. Tyndall, M.B., Numerical modelling of shocks in gases and metals, Materials Research Laboratory Report, in preparation, 1989.

9. Zalesak, S.T., Fully Multidimensional Flux-Corrected Transport Algorithms for Fluids, *J.Comput.Phys*, vol. 31, pp. 335-362, 1979.

Computation of the Reduced Navier-Stokes Equations with Multigrid Acceleration

Z. ZHU and C. A. J. FLETCHER
Department of Mechanical Engineering
University of Sydney
Sydney, NSW 2006, Australia

INTRODUCTION

For viscous flows with a dominant flow direction it is often possible to cast the governing equations in a form where specific terms can be neglected, or the governing equations can be solved in such a sequence as to benefit greatly in convergence rate and stability. For steady flows, this raises the possibility that the mathematical character of the governing equations may change from elliptic to non–elliptic or slightly elliptic with respect to the dominant flow direction. If this occurs, stable computational algorithms can be constructed that permit the solution to be obtained in a downstream–march type procedure (Fletcher and Zhu, 1988; Briley, 1974).

For some forms of reduced Navier–Stokes (RNS) equations, the equations are non–elliptic and stable solutions can be obtained efficiently in a single spatial march in the approximate downstream flow direction. For slightly elliptic flows computational strategies based on repeated downstream marches are possible and effective (Fletcher and Zhu, 1988). The reduced form of the original governing equations can be used to obtain an approximate solution very efficiently and this solution serves as initial data for an iterative solution of the actual governing equations. If the ellipticity of the original governing equations is not too large, it is still possible to use an equivalent non–elliptic algorithm to obtain a solution. In this way, more efficient overall iterative techniques can be constructed.

In this paper, a repeated marching algorithm is described which is based on the use of the axial momentum equation to obtain the pressure distribution. Subsequently, in two dimensions, the transverse velocity component is obtained from the transverse momentum equation and the axial velocity component is chosen to satisfy continuity. The multigrid technique is used to accelerate the overall convergence of the algorithm. The incompressible RNS equations are presented in general tensor notation in Section 2 and discretised to construct a repeated marching algorithm in Section 3. Section 4 details the multigrid technique scheme. In Section 5, results are presented for a problem of suddenly expanding flow with large recirculating areas and the convergence histories of different schemes are compared.

INCOMPRESSIBLE REDUCED NAVIER–STOKES (RNS) EQUATIONS

For incompressible steady flow, the non–dimensional Navier–Stokes and continuity equations can be expressed in general tensor notation as:

$$(V^k V^i)_{,k} = -p_{,k}g^{ki} + \{\nu_t(V^i_{,m}g^{mk} + V^k_{,m}g^{mi})\}_{,k}, \tag{1}$$

$$V^k_{,k} = 0. \tag{2}$$

The boundary conditions are: at the wall surface: $V^i = 0$, $i = 1,2$; at the duct inlet: $V^1 = U^1$, and $V^2 = U^2$; and at the duct outlet: $p = p_{out}$. Here, V^i are the contravariant velocity components, p is the pressure, ν_t is the total eddy viscosity, which includes both the viscous and the turbulence effects.

The tensorial differentiation in eq. (1) is evaluated to express the equation in strong conservation form (Fletcher and Zhu, 1988). Eqs. (1) and (2) become

$$\frac{\partial g^{\frac{1}{2}} V^k V^i}{\partial x^k} = -\frac{\partial p}{\partial x^k} g^{\frac{1}{2}} g^{ki} + \frac{\partial}{\partial x^k}\left[\nu_t \alpha_r^i \frac{\partial g^{\frac{1}{2}} V^s g^{jm} \alpha_m^k \alpha_s^r}{\partial x^j}\right] + \frac{\partial \nu_t}{\partial x^k}\frac{\partial g^{\frac{1}{2}} V^m g^{jr} \alpha_r^i \alpha_m^k}{\partial x^j} \quad (3)$$

$$\frac{1}{g^{\frac{1}{2}}}\frac{\partial g^{\frac{1}{2}} V^k}{\partial x^k} = 0 \quad (4)$$

For steady flow in a duct the diffusion terms in the stream direction are sufficiently small in comparison with the transverse diffusion terms, that they can be neglected from the axial and transverse momentum equations. The generalised coordinates in the present formulation are constructed so that x^1 is in the approximate downstream direction. With the neglect of the diffusive terms, the resulting RNS equations, in place of eq. (3), become

$$\frac{\partial g^{\frac{1}{2}} V^k V^i}{\partial x^k} = -\frac{\partial p}{\partial x^k} g^{\frac{1}{2}} g^{ki} + \frac{\partial}{\partial x^k}\left[\nu_t \alpha_r^i \frac{\partial g^{\frac{1}{2}} V^s g^{jm} \alpha_m^k \alpha_s^r}{\partial x^j}\right]_{k=j\neq 1} + \frac{\partial \nu_t}{\partial x^k}\frac{\partial g^{\frac{1}{2}} V^m g^{jr} \alpha_r^i \alpha_m^k}{\partial x^j} \quad (5)$$

The above equations can be solved using a marching algorithm. As for slightly parabolic flow, repeated marching in the downstream direction must be carried out to allow for the elliptic pressure influence. But if the elliptic effects are large or some recirculating areas exist, a large number of the repeated downstream marches has to be done to achieve a reasonable convergence rate. In this paper, a pressure Poisson equation derived from the momentum equations is solved for the whole domain after each downstream march. This has proved to be very effective in accelerating the convergence rate of the algorithm without causing much extra computational work. Such an algorithm is described in the next section.

INTERNAL FLOW SOLUTION ALGORITHM

A two–level fully implicit scheme (Fletcher, 1988) is employed here for marching in the axial (x^1) coordinate direction, and three–point central differencing is used in the transverse (x^2) direction. If n and j are the grid locations in the x^1 and x^2 directions respectively, and if the velocities and pressures at the nth step in the marching direction, $[V^1]_j^n$, $[V^2]_j^n, p_j^n, j = 1,...,$ NNY, are all known, eq. (5), after discretisation and rearrangement, becomes:

$$A[g^{\frac{1}{2}} V^r \alpha_r^i]_{j-1}^{n+1} + B[g^{\frac{1}{2}} V^r \alpha_r]_j^{n+1} + C[g^{\frac{1}{2}} V^r \alpha_r^i]_{j+1}^{n+1} + = -\frac{Dp}{Dx^k} g^{\frac{1}{2}} g^{ki} + [S^i]_j^n, i = 1,2, \quad (6)$$

where $D_p = [p_j^{n+2}]^{l-1} - [p^{n+1}]^l$, $Dx^k = [x^k]^{n+1} - [x^k]^n$, and l is the current downstream march number. As A,B,C are also functions of $[V^1]_j^{n+1}, [V^2]_j^{n+1}, j = 1,...,$NNY, the equations are nonlinear. The continuity equation becomes:

$$\frac{1}{g^{\frac{1}{2}}}\frac{[g^{\frac{1}{2}} V^1]_j^{n+1} - [g^{\frac{1}{2}} V^1]_j^n}{Dx^1} + \frac{1}{g^{\frac{1}{2}}}\frac{[g^{\frac{1}{2}} V^2]_{j+1}^{n+1} - [g^{\frac{1}{2}} V^2]_{j-1}^{n+1}}{2Dx^2} = 0 \quad (7)$$

As the coordinate system is non–orthogonal, $\alpha_r^i \neq O$, in eq. (6), which makes V^1 and V^2 interact with each other very strongly.

Knowing that the RNS equations are slightly elliptic, we can now solve eqs. (6), (7) by the following procedure:

Step 1: The transverse momentum equation is solved for $[V^2]_j^{n+1}$:

$$A^n[g^{\frac{1}{2}}V^2\alpha_2^2]_{j-1}^{n+1} + B^n[g^{\frac{1}{2}}V^2\alpha_2^2]_j^{n+1} + C^n[g^{\frac{1}{2}}V^2\alpha_2^2]_{j+1}^{n+1} =$$

$$-\frac{Dp}{Dx^k}g^{\frac{1}{2}}g^{k2} - [S^i]_j^n - \{A[g^{\frac{1}{2}}V^1\alpha_1^2]_{j-1}^{n+1} + B[g^{\frac{1}{2}}V^1\alpha_1^2]_j^{n+1} + C[g^{\frac{1}{2}}V^1\alpha_1^2]_{j+1}^{n+1}\} \qquad (8)$$

After linearisation, eq. (8) is tridiagonal in the transverse direction. Terms on the right hand side of eq. (8) are evaluated from the most recent values.

Step 2: The continuity equation is solved for $[V^1]_j^{n+1}$

$$\frac{1}{g^{\frac{1}{2}}}\frac{[g^{\frac{1}{2}}V^1]_j^{n+1} - [g^{\frac{1}{2}}V^1]_j^n}{Dx^1} = -\frac{1}{g^{\frac{1}{2}}}\frac{[g^{\frac{1}{2}}V^2]_{j+1}^{n+1} - [g^{\frac{1}{2}}V^2]_{j-1}^{n+1}}{2Dx^2} \qquad (9)$$

Step 3: The values of V^1 and V^2 are substituted into eq. (6). Because of the sequential nature of the algorithm, eq. (6) cannot be satisfied exactly. The contravariant residuals for eqs. (6), which are denoted as R^i, $i = 1,2$, are given by:

$$R^i = -\frac{Dp}{Dx^k}g^{ki} + \frac{1}{g^{\frac{1}{2}}}\{A[g^{\frac{1}{2}}V^r\alpha_r^i]_{j-1}^{n+1} - B[g^{\frac{1}{2}}V^r\alpha_r^i]_j^{n+1} - C[g^{\frac{1}{2}}V^r\alpha_r^i]_{j+1}^{n+1}\} \qquad (10)$$

Step 4: It is assumed that the contravariant residuals can be reduced to zero by introducing Δp as a correction to p, that is:

$$\frac{D\Delta p}{Dx^i} = R^k g_{ki} \qquad (11)$$

In this paper, p^{n+1} is updated using the axial pressure gradient ($i = 1$) component only. In eq. (11) $D\Delta p$ is discretised as a forward difference to account for the upstream effect, and Δp_j^{n+2} is set to zero, thus.

$$\Delta p_j^{n+1} = -Dx^1 \cdot R^k g_{k1} \qquad (12)$$

If Δp_j^{n+1} is smaller than a predefined small amount, step 5 is executed. Otherwise, with the newly obtained p and V^k, another iteration at this axial station is carried out, i.e. steps 1 to 4 are repeated.

Step 5: The above iterative algorithm, steps 1 to 4, is repeated at each downstream location (increasing n) until the outlet is reached.

Step 6: After finishing each sweep, residuals of the two momentum equations are calculated once again by eq. (10), and the following Poisson equation is solved by

375

an Approximate Factorization Iterative method (AFI) to give a further pressure correction p^c:

$$\frac{1}{g^{\frac{1}{2}}}\frac{\partial g^{\frac{1}{2}}p^c_{,k}g^{ki}}{\partial x^i} = R^i_{,i} \tag{13}$$

The algorithm is repeated for successive downstream marches until the solution no longer changes.

MULTIGRID ACCELERATION

It is well known that the multigrid technique is very effective in accelerating convergence when a typical iterative method, e.g. ADI, SOR or AFI, (Fletcher, 1988) fails to efficiently reduce residuals with dominant long wavelengths, and that it is also very effective in conveying information very quickly through large scale domain. It is precisely these characteristics of the multigrid technique that are needed to enhance the overall performance of downstream–marching algorithms of RNS equations. In this paper, both linear (CS) and nonlinear (FAS) multigrid techniques (Brandt, 1977) are applied to the algorithm described in Sect. 2 and 3.

At each mesh level, the procedure to compute velocities and pressure or their corrections (in the case of the CS scheme) is exactly the same as the procedure described in Sect. 3, except for coarser meshes, where the momentum equations have extra terms on the right hand side, and where in the case of CS scheme, the convective nonlinear terms are linearised.

The symbol l represents mesh level, the finest mesh level being $l = 1$, at which the solution is regarded as the solution of the original physical problem, and any one level coarser than l is $l = l + 1$. For the nonlinear multigrid FAS scheme at mesh level $l > 1$, equations corresponding to eq. (1) have two extra terms on the right hand side:

$$([V^k]^l)_{,k} = -[p]^l_{,k}\, g^{ki} + \{\nu_t([V^i]^l_{,m}\, g^{mk} + [V^k]^l_{,m}\, g^{mi})\}_{,k}$$

$$+ R^{*i} + (V^{*k}V^{*i})_{,k} - [-p^*_{,k}\, g^{ki} + \{\nu_t(V^{*i}_{,m}\, g^{mk} + V^{*k}_{,m}\, g^{mi})\}_{,k}] \tag{14}$$

Here, $[*]^l$ denotes the variable is defined at mesh level l and the subscript l should not be confused with normal tensorial notation and rules. R^{*i} is the contravariant residual injected from level $l-1$, i.e. $R^{*1} = I^l_{l-1}[R^i]^{l-1}$. Here $[R^i]^{l-1}$ is given by eq. (10) calculated at level $l-1$, and I^l_{l-1} is an injection function which interpolates variables from mesh level $l-1$ to l. Similarly, $V^{*1} = I^l_{l-1}[V^i]^{l-1}$ are velocities injected from velocities at level $l-1$, and $p^* = I^l_{l-1}[p]^{l-1}$.

The solution $[V^i]^l, [p]^l$, by solving the above equations and the continuity equation is then interpolated back as a correction to level $l-1$.:

$$[V^i]^{l-1} = [V^i]^{l-1} + P^{l-1}_l\{[V^i]^l - V^{*1}\} \qquad [p]^{l-1} = [p]^{l-1} + P^{l-1}_l\{[p]^l - p^*\}.$$

Here, P^{l-1}_l extrapolates variables from mesh level l to $l-1$.

For the linear CS scheme, correction equations for mesh level $l > 1$, corresponding to eq. (1), can be expressed as:

$$(U^k[V^i]^l)_{,k} = -[p]^l_{,k}\, g^{ki} + \{\nu_t([V^i]^l_{,m}\, g^{mk} + [V^k]^l_{,m}\, g^{mi})\}_{,k} + R^{*i} \tag{15}$$

Here, the $U^i = [V^i]^{l=1}$ are velocities at the finest mesh, $R^{*i} = I^l_{l-1}[R^i]^{l-1}$, $[R^i]^{l-1}$ is given by eq. (10) calculated at mesh level $l-1$. The solution at level l of the above equations and the continuity equation, by the procedure described in Sect. 3, is used as a direct correction to mesh level $l-1$:

$$[V^i]^{l-1} = [V^i]^{l-1} + P^{l-1}_l[V^i]^l \qquad [p]^{l-1} = [p]^{l-1} + P^{l-1}_l[p]^l.$$

In this paper, at each mesh level, the convergence rate is determined by the ratio of RMS of momentum equations' residuals at two consecutive marches. If the rate is larger than a certain predefined value, the calculation continues at this level until full convergence is reached and the solution is used to correct the solution at level $l-1$; if the convergence rate is smaller than the predefined value, calculation at this l level stops and computation at mesh level $l+1$ starts. So, the multigrid technique applied here is actually a type of $W-$ cycle (Brandt, 1977).

COMPUTATION AND DISCUSSION

The test problem (Fig. 1) of recirculating flow in an expanding duct is used to validate the present method. In the present formulation, lengths are nondimensionalised by D, the inlet depth of the duct, and velocities by U_0, the average velocity at the inlet; so that the Reynolds number, $Re = DU_0/\nu$. In this test case, $Re = 50$. For all the figures, X,Y correspond to x^1, x^2 respctively, and $U = [V^1\vec{g}_1 + V^2\vec{g}_2]\cdot\vec{1}_x$ in Sect. 3. Details of the test case's geometric data are given by Zhu and Fletcher, 1989.

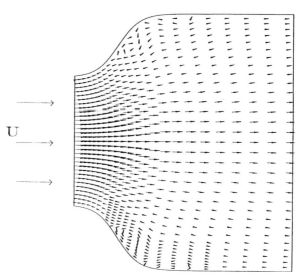

U

FIGURE 1. Velocity vectors for a 65 x 65 grid.

Fig. 1 shows the velocity vectors of the flow passing through the expanding duct. The flow is symmetric and grid points are clustered near the entrance and near the walls. In Fig. 1, a very big separating area along the lower wall, consisting of two recirculating bubbles, is clearly shown by the velocity vectors. But only one long separation bubble is predicted by Burns and Wilkes 1987. The discrepancy in the structure of the recirculation is probably due to neglecting the streamwise diffusive term in the momentum equations. The predicted vorticity and pressure along the lower wall show excellent agreement (Figs. 2,3) with the computed results given by Burns and Wilkes 1987, confirming the accuracy of the present algorithm. But some difference can be detected around the recirculating area, with the pressure having the more noticeable discrepancy. If the streamwise diffusive term is not neglected (Zhu and Fletcher, 1989), the present method produces consistent result with that of Burns and Wilkes, 1987.

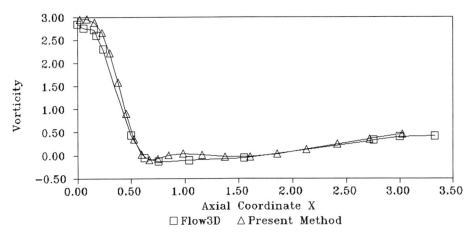

FIGURE 2.   Vorticity along the wall.

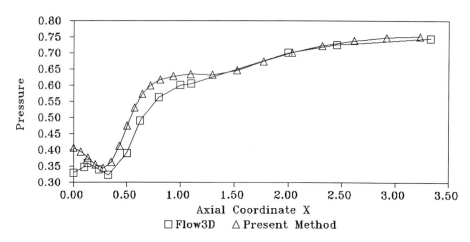

FIGURE 3.   Pressure along the wall.

Fig. 4 shows the effects of different multigrid schemes on the convergence rate of the algorithm. It can be seen that the multigrid technique can accelerate the convergence rate by at least a factor of 2. The linear CS scheme has almost the same effect as the nonlinear FAS scheme, even though the present problem is nonlinear. From our experience, the CS scheme is not as stable as the FAS scheme when applied to nonlinear problems, and this is evidenced in Fig. 4 by the initial increase in the RMS of the CS scheme. Figs. 5,6 show the effects on convergence rate when different level multigrid are implemented in the CS and FAS schemes respectively. For both schemes, 2–level multigrid doesn't produce the fastest convergence rate, and 3,4–levels seem to have the same performance. Since more mesh levels mean more computational work and too few meshes can often cause divergence, 3–level multigrid is suggested to be the proper choice.

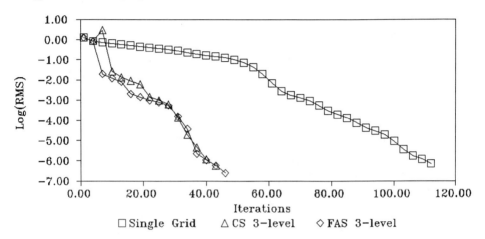

FIGURE 4.   Convergence Rate.

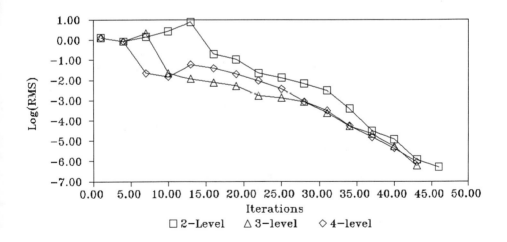

FIGURE 5.   CS Convergence Rate.

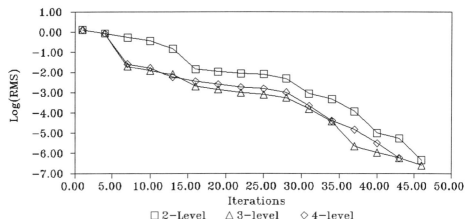

FIGURE 6. FAS Convergence Rate

CONCLUSION

A robust and accurate downstream—marching algorithm has been developed to predict complex internal flows with difficult geometric boundaries. Comparison with benchmark solution indicates that the present algorithm produces accurate predictions even when large areas of separation are present. The use of multigrid acceleration is shown to be effective for this class of algorithms with a 3–level FAS W — cycle producing the optimal performance in convergence and stability.

ACKNOWLEDGEMENT

The authors are grateful to the Australian Research Council for their continuing support of this research work.

REFERENCES

1.   A. Brandt "Multi—Level Adaptive Solutions to Boundary—Value Problems", *Mathematics of Computation*, Vol. 31, No. 138. April, pp. 333–390, 1977

2.   W.R. Briley "Numerical Method for Predicting Three—Dimensional Steady Viscous Flow in Ducts", *J. Comp. Phys.*, Vol. 14, pp. 8–28, 1974

3.   A.D. Burns and N.S. Wilkes "A Finite Difference Method for the Computation of Fluid Flows in Complex Three Dimensional Geometries", Harwell Laboratory Report: AERE R 12342, July, 1987

4.   C.A.J. Fletcher (1988): *Computational Techniques for Fluid Dynamics*, Vol. 2, Springer—Verlag, Heidelberg, Chap. 16.

5.   C.A.J. Fletcher and Z. Zhu "Efficient Internal Flow Solver Using Reduced Navier—Stokes Equations", Proc. Int. Conf. on Computational Methods in Flow Analysis (ed. M. Kawahara), Okayama, pp. 304–311, 1988

6.   Z. Zhu and C.A.J. Fletcher "An Efficient Repeated Marching Algorithm for Viscous Internal Flows with Axial Recirculation", in preparation, 1989

FINITE ELEMENT METHODS AND APPLICATIONS

An Elasto-Hydrodynamic Study of Journal Bearing

N. O. FREUND and A. K. TIEU
Department of Mechanical Engineering
University of Wollongong
NSW 2500, Australia

INTRODUCTION

The hydrodynamic analysis is based on the laminar Reynolds equation which yields the pressure distribution. This is coupled to the elasticity analysis which gives the bearing housing deflection and in turn modifies the lubricating film thickness in the Reynolds equation. The above steps are repeated until convergence is achieved. The effects of elastic deformation have been studied by other authors such as Huebner (1973) and Conaway & Lee (1975) but they have not considered using deformation as a means of improving load capacity.

The bearing is specially modified to include an under cut, which has the potential to increase the load capacity of the journal bearing. The undercut deflects due to the hydrodynamic pressure and therefore modifies the convergent lubricating film. This acts to increase the flow at inlet. The extent of deflection can be controlled by modifying the shape and thickness of the under cut, as shown in Fig. 1.

A CAD package, Autocad is used to define the geometry of the finite element model and later display the results. This allows journal bearings of arbitrary size to be modelled. The FEM study has been carried out on a 386 personal computer, using Fortran.

NOTATION

$[B]$	Strain displacement matrix
$[C]$	Material matrix
F_0, F_1, F_2	Viscosity variables
$\vec{f}^s, \vec{F}_i$	Surface & concentrated load vector respectively
G	Viscosity variable
h	Film thickness

H_s	Shape function at element surface
J	Jacobian
N	Shape function
$[K_P]$	Stiffness matrices
P	Pressure
r,s,t	Local coordinate system for isoparametric elements
$\{R\},\{R_c\},\{R_s\}$	Resultant, concentrated & surface load vector
S_P	Surface where pressures are known
T,T_0	Temperature of oil & reference oil temperature
$\vec{V}_1,U_1,V_1$	Film velocity vectors
$\{u\}$	Displacement array in x,y,z directions
$\vec{u}_{cv}^s,\vec{u}_{iv}^s$	Surface displacement vectors: continuous & discrete
$\Delta\overline{U},\Delta U_x,\Delta U_y$	Change in film velocity vectors
(x,y,z)	Global coordinate system
β	Viscosity coefficient
$\vec{\epsilon}_v$	Virtual displacement derivative vector
μ,μ_0	Oil viscosity & its reference point at T_0
$\vec{\tau}$	Stress vector
Ω	Surface to integrate

NUMERICAL SIMULATION

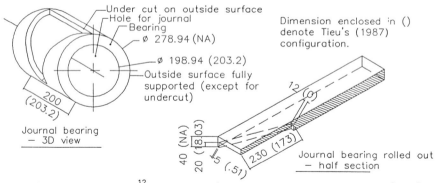

Young's modulus $.103(10)^{12}$ [Pa], Poison's ratio .3, Bearing speed 1000 [RPM], Oil viscosity 60.7 [cP], Eccentricity 0.3, Radial clearance .0762 [mm].

FIGURE 1.   Physical model of journal bearing.

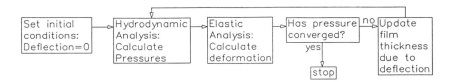

FIGURE 2.   Flow chart of FEM solution procedure.

Numerical simulation is achieved by coupling the hydrodynamic
and elastic analysis together by an iterative solution as shown
in Fig. 2.

Generalized Reynolds Equation for pressure

If body forces, diffusion flow and time varying film thickness
are not considered, the generalized Reynolds equation of
pressure which takes into account three-dimensional viscosity
variations in the film can be formed:

$$-\nabla \cdot G \nabla P = \nabla \cdot \left(h \vec{V}_1 \right) + \nabla \cdot \left(\Delta \vec{U} F_2 \right) \qquad \ldots (1)$$

For the Reynolds equation to be solved, pressure constraints
must be specified on the boundary (boundary flow constraints
not considered) ie, $P = P(x,y)$ on S_p

For the thermo-hydrodynamic case, it can be shown (Huebner
(1982),Tieu (1975)) that the pressure distribution that
satisfies equ. (1) and boundary condition also minimizes the
functional:

$$I(P) = \int_\Omega \left[-\frac{1}{2} G \nabla P \cdot \nabla P - \left(h \vec{V}_1 \right) \cdot \nabla P - \left(\frac{\Delta \vec{U}}{F_0} \int_0^h \int_0^z \frac{dz}{\mu} dz \right) \cdot \nabla P \right] d\Omega \qquad \ldots (2)$$

By discretizing the solution domain with isoparametric finite
elements and applying: $\partial I(P)/\partial P_k = 0$ to all pressure nodes the
following finite element equations result:

$$\left[K_P \right] \{ P \} = - \left[K_{U_1} \right] \{ U_1 \} - \left[K_{V_1} \right] \{ V_1 \} - \left[K_{\Delta U_x} \right] \{ \Delta U_x \} - \left[K_{\Delta U_y} \right] \{ \Delta U_y \} \qquad \ldots (3)$$

$$K_{P_{i,j}} = - \int_{r=-1}^{+1} \int_{s=-1}^{+1} G \left(\frac{\partial N_i}{\partial x} \frac{\partial N_j}{\partial x} + \frac{\partial N_i}{\partial y} \frac{\partial N_j}{\partial y} \right) |J| dr ds \qquad \ldots (4)$$

$$K_{U_{1_{i,j}}} = \int_{r=-1}^{+1} \int_{s=-1}^{+1} h \frac{\partial N_i}{\partial x} N_j |J| dr ds \quad , \quad K_{V_{1_{i,j}}} = \int_{r=-1}^{+1} \int_{s=-1}^{+1} h \frac{\partial N_i}{\partial y} N_j |J| dr ds \qquad \begin{matrix} \ldots (5) \\ \ldots (6) \end{matrix}$$

$$K_{\Delta U_{x_{i,j}}} = \int_{r=-1}^{+1} \int_{s=-1}^{+1} F_2 \frac{\partial N_i}{\partial x} N_j |J| dr ds \quad , \quad K_{\Delta U_{y_{i,j}}} = \int_{r=-1}^{+1} \int_{s=-1}^{+1} F_2 \frac{\partial N_i}{\partial y} N_j |J| dr ds \qquad \begin{matrix} \ldots (7) \\ \ldots (8) \end{matrix}$$

the variables are typically discretized by:

$$h = \sum_{n=1}^{N} N_n h_n \quad , \quad P = \sum_{n=1}^{N} N_n P_n \quad \text{etc.} \qquad \ldots (9)$$

Then the general form of the viscosity variables can be written in the following manner along with their isothermal cases

$$F_0 = \int_0^h \frac{dz}{\mu} = \frac{h}{\mu} \quad , \quad F_1 = \int_0^h z \frac{dz}{\mu} = \frac{h^2}{2\mu} \quad , \quad F_2 = \frac{1}{F_0} \int_0^h \int_0^z \frac{dz}{\mu} dz = \frac{h}{2} \begin{matrix} \ldots (10) \\ \ldots (11) \\ \ldots (12) \end{matrix}$$

$$\mu = \mu_0 \exp(-\beta(T - T_0)) \quad , \quad G = \int_0^h \int_0^z \frac{z}{\mu} dz dz - \frac{F_1}{F_0} \int_0^h \int_0^z \frac{dz}{\mu} dz = -\frac{h^3}{12\mu} \begin{matrix} \ldots (13) \\ \ldots (14) \end{matrix}$$

Elasticity Equations

Elasticity equations presented are based on the principle of virtual displacements. Since the total internal virtual work is equal to the total external virtual work this gives:

$$\int_V \vec{\epsilon}_v^T \vec{\tau} \, dV = \int_S \vec{u}_{cv}^{ST} \vec{f}^s \, dS + \sum_i \vec{u}_{iv}^T \vec{F}_i \qquad \ldots (15)$$

It can be shown that by using Hook's Law , equ. (15) and the principle of virtual displacements that the following finite element equations result:

$$[K]\{u\} = \{R\} \qquad \ldots (16)$$

$$K = \sum \int_V [B]^T [C][B] dV \quad , \quad \{R\} = \{R_c\} + \{R_s\} \quad , \quad \{R_s\} = \sum \int_S [H^s]^T \{f^s\} dS \quad \begin{matrix} \ldots (17) \\ \ldots (18) \\ \ldots (19) \end{matrix}$$

Three-dimensional 20 node isoparametric elements are used for elasticity analysis, which can easily be coupled to the 8 node hydrodynamic elements.

Integration

To obtain convergent results integration orders of 3x3x3 and 3x3 are chosen for the 20 node elastic brick elements and 8 node hydrodynamic elements respectively, as suggested by Bathe (1982).

Solving Matrix equations

Matrix equations (16) and (3) are solved by a variant of the Gaussian elimination method. Since the stiffness matrices can get extremely large (1.05 Mbytes for elasticity analysis), a compaction scheme is necessary. The sky line scheme (Bathe (1982)) is chosen, which takes advantage of elements in the stiffness matrix which are initially zero and remain zero during the Gaussian elimination.

Finite Element Modelling Considerations

To enable comparisons, the journal bearing modelled closely resembles the one used by Tieu (1987). The basic dimensions of the journal bearing are shown in Fig. 1. To save on incore computer storage requirements, symmetry conditions are taken advantage of, and therefore only one half of the journal bearing is modelled.

Keeping in mind the FEM model is applied on a personal computer the mesh density is kept to a minimum. Further mesh refinements can be made if a mainframe computer is employed. The geometry is generated manually with the aid of a CAD package. Information entered for the elasticity analysis includes node position, node number, degree of freedom in x,y,z directions, element number, element node list, & material set number. The model is drawn as a wire mesh with node and element data points, referred as blocks. These blocks contain textural information which forms a database in the CAD system. To aid in this data entry, Lisp (from AUTOCAD) routines are used to automate node and element numbering and entering element node lists. This data can be exported from Autocad and transformed into databases used by the FEM routines.

Elasticity Analysis. The 3-dimensional elasticity model shown in Fig. 3 below, is used as a master geometric model. Once the elasticity FEM database is created, it can be used to automatically generate the hydrodynamic FEM database. The master model is described as a rolled out journal bearing for ease of manual definition, and is transformed by Fortran routines to the final curved model shown in Fig. 3. The symmetry condition is imposed by disallowing axial displacement on the line of symmetry. The outside surface of the bearing is not allowed to displace at all and all other portions of the bearing are free.

The statistics for the elasticity model is shown below Table 1.

Hydrodynamic analysis. The hydrodynamic FEM mesh is shown below in Fig. 4. Pressures are specified as zero on all external boundaries. The symmetry condition is imposed by

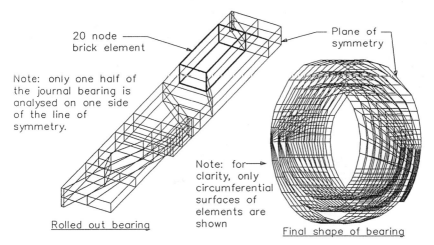

20 node brick element

Plane of symmetry

Note: only one half of the journal bearing is analysed on one side of the line of symmetry.

Note: for clarity, only circumferential surfaces of elements are shown

Rolled out bearing

Final shape of bearing

FIGURE 3. Elasticity FEM model.

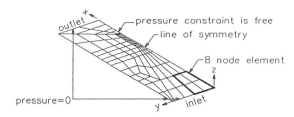

FIGURE 4. Hydrodynamic FEM model.
TABLE 1. FEM model statistics

Description	Elasticity	Hydrodynamic
Number of nodes	392	88
Number of elements	54	23
Degrees of freedom	830	67
Size of compacted stiffness matrix	1 Mb	5 kb
Run time for calculating & solving stiffness matrix	29 min	12 s

having a free pressure boundary on the line of symmetry.

RESULTS AND DISCUSSION

If the bearing under consideration has unsupported regions,
elasticity analysis must be carried out when the deflections
become comparable to the film thickness. This occurs for the
bearing under investigation in the region of the under cut as
shown below in Fig. 6. The deformed shape of the bearing is
shown in Fig. 5. It is interesting to note the negative
deflection of bearing at the end of the under cut. Further
investigation is warranted to see if this persists at higher
bearing loads and how the under cut shape can be modified to
reduce this.

The addition of an under cut to this journal bearing shows
promise for improving its load capacity. For an eccentricity
ratio of 0.3 and radial clearance of 76.2 μm the load capacity
without bearing deflection is 143,714 N whilst for the
deflected case it rises to 158,190 N. For this bearing
configuration the increase is not substantial. However it is
hoped that by altering the position, extent and thickness of
the under cut that the load capacity can be improved. Another
parameter, as suggested by Tieu (1987) that can be altered is
the angle of the undercut, and if increased will increase the
load capacity.

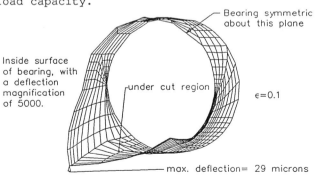

FIGURE 5. Deformed journal bearing under Hydrodynamic loading.

388

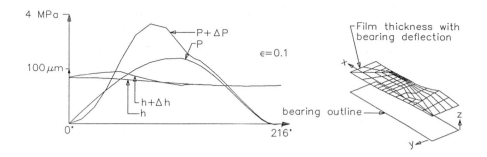

FIGURE 6. Film thickness and pressure on center line of
journal bearing & film thickness overall.

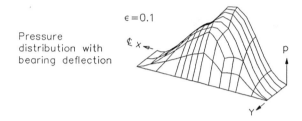

FIGURE 7. Pressure distribution of journal bearing.

By decreasing the eccentricity ratio to 0.1 a marked increase
in bearing capacity results by including the under cut, from 45
kN to 58 kN. The maximum deflection for this case is 15 μm.

The FEM routines have been tested by comparison to known
solutions from problems that can easily be solved by analytical
means. The elasticity analysis was checked by examining a bar
under tension, which yielded perfect correlation. The
hydrodynamic analysis was tested by modelling a long bearing
and comparing the results with the analytical formulae. The two
pressure distributions followed each other closely. In addition
the symmetry conditions of zero gradients of deflection and
pressure normal to the plane of symmetry were satisfied, as
shown in Fig. 6 & Fig. 7.

It can be seen from Fig. 6 that there exists a small region of
negative pressure which indicates a cavitation region. Since
this is small in comparison with the maximum pressure it is not
expected to affect the results adversely. If this negative
region posed a problem the FEM mesh would be altered to allow
exit pressure boundary conditions to be moved upstream to
produce a positive pressure distribution throughout.

As a comparison, investigations carried out by Tieu (1987) with
approximately the same bearing dimensions and operating
conditions, gave a maximum bearing deflection of 42 μm,
whereas this investigation yielded 33 μm. The improvement in

389

bearing load capacity with the introduction of the undercut for this investigation is 0.101, compared to 0.31 from the work of Tieu (1987). These differences can be attributed to the differing boundary conditions for the elasticity analysis at the bearing inlet. Tieu (1987) used a free inlet edge whilst this investigation uses a fixed edge.

CONCLUSIONS

It has been shown that for studying the performance of journal bearings it is necessary to include the deflection of the bearing housing. If this deflection is controlled by the introduction of an under cut the load capacity of the journal bearing can be enhanced.

Employing a CAD system proves a useful tool for manual definition of FEM meshes, which can then be transferred to FEM databases for analysis. The CAD system also provides an efficient means of displaying the results.

Scope for further work is seen to be in the area of altering bearing and undercut dimensions to achieve maximum journal bearing performance. In addition thermal effects on the bearing and fluid could also be included.

REFERENCES

1. Conway, H. D., and Lee, H. C., The Analysis of the Lubrication of a Flexible Journal Bearing, *Journal of Lubrication Technology*, pp. 599-604, 1975.

2. Bathe, K., *Finite Element Procedures in Engineering Analysis*, Prentice-Hall, Englewood Cliffs New Jersey, 1982.

3. Huebner, K., Application of Finite Element Method to Thermohydrodynamic Lubrication, *International Journal for Numerical Methods in Engineering*, vol. 8, pp. 139-165, 1974.

4. Huebner, K., *The Finite Element Method for Engineers*, John Wiley & Sons, Inc., 1982.

5. Tieu, A. K., A Thermo-Elasto-Hydrodynamic Analysis of Journal Bearings, *5th Int. Conf. in Aust. on Finite Element Methods Melb.*, pp. 169-174, 1987.

6. Tieu, A. K., A Numerical Simulation of Finite-Width Thrust Bearings, taking into account viscosity variation with temperature and pressure, vol. 17, no. 1, pp. 1-10, 1975.

Partially Orthogonal Series Approximations in Field Problems

B. W. GOLLEY and J. PETROLITO
University College
Australian Defence Force Academy
Canberra, ACT, Australia

INTRODUCTION

The finite strip method is a specialised finite element procedure which has computational advantages for solving a restricted class of problems. The method has its origins in structural mechanics, and was introduced by Cheung (1968) for the static analysis of rectangular plates with two opposite simply supported edges. The approximating functions within each strip were chosen as products of polynomials and truncated sine series, which automatically satisfied the simple support boundary conditions. The resulting stiffness equations uncoupled into smaller systems of equations with narrow bandwidths, which has advantages in both core requirements and solution times compared with finite element methods. With other boundary conditions, the truncated sine series were replaced by vibrating beam functions. In these cases, the uncoupling no longer occurs, and much of the computational efficiency of the method is lost. Most structural applications, which now embrace dynamic and stability analysis of plates, folded plates and shells emphasise the simply supported case where uncoupling occurs.

The finite strip method has also been applied to transient heat conduction analysis of rectangular plates (Chakrabarti, 1980). The examples considered in this thermal case had constant zero temperatures on two opposite plate boundaries, which leads to an uncoupling of equations similar to that occurring in the simply supported structural case. With other boundary conditions, uncoupling does not occur.

The authors have recently developed a technique called the finite strip-element method (Golley et al, 1987), in which displacement functions are combinations of those used with finite elements and those used with simply supported finite strips. The method has been applied to the static analysis of thick and thin plates, for which the following advantages are noted.

1. Stiffness equations partially uncouple, which is computationally efficient.

2. Displacement functions are very simple, permitting exact integration of stiffness coefficients, which are themselves very simple expressions.

391

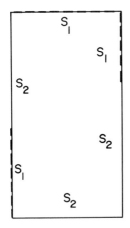

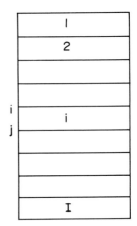

FIGURE 1. Typical problem.   FIGURE 2. Strip-element division.

3.   Mixed boundary conditions on two opposite boundaries of the rectangular domains are easily treated.

In this paper, we consider the application of the method to the solution of Poisson's equation, and discuss the computational advantages of the method compared with other techniques.

OUTLINE OF METHOD

We consider determining a two-dimensional function $\phi(x, y)$ which satisfies the equation

$$\nabla^2 \phi = q \tag{1}$$

in a domain A subject to the boundary conditions

$$\phi = \overline{\phi} \tag{2}$$

on portion S_1 of the boundary and

$$\phi_{,n} = 0 \tag{3}$$

on portion S_2 of the boundary. In Eqs. 1, 2 and 3, q is a specified function of x and y, and n is the outward normal coordinate to the boundary. The procedure may also be applied to two- and three-dimensional domains involving non-homogeneous materials and to cases where non zero normal derivatives are specified on the boundary but the above case is considered for brevity.

The alternative variational formulation is to determine ϕ to minimise the functional

$$\chi = \int_A \{\frac{1}{2}(\phi_{,x}^2 + \phi_{,y}^2) - q\phi\} \, dA \tag{4}$$

subject to ϕ satisfying Eq. 2 on S_1. We consider the determination of ϕ in a rectangular domain, as shown in Fig. 1. Uniform boundary conditions on two opposite sides are necessary, but mixed boundary conditions on the other two sides may prevail.

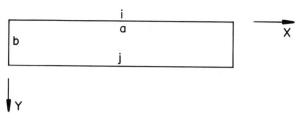

FIGURE 3. Typical strip-element.

The domain is divided into I finite strip elements by $I + 1$ nodal lines as shown in Fig. 2. A typical strip-element, ij of dimensions a and b is shown in Fig. 3, together with the local coordinates x and y.

On nodal line i, ϕ is approximated by

$$\phi^i(\xi) = \sum_{m=1}^{M} \Phi_m^i \sin \alpha_m \xi + \sum_{m=1}^{2} \overline{\Phi}_m^i P_m(\xi) \tag{5}$$

where $\xi = x/a$, $\alpha_m = m\pi$, $P_1(\xi) = 1 - \xi$ and $P_2(\xi) = \xi$. In the simplest of strip elements, ϕ is interpolated linearly in the y direction although higher order interpolation could be adopted by introducing internal nodal lines. Hence in the linear case, ϕ is approximated in strip-element ij by

$$\phi(\xi, \eta) = \phi^i(\xi)P_1(\eta) + \phi^j(\xi)P_2(\eta) \tag{6}$$

where $\eta = y/b$. The approximating function is C_0 continuous in A.

Substituting $\phi(\xi, \eta)$ from Eq. 6 into Eq. 4, integrating over the area of the strip-element and differentiating with respect to the function parameters $\Phi_1^i, \Phi_2^i \ldots \overline{\Phi}_2^j$ we obtain

$$\partial \chi / \partial \left\{ \begin{array}{c} \boldsymbol{\Phi}_1 \\ \boldsymbol{\Phi}_2 \\ \vdots \\ \boldsymbol{\Phi}_m \\ \vdots \\ \boldsymbol{\Phi}_M \\ \overline{\boldsymbol{\Phi}} \end{array} \right\} = \begin{bmatrix} \mathbf{k}_{11} & 0 & \cdots & 0 & \cdots & 0 & \overline{\mathbf{k}}_1 \\ 0 & \mathbf{k}_{22} & \cdots & 0 & \cdots & 0 & \overline{\mathbf{k}}_2 \\ \vdots & \vdots & \ddots & \vdots & & \vdots & \vdots \\ 0 & 0 & \cdots & \mathbf{k}_{mm} & \cdots & 0 & \overline{\mathbf{k}}_m \\ \vdots & \vdots & & \vdots & \ddots & \vdots & \vdots \\ 0 & 0 & \cdots & 0 & \cdots & \mathbf{k}_{MM} & \overline{\mathbf{k}}_M \\ \overline{\mathbf{k}}_1^T & \overline{\mathbf{k}}_2^T & \cdots & \overline{\mathbf{k}}_m^T & \cdots & \overline{\mathbf{k}}_M^T & \overline{\overline{\mathbf{k}}} \end{bmatrix} \left\{ \begin{array}{c} \boldsymbol{\Phi}_1 \\ \boldsymbol{\Phi}_2 \\ \vdots \\ \boldsymbol{\Phi}_m \\ \vdots \\ \boldsymbol{\Phi}_M \\ \overline{\boldsymbol{\Phi}} \end{array} \right\} = \left\{ \begin{array}{c} \mathbf{q}_1 \\ \mathbf{q}_2 \\ \vdots \\ \mathbf{q}_m \\ \vdots \\ \mathbf{q}_M \\ \overline{\mathbf{q}} \end{array} \right\} \tag{7}$$

where

$$\boldsymbol{\Phi}_m = [\Phi_m^i \; \Phi_m^j]^T \tag{8}$$

$$\overline{\boldsymbol{\Phi}} = [\overline{\Phi}_1^i \overline{\Phi}_2^i \overline{\Phi}_1^j \overline{\Phi}_2^j]^T \tag{9}$$

$$\mathbf{k}_{mm} = \frac{a}{2b} \begin{bmatrix} 1 & -1 \\ -1 & 1 \end{bmatrix} + \frac{\alpha_m^2 b}{12a} \begin{bmatrix} 2 & 1 \\ 1 & 2 \end{bmatrix} \tag{10}$$

$$\overline{\mathbf{k}}_m = \frac{a}{\alpha_m b} \begin{bmatrix} 1 & -(-1)^m & -1 & (-1)^m \\ -1 & (-1)^m & 1 & -(-1)^m \end{bmatrix} \tag{11}$$

and

$$\overline{\overline{\mathbf{k}}} = \frac{1}{6} \begin{bmatrix} 2(\frac{a}{b} + \frac{b}{a}) & (\frac{a}{b} - 2\frac{b}{a}) & (-2\frac{a}{b} + \frac{b}{a}) & -(\frac{a}{b} + \frac{b}{a}) \\ (\frac{a}{b} - 2\frac{b}{a}) & 2(\frac{a}{b} + \frac{b}{a}) & -(\frac{a}{b} + \frac{b}{a}) & (-2\frac{a}{b} + \frac{b}{a}) \\ (-2\frac{a}{b} + \frac{b}{a}) & -(\frac{a}{b} + \frac{b}{a}) & 2(\frac{a}{b} + \frac{b}{a}) & (\frac{a}{b} - 2\frac{b}{a}) \\ -(\frac{a}{b} + \frac{b}{a}) & (-2\frac{a}{b} + \frac{b}{a}) & (\frac{a}{b} - 2\frac{b}{a}) & 2(\frac{a}{b} + \frac{b}{a}) \end{bmatrix} \tag{12}$$

The vectors $\mathbf{q}_m$ and $\overline{\mathbf{q}}$ depend on $q(x,y)$. In solving the torsion problem of homogeneous prisms using Prandtl's stress function, q is taken as minus 2 and for this case

$$\mathbf{q}_m = \frac{ab(1 - (-1)^m)}{\alpha_m} \begin{Bmatrix} 1 \\ 1 \end{Bmatrix} \tag{13}$$

and

$$\overline{\mathbf{q}} = \frac{ab}{2} \begin{Bmatrix} 1 \\ 1 \\ 1 \\ 1 \end{Bmatrix} \tag{14}$$

The null submatrices $\mathbf{k}_{mn}$, $(m \neq n)$ are due to orthogonality conditions arising from the choice of shape functions. The submatrices $\mathbf{k}_{mm}$ and $\mathbf{q}_m$ are the same as those obtained using the finite strip method (Chakrabarti, 1980), while the submatrices $\overline{\overline{\mathbf{k}}}$ and $\overline{\mathbf{q}}$ are the same as those obtained using bilinear displacement functions with the finite element method.

Assembling strip-element matrices using standard procedures and modifying the equations to incorporate known coefficients leads to the global equations

$$\begin{bmatrix} \mathbf{K}_{11} & 0 & \cdots & 0 & \cdots & 0 & \overline{\mathbf{K}}_1 \\ 0 & \mathbf{K}_{22} & \cdots & 0 & \cdots & 0 & \overline{\mathbf{K}}_2 \\ \vdots & \vdots & \ddots & \vdots & & \vdots & \vdots \\ 0 & 0 & \cdots & \mathbf{K}_{mm} & \cdots & 0 & \overline{\mathbf{K}}_m \\ \vdots & \vdots & & \vdots & \ddots & \vdots & \vdots \\ 0 & 0 & \cdots & 0 & \cdots & \mathbf{K}_{MM} & \overline{\mathbf{K}}_M \\ \overline{\mathbf{K}}_1^T & \overline{\mathbf{K}}_2^T & \cdots & \overline{\mathbf{K}}_m^T & \cdots & \overline{\mathbf{K}}_M^T & \overline{\overline{\mathbf{K}}} \end{bmatrix} \begin{Bmatrix} \boldsymbol{\Phi}_1^g \\ \boldsymbol{\Phi}_2^g \\ \vdots \\ \boldsymbol{\Phi}_m^g \\ \vdots \\ \boldsymbol{\Phi}_M^g \\ \overline{\boldsymbol{\Phi}}^g \end{Bmatrix} = \begin{Bmatrix} \mathbf{Q}_1 \\ \mathbf{Q}_2 \\ \vdots \\ \mathbf{Q}_m \\ \vdots \\ \mathbf{Q}_M \\ \overline{\mathbf{Q}} \end{Bmatrix} \tag{15}$$

where $\boldsymbol{\Phi}_m^g$, $(m = 1, 2 \ldots M)$ and $\overline{\boldsymbol{\Phi}}^g$ are vectors of global coefficients. The submatrices $\mathbf{K}_{mm}$ are symmetrical and have a half bandwidth of two and may be expressed in the form

$$\mathbf{K}_{mm} = \mathbf{K}_1 + \alpha_m^2 \mathbf{K}_2 \tag{16}$$

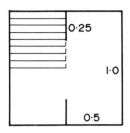

FIGURE 4. Cracked prism cross section.

where $\mathbf{K}_1$ and $\mathbf{K}_2$ are independent of m.

The submatrices $\overline{\mathbf{K}}_m$ are of the form

$$\overline{\mathbf{K}}_m = \frac{1}{\alpha_m}\overline{\mathbf{K}}^o \quad (m \text{ odd})$$

$$= \frac{1}{\alpha_m}\overline{\mathbf{K}}^e \quad (m \text{ even}) \tag{17}$$

where $\overline{\mathbf{K}}^o$ and $\overline{\mathbf{K}}^e$ are independent of m.

From the mth block of equations in Eq. 15, namely

$$\mathbf{K}_{mm}\boldsymbol{\Phi}_m^g + \overline{\mathbf{K}}_m\overline{\boldsymbol{\Phi}}^g = \mathbf{Q}_m \tag{18}$$

$\boldsymbol{\Phi}_m^g$ may be eliminated using Gaussian elimination, making use of the symmetry and narrow banding of $\mathbf{K}_{mm}$. The matrices $\mathbf{K}_{mm}$ and $\overline{\mathbf{K}}_m$ occurring for any value of m are simply generated using Eqs. 16 and 17. Repeating this procedure for $m = 1, 2, \ldots, M$ leads to the reduced equations

$$\overline{\overline{\mathbf{K}}}^*\overline{\boldsymbol{\Phi}}^g = \overline{\mathbf{Q}}^* \tag{19}$$

which may be solved for $\overline{\boldsymbol{\Phi}}^g$. As a consequence of the elimination procedure, $\overline{\overline{\mathbf{K}}}^*$ is not banded. Any number of trigonometric terms, M, may be taken without increasing computer core requirements. Convergence with increasing M is therefore readily assessed.

EXAMPLES

A prism with a square cross section of unit side length and a cracked prism with the cross section shown in Fig. 4 were analysed by dividing a quarter of the section into 4, 8, 16 and 32 strip-elements. The stress function at the centres of the prisms obtained using the proposed method are summarised in Tables 1 and 2 for various values of M. The exact value for the square prism is 0.14734, while an approximate value of 0.01037 for the cracked prism was obtained by the alternative procedure of dividing a quarter of the section into two macro elements (Golley, 1976).

The approximate values obtained for the square prism converge very rapidly with increasing M to values very close to the exact value. Convergence is slower with the cracked prism, which has stress singularities at the crack tips. Solutions converge with increasing numbers of strip-elements and increasing numbers of trigonometric terms to a value close to that obtained using macro elements.

TABLE 1. ϕ at Centre of Square Prism. (Exact = 0.14734)

	Number of strips			
M	1	2	4	8
4	0.14830	0.14670	0.14743	0.14739
8	0.14827	0.14757	0.14740	0.14736
16	0.14826	0.14757	0.14740	0.14736

TABLE 2. ϕ at Centre of Cracked Prism. (Macro Element Solution = 0.10327)

	Number of strips			
M	4	8	16	32
4	0.09868	0.9994	0.10020	0.10028
8	0.09833	0.10083	0.10147	0.10170
16	0.09819	0.10104	0.10198	0.10240
32	0.09819	0.10107	0.10212	0.10267
64	0.09819	0.10107	0.10214	0.10275
128	0.09819	0.10107	0.10215	0.10276
256	0.09819	0.10107	0.10215	0.10276

CONCLUSIONS

The use of shape functions which are partially orthogonal is computationally efficient for solving a class of second order problems. The method discussed is a combination of the finite element and finite strip methods and contains some advantages of both techniques. In common with the finite element method, a variety of mixed boundary conditions may be treated, while in common with the finite strip method as originally applied to simply supported structures, any number of trigonometric terms may be taken without requiring increased computer capacity.

REFERENCES

1. Chakrabarti, S., Heat Conduction in Plates by Finite Strip Method, *J. Engineering Mechanics Division, ASCE*, vol. 106, no. EM2, pp. 233-244, 1980.

2. Cheung, Y. K., Finite Strip Method in the Analysis of Elastic Plates with Two Opposite Ends Simply Supported, *Proc. Instn. Civ. Engrs,* vol. 40, pp 1-7, 1968.

3. Golley, B. W., Grice, W. A. and Petrolito, J., Plate Bending Analysis using Finite Strip-Elements, *J. Structural Engineering, ASCE*, vol. 113, pp. 1282-1296, 1987.

4. Golley, B. W., Rectangular and Right Isosceles Triangular Elements for Torsion Analysis, *Proc. Int. Conf. on Finite Elements in Engineering*, eds Y. K. Cheung and S. G. Hutton, pp. 33.1-33.15, Adelaide, 1976.

An Algorithm for Modeling Crack Growth and Predicting Rupture in Post-Yield Fracture Mechanics

E. S. HWANG, C. PATTERSON, and D. W. KELLY
School of Mechanical and Industrial Engineering
University of New South Wales
Kensington, NSW, 2033, Australia

INTRODUCTION

The use of the finite element method to model crack growth initiation and stable crack growth in elastic-plastic fracture has been given much attention. The modelling process must not only study the selection of criteria for crack growth initiation and stable crack growth, but also develop a stable algorithm for extending the crack across the finite element mesh.

A large number of fracture criteria such as Crack Opening Displacement[COD] described by Wells (1961), the J-integral of Rice (1968) and deLorenzi (1982) and crack tip node force methods of Kobayashi et al (1973), Light and Luxmoore (1977) and Evans et al (1980) have been developed. Kanninen et al (1979) studied several of these for Al-2219-T87 aluminium alloy and A533-B steel materials using experimentally obtained load line displacement data in their finite element model. This work indicated that, of the parameters examined, crack tip node force, work done in separating the crack faces, energy release rate based on a computational process zone and crack tip opening angle are constant during stable crack growth. They therefore provide possible criteria for predicting the growth of the crack.

Algorithms for modelling the crack growth using finite elements have also been described in the literature. These are based on uncoupling nodal points ahead of the crack tip and releasing the nodal forces. Kobayashi (1973) and Light and Luxmoore (1977) applied reaction forces to the crack tip nodes on the newly created degrees of freedom and reduced these reaction forces zero to grow the crack. Anderson (1974) and Newman (1977) added an infinite stiffness to the crack tip nodes and reduced this additional stiffness to zero. Shih et al (1976) modelled crack growth by shifting the nodes.

The energy release rate at the crack tip is the limiting value of the change in strain energy for a crack growth increment δa divided by the growth increment. The limit is taken as the growth increment goes to zero. The energy release rate is then a valid criterion for crack growth initiation when the region near the crack tip is considered to behave elastically. In a previous report (Kelly et al (1988)), the authors confirmed the widely held view that crack tip energy release tends to zero as δa goes to zero when post-yield effects are correctly modelled and the limiting work hardening slope is zero. If the limiting work hardening slope is not zero, however, the crack tip energy release rate drops sharply as δa becomes small in a way which makes uncertain any conclusion about whether a finite value of energy release rate is achieved as δa goes to zero.

In this paper, we have extended the investigation to the holding back forces $F^{\delta a}$ and energy release rate $G^{\delta a}$ for a finite δa when the limiting work hardening slope is positively finite, and found that the forces $F^{\delta a}$ are constant during the crack growth confirming the results produced by Kanninen et al (1979). We have also found that a combination of energy release rate and appearance of rupture stress in the immediate vicinity of the crack tip appears to control the onset of rupture.

In the following section, the finite element algorithm is described. It simulates crack growth by determining the holding back forces $F^{\delta a}$ for a finite δa at the crack tip prior to crack growth and releasing these forces to extend the crack. The criteria for the crack growth and final rupture include a combination of the magnitude of the holding back forces, the energy release rate and the value of the effective stress in the plastic process zone ahead of the crack tip. The plastic process zone spreads over a large area as the crack grows. To test the crack growth criteria, a complete experimental load and crack growth history has been followed by imposing growth of the crack to match experimental results.

An important result from the work by Kelly et al (1988) is the recognition of the presence of an $r^{-\frac{1}{2}}$ stress/strain singularity at the crack tip inside the plastic process zone when the work hardening slope is not zero - leading to discussions about the influence of the design of the finite element mesh on the crack growth criteria. It is apparent that the finite element mesh must be sufficiently fine near the crack tip to model this singularity and any crack growth increment used to determine crack growth parameters must be small compared to the zone of this singularity.

Numerical experiments in Section 3 consider a centre cracked panel of aluminium alloy Al-2219-T87 for which experimental data has been published by Kanninen et al (1979). The experiments show that the reactive holding back forces and energy release rate are geometry and mesh size independent for a suitably determined δa. The results also confirm the conclusion of Kanninen et al (1979) that holding back forces are approximately constant for a suitably determined δa during the crack growth. Further it has been found that the magnitudes of critical energy release rate for δa are monotonically increasing in the stable crack growth regime and rapidly increasing in the rapid fracture regime.

THE FINITE ELEMENT ALGORITHM

The computer program used in the numerical experiments is the elastic-plastic small strain plasticity program given in Owen and Fawkes (1983), modified to permit the crack tip extension by releasing the holding back reactive nodal forces at the crack tip, and containing restart procedures to allow increments of the crack growth to test for energy release during monotonic loading. For a symmetric specimen, with the crack lying along the plane of symmetry, nodes at the ahead of the crack tip are restrained by zero normal prescribed displacement boundary conditions. The program allows the reactive nodal forces to be evaluated at these restrained nodes. These nodal reactive forces are recorded and then simultaneously reduced to zero in a series of increments. As the crack opens the crack tip energy release is calculated as

$$\Delta U = \sum_{i=1}^{n} \int F_i^{\delta a} \, dv_i \tag{1}$$

where $F_i^{\delta a}$ are the reactive holding back forces and v is the displacement perpendicular to the crack line as the crack opens.

Modelling the Growth Increment

The algorithm has been supplemented by a restart facility which allows the holding back forces and energy release to be calculated for a given δa after an elastic-plastic loading of the cracked specimen. If crack growth is not indicated by the growth criteria, further loading of the specimen is carried out at the original crack length. The quadratic 8 node isoparametric element has been used in all numerical experiments with the crack extension across an even number of nodes simultaneously so that crack grows along full element sides. For computational accuracy the holding back forces are released in 10 increments to simulate the crack growth and computational iterations have been performed to obtain a specified percentage of convergence, that is, the ratio of residual forces over applied forces is less than 0.0001 % .

Criteria for Growth and Rupture

For the crack growth initiation and stable growth criteria, the $F^{\delta a}$ have been divided by δa to provide an average force at the crack tip and $G^{\delta a}$ per unit thickness t has been calculated from the following equation.

$$G^{\delta a} = \frac{\Delta U}{\delta a \; t} \tag{2}$$

The paper by Kanninen et al (1979) concluded that reactive holding back forces are constant during the crack growth and that the energy release rate had to be confined to that occuring in a small process zone near the crack tip. However, a criterion for the prediction of final rupture was not reported. The authors have extended this investigation to include the full energy release rate in the search for a criterion capable of predicting the final rupture of the specimen.

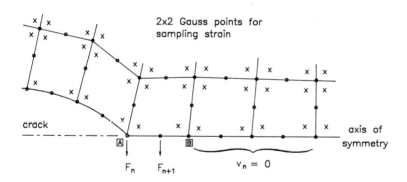

F_n and F_{n+1} are reactive forces which are reduced simultaneously to zero over 10 increments to advance the crack tip from A to B

FIGURE 1. Crack advancement

NUMERICAL IMPLEMENTATION

The numerical experiments for the centre cracked panel and compact tension specimens of Al-2219-T87 aluminium alloy have been carried out under the plane stress and Mode I loading conditions. One configuration of centre cracked panel is shown in Figure 2(a) and the results will be reported here. The finite element mesh of eight node serendipity elements is shown in Figure 2(b).

Crack Growth Initiation

Reactive holding back forces $F_0^{\delta a}$ and energy release rate $G_0^{\delta a}$ have been calculated under the applied load at the onset of crack growth for δa varying from 0.005 millimeter (one element edge in a fine mesh) to 3.5 millimeters (several element edges in a coarse mesh). Both parameters have been shown to have the same value at the onset of crack growth for all the specimens considered.

Modelling Crack Growth

The load versus crack growth Δa history has been modelled with full elastic-plastic analysis. For computational accuracy, small external load increments have been carried out and the 0.0001 percentage convergence described above has been enforced. Figure 3 shows that the plastic process zone spreads over a large area and finally reaches the remaining ligament of structure as the crack propagates.

The reactive holding back forces were read from the reaction forces at the crack tip as the crack grew. Figure 4 shows that the critical values of $F^{\delta a}$ are constant, within a few percent, during the crack growth for a suitably determined δa which is based on the zone of influence of the $r^{-\frac{1}{2}}$ stress/strain singularity. This result is quite similar to result produced by Kanninen et al (1979).

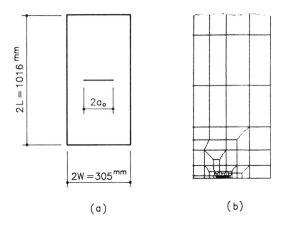

(a)   (b)

FIGURE 2. Configuration of a centre cracked panel with $2a_0 = 102\,^{mm}$ and finite element mesh for one quadrant. This panel is $6.35\,^{mm}$ thick $2219 - T\,87$ aluminum alloy

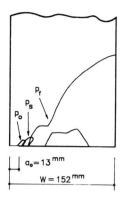

$a_o = 13^{mm}$

$W = 152^{mm}$

P_o : load at the onset of crack growth
P_f : load at the rapid fracture
$$P_o < P_s < P_f$$

FIGURE 3.   Plastic zone as the crack grows

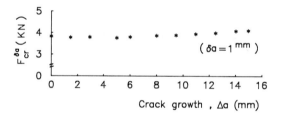

FIGURE 4.   Critical values of reactive holding back forces as the crack grows
($2a_o = 102^{mm}$)

Final Rupture

The crack tip energy release increases as the crack propagates. Figure 5 shows that the values of $G_{cr}^{\delta a}$ are monotonically increasing in the stable crack growth regime and rapidly increasing in the final rupture regime.

The Von Mises effective stress ahead of the crack tip is plotted against the elastic-plastic loading in Figure 6. The plot shows that the effective stress $\bar{\sigma}$ ahead of the crack tip increases as the applied load increases. At the applied load at which rupture occured in the experiments, the effective stresses at the crack tip departs from the $r^{-\frac{1}{2}}$ singularity and exceeds the rupture stress σ_r.

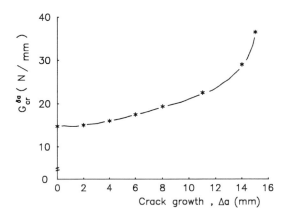

FIGURE 5. Critical values of energy release rate as the crack grows ($2a_0 = 102^{mm}$)

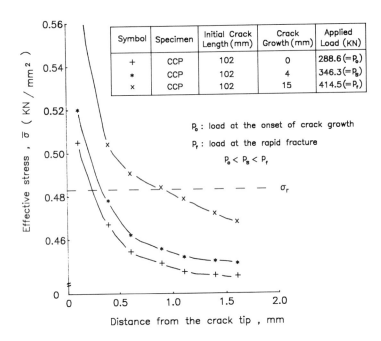

Symbol	Specimen	Initial Crack Length (mm)	Crack Growth (mm)	Applied Load (KN)
+	CCP	102	0	288.6 (= P_0)
*	CCP	102	4	346.3 (= P_s)
x	CCP	102	15	414.5 (= P_r)

P_0 : load at the onset of crack growth

P_r : load at the rapid fracture

$$P_0 < P_s < P_r$$

FIGURE 6. Effective stresses ahead of the crack tip as the crack grows

PROPOSED ALGORITHM FOR CRACK GROWTH

From the foregoing results, the following algorithm can be proposed.

1. If the value of $F^{\delta a}$ and/or $G^{\delta a}$ under an external load does not exceed the critical value for the onset of crack growth, increase the load.

2. When the $F^{\delta a}$ and/or $G^{\delta a}$ exceeds the critical value, allow the crack grow by a small amount of crack extension Δa_1 .

3. If $F^{\delta a}$ is less than the $F^{\delta a}_{cr}$, increase the applied load P until $F^{\delta a}$ equals $F^{\delta a}_{cr}$.

In this manner, allow the crack to grow calculating the $G^{\delta a}_{cr}$ under the same load level and geometry. If $G^{\delta a}_{cr}$ increases slowly compared with other values and the effective stress ahead of the crack tip is below the rupture stress, then the crack is still in the stable crack growth regime. When, however, these criteria are not met or $F^{\delta a}$ does not become less than $F^{\delta a}_{cr}$ as the crack grows, then rupture has occurred.

CONCLUDING DISCUSSION

Holding back forces and energy release rate, limited to growth increments well inside the zone of influence of the $r^{-\frac{1}{2}}$ stress/strain singularity, have been confirmed as possible criteria in elastic-plastic fracture with finite limiting work hardening slope. The following conclusions can be drawn.

1. Reactive holding back forces are approximately constant during the crack growth.

2. Energy release rate is monotonically increasing in the stable crack growth regime and rapidly increasing in the rapid fracture regime.

3. For thee materia tested the plastic process zone spreads over a large area and finally reaches the remaining ligament of structure as the crack grows.

These conclusions have been confirmed by application to cracks of different length in the centre cracked panel and to a test using a compact tension specimen.

The algorithm is now to be applied to the different materials given in Newman (1985) where the criteria will be used to predict the crack growth history and final rupture.

REFERENCES

1. Anderson, H., Finite Element Treatment of a Uniformly Moving Elastic-Plastic Crack Tip, *J. of Mech. Phys. Solids*, Vol. 22, pp. 285-308, 1974.

2. Evans, W. T., Light, M. F., and Luxmoore, A. R., An Experimental and Finite Element Investigation of Fracture in Aluminium Thin Plates, *J. of Mech. Phys. Solids*, Vol. 28, pp. 167-189, 1980.

3. Kanninen, M. F., Rybicki, E. F., Stonesifer, R. B., Broek, D, Rosenfield, A. R., Marschall, C. W., and Hahn, G. T., Elastic-Plastic Fracture Mechanics for Two-Dimensional Stable Crack Growth and Instability Problems, *ASTM STP 668*, pp. 121-150, 1979.

4. Kelly, D. W., Patterson, C., Hwang, E. S., and Leong, S. S., Singular Stress Fields and Crack Tip Energy Release Rates For Mode I Cracks in Strain Hardening Plasticity, *Internal Report 1988 /am/1*, School of Mechanical and Industrial Engineering, University of New South Wales, Sydney, Australia, 1988.

5. Kobayashi, A. S., Chiu, S. T., and Beeuwkes, R., A Numerical and Experimental Investigation on the Use of J-integral, *Engineering Fracture Mechanics*, Vol. 5, pp. 293-305, 1973.

6. Light, M. F., and Luxmoore, A. R., Crack Extension Forces in Elasto-Plastic Stress Fields, *J. of Strain Analysis*, Vol. 12, No. 4, pp. 305-309, 1977.

7. deLorenzi, H. G., On the Energy Release Rate and J-integral for 3D Crack Configurations, *Int. J. of Fracture,* Vol. 19, pp. 183-193, 1982.

8. Newman, J. C. Jr., Finite Element Analysis of Crack Growth Under Monotonic and Cyclic Loading, *ASTM STP 637*, pp. 56-80, 1977.

9. Newman, J. C. Jr., An Evaluation of Fracture Analysis Methods, *EPFM Technology, ASTM STP 896*, pp. 5-96, 1985.

10. Owen, D. R. J., and Fawkes, A. J., *Engineering Fracture Mechanics, Numerical Methods and Applications*, Pineridge Press, 1983.

11. Rice, J. R., In Fracture, *An Advanced Treatise*, edited by Liebowitz, H., Academic Press, Vol. 2, pp. 191-311, 1968.

12. Shih, C. F., deLorenzi, H. G., and German, M. D., Crack Extension Modelling with Singular Quadratic Isoparametric Elements, *Int. J. of Fracture*, Vol. 12, pp. 647-651, 1976.

13. Wells, A. A., Unstable Crack Propagation in Metals, Cleavage and Fast Fracture, *The Crack Propagation Symposium*, Cranfield, 1961.

Automatic Detail Removal of CAD Models for Finite Element Analysis

Y. C. LAM and W. SIEW
Department of Mechanical Engineering
Monash University
Clayton, Victoria 3168, Australia

K. HO-LE
Abak Computing Pty. Ltd.
6, Winifred Crescent
Glen Iris, Victoria 3146, Australia

INTRODUCTION

The finite element (FE) method is widely used in engineering analysis and design. The method consists of three major phases : model preparation or pre-processing, analysis and post-processing. Model preparation is the most labour intensive and error prone of the three phases. In this phase, the object to be analysed is simplified into a geometrical model which is then discretized into smaller blocks called the finite elements. This model is called the finite element model and the collection of such small blocks is known as the finite element mesh. From the time when the concept of FE method was introduced in the 1950's, through to the 1970's, model preparation was done manually. The task would be tedious when the object to be analysed has complex geometry, or irregular boundaries, or is non-symmetrical.

Automatic mesh generators had been developed since the early 1980's [Nguyen (1982), Yerry and Shephard (1983, 84), Wördenweber (1984), Kela et. al. (1986), Shephard et. al. (1986), Ho-Le (1987)] to perform the discretization using the computer. The algorithms used in these mesh generators were based on different techniques such as triangulation (Wördenweber), modified quadtree and octree methods (Yerry and Shephard), topologically based element removal method (Shephard et. al.), recursive spatial subdivision (Kela et. al.) and grid-based method (Ho-Le).

The advent of computer-aided design (CAD) systems enables the design of engineering products to be done easily and quickly. The computer representation of the product geometry is called the CAD model. This model acts as the basis for subsequent analyses and manufacturing processes. With the help of CAD facilities and the implementation of automatic mesh generators in FE systems, the effort in preparing an FE model is greatly decreased. However, the initial step of converting the CAD model into a FE model still requires significant effort, especially for geometries of high complexity [Butlin (1983), Revelli (1985)]. These geometries usually contain too many details such as edges, holes, and fillets that are very small compared to the overall dimension of the object. As a result, an excessive number of elements will be generated by the mesh generator; not to mention the limited capability of some mesh generators in handling large amounts of fine detail.

This paper examines some of the issues involved in simplifying the CAD model prior to mesh generation using the technique of *automatic detail removal* (ADR). A preliminary two-dimensional (2-D) ADR algorithm for straight edges will be presented. This algorithm interfaces with the automatic mesh generator developed by Ho-Le (1987) which will be briefly described.

A GRID-BASED AUTOMATIC MESH GENERATOR

This 2-D and 3-D mesh generator is a result of a combined research effort between the Department of Mechanical Engineering, Monash University and the Division of Manufacturing Technology, CSIRO. The mesh generator utilizes the grid-based method to place a mesh on an object. The 3-D mesh generator will not be discussed since only the 2-D mesh generator is of interest here.

The input to this program consists of two parts, the mesh density specifications and the object specifications. The object is specified as a set of "loops" forming the boundary of the object, where a loop is an ordered circular list of vertices, surrounding the material on the left hand side. Due to this specification, the mesh generator is able to handle multiply- connected region (for example, objects with holes).

Figure 1 shows that multiply-connected regions do not need to be represented with a cut as required by some other mesh generators. It also demonstrated the capability of the mesh generator in placing elements around short edges. Mesh refinement around regions of interest is shown in Fig. 2.

AUTOMATIC DETAIL REMOVAL

From CAD Model to FE Model

Although the CAD model can be created quickly and accurately, it is very often not suitable for meshing because it contains too many details. Thus the FE model usually has to be reconstructed from scratch. This does not only involve a duplicated effort in model preparation, but also the possible introduction of errors when the model is rebuilt. It is very common that after the initial analysis of the overall model, certain critical regions are to be re-analysed with more detail added in, or changes may be made to the boundary conditions. Re-analysing the critical regions will require the construction of sub-models; and the changes to the boundary conditions may require the existing model to be modified in order to represent the new boundary conditions adequately. Thus it is desired that all these models are based on a common source,

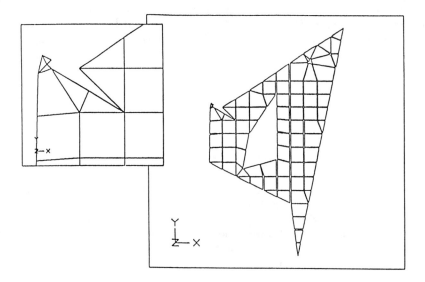

FIGURE 1. An object meshed with quadrilateral and triangular elements. The elements have been shrunk to show the small hole at the top right and a magnification of the elements around the short edges at the left hand projection is given.

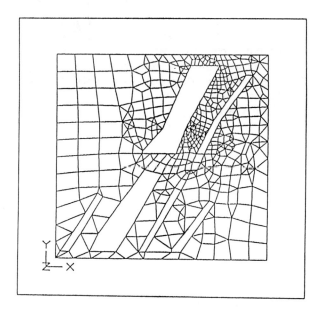

FIGURE 2. Mesh refinement around some specified points of interest. The resulting mesh is non-uniform.

407

that is the CAD model, such that model preparation can be made more efficient and the errors reduced. In order for this to be feasible, excessive detail from the CAD model must be identified and removed. Manual detail removal is inefficient and error prone [Butlin (1983)]. ADR will eliminate this problem.

ADR – The Algorithm

The basic function of ADR is to simplify the CAD model to produce a geometrical model, which may then be meshed manually or by an automatic mesh generator. The algorithm to be presented is based upon the criterion of a prescribed minimum length l. A short edge is then defined as an edge with length less than l which is a detail to be removed. The model or geometry is assumed to be made up of at least one polygon. As in the automatic mesh generator mentioned earlier, the edges of each polygon are connected together to form a loop. (A loop is an ordered circular list of edges.)

There are three major steps in the algorithm :

Step 1 — Joining a series of short edges. In this step, consecutive short edges are identified and joined together to form one edge. These consecutive edges are detected by traversing the loops starting from the first edge. The lengths of an edge and the edge following it are compared against l. If both lengths fall below l, they will be joined together to form a new edge, starting from the first end-point of the first edge and terminating at the second end-point of the other edge, as shown in Fig. 3. The length of this new edge is calculated; if it is found to be less than l still, the procedure of comparing two edges at one time with length l will be repeated. Thus, a series of edges shorter than l will be joined together until an edge of at least l in length is formed or when the next edge in the loop is longer than l. Figure 4 shows how a series of short edges are joined to form a new edge.

Each time before two edges are joined, a check is made to ensure that the resulting edge will not affect the validity of the model. This is done by checking to see if the

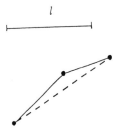

FIGURE 3. Connecting two edges (solid lines) to form one edge (broken line).

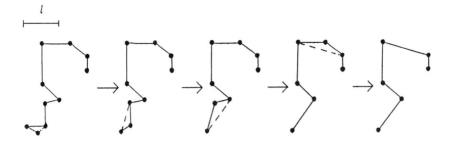

FIGURE 4. Joining several short edges together.

resulting edge interferes/intersects with other edges of the model. Such interference can occur when a concave and a convex surface are close to each other. The check stops once interference is detected.

Whenever there is interference, the two edges will not be joined. Instead, one of the short edge will be joined to the long edge next to it, where there is no interference. This will also eliminate the occurrence of two short edges next to each other.

In general, joining a series of short edges will produce a longer edge. The procedure also reduces the number of edges in the model. The database will be modified to reflect the change, discarding any unnecessary data to economize on memory space.

Step 2 — Elimination of a single short edge. At the completion of the above step, it is possible to have a short edge between two long edges in the CAD model. This short edge can be removed in a number of ways, depending on the position of this short edge relative to the two long edges :

Merging collinear edges — As a result of joining edges, a new edge may be formed such that it is in line (collinear) or almost in line (to a specified tolerance) with the edges connected to it. Here, near-collinearity will be treated as collinear. When a short edge is found to be collinear with the edge next to it, the edges will be merged to form one long edge. The merging is basically of the same concept as joining the edges which was mentioned in the earlier section.

Z-shape elimination — This method is named thus owing to the geometrical shape when the long edges lie on the opposite sides of the short edge as shown in Fig. 5. The easiest way to eliminate the short edge and at the same time minimizing local changes is to move one end-point of each of the long edge to the mid-point of the short edge (Fig. 5(a)). This is possible only if interference will not happen. If interference results, one end-point of the short edge will be deleted such that only one long edge need to be changed (Fig. 5(b)).

U-shape elimination — When the long edges on either sides of the short edge are parallel or near-parallel to each other, the formation looks like the letter U. Edges that are close to being parallel within a specified tolerance will be considered as

409

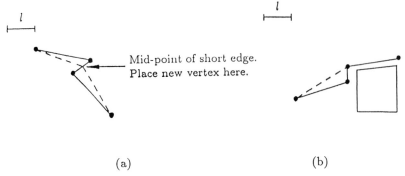

(a)   (b)

FIGURE 5. Removing a short edge by displacing some end-points.

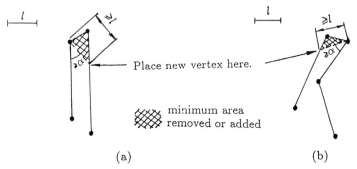

(a)   (b)

FIGURE 6. Replacing a short edge with an edge of length l.

parallel here. The least area change will occur when a triangular section is removed[2] from the "top" as shown in Fig. 6, subjected to the requirement of replacing the short edge with an edge of length l, inclined at an angle α. The angle condition is specified such that the elements subsequently formed will be well shaped, for example α could be set to a minimum value of 30°. When the edges are in the formation shown in Fig. 6(b), the requirement of least area change under the constraints will be satisfied by U-shape elimination as well.

V-shape elimination — This is the third group of the edges formation where the long edges diverge away from the short edge (Fig. 7). If they formed an angle which is α or greater, the edges are extended, thus eliminating the short edge as shown in Fig. 7(a); otherwise, a triangular or quadrilateral piece will be removed (or added) such that it conforms to the angle condition and the length requirement,with minimum area change (Fig. 7(b)).

Step 3 — Elimination of degenerate loops. When all short edges have been eliminated, the model will be left with loops with all long edges, loops with only one edge (a line) or some empty loops (points). These lines and points must be removed.

---
[2]
The area is removed for a convex surface, the same area will be added if the surface is concave.

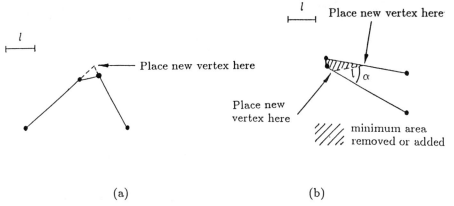

(a)   (b)

FIGURE 7. The long edges can be extended as in (a) or lengthening the short edge by removing (or adding) material to satisfy the length l and angle α constraints.

In the first two steps, degenerate loops will be removed when met during the joining or short edge elimination procedures. For example, the joining may end with a single short edge which will be reduced to a point, or stops when a single long edge is formed.

A Reduced CAD Model

Figure 8(a) is an example of a CAD model. On applying ADR to this model, the simplified form is shown in Fig. 8(b).

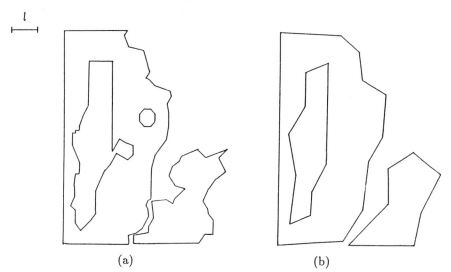

(a)   (b)

FIGURE 8. A simpler model resulting from the use of ADR.

ACKNOWLEDGEMENT

The authors wish to acknowledge the financial support of the Harold Armstrong Memorial Fund.

REFERENCES

1. Butlin, G., CAD/FEM Interfacing, *Effective CADCAM, I. Mech. E. Conference Publicati9ons 1989-7*, pp. 95 – 104, Mechanical Engineering Publications Limited for The Institution of Mechanical Engineers, London, 1983.

2. Ho-Le, K., *A Grid-based Method for the Automatic Generation of Two- and Three-dimensional Finite Element Meshes*, PhD thesis, Monash University, Australia, 1987.

3. Kela, A., Perucchio, R. and Voelcker, H., Toward Automatic finite Element Analysis, *Computers in Mechanical Engineering*, vol. 5, no. 1, pp. 57 – 71, 1986.

4. Nguyen-Van-Phai, Automatic Mesh Generation With Tetrahedron Elements, *International Journal for Numerical Methods in Engineering*, vol. 18, no. 2, pp. 273 – 289, 1982.

5. Revelli, V. D., Getting CAD and FE to Cooperate, *Computers in Mechanical Engineering*, vol. 4, no. 2, pp. 39 – 44, 1985.

6. Shephard, M. S., Grice, K. R. and Georges, M. K., Some Recent Advances in Automatic Mesh Generation, *Proc. Session in Conjunction with the ASCE National Convention : modern Methods for Automating Finite Element Mesh Generation*, ed. Kenneth Baldwin, pp. 1 – 18, The American Society of Civil Engineers, 1986.

7. Wördenweber, B., Finite Element Mesh Generation, *Computer-Aided Design*, vol. 16, no. 5, pp. 285 – 291, 1984.

8. Yerry, M. A. and Shephard, M. S., A Modified Quadtree Approach to Finite Element Mesh Generation, *IEEE Computer Graphics and Applications*, vol. 3, no. 1, pp. 39 – 46, 1983.

9. Yerry, M. A. and Shephard M. S., Automatic Three-Dimensional Mesh Generation by the Modified-Octree Technique, *International Journal for Numerical Methods in Engineering*, vol. 20, no. 11, pp. 1965 –1990, 1984.

Multilevel Substructuring for Finite Element Analysis

LIU XIAO-LIN and Y. C LAM
Department of Mechanical Engineering
Monash University
Clayton, Victoria 3168, Australia

INTRODUCTION

The finite element method has the ability to provide relatively accurate analyses for various categories of practical design problems, such as stress analysis, heat transfer, fluid mechanics, eletro-magnetical fields and others. Thus, it has become one of the most popular methods now available for getting numerical solutions. However, the size of a problem that can be solved using the finite element method is limited by the available computing resource. On the other hand, even with a large scale computer, the computational cost would become prohibitively high if the problem size is very large. In those cases, some form of substructuring which partitions the structure into segments has to be performed for the purpose of the analysis [Dodds and Lopez (1980), Furuike (1972), Hitchings and Balasubramaniam (1984), Kela et. al. (1986), Noor et. al. (1978), Peterson and Popov (1977)].

This paper presents a new technique of substructuring which is based on the regular mesh and the use of a special numbering-addressing system. It is shown that the present technique is of advantages in terms of the computational efficiency and the core memory requirement for large size problems.

METHOD OF ANALYSIS

The Concept of Substructuring

Basically, substructuring contains the following major stages:

i) DISCRETIZATION. Divide the structure into substrucrures which are interfaced by interior boundaries.

Liu Xiao-Lin is visiting, on leave from Laboratory of Metalworking Technology, General Research Institute for Non-Ferrous Metals, Beijing, PRC.

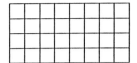

a) Original mesh.

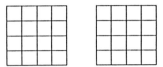

b) divided into two substructures.

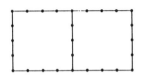

c) internal d.o.f condensed and assembled

FIGURE 1. Two dimensional region illustrating the concept of substructuring.

ii) FORWARD CONDENSATION. For each substructure, obtain the boundary stiffness matrix and the equivalent loading vector by static condensation.

iii) ASSEMBLY. Assemble the substructure matrices and vectors into the global equations, and solve for the boundary displacements.

iv) BACKWARD SUBSTITUTION. Using the boundary displacements, compute the internal displacements for each substructure, and strains and stresses for each element.

As shown in Fig. 1, the original structure is divided into two substructures. The equilibrium equation for a subustructure can be written as:

$$\begin{bmatrix} K_{ii} & K_{ib} \\ K_{bi} & K_{bb} \end{bmatrix} \begin{Bmatrix} U_i \\ U_b \end{Bmatrix} = \begin{Bmatrix} F_i \\ F_b \end{Bmatrix} \tag{1}$$

Where the subscripts i and b stand for the internal d.o.f. (degrees of freedom) and the boundary d.o.f. respectively.

Expanding the above equation gives:

$$\begin{cases} K_{ii}U_i + K_{ib}U_b = F_i \\ K_{bi}U_i + K_{bb}U_b = F_b \end{cases} \tag{2}$$

414

Solving for the internal d.o.f. in terms of the boundary ones leads to:

$$\begin{cases} U_i = K_{ii}^{-1}(F_i - K_{ib}U_b) \\ (K_{bb} - K_{bi}K_{ii}^{-1}K_{ib})U_b = F_b - K_{bi}K_{ii}^{-1}F_i \end{cases} \tag{3}$$

Let

$$K_1 = K_{bi}K_{ii}^{-1} \tag{4}$$

We have

$$\begin{cases} K_b = K_{bb} - K_1 K_{ib} \\ F_{b1} = F_b - K_1 F_i \\ K_b U_b = F_{b1} \end{cases} \tag{5}$$

The substructure boundary stiffness matrix, K_b and the equivalent loading vector, F_{b1} are to be assembled into the global equations the same way as an element. Therefore, a substructure with its internal d.o.f being condensed is often referred as a superelement.

Solving the global equations, we get U_b and then from eqn (3) U_i can be calculated for each substructure.

Multi-level Substructuring Based on Regular Mesh

For simplicity, let us limit our discussion on two dimensional cases. The regular mesh that we use to carry out the multi-level substructuring is the $N \times N$ nodes finite element mesh which divides a rectangular area into $(N-1) \times (N-1)$ identical rectangular elements. In solid mechanics, a domain of interest can often be modelled by a combination of the regular mesh areas and some boundary elements of irregular shapes. There are two advantages in using the regular mesh discretization. Firstly, the stiffnesses of the elements within a regular mesh area need to be analyzed only once since they are all the same. Secondly, the regularity would enable an efficient recursive calculation scheme of condensation and backward substitution.

If we consider each of the regular mesh areas as a substructure, the higher level substructure boundary stiffness matrices can be calculated by condensation of the lower level ones. In the recursive procedure, two series of patterns of boundary stiffness matrices are obtained depending if the number of nodes along the boundary is odd or even, see Fig. 2.

Suppose the hightest level of the substructures is M, the recursive calculation equations for the forward condensation and the backward substitution respectively are:

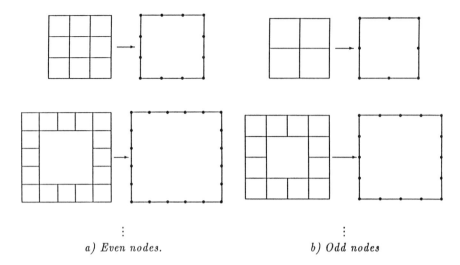

a) Even nodes.   *b) Odd nodes*

FIGURE 2. Two series of patterns of superelements.

$$
\begin{cases}
K_{ii}^{(j)} = K_b^{(j-1)} \\
K_1^{(j)} = K_{bi}^{(j)}(K_{ii}^{(j)})^{-1} \\
K_b^{(j)} = K_{bb}^{(j)} - K_1^{(j)} K_{ib}^{(j)} \\
F_{b1}^{(j)} = F_b^{(j)} - K_1^{(j)} F_{b1}^{(j-1)} \\
\quad (j = 1, 2, 3, \ldots, M)
\end{cases}
\tag{6}
$$

$$
U_i^{(j)} = U_b^{(j-1)} = (K_{ii}^{(j)})^{-1}(F_i^{(j)} - K_{ib}^{(j)} U_b^{(j)})
$$
$$
(j = M, M-1, M-2, \ldots, 1)
\tag{7}
$$

By using a special numbering-addressing system which takes advantage of the physical symmetry of the regular mesh, the stiffness matrix of each level can be made
symmetrical with respect to both diagonals. Therefore, only about one quarter of it
need to be processed during the forward and the backward procedures.

TEST EXAMPLES

Based on the above concept, a two dimensional finite element analysis program,
"FEMSUB", has been implemented in the University's VAX network. A few examples
have been tested to check its validity, computational efficiency and core memory
requirement for the global stiffness matrix. Two other programs have been used for
comparison. The first one is the well established package, "SAP2". The second one,
"FEM", is a conventional two dimensional finite element analysis program which uses
the same routines as "FEMSUB" except subroutine "substructure".

416

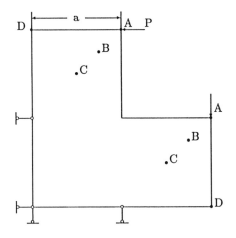

Mesh Density : 108 *elements*, 133 *nodes.*
Thickness $= 1mm$, $a = 240mm$, $P = 10KN$
$\qquad E = 100KN/mm^2$, $\nu = 0.25$

FIGURE 3. L- shaped plate.

Example 1 is a L-shaped plate loaded symmetrically by the concentrated forces at points A, B and C respectively, as shown in Fig. 3. The plate is analyzed using three identical substructures by "FEMSUB" and using the same mesh density by "FEM" and "SAP2". The results from "FEM" and "FEMSUB" are symmetrical and in good agreement with those from "SAP2". Table 1 shows the calculated displacements at the typical points D by the programs. The slight difference between them can be attributed to that no double precision varibles have been used in "FEMSUB" and "FEM".

Table 2 lists the occupied array size of the global stiffness matrix and the total running CPU time for each of the loading cases. Since the global stiffness matrix takes up most of the data core memory and "FEMSUB" shares the same assembler and solver with "FEM", the table indicates that both the core memory requirement and the computional time are less by using "FEMSUB" as compared with "FEM".

The second example is a cantilever beam which is discretized by the regular mesh of different density and analyzed using different number of substructures (Fig. 4).

TABLE 1. Calculated Displacements at Point D (mm)

Program	Loading Points		
	A	B	C
SAP2	-0.5328,-0.9915	-0.4416,-0.8377	-0.2343,-0.4777
FEM	-0.5312,-0.9906	-0.4387,-0.8358	-0.2306,-0.4742
FEMSUB	-0.5312,-0.9906	-0.4387,-0.8358	-0.2306,-0.4742

TABLE 2. Comparison of Memory and CPU Time

Program	Global K size	CPU Time (sec.)		
		A	B	C
FEM	6111	8.07	8.07	8.07
FEMSUB	4066	7.72	7.79	7.81

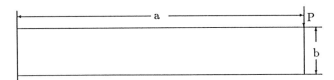

$a/b = 6$

Mesh Density : $A = 100$ *elements*, $B = 400$ *elements*.

FIGURE 4. Cantilever beam.

The resulted memory requirement and computational efficiency are shown in Table 3 and 4, respectively. From the tables, it is obvious that with an increase in the number of degrees of freedom, the saving in memory and CPU time would become more significant.

Although the use of higher level substructure definitely results in less memory requirement, its effect on the computational efficiency is not straightforward. As shown in Table 4, computation is more efficient by using a larger number of the same substructure of relatively lower level as compared with smaller number of higher level substructure. This is probably because less time is needed for the condensation in the lower level case.

TABLE 3. K Size and CPU Time while Using 4 Substructures

Program	Mesh Density	K Size	CPU Time (sec.)
FEM	A	3678	6.10
	B	22353	51.77
FEMSUB	A	3046	5.55
	B	12201	35.60

TABLE 4. K Size and CPU Time while Using Different Number of Substructures (Mesh Density B)

Number of Substructures	K Size	CPU Time (sec.)
0	22353	51.77
4	12201	35.60
16	15457	27.29

CONCLUSION

A new technique of multi-level substructuring has been developed for finite element analysis. A number of test examples have been conducted using the two dimensional finite element programs based on the new technique and the conventional one, respectively. Comparison of the test results shows that the newly developed technique can result in saving of both the core memory and the computational time, hence the total computational cost. This saving becomes more significant for the problems of larger size .

More numerical tests have to be carried out to investigate the exact effect of the substructure level on the over-all computational efficiency.

This multi-level substructuring technique is not restricted to problems with regular boundary only. For problems with irregular boundry, the substructures can be considered as superelements which can be connected to other non-uniform elements which define the boundary.

ACKNOWLEDGEMENT

The project is financially supported by the Chinese Education Commission and the Monash Research Grant.

REFERENCES

1. Dodds Jr., R. H. and Lopez, L. A., Substructuring in Linear and Nonlinear Analysis, *International Journal for Numerical Methods in Engineering*, vol. 15, pp. 583-597, 1980.

2. Furuike, T., Computerized Multiple Level Substructuring Analysis, *Computers and Structures*, vol. 2, pp. 1063-1073, 1972.

3. Hintchings, D. and Balasubramaniam, K., The Cholesky Method in Substructuring with an Application to Fracture Mechanics, *Computers and Structures*, vol. 18, pp. 417-424, 1984.

4. Kela, A., Perucchio, R. and Voelcker, H., Toward Automatic Finite Element Analysis, *Computers in Mechanical Engineering*, pp. 57-71, 1986.

5. Noor, A. K., Kamel, H. A. and Fulton, R. E., Substructuring Techniques – Status and Projections, *Computers and Structures*, vol. 8, pp. 621-632, 1978.

6. Petersson H. and Popov, E. P., Substructuring and Equation System Solutions in Finite Element Analysis,*Computers and Structures*, vol. 7, pp. 197-206, 1977. pp. 197-206.

A Solenoidal Element for Inherent Mass Conservation

A. N. F. MACK
School of Mechanical Engineering
University of Technology, Sydney
P.O. Box 123
Broadway, NSW, 2007, Australia

INTRODUCTION

In connection with finite element solutions to viscous incompressible
flows, an idea which has received scant attention is the solenoidal
approach. The attraction of this approach is the uncoupling of the
pressure from the solution. The velocity components are determined
solely from their values at a previous iteration. The difficulty with
the approach lies in the construction of an admissible element, one which
is solenoidal as it stands on its own.

The present paper promotes one such element, in fact, the first known
practical example. The triangle suggests itself as the shape of the
polygon. The solution variables, then, are representable by complete
quartics. Of the 30 equations, 6 are supplied by the imposition of the
solenoidal constraint at special integration points, the remainder are
obtained from the assignment of the nodal values in the usual manner. Of
utmost importance is the location of the integration points.

MATHEMATICAL FORMULATION

Consider the plane laminar flow of an incompressible Newtonian fluid
whose viscosity is constant. Without jeopardy to the generality of the
arguments, it is instructive to restrict discussion to creeping motion,
the differential equations for which take the form

$$u_x + v_y = 0 \tag{1}$$

$$\nabla^2 u - p_x = 0$$
$$\nabla^2 v - p_y = 0 \tag{2}$$

where p is the pressure with respect to some datum, u is the velocity in
the x-direction, v is the velocity in the y-direction, all of which have
been non-dimensionalised.

The finite element method operates, not on the differential equations, but on an integral formulation. This can be obtained from an inner product with the arbitrary variations δp, δu, δv so that

$$\int_A \delta p(u_x + v_y) \, dA + \int_A \delta u(\nabla^2 u - p_x) \, dA + \int_A \delta v(\nabla^2 v - p_y) \, dA = 0 \qquad (3)$$

where A is the integration domain. With the aid of Green's theorem,

$$-\int_A [\delta(u_x + v_y)p + \delta p(u_x + v_y)] \, dA + \int_A (\delta u_x u_x + \delta v_x v_x + \delta u_y u_y + \delta v_y v_y) \, dA$$

$$= \int_1 [\delta u(u_n - \alpha p) + \delta v(v_n - \beta p)] \, dl \qquad (4)$$

where α, β are the direction cosines for the normal n to the boundary 1.

Although often not formulated as such, the popular finite element approaches all follow this path. However, the solenoidal approach now takes the important step which recognises the advantage to be gained from the imposition, at an element level, of velocity components with zero divergence. If this can be achieved, then

$$\int_A (\delta u_x u_x + \delta v_x v_x + \delta u_y u_y + \delta v_y v_y) \, dA$$

$$= \int_1 [\delta u(u_n - \alpha p) + \delta v(v_n - \beta p)] \, dl \qquad (5)$$

wherein the pressure is eliminated from the solution. Space precludes an explanation of the transformation from this integral formulation to an algebraic formulation, but this is covered in detail in Mack (1983).

ELEMENT CONSTRUCTION

The difficulty, of course, centres on the derivation of appropriate interpolation functions. To date, elements for the implementation of the solenoidal approach have proved somewhat elusive. For a start, those which are not inherently solenoidal, that is, not solenoidal an entities by themselves, can be discounted. Elements have been suggested before, in particular by Fortin (1972), but always insufficient development is demonstrated to invite speculation on the aptness of the proposals. The element discussed here, for the first time in the published literature, is that of Mack (1983).

The triangle suggests itself as the shape of the polygon. The reason for this is the ease with which triangles can be arranged to accommodate complicated geometries, without the need for a coordinate transformation.

Added to this is the manner in which triangles can be guaranteed to produce isotropic expansions. This permits the velocity components to be approximated by complete polynomials, where both functions receive like treatment. Now, in the integral formulation, the derivatives do not exceed first order. Thus, the functions, as against their derivatives, are suitable nodal parameters. With scope for the imposition of the solenoidal constraint, the simplest possible functions are quartics.

A conforming element is defined by the assignment of the components to 12 special nodes. A solenoidal element is defined by the imposition of the constraint at 10 special points. But, with only 30 unknowns, it is not possible to construct a conforming element which is pointwise solenoidal. A compromise, however, is to impose the constraint in an average sense, a relaxation which is consistent with the transformation to the weaker statement in the integral formulation. This is achieved through the use of 6 points, at each of which Eq.(1) is satisfied, so that

$$\int_{Ae} (u_x + v_y) \, dA = 0. \tag{6}$$

For 6 points to integrate a cubic polynomial, a Gauss scheme is wanted. To avoid linear dependence, the points need to be positioned off the medians of the triangle. An arbitrary point, then, is definable by the area coordinates σ, τ for which, together with $1 - \sigma - \tau$, there are 6 permutations. Now suppose

$$\mu = \frac{\sigma}{\tau} \tag{7}$$

where

$$0 < \mu < 1. \tag{8}$$

A quadratic polynomial is integrated exactly when

$$\sigma = \frac{\mu}{2(1 + \mu + \sqrt{\mu})}$$
$$\tag{9}$$
$$\tau = \frac{1}{2(1 + \mu + \sqrt{\mu})}$$

for all values of μ. One value of μ does likewise for a cubic polynomial, as shown by Reddy (1978).

Nevertheless, considerable care is required in the selection of μ. Figure 1 presents a measure of the linear dependence as a function of μ. Figure 2 presents a measure of the integration error as a function of μ. An unfortunate, but not unexpected, result is the coincidence of the critical values. Not unexpected because Eq.(6) implies

$$\int_{le} (\alpha u + \beta v) \, dl = 0. \tag{10}$$

423

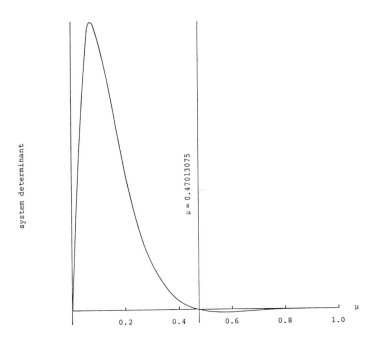

FIGURE 1. Measure of linear dependence as function of μ.

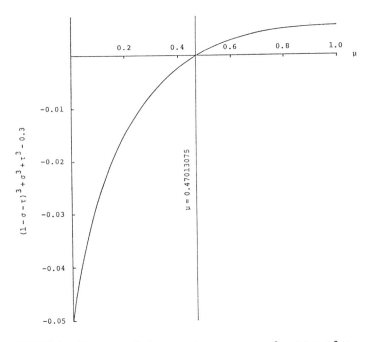

FIGURE 2. Measure of integration error as function of μ.

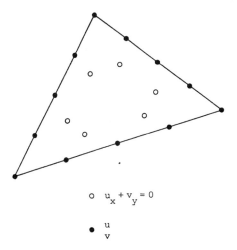

$$\circ \quad u_x + v_y = 0$$

$$\bullet \quad \begin{matrix} u \\ v \end{matrix}$$

FIGURE 3. Solenoidal element configuration.

Thus, not only can it not be exactly pointwise solenoidal, but also a
conforming element cannot even be exactly elementwise solenoidal. A
response to this dilemma, however, is to set μ equal to 0.4, the cost of
which is a maximum error of just 1% in the integration of the cubic
terms. The actual element configuration is displayed in Figure 3.

An element's performance is assessed on its imitation of the continuum.
There, a balance exists of one constraint equation for two component
equations. An ideal element is one which produces the same ratio over an
infinite domain. An element, therefore, is defective to the extent that
this does not occur. With a ratio of 0.6, the solenoidal element, as it
happens, comes close to the optimum, with the preferred bias.

SAMPLE APPLICATION

Removing the creeping flow restriction, the element can be used for the
solution of viscous flows with inertia effects, a code for which is given
in Mack (1984) and its subsequent revisions. An example of such an
application is the flow in a tunnel with a step, with variable base
bleed. This flow is characterised by the formation of a vortex in the
corner behind the step. Base bleed detaches the vortex from the face of
the step.

The problem statement is defined in Figure 4. Figure 5 to Figure 9
present results for a Reynolds number of 100, the characteristic length
for which is the step height, the characteristic speed for which is the
maximum velocity. In spite of the limitations of the discretisation, the
expected behaviour is clearly discernible.

CONCLUSIONS

In summary, then, what are the features of the element. It is

425

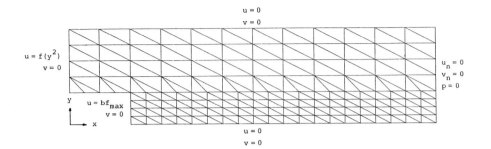

FIGURE 4. Problem statement for step flow.

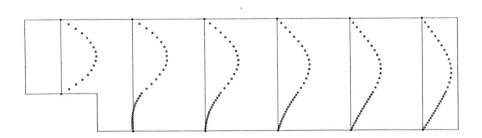

FIGURE 5. Velocity profiles for step flow at Re = 100, b = 0.

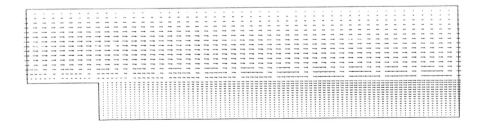

FIGURE 6. Velocity vectors for step flow at Re = 100, b = 0.

426

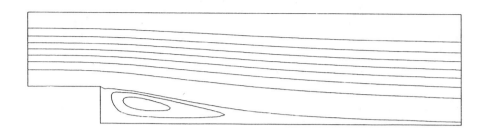

FIGURE 7. Streamlines for step flow at Re = 100, b = 0.

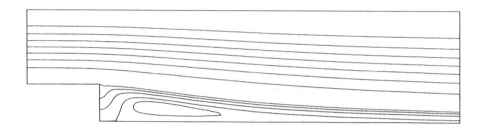

FIGURE 8. Streamlines for step flow at Re = 100, b = 0.025.

FIGURE 9. Streamlines for step flow at Re = 100, b = 0.050.

conforming. Due to the imposition of the external constraint, it has cross-coupled interpolation functions. To the knowledge of the author, this is unique. The solenoidal constraint is satisfied at 6 locations. As well, it is satisfied on an average over the element. Ignoring the small integration error, the consequences of such an approximation are these. For constant pressure, the solenoidal approach is exact within the bounds of the interpolation functions. Otherwise, it becomes exact in the limit as the size of the element tends to zero. In what is a significant test, the sample application demonstrates the worth of the approach, and of the element in particular. Moreover, their use is applicable, not just with viscous incompressible flows, but wherever the governing equations are of a similar form to the Eq.1-Eq.2 pairing.

REFERENCES

1. Fortin, M., Approximation des Fonctions a Divergence Nulle par la Methode des Elements Finis, *Proceedings of the Third International Conference on Numerical Methods in Fluid Mechanics, Paris,* pp. 99-103, 1972.

2. Mack, A.N.F., A Solenoidal Approach to Viscous Flow Simulation, Thesis, Doctor of Philosophy, University of Sydney, 1983.

3. Mack, A.N.F., The SOLFEM Code - A User Reference, Technical Report, Charles Kolling Research Laboratory, University of Sydney, 1984.

4. Reddy, C.T., Improved Three Point Integration Schemes for Triangular Finite Elements, *International Journal for Numerical Methods in Engineering,* vol. 12, pp. 1890-1896, 1978.

Finite Element Method for Resistive Plasma Stability Studies in Cylindrical Geometry

A. PLETZER and R. L. DEWAR
Department of Theoretical Physics and Plasma Research Laboratory
Research School of Physical Sciences
The Australian National University
G.P.O. Box 4
Canberra ACT 2601, Australia

INTRODUCTION

For a weakly resistive plasma, the linear stability of a magnetic configuration can be studied by using the method of asymptotic matching (Coppi, Greene & Johnson 1966). This consists in separating the plasma into two distinct classes of region called inner and outer layers. The resistivity plays an important role only in thin inner layers located at 'rational surfaces' where the magnetic field lines close on themselves after a finite number of circuits. A similar situation occurs at the critical layer in shear flow instabilities in weakly viscous fluids.
In the outer layers, on the other hand, the resistivity has negligible effect and the ideal marginal MHD equations are taken to be satisfied. The displacement of the plasma from its equilibrium position is the solution of an Euler-Lagrange equation possessing regular singular points at the magnetic axis and at rational surfaces. It is the problem of solving the outer layer equations and extracting the requisite asymptotic information as a rational surface is approached which is the subject of this paper.

A finite element method has been developed by Miller & Dewar (1986) to obtain the solution of the Euler-Lagrange equation and, by applying a 'generalised Green's function method', the asymptotic information needed for matching to the inner layer. The generalised Green's function method utilises an integral expression to evaluate the ratio of recessive and dominant solutions and is thus less sensitive to inaccuracies than a method (Dewar & Grimm 1984) based on localised behaviour. Miller & Dewar (1986) treat only an idealised, one sided problem. This paper represents a first step in generalising and applying their method to realistic geometries used for fusion plasmas. We treat a cylindrical plasma with periodic boundary conditions in the axial (z) direction, which models the topology of a toroidal plasma, such as that in a tokamak, but keeps the problem one dimensional.

THEORETICAL BACKGROUND

According to Newcomb (1960), the zero frequency displacements, normal to the magnetic flux surfaces ψ,

$$\xi \equiv \vec{\xi} \cdot \vec{\nabla}\psi \tag{1}$$

of the plasma from its equilibrium position, satisfy the Euler-Lagrange equation

$$L(\xi) \equiv \frac{d}{d\psi}\left(f(\psi) \, \frac{d\,\xi(\psi)}{d\psi} \right) - g(\psi) \, \xi(\psi) = 0 \tag{2}$$

valid for a cylindrical plasma of circular section. Here, ψ is the poloidal flux of the magnetic field

$$\vec{B} = \left(h\,\vec{\nabla}z - q(\psi)\,\vec{\nabla}\theta \right) \times \vec{\nabla}\psi \tag{3}$$

expressed in the straight field line system of coordinates (ψ, θ, z) (Dewar, Monticello & Sy 1984). In expression (3), $2\pi/h$ is the axial periodicity length and the poloidal angle θ is the geometric angle about the cylindrical axis. In Eq. (2), $f(\psi)$ and $g(\psi)$ are given by

$$f(\psi) = -\,(m - nq)$$

$$g(\psi) = K + Q^2 G - \frac{d}{d\psi}(i\,(m - nq)\,QG) \tag{4}$$

for modes $\xi(\psi, \theta, z) = \xi(\psi)\,\exp(i\,m\,\theta - i\,h\,n\,z)$. The poloidal and axial mode numbers m and n are integers. In the above expression, G, K and Q depend on the magnetic configuration, that is on equilibrium quantities such as the the pressure $p(\psi)$, the Jacobian of the system of coordinates $\mathcal{J}(\psi)$ and the current density in the z direction $j_z(\psi)$. Without entering into the details (Pletzer & Dewar 1988) of the definitions of G, K and Q, let us simply give their asymptotic behaviour as the $\psi\to0$ limit

$$f \sim -\,(m - nq)^2 \; \frac{2h\,\psi}{m^2 + \dfrac{4h^3 n^2}{j_z(0)}\,\psi} \;,$$

$$g \sim -\,(m - nq)^2 \frac{h}{2\psi} \qquad\qquad \text{for } m \neq 0$$

$$\sim O(\psi^0) \qquad\qquad\qquad\qquad \text{for } m = 0. \tag{5}$$

From (4) and (5), we see that we have two types of singular surface, one at the axis, $\psi=0$, and one at $\psi=\psi_s$, a rational surface for which $m-nq(\psi_s)$ vanishes. To leading order, (2) reduces near $\psi=0$ to

$$\frac{d}{d\psi}\left(4\,\psi\,\frac{d\xi}{d\psi} \right) - \frac{m^2}{\psi}\,\xi = 0 \qquad\qquad \text{for } m \neq 0, \tag{6}$$

whence we obtain the asymptotic behaviour $\xi \sim \psi^{\pm m/2}$ near $\psi=0$. For
m=0 [provided q(0)≠0], $\xi \sim \psi$ or ψ^0. Regularity of ξ_r at $\psi=0$
selects only the solution $\sim \psi^{+m/2}$ as physically admissible for
m≠0, or $\sim \psi$ for m=0.

The second type of regular singularity appears when the safety
factor $q=q(\psi_s)\equiv m/n$. (For simplicity we assume q to be monotonic
so that there is at most one ψ_s for given m and n.)   Let us
expand $f(\psi)$ and $g(\psi)$ in Taylor series around $\psi=\psi_s$

$$f(\psi) = \sum_{k=0}^{\infty} f_k \ (\psi-\psi_s)^{k+2}$$

$$g(\psi) = \sum_{k=0}^{\infty} g_k \ (\psi-\psi_s)^k \tag{7}$$

and use the more general Frobenius expansion for the solution

$$\xi = \sum_{k=0}^{\infty} \xi_k |\psi-\psi_s|^\alpha \ (\psi-\psi_s)^k \tag{8}$$

where α is real. We obtain after introducing expansions (7) and
(8) into Eq.(2) the recurrence formula

$$\left(f_0(\alpha+k)(\alpha+k+1)-g_0\right)\xi_k = -\sum_{k'=0}^{k-1}\left((\alpha+k')(\alpha+k+1)f_{k-k'} - g_{k-k'}\right)\xi_{k'}. \tag{9}$$

At the lowest order (k=0), expression (9) reduces to the
indicial equation

$$f_0 \alpha \ (\alpha + 1) - g_0 = 0 \tag{10}$$

whose roots are

$$\alpha^{(s,b)} = -\frac{1}{2} \pm \sqrt{\frac{1}{4} + \frac{g_0}{f_0}} \equiv -\frac{1}{2} \pm \mu. \tag{11}$$

The minus sign corresponds to the dominant (big) and the plus
sign to the recessive (small) solution. The superscripts (s) and
(b) denote the small and big solution respectively. For
simplicity we assume that μ is not close to an integer or half
integer. (A way to handle the problems arising in such cases was
suggested by Miller & Dewar 1986). At higher orders k=1,2,3...,
Eq.(9) becomes

$$\xi_k^{(b)} = -\frac{\sum_{k'=0}^{k-1}\left((\alpha^{(b)}+k')(\alpha^{(b)}+k+1)f_{k-k'} - g_{k-k'}\right)\xi_{k'}^{(b)}}{k(k-2\mu)} \tag{12}$$

for the Frobenius coefficients ξ_k of the big solution, and

$$\xi_k^{(s)} = - \frac{\sum_{k'=0}^{k-1}\left(\left(\alpha^{(s)}+k'\right)\left(\alpha^{(s)}+k+1\right)f_{k-k'} - g_{k-k'}\right)\xi_{k'}^{(s)}}{k(k+2\mu)} \tag{13}$$

for the coefficients of the small solution. We normalise the solutions by defining the leading order coefficients to be unity, i.e. $\xi_0^{(s)} \equiv \xi_0^{(b)} \equiv 1$. The general solution of Eq. (2) on one side or other of the rational surface is

$$\xi(\psi) = \text{const.}\left(\Delta\,\xi^{(s)}(\psi) + \xi^{(b)}(\psi)\right), \tag{14}$$

where Δ is the ratio between the small and big solutions, $\xi^{(s)}$ and $\xi^{(b)}$.

Owing to the regular singular point at ψ_s, the weak solution admits jumps in the two arbitrary constants in Eq. (14). The Δ coefficient applying for $\psi<\psi_s$, Δ_-, is determined from the regularity condition at $\psi=0$. The Δ coefficient for $\psi>\psi_s$, Δ_+, is determined from the boundary condition $\xi=0$ at $\psi=\psi_{max}$ (where ψ_{max} is the value of ψ at the plasma boundary) in the case of a perfectly conducting wall, or a Neuman boundary condition in the case of vacuum surrounding the plasma. The physical solution is obtained by using the information embodied in $\Delta_\pm$ as boundary conditions for the inner layer solution.

NUMERICAL APPROXIMATION OF $\Delta_\pm$

Since the solutions on either side of the rational surface are completely decoupled in this one dimensional problem we can subdivide the domain of ψ in Eq. (2) into $0<\psi<\psi_s$ and $\psi_s<\psi<\psi_{max}$ to apply the numerical analysis developed by Miller & Dewar (1986) for the one-sided problem, with only the slight complication brought about by the singularity at $\psi=0$.

The essence of this theory is to decompose ξ into a 'finite energy part' $\xi_0(\psi)$, lying within a suitably defined Hilbert space, and an 'infinite energy part' $\xi_1(\psi)$ containing the non-square-integrable singularity of the big solution. The finite energy part can then be approximated by finite elements in the normal way, while $\xi_1(\psi)$ appears in a forcing term.

The finite energy parts $\xi_0^\pm(\psi)$ of the solutions $\xi^\pm$ with support on the intervals $0<\psi<\psi_s$ (ξ^-) and $\psi_s<\psi<\psi_{max}$ (ξ^+) are defined by

$$\xi_0^\pm(\psi) \equiv \xi^\pm(\psi) - \xi_1^\pm(\psi)$$

$$\xi_0^+(\psi) \sim \Delta_+\,|\psi-\psi_s|^{-\frac{1}{2}+\mu} \qquad \text{as } \psi\to\psi_s+$$

$$\xi_0^-(\psi) \sim \Delta_-\,|\psi-\psi_s|^{-\frac{1}{2}+\mu} \qquad \text{as } \psi\to\psi_s- . \tag{15}$$

In (15), $\xi_1^\pm(\psi)$ is a function chosen to approximate the big solution (12) in the vicinity of the rational surface up to order N

$$\xi_1^\pm(\psi) = S^\pm(\psi) \, | \, \psi - \psi_s |^{-\frac{1}{2}-\mu} \sum_{k=0}^{N} \xi_{k}^{(b)} \, (\psi - \psi_s)^k, \tag{16}$$

where the shape factor $S^-(\psi)$ has been introduced in order to make ξ_1^- behave like the physical solution near $\psi=0$,

$$S^-(\psi) \equiv \left[1 - \left(\frac{\psi_s - \psi}{\psi_s} \right)^{N+1} \right]^{\frac{m}{2}} H(\psi_s - \psi) \, , \tag{17}$$

where $H(.)$ is the unit step function. (For simplicity we shall always assume $m \neq 0$, the handling of the case $m=0$ being obvious.) We take $S^+(\psi)$ to be simply $H(\psi-\psi_s)$, though it could be used to make $\xi_1^+(\psi_{max})=0$ if desired. The Frobenius expansion (16) is truncated at $N \geq 2\mu+1$ in order to guarantee the asymptotic small behavior (15), i.e. so that corrections to the big solution do not overwhelm the leading order behaviour of the small solution.

Now the equations to solve become

$$L(\xi_0^\pm) = -L(\xi_1^\pm) \tag{18}$$

with boundary conditions $\xi_0^+(\psi_{max}) = -\xi_1^+(\psi_{max})$, in the case of a wall on the plasma, and $\xi_0^-(\psi) \sim a_0 \psi^{m/2}$ at the origin with undetermined coefficient a_0. The right hand sides $L(\xi_1^\pm)$ act as known forcing terms, with the responses $\xi_0^\pm$ to be determined using the finite element expansion

$$\xi_0(\psi) = \sum_{i=1}^{M} \Xi_i \, e_i(\psi) \tag{19}$$

where the basis functions $e_i(\psi)$ are piecewise linear (Fig. 1), except for the elements with central node points $i=1, s-1, s+1$, near the singular points, where we take

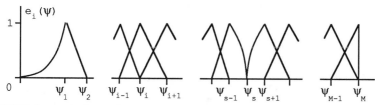

FIGURE 1. Finite elements used as basis functions for expanding $\xi_0^\pm$ showing the use of fractional power dependence near the singular points $\psi=0$ and $\psi=\psi_s$. The last node is at $\psi_M \equiv \psi_{max}$. Note that there is no element $e_s(\psi)$.

$$e_1(\psi) = \left(\frac{\psi}{\psi_1}\right)^{\frac{m}{2}} \qquad \text{for } 0 < \psi < \psi_1 \ ,$$

$$e_{s-1}(\psi) = \left(\frac{\psi_s - \psi}{\psi_s - \psi_{s-1}}\right)^{-\frac{1}{2}+\mu} \qquad \text{for } \psi_{s-1} < \psi < \psi_s \ ,$$

$$e_{s+1}(\psi) = \left(\frac{\psi - \psi_s}{\psi_{s+1} - \psi_s}\right)^{-\frac{1}{2}+\mu} \qquad \text{for } \psi_s < \psi < \psi_{s+1} \ . \tag{20}$$

One way to compute the ratios Δ_+ and Δ_- is by estimating

$$\Delta_\pm = \lim_{\psi \to \psi_s \pm} \frac{\xi_0^\pm(\psi) \ |\psi - \psi_s|^{-2\mu}}{\xi_1^\pm(\psi)} \tag{21}$$

which is called the direct method, and has the advantage of simplicity. In contrast to the direct method, the generalised Green's function method expresses the ratio in terms of an integral rather than the values, inaccurately determined, of the solution near the singular point. By multiplying Eq.(18) with a test function ϕ^-, and integrating by parts twice over the interval $0 \leq \psi < \psi_s$, we obtain

$$\int_0^{\psi_s^-} d\psi \, L(\phi^-)\xi_0^- + \left(f\frac{d\xi_0^-}{d\psi}\phi^- - f\xi_0^- \frac{d\phi^-}{d\psi}\right)\Bigg|_{\psi=0}^{\psi=\psi_s^-} = -\int_0^{\psi_s^-} d\psi \, L(\xi_1^-)\phi^- \ . \tag{22}$$

If now we choose ϕ^- such that

$$\phi^-(\psi) \propto \psi^{\frac{m}{2}} \qquad \text{as } \psi \to 0$$

$$\phi^-(\psi) \sim |\psi - \psi_s|^{-\frac{1}{2}-\mu} \qquad \text{as } \psi \to \psi_s^- \tag{23}$$

in order that the endpoint contribution vanishes while the rational surface contribution is finite, Eq.(22) reduces to

$$2\mu\Delta_- = \int_0^{\psi_s^-} d\psi \left(L(\phi^-)\xi_0^- + L(\xi_1^-)\phi^-\right) \ . \tag{24}$$

On the right-hand side of the rational surface, a development, in all points similar but with the requirement $\phi^+(\psi_{max}) = 0$, leads to

$$2\mu\Delta_+ = \int_{\psi_s^+}^{\psi_{max}} d\psi \left(L(\phi^+)\xi_0^+ + L(\xi_1^+)\phi^+\right) + \left[f\xi_0^+ \frac{d\phi^+}{d\psi}\right]_{\psi=\psi_{max}} \ . \tag{25}$$

The functions $\phi^\pm$ are called generalised Green's functions because they provide kernels which extract the desired information when applied to the finite element solution.

TEST CASE

We choose the following equilibrium characterised by the pressure $p(\psi)$ and the safety factor $q(\psi)$ profiles

$$p(\psi) = \frac{p_0 \left[1 - \left(\frac{\psi}{\psi_{max}} \right)^2 \right]}{\left[1 + \left(\frac{\psi}{\psi_{max}} \right)^2 \right]^2} \tag{26}$$

$$q(\psi) = q_0 \left[1 + \left(\frac{\psi}{\psi_{max}} \right)^2 \left(\frac{q_a}{q_0} - 1 \right) \right] \tag{27}$$

with ψ_{max} normalised to 1 and $p_0 = 0.25$. By setting $q_0 = q(\psi=0) = 1$ and $q_a = q(\psi_{max}) = 3$, the rational surface is localised at $\psi = \psi_{max}/2$ for the $m=3$ and $n=2$ mode. The corresponding Frobenius exponents of this equilibrium are $\alpha^{(b)} = -0.879$ and $\alpha^{(s)} = -0.121$.

In order to improve the convergence, the width of the singular elements e_1, e_{s-1} and e_{s+1} as well as the neighbouring elements are scaled, following Miller and Dewar (1986), to the number of mesh points $M_\pm$ on the left, resp.right, hand side of ψ_s. In particular we have

$$\psi_1 = 0.007 \ (M_-/256)^{-\frac{2}{m}} \ \psi_s$$

$$\psi_s - \psi_{s-1} = c_{mesh} \ (M_-/256)^{1-\frac{1}{\mu}} \ (\psi_s - \psi_1)$$

$$\psi_{s+1} - \psi_s = c_{mesh} \ (M_+/256)^{1-\frac{1}{\mu}} \ (\psi_{max} - \psi_s) \ , \tag{28}$$

the value 0.007 being imposed by the accuracy of the equilibrium solving part of the code. A grading mesh $\psi_{s\pm i} = \psi_s \pm (i/M_\pm)^{1/\mu}$ ($i=1,2,3...$) is centred around the rational surface for mesh nodes in the interval $\psi_s - x_{grad} < \psi_{s\pm i} < \psi_s + x_{grad}$.

Figure 2 compares the convergence of the direct method (d- and d+) with the generalised Green's function (GGF) methods (g-, gopt-, g+ and gopt+) for $0 < \psi < \psi_s$ (minus sign) and $\psi_s < \psi \leq 1$ (plus sign). The distinction between g and gopt is due to a different choice of Green functions $\phi^{\pm}$. For the "g" method we take $\phi^{\pm} = \xi_1^{\pm}$ while for "gopt" $\phi^{\pm} = \xi = \xi_0^{\pm} + \xi_1^{\pm}$ is chosen. The limit in Eq. (21) of the direct method is approximated by taking $\psi = \psi_{s\pm 1}$.

We note first the superiority of the GGF method in terms of convergence (error $\propto M_\pm^{-2}$ for gopt-,g+ and gopt+) and accuracy. The relatively important error of g- can be attributed to the $L(\phi^-)\xi_0^-$ term in Eq. (24). In the left hand side region, the integral of the latter is almost exactly balanced by the integral of $L(\xi_1^-)\phi^-$, leading to a larger relative error of Δ_-.

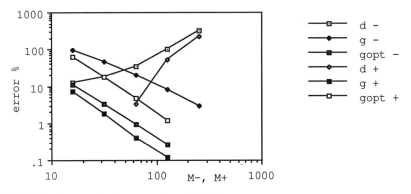

FIGURE 2. cmesh=1. xgrad = 0.025

ACKNOWLEDGMENT

The calculations upon which this work is based where carried out using the Fujitsu VP100 of the Australian National University Supercomputer Facility.

CONCLUSION

The generalised Green's function method has been applied to a test equilibrium. The matching conditions Δ_- and Δ_+ obtained by this method have proved to be more accurate. The direct method does not converge for a negative Frobenius exponent $\alpha^{(s)} = -1/2 + \mu$. A mesh grading in $1/\mu$ around the rational surface ensures the $M_\pm^{-2}$ convergence of the GGF method.

REFERENCES

1. Coppi, B., Greene, J.M. & Johnson, J.L., Resistive Instabilities in a Diffuse Linear Pinch, *Nucl. Fusion*, vol.6, pp.101-117, 1966.

2. Dewar, R.L. & Grimm, R.G., Singular Finite Element Methods in Plasma Stability Computations - a Simple Model, in *Computational Techniques and Applications: CTAC*-83, eds. J. Noye & C. Fletcher, pp.730-740, North Holland, Amsterdam, 1984.

3. Dewar, R.L., Monticello, D.A. & Sy, W.N.-C., Magnetic Coordinates for Equilibria with a Continuous Symmetry, *Phys. Fluids*, vol.27, pp.1723-1732, 1984.

4. Miller, A.D. & Dewar, R.L., Galerkin Method for Differential Equations with Regular Singular Points, *J. Comput. Phys.*, vol.66, pp.356-390, 1986.

5. Newcomb, W.A., Hydromagnetic Stability of a Diffuse Linear Pinch, *Ann. Phys.*, vol. 10, pp.232-267, 1960.

6. Pletzer, A. & Dewar, R.L., Reduction of the Generalised Newcomb Equation to Newcomb's form by using a Canonical Transformation, ANU Plasma Research Lab. Report ANU/TPP88/05, December 1988.

A Flux-Split Solution Procedure
for the Euler Equations

H. S. PORDAL, P. K. KHOSLA, and S. G. RUBIN
Departement of Aerospace Engineering and Engineering Mechanics
University of Cincinnati
Cincinnati, Ohio 45221, USA

INTRODUCTION

Present day need is for fast and accurate prediction of flow behaviour over complex geometries. Toward this goal, steady and unsteady flows over aerodynamic geometries of interest are predicted by a flux splitting procedure. The present study concerns the computation of inviscid steady and unsteady internal and external flows. The governing equations are written in general curvilinear co-ordinates and discretized using a flux vector splitting imbedded in a global pressure relaxation technique developed by Rubin (1986) and Khosla (1983). The global pressure relaxation procedure considered here was originally developed for the solution of the asymptotic Reduced Navier Stokes equations (RNS), and applied for flows with strong viscous/inviscid interaction. However, the same methodology is also applicable for the inviscid Euler equations. The boundary value nature of these equations is associated with the subsonic elliptic streamwise acoustic flux splitting in the axial or mainstream direction. The basic formulation remains the same for viscous and inviscid subsonic, transonic and supersonic flows, see Rubin (1986,1982,1983), Khosla (1983,1987) and Ramakrishnan (1984), and adapted to three dimensional flow computations by Cohen and Khosla (1989), Himansu et al (1989). Ramakrishnan and Rubin (1986,1987) have extended the global relaxation process for low subsonic unsteady flows at large Reynolds number. Significant features of this procedure are that it is independent of the mach number and Reynolds number and does not require the addition of artificial viscosity, other than that associated with the discretization. This is minimized on fine grids, see Rubin and Himansu (1989).

The first part of the present study deals with the computation of two dimensional external flow for a range of mach numbers. Oscillation free strong normal and oblique shocks are captured over three mesh points. In the second part, the applicability of the flux vector relaxation technique for unsteady flows involving moving shocks is demonstrated. For these computations, the relaxation procedure is replaced with a time consistent direct sparse matrix solver. Bender (1988) has demonstrated the applicability and efficacy of direct sparse matrix solvers and Newton iteration for the numerical solution of fluid problems. The choice of the direct solver is dictated by considerations of stability, robustness, accuracy and time consistency. For steady computations, the solution technique should permit large time increments and have strong convergence properties; whereas, for transient flows, time consistency plays a major role. Implicit consistent iterative procedures usually do not have strong convergence properties and may require added transient or steady state artificial viscosity see Khosla and Rubin (1987). To retain the simplicity and robustness of the time marching procedure a direct solver is applied. Added artificial viscosity is not required.

GOVERNING EQUATIONS

The conservation form of the Euler equations are written in general non-orthogonal curvilinear co-ordinates (ξ, η). This form of the equations is capable of adapting to any type of grid generator.

Continuity Equation

$$(\rho g)_\tau + (\rho g U)_\xi + (\rho g V)_\eta = 0.$$

X Momentum Equation

$$[\rho g(U X_\xi + V X_\eta)]_\tau + (\rho g U^2 X_\xi)_\xi + (\rho g U V X_\eta)_\xi + (\rho g U V X_\eta)_\eta + (\rho g V^2 X_\eta)_\eta +$$
$$(P Y_\eta)_\xi - (P Y_\xi)_\eta = 0 \tag{1}$$

Y Momentum Equation

$$[\rho g(V Y_\eta + U Y_\xi)]_\tau + (\rho g U^2 Y_\xi)_\xi + (\rho g U V Y_\eta)_\xi + (\rho g U V Y_\xi)_\eta + (\rho g V^2 Y_\eta)_\eta$$
$$+ (P X_\xi)_\eta - (P X_\eta)_\xi = 0.$$

Energy Equation

$$(\rho g h)_\tau + (\rho g h U)_\xi + (\rho g h V)_\eta = \{(\gamma-1)M_\infty^2 / [1+.5(\gamma-1)M_\infty^2]\}(P g)_\tau$$

Equation of State

$$P + (\gamma-1)\rho V^2/(2\gamma) = (\gamma-1)[1/((\gamma-1)M_\infty^2)) + .5] (\rho h/\gamma)$$

U and V are the contravariant velocity components in the ξ and η directions respectively; ρ is the density; P is the pressure; h is the total enthalpy; γ is the ratio of specific heats and M_∞ is the free stream mach number. The quantities X_ξ, X_η, Y_ξ, Y_η are metrics associated with the co-ordinate transformation; g is the jacobian of transformation and τ is time.

Distances have been normalized with respect to a reference chord or channel width, the velocities, density, temperature, total enthalpy are non-dimensionalized with respect to the corresponding freestream values; the pressure is non-dimensionalized with respect to twice the free stream dynamic pressure.

BOUNDARY CONDITIONS

At the inflow U, V, ρ, P and h are all prescribed. At the outflow, for external flow, the negative eigenvalue fluxes are neglected. For internal flow computations, the back pressure is specified at the outflow.
Far from the surface, uniform flow conditions are imposed and at the surface, zero normal velocity or injection is specified.

DISCRETIZATION

The difference scheme has previously been discussed in earlier references by Rubin (1986), Khosla (1983) and Ramakrishnan (1984). All ξ derivatives (except the P_ξ term) are backward differenced and all η derivatives are central differenced. Central differencing for the normal convective terms (η derivatives) in the x momentum equation, works quite well for normal shocks, but leads to oscillations ahead of strong oblique shocks. This problem is resolved by the following averaging:

$\alpha\{$ (x momentum)$_{\text{centered at } j-1/2}$ + (x momentum)$_{\text{centered at } j+1/2}$ $\}$ +

$(1-\alpha)\{$ (x momentum)$_{\text{centered at } j}$ $\}$ = x momentum discretization.

The x momentum equation written at $j\pm1/2$ is a two point equation and eliminates oscillations in the y direction across oblique shocks. The parameter α is chosen such that the monotonicity of the solution is assured see Li (1983). The axial pressure gradient term is written as:

$P_\xi = (\omega_{i-1/2})(P_i-P_{i-1})/\Delta\xi_i$ + $(1.-\omega_{i+1/2})(P_{i+1}-P_i)/\Delta\xi_{i+1}$

The parameter ω is computed as follows:
For unsteady flows where a differential form of the energy equation is employed, see flux vector analysis by Rubin (1988).

$\omega = M_\xi^2$ for $M_\xi < 1.$
$\omega = 1.$ for $M_\xi > 1.$

For computations where a constant stagnation enthalpy condition is used, in lieu of the differential energy equation, the parameter ω is given by the following relation first obtained by Vigneron (1978) and is also recovered in the analysis by Rubin (1988).

$\omega = \gamma M_\xi^2/(1.+(\gamma-1.)M_\xi^2)$ for $M_\xi < 1.$
$\omega = 1.$ for $M_\xi > 1.$

This flux form of the streamwise pressure gradient term is capable of capturing very sharp normal shocks, e.g. three grid points. It should be noted that the flux splitting is employed only in the main flow or ξ direction. In the normal, and/or secondary flow direction, central differencing is applied. The resulting scheme has minimal artificial viscosity. For blunt nosed bodies, the present version of flux vector splitting should be employed in both co-ordinate directions, this is currently under investigation.

For external flows, the free stream pressure is specified as the far field condition on pressure; However, for internal flows we have a zero normal velocity condition at both boundaries (j=1, and j=jmax). A pressure boundary condition does not exist. In the present study, the zero injection condition is directly imposed at both boundaries and the wall pressures are computed directly. This is accomplished by regrouping the equations near the boundary. More details on applications to internal flows are covered by Reddi (1988).

SOLUTION PROCEDURE

The discretized equations are quasilinearized using Newton's method and written in a nine point star in delta form.

$$A_{ij}\delta\phi_{ij-1} + B_{ij}\delta\phi_{ij} + C_{ij}\delta\phi_{ij+1} + D_{ij}\delta\phi_{i-1j} + E_{ij}\delta\phi_{i+1j} + AM_{ij}\delta\phi_{i-1j-1} +$$
$$CM_{ij}\delta\phi_{i-1j+1} + EM_{ij}\delta\phi_{i+1j-1} + EP_{ij}\delta\phi_{i+1j+1} = G_{ij} \qquad (2)$$

where $\delta\phi$ is the solution vector and the coefficients A_{ij} ... EM_{ij} are (5x5) matrices and G_{ij} is a (5x1) matrix . For steady supersonic flows E_{ij}, EM_{ij} and EP_{ij} are zero and the method reduces to a standard initial value technique.

For unsteady flow, the system (2) is solved with a sparse matrix direct solver. The Yale Sparse Matrix Package (YSMP), developed by Eisenstadt (1977) and modified for coupled systems and the boundary conditions detailed previously is applied here. This is an efficient solver as it stores only non-zero elements, and reorders equations to minimize fill-in during LU decomposition. This solver has previously been successfully employed for steady subsonic and transonic flow computations by Bender (1988).

In order to capture the correct transient behaviour the complete flow field is computed at each time increment. The application of a direct solver can be extremely resource and time intensive therefore a pseudo iterative technique developed by Himansu et al (1989) for three dimensional flows is used. If the coefficient matrices at several consecutive time levels are sufficiently similar, then the LU decomposition at the first time level is stored and used in the back-substitution step for the subsequent time levels. LU decompositions are only performed at time levels where the previous LU decomposition is inadequate for convergence. The procedure is further supplemented by a correction procedure due to Hotteling, as described by Feddeeva (1959), this allows for changes in the jacobian at the different time levels.

RESULTS

Steady Flow Computations

Steady flow over a 10 % thick parabolic airfoil is computed for various mach numbers varying from almost incompressible (M_∞ = .01) to high

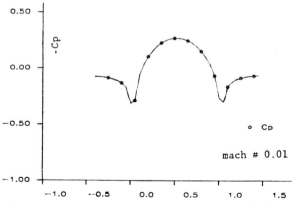

FIGURE 1A. Cp for 10% Parabolic Arc Airfoil

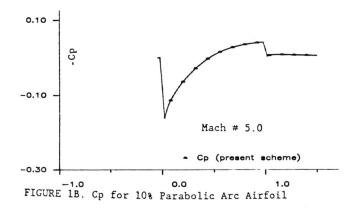

FIGURE 1B. Cp for 10% Parabolic Arc Airfoil

supersonic (M_∞ = 5.). The results are presented in figures.1a-b. It must be noted that the same Euler code has been used to cover the entire mach number range.

Unsteady Flow Computations

The unsteady formulation is used to compute the flow in an engine inlet. At supersonic flight mach numbers the performance of an aircraft inlet is affected by the nature and location of the terminal shock in the inlet/diffuser. The shock pattern and the location in the inlet are critical to the performance and stability of the diffuser flow. For certain flight conditions and diffuser design a shock can move ahead of the cowling so that inlet unstart occurs. This causes a sharp reduction in mass flow and pressure recovery and an associated large drag increase. The shock can be swallowed (restart) by increasing the mach

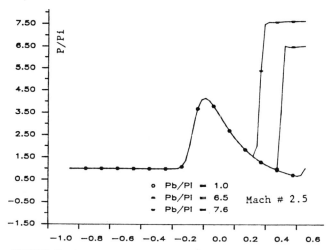

FIGURE 2. Pressure Ratio (P/Pi)

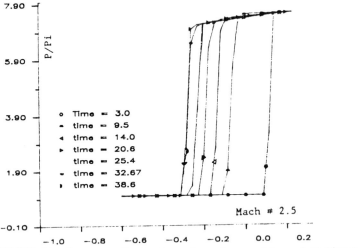

FIGURE 3A. Pressure Variation Along Center Line of Diffuser

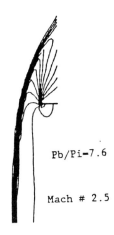

Pb/Pi=7.6

Mach # 2.5

FIGURE 3B. Pressure Contours (unstarted diffuser)

number or by increasing the throat area, the performance of the inlet is then optimized. The unstart and restart of a supersonic inlet are investigated herein.

As a first step, a converging diverging geometry is considered. Supersonic flow through a diffuser geometry was computed for a free stream mach number of 2.5 and several back pressure ratios. The terminal shock location depends on the back pressure. As the back pressure is increased the shock moves towards the throat (fig.2). For a sufficiently large back pressure the mass flow behind the normal shock would be larger than the maximum allowable mass for the given throat area. As a result, the shock would be expelled out of the inlet. This results in spillage of the excess mass and unstart of the inlet. To understand this phenomena a simple geometry is investigated. Supersonic flow between two

442

parallel plates with a sufficiently high back pressure is computed. A normal shock exists at the exit of the geometry. Since the back pressure is sufficiently high, the mass flow behind the shock is greater than allowable and, as a result the normal shock moves towards the inlet. The inlet eventually unstarts and the shock appears as a curved bow shock. Figure 3a depicts the shock location with time. For a given mach number and geometry the shock stand off distance depends on the back pressure. Figure 3b shows the steady state pressure contours. We clearly see a curved bow shock standing at a distance from the inlet.

The code is currently being modified to include the viscous RNS terms. The effects of boundary layer interaction on shock movement and inlet unstart are to be considered.

ACKNOWLEDGEMENT

The work has been performed for the NASA Lewis Research Center (T.Benson, Technical Monitor) under Grant No. NAG 3-716 The computations have been performed on the Cray X-MP/28 at the Ohio Supercomputer center.

REFERENCES

1. Rubin, S.G., and Reddy, D.R., Analysis of Global Pressure Relaxation for Flows with Strong Interaction and Seperation, *Computers and Fluids*, vol.11, no.4, pp. 281-306, 1986.

2. Khosla, P.K., and Lai, H.T., Global PNS Solutions for Subsonic Strong Interaction Flow over a Cone-Cylinder Boat-Tail Configuration, *Computers and Fluids*, vol.11, no.4, pp. 325-339, 1983.

3. Rubin, S.G., A Review of Marching Procedures for PNS Equations, *Proceedings of a Symposium on Numerical and Physical Aspects of Aerodynamic Flows*, Springer-Verlag, CA, pp. 171-186, 1982.

4. Rubin, S.G., and Reddy, D.R., Global PNS Solution for Laminar and Turbulent flows, *6th Computational Fluid Dynamics Conference*, AIAA 83-1911, Denver, MA, 1983.

5. Khosla, P.K., and Lai, H.T., Global Relaxation Procedure for Compressible Solutions of the Steady-State Euler Equations, *Computers and Fluids*, vol.15, no.2, pp. 215-218, 1987.

6. Ramakrishnan, S.V., and Rubin, S.G., Global Pressure Relaxation for Steady, Compressible, Laminar, Two Dimensional Flows with Full Pressure Coupling and Shock Waves., Report AFL 84-100, University of Cincinnati, 1984.

7. Cohen, R., and Khosla, P.K., Three Dimensional Reduced Navier Stokes Solutions for Subsonic Separated and Non Separated Flows, using a Global Pressure Relaxation Procedure, Accepted for publication, *International Journal for Numerical Methods in Fluids*, 1989.

8. Himansu, A., Khosla, P.K., and Rubin, S.G., Three Dimensional Recirculating Flows, 27th Aerospace Sciences Meeting, AIAA 89-0552, Reno, Nevada, 1989.

9. Ramakrishnan, S.V., and Rubin, S.G., Numerical Solution of Unsteady Compressible Reduced Navier Stokes Equations, 24th Aerospace Sciences Meeting, AIAA 86-0205, Reno, Nevada, 1986.

10. Ramakrishnan, S.V., and Rubin, S.G., Time Consistent Pressure Relaxation Procedure for Compressible Reduced Navier Stokes Equations, *AIAA journal,* vol 25, no.7, pp. 905-913, 1987.

11. Rubin, S.G., and Himansu, A., Convergence Properties of High Reynolds Number Separated Flow, Accepted for publication, *International Journal for Numerical Methods in Fluids*, 1989.

12. Bender, E.E., The use of Direct Sparse Matrix Solver and Newton Iteration for the Numerical Solution of Fluid Flow, Ph'd thesis, University of Cincinnati, 1988.

13. Khosla, P.K., and Rubin, S.G., Consistent Strongly Implicit Iterative Procedure for Two Dimensional Unsteady and Three Dimensional Space-Marching Flow Calculations, *Computers and Fluids*, vol.15, no.4, pp. 361-377, 1987.

14. Li, Z.C., Zhan, L.J., Wang, H.L., Difference Methods of Flow in Branch Channel, *Journal of Hydraulic Engineering*, vol.109, no.3, pp. 424-447, 1983.

15. Rubin, S.G, Reduced Navier Stokes/Euler Pressure Relaxation and Flux Vector Splitting, *Computers and Fluids*, vol.16, no.4, pp. 285-290, 1988.

16. Vigneron, Y., Calculation of Supersonic Viscous Flow over Delta Wings with Sharp Leading Edges, AIAA 78-11377, Seattle, Washington, 1978.

17. Reddy, D.R., and Rubin, S.G., Consistent Boundary Condition for Reduced Navier Stokes Scheme Applied to Three Dimensional Internal Viscous Flows, 26th Aerospace Sciences Meeting, AIAA 88-0714, Reno, Nevada, 1988.

18. Eisenstadt, S.G., Gursky, M.C., Schultz, M.H., and Sherman, A.H., Yale sparse matrix package II. The Non-Symmetric Codes, Report 114, Yale University Department of Computer Science, 1977.

19. Feddeeva, V.N., *Computational Methods of Linear Algebra*, 2d-ed., pp. 99-109, Dover publication, New York, 1959.

444

Numerical Modeling of the Flow of Granular Solids through Silos

L. C. SCHMIDT and Y. H. WU
Department of Civil and Mining Engineering
University of Wollongong
Wollongong, NSW 2500, Australia

INTRODUCTION

The storage, transport and handling of bulk materials are a major interest to primary industries such as agriculture, mining, and refractory material and cement manufacture and storage. The discharge rate of bulk solids from storage silos and the structural stability of silos are two important aspects to be considered in silo design.

In addition to experimental and analytical studies such as those by Arnold et.al. (1978) and Jenike and Johanson (1968), the finite element method has been used to predict the discharge rate of bulk solids from silos and the structural response of silos during discharge of materials, such as those methods due to Haussler and Eibl (1984), Runesson and Nilsson (1986), and Schmidt and Wu (1989). However, in the models mentioned the interaction of flowing solids with the flexible walls of silos is ignored and a rigid wall condition is assumed. Moreover, these models are not capable of predicting the internal stress resultants in silo walls.

In this paper shell theory is employed for the analysis of a silo wall. Both the internal forces and the bending moments in the silo wall are considered in the analysis. The bulk material domain is divided into a finite number of quadrilateral elements, and the silo wall is discretized into a number of shell elements. A numerical procedure which couples the finite element formulae for the motion of bulk solids with the finite element formulae for the motion of the shell has been developed and presented in the paper.

CONSTITUTIVE MODEL OF BULK SOLIDS

As bulk granular materials may in some aspects be regarded as solid and in others as a fluid, an incremental viscoplastic constitutive law is used to describe the material behaviour, which is in the form of

$$\dot{\underline{\sigma}}(\underline{d}, \dot{\underline{d}}) = \dot{\underline{\sigma}}_s(\underline{d}) + \dot{\underline{\sigma}}_v(\dot{\underline{d}}) \tag{1}$$

where

$$\dot{\underline{\sigma}} = \frac{\partial \underline{\sigma}}{\partial t} + \nabla \underline{\sigma} \cdot \underline{v} + \underline{\sigma}\,\underline{w} - \underline{w}\,\underline{\sigma}$$

445

$\dot{\underline{\sigma}}_s$ is the rate independent part of stress rate, which is associated with the material elastic-plastic behaviour. A Lade (1977) elastic-plastic model with two yield surfaces is employed to relate the stress rate with strain rate, i.e.

$$\dot{\underline{\sigma}}_s = \underline{H}\,\underline{d} \tag{2}$$

$\dot{\underline{\sigma}}_v$ is the rate dependent part of the stress rate, which is associated with the viscous behaviour of the flow of bulk materials. A rate dependent constitutive law proposed by Haussler and Eibl is employed for the computation of this part of the stress rate, i.e.,

$$\dot{\underline{\sigma}}_v = \underline{G}\,\dot{\underline{d}} \tag{3}$$

with

$$\dot{\underline{d}} = \frac{\partial \underline{d}}{\partial t} + \nabla \underline{d}.\ \underline{v} + \underline{d}\ \underline{w} - \underline{w}\ \underline{d}$$

Regarding eq.(1) to eq.(3), the stress rate in an Eulerian frame of reference can be determined as

$$\frac{\partial \underline{\sigma}}{\partial t} = \underline{H}\,\underline{d} + \underline{G}\,\frac{\partial \underline{d}}{\partial t} - (\underline{\sigma}\,\underline{w} - \underline{w}\,\underline{\sigma} + \nabla\underline{\sigma}\,\underline{v}) + \underline{G}\,(\underline{d}\,\underline{w} - \underline{w}\,\underline{d} + \nabla\underline{d}\,\underline{v}) \tag{4}$$

FINITE ELEMENT FORMULATION

The analysis deals with the axisymmetric filling and discharge of the silo, which is a common practice.

The composite system of the bulk solids and the silo wall is treated as a continuum discretized by a number of finite elements.

The bulk solid media is replaced by a number of axisymmetric quadrilateral ring elements connected at a finite number of circumferential joints or nodes. As the behaviour of the bulk material and the bin wall is axisymmetric, the displacements considered are those components in the radial (r) and longitudinal (z) directions only. Tangential displacements do not exist and the strains and stresses do not vary in the tangential direction. Thus the axisymmetric body of the bulk material can be viewed mathematically as two-dimensional in nature, as shown in Fig.1.

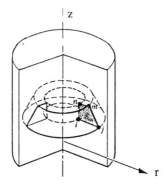

FIGURE 1. Element of an axisymmetric solid

The finite element formulae for the motion of bulk materials with large deformations can be obtained in an Eulerian frame of reference by using the virtual work principle and a finite element discretized procedure (Haussler and Eibl, 1984) as

$$\underline{M}^g \, \dot{\underline{a}} + \underline{M}_c \underline{a} + \underline{C} \, (\underline{a} - \underline{a}^o) + \Delta t \, \underline{K}^g \underline{a} = \underline{F}^g + \underline{F}^g_{gs} \qquad \text{or}$$

$$\underline{\Psi}^g (\underline{a}) = \underline{F}^g + \underline{F}^g_{gs} - \frac{1}{\Delta t} \underline{M}^g (\underline{a} - \underline{a}^o) - \underline{M}_c \underline{a} - \underline{C} \, (\underline{a} - \underline{a}^o) - \Delta t \, \underline{K}^g \underline{a} \qquad (5)$$

where $\underline{a}$ = nodal velocity, $\underline{M}^g$ = mass matrix, $\underline{M}_c$ = a matrix due to the geometrical nonlinearity, $\underline{C}$ = viscous matrix, $\underline{K}$ = stiffness matrix, $\underline{F}^g$ = nodal forces due to the gravity of the bulk granular material, $\underline{F}^g_{gs}$ = force due to the friction between the bulk granular material and the silo wall.

At an instant of time t^{n+1}, solution of eq.(5) may be found by a modified Newton-Raphson Scheme:

$$\underline{A}^g \left(\overset{i+1}{\underline{a}}{}^{n+1} - \overset{i}{\underline{a}}{}^{n+1} \right) = \underline{\Psi}^g (\overset{i}{\underline{a}}{}^{n+1}) \qquad (6)$$

with

$$\underline{A}^g = \frac{1}{\Delta t} \underline{M}^g + \underline{M}^n_c + \underline{C} + \Delta t \, \underline{K}^{gn}$$

The silo wall is divided into a series of conical frustums connected at their edges. As the shell and the loading are axisymmetric, the displacement of a point on the shell is uniquely determined by two components u and w in the longitudinal tangential and normal directions respectively. Lateral tangential displacement does not exist and the internal stress resultants (N_s, N_θ, M_s and M_θ) shown in Fig.2 do not vary in the lateral tangential direction. Thus the element becomes one dimensional, as shown in fig. 3.

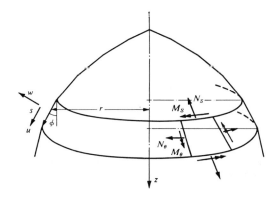

FIGURE 2. Axisymmetric shell and displacements and stress resultants (after Zienkiewicz, 1979)

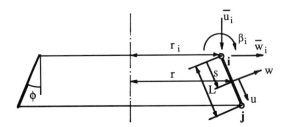

FIGURE 3. Element of an axisymmetric shell (after Zienkiewcz, 1979)

The displacement model within the element is assumed to be

$$u = \alpha_1 + \alpha_2 s \tag{7a}$$

$$w = \alpha_3 + \alpha_4 s + \alpha_5 s^2 + \alpha_6 s^3 \tag{7b}$$

The details of the derivation of the element stiffness matrix, the strain - displacement matrix and the stress-strain matrix have been presented by Zienkiewicz (1977), and the finite element formulae for the dynamic problem can be expressed in an incremental form as

$$\underline{M}^s \dot{\underline{a}} + \Delta t \, \underline{K}^s \underline{a} = \underline{F}^s + \underline{F}^s_{gs} \qquad \text{or}$$

$$\underline{\psi}^s (\underline{a}) = \underline{F}^s + \underline{F}^s_{gs} - \frac{1}{\Delta t} \underline{M}^s (\underline{a} - \underline{a}^o) - \Delta t \, \underline{K}^s \underline{a} \tag{8}$$

By using the same procedure as that for the solution of eq.(5), the solution of eq.(8) can be obtained from:

$$\underline{A}^s \left({}^{i+1}\underline{a}^{n+1} - {}^{i}\underline{a}^{n+1} \right) = \underline{\psi}^s ({}^{i}\underline{a}^{n+1}) \tag{9}$$

with

$$A^s = \frac{1}{\Delta t} \underline{M}^s + \Delta t \, \underline{K}^{sn}$$

The motion of the bulk granular material and the motion of the silo structure are coupled by the compatibility requirements on the interface between the two materials:
(1) displacements normal to the interface are equal;
(2) sliding along the interface is allowed, however, friction forces between the two materials have to be considered as external loads and applied to the boundaries of both materials.

To implement the compatibility condition of displacements, the equations of motion for the domain of the flowing granular material are partitioned into boundary velocity components normal to the interface, denoted by subscript B, and all other components denoted by subscript G. In the same way, the equation for the motion of the silo wall is partitioned into the displacement components normal to the interface, denoted by subscript B, and all other components, denoted by subscripts S. Therefore, each node in the bulk material domain has two degrees of freedom, three degrees of freedom in the shell domain, and four degrees of freedom in the interface.

The combination of partitioned equations (6) and (9) gives the coupled equations of motion for the bulk solids - silo system:

$$
\begin{pmatrix}
\underset{\sim}{A}{}^{g}_{GG} & \underset{\sim}{A}{}^{g}_{GB} & 0 \\
\underset{\sim}{A}{}^{g}_{BG} & \underset{\sim}{A}{}^{g}_{BB}+\underset{\sim}{A}{}^{s}_{BB} & \underset{\sim}{A}{}^{s}_{BS} \\
0 & \underset{\sim}{A}{}^{s}_{SB} & \underset{\sim}{A}{}^{s}_{SS}
\end{pmatrix}
\begin{pmatrix}
{}^{i+1}\underset{\sim}{a}{}^{n+1}_{G} - {}^{i}\underset{\sim}{a}{}^{n+1}_{G} \\
{}^{i+1}\underset{\sim}{a}{}^{n+1}_{B} - {}^{i}\underset{\sim}{a}{}^{n+1}_{B} \\
{}^{i+1}\underset{\sim}{a}{}^{n+1}_{s} - {}^{i}\underset{\sim}{a}{}^{n+1}_{s}
\end{pmatrix}
=
\begin{pmatrix}
{}^{i}\underset{\sim}{\psi}{}^{g(n+1)}_{G} \\
{}^{i}\underset{\sim}{\psi}{}^{g(n+1)}_{B} + {}^{i}\underset{\sim}{\psi}{}^{s(n+1)}_{B} \\
{}^{i}\underset{\sim}{\psi}{}^{s(n+1)}_{s}
\end{pmatrix}
\qquad (10)
$$

From eq.(10), the velocity field can be obtained by using an iteration procedure. The iteration starts with ${}^{o}\underset{\sim}{a}{}^{n+1} = 0$.

Once the nodal velocities are obtained, the stresses in the bulk solid domain and the stress resultants in the silo wall can be computed by the strain rate - velocity relations and the constitutive laws adopted.

EXAMPLE

A typical example is presented to show the application of the method.

Consider an axisymmetric hopper-silo combination. Fig.4 show the finite element mesh adopted for half of the hopper-silo cross section through the symmetrical axis. The thickness of the silo wall is 4 mm, the Young's modulus of the silo material E=210000 MPa, and Poisson's ratio $v=0.25$. The friction angle of the silo wall is $\phi_w=15°$.

The bulk material adopted has the properties of density $\rho=1600$ kg/m³, and elastic-plastic properties as those given in Lade's paper (1977).

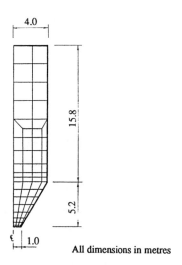

FIGURE 4.   Finite element mesh

449

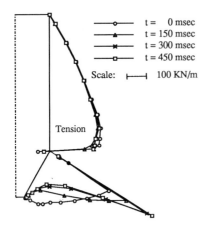

FIGURE 5.   Circumferential force in silo wall

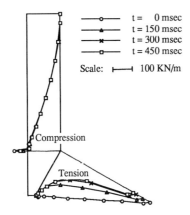

FIGURE 6.   Meridional force in silo wall

Fig.5 shows the distributions of the circumferential forces in the wall of the silo at several instants of time during discharge. There is a decrease of the circumferential forces near the outlet during the initial phase of discharge. However, a strong increase of the circumferential forces occurs near the transition area between the hopper and the bin wall.

Fig.6 shows the distributions of the meridional forces in the silo wall at several different instants of discharge. It is found that the hopper wall is subjected to tensile forces, however, the bin wall is subjected to compressive forces. In addition, it is noted that the filling condition is the critical condition for the meridional forces in the hopper.

Fig.7 shows the vertical velocity distributions on the centre line of the silo at some instants during discharge of the bulk material. It is clear that there is a uniform increase of velocity in the bin area. Large increases of velocity occur in the region from the outlet to the transition point.

450

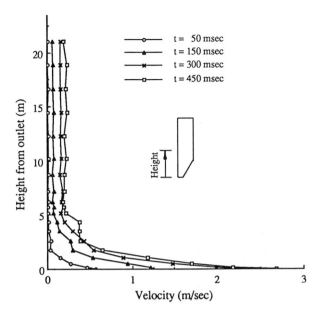

FIGURE 7. Velocity distributions on centre line at various instants

CONCLUSIONS

A numerical procedure is presented that can simulate the flow of bulk materials from silos. The method can provide the transient stress resultants in the walls of silos and the transient velocity fields in the domain of bulk materials.

NOMENCLATURE

$\underline{A}$	eq.6;
$\underline{C}$	viscous matrix;
$\underline{H}$	elastic-plastic matrix;
$\underline{K}$	stiffness matrix;
$\underline{M}$	mass matrix;
$\underline{a}$	nodal velocity;
$\underline{d}$	stain rate;
t	time;
$\underline{v}$	velocity;
$\underline{\sigma}$	stress tensor; and
ϕ_w	angle of friction between solids and silo wall.

Superscripts

g	granular solids;
s	structure;
n, n+1	time steps;
i, i+1	iterative cycles; and
(˙)	differential with respect to time.

Subscripts

B	interface between solids and structure;
G	granular solids;
S	structure; and
o	initial.

REFERENCES

Arnold, P.C., McLean, A.G., and Roberts, A.W., *Bulk Solids: Storage, Flow and Handling*, Tunra Limited, The University of Newcastle, Australia, 1978.

Haussler,U., and Eibl, J., Numerical Investigation on Discharging Silos, *J. Engrg. Mech.*, ASCE, Vol.110, No.6, pp.957-971, 1984.

Jenike, A.W., and Johanson, J.R., Bin Loads, *J. Struct. Div.*, ASCE, Vol.94, No.4, pp.1011-1041, 1968.

Lade, P.V., Elasto-plastic Stress Strain theory for Cohesionless Soil with Curved Yield Surface, *International Journal of Solids and Structures*, Vol.13, pp.1019-1035, 1977.

Runesson, K., and Nilsson, L., Finite Element Modelling of the Gravitational Flow of a Granular Material, *International Journal of Bulk solids Handling*, 6(5), pp.877-884, 1986.

Schmidt, L.C. and Wu, Y.H., Prediction of Dynamic Pressures on silos, *International Journal of Bulk Solids Handing,* August, 1989.

Zienkiewicz, O.C., *The Finite Element Method*, 3rd ed., McGraw-Hill, New Delhi, 1979.

The Effects of Elastic Deformation
on the Dynamic Stability
of Journal Bearings

L. H. WANG and A. K. TIEU
Department of Mechanical Engineering
The University of Wollongong
Wollongong, 2500, Australia

INTRODUCTION

In high speed machinery, it was noted that rotors supported by an oil film lubricated bearings sometimes experienced large amplitudes of vibration, having a whirl frequency close to half the rotation speed of the machine. Their instability problems have been investigated by many researchers. Newkirk and Taylor (1925), were the first to report violent whipping occurring at speeds equal to twice the first critical frequency of the shaft, and it persisted at higher rotation speeds. Hagg (1946) and Cameron (1953) explained how the whirling frequency has to be equal to half the journal speed, in order to satisfy the continuity of flow in a journal bearing under whirl conditions. Orbeck (1961), Badgly and Booker (1969), Akers et al (1971) and Dostal et al (1974) derived the stability criteria, and the onset of stability of the rotor. Lund (1980) provided a means for evaluating the sensitivity of the critical speeds of a system to change in model elements. He used the energy distribution in the system and utilized a first-order Eigenvalue-vector perturbation technique. Majumdar et al. (1988) carried out an investigation which dealt with the stability characteritics of oil film journal bearings, including the effect of elastic distortion in the bearing liner.

In this paper, the finite element method was used to solve the Reynolds equation for the oil film pressure distribution of plain journal bearings.The dynamic characteristics and instability are discussed when an arbitrary elastic deformation of the bearing is considered.

NOTATION

$B_{XX}, B_{XY}, B_{YX}, B_{YY}$ = damping coefficients; dimensionless coefficients, $B_{ij} = B_{ij}c\omega/WS$
c = radial clearance.
C_1, C_2 = coefficients of solution of characteristic equation.
e = eccentricity.
$h, \Delta h$ = oil film thickness, deformation of bearing
$K_{XX}, K_{XY}, K_{YX}, K_{YY}$ = stiffness coefficients; dimensionless coefficients, $K_{ij} = K_{ij}c/WS$
L_c = bearing circumferential length
M = mass of external static load.
p = oil film pressure, p = non-dimensional oil film pressure, $p = pc/\eta U$.
R = journal radius.
U = tangential velocity of any arbitrary point on the journal surface.
W = static load.
δ = oil film extent
ε = eccentricity ratio = e/c.
θ = angular coordinate for bearing.

η	= lubricant viscosity.
ϕ	= attitude angle.
ω	= angular speed.
ω_c	= stability parameter, $\omega_c = Mc\omega^2/F$.
γ	= angular position of journal center in the clearance circle
λ	= solution of characteristic equation of motion equation.

STATIC SOLUTION

Basic Equation

The basic equation for this problem is the Reynolds equation which governs the oil film pressure distribution in journal bearings, and it has the following dimensionless form :

$$\frac{c}{L_c}\left[\frac{\partial}{\partial \bar{x}}\left(\frac{\bar{h}^3}{12}\frac{\partial \bar{p}}{\partial \bar{x}}\right) + \frac{\partial}{\partial \bar{y}}\left(\frac{\bar{h}^3}{12}\frac{\partial \bar{p}}{\partial \bar{y}}\right)\right] = \frac{1}{2}\frac{\partial \bar{h}}{\partial \bar{x}} + \frac{L_c}{c}\bar{\dot{h}} \qquad (1)$$

where

$$\bar{x} = \frac{x}{L_t}, \quad \bar{y} = \frac{y}{L_c}, \quad \bar{h} = \frac{h}{c} = 1 + \varepsilon \cos\theta \qquad \bar{p} = \frac{pc}{\eta U}, \quad \bar{\dot{h}} = \left(\frac{dh}{dt}\right)/U \qquad (2)$$

Solution of Reynolds Equation:

A variational principle was employed to define a variational equivalent of equation (1) as shown in equation (3)

$$I = \int\int\left\{\frac{c}{L_c}\left[\frac{\bar{h}^3}{24}\left(\frac{\partial \bar{p}}{\partial \bar{x}}\right)^2 + \frac{\bar{h}^3}{24}\left(\frac{\partial \bar{p}}{\partial \bar{y}}\right)^2\right] - \left[\frac{\bar{h}}{2}\frac{\partial \bar{p}}{\partial \bar{x}} + \frac{L_c\bar{\dot{h}}}{c}\bar{p}\right]\right\}d\bar{A} \qquad (3)$$

And according to the principle, when the minimum of I is attained, that is when $\partial I/\partial p = 0$, the functional (3) will govern a stable solution of equation (1).

In the finite element method, more than 200 rectangular elements were used to cover the pressure area in the bush as shown in Figs. 1 and 2. The non-dimensional pressure p and oil film thickness h are expressed in terms of the nodal values as followed :

$$\bar{p} = \sum_{j=1}^{4} N_j \bar{p}_j \quad ; \quad \bar{h} = \sum_{j=1}^{4} N_j h_j \qquad (4)$$

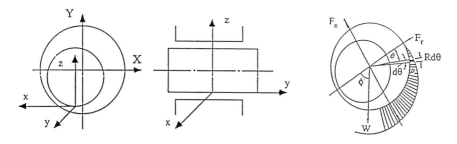

FIGURE 1. Bearing Coordinate system   FIGURE 2. Oil Film Geometry

Substituting expressions (4) into the expression $\{\partial I/\partial P = 0\}$

$$\iint \frac{c}{12L_c}\left(\sum_{j=1}^{4} N_j \bar{h}_j\right)^3 \left[\frac{\partial N_i}{\partial \bar{x}}\sum_{j=1}^{4}\frac{\partial N_j}{\partial \bar{x}}\bar{p}_j + \frac{\partial N_i}{\partial \bar{y}}\sum_{j=1}^{4}\frac{\partial N_j}{\partial \bar{y}}\bar{p}_j\right]d\bar{A} = \iint \left[\frac{1}{2}\left(\sum_{j=1}^{4} N_j \bar{h}_j\right)\frac{\partial N_i}{\partial \bar{x}} + \frac{L_c}{c}\left(\sum_{j=1}^{4} N_j \bar{h}_j\right)N_i\right]d\bar{A}$$

and in matrix form $A \times P = B$ \hfill (5)

where
$$P = [\ \bar{p}_1,\ \bar{p}_2,\ \bar{p}_3,\ \bar{p}_4\ ]$$

$$B = [\ b_1,\ b_2,\ b_3,\ b_4\ ] \hfill (6)$$

$$a_{ik} = \iint \frac{c}{12L_c}\left(\sum_{j=1}^{4} N_j\ \bar{h}_j\right)^3 \left[\frac{\partial N_i}{\partial \bar{x}}\frac{\partial N_k}{\partial \bar{x}} + \frac{\partial N_i}{\partial \bar{y}}\frac{\partial N_k}{\partial \bar{y}}\right]d\bar{A}$$

$$b_i = \iint \left[\frac{1}{2}\left(\sum_{j=1}^{4} N_j\ \bar{h}_j\right)\frac{\partial N_i}{\partial \bar{x}} + \frac{L_c}{c}\left(\sum_{j=1}^{4} N_j\ \bar{h}_j\right)N_i\right]d\bar{A}$$

Gauss integration method was employed to determine a_{ik} and b_i. After the matrix A and the column B in each element was established, all of these were assembled together by the principle of finite element method to compose the (n x n) order coefficient matrix and n order right hand side column. Then this system of simultaneous linear equation was solved by the Gauss elimination method to yield the oil film pressure distribution such as those for two typical eccentricity ratio shown in Fig.3.

DYNAMIC CHARACTERISTICS AND STABILITY

Bearing Dynamic Characteristics

The dynamic characteristics of journal bearings is described by four stiffness and four damping coefficients which are defined in equation (7).

$$K_{XX} = \frac{\partial F_X}{\partial X},\ K_{XY} = \frac{\partial F_X}{\partial Y},\ K_{YX} = \frac{\partial F_Y}{\partial X},\ K_{YY} = \frac{\partial F_Y}{\partial Y} \hfill (7)$$

$$B_{XX} = \frac{\partial F_X}{\partial \dot{X}},\ B_{XY} = \frac{\partial F_X}{\partial \dot{Y}},\ B_{YX} = \frac{\partial F_Y}{\partial \dot{X}},\ B_{YY} = \frac{\partial F_Y}{\partial \dot{Y}}$$

F_X and F_Y are the oil film forces acted on the shaft and they are determined by :

$$F_X = \left(\int_0^{R\delta}\int_{-L/2}^{+L/2} p\cos\theta dxdy\right)\sin\phi - \left(\int_0^{R\delta}\int_{-L/2}^{+L/2} p\sin\theta dxdy\right)\cos\phi \hfill (8)$$

$$F_Y = \left(\int_0^{R\delta}\int_{-L/2}^{+L/2} p\cos\theta dxdy\right)\cos\phi + \left(\int_0^{R\delta}\int_{-L/2}^{+L/2} p\sin\theta dxdy\right)\sin\phi$$

$\varepsilon = 0.7$ $\varepsilon = 0.8$

FIGURE 3. Oil Film Pressure Distribution for Different Eccentricity Ratio

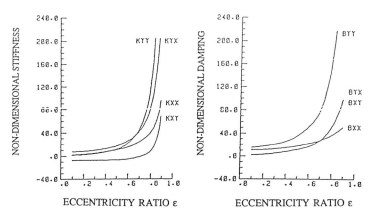

FIGURE 4. Oil Film Coefficients in Non-Deformation Case (L/D=1.0)

These equations can be calculated by means of perturbation method in which a small displacement Δx and Δy are given to the shaft respectively and the oil film forces in these cases can be obtained, therefore by comparing these results with those at equilibrium position and using equation (7), four stiffness coefficients can be evaluated. By the same way, a small Δx and Δy are given to the shaft respectively in the calculation of four damping coefficients.

From this method, some coefficients results for different values of L/D in the case of nondeformation have been obtained as shown in Fig.4. Good agreement was obtained with other results [9] by the finite difference method.

Motion Analysis of journal

In the simulation of journal bearing dynamics, the equations of motion have been established for this vibration system at an equilibrium position, and the properties of stiffness and damping of the oil film are regarded as linear in a small area around this position. The analysis of forces on the journal is shown in Fig.5.

$$M\ddot{X} = F\sin\omega t - (P_x + \Delta P_x)\sin(\gamma + \phi) - (P_y + \Delta P_y)\cos(\gamma + \phi) \tag{9}$$

$$M\ddot{Y} = F\cos\omega t - W - (P_y + \Delta P_y)\sin(\gamma + \phi) + (P_x + \Delta P_x)\cos(\gamma + \phi)$$

where

$$P_x = F_n \sin\phi + F_r \cos\phi \qquad\qquad P_y = -F_n\cos\phi + F_r\sin\phi \tag{10}$$

$$\Delta P_x = K_{xx}x + K_{xy}y + B_{xx}\dot{x} + B_{xy}\dot{y} \qquad \Delta P_y = K_{yx}x + K_{yy}y + B_{yx}\dot{x} + B_{yy}\dot{y} \tag{11}$$

This is a non-linear vibration problem because the stiffness and damping coefficients are the functions of the position and motion state of the journal center. Here the fourth-order Runge-Kutta method for solving linear differential equation has been employed to do this work.

For example, if an arbitrary position 1 is a starting point of the journal center, the first thing to be done is to determine the position of this point in the clearance circle which is determined by the eccentricity e_1 and the angle γ_1 as shown in Fig.5. And according to these two parameters, the eight oil film coefficients and the static reaction forces acted on the journal by the oil film can be determined. In this time, the origin of X-Y coordinate system is located at

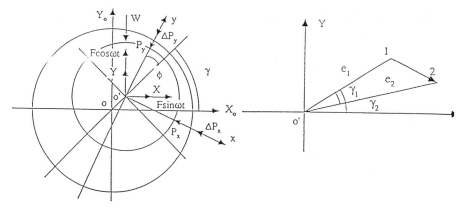

FIGURE 5. Analysis of Forces on Journal in Motion

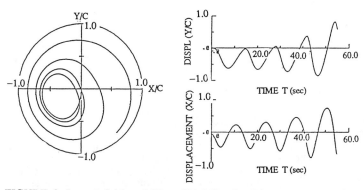

FIGURE 6. Journal Orbit and X- and Y- Vibration Displacements (ε=0.4)

this point and equation (9) describes the motion of journal in small area around position 1. If the coefficients and the static reaction forces are regarded as constant in this small area, equation (9) is a general differential equation which can be solved by using Runge-Kutta method directly. In this way, a new position 2 is obtained.

At the new position, the new coefficients and static reaction forces are determined first according to new position parameters c_2 and γ_2. The above work is repeated again to obtain next position and so on until the orbit of journal center is determined. A typical journal orbit and its x,y vibration displacement are shown in Fig.6.

Stability Analysis

In this paper, a linear method is used to investigate the stability of oil film-journal vibration system by analysing motion equation at every equilibrium position. Here the homogeneous equations of motion equation have the following form:

$$M\ddot{X} + K_{XX}X + K_{XY}Y + B_{XX}\dot{X} + B_{XY}\dot{Y} = 0 \qquad (12)$$
$$M\ddot{Y} + K_{YX}X + K_{YY}Y + B_{YX}\dot{X} + B_{YY}\dot{Y} = 0$$

In the above equations, the stiffness and damping coefficients are kept constant about the equilibrium position. Therefore in this case, a general solution of the following form can be assumed,

457

$$X = C_1 e^{\lambda t}, \quad \dot{X} = C_1 \lambda e^{\lambda t} = \lambda X, \quad \ddot{X} = C_1 \lambda^2 e^{\lambda t}$$
$$Y = C_2 e^{\lambda t}, \quad \dot{Y} = C_2 \lambda e^{\lambda t} = \lambda Y, \quad \ddot{Y} = C_2 \lambda^2 e^{\lambda t}$$

Substituting these solutions into equation (12) and assuming : $\quad \dfrac{K_{ij}}{M} = k_{ij}, \quad \dfrac{B_{ij}}{M} = b_{ij}$

then the following quadratics are obtained :

$$(\lambda^2 + k_{XX} + b_{XX}\lambda)X + (k_{XY} + b_{XY}\lambda)Y = 0 \tag{13}$$
$$(k_{YX} + b_{YX}\lambda)X + (\lambda^2 + k_{YY} + b_{YY}\lambda)Y = 0$$

Its characteristic equation is,

$$\lambda^4 + A_1\lambda^3 + A_2\lambda^2 + A_3\lambda + A_4 = 0 \tag{14}$$

where
$$A_1 = b_{YY} + b_{XX}$$

$$A_2 = k_{YY} + k_{XX} + b_{XX}b_{YY} - b_{YX}b_{XY}$$

$$A_3 = k_{YY}b_{XX} + k_{XX}b_{YY} - b_{XY}k_{YX} - b_{YX}k_{XY}$$

$$A_4 = k_{XX}k_{YY} - k_{XY}k_{YX}$$

According to the stability principle of solution of differential equation, the condition of stability is that the real parts of characteristic equation (14) are equal to or less than zero. Therefore the Routh criteria is employed to determine the onset of instability from the fourth-order equation (14). When the absolute of a maximum real root among the four real roots of equation (14) is equal to or less than 10^{-4}, the vibration system was thought to be in the critical state. The above work is repeated at each equilibrium position, so an instability boundary line can be determined.

For the convenience in application to journal bearings having the same geometry but different clearance and different rotating speed, a stability parameter was introduced which is defined in the form $\omega_c = Mc\omega^2/F$. The instability results of different values of L/D are shown in Fig.7.

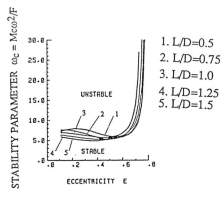

1. L/D=0.5
2. L/D=0.75
3. L/D=1.0
4. L/D=1.25
5. L/D=1.5

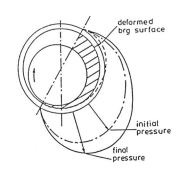

FIGURE 7. Instability Curves in Non-
Deformation Cases

FIGURE 8. Bearing Surface Deformation

458

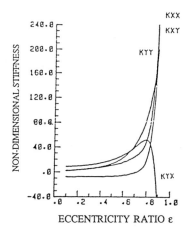 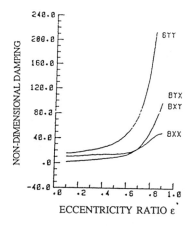

FIGURE 9. Oil Film Coefficients in Deformation Case

EFFECTS OF ELASTIC DEFORMATION ON BEARING DYNAMIC CHARACTERISTICS AND STABILITY

When a journal bearing is subjected to an external load in operation, oil film pressure is produced by hydrodynamic action to balance the external load. At the same time, the oil film pressure can cause the elastic deformation of bearing which in turn modifies the oil film profile and improves the load capacity of the journal and thrust bearings [11]. These deformation can influence the dynamic characteristics and stability of journal bearing.

An arbitrary elastic deformation Δh from Tieu (1986), shown in Fig.8, is introduced to modify the oil film profile as shown in the following equation: $h = c + e \cos \theta + \Delta h$

When the modified oil film thickness is introduced into equation (1), the above computational technique as described in previous sections is again repeated. The analysis results about dynamic characteristics and stability of elastically deformed journal bearing are shown in Fig.9 and Fig.10.

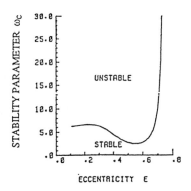

FIGURE 10. Instability Curve for Deformation Case

459

CONCLUSIONS

In this paper, finite element method was used to solve the Reynolds equation.The perturbation method was employed to determine the oil film force coefficients and although this is a linear method, the results obtained by it have shown to produce good agreement with other theoretical and experimental results elsewhere.

The effects of elastic deformation on the bearing stiffness characteristics are more obvious when the eccentricity ratio ε is greater than 0.6. The larger is the deformation of bearing surface, the more the stiffness coefficients K_{yx} and K_{yy} decrease, and the more the coefficients K_{xx} and K_{xy} increase. The effects on damping coefficients are not significant. The bearing stability deteriorates further with the expansion of the unstable region for $0.65 > \varepsilon > 0.3$, but the stability improves for $0.7 > \varepsilon > 0.65$.

ACKNOWLEDGEMENT

This work is funded by an Australian Electrical Research Board grant supervised by Mr A.Gibson of Electricity Commission of NSW.

REFERENCES

1. Akers,A., Michaelson,S. and Cameron,A., Stability Contours for a Whirling Finite Journal Bearing , *Journal of Lubrication Technology*, Vol.93, p.177, 1971.

2. Badgley,R.H. and Booker,J.F., Turborotor Instability: Effect of Initial Transients on Plain Motion , *Journal of Lubrication Technology*, Vol.91, P.625,1969.

3. Cameron,A., Oil Whirl in Bearings , *Engineering*, Vol.179, pp237-239, 1955.

4. Dostal,M., Roberts,J. and Holms,R., Stability Control of Flexible Shafts Supported on Oil-film Bearings ,*Journal of Sound and Vibration*,Vol.35, No.3, pp.361-377, 1974.

5. Hagg,A., The Influence of Oil-film Journal Bearings on the Stability of Rotating Machines , *Journal of Application Mechanics*, Vol.13, pp.A-211, 1946.

6. Newkirk,B. and Taylor,M., Shaft Whipping due to Oil Action in Journal Bearings , *General Electric Review*, Vol.28, pp.559 , 1925.

7. Lund,J.W., Sensitivity of the Critical Speeds of a Rotor to Changes in the Design , *Journal of Mechanical Design* , Vol.102, pp.115-121,1980.

8. Majumdar,B.C., Brewe,D.E. and Khonsari,M.M, Stability of a Rigid Rotor Supported on Flexible Oil Journal Bearings , *Journal of Tribology*, Vol.110, pp.181-187, 1988.

9. Makdissy,J., A Theoretical and Experimental Study into the Dynamic Characteristics of a Hydrodynamic Oil Film Bearing , *Ph.D Thesis*, Cranfield Institute of Technology, 1980.

10. Orbeck,F., Stability Criterion for Oil Whip of a Rotor in Journal Bearings , *The Engineer*, Vol.212, p.303, 1961.

11. Tieu,A.K. A Thermo-Elasto-Hydrodynamic Analysis of Journal Bearings , *5th Int. Conf. in Australia on Finite Element Methods*, Melbourne 1987 .

Finite Element Analysis of Free Surface Flows

DER-LIANG YOUNG and YU-SUN LIU
Department of Civil Engineering
National Taiwan University
Taipei, Taiwan, 10764, ROC

INTRODUCTION

The inclusion of the effects of free surface into the Navier-Stokes equations of incompressible viscous flows is a very annoying task in the realm of computational fluid dynamics. Therefore, this field has attracted tremendous attention by and by and becomes a very interesting research topic. Traditionally, there are three different methodologies adopted to simulate the dynamics of an incompressible fluid with a free surface. They are the Lagrangian, Eulerian, and arbitrary Lagrangian-Eulerian coordinates. Each approach claims its advantages or drawbacks in dealing with numerical simulations. As evolving from the development of numerical analysis, the finite difference scheme comes first, and then follows by the finite element or boundary element methods. For example, in the Lagrangian description, Hirt et al(1970) discretized the system by the finite difference while Ramaswamy and Kawahara(1987a) adopted the finite element. In the Eulerian description, Harlow and Welch(1965) used the finite difference, on the other hand Washizu et al(1984) employed the finite element. As far as arbitrary Lagrangian-Eulerian method is concerned, Hirt et al(1974) used the finite difference while finite element was adopted by Ramaswamy and Kawahara(1987b).

In this study, Eulerian finite element method is chosen by considering the merit of generality and simplicity of describing the flow dynamics except on the free surface. To circumvent the difficulties of modeling free surfaces, which vary with time in a manner not known a priori, a boundary fitted coordinate transformation is introduced, so that the irregular and time-dependent free surface can be mapped into a fixed flat surface for easy computation by the finite element algorithms. Other techniques to cope with the free surface via the Eulerian approach, such as the concept of a fractional volume of fluid (VOF)(Hirt and Nichols, 1981), or a surface height method(Hirt et al 1975), or interface tracking technique(McMaster and Gong, 1979) can also be found from the literature.

461

For a time-dependent two-dimensional viscous incompressible flow of fluid with a free surface, the governing equations are the continuity and Navier-Stokes equations, as represented by the following dimensionless form:

$$u_t + uu_x + vu_y = -p_x + R^{-1}(u_{xx} + u_{yy}) \tag{1}$$

$$v_t + uv_x + vv_y = -p_y + R^{-1}(v_{xx} + v_{yy}) - F^{-2} \tag{2}$$

$$u_x + v_y = 0 \tag{3}$$

where x and y are Cartesian coordinates, t is time, u is x-direction velocity, v is y-direction velocity, p is the pressure, R is the Reynolds number and defined by $R = VL/\nu$ with characteristic length L, characteristic velocity V and kinematic viscosity of fluid ν, and F is the Froude number and defined by $F = V/(gL)^{1/2}$ with acceleration of gravity g.

Some boundary conditions on solid and free surface boundaries must be imposed, they are reiterated as follows. For solid boundaries, either non-slip or free-slip boundary condition may be imposed. As far as free surface boundaries are concerned, they must satisfy both the kinematic as well as dynamic free surface boundary conditions (Washizu et al, 1984).

A boundary fitted coordinates transformation is introduced to circumvent the moving free surface boundaries by the following:

$$\tilde{x} = x \tag{4}$$

$$\tilde{y} = [y + d(x)]/[h(x,t)] - 1 \tag{5}$$

$$\tilde{t} = t \tag{6}$$

Where h is the height of free surface measuring from basin bottom elevation, d is the height of undisturbed water depth of the basin. After this coordinate system transformation is undertaken, the complex domain of both free surface and solid boundaries are mapped into a simple geometric domain. In addition, the governing equations of Eq.1 to 3 are transformed to the following new equations (for convenience, the symbol " $\sim$ " is dropped) :

$$u_t + C0u_y + u(u_x + C1u_y) + C2vu_y = -p_x - C1p_y + R^{-1}[u_{xx}$$
$$+ 2C1u_{xy} + (C1^2 + C2^2)u_{yy} + (C1_x + C1C1_y)u_y] \tag{7}$$

$$v_t + C0v_y + u(v_x + C1v_y) + C2vv_y = -C2p_y - F^{-2} + R^{-1}[v_{xx}$$
$$+ 2C1v_{xy} + (C1^2 + C2^2)v_{yy} + (C1_x + C1C1_y)v_y] \tag{8}$$

$$u_x + C1u_y + C2v_y = 0 \tag{9}$$

Where coefficients C0, C1 and C2 are related to the basin geometry, and free surface. Similar transformations can be obtained for the corresponding kinematic as well as dynamic boundary conditions. This boundary fitted coordinate transformation technique is also used in the finite difference discretization(Hung and Wang, 1987). However, it is noticed that much complex transformed equations(including the x coordinate) appear in their finite difference scheme.

FINITE ELEMENT ANALYSIS

Standard Galerkin finite element analysis (Zienkiewicz, 1977) is introduced to discretize the transformed governing equations of Eq. 7 to 9 by using the mixed shape functions of quadratic triangular element for velocity and linear triangular element for pressure. The time is discretized by an implicit finite difference scheme. Therefore

$$\{Fu\} = [S00](\{u\}^n - \{u\}^{n-1})/\Delta t + \theta\{Fx\}^n + (1-\theta)\{Fx\}^{n-1}$$

$$+([S01] + C1[S02])\{p\} - \{Rx\} = \{0\} \tag{10}$$

$$\{Fv\} = [S00](\{v\}^n - \{v\}^{n-1})/\Delta t + \theta\{Fy\}^n + (1-\theta)\{Fy\}^{n-1}$$

$$+ C2[S02]\{p\} - \{Ry\} = \{0\} \tag{11}$$

$$\{Fp\} = ([T01] + C1[T02]) \{u\} + C2 [T02]\{v\} = \{0\} \tag{12}$$

where [S00], {Fx}, {Fy}.... are all coefficient matrices, while {Rx}, {Ry} are the loading vectors, and can be obtained by conventional numerical integration procedures. θ is a time integrating factor and lies between 0 and 1. A pressure-velocity correction scheme was proposed to solve the coupling system of Eq. 10 to 12. That is, velocity and pressure fields are solved separably by the iterative procedures. The process is described as follows. Firstly a pressure field is assumed known at the beginning, so that Eq. 10 and 11 of momentum equations can be used to calculate the velocity components by the modified Newton-Raphson method since the system is nonlinear, namely

$$[J]\{ \Delta u \}_i^n = \{ - Res\{Fu\}_{i-1} \} \tag{13}$$

$$[J]\{ \Delta v \}_i^n = \{ - Res\{Fv\}_{i-1} \} \tag{14}$$

Noticing that the correction of velocity component does not need the coupling of u and v and is solved separately.

Secondarily, the new trial velocity fields are substituted into the Eq. 12 of continuity equation. If the velocity components satisfy the continuity equation, then we have found the desirable velocity and pressure distribution. On the other hand a pressure correction process is now indispensable, and the computational cycle is repeated until the criteria of convergence of both the velocity and pressure are reached.

There are various alternatives in the pressure correction process(Young, 1989). However, in this study a modified algorithm of semi-implicit method for pressure-linked equation(SIMPLE)(Patankar and Spalding, 1972) is selected. In general, the correction of pressure can be formulated as

$$[K]\{ \Delta p \}_i^n = [C11]\{u\}_{i-1} + [C22]\{v\}_{i-1} \tag{15}$$

Calculation of the elevation of the free surface is achieved by
incorporating kinematic boundary condition into the global system. A
predictor-corrector method of finite difference scheme was adopted in
this analysis. The Euler method of numerical discretization was found to
give very efficient and stable solutions of the corresponding free
surfaces.

MODEL APPLICATIONS

To test the feasibility of the numerical model, three case studies will
be made. They are (1) propagation of a moving bore due to an impulse
force, (2) wind-driven circulation of a cavity flow with and without the
effects of free surface, and (3) the viscous effects on the damping of
the free oscillation of sloshing in a tank.

When a horizontal column of water is pushed into a rigid and vertical
wall, a moving bore is generated which runs away from the wall. This
undular bore evolution was adopted by Nichols et al (1980) to test their
SOLA-VOF Codes : a solution algorithm for transient fluid flow with
multiple free boundaries. The same problem was also simulated by using
the present model. Fig. 1 shows the distributions of velocities and

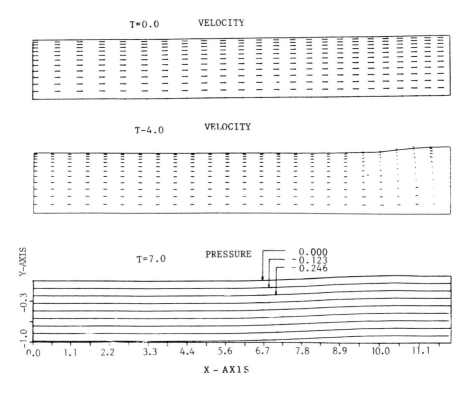

FIGURE 1. Wave propagation of a moving bore due to an impulse force

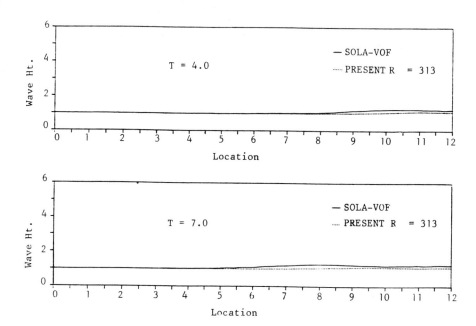

FIGURE 2.   Comparison of wave height of a moving bore due to   an   impulse
force

free-surface   configuration for the undular bore propagation. Times   are
0.0,   4.0,   and 7.0 respectively.   The right and bottom walls   are   rigid
free-slip boundaries.   A continuative boundary condition is taken at   the
left boundary which allows a continuous input of fluid and   prevents   any
waves   from   being   generated   there.   In   general,   the   two   different
approaches yield very similar results except the amplitude of the undular
bore of present studies is lower than that of SOLA-VOF model as   depicted
in   Fig.   2.   The discrepancy is attributed to the   viscous   damping   of
present study, since SOLA-VOF considers an ideal fluid (which corresponds
to taking R   approaching infinity) while R   = 313 is chosen in our   case.
The deviation of bore height will be improved for both approaches, when R
is increased.   For example, improvement is made when R   is set equal   to
1000.

In the classic wind-driven lake circulation analysis, the effects of free
surface   are   considered   to   play   a minor   role   so   that   a   rigid-lid
approximation   can   be applied (Liggett and   Hadjitheodorou,   1969).   To
simplify   the   calculation,   a   square cavity   flow   with   and   without
considering the influences of free surface is undertaken.   Fig. 3 depicts
the response of the velocity and pressure distributions for R   = 10 under
the   shear driven flow with the consideration of free surfaces.   At   the
beginning,   the flow circulation pattern is quite different from that   of
the rigid-lid approximation which ignores the free surface.   However,   as
the   evolution of the circulation gyre develops, the effects of the   free

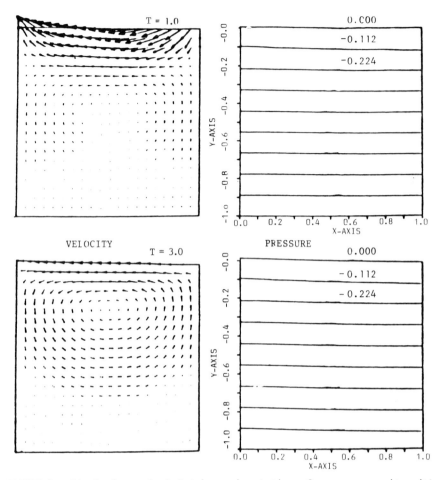

FIGURE 3. Simulation of wind-driven circulation of a square cavity with
free surface

surfaces are damped gradually by the viscosity dissipation, so that not
much difference is detected on the free surfaces in the long run.   The
major differences are the wiggling surface with vertical velocity when
the free surface is taken into consideration.

The final example is the simulation of the viscous effects on the damping
of the free oscillation of a sloshing tank from a sudden movement of free
surface. In this problem, the fluid initially at rest in a rigid, open,
rectangular container, is impulsively set in motion by tilting the tank.
Fig. 4 shows the time histories of the free surface displacement at both
sides of walls under a free oscillation. The decreased amplitudes of
the surface elevation are damped by the viscosity, as manifestly
delineated from Fig. 4. The oscillating frequency of the sloshing tank
is found almost invariant. These two characteristics are found to be
very consistent to the salient features of physics, as also examined by
other investigators, such as Washizu et al(1984), Ramaswamy and
Kawahaza(1987b) among others.

466

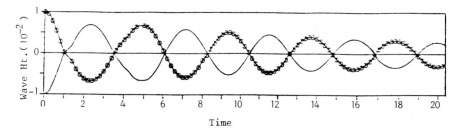

FIGURE 4. Viscous effects on the damping of the free oscillation of a sloshing tank

CONCLUSION

This model was used to investigate some transient two-dimensional viscous incompressible free surface flow fields, such as wave propagation of a moving bore due to an impulse force; wind-driven circulation of a cavity flow with and without the effects of free surface; as well as the vicous effects on the damping of the free oscillation of a sloshing tank. It is demonstrated that the present methodology is a very efficient way to deal with the general response of viscous free surface flows for incompressible fluids, which are of considerable importance to the industrial applications as well as academic interests.

REFERENCES

1. Harlow, F.H., Welch, J.E., Numerical Calculation of Time-dependent Viscous Incompressible Flow of Fluid with a Free Surface, *Phys. Fluids*, Vol. 12, pp. 2182-2189, 1965.

2. Hirt, C.W., Cook, J.L., and Bulter, T.D., A Lagrangian Method for Calculating the Dynamics of an Incompressible Fluid with Free Surface, *J. Comp. Phys.*, Vol. 5, pp. 103-124, 1970.

3. Hirt, C.W., Amsden, A.A., and Cook, J.L., An Arbitrary Lagrangian-Eulerian Computing Method for all Flow Speeds, *J. Comp. Phys.*, Vol. 14, pp. 227-253, 1974.

4. Hirt, C.W., Nichols, B.D., and Romero, N.C., SOLA-A Numerical Solution Algorithm for Transient Fluid Flows, Los Alamos Scientific Laboratory report LA-5852 April 1975.

5. Hirt, C.W., and Nichols, B.D., Volume of Fluid(VOF) Method for the Dynamics of Free Boundaries, *J. Comp. Phys.*, Vol. 39, pp. 201-225, 1981.

6. Hung, T.K., and Wang, M.H., Nonlinear Hydrodynamic Pressure on Rigid Dam Motion, *J. of Engineering Mechanics, ASCE*, Vol. 113, No. 4, pp.482-499, 1987.

7. Liggett, J.A., and Hadjitheodorou, c., Circulation in Shallow

Homogeneous Lakes, *J. of the Hydraulics Division, ASCE*, Vol. 95, No. HY2, pp. 609-620, 1969.

8. McMaster, W.H., and Gong, E.Y., PELE-IC User's Manual, Lawrence Livermore Laboratory report UCRL-52609, May 1979.

9. Nichols, B.D., Hirt, C.W., and Hotchkiss, R.S., SOLA-VOF : A Solution Algorithm for Transient Fluid Flow with Multiple Free Boundaries, Los Alamos Scientific Laboratory, Report, LA-8355, August 1980.

10. Patankar, S.V., and Spalding, D.B., A Calculation Procedure for Heat, Mass and Momentum Transfer in Three-dimensional Parabolic Flows, *Int. J. Heat Mass Transfer*, Vol. 15, pp. 1787-1806, 1972.

11. Ramaswamy, B., and Kawahara, M., Lagrangian Finite Element Analysis Applied to Viscous Free Surface Fluid Flow, *Int. J. Numer. Methods Fluids*, Vol. 7, pp. 953-984, 1987a.

12. Ramaswamy, B., and Kawahara, M., Arbitrary Lagrangian-Eulerian Finite Element Method for Unsteady, Convective, Incompressible Viscous Free Surface Fluid Flow, *Int. J. Numer. Methods Fluids*, Vol. 7, pp. 1053-1057, 1987b.

13. Washizu, K., Nakayama, T., Ikegawa, M., Tanaka, Y., and Takoshi, A., Some Finite Element Techniques for the Analysis of Nonlinear Sloshing Problems, in *Finite Elements in Fluids*, Vol. 5, pp. 357-376, Wiley-Interscience, N.Y. 1984.

14. Young, D.L., Finite Element Analysis of Computational Fluid Dynamics, In Press, *Proc. of the National Scierce Council, Part A: Physical Science and Engineering, ROC.* 1989.

15. Zienkiewicz, O.C., *The Finite Element Method*, 3rd ed. McGraw-Hill, London, 1977.

INDUSTRIAL MATHEMATICS

Numerical Simulation of Two-Phase Flow Arising from Bottom Injection of Gas into Liquid

M. R. DAVIDSON
CSIRO Division of Mineral and Process Engineering
Lucas Heights Research Laboratories
Lucas Heights, NSW 2232, Australia

INTRODUCTION

Submerged injection of gas through an orifice in the containing vessel is widely used in smelting and refining processes for stirring liquid metals and slags to promote mixing and chemical reaction. In bottom gas injection, a plume of rising gas bubbles spreads out from the orifice and sets up recirculating flow patterns in the liquid. Early models of such two-phase flows (e.g. Szekely et al., 1976; Grevet et al., 1982; Muzumdar and Guthrie, 1985), based on a mixture formulation, achieved reasonable success in predicting the liquid circulation by specifying the void fraction and plume velocity distributions *a-priori* . Subsequent use of two-fluid models, to calculate both the void fraction and the liquid circulation simultaneously, yielded adequate predictions of the bath circulation (Cross et al., 1984) but was unable to correctly predict the spreading of the plume when the drag force was the only phase interaction effect included. Schwarz (1989) overcame this deficiency by adding diffusion-like terms to the mass balance equations to model the effect of the phase interaction forces which cause the plume to spread. He demonstrated acceptable agreement with air-water experimental data of Grevet et al. (1982) who measured liquid velocities, and the data of Castillejos and Brimacombe (1986) who measured the void fraction and gas velocity in the plume. Davidson (1989a,b) took a more fundamental approach to modelling the flow by attempting to include the important phase interaction forces directly into the momentum equations without the addition of terms to the mass balance equations. He showed that inclusion of the particle lift force, which arises from the relative motion of the phases in cross-stream velocity gradients, together with a diffusive force which models the turbulent dispersion of bubbles, could account for the gas distribution in the plume and the centre-line gas velocity away from the orifice. The flow equations were solved numerically by finite differences using a modified form of the transient, two-fluid flow program K-FIX (Rivard and Torrey, 1977).

So far Davidson's model has only been tested against a single experiment of Castillejos and Brimacombe and further validation is desirable. Here the model is summarised, and results of additional simulations of the Castillejos/Brimacombe and Grevet et al. experiments (the same ones simulated by Schwarz, 1989) are reported. The predicted steady state void fraction and velocity at different flow rates are compared with the corresponding experimental data, and with the results obtained by Schwarz (1989). Both models are shown to achieve similar agreement with the data despite major differences in formulation. Two such differences are highlighted and their apparent unimportance is discussed.

THE MODEL EQUATIONS

The equations governing the transient motion of each phase (assumed to be incompressible) have the form (Drew, 1983)

$$\frac{\partial \alpha_k}{\partial t} + \nabla \cdot (\alpha_k \mathbf{u}_k) = 0 \tag{1}$$

471

$$\frac{\partial(\alpha_k\rho_k\mathbf{u}_k)}{\partial t} + \nabla\cdot(\alpha_k\rho_k\mathbf{u}_k\mathbf{u}_k) = -\alpha_k\nabla p_k - \alpha_k\rho_k g\widehat{\mathbf{z}} + (p_{ki} - p_k)\nabla\alpha_k + \mathbf{M}_k + \nabla\cdot(\alpha_k\tau_k) \qquad (2)$$

with

$$\alpha_G + \alpha_L = 1 \qquad (3)$$

and

$$\mathbf{M}_L = -\mathbf{M}_G \qquad (4)$$

where the subscript k = G or L to denote the gas or liquid phases, and ρ, α, $\mathbf{u}$, p, $\mathbf{M}$, τ, g, $\widehat{\mathbf{z}}$ denote density, volume fraction, velocity, pressure, phase interaction force, viscous stress tensor, gravitational acceleration, and the upward vertical unit vector, respectively. Davidson (1989a) simplified the viscous term by setting

$$\nabla\cdot(\alpha_k\tau_k) = \nabla\cdot(\alpha_k\mu_{kE}\nabla\mathbf{u}_k) \qquad (5)$$

for two-dimensional axi-symmetric flow, where μ_{kE} is the eddy viscosity of phase k, and the turbulent kinematic viscosities (μ_{kE}/ρ_k, k = G, L) are taken to be equal and uniform throughout the bath.

The pressure p_{ki} represents the average pressure of phase k at microscopic phase interfaces. In the spirit of Pauchon and Banerjee (1986), Davidson (1989a) set

$$p_{Li} = p_{Gi} = p_G \qquad (6)$$

and

$$p_{Li} - p_L = -\frac{1}{4}\alpha_L\rho_L|\mathbf{u}_G - \mathbf{u}_L|^2 \qquad (7)$$

Equation (6) is a good approximation for gas bubbles when surface tension is neglected, and equation (7) reduces to the corresponding expression for potential flow around an isolated sphere of constant radius when $\alpha_L = 1$.

Based on the work of Drew et al. (1979), Drew (1983) and Drew and Lahey (1987), the interphase force terms comprising $\mathbf{M}_G$ in the Davidson model are given below. As in equation (7), expressions valid for low void fraction are multiplied by α_L to ensure the correct limit when only gas is present.

interfacial drag force:

$$K(\mathbf{u}_L - \mathbf{u}_G) \qquad (8)$$

where (following Cook and Harlow, 1983)

$$K = \frac{3}{8}C_D\alpha_G\alpha_L\rho_L\frac{|\mathbf{u}_G - \mathbf{u}_L|}{r_b} \qquad (9)$$

and C_D is the drag coefficient for a single bubble of radius r_b.

virtual mass force :

$$\alpha_G\alpha_L\rho_L C_m\left(\frac{\partial\mathbf{u}_L}{\partial t} + \mathbf{u}_G\cdot\nabla\mathbf{u}_L - \frac{\partial\mathbf{u}_G}{\partial t} - \mathbf{u}_L\cdot\nabla\mathbf{u}_G + (1 - \lambda)(\mathbf{u}_G - \mathbf{u}_L)\cdot\nabla(\mathbf{u}_G - \mathbf{u}_L)\right) \qquad (10)$$

where $C_m \geq 1/2$ (the value for a single isolated sphere) is the virtual mass coefficient of bubbles in a mixture which is predominately liquid, and λ is a parameter in the range (0, 2). The virtual mass force arises because the acceleration of gas bubbles must overcome the inertia of displaced fluid, and is expressed here in the frame indifferent form of Drew et al (1979). Virtual mass terms do not significantly affect the final result in most cases since they are only important at high frequency or rapid acceleration. However, such terms are included since the effect on the short wavelengths is to prevent their rapid growth and the consequent onset of instability.

lift force:

$$- \alpha_G \alpha_L \rho_L (u_G - u_L) \cdot \frac{1}{2} \left(\nabla u_m + (\nabla u_m)^T \right) \tag{11}$$

where gradients of the mixture velocity

$$u_m = \alpha_G u_G + \alpha_L u_L \tag{12}$$

are used, rather than those of the liquid velocity, to ensure stability for all $C_m \geq 1/2$; this matter is discussed by Davidson (1989a). The lift force arises from the relative motion of the phases in cross-stream velocity gradients. It is an inertial effect, which has been observed and analysed for Poiseuille flow in tubes, whereby a spherical particle travelling faster (or slower) than the local fluid velocity migrates away from (or towards) the faster moving fluid at the centre-line. Further, Drew and Lahey (1987) have shown that the inclusion of the lift force in their generalised expressions for the phase interaction force is necessary for the combination to reduce to that calculated for a single sphere. Note that Drew and Lahey (1987) recently redefined the virtual mass and lift terms keeping the sum of these terms unchanged; however, equations (10) and (11) are based on their earlier partitioning.

diffusive interfacial force:

$$-K \frac{D_{eff}}{\alpha_G} \nabla \alpha_G = -\frac{3}{8} C_D \alpha_L \rho_L \frac{|u_G - u_L|}{r_b} D_{eff} \nabla \alpha_G \tag{13}$$

substituting equation (9) for K. The effective bubble dispersion coefficient D_{eff} is set equal to the (assumed) uniform turbulent kinematic viscosity. This term models the turbulent dispersion of bubbles along a gradient in void fraction (Davidson, 1989b). Fluid eddies oscillate unevenly in such a gradient, due to variations in instantaneous drag, resulting in the ratchet-like motion of entrained bubbles away from the region of high void fraction (Lee and Weisler, 1987).

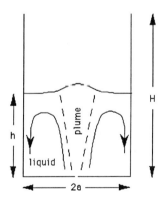

FIGURE 1. Schematic representation of the bath.

THE NUMERICAL METHOD

The model 2-D equations for the gas and liquid phases are solved numerically for the velocities ($\mathbf{u}_k = (U_k, V_k)$, $k = G, L$), pressure and void fraction on a uniform mesh in cylindrical coordinates (r, z). The K-FIX computer code (Rivard and Torrey, 1977) for calculating transient, two-fluid flow is used after adaption to include terms in the momentum equations involving void fraction gradient, virtual mass, particle lift and diffusion force. The virtual mass force was incorporated by following Cook and Harlow (1983).

The program K-FIX is based on an implicit multifield solution method (Harlow and Amsden, 1975) which reduces to the ICE technique (Harlow and Amsden, 1971) for single phase flow. The method uses a staggered mesh with pressure and void fraction defined at the centre of computational cells, and with normal velocities at cell walls defined at their mid-points. Donor cell differencing of the advection terms is used and, at each time step, the pressure is adjusted to ensure conservation of mass by a point relaxation technique.

The computation domain is cylindrical (shown schematically in Figure 1) with radius a and height H having an initial depth h of liquid. The flow is taken to be axisymmetric since surface waves and plume precession were damped by a baffle in the Castillejos and Brimacombe (1986) experiments, and do not occur in the Grevet et al. (1982) experiments at the low flow rates considered. In the present work, free slip of both gas and liquid is assumed at solid boundaries; of course, the normal velocity of each phase and the normal gradient of void fraction are set to zero there also. At the top boundary ($z = H$) we set $V_L = 0$ (to prevent liquid from passing through) and

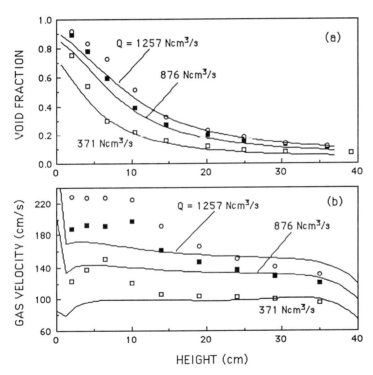

FIGURE 2. Predicted centre-line void fraction (a) and gas velocity (b) compared with corresponding data of Castillejos and Brimacombe (1986) at three different gas flow rates (o 1257 Ncm³/s; ■ 876 Ncm³/s; □ 371 Ncm³/s) when the orifice diameter is 0.635 cm and initial depth of liquid is 40 cm.

$\partial V_G/\partial z = \partial \alpha_G/\partial z = 0$. To ensure that the correct gas volume and momentum is injected into the liquid, gas is assumed to enter through an apparent orifice (centrally located at the bottom) of diameter equal to the width of bubbles at formation, with the void fraction less than 1 immediately behind the orifice to account for bubble frequency. Further details are given in Davidson (1989a).

DISCUSSION OF RESULTS

Davidson (1989a,b) simulated only a single experiment of Castillejos and Brimacombe (1986) and compared predicted and experimental values of steady state void fraction and gas velocity. Figure 2 shows the results of such comparisons at the centre-line for the three flow rates (Q = 371, 876, 1257 Ncm^3/s where N denotes NTP conditions) considered by Castillejos and Brimacombe when the (actual) orifice diameter is 0.635 cm and the initial depth (h) of liquid is 40 cm. Case Q = 876 Ncm^3/s was considered previously by Davidson. At each flow rate in Figure 2a, the predicted centre-line void fraction agrees closely with the experimental data over most of the bath depth. Differences are most pronounced within 10 cm of the orifice where predicted void fractions are too low. In Figure 2b, agreement between predicted and experimental centre-line gas velocities is achieved only in the upper half of the bath with best results occurring at the lowest flow rate. In the lower half of the bath, predicted gas velocities are up to 20 - 30 per cent too low.

A distinguishing feature of the calculated velocities in Figure 2b is their near uniformity with depth away from the orifice and the free surface. The fall in predicted velocity very close to the orifice occurs because of the choice (described in Davidson, 1989a) of initial velocity for bubbles formed at

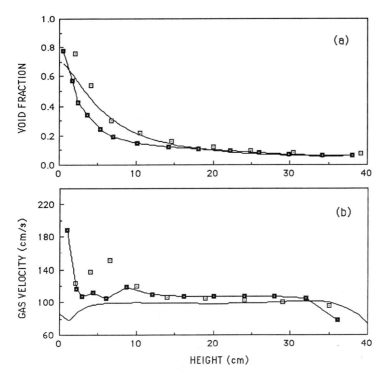

FIGURE 3. Predicted centre-line void fraction (a) and gas velocity (b) compared with corresponding data (□) of Castillejos and Brimacombe (1986) and the predictions (—■—) of Schwarz (1989) for the flow rate 371 Ncm^3/s when the orifice diameter is 0.635 cm and initial depth of liquid is 40 cm.

the orifice; a lower initial velocity has little effect of the predicted velocity curves other than to reduce the initial fall. The drop in gas velocity as the free surface is aproached occurs because of an increase in plume cross-sectional area combined with an almost uniform void fraction in this region.

Figure 3 compares the centre-line void fraction and gas velocity predictions for the case $Q = 371$ Ncm^3/s above with those of Schwarz (1989), derived using the commercially available fluid flow computer program PHOENICS (Rosten et al., 1983). Both models agree with the experimental data in the upper half of the bath. Nearer the orifice, the present model gives better results for the void fraction whereas the Schwarz model yields better results there for the gas velocity. However, when the effective orifice diameter in the Schwarz model is set approximately equal to the bubble diameter (as in the present work), the two models predict gas velocities which are similarly too low near the orifice. In practice, accurate modelling of the region near the orifice cannot be expected without considerable mesh refinement (especially at high flow rates) and perhaps inclusion in the model of factors such as bubble breakup and the effect of forced flow conditions on the drag force.

Two major differences between the two models are (i) the treatment of the phase interaction terms asociated with the spreading of the plume, and (ii) the modelling of the free surface. These differences are now described below.

(i) Schwarz (1989) includes the more commonly considered terms (pressure gradient, gravity, viscous drag, turbulent shear stress) in the momentum equations, but adds diffusion-like terms to the mass balance equations. In the present model the traditional form of the mass balance equations (in a two-fluid formulation) is retained but additional force terms, associated with particle lift and turbulent dispersion, are added; two other forces included to ensure stability (the virtual mass force and the force due to microscopic - bulk presure differences) are expected to have little affect on the final results.

(ii) The present model includes both the liquid and gas top space in the calculation domain; it thus accounts for the free surface deformation and the presence of liquid in the top space. Schwarz (1989) takes the top boundary surface to be flat and excludes the gas top space from the calculation domain.

Figure 4 compares the magnitudes of the liquid velocity in the vessel, predicted at various heights above a nozzle through which gas enters at a velocity of 320 cm/s, with the corresponding data of Grevet et al. (1982), and with predicted values presented by Schwarz (1989). For the given orifice diameter (1.27 cm) the operating flow rate is 405 cm^3/s. The present model succeeds in reproducing the major trends in the experimental liquid velocity data at all heights. At $z/h = 0.3$ the predicted strength of the recirculating velocity at the wall is too low and, at $z/h = 0.68$ the calculated velocity minimum is too high. The Schwarz model fits the data more closely at $z/h = 0.3$ whereas the present model is better at $z/h = 0.1$. At $z/h = 0.68$ the level of agreement with the data is similar for both models. Results for the remaining heights shown in Figure 4 are not given in Schwarz (1989).

CONCLUDING REMARKS

The model described by Davidson (1989a,b) is shown to adequately predict the void fraction and velocity in experiments by Castillejos and Brimacombe (1986) and Grevet et al. (1982), except for the gas velocity in a region near the orifice. The results reinforce the notion (initially exemplified by the simulation of a single experiment) that the particle lift and diffusion interfacial forces play an important role during bottom gas injection. Results are also compared with those of Schwarz (1989), who has previously simulated the same experiments, and are found to be similar. The success of both models in simulating these experiments occurs despite some major differences in formulation. Although explicit consideration of the forces responsible for the spreading of the plume does not improve on the overall fit with the above data achieved by modelling such effects by mass diffusion terms, the more fundamental approach has the potential for greater generality.

The validity of both models also shows that, at these low flow rates for which the free surface deformation occurs as a central upwelling rather than a fountain, including the gas top space in the calculation results in little improvement over a model which ignores the top space and assumes a flat top boundary (this has been confirmed by solving the present model equations in a geometry without a top space). This is convenient since the computational effort required is much reduced when the top space is omitted. However, at high flow rates it seems likely that inclusion of the gas top space, to account for the vertical liquid flows therein, will be necessary to achieve correct recirculation in the is bath. Recent efforts by Schwarz et al. (1988) to model the effect of increasing flow rate suggest this

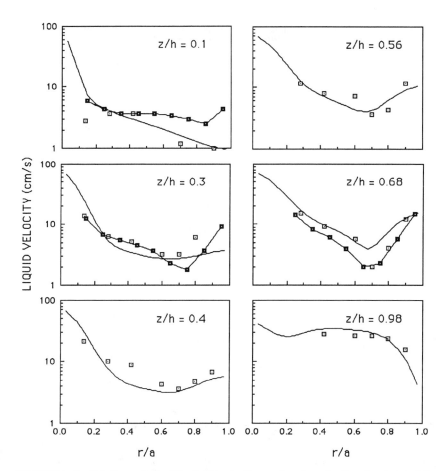

FIGURE 4. Predicted magnitude of the liquid velocity, as a function of radius at various heights above the orifice, compared with corresponding data (□) of Grevet et al. (1982) and the predictions (─■─) of Schwarz (1989) for a nozzle gas velocity of 320 cm/s.

is so, although discretization errors in the plume and errors in modelling the turbulence are complicating factors.

REFERENCES

1. Castillejos, A.H. and Brimacombe, J.K., Structure of Turbulent Gas-Liquid Plumes in Vertically Injected Jets, *SCANINJECT IV, Proc. 4th Int. Conf. on Injection Metallurgy, Lulea, Sweden,* pp. 16:1-16:34, 1986.

2. Cook, T.L. and Harlow, F.H., VORT: A Computer Code for Bubbly Two-Phase Flow, Los Alamos National Laboratory Rept. LA-10021-MS, 1983.

3. Cross, M., Markatos, N.C. and Aldham, C., Gas Injection in Ladle Processing, *CONTROL '84, Proc. 1st Int. Symp. on Automatic Control in Mineral Processing and Process Metallurgy, AIME -SME* pp. 291-297, 1984.

4. Davidson, M.R., Numerical Calculations of Two-phase Flow in a Liquid with Bottom Gas Injection: The Central Plume, *Appl. Math. Modelling,* submitted, 1989a.

5. Davidson, M.R., On Numerical Simulation of the Central Plume Arising from Bottom Injection of Gas into a Liquid, *Appl. Math. Modelling,* to be submitted, 1989b.

6. Drew, D.A., Cheng, L. and Lahey, R.T., The Analysis of Virtual Mass Effects in Two-Phase Flow, *Int. J. Multiphase Flow*, vol. 5, pp. 233-242, 1979.

7. Drew, D.A., Mathematical Modelling of Two-Phase Flow, *Ann. Rev. Fluid Mech.*, vol. 15, pp. 261-291, 1983.

8. Drew, D.A. and Lahey, R.T., The Virtual Mass and Lift Force on a Sphere in Rotating and Straining Inviscid Flow, *Int. J. Multiphase Flow*, vol. 13, pp. 113-121, 1987.

9. Grevet, J.H., Szekely, J. and El-Kaddah, N., An Experimental and Theoretical Study of Gas Bubble Driven Circulation Systems, *Int. J. Heat Mass Transfer*, vol. 25, pp. 487-497, 1982.

10. Harlow, F.H. and Amsden, A.A., A Numerical Fluid Dynamics Calculation Method for All Flow Speeds, *J. Comput. Phys.*, vol. 8, pp. 197-213, 1971.

11. Harlow, F.H. and Amsden, A.A., Numerical Calculation of Multiphase Fluid Flow, *J. Comput. Phys.*, vol. 17, pp. 19-52, 1975.

12 Lee, S.L. and Weisler, M.A., Theory on Transverse Migration of Particles in a Turbulent Two-Phase Suspension Flow due to Turbulent Diffusion - I, *Int. J. Multiphase Flow*, vol. 13, pp. 99-111, 1987.

13. Mazumdar, D. and Guthrie, R.I.L., Hydrodynamic Modelling of Some Gas Injection Procedures in Ladle Metallurgy Operations, *Metall. Trans.*, vol. 16B, pp. 83-90, 1985.

14. Pauchon, C. and Banerjee, S., Interphase Momentum Interaction Effects in the Averaged Multifield Model. Part 1: Void Propagation in Bubbly Flows, *Int. J. Multiphase Flow*, vol. 12, pp. 559-573, 1986.

15. Rivard, W.C. and Torrey, M.D., K-FIX: A Computer Program for Transient, Two-Dimensional, Two Fluid Flow, Los Alamos Scientific Laboratory, LA-NUREG-6623, 1977.

16. Rosten, H.I., Spalding, D.B., and Tatchell, D.G., PHOENICS: A General-Purpose Program for Fluid-Flow, Heat-Transfer and Chemical-Reaction Processes, *Proc. 3rd Int. Conf. on Engineering Software*, pp. 639-655, 1983.

17. Schwarz, M.P., Zughbi, H.D., White, B.F., and Taylor, R.N., Flow Visualization and Numerical Simulation of Liquid Bath Agitation by Bottom Blowing, *CHEMECA 88, Proc. 16th Austral. Chem. Eng. Conf.*, pp. 537-542, 1988.

18. Schwarz, M.P., Numerical Simulation of Gas-Liquid Plumes Occurring in Injection Processes, *J. Fluids Eng.*, submitted, 1989.

19. Szekely, J., Wang, H.J., and Kiser, K.M., Flow pattern velocity and turbulent energy measurements and predictions in a water model of an argon-stirred ladle, *Metall. Trans.*, vol. 7B, pp. 287-295, 1976.

Automated Model Building
in Mineral Processing

T. KOJOVIC and W. J. WHITEN
Julius Kruttschnitt Mineral Research Centre
Isles Road, Indooroopilly
Brisbane, Qld 4068, Australia

INTRODUCTION

Although a physical model is the ultimate goal of any modelling study, most agree that in mineral processing operations the underlying physical theory is still not fully understood, and such a model can rarely be achieved. In this regard, a model based on sound physical concepts, which is then empirically adjusted to represent the real operation, presents the most realistic solution. The development of mathematical models using this approach is largely dependent on parameter estimation and regression techniques.

There is much literature on general model building procedures (eg. Osborne and Watts,1977; Draper and Smith,1981; and Tucker,1985), but little on methods suitable for constructing physical/empirical models of mineral processing units. Whiten's (1971) procedure is the most appropriate (Kojovic,1988) and, unlike the other methods, is comprehensive and explicitly stated. In this procedure, models are developed by using nonlinear least-squares methods to calculate the parameters of the model for each of the individual tests and then by using regression techniques to predict the parameter values over all of the tests. This model building procedure has been successfully applied in developing models of cone crushers and vibrating screens that are still used today. In Whiten's procedure, however, there are still many points which require manual intervention by the model builder, and there is accordingly potential for improvement.

In the context of this work, a model (which is always multidimensional and usually both complex and nonlinear) is used to predict the steady state output from the unit given a description of the unit model parameters and the unit input. It is generally impossible to derive an accurate model from first principles, but there may be some physical theory available to create a structure. This structure may be as simple as a partition curve, or can be based on a detailed physical model.

The data for model building typically consists of a series of tests each giving a description of input and output from the unit plus the values of the operating parameters for each test. Model building at the JKMRC is typically done by collecting data in the form of multiple tests on industrial or pilot plant units and using the data to construct and verify simulation models. This construction and verification may take several months to complete.

REQUIREMENTS FOR AUTOMATED MODEL BUILDING

The literature suggests that model building is largely an art requiring the model builder to make many manual decisions and investigate many alternatives. The aim of an automated model builder was to reduce the number of manual decisions to a minimum and to present only relevant information to the user. Where possible, decisions need to be made automatically and alternatives evaluated for the user.

Clearly this was a very large project that could not be completed without a long series of improvements and upgrades. A subset of limited but significant aims was therefore identified. The literature shows that Whiten (1971) has gone furthest in identifying the steps required for this type of model building. However, he did not address the problems of automated decision making.

It was assumed that the data consists of multiple tests which give input, operating conditions and output of a processing unit with the same values measured for each test. The following subset of Whiten's steps was chosen to form the basis of an automated procedure:

1. Model form specification,
2. Model parameter calculation for each individual test by nonlinear least-squares,
3. Calculation of a small number of possible linear relations for each of the parameters,
4. Nonlinear fitting of coefficients in alternative sets of parameter relations to the complete set of original data values,
5. Evaluation of the alternative nonlinear fits to the data and presentation to the user.

These steps cover the major thrust of Whiten's procedure but follow a more direct approach, thus reducing the number and types of empirical decisions required.

It was clear that various quality criteria would be needed in this project and for uniformity all criteria, including those originating from approximate statistical tests, were converted to a 0 to 100 scale with 100 being the best possible.

DETAILED MODEL BUILDING STEPS

1. Model Form

A model building control program has been developed according to the requirements discussed above. The automated model building techniques have been applied successfully to develop and evaluate models of the form:

$$D \text{ is predicted by } F[\, I, P(C)\,]$$

where
	D	= unit outputs, data vector for each test.
	F	= nonlinear vector function
	I	= unit inputs, data vector for each test.
	P	= model parameter values, vector.
	C	= operating conditions (generally calculated from I), vector.

The actual form of F is up to the user, while P(C) is determined automatically to complete the model. A uniform and flexible structure to interface to the model building program was developed so that the user can specify the model structure once only, rather than as continuing variations as in Whiten (1971). This interface consists of five subroutines which the user is required to write in Fortran prior to using the model builder:

INPUTV	-	data input. INPUTV is called to input the data file so that it is available for use by the other subroutines.
ERRORU	-	error calculation. ERRORU subroutine allows the user to specify the structure of the model using only the model parameters to be used in the objective error function.
PRINT	-	results output. The PRINT subroutine is user specific and enables the user to print information from the main data base and also from the model calculations.
COND	-	operating condition value calculation. The COND subroutine defines the operating conditions that make up the proposed model structure. This structure was chosen for its simplicity and flexibility in specifying different model forms.
PVALUE	-	initial parameter value estimates. The subroutine PVALUE has been incorporated in the interface to optionally provide estimates of the parameter values for each data set using the information contained in the main data base.

2. Parameter Calculation

The first stage in the calculation is to determine the corresponding value of P for each available test. This is done with a nonlinear least squares fitting program that determines the best fit of F(I,P) to data values D.

The majority of nonlinear parameter estimation problems have been formulated using a least-squares approach. Rice et al (1979) claim that these problems make up most of the unconstrained problems that can be solved numerically. They believe that methods tailored to nonlinear least-squares are especially effective, but their exploitation has lagged behind that of methods for the general unconstrained problem.

Levenberg-Marquardt's approach is the most popular because of its general reliability in solving nonlinear least-squares problems (Levenberg,1944; Marquardt, 1963). The method will proceed steadily, if slowly, towards lower sum of squares values even if presented with very poorly scaled functions (Nash and Walker-Smith, 1987). For reasonably scaled problems with good initial parameter estimates, Marquardt's method is very efficient. For most problems, direct solution of the normal (or modified normal) equations is considered to be adequate and computationally less time consuming than orthogonal transformations (Branham, 1987).

The nonlinear fitting program used implements a Levenberg-Marquardt least-squares method using numerical derivatives with some control of derivative accuracy. It also allows the errors (D-F[..]) to be presented in blocks, corresponding to the natural structure of the data. This means storage for a possibly very large number of derivative values is not required.

This algorithm has been well tested at the JKMRC and has the reputation of being robust. However, it is similar to other nonlinear least-squares programs in that convergence may be slow near the minimum and it may have problems getting an accurate minimum for near-singular problems.

The program can be used as a stand alone program, but includes control parameters for automatic model building. These include the separation of control and data files and the generation of files for subsequent model building stages. The run time varies depending on the size of problem (number of data sets, parameters, and errors), but typically is less than 10 minutes on a PC-AT. Adequate starting values are a problem and supplying suitable values was found to reduce run time.

3. Linear Parameter Relations

The next stage in the model building is to determine a set of tentative linear relations for P(C). While linear regression is well defined, the choice of best equations is not. The literature on selecting the 'best' linear regression equation presents no clear solutions for this problem. The situation is best summarised by a quotation from Draper and Smith (1981) on the problem of selecting the best regression equation: "there is no unique statistical procedure for doing this."

Several authors have recently found the same. Healy (1987) concluded that "the problem of determining the importance of different predictors is an extremely tricky one; it is usually ill-defined and there is no universal text book solution."

Several procedures are available for the selection of a subset of independent variables in fitting a multiple regression equation to a set of experimental data. It was found that none of these methods can claim to achieve the optimum solution (Kojovic,1988). In fact, often there is no unique best solution, but instead a small number of possible alternatives worthy of further consideration.

The approach developed by the authors is to generate all possible equations and to rank them in an approximate order of utility using a criterion value. This ensures that all relevant equations are considered. This criterion takes into account the accuracy, the significance of terms in the equation, the complexity of the equation, and also user preferences.

Because the number of calculations increases exponentially with the number of variables in the regression, it was important to have an efficient algorithm for performing the necessary computations. An algorithm based on the Cholesky reduction of the normal matrix was chosen, as this is recognised to be the most efficient way of solving least-squares problems.

This procedure is different to that of Schatzoff et al. (1968) who use a variation of the pivotal inversion technique to a particular row and column of the normal matrix A ($=X^TX=LL^T$) to either introduce the variable corresponding to that row and column into the regression equation, or remove it if it was already in the equation. In the authors' algorithm, only one row is added to the lower triangular matrix L for each new regression equation to be fitted. Furthermore, this technique updates the triangular matrix from the original normal matrix elements and is unlikely to suffer from problems with rounding errors usually associated with pivotal inversion methods.

The problem of selecting the equations which are worthy of further consideration was approached with the objective of developing a criterion value for each subset equation which could be used to sort the equations in order of highest to lowest criterion values. The philosophy behind this criterion is:

1. Equations selected must be accurate (ie. high (1-F-test) value).
2. Equations selected must have significant terms (ie. the minimum (1-t-test) over all coefficients must be high).
3. Other requirements being equal, simpler equations are preferred, hence a factor that gives some preference to simpler equations is required. A preference of 0.9 per additional term (ie. 0.9^{nt-1}) is used.
4. The user (or high level program) needs to be able to specify a preference for certain terms. This may be to bias towards the more physically meaningful, simpler to use, or an external indication (eg. from the model evaluation program) of usefulness.

These four requirements were combined into a single criterion value so that the equations can be put into an order of preference. The four requirements can be trivially converted to the 0 to 100 scale, and then a product of the values is used to create an overall criterion. Fortunately an exact rating is not necessary as the program can present all the equations with high criterion values for further consideration.

The ranking system provided by the criterion value has been proven to be successful in many regression applications encountered recently at the JKMRC. It is beyond the scope of this paper to describe case studies but, it generally gives better results than stepwise regression and Mallow's Cp statistic (Mallows,1973), both of which have a close relationship to the R^2 statistic and result in a selection that is very dependent on the accuracy of the equation.

The run time for the linear regression program is always quite short on a PC-AT. The maximum run time for 10 independent variables is about 3 minutes, with simpler cases running in a few seconds.

4. Nonlinear Fitting over All Data

The various alternative linear relations are inserted into the model and the coefficients in these relations are determined in a nonlinear fit over all the available data (ie., the coefficients in $P(C)$ are found to give a best fit of the functions F_i $(I,P(C))$ to the data D_i over all data sets i).

As in step 2, this is adequately covered by standard techniques. Good initial estimates are available from the linear equations found for the parameters in step 3. However, the amount of data and number of coefficients to fit have both increased significantly and also fits for each of the various combinations of possible linear combinations must be performed.

Nonlinear fitting of coefficients to the complete set of original data values is an important step in the automated model building procedure because of the following:

1. Errors are more likely to be normally distributed.
2. Data values are more likely to have reliable standard deviations, which can be determined externally.
3. Parameter values appear to have a distribution with an extreme tail, making it possible for them to have incorrect values, and they often can be changed significantly without affecting the overall error very much.
4. Parameters are often interrelated and do not behave independently.

The equations considered best (both manually and by automatic criterion) in step 3 seldom give the best overall model. Examining parameter relations from step 3 is often useful but the model quality ultimately needs to be related to the original data. The subroutines used in step 2 to access the data and provide error values for nonlinear fitting of individual tests also provide the required information to enable coefficients to be fitted over all the data.

This step is by far the most computationally intensive and run times on a PC-AT range from a few minutes to several hours for complex models with extensive data.

5. Evaluation of Nonlinear Models

Model validation has been found to be one of the most important steps in the development of simulation models (Wilson, 1977; Snee, 1977; Hoover and Perry, 1984; Balci and Sargent, 1984; Sargent, 1985), but little guidance is available on what techniques should be applied. This is best summarised by Sargent (1985): "Unfortunately, there is no set of specific tests that can be easily applied to determine the validity of the model. Furthermore, no algorithm exists to determine what techniques or procedures should be used."

According to Sargent (1985), these techniques can be used subjectively or objectively, that is, using some type of statistical test or procedure (hypothesis tests, goodness-of-fit tests, tests of normality, and confidence intervals). A combination of techniques is normally recommended. The methods of validation which are commonly used are:

1. comparison of model predictions to the real system data (statistical, graphical, or ad hoc), and
2. face validity, where a knowledgeable person decides if the model is reasonable.

In the development of an automated model builder, there was a need for a method which not only identifies problems with the model, but also indicates how to improve the model (to provide feedback to the decision making steps).

Whiten's (1971) procedure for evaluating the quality of the model is the only technique which may suggest how the model may be improved if found to be invalid. His model evaluation is based on comparison of the model predictions and the data, using regression techniques that test for, and identify, non-random behaviour in the error terms, and the input terms that predict the non-random behaviour. However, in his original paper, Whiten gave no clues on how to quantitatively combine information from each equation or from different errors into a form useful in step 4 or for compact presentation to the user.

Once a set of parameter relations have been determined, the model is evaluated by looking for non-random behaviour in the error terms (D-F[..])/s, ie. observed-predicted output values normalised by an error estimate s. The method selected uses multiple linear regression to determine if the operating conditions can predict any of the error terms. It has been found that the errors that can be predicted can usually be predicted by several equations. This clearly makes the specification of the result from evaluating the model quality more open and complex.

If only random behaviour is found the model has explained all the systematic behaviour of the available data and thus can be considered satisfactory at least until unsatisfactory behaviour is demonstrated. It was decided that the model evaluation results should indicate two things :

1. Which error terms are biased or predictable using the operating conditions, and
2. Which operating conditions should be used to improve the model.

The initial part of the output was meant to address the first question above. The simplest and most appropriate information was considered to be the mean and standard deviation of each error term, the standard deviation of the best regression equation for that error term, and a criterion value based on the F-distribution indicating how much different the standard deviations are.

The second part of the output is intended to give a summary of the relative importance of the operating conditions in being able to predict the error terms and so improve the model. The information required to do this was not easy to identify, and three possible criteria were examined. The criteria or ranking selected was based on all the regression results and considers the benefit of each operating condition in improving the regression, and its significance in the regression.

The program uses an algorithm similar to that of step 2 to systematically and efficiently perform multiple regressions. Experience has indicated that this gives a comprehensive evaluation of models. It has the advantage of being both quantitative and uniform. Alternative models can be easily and quickly compared.

The run time for the model evaluation program is reasonably short on a PC-AT; it varies with the number of error terms and operating conditions. The maximum run time is about 5 minutes, with simpler cases running in less than a minute.

CONCLUSIONS

A control program coordinates steps 2 to 5 described above to provide an integrated package which can run without manual intervention. Details of the data flow in this package are given in Figure 1. Once the necessary base model form has been provided by the user, the program executes the following steps:

1. Model parameter estimation for each data set using nonlinear fitting program (NLFMB).

2. Relationships between the model parameters and the operating conditions, as specified the user, are determined by linear regression program (LREGMB).

3. For each highly rated combination of equations (from step 2) for the model parameters

    a. Nonlinear fit the coefficients to all the data (ie. over all data sets).
    b. Model evaluation of the nonlinear fit in 3a using program (MVALMB).

4. Rating of the models evaluated in 3b in order of an acceptability criterion.

The overall model building time is dominated by steps 3 and 4 which are repeated for each highly rated combination of model parameter equations. Run times on a PC-AT range from less than an hour to a few days for the complex models with extensive data.

The construction of an automated model builder cannot be claimed to be complete until it can automatically create a model, of any unit, from a minimum (perhaps zero) amount of data. This aim can be contrasted with the comment of Draper and Smith (1981): "The important point to remember is that the screening of variables should never be left to the sole discretion of any statistical procedure."

This paper has briefly described the techniques that make an automated model builder possible for models of a particular type. In a demonstration of its use in modelling particle size separators, the automated model builder gave good results for a simple cone classifier unit. For the more complex hydrocyclone, there was an improvement of a factor of 2 in the accuracy of the existing models, but not a definitive result in the form of a single universal hydrocyclone model. This is no doubt related to the greater number of variables in the hydrocyclone model.

The automated model building program has also been used by Baguley (1988) in developing models of dense medium drum separators for his MEng.Sc thesis, Scott (1988) in completing his Ph.D work on modelling dense medium cyclones, and Andersen (1989) in his MEng.Sc work on modelling crushers. Individually, the linear regression and nonlinear fitting sections have been used extensively in numerous other applications.

The automated model builder does not replace the initial intellectual requirements in model building and the model quality and robustness depends critically on the users ability to specify an appropriate base model form and corresponding operating conditions. However for models where a limited range is satisfactory, a simple base is often very adequate. Once the base has been established and perhaps some preferences indicated, the automated model builder does integrate and automate the model completion and evaluation in a manner that justifies its name.

Although the model building run time may extend into a few days, it is a vast improvement over previous practice which typically took several months to produce a model of lower quality. It allows a more comprehensive investigation of a wider range of options in the model form, while freeing the model builder of most of the tedious aspects of the model construction.

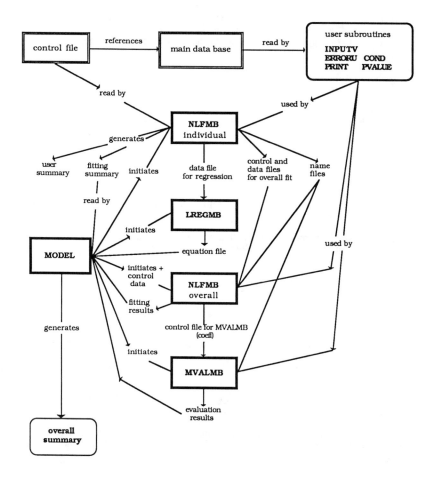

FIGURE 1 : Data flow in model building

MODEL - control program   **LREGMB** - multiple linera regression
NLFMB - nonlinear fitting program   **MVALMB** - model evaluation program

The work described in this paper has clearly demonstrated that automatic model building is practical and of great potential value. The aim of this work was to demonstrate a complete system and this has, by necessity, required that many of the sub-sections are not yet optimum.

The program and techniques used in this model builder are currently being improved and extended as part of a Australian Research Council Project. We believe that this work will result in several significant improvements.

REFERENCES

1. Andersen, J. Development of a Cone Crusher Model, M.Eng.Sc. Thesis, University of Queensland, 1988.

2. Baguley, P. Modelling and Simulation of Dense Medium Drum Separators, M.Eng.Sc. Thesis, University of Queensland, 1988.

3. Balci, O. and Sargent, R.G. A Bibliography on the Credibility, Assessment and Validation of Simulation and Mathematical Models, *Simuletter*, vol.15, pp. 15-27, 1984.

4. Branham, R.L., Jr. Are Orthogonal Transformations Worthwhile for Least Squares Problems? *ACM SIGNUM Newsletter*, vol.22, pp.14-19, 1987.

5. Draper, N.R. and Smith H. *Applied Regression Analysis*, New York: John Wiley & Sons Inc., 1981.

6. Healy, J.R. Interpreting the Results of Multiple Regression, *Communications of the ACM*, vol.30, p. 168, 1987.

7. Hoover, S.V. and Perry, R.F. Validation of Simulation Models: The Weak/Missing Link, Proceedings of the 1984 Winter Simulation Conference, edited by Sheppard, Pooch and Pegden, Dallas, Texas, pp.75-80, 1984.

8. Kojovic, T. The Development and Application of MODEL - An Automated Model Builder for Mineral Processing, Ph.D Thesis, University of Queensland, 1988.

9. Levenberg, K. A Method for the Solution of Certain Non-Linear Problems in Least Squares, *Quarterly of Applied Mathematics*, vol.2, pp. 164-168, 1944.

10. Mallows, C.L. Some Comments on C_p, *Technometrics*, vol.15, pp. 661-675, 1973.

11. Marquardt, D.W. An Algorithm for Least-Squares Estimation of Nonlinear Parameters, *SIAM Journal of Applied Mathematics*, vol.11, pp. 431-441, 1963.

12. Nash, J.C. and Walker-Smith, M. *Nonlinear Parameter Estimation - An Integrated System in BASIC*, New York: Marcel Dekker Inc., 1987.

13. Osborne, M.R. and Watts, R.O. Model Construction and Implementation, *Simulation and Modelling*, edited by Osborne and Watts, Brisbane: University of Queensland Press, pp.3-29, 1977.

14. Rice, J.R. et al. Numerical Computation - its nature and research directions, *SIGNUM Newsletter*, Special Issue, 1979.

15. Scott, I.A. A Dense Medium Cyclone Model based on the Pivot Phenomenon, Ph.D Thesis, University of Queensland, 1988.

16. Sargent, R.G. An Expository on Verification and Validation of Simulation Models, Proceedings of the 1985 Winter Simulation Conference, pp.15-22, 1985.

17. Schatzoff, M., Tsao, R. and Fienberg, S. Efficient Calculation of All Possible Regressions, *Technometrics*, vol.10, pp.769-779, 1968.

18. Snee, R.D. Validation of Regression Models: Methods and Examples, *Technometrics*, vol.19, pp. 415-428, 1977.

19. Tucker, P. Development of a Simulation Package for Flowsheet Design in the Mineral Processing Industries, Paper presented at the seminar Datorer inom mineraltekniken, Luleå, Sweden, November, pp.114-136, 1985.

20. Whiten, W.J. Model Building Techniques for Mineral Treatment Processes, Symposium on Automatic Control Systems in Mineral Processing Plants, Brisbane, pp.129-148, 1971.

21. Wilson, S.R. The Use of Statistics in Model Validation, *Simulation and Modelling*, edited by Osborne and Watts, Brisbane: University of Queensland Press, pp.30-42, 1977.

Mathematics Applied to Subsea Hydraulic Control Systems

P. J. OWEN and A. A. INAYAT-HUSSAIN
BHP Melbourne Research Laboratories
P.O. Box 264
Clayton, Victoria 3168, Australia

E. H. VAN LEEUWEN
BHP Aerospace and Electronics
P.O. Box 264
Clayton, Victoria 3168, Australia

C. J. SIMMONS
BHP Petroleum
G.P.O. Box 1911R
Melbourne, Victoria 3001, Australia

INTRODUCTION

The subsea controls model was developed to enable prediction of
the dynamic response of the hydraulic system used to control oil
production from subsea well-heads. The dynamic response is
determined principally by the shut-in times of the valves (which
control the oil flow) and the magnitudes of the pressure
transients in the hydraulic fluid; these quantities also provide
a measure of the performance of the system. Figure 1 shows a
typical circuit of the control system. The accumulators, valves
and actuators are located on the subsea Christmas tree which can

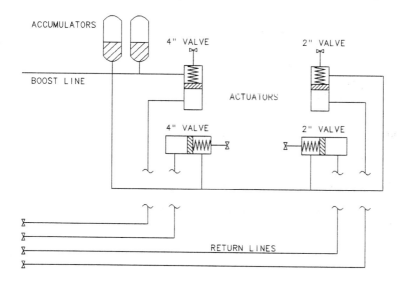

FIGURE 1. Subsea hydraulic control system.

be at a distance of a few kilometres from the control facility,
e.g., a ship. To shut the tree valves, the hydraulic fluid in
the return lines (which is at a higher pressure than the boost
line) is vented at the control facility. The long length of the
hydraulic control umbilicals (i.e., the boost and return lines
on Figure 1) makes it difficult to assess at the design stage
the performance of the hydraulic system. Experimental
determination of the performance of the system is not
economically feasible, and in any case, it does not yield
complete information on the essential parameters affecting the
performance.

The subsea controls model enables accurate simulations of the
real system, thereby clarifying, e.g., the relative importance
of the accumulators. This can assist in the design of control
systems which are of low cost and high reliability, and preclude
the need to use more complicated control systems, such as,
hydraulic sequenced, electronic and multiplexed systems. The
model has already been applied successfully to Jabiru and
Challis field production projects in the Timor sea.

MODEL

The fluid motion during a valve shut-in operation is unsteady
and is accompanied by changes in the cross-sectional area of the
umbilicals. This motion is effectively one-dimensional (i.e.,
along the length of the umbilical) and it can be adequately
described by the mass and momentum conservation equations (Wylie
and Streeter, 1983):

$$b(p) \frac{\partial p}{\partial t} + \frac{\partial Q}{\partial x} = 0 , \tag{1}$$

$$\frac{\partial Q}{\partial t} + \frac{2 Q}{\rho A} \frac{\partial Q}{\partial x} + [A - \frac{Q^2 b(p)}{\rho^2 A^2}] \frac{\partial p}{\partial x} = - \frac{Q|Q| \lambda}{2 \rho A d} , \tag{2}$$

where $b(p) \equiv \partial (\rho A) / \partial p$ and where the variables with their
respective units given in parentheses are defined thus: t is the
time variable (s), x is the distance along an umbilical (m), p
is the fluid pressure (Pa), Q is the fluid mass flow (kg s^{-1}), ρ
is the density of the hydraulic fluid (kg m^{-3}), λ is the friction
factor, d is the internal diameter of an umbilical (m), and A
is the cross-sectional area of an umbilical (m^2). The functional
form of the coefficient $A = A(p)$ can be determined from the
published or measured volumetric expansion data for the
umbilicals. This coefficient is usually a highly nonlinear
function of the pressure (for details, see Inayat-Hussain et.
al., 1987).

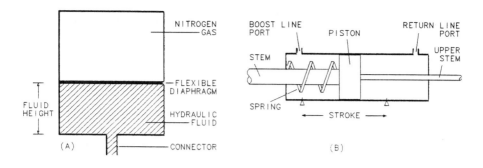

FIGURE 2.   (A) an accumulator and (B) an actuator.

A full description of the dynamic response of the hydraulic
system involves coupling the fluid flow in the umbilicals and in
the accumulators (Fig. 2A) with the mechanical motion of the
actuators (Fig. 2B).   This coupling is achieved through the
physical boundary conditions, namely, the continuity of the mass
flows and pressures.   It is interesting to note that the mass of
the piston (and stem assembly) in the actuators is not required
in the description of the piston motion.   This can be seen from
scaling of the appropriate force balance equation which shows
that the dominant contributions to the force balance come from
the fluid pressures and the spring force.

The specification of the model is completed by prescribing the
initial conditions, which are the boost and return line
operating pressures (which determine the initial volume of fluid
in the accumulators) and the initial positions of the pistons in
the actuators.   These considerations yield an initial boundary
value problem for the hydraulic control system.   Efficient
finite-difference software has been developed for the numerical
solution of this problem (Inayat-Hussain et. al., 1987).   The
development of a dynamic visual display for this software is
currently in progress.

RESULTS

Numerical calculations based on the model display the dependence
of the valve shut-in times on the operating pressures, the
length and diameter of the umbilicals, and on the numbers and
sizes of the accumulators and valve actuators.   The calculations
also yield the fluid pressures and mass flows in the boost and
return lines, height of fluid in the accumulators, as well as
the positions and velocities of the actuator pistons at any
given time.   The values of the mass flows and pressures can be
used to specify the return pipes which drain the high pressure
fluid from the return lines during a shut-in operation.

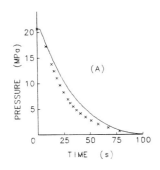
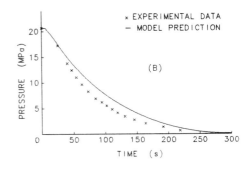

FIGURE 3. Decay of pressure in (A) a 2.8 km umbilical and in
(B) a 5.7 km umbilical.

Thus, for BHP's current design of subsea control systems, the
numerical calculations have indeed revealed the essential
parameters affecting its performance, and in particular
elucidated the subtle role of the accumulators. This has led to
simple rules of thumb for the design of subsea completions;
e.g., doubling the length of the hose effectively doubles the
shut-in times of the tree valves, and the pressure absorbing
effect of the accumulators can be effectively achieved by
marginally increasing the diameter of the boost line.
Significant reductions in the complexity (but with no adverse
effects on the performance) of the subsea control systems have
resulted from applications of our model. Indeed, in one
particular application, the accumulators have been shown to be
totally redundant.

Results of the numerical calculations have shown good agreement
with data obtained from experiments on a single umbilical as
well as from land tests for a particular subsea Christmas tree.

Figure 3A shows the curves for the decay of the pressure at the
closed end of a 2.8 km umbilical which was vented through a
valve with a discharge coefficient of 1 U.S. gal/(min psi$^{1/2}$);
the corresponding curves for a 5.7 km umbilical are shown in
Fig. 3B. The measured and calculated results are in reasonable
agreement. Calculated piston positions and velocities for one
of BHP Petroleum's hydraulic control systems are shown
respectively in Figs. 4A and 4B, where it is worth noting that
the sudden change in the velocities at approximately 22s and 33s
for the 2-inch and 4-inch valves respectively is due to a gate
drag which comes into play over the last 12 mm of the piston
strokes. The valve shut-in times (measured from the moment the
pistons start to move to when they stop) for this control system
are given in Table 1. Again, there is good agreement between
the measured and the calculated results.

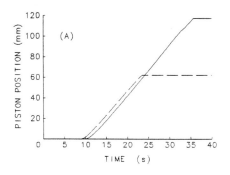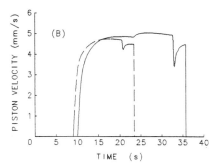

FIGURE 4. (A) positions and (B) velocities of the 2-inch
actuator piston (broken line) and the 4-inch actuator piston
(solid line).

TABLE 1. Valve shut-in times in seconds.

Valve type	Measured	Calculated
4-inch	29	26
4-inch	22	26
2-inch	12	14
2-inch	13	14
2-inch	13	14

The success of the subsea controls model in predicting
accurately the data from land tests has already prompted its
application to a feasibility study of a piloted hydraulic system
for an extensive new development offshore Australia and to
determining whether a particular field should be developed with
a purely hydraulic system or an electrohydraulic system. Our
subsea controls model will serve as a powerful tool for the
future design of hydraulic control systems for oil production
from remote subsea wells.

REFERENCES

1. Inayat-Hussain, A.A., Owen, P.J., van Leeuwen, E.H., and
    Simmons, C.J., Dynamics of the Hydraulic Control System for
    Christmas Tree Valves - a Mathematical Model, *BHP Melbourne
    Res. Labs. Rept.*, no. MRL/PG18/87/001, pp. 1 - 39, 1987.

2. Wylie, E.B., and Streeter, V.L., *Fluid Transients*,
    Corrected edition, FEB Press, Michigan, 1983.

Mathematical Analysis of Free-Surface Flow through a Slot

N. V. QUY
Research and Technology Centre
Coated Products Division
BHP Steel International Group
P.O. Box 77
Port Kembla NSW 2505, Australia

INTRODUCTION

Pre-metered processes have been increasingly favoured as a means for producing thin films of liquid because the thickness of the produced films is dependent solely on the flow-rate of the pre-metered feed stream (Ruschak 1985). Figure 1 shows the diagram of one of such processes which was studied by Ruschak (1976) and Higgens and Scriven (1980). Liquid at a pre-determined flow-rate flows downward and onto a moving substrate between two free menisci: upstream is a meniscus which bridges the gap between the substrate and the confining solid, and downstream is the free surface of the deposited film. Although the variation of physical properties such as viscosity and density with temperature, pressure, and time does not influence the film thickness, the operating substrate speed and, to a lesser extent, the physical properties influence the stability of the liquid film. The stability problem can be solved analytically for very thin films. The solution was developed first by Ruschak (1976) for a simple geometry and extended by Higgins and Scriven to more general geometries. They showed that for a given film thickness, the operation was stable only if the pressure difference between the upstream and downstream menisci was greater than a lower bound and smaller than an upper bound. The pressure bounds varied with the film thickness and the operation was always unstable if the thickness was less than a lower limit. Since the substrate speed was directly proportional to the film thickness for a constant liquid feed-rate, the stable substrate speeds were therefore also bounded by the pressure bounds. Tallmadge el al. (1979) experimentally showed that Ruschak's theory was satisfactory only for large slot sizes but over-predicted the maximum substrate speeds by a factor of about two for small slot sizes. The unexpected experimental result may be explained by noting that, for a constant film thickness, the smaller slot sizes corresponded with the smaller ratios between the slot size and the film thickness. Consequently, Ruschak's assumption of very thin films was not satisfied in the case of the smaller slot sizes.

Ruschak's theory is to be extended to be applicable to thick films as well as thin films. As an analytical solution is not possible for thick films, the finite element method was used to evaluate the stable operating regions for the substrate speeds and to determine the range of pressure difference between the upstream and downstream menisci. Saito and Scriven (1981) reported finite element solutions for the slot coating process; however, they did not evaluate the pressure bounds for stable operations.

THEORY AND COMPUTATIONAL RESULTS

The process can be divided (Silliman 1979) into three independent zones: the downstream, upstream and the interjacent zones. The downstream zone begins at about one slot width upstream of the exit and terminates at the downstream meniscus. The upstream zone begins at the upstream meniscus and has the length of about one slot width. The interjacent zone is situated between the other zones and is fully developed. Each zone can be calculated independently provided that the variables at the zonal interfaces match each other.

Dimensionless Variables

Referring to Fig. 1, let the regions just inside and outside the upstream meniscus be Locations 1 and 0 respectively, the intersection between the downstream zone and the interjacent zone the Location 2, the intersection between the interjacent zone and the upstream zone Location 3, and the gas region just outside the upstream interface Location b. Let the substrate speed be U^*, the size of the exit gap D_0^*, the size of the upstream gap D_b^*, the asymptotic film thickness d^*, the fluid pressure p^*, the velocity in the x-direction u^*, the velocity in the y-direction v^*, the liquid viscosity μ, the surface tension σ and the liquid density ρ. It is assumed that the exit and the upstream gaps are equal; an extension for unequal gap sizes is trivial.

In the system being considered, the characteristic distance is D_0^* and the characteristic velocity is U^*. The dimensional coordinates, flow velocities, and pressure can be expressed in dimensionless forms x, y, u, v, and p as shown in Eq. 1 with Superscript (*) denoting a dimensional variable and the same symbol without the superscript representing its non-dimensional value. Variables are combined in two dimensionless groups: the capillary number, Ca, and film reduction number, Fm; their definitions are given in Eq. 2. The film reduction number represents the fraction of reduction of the film from the size of the exit gap; the smaller is this number, the greater is the reduction.

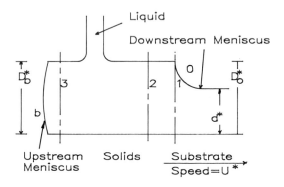

FIGURE 1. Process diagram.

$$x = \frac{x^*}{D_0^*} \quad y = \frac{y^*}{D_0^*} \quad u = \frac{u^*}{U^*} \quad v = \frac{v^*}{U^*} \quad p = \frac{p^*}{\mu U^*/D_0^*} \tag{1}$$

$$Ca = \frac{\mu U^*}{\sigma} \quad Fm = \frac{d^*}{D_0^*} \tag{2}$$

Interjacent Zone

The flow is nearly rectilinear in the interjacent zone and the velocity and pressure distributions can be derived analytically from lubrication theory. In general, the flow consists of two components: the Couette contribution imposed by the relative motion between the substrate and the stationary surface and the Poiseuille contribution by the fluid flowing between two solid surfaces. The velocity distribution, the Couette contribution towards the pressure drop, ∂p_c, and the the Poiseuille contribution towards the pressure drop, ∂p_p, are given by Eq. 3 with D being the clearance at location x.

$$u = 1 + (6Fm - 4)y - (6Fm - 3)y^2 \;\; ; \;\; \frac{\partial p_c}{\partial x} = \frac{6}{(D/D_0^*)^2} \;\; ; \;\; \frac{\partial p_p}{\partial x} = -\frac{12Fm}{(D/D_0^*)^3} \tag{3}$$

The total pressure drop in the interjacent zone is calculated by integrating Eq. 3.

Downstream Zone

Theory for thin films. Ruschak (1976) considered the case where the film thickness was small, the Bond number $[g(d^*)^2\rho/\sigma]$ less than unity. From geometry, the radius of the gas and fluid interface, r_0^*, is bounded by the inequality shown in Eq. 4 for a very thin film and Eq. 5 for a thicker film:

$$\frac{D_0^*}{2} \;\; \leq \;\; r_0^* \;\; < \;\; \infty \tag{4}$$

$$\frac{(D_0^* - d^*)}{2} \;\; \leq \;\; r_0^* \;\; < \;\; \infty \tag{5}$$

Since the difference between the pressures just inside, p_1^*, and just outside, p_0^*, is:

$$p_0^* - p_1^* = \frac{\sigma}{r_0^*} \tag{6}$$

the bounds on the pressure difference are given by Eq. 7:

495

$$\frac{2\sigma}{(D_0^* - d^*)} \geq p_0^* - p_1^* > 0 \tag{7}$$

Equation 7 can easily be changed to a non-dimensional form as shown in Eq. 8:

$$\frac{2}{Ca(1 - Fm)} \geq p_0 - p_1 > 0 \tag{8}$$

Ruschak also proved that the pressure difference for very thin films satisfied Eq. 9:

$$p_0 - p_1 = \frac{1.34}{FmCa^{\frac{1}{3}}} \tag{9}$$

Equations 8 and 9 formed the bounds for the pressure drop across the downstream interface for a thin film. By substituting Eq. 9 into Eq. 8, the relationship between the capillary number and the film reduction number at the onset of instability, Fm_{min}, can be deduced:

$$\frac{Fm_{min}}{(1 - Fm_{min})} = 0.67Ca^{\frac{2}{3}} \tag{10}$$

Finite element solution. To investigate the applicability of Ruschak's model to thick films, a detailed finite element analysis of the downstream zone has been performed. Kistler's method (1983) for solving problems involving free surface flows is used. The ranges of dimensionless numbers used in the calculation are: the capillary number between 0.1 and 5.6, the film reduction number between 0.15 and 0.30. The effect of the dimensionless numbers has been investigated by varying one number while keeping the other numbers constant.

The effect of the capillary number on the shape of the downstream meniscus is shown in Fig. 2. At low capillary number, the interface is completely outside the slot. As the number increases, the surface moves towards the slot and then enters the slot. Eventually, the interface breaks down which is equivalent to the violation of Ruschak's criterion for stable film, Eq. 10. In this work, the stability limit is chosen when the interface penetrates 20% of the slot.

The effect of the film reduction number on the shape of the gas and fluid interface is shown in Fig. 3. At high film reduction numbers, the interface is completely outside the slot. As the number decreases, the surface moves towards the slot and then enters the slot.

Because the pressures along the interface are not constant, a direct comparison with Ruschak's equations for thin films cannot be performed. However, an equivalent pressure at the interface can be calculated. Let p_2 be the fluid presure at Location 2, the boundary between the downstream and the interjacent zones, Δp_c be the

Couette pressure drop and Δp_p be the Poiseulle pressure drop between Location 2 and Location 1. Then the equivalent interface pressure, p_1^e is given by Eq. 11.

$$p_1^e = p_2 + \Delta p_c + \Delta p_p \tag{11}$$

Since the pressure at Location 2 is known from the finite element solution and Δp_c and Δp_p are calculable from Eq. 3, the equivalent pressure at the interface can be evaluated. The computed pressure drop across the interface, $p_0 - p_1$, is shown in Fig. 4. For a given film reduction number, the dimensionless pressure drop decreases as the capillary number increases. When the capillary number reaches a limit, the film becomes unstable. The relationship between the pressure drop and the capillary number at the onset of instability is closely approximated by Ruschak's stability criterion at the upstream interface (Eq. 8). It should be noted that the relationship is hyperbolic: the variation in pressure drop is large at low capillary number and small at high capillary number. Although the Ruschak's stability criterion is satisfactory for a thick film, the formula for evaluating the pressure drop, Eq. 9, does not match the finite element results for high film reduction number. Consequently, the analytical relationship between the capillary number and the film reduction number at the stability limit, Eq. 10, can only be used for films having a film reduction number smaller than 0.25 (see Fig. 5). It is interesting to note the sudden jump in values of capillary number as the film reduction number increases above 0.25. This result is caused by the hyperbolic nature of Ruschak's stability criterion.

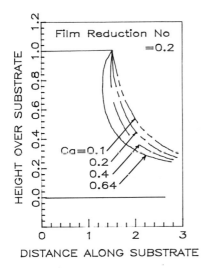

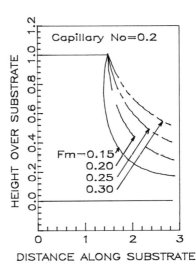

FIGURE 2. Effect of capillary number on downstream meniscus.

FIGURE 3. Effect of film reduction number on downstream meniscus.

Upstream Zone

Higgins and Scriven discussed the evaluation of the bounds on the capillary pressure drop across the downstream meniscus for many different geometries and for pinned (when the surface is fixed to the edge of the corner of the die) and free surfaces. Although their derivation was basically correct, they failed to distinguish two distinctly different situations for convex and concave surfaces, i.e. when the contact angle was greater or less than 90^0. To illustrate this point, let us consider the case when the upstream surface is pinned. If the surface is convex, the bounds on the radius of the surface, r_b^*, can be determined from geometry according to Eq. 12 with θ being the contact angle between the substrate and the fluid.

$$\frac{D_b^*}{1 - cos\theta} \leq r_b^* < \infty \tag{12}$$

Consequently, the bounds on the capillary pressure are given in Eq. 13:

$$\frac{\sigma(1 - cos\theta)}{D_b^*} \geq p_3^* - p_b^* > 0 \tag{13}$$

On the other hand, if the surface is concave, the bounds on radius and pressure are given in Eq. 14 and 15:

$$\frac{D_b^*}{1 + cos\theta} \leq r_b^* < \infty \tag{14}$$

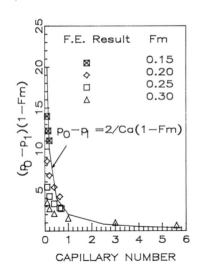

FIGURE 4. Variation of downstream pressure drop.

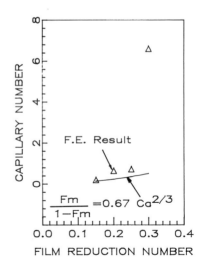

FIGURE 5. Condition at onset of instability.

$$\frac{-\sigma(1+cos\theta)}{D_b^*} \leq p_3^* - p_b^* < 0 \tag{15}$$

Higgins combined Eq. 11 with Eq. 14 and Eq. 12 with Eq. 15. However, because a surface can only be either convex or concave, the combination is meaningless.

Let us consider convex surfaces only, the treatment for concave surfaces is similar. The dimensionless form for the pressure bounds, Eq. 13, is given in Eq. 16:

$$\frac{(1-cos\theta)}{Ca} \geq p_3 - p_b > 0 \tag{16}$$

Total Pressure Difference Between Menisci

The pressure difference between the upstream gas and downstream gas can be evaluated by adding the pressure drops in the upstream, interjacent and downstream zones. As the upstream and downstream gas pressures are bounded, the total pressure difference is also bounded. By adding Equations 8,3, and 16, the bounds for the pressure difference are given in Eq. 17:

$$\frac{2}{Ca(1-Fm)} + \frac{(1-cos\theta)}{Ca} + \Delta p_{1-3} \geq p_0 - p_b > \Delta p_{1-3} \tag{17}$$

where Δp_{1-3} is the pressure drop in the interjacent zone and can be evaluated by integrating Eq. 3.

The general shape of the bounds are shown in Fig. 6. In the dimensionless form, the upper bound is a hyperbolic curve and the lower bound is approximately a straight line. Figure 6 indicates that for a given capillary number and if the pressure difference is within the bounds, the process is stable for some film reduction numbers.

When the film reduction number is specified, the pressure drop across the downstream surface can be evaluated, either from Eq. 9 for thin films or Fig. 4 for thick films. Therefore the pressure difference is only bounded by the pressure bound in the downstream surface. For example, the bounds for thin films are given in Eq. 18:

$$\frac{1.34}{FmCa^{1/3}} + \frac{(1-cos\theta)}{Ca} + \Delta p_{1-3} \geq p_0 - p_b > \frac{1.34}{FmCa^{1/3}} + \Delta p_{1-3} \tag{18}$$

Equation 18 is shown as the shaded area in Fig. 6. Note that the capillary number Ca is limited by Eq. 10.

CONCLUSIONS

(a)   A pre-metered process for producing thin and thick films of liquid was investigated using an analytical method in conjunction with the finite element method.

(b)   The conditions for a stable coating operation were determined.

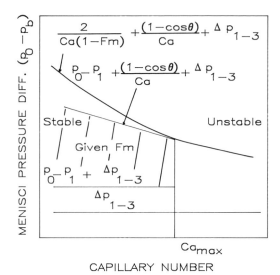

CAPILLARY NUMBER

FIGURE 6. Pressure operating bounds for upstream and downstream menisci.

ACKNOWLEDGEMENTS

The author wishes to thank the management of BHP Steel Coated Products Division for permission to publish this paper.

REFERENCES

1.   Higgens, B.G. and Scriven, L.E., Capillary Pressure and Viscous Pressure Drop Set Bounds on Coating Bead Operability, *Chem. Eng. Sci.*, vol. 35, pp. 673–682, 1980.

2.   Kistler, S.F., *The Fluid Mechanics of Curtain Coating and Related Viscous Free Surface Flows with Contact Lines*, Ph.D. Thesis, University of Minnesota, 1983.

3.   Ruschak, K.J., Limiting Flow in a Pre-Metered Coating Device, *Chem. Eng. Sci.*, vol. 13, pp. 1057–1060, 1976.

4.   Ruschak, K.J., Coating Flows, *Ann. Rev. Fluid. Mech.*, vol. 17, pp. 65–89, 1985.

5.   Saito, H. and Scriven, L.E., Study of Coating Flow by the Finite Element Method, *J. Comput. Phys.*, vol. 42, pp. 53–76, 1981.

6.   Silliman, W.J., *Viscous Film Flow with Contact Lines*, Ph.D. Thesis, University of Minnesota, 1979.

7.   Tallmadge, J.A., Weinberger, C.B., and Faust, H.L., Bead Coating Instability: A Comparison of Speed Limit Data and Theory, *AIChE J.*, vol. 25, pp. 1065–1072, 1979.

Modeling of Batch and Continuous Sedimentation of Suspensions

S. J. SPENCER, D. R. JENKINS, and N. G. BARTON
CSIRO Division of Mathematics and Statistics
P.O. Box 218
Lindfield, NSW 2070, Australia

INTRODUCTION

Sedimentation has been in practical use for thousands of years in batch separation processes. Though in use for almost a hundred years, continuous sedimentation is still subject to improvements in the practical design of thickeners and the theory of continuous thickening.

There is a vast range of applications for the sedimentation process. These include potable water treatment, food processing, industrial water reclamation, leaching and solvent extraction in mineral separation, gravity thickeners, blood fractionation, paint technology and hydraulic transport of mineral slurries. New work on sedimentation applications includes highly efficient flocculation agents, high rate thickeners, solids-contact clarifiers for recirculation of mixtures and deep compression zone thickeners for maximising underflow sediment concentration (see review by Dahlstrom and Emmett(1983)).

Coe and Clevenger(1916) were the first to begin semi-empirical studies of the theory of sedimentation under gravity by establishing the existence of four zones of settling in batch sedimentation of metallurgical slimes. These are (with increasing depth in the sedimentation vessel) an upper layer of clear liquid from which all particles have already fallen, a layer of sedimenting particles at constant density and settling rate, a layer of continuously increasing density with depth and finally a compression zone at the bottom of the sedimentation where the sediment flocs or individual particles are in contact but still compressing downwards under the overlying weight of particles. Free settling of particles above the compression zone is seen as the result of gravitational and hydrodynamic drag forces and purely a function of local sediment concentration while in the compression zone the sedimentation velocity is a function of both concentration and depth. Liquid is squeezed from the interparticle volume and, for flocculated systems, from the aggregations of particles themselves, through channels or pore structures of varying size.

Kynch(1952) developed a mathematical model for batch sedimentation based on the assumption that settling rate depends only on local particle concentration. This analysis was the first real attempt at a rigorous though limited quantitative theoretical model for batch sedimentation. Settling rate is seen as being dependent on a power law of local particle concentration in the "free settling" zone. Recognising that compaction of sediments at higher densities occurs at more appreciable rates than suggested by free settling, Kynch envisaged a compression zone where the settling rate is due to Darcy type flow of the liquid through the porous media of compacting sediment.

Study of flocculated suspensions led Michaels and Bolger(1962) to a model assuming

501

compaction of sediments due to gravity, drag forces and stresses transmitted through particles in contact. Tiller(1981) extended Kynch theory to regions of compacting sediment by recognising the different physical mechanisms of free settling and the compression zone. Another model of sedimentation based on Darcy's law for flow in porous media was formulated by Blake et al.(1979).

Auzerais et al.(1988) solved both momentum and continuity equations to obtain accurate concentration profiles across regions of rapid variation of concentration for batch settling. Huang and Somasundaran(1988) used stochastic theory of particle settling based on gravitational forces and random Brownian motion of particles due to collision with other particles.

Park et al.(1983) and Fryer and Uhlherr(1980) have made a steady state analysis of continuous sedimentation, for various forms of sedimentation velocity by including the bulk fluid velocity caused by the system input and output in their calculations. To date though, continuous thickeners are largely based on the semi-empirical results and techniques of Coe and Clevenger.

Optimisation of thickener design is achieved by determining settling rates either in a series of batch tests with different initial concentrations (Coe and Clevenger(1916)), or by analysis from a single batch test (Talmage and Fitch(1955)). The aim of the optimisation is to minimise the area of settling necessary for a given feed rate and concentration while ensuring clear fluid (or very low sediment concentration) is removed from the top, and also to maximise feed rate and underflow sediment concentration for a steady state with clear supernatant fluid.

There is a need for a simple, flexible solution to the sedimentation problem, that is quantitatively accurate over all flow regimes in both flocculated and simple non-colloidal suspensions. Kynch theory is still one of the simplest ways of viewing the problem, provided suitable extensions can be made for continuous flow and flocculation effects.

Due to the complex nature of continuous flow problems, use of a mathematical model is imperative in order to take into account effects such as feed rate, location and concentration; underflow rate and concentration; effective thickener area; addition of flocculant; overflow clarity; recycling of underflow. Such a broad scope of parameters effectively rules out an empirical approach to optimisation of the process. As industrial sedimentation is a process whose efficiency is often crucial for plant operation, accurate and simple mathematical modelling is therefore economically very desirable.

In this paper we present a model of batch and continuous sedimentation based on the Kynch theory. The method of characteristics is used to track the progress of discontinuities in sediment concentration and results are presented for batch sedimentation. The various steady states which may occur in continuous settling are presented and we conclude with a description of the proposed method for solution of the transient continuous sedimentation problem.

BATCH SEDIMENTATION

In Kynch theory, the settling rate can be determined entirely by a continuity equation which is a quasi-linear, first order P.D.E., thus ignoring the complications inherent in also solving a momentum equation. In the simplest approximation a one-dimensional analysis is sufficient with particles falling vertically under gravity, which leads to the continuity equation

$$\frac{\partial C_s}{\partial t} + \frac{\partial (v_s C_s)}{\partial x} = 0 \tag{1}$$

where C_s is the local sediment concentration, $v_s(C_s)$ is the local sediment velocity in a column. Equation (1) leads to

$$\frac{\partial C_s}{\partial t} + \gamma(C_s)\frac{\partial C_s}{\partial x} = 0 \tag{2}$$

where

$$\gamma(C_s) = \frac{d(C_s v(C_s))}{dC_s} \tag{3}$$

Solution of equation (2) is achieved by the Method of Characteristics. The characteristics are straight lines of slope $\gamma(C_s)$, along which C_s is constant. The shape of the flux curve $v_s C_s$ against C_s is crucial in determining the behaviour of the sediment.

A parametric form is proposed for the sedimentation velocity that takes into account both free settling and a compression zone at higher concentrations.

$$v_s = v_0(C_m - C_s)^q + v_1 C_s(C_m - C_s) \tag{4}$$

where v_s is the sedimentation velocity, v_0 is the superficial velocity due to free settling, C_m is the maximum concentration of sediment possible, q is a positive constant and v_1 is the superficial velocity due to Darcy type flow of liquid through sediment particles in contact. This form allows for a wide variety of behaviour which is compatible with experiments and previous analyses (Kynch(1952), Auzerais et al.(1988), Fryer and Uhlherr(1980), Park et al.(1983)).

The Method of Characteristics may allow a multiplicity of solutions where characteristics intersect. This is indicative of shock fronts which are regions of rapidly varying concentration separating continuous regions of very different concentration. A shock propagates through the sediment at a speed (Carrier and Pearson(1976))

$$U = \frac{C_s^+ v_s(C_s^+) - C_s^- v_s(C_s^-)}{C_s^+ - C_s^-} \tag{5}$$

where C_s^+ is the concentration just above the shock and C_s^- is the concentration just below it. These concentrations can be determined by using a Newton-Raphson method on the equation of intersection of the characteristics, given their initial position and the current time and position.

A numerical scheme has been developed which allows shocks to be tracked over time until a suspension is fully compacted. Favourable comparisons of results of this method with experiments on aqueous calcium carbonate suspensions have been presented elsewhere (Barton et al (1988)). They assumed a parametric form for the sedimentation velocity similar to (4), but with $v_1 = 0$.

Fig 1(a) shows a plot of the flux curve for a sedimentation velocity derived from eq (4) with $v_1 \neq 0$, and the resulting transient settling behaviour is evident from the shock diagram in fig 1(b).

There are four different shocks evident in fig 1(b), comprising

1. Falling free surface which is a front dividing clear, supernatant liquid from the sediment at initial concentration. This shock falls from the top of the column at a constant rate equal to the sedimentation velocity of the initial concentration.

2. An intermediate shock propagating upwards from the bottom of the column. This separates the region of constant, initial concentration from a region of increasing concentration nearer the bottom of the column.

3. A shock to maximum incompressible concentration propagating upwards from the bottom. This occurs in the so called compression zone where sedimentation is dominated by Darcy type flow.

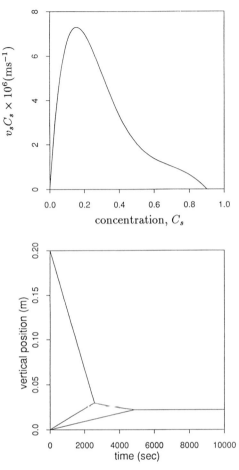

FIGURE 1. (a) Graph of $v_s C_s$ versus C_s, called a flux curve, derived from the parametric form of equation (4) with $v_0 = -0.2$, $v_1 = -0.01$, $q = 5$ and $C_m = 0.9$, where all velocities are in mm/s. (b) Location of shocks in a sedimentation column, for the above flux curve. The initial concentration in the column was $C_s = 0.1$.

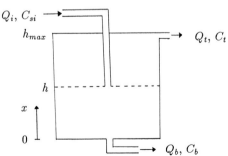

FIGURE 2. Schematic of a continuous flow sedimentation column of cross-sectional area A.

4. Falling free surface separating the clear, supernatant liquid later in time from the underlying zone of increasing concentration with depth. This is the so-called "falling rate period" where the descent of the top of the suspension slows down in accordance with the increased concentration immediately beneath the shock front (Auzerais et al.(1988) and Fitch(1971)).

CONTINUOUS SEDIMENTATION

Thickening tanks are modelled as a single feed at height h in a tank of constant cross-sectional area A, with an underflow removing concentrated solution at the bottom and an overflow removing liquid from the top at height h_{max}, as shown in fig 2.

Total volume is conserved in the system, as is the volume of the sediment. Hence

$$Q_i = Q_t + Q_b \tag{6}$$

at all times, and at steady states

$$C_{si}Q_i = C_t Q_t + C_b Q_b \tag{7}$$

Three dimensional effects will occur near the feed and outlets but to first order they are ignored. The concentration weighted bulk flow velocity V is defined as

$$V = u_s C_s + u_w(1 - C_s) \tag{8}$$

where u_s, u_w are the total velocities of the separate phases. The velocity v_s is therefore the sedimentation speed relative to the bulk fluid motion. By modelling the feed as a point source at $x = h$

$$q_i = Q_i \delta(x - h) \tag{9}$$

we obtain

$$V(x) = \frac{-Q_b + Q_i H(x - h)}{A} = -V_b + V_i H(x - h) \tag{10}$$

where $H(x - h)$ is the Heaviside step function. Kynch theory may be modified for continuous settling, by including bulk flow and feed injection terms in the equation of continuity (1), which then becomes

$$\frac{\partial C_s}{\partial t} + \frac{\partial}{\partial x}[C_s(v_s + V)] = \frac{q_i}{A}C_{si} \tag{11}$$

or equivalently,

$$\frac{\partial C_s}{\partial t} + (\gamma + V)\frac{\partial C_s}{\partial x} = \frac{Q_i}{A}\delta(x - h)(C_{si} - C_s) \tag{12}$$

Steady States

At steady state eq (11) becomes

$$\frac{\partial}{\partial x}[C_s(v_s + V)] = V_i C_{si}\delta(x - h) \tag{13}$$

which upon integration gives

$$\Phi = [v_s(x) + V(x)]C_s(x) = V_i C_{si} H(x - h) + [v_s(0) + V(0)]C_s(0) \tag{14}$$

where Φ is the flux corrected for the bulk motion. Fig 3 shows a graph of the corrected flux based on the parametric form of eq (4), with $v_0 = -0.1$, $v_1 = 0$, $q = 5$, $V_i = 0.04$ and $V_b = -0.02$, where all velocities are expressed in mm/s.

505

Below the injection point, $H = 0$, so eq (14) becomes

$$[v_s(x) - V_b]C_s(x) = [v_s(0) - V_b]C_s(0) = \Phi_b, \text{ a constant.} \tag{15}$$

Above the injection point, $H = 1$, so eq (14) becomes

$$\Phi_a = [v_s(x) + (V_i - V_b)]C_s(x) = V_iC_{si} + [v_s(0) - V_b]C_s(0) = \Phi_b + V_iC_{si} \tag{16}$$

The corrected flux above the feed is therefore also a constant. For a clear overflow, Φ_a must be zero. Three possible steady states may exist, depending on the value of the feed concentration, C_{si}:

1. The concentration throughout the column equals the feed concentration, C_{i1} (overflow case). The concentration at the top outlet is C_{t1} and at the bottom outlet is C_{b1}, as shown in fig 3.

2. A jump from zero concentration above the injection point to concentration C_2 below the injection point (underflow case). The concentration at the bottom outlet is C_{b2}, as shown in fig 3.

3. A jump from zero concentration to the concentration C_{i3} at some height above the injection point. (limiting load case). The concentration is C_{i3} throughout the rest of the column and the concentration in the bottom outlet is C_{b3}, as shown in fig 3.

Underflow occurs for concentrations less than the critical value C_{i3}, where the upward bulk flow of the liquid above the feed is exceeded in magnitude by the downward sedimentation velocity. This means all the sediment falls to the feed level and below, leaving the region above clear. Overflow occurs for concentrations greater than C_{i3}, where bulk flow upwards is sufficient to carry sediment to the top of the column.

Knowing the feed sediment concentration then allows graphical solution of the problem in terms of possible discontinuities and concentrations in the sedimentation column and at the outlets. Consequently it is straightforward to determine if the system is underloaded, overloaded or critically loaded for steady states. More complicated steady state behaviour is possible, depending on the nature of the flux curve (Fryer and Uhlherr(1980) and Park et al.(1983)).

What cannot be determined from a steady state analysis is the time for the system to reach equilibrium and the position throughout the column of the various possible discontinuities. These considerations may be extremely important for practical applications of sedimentation theory. In order to determine this information a transient analysis must be performed.

Continuous Sedimentation Transient Solution

Solution of the time-dependent continuous flow problem may be achieved by the Method of Characteristics. In this case, characteristics are straight lines of constant concentration until they cross the feed height, where the concentration may change due to the injection term. As the feed is idealised as a point source, the concentration and hence the slope of a given characteristic simply changes to another value at which it remains and hence the characteristic is again a straight line. This analysis once again allows for the possibility of shocks, dependent on the shape of the flux curve corrected for the bulk flow (see Fryer and Uhlherr(1980) and Park et al.(1983)). In fact, the transition from one steady state to another occurs by the development and propagation of shocks. Work is currently in progress to develop software that tracks the position of shocks and hence enables an examination of the development of steady states as well as the transition to a new steady state when one of the parameters (such as the feed concentration) is varied.

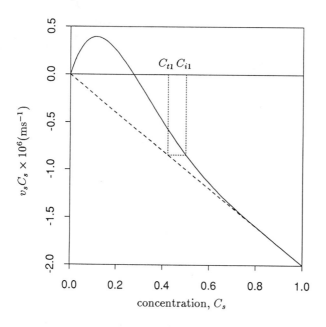

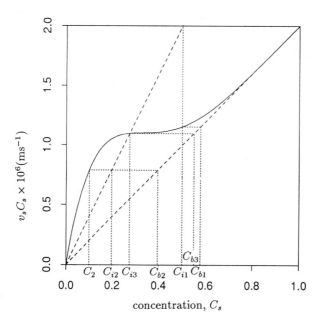

FIGURE 3. Corrected flux plots for continuous sedimentation, for the parts of the column (a) above the feed level and (b) below the feed level.

DISCUSSION

In this paper we have described the basis of a simple approach to the mathematical modelling of batch and continuous sedimentation. The ultimate aim is to produce appropriate software for the analysis and control of operating sedimentation systems. It is obvious that further development is necessary, in order to include effects such as the addition of flocculants, before such a task can be completed. Nevertheless, the paper illustrates the amount of information and understanding of the process that can be obtained quite rapidly by means of mathematical modelling. This is a significant benefit to the analysis of sedimenting systems, because the timescale involved in experimentation may be days or even weeks, due to the extremely slow settling rates of some systems. Consequently, appropriate mathematical modelling is very cost-effective. The model allows rapid analysis of a wide range of operating regimes.

The benefits of mathematical modelling to process control in continuous sedimentation systems is also important. The work described here should provide information about the transition from one steady state to a new steady state due to change of input, including the time necessary to achieve a steady state. There is no other way to obtain such useful information.

REFERENCES

1. Auzerais, F.M., Jackson, R. & Russel, W.B. The resolution of shocks and the effects of compressible sediments in transient settling. *J. Fluid Mech.*, vol. 195, pp. 437-462, 1988.

2. Barton, N.G., Jenkins, D.R., Collings, A.F. & Hung, W.T. Sedimentation in concentrated suspensions. *Xth Int. Congress on Rheology, Sydney*, pp. 173-176, 1988.

3. Blake, J.R., Colombera, P.M. & Knight, J.H. A One-Dimensional Model of Sedimentation Using Darcy's Law. *Separation Science and Technology*, vol. 14, no. 4, pp. 291-304, 1979.

4. Carrier, A.F. & Pearson, C.E. *Partial Differential Equations - Theory and Technique.* Academic Press, 1976.

5. Coe, H.S. & Clevenger, G.H. Methods for determining the capacities of slime-settling tanks. *Trans. AIME*, vol. 55, pp. 356-384, 1916.

6. Dahlstrom, D.A. & Emmett, R.C. Recent Developments in Gravitational Sedimentation. *Proc. - 3rd Pacific Chem. Eng. Congress*, vol. 1, pp. 92-99, 1983.

7. Fitch, B. Batch Tests Predict Thickener Performance. *Chemical Engineering* August 23, pp. 83-88, 1971.

8. Fryer, C. & Uhlherr, P.H.T. Continuous Thickeners - Behaviour and Design. *Eighth Aust. Chem. Eng. Conf., Melbourne*, pp. 80-84, 1980.

9. Huang, Y.B. & Somasundaran, P. Discrete modelling of sedimentation. *Physical Review A*, vol. 38, no. 12, pp. 6373-6376, 1988.

10. Kynch, G.J. A theory of sedimentation. *Trans. Faraday Soc.*, vol. 48, pp. 166-176, 1952.

11. Michaels, A.S. & Bolger, J.C. Settling rates and sediment volumes of flocculated kaolin suspensions. *Ind. Engng. Chem. Fundam.*, vol. 1, pp. 24-33, 1962.

12. Park, K.H., Andrews, J.R.G. & Uhlherr, P.H.T. Batch sedimentation and thickener behaviour. *Proc. - 3rd Pacific Chem. Eng. Congress*, vol. 1, pp. 100-105, 1983.

13. Talmage, W.P. & Fitch, E.B. Determining Thickener Unit Areas. *Ind. Eng. Chem.*, vol. 47, pp. 38-41, 1955.

14. Tiller, F.M. Revision of Kynch sedimentation theory. *AICHE J.*, vol. 27, pp. 823-829, 1981.

Inverse Problems in Machine Vision

D. SUTER
Department of Computer Science
La Trobe University
Bundoora 3083, Australia

INTRODUCTION

The theory of regularization has been applied to many real world problems that are inverse problems (Tikhonov and Goncharsky, 1987). Included amongst these problems are some involving restoration of an image suffering from some degradation. Recently, the theory of regularization has been promoted as the correct theory for formulation of problems in vision (Bertero et al, 1988).

The regularization procedure usually results in the formulation of the problem as a minimization of a functional. This functional cannot generally be solved analytically and thus one needs to develop computational techniques for efficient solution. This paper reviews the regularization based approaches to problems in low level vision: from the problem formulation to the algorithm specification. The second section discusses the regularization formulation, the third section illustrates the different routes from formulation to algorithm specification. The fourth section summarises particular implementation approaches including current work on a transputer based hypercube. The final section provides an illustration using the shape from shading problem.

INVERSE PROBLEMS AND REGULARIZATION

The promotion of the treatment of vision problems as ones falling within the bounds of regularization theory was due to the work of Bertero, Poggio, Torre and Marroquin (see (Bertero et al., 1988) for a review). It seems natural to regard vision as a collection of inverse problems. We receive sensed data by the pattern of light falling upon the retina; this pattern of light is determined by the characteristics of the objects within the scene, and it is generally a subset of these characteristics that one wishes to recover. We will see how many of the problems of low level vision can be formulated as instances of a model regularization formulation.

The Model Problem

Let us denote by f the set of values we wish to recover (it can be viewed as a function $f(x, y)$ giving the depth to the surface point imaged at retinal coordinates (x, y), or it can represent other more complex features), and let g be the sensed data. It is usual to consider that $g = Af + \epsilon$; where A represents the known operator relating the real world features to the sensed data, and ϵ represents the unknown "distortions" such as sensor noise (the latter is usually assumed to have a known statistical distribution). If A is invertible, and the noise is negligible, we could try to simply invert the above equation, however, such inverse problems are often ill-posed in the sense that they violate at least one of the three conditions for a well-posed problem:

- the solution must exist

- the solution must depend continuously on the data

- the solution must be unique.

A common regularization approach replaces the original problem with a series of closely related problems of finding f minimizing:
$$\Phi_\lambda(f) = \|Af - g\|_X + \lambda\|Cf\|_Y.$$
Intuitively, the first term penalizes solutions that are less faithful to the data whilst the last term imposes some suitability criteria which may embody prior knowledge of the form of the solution (usually it is a smoothness criteria). The choice of λ can be thought of as balancing fidelity to data with the constraints due to expectations on the form of the solution. To fully specify the problem we have to specify the norms $\|.\|_X, \|.\|_Y$, the smoothness operator C, and the *regularization parameter* λ. Ideally we would like to be able to prove analytically that the above three conditions are met for our reformulated problem. In practice this is very difficult for the complex functionals generally used in vision problems. Furthermore, we would like to be able to derive analytically solutions or, if not, derive efficient methods for approximation of the solution. Rarely in the problems in vision can one derive analytic solutions, we discuss some methods for deriving approximation methods in subsequent sections of this paper.

Stochastic Formulation

The stochastic formulations stem from the pioneering work of Geman and Geman on reconstruction of images (Geman and Geman, 1984). It is possible to use this approach for bayesian regularization of problems in vision (Marroquin, 1985). In general one maps the problem onto a Markov Random Field (MRF) model where the probability of a solution f (given the data g) is given by:
$$P(f|g) = \tfrac{1}{Z}exp\big(\tfrac{(-\|Af-g\|_X)}{kT}\big)exp\big(\tfrac{-(\|Cf\|_Y)}{kT_0}\big),$$
where the first factor is a normalisation factor (partition function in statistical mechanics) and the two exponentials represent probability factors for the data g given the solution f, and the apriori probability of f, respectively.

If one takes the logarithm of the above equation one realises that maximization of the above probability distribution is equivalent to minimization of a functional of the same form as our previous deterministic formulation. However, the statistical formulation is more general, and the recognition that the functional is usually one with many extrema, allows one to use the statistical mechanics models (such as simulated annealing) to follow the evolution of the above MRF to avoid being trapped in local minima of the functional. There are many possible variations one can use in solving problems using such a stochastic formulation (Suter and Cohen, 1988).

Choice of Parameters

The formulations outlined above involve the estimation of a single parameter λ (although this parameter can be made spatially varying and thus, in the discrete case one has a different parameter at each image position). In areas such as smooth approximation to contaminated data, and image reconstruction, the choice of λ has been examined in some detail. In these problem domains the parameter λ reflects the confidence in our data and is related to the noise level. If λ is too large then the solution will be too smooth, if λ is too small the problem may become unstable to small amounts of noise.

One of the most often promoted methods involves cross-validatory techniques. Generalised Cross Validation minimises a quadratic risk criterion $G(\lambda)$. Thompson et al(Thompson et al, 1988a) (Thompson et al, 1988b) investigate experimentally several methods of choosing the regularization parameter for a image restoration problem, including cross-validation. They found that there were some cases where either $G(\lambda)$ had multiple minima, no minima for $\lambda > 0$, or an ill-defined minima (making practical solution for the minimum difficult). Furthermore, they report some instances where the value of λ_{GCV} so chosen leads to an unsatisfactory solution to the regularization problem.

Accuracy of Formulation
The formulation outlined above seems plausible; however, there are many rather arbitrarily chosen features of a complete formulation especially when one tries to allow the possibility of the reconstructed function having a few discontinuities (the number and location of which is not known). Current proposals for incorporating such "controlled continuity" (Geman and Geman 1984)(Terzopoulos 1988) introduce many parameters that have to be estimated; as well as making the functional very complex and nonlinear.

Incorporating Hard Constraints
Accuracy of the formulation is a major concern with the penalty function based regularization methods: the solution will only meet the constraint $Af = g$ if λ approaches zero. This can be a particular problem when the constraints we wish to impose are *hard* constraints. In optimization theory the classic way to incorporate constraints is to form the Lagrangian and to use Lagrange multipliers. The Lagrangian formulation turns a constrained functional minimization problem into an unconstrained search for a saddlepoint ((Becker et al, 1981) Vol. II p 111). Horn and Brooks (Horn and Brooks, 1986) tried to use the Lagrangian formulation in the shape from shading problem without much success.

Recently, the author has investigated a mathematical programming approach based upon the independent proposals of Snyman (Snyman, 1988) and Platt (Platt, 1987). Essentially, one performs gradient ascent on the Lagrange multipliers and descent on the primary variable of the problem. The author has investigated applying this method to several problems in vision (Suter 1989); numerical convergence and stability can be a problem but implementation in analog for remains a distinct possibility (Umminger and DeWeerth, 1989).

ALGORITHM FORMULATION
Usually one formulates the problem in a continuous setting, as outlined above. One can discretize this formulation directly, or one can derive the corresponding Euler-Lagrange equations first and then discretize.

There are other alternatives such as to recognise the problem as discrete and formulate directly as a discrete functional (using finite elements or using finite difference analogs). Furthermore, one can try to solve the discretised functional minimization directly, without proceeding to a Euler-Lagrange formulation. For example, one can perform a simulated annealing based stochastic search, or one can try dynamic programming type approaches. The latter of these has found limited application in vision as it becomes intractable for 2-d problems; it has, however, been used on 1-d versions of the model problems to derive exact solutions for comparison purposes (Blake, 1989) (Marroquin, 1985).

Discretization

One can either adopt a Finite Difference or a Finite Element approach to discretization of the functional. The application of finite element methods to visual problems was pioneered by Terzopoulos in his approach to surface interpolation to regularise the reconstruction of surfaces from feature based stereo data (Terzopoulos, 1988). The advantages of this method over finite difference formulations are that the method naturally deals with irregular nodal spacing, and that the supporting theory enables more accurate approximations to be derived in a rather elegant way. A discussion of the theoretical properties of this method is beyond the scope of this paper: the reader will find that books (Becker et al, 1981) provide an extensive coverage of the theory.

IMPLEMENTATIONS

If we take an approach based upon the Euler-Lagrange equations we then have a system of coupled algebraic equations. If the system of algebraic equations is linear, we can be cast it in the following matrix equation form: $Bf = d$. We can often then solve this system of equations using fairly standard direct or iterative methods. If the system is nonlinear, we generally must opt for iterative methods, and less is known of how to derive such methods with guaranteed stability and convergence speed.

We can avoid the nonlinear Euler-Lagrange equations by trying one of the deterministic or stochastic search methods directly on the discretized functional. The latter methods are preferable as they can avoid being trapped in local minima.

In any method we have an enormous number of variables to manipulate and parallel implementations are generally to be preferred.

Matrix Equation Solution

For the linear cases we can exploit standard matrix inversion techniques using either direct or iterative methods. Our system is typically very large and sparse, this generally suggests iterative methods since the matrix need not be stored and manipulated in full. Fine grained iterative methods also suggest cooperative methods akin to those presumed to operate within the human visual system. They also have the advantage that, in some cases, partial solutions can be adequate for certain purposes - especially when processing time is crucial (e.g. obstacle avoidance without full recognition of the nature of the obstacle). Iterative methods can be improved using multigrid acceleration.

Stochastic Methods

Many of the stochastic methods are version of simulated annealing using either the Metropolis, the Heat Bath, or the Gibbs Sampler approach (Marroquin, 1985). One can also use microcanonical approaches that may have computational advantages in some cases (Barnard, 1988). These methods have also inspired deterministic "mean field" approaches. The processing time is generally far too long for standard hardware but well suited to parallel fine grained processors (e.g. connection machine) or to analog implementations (where the iteration count becomes the time constant of the circuit). For the complex multi-modal functionals constructed in vision, such methods promise efficient solution methods that avoid the problems of local minima.

Parallel Implementation
One expects that the translational invariance symmetry inherent in the problem will be reflected in a parallel formulation of the solution where individual processors can carry out identical processing on subimages. This type of parallelism is usually termed "data parallelism" or "geometric parallelism". The computational demands of the algorithms suggested can be of the order of hours of processing time even for images as small as 64×64. Since computation is proportional to the square of the linear dimension of an image a more realistic size of 256×256 image requires 16 times more processing time. A "simple" solution is to divide the image into 16 parts and have a processor such as a transputer process each subimage.

Work has commenced on implementing such algorithms on an 8 transputer array organised in a cube. This system uses the *Express* operating system devised for hypercubes at Caltech (Fox et al, 1988) as it provides support for automatic problem decomposition and message routing, this removes many of the problems of developing a flexible yet easy to use and program array of transputers (Suter et al, 1989).

Analog Implementation
It has been suggested (Mead 1989) that analog methods of solution of these problems are the only feasible alternatives for real time performance; such analog implementations are usually variations on a resistive network incorporating op-amps. These analog networks are thought to be close models of neural network apparatus in biological vision. It is interesting to note that the multilayered network proposed by Harris (Harris 1989) is analogous to the mixed finite element ((Becker et al, 1989) Vol. II) approach for approximating a high order problem by several coupled lower order problems.

RECOVERY OF SHAPE
If we look at Fig. 1 we usually have no trouble in recognising a 3-d spherical shape. Indeed, this synthetic image was generated using a shading model to produce precisely this impression. This reconstruction of 3-d information from 2-d data, used by our visual system, is a remarkable achievement which we hope to understand. In this section we will first introduce the image model used to generate the image, then we will discuss the inverse problem of reconstructing the 3-d shape.

Lambertian Shading Image Model
The lambertian shading model is a common and simple model used in computer graphics. We essentially assume that the surface is perfectly matt, reflecting equally in all directions away from the surface. For this model we find the the radiance R leaving the surface in any direction is related to the irradiance I, the surface normal n and the surface albedo A by (Horn, 1986)(Chapter 10): $R = A(\vec{n} \bullet \vec{I})$. Given the description of a surface that allows the normal to be derived, the forward problem of computing the image is rather straightforward. We are interested here in the inverse problem which will be discussed next.

Inverse Problem - Recovery of Shape
We have already noted that Fig. 1 usually evokes a perception of a sphere: we have somehow solved the inverse problem. This inverse problem is not simple and can easily be shown to be ill-posed. In particular the perceived spherical shape is not

FIGURE 1 (left) and 2 (right) - Lambertian Spheres

the only solution - the solutions are many fold. For example, the light distribution could be due entirely to albedo changes rather than surface changes (Indeed for this **image** it actually is!). There are infact two possible spherical surfaces consistent with the data (concave up or down). This is typical of many problems in vision, the first formulation is underconstrained; we usually try to solve the problem by adding in constraints related to assumptions and heuristics that we believe the visual system uses in arriving at a stable and (usually) accurate interpretation. Many previously proposed solutions use constraints of known surface orientation along a curve (such as the occluding boundary of the "edges" of the sphere): these, of course, beg the question of how one recognises an occluding boundary. Furthermore, most schemes assume constant albedo and smooth surfaces free of any discontinuities. An excellent reference is (Horn and Borrks, 1986).

On the face of it, one could suggest that, since we see a sphere, we assume the albedo and irradiance is constant; changes are due to orientation. However, this assumption still leaves two interpretations valid for spherical surfaces and would rule out the usual interpretation of Fig. 2: usually we see a "pool ball" type shape where the dark band is seen as a surface marking (albedo change) on the sphere. It then seems plausible that we insist that radiance changes are due to orientation changes (constant albedo) - unless allowing some albedo changes at a few contours (set of measure zero) leads to a more simple and smooth underlying surface. Thus we must provide a formulation that allows for some (isolated) discontinuities. We can also postulate that we prefer convex to concave interpretations. This latter assumption could be supported by the fact that we usually find that objects have basically convex surfaces.

Problem Formulation

We now proceed to derive a functional that expresses, we hope, the desired characteristics of our solution. Our first concern is, of course, that the resultant extracted shape is consistent with our data (using the lighting model). We adopt the fairly standard "brightness error criteria" (discretizing immediately):

$$\sum_x \sum_y (E(x,y) - R(x,y,\vec{n}_{xy}))^2$$

should be minimum. We have already mentioned that our chosen model for the reflectance function R will be the lambertian function $A(x,y)\vec{I} \bullet \vec{n}_{xy}$. We have also noted that this is still an ill-posed problem since there still exist an infinite number of solutions even given E and I. We want the solutions to be smooth so we add on a regularising term requiring the surface to be a smooth as possible. There are many ways to characterise smoothness, since we have chosen the representation of the surface to be in terms of surface normals we choose:

$$-\sum_x \sum_y \sum_{lk \in N(xy)} \vec{n}_{xy} \bullet \vec{n}_{kl}$$

where $N(x,y)$ is the usual NEWS neighborhood of (x,y). One can see we are using a Heisenberg Spin magnet model for the smoothness criteria. The symmetry of the dot product gives us the two possible interpretations (concave or convex) having equal energy. In order to impose preference for convex surfaces we need to add an extra field to break this symmetry. Alternatively, we can impose boundary constraints (such as the direction of the surface normals on the occluding edge of the sphere). The solutions to these functionals tend to be too smooth, the smoothness constraint is applied even across the boundaries. Furthermore, we want to favour constant albedo in almost all cases. A more complex functional to allow discontinuities and to favour constant albedo is:

$$\sum_x \sum_y ((E(x,y) - A(x,y)\vec{I} \bullet \vec{n}_{xy})^2 - \sum_{lk \in N(xy)} (\delta(xy,kl)\vec{n}_{xy} \bullet \vec{n}_{kl} + \alpha(xy,kl)A_{xy} \bullet A_{xy})) + Vc(\alpha) + Vc(\delta)$$

where α and δ correspond to line elements (Geman and Geman, 1984) that turn of the constraints of smoothness in surface normal and albedo (respectively) at certain sites. We encourage piecewise constant albedo function by using a Potts model type interaction $A_{xy} \bullet A_{jk}$. The last two terms are functionals on the line elements to discourage their number and encourage such things as their continuity.

CONCLUSION

This paper has outlined the basic regularization approach to vision. Included is a discussion of the various methods used to proceed from this formulation; first to an algorithm, and then an implementation. Finally, an illustration was provided using the example of recovering shape from shading. The functional formulated for this problem is unique and is currently being studied using a number of algorithms implemented on a transputer hypercube array.

REFERENCES

1. Barnard, S.T., Stochastic Stereo Matching over Scale, *Proceedings of DARPA Image Understanding Workshop, Cambridge, MA*, pp. 769-778, 1988.

2. Becker, E. B., Carey, G.F., and Oden, J. T., *Finite Elements: A First Course*, Vol I,II,IV, Prentice-Hall, Englewood Cliffs, NJ, 1981.

3. Bertero, M., Poggio, T. A., and Torre, V., Ill-Posed Problems in Early Vision, *Proc. IEEE*, Vol. 76, No. 8, pp. 869-889, 1988.

4. Blake, A., Comparison of the Efficiency of Deterministic and Stochastic Algorithms for Visual Reconstruction, *IEEE Trans. on Pattern Analysis and Machine Intelligence*, Vol. 11, No. 1, pp. 2-12, 1989.

Development of Adaptive Noise Cancellation for the Detection of Damaged Bearing Signals

C. C. TAN
Queensland University of Technology
Faculty of Engineering
P.O. Box 2434
Brisbane, Qld 4001, Australia

INTRODUCTION

Application of computers to the detection of bearing failure signal not only provides programming flexibility allowing a large number of tests to be performed for a range of parameters but also provides high speed data acquisition. It also has the advantage of allowing a large quantity of data to be stored and rapidly recalled and presented in a convenient format. In bearing failure detection, statistical techniques such as signal averaging, correlation test and frequency spectrum analysis are often performed by computer. These methods have certain drawbacks as they require a relatively large signal-to-noise ratio for the techniques to be effective. Unfortunately in most machinery vibration early detection of incipient bearing failure is often made difficult due to the masking of heavy background vibration (noise) generated by other machine components.

A method used to remove unwanted noise is the application of adaptive filter. An early version of the adaptive filter developed by Widrow (1957) and later he and Hoff (1960) developed the least-mean square (LMS) adaptive algorithm. This algorithm is used to update the parameters within the adaptive filter so as to achieve optimum minimum error at its output. The filter adjusts its parameters automatically according to the noise source. This requires little or no prior knowledge of the noise and signals characteristics. Its application to the detection of desired signals corrupted by noise in numerous applications is presented by Widrow et.al (1975). Its application for the detection of incipient bearing defect signal is demonstrated by Tan and Dawson (1983,1987).

In this paper, the design and development of computer software used for the detection of artifically seeded bearing signals corrupted by background noise is presented. The work involved data acquisition, design of adaptive algorithm and presentation of results at computer outputs. The developed

517

software program was tested on a computer simulation test involving artificially generated noise and signals, and practical tests with data from a bearing test-rig.

ADAPTIVE NOISE CANCELLATION CONCEPT

The basic concept of the adaptive noise cancellation (ANC) technique is shown in Figure 1. The reference noise (n_1) which is uncorrelated to the desired signal (s) is adaptively filtered to produce an output that is as close a replica as possible of primary noise (n_0). The filter output is subtracted from the primary input to produce the system output.

The adaptive filter shown in Figure 2 is employed to adjust and minimise the system output (e) and consequently the filter output (y) is adaptively filtered to obtain the best least squares estimate of the primary noise (n_0).

Assuming that the input signals to the combiner of the set occurring simultaneously in the time denoted by the subscript i, then the weighted output y_i is given by,

$$y_i = X_i^T.W \tag{1}$$

where x_i is the input signal vector and W the weight vector.

The adaptive algorithm is employed to adjust the weights of the adaptive linear combiner to obtain minimum mean-square error. Assuming that the input signals and the desired response d_i are statistically stationary and that the weights are fixed, the error e_i is defined as the difference between the desired response d_i and the output response y_i and is given by

$$e_i = d_i - x_i^T.W \tag{2}$$

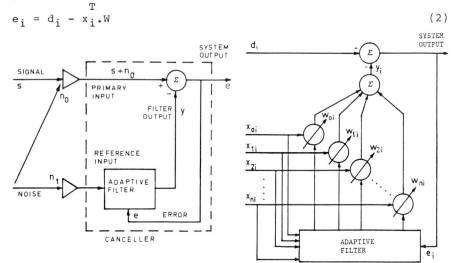

FIGURE 1.   ANC Concept   FIGURE 2.   LMS Adaptive Filter

518

Squaring the error e_i and taking the expected value of the error squared, gives the mean-square error,

$$E[e_i{}^2] = E[d_i{}^2] - 2E[d_i{}^T x_i]W + W^T E[X_i{}^T X_i]W \qquad (3)$$

Where $E[d_i{}^T x_i]$ is the cross-correlation between the desired response (scalar) and the input signal x_i, and $E[x_i x_i]$ is the input signal x_i correlation matrix.

The above expression is a quadratic function of weights and the optimum set of weights, W, may be obtained by differentiating the mean-square error function, $E[e_i{}^2]$. The technique for estimating a set of weights is given by Widrow (1959) namely,

$$W_{i+1} = W_i + 2ue_i X_i \qquad (4)$$

With the initial estimate of the set of weights, the algorithm will converge in the mean and will remain stable as long as the convergency factor u is less than the reciprocal of the largest eigen value λ_{max} of the input matrix and greater than zero.

INSTRUMENTATION AND SIMULATION TESTS

The formation of primary input for the canceller involves computer generation of desired periodic signals (Figure 3) and storing of a seeded bearing signal (Figure 4) obtained from an accelerometer fixed on the bearing test rig. These signals are then corrupted by computer generated random and sinusoidal noise to produce the primary input. The reference input consists of only random noise or sinusoidal signal.

The instrumentation for data acquisition and analysis is shown in Figure 5. The analogue data is amplified and conditioned before storing in a micro-computer using Dash 16 A/D board

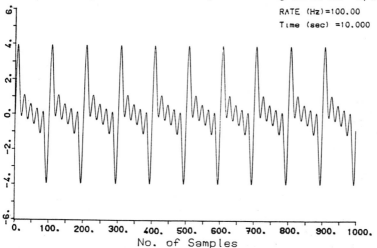

FIGURE 3. Computer Generated Signal

519

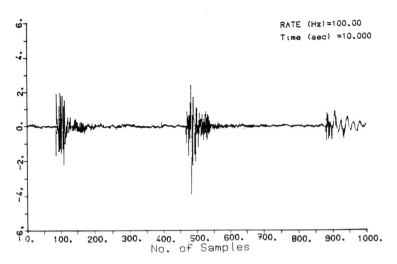

RATE (Hz)=100.00
Time (sec) =10.000

No. of Samples

FIGURE 4.   Unmasked Bearing Signal

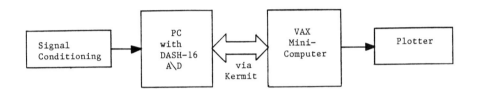

| Signal Conditioning | PC with DASH-16 A\D | via Kermit | VAX Mini-Computer | Plotter |

FIGURE 5.   Instrumentation

with a sampling rate of 25 kHz on two channels of data simultaneously. The digitised data is subsequently transferred to a VAX-11 mini-computer via Kermit. Analysis of data using the ANC technique is then performed on the mini-computer and output to a plotter.

The analysis technique to remove corrupting noise is shown in Figure 6. The simulated signals and noise or digitised practical data stored in the disc is recalled and displayed in analogue form. If the adaptive filtering is not in process, the program will evaluate the initial weight vector and convergency factor. These values are then used in the adaptive filter to form an estimate of the noise component using LMS procedure shown in Figure 7. At the end of the cancellation process the output signal is presented on the terminal and hard copy obtained via the plotter.

DISCUSSIONS

Samples of desired signal generated by computer and measurement from test-rig are respectively shown in Figure 3 and 4. When masked by computer generated random noise and sinusoidal signal of 50Hz (reference signal of Figure 8(a)-10(a), the desired signals are no longer visible as shown in primary signals of the above figures.

520

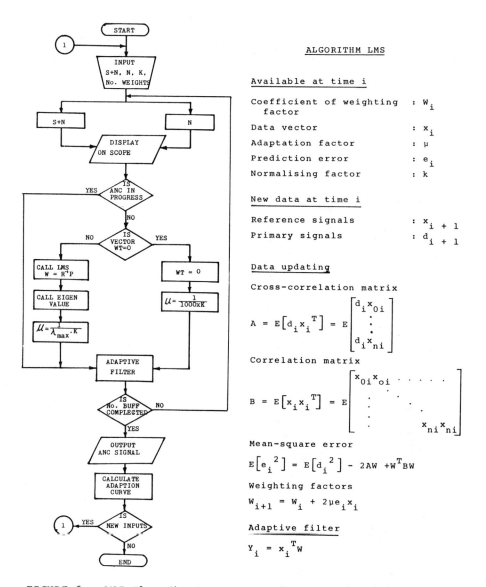

ALGORITHM LMS

Available at time i

Coefficient of weighting factor	: W_i
Data vector	: x_i
Adaptation factor	: μ
Prediction error	: e_i
Normalising factor	: k

New data at time i

Reference signals	: x_{i+1}
Primary signals	: d_{i+1}

Data updating

Cross-correlation matrix

$$A = E\left[d_i x_i^T\right] = E\begin{bmatrix} d_i x_{0i} \\ \cdot \\ \cdot \\ \cdot \\ d_i x_{ni} \end{bmatrix}$$

Correlation matrix

$$B = E\left[x_i x_i^T\right] = E\begin{bmatrix} x_{0i}x_{0i} & \cdots \\ \cdot & \cdot \\ \cdot & \cdot \\ \cdot & x_{ni}x_{ni} \end{bmatrix}$$

Mean-square error

$$E\left[e_i^2\right] = E\left[d_i^2\right] - 2AW + W^T BW$$

Weighting factors

$$W_{i+1} = W_i + 2\mu e_i x_i$$

Adaptive filter

$$Y_i = x_i^T W$$

FIGURE 6. ANC Flow Chart.   FIGURE 7. LMS Adaptive Algorithm

Computer simulation on the ANC technique was tested with a series of weight numbers (1 to 16) and convergency factor ($0.01 \times 10^{-3} - 1.0 \times 10^{-3}$) in order to determine the optimum weight number and convergency factor. The initial set of weights for the updating process were detemined using the LMS algorithm. The choice of the initial set of weights does not affect the final results in the adaption process since the adaptive algorithm will update the weights accordingly so as to achieve minimum mean-square error. However, the choice of the rate of convergence can affect the initial weight choice.

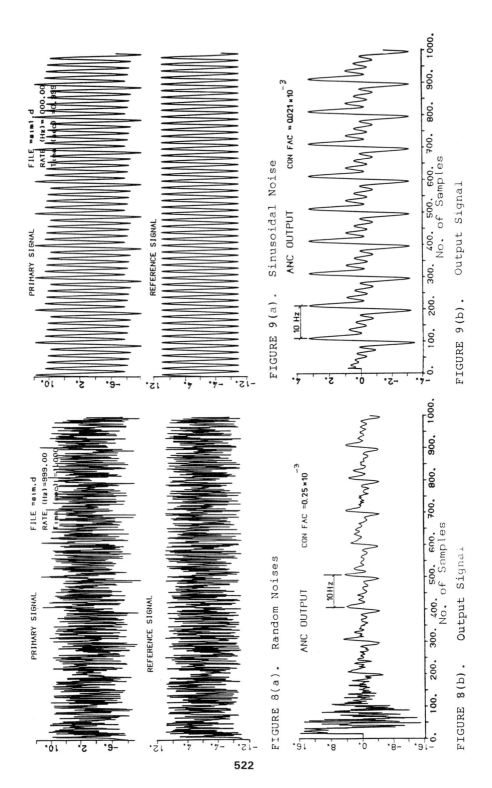

FIGURE 8(a). Random Noises

FIGURE 8(b). Output Signal

FIGURE 9(a). Sinusoidal Noise

FIGURE 9(b). Output Signal

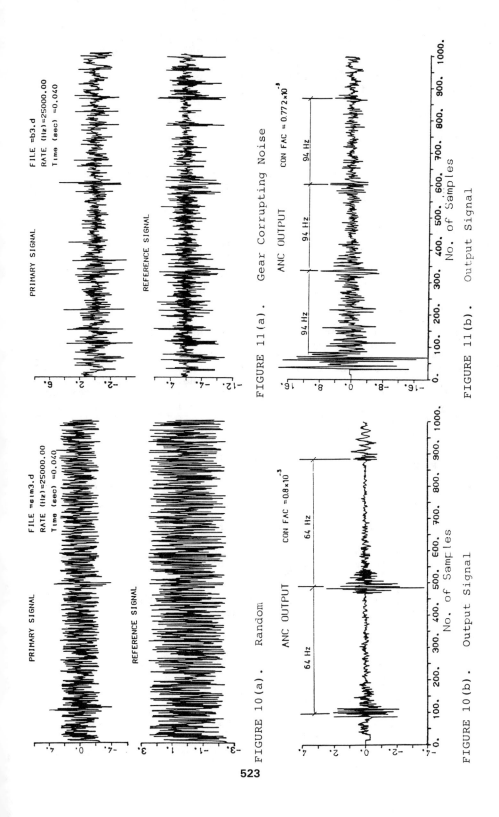

PRIMARY SIGNAL

FILE =sim3.d
RATE (Hz)=25000.00
Time (sec) =0.040

REFERENCE SIGNAL

FIGURE 10(a). Random

ANC OUTPUT

CON FAC =0.8×10⁻³

64 Hz

64 Hz

No. of Samples

FIGURE 10(b). Output Signal

PRIMARY SIGNAL

FILE =b3.d
RATE (Hz)=25000.00
Time (sec) =0.040

REFERENCE SIGNAL

FIGURE 11(a). Gear Corrupting Noise

ANC OUTPUT

CON FAC = 0.772×10⁻³

94 Hz

94 Hz

94 Hz

94 Hz

No. of Samples

FIGURE 11(b). Output Signal

523

With 16 weights and varying convergency factors (u) the ANC outputs after the cancellation process clearly exhibit the original simulated signals, as shown in Figures 8(b)-10(b). With random and sinusoidal noise on generated desired signal the values of u were 0.25 x 10^{-3} and 0.021 x 10^{-3} respectively. With random noise on seeded bearing signals a much higher value of u (0.8 x 10^{-3}) was applied. The effect of u on the cancellation of correlated noise plays a very important role in the technique and is not within the scope of this paper.

To further illustrate the effectiveness of the technique, data from a bearing test-rig with corrupting gear noise was obtained and shown in Figure 11. The primary and reference signals were recorded at two different positions to comply with the ANC theory. Both the primary and reference signals show no sign of a bearing damaged signal. Feeding these two signals into the canceller and using the above weight number and a convergency factor of 0.772 x 10^{-3} the output signal clearly shows the spall type bearing defeat occurring at 94 Hz which corresponds to a damage on the bearing outer ring.

CONCLUSIONS

The developed algorithm to perform the ANC technique using computed generated data and practical data for the detection of seeded bearing signal is successfully demonstrated. The results show that a different value of u needs to be evaluated for a different set of inputs. With the weight number fixed at sixteen and the above range of u, satisfactory convergence was achieved. The results from these tests are very promising and with proper design of sensors for the primary and referency signals, the technique could be applied to detect machine failure.

REFERENCES

1.   Tan, C.C. & Dawson, B., in *Application of Adaptive Noise Cancellation to the Condition Monitoring of Rolling Element Bearings*, ed. D.W. Butcher, Chapter 12, pp. 167-181, Ellis Horwood Ltd., Chichester, U.K., 1983.

2.   Tan, C.C., Adaptive Noise Cancellation Approach for condition Monitoring of Gear-box Bearings, *Proc. of IEAust, Int'l Tribology Conf.*, *Melbourne*, No.87/18, pp.360-365, Nov.1987.

3.   Widrow, B., Adaptive Samples - Data Systems - A Statistical Theory of Adaption, *1959 Wescon Conv. Rec., Inst. of Radio Engineers*, Pt. 4., pp. 74-85, 1959.

4.   Widrow, B. & Hoff, Jr. M.E., Adaptive Switching Circuits, *1960 Wescon Conv., Rec., Inst. of Radio Engineers*, Pt. 4., pp.96-104, 1960.

5.   Widrow,B., McCool,J., Williams,C., Kaunitz,J., Hearn,R., Zeidler,J., Dong,E., and Goodlin R., Adaptive Noise Cancelling: Principles and Applications, *Proc. IEEE*, Vol.63, No. 12, pp. 1692-1716, Dec. 1975.

Numerical Comparison of Pressure Distributions for Nonconforming Line Contact between Circular Elastic Cylinders

A. A. TORDESILLAS and J. M. HILL
Department of Mathematics
University of Wollongong
Wollongong, NSW, 2500, Australia

INTRODUCTION

Contact problems involving circular elastic cylinders are of considerable practical and theoretical interest and arise in many industrial situations such as the coating, textile, paper and steel industries. Our work is specifically concerned with their application in the roller coating process which is the painting technique used in the coil coating industry. A typical roller coater configuration is the forward roller coating system (see Figure 1). The bottom rubber-covered roller is half immersed in a pan of paint and the top steel roller which carries the strip to be coated is pressed against it. As the bottom roller rotates, paint is drawn up through a narrow gap. The flow emerging downstream splits into two surface layers. The layer remaining on the bottom roller is returned to the pan while the layer attached to the metal strip forms the final coated film. The film thickness is metered by the entire paint flow in the gap between the two rollers. Of major interest is the determination of the film thickness and the pressure distribution in the contact zone.

This process is modelled assuming two heavily loaded cylinders in plane contact lubricated such that one or both are deformed under the action of fluid pressure and with the following assumptions for the contact zone:

 (i) the lubricant (paint) is a Newtonian fluid subject to a steady two dimensional laminar flow,

 (ii) the film thickness is vanishingly small compared to the radii of the cylinders,

 (iii) the lubricant pressure is constant across the film thickness,

 (iv) inertial and side leakage effects are neglected,

 (v) the cylinders are long compared to the width of the contact region.

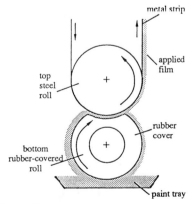

FIGURE 1: Forward roller coating system

Based on these assumptions, the theory of elastohydrodynamic lubrication (EHL) is applied to analyze the contact zone. In EHL flows the elastic deformation of the surfaces through which the fluid flows may become significant. Accordingly we need to relate the hydrodynamic lubrication action with the elastic deformation and in order to obtain solutions for the pressure distribution and lubricant film profile, the Reynold's equation and the elasticity equations must be solved simultaneously (see Dowson and Higginson (1977) p.65). Grubin (1949), based on a number of engineering approximations, proposed the first solution of this EHL problem. Grubin isolated the surface deformation aspect of the problem by assuming that the shape of the deformed rolling surfaces is simply that which would occur under dry static contact. Thus, the lubricant is presumed to simply separate the surfaces by some uniform amount which is unknown *a priori*. Hooke (1977) confirmed this, especially for highly loaded contact where the elastic deformation is several orders of magnitude larger than the film thickness. Accordingly a successful analysis of the EHL problem hinges on an accurate determination of the associated dry static contact problem.

The subject of contact mechanics began when Hertz (1882) presented the first theory of elastic contact which has stood the test of time with remarkable endurance (see Gladwell (1980) and Johnson (1984)). Hertz theory, based on classical linear elasticity, assumes no slipping between surfaces which are approximated by quadratic terms only, that the contact area is small in comparison to the size of the bodies and any curvature of the surface is completely neglected. The latter assumption means that the linear elastic Flamant solution for a concentrated load acting on an infinite half-space can be exploited to yield a singular integral equation involving the finite Hilbert transform. This integral equation can be solved to give the pressure distribution

$$p(z) = p_{\max} \left[1 - \left(\frac{z}{h} \right)^2 \right]^{1/2}, \tag{1}$$

where $p_{\max}$ is the maximum pressure at the centre of the contact region, z is the distance from the centre and $2h$ is the contact width. Furthermore, for two elastic cylinders in contact, Hertz theory predicts that the maximum pressure and half contact width are determined respectively by

$$p_{\max} = \frac{2F}{\pi \ell h}, \quad h = \left[\frac{4Fab}{\pi \ell (a+b)} \left(\frac{1 - \sigma_1^2}{E_1} + \frac{1 - \sigma_2^2}{E_2} \right) \right]^{1/2}, \tag{2}$$

where E and σ denote Young's modulus and Poisson's ratio respectively, a and b denote the radii of the cylinders, F is the total load pressing the two cylinders together and ℓ is the length of the cylinders.

Although in many cases (1) provides a remarkably robust approximation, Parish (1958) reported non-Hertzian pressure distributions for experiments involving contact between rubber covered and rigid circular cylindrical rollers. The experiments performed by Parish (1958) were conducted on extremely thin rubber covered rollers, which could possibly explain the departures from the Hertzian theory, except that the modification of the Hertz theory due to Hannah (1951) also does not account for the experimental data. Thus, although (1) can be viewed as a reasonably accurate approximation, nevertheless the problem of genuinely incorporating curvature effects for contacting rolling cylinders is still of considerable practical and theoretical interest.

The problem of incorporating curvature effects was first attempted by Loo (1958) who considered the symmetrical contact of a central elastic cylinder by two identical elastic cylinders. For this situation, the well known linear elastic solution for a cylinder under equal and opposite concentrated loads at the boundary (see Figure 2 and Muskhelishvili (1963), p.147) can be utilized to obtain a singular integral equation for which Loo (1958) gave an approximate solution. This approximate solution is essentially the Hertz approximation but with different constants. Sternberg and Turteltaub (1972) successfully incorporated curvature effects for the problem of compression of an elastic roller, compressed between two rigid flat parallel plates. These authors give a remarkably simple closed expression for the pressure distribution for this problem, namely

$$p(\theta) = \frac{E(\sin^2 \theta_0 - \sin^2 \theta)^{1/2}}{(1 - \sigma^2)(1 + \cos^2 \theta_0)} \quad |\theta| \le \theta_0, \tag{3}$$

where the contact angle θ_0 is determined by the condition that the total force exerted by the horizontal pressure is the prescribed applied force $P = F/\ell$, thus

$$\int_{-\theta_0}^{\theta_0} p(\theta) a d\theta = P, \tag{4}$$

which in this case, can be expressed in terms of elliptic integrals.

In two recent papers (Hill and Tordesillas (1989) and (1990)) the present authors propose two distinct models for the compression of a central elastic cylinder by two symmetrically placed identical elastic cylinders or identical elastic parallel plates. The first model (Hill and Tordesillas (1989)) is precisely that proposed by Loo (1958) except that his singular integral equation is solved exactly and includes the closed expression (3) as a special case. This model assumes that only the horizontal circumferential displacement is prescribed in the contact region and accordingly reduces to a single singular integral equation. However for this model some slip in the vertical direction is allowed. The second model proposed (Hill and Tordesillas (1990)) assumes fully adhesive contact so that in the contact region there is a prescribed horizontal displacement and zero vertical displacement. The latter model involves the solution of a coupled system of singular integral equations and two possible analytical approximations are presented. The purpose of this paper is to bring these two models together to present a full numerical comparison with the Hertz theory which is done only for the case of compression by rigid compressors. Details for parallel plates and elastic compressors can be found in the two papers. The two models are briefly summarized in the following two sections and numerical results are given in the final section of the paper.

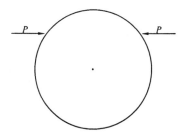

FIGURE 2: Circular cylinder under equal and opposite forces applied at the boundary

SYMMETRICAL CONTACT OF CIRCULAR ELASTIC CYLINDERS BY RIGID COMPRESSORS

We consider the case of an elastic cylinder of radius a in symmetrical contact between two rigid compressors (see Figure 3) described by $X = \pm f(Y)$ which can be of any shape provided they are themselves symmetric. In the contact region $-\theta_0 \le \theta \le \theta_0$ we assume that circumferential points of the cylinder undergo a horizontal displacement of magnitude

$$\epsilon - \{a(1 - \cos\theta) + [f(a\sin\theta) - a]\}, \tag{5}$$

towards the centre of the elastic cylinder, where ϵ is assumed to be the rigid displacement of each cylinder towards the central cylinder. Following Loo (1958) we can show, by considering the horizontal displacement of the circumferential points of the cylinder and using the classical linear elastic solution for two equal and opposite horizontal concentrated loads, the problem may be reduced to the integral equation

$$\frac{2(1 - \sigma^2)}{\pi E} a \int_{-\theta_0}^{\theta_0} p(\psi) \left\{ \log\tan\frac{|\theta - \psi|}{2} + \cos\theta \cos\psi \right\} d\psi = f(a\sin\theta) - a\cos\theta - \epsilon, \tag{6}$$

where $p(\theta)$ is an even function and denotes the unknown horizontal pressure distribution in the contact region. By a sequence of transformations, Hill and Tordesillas (1989) simplify this integral equation to one involving the standard finite Hilbert transform or the well known airfoil equation,

$$\frac{1}{\pi} \int_{-1}^{1} \frac{\phi(y)}{(y - x)} dy = -g(x), \tag{7}$$

where the following variables and parameters have been used

$$z = \sin\theta, \quad \xi = \sin\psi, \quad x = z/z_0, \quad y = \xi/z_0,$$

$$p^*(z) = p(\sin^{-1}(z)), \quad \phi(x) = p^*(xz_0), \tag{8}$$

$$g(x) = \left\{ \frac{Ez_0}{2(1 - \sigma^2)} + \frac{z_0^2}{\pi} \int_{-1}^{1} \phi(y) dy \right\} x + \frac{E}{2(1 - \sigma^2)} f'(az_0 x)(1 - z_0^2 x^2)^{1/2}.$$

The unknown horizontal pressure distribution is essentially $\phi(x)$ and the known function $g(x)$ contains information describing the rigid compressors. The exact solution of (7) is itself expressed in terms of the finite Hilbert transform and is given by,

$$(1 - x^2)^{1/2}\phi(x) = \frac{1}{\pi} \int_{-1}^{1} \phi(y) dy + \frac{1}{\pi} \int_{-1}^{1} \frac{(1 - y^2)^{1/2} g(y)}{(y - x)} dy, \tag{9}$$

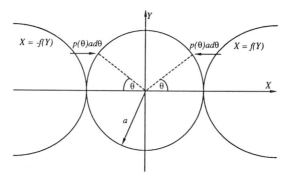

FIGURE 3: Symmetrical rigid compression of a circular elastic cylinder. (The rigid compressors are assumed given by $X = \pm f(Y)$ where f is even and $f(0) = a$.)

which, in some cases, can be evaluated with relative ease depending on the form of $g(x)$. For example, for rigid parallel plates we may readily deduce the expression (3). Similarly for rigid compressing circular cylinders of the same radii as the central elastic cylinder we may deduce

$$p(\theta) = \frac{2E(\sin^2\theta_0 - \sin^2\theta)^{1/2}}{(1-\sigma^2)(1+\cos^2\theta_0)} \quad |\theta| \le \theta_0, \tag{10}$$

where θ_0 is again determined by the condition (4). For the general case when the radii are distinct we may obtain the approximate expression

$$\phi(x) \simeq \frac{Ez_0(1-x^2)^{1/2}}{(1-\sigma^2)(2-z_0^2)} \left(\frac{a}{b}+1\right) \left\{ 1 + \frac{z_0^2}{16}\frac{a}{b}\left(\frac{a}{b}-1\right)[z_0^2+4+4x^2(2-z_0^2)] \right\}, \tag{11}$$

where b denotes the radius of the rigid cylindrical compressors and by evaluating the integral in (9) numerically, this approximation can be shown to be reasonably accurate especially if $a/b < 1$.

SYMMETRICAL ADHESIVE CONTACT OF CIRCULAR ELASTIC CYLINDERS BY RIGID COMPRESSORS

If the coefficient of friction between the contacting bodies is large then there is no relative displacement of the contacting surfaces, that is, there is no interfacial slip. For the prescribed horizontal circumferential displacement (5) and zero vertical circumferential displacement Hill and Tordesillas (1990) utilize the linear elastic solution for four pairs of concentrated loads acting on the boundary of a circular elastic cylinder (see Figure 4) to deduce coupled singular integral equations for the horizontal and vertical pressures $p(\theta)$ and $q(\theta)$, thus

$$\frac{1}{\pi}\int_{-1}^{1} \frac{\Phi(y)dy}{(y-x)} - \beta\Psi(x) = -Bx - \frac{E}{2(1-\sigma^2)}f'(az_0x)(1-z_0^2x^2)^{1/2},$$

$$\frac{1}{\pi}\int_{-1}^{1} \frac{\Psi(y)dy}{(y-x)(1-z_0^2y^2)^{1/2}} + \beta\frac{\Phi(x)}{(1-z_0^2x^2)^{1/2}} = A, \tag{12}$$

where the constants A and B are defined by

$$A = \frac{z_0}{\pi}\int_{-1}^{1}\left\{\Phi(y) + \Psi(y)\frac{yz_0}{(1-z_0^2y^2)^{1/2}}\right\}dy, \quad B = Az_0 + \frac{Ez_0}{2(1-\sigma^2)}, \tag{13}$$

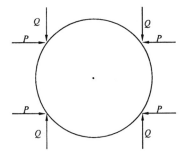

FIGURE 4: Circular cylinder under four pairs of equal and opposite concentrated forces applied at the boundary

and the variables and functions introduced are defined in the following manner,

$$z = \sin\theta, \quad \xi = \sin\psi, \quad x = z/z_0, \quad y = \xi/z_0,$$

$$p^*(z) = p(\sin^{-1}(z)), \quad q^*(z) = q(\sin^{-1}(z)), \quad \Phi(x) = p^*(xz_0), \quad \Psi(x) = q^*(xz_0), \tag{14}$$

and moreover $p(\theta)$ and $q(\theta)$ are assumed to be even and odd functions respectively.

Two approximate solutions can be given for the coupled system (12). The first (subscripted 0) stems from the fact that the coupled system is exactly solvable when z_0 is set to zero in the left-hand side of (12) while the second (subscripted 1) arises from a consistent small contact angle approximation. In both approximate schemes, the integral equations can be combined to give a singular integral equation of the second kind having the general form

$$\phi(x) + \frac{\lambda}{\pi} \int_{-1}^{1} \frac{\phi(y)}{(x-y)} dy = g(x), \tag{15}$$

where $\phi(x) = \Phi(x) + i\Psi(x)$. The complex function $g(x)$ is known and $\lambda = i/\beta$ is a parameter whose value depends upon the Poisson's ratio and it is required to find the function $\phi(x)$ subject to the boundary conditions $\phi(\pm 1) = 0$. The method of solving (15) is discussed in detail in Hill and Tordesillas (1990) and the final general solution takes the form

$$\phi(x) = \left(\frac{1-x^2}{1-\beta^2}\right)^{1/2} \left(\frac{1+x}{1-x}\right)^{i\delta} \{(a_0 - a_2 x^2) + ib_0 x\}, \tag{16}$$

where $\cot(i\pi\delta) = \lambda$ and where a_0, b_0 and a_2 are various constants distinct for each approximation and the type of rigid compressors considered. Thus the required solution of the original coupled system becomes

$$\Phi(x) = \left(\frac{1-x^2}{1-\beta^2}\right)^{1/2} \left\{(a_0 - a_2 x^2)\cos\left[\delta\log\left(\frac{1+x}{1-x}\right)\right] - b_0 x \sin\left[\delta\log\left(\frac{1+x}{1-x}\right)\right]\right\},$$

$$\Psi(x) = \left(\frac{1-x^2}{1-\beta^2}\right)^{1/2} \left\{b_0 x \cos\left[\delta\log\left(\frac{1+x}{1-x}\right)\right] + (a_0 - a_2 x^2)\sin\left[\delta\log\left(\frac{1+x}{1-x}\right)\right]\right\}. \tag{17}$$

Thus for example, for compression by rigid circular cylinders of radius b the approximation subscripted by zero takes the form (17) where the various constants are given by

$$a_0 = \frac{(k+1)Ez_0}{2(1-\sigma^2)(1+2\delta z_0)} \left\{1 - k(k-1)z_0^2\left[\frac{\delta z_0(1+4\delta^2)}{3} - \frac{(1-4\delta^2)}{4}\right]\right\},$$

$$a_2 = -\frac{k(k^2-1)Ez_0^3}{4(1-\sigma^2)}, \quad b_0 = -\frac{k(k^2-1)E\delta z_0^3}{2(1-\sigma^2)}, \tag{18}$$

where $k = a/b$, $\beta = (1-2\sigma)/2(1-\sigma)$ and the constant δ is given explicitly by

$$\delta = -\frac{1}{2\pi}\log\left(\frac{1+\beta}{1-\beta}\right) = -\frac{1}{2\pi}\log(3-4\sigma). \tag{19}$$

Details for other approximations can be found in Hill and Tordesillas (1990).

NUMERICAL RESULTS

Figures 5(a) and 5(b) show the non-dimensional load variation with contact angle for the various models for Poisson ratios of $\sigma = 0.45$ and 0.5 respectively and $a/b = 0.2, 1.0$ and 3.0. For a fixed a/b, the closeness of these curves is clearly evident for small contact angles. However, both our models deviate substantially from the Hertz estimate at increasing values of the contact angle and

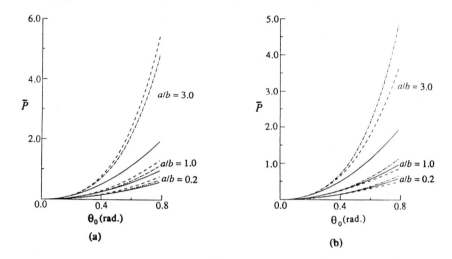

FIGURE 5: Variation with contact angle θ_0 of the non-dimensional force per unit length of cylinder, $\overline{P} = \frac{P(1-\sigma^2)}{aE}$. Hertz ——; first model - • - • -; second model - - -.
(a) $\sigma = 0.45$; (b) $\sigma = 0.50$.

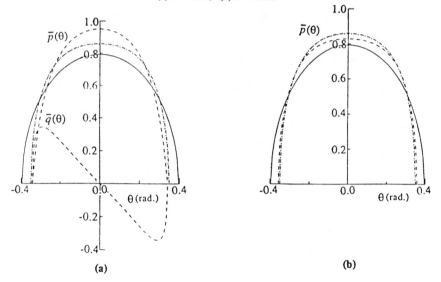

FIGURE 6: Variation with angle θ of the non-dimensional horizontal and vertical pressures, $\overline{p}(\theta) = \frac{(1-\sigma^2)}{E}p(\theta)$ and $\overline{q}(\theta) = \frac{(1-\sigma^2)}{E}q(\theta)$, respectively. Hertz ——; first model - • - • -; second model - - -. (a) $\sigma = 0.45, a/b = 3.0, \overline{P} = 0.5$; (b) $\sigma = 0.50, a/b = 3.0, \overline{P} = 0.5$.

a/b. For large a/b, we would expect both our models (which take into account the curvature of the bodies) to provide a more accurate estimate. For small values of a/b the curves are similar. However, as anticipated the curves for both our models deviate rapidly from Hertz for large values of a/b, that is, where the cylindrical compressor is much smaller than the elastic cylinder. In this case, the curvature of the compressor becomes significant as the contact area grows and Hertz' assumptions become inappropriate. Figures 6(a) and 6(b) show the corresponding pressure distributions in the contact region resulting from a fixed load of $\overline{P} = 0.5$ applied to cylinders with $a/b = 3.0$ and Poisson's ratio $\sigma = 0.45$ and 0.5 respectively. Consistent with Figures 5(a) and 5(b), our models predict a smaller contact angle than Hertz at the expense of a higher maximum pressure. Moreover, in the second model the vertical pressure vanishes when the elastic cylinder is incompressible ($\sigma = 0.5$) and generally our results indicate that the effect of the vertical pressure on the horizontal pressure is not significant.

ACKNOWLEDGEMENT The authors are grateful to Mark Davies and Keith Enever of B.H.P. Coated Products Division, Port Kembla, for many helpful discussions. They are also indebted to Traci Carse for her careful typing of this paper.

REFERENCES

1. Dowson, D. and Higginson, G.R., *Elastohydrodynamic Lubrication,* 2nd ed., Pergamon, Oxford, 1977.
2. Gladwell, G.M.L., *Contact problems in the classical theory of elasticity,* Sijthoff and Noordhoff, London, 1980.
3. Grubin, A.N., *Central Scientific Research Institute for Technology and Mechanical Engineering, Book No. 30,* Moscow, D.S.I.R. Translation No. 377, pp. 115-166, 1949.
4. Hannah, M., Contact stress and deformation in a thin elastic layer. *Quart. J. Mech. appl. Math.* vol. 4, pp. 94-105, 1951.
5. Hertz, H., On the contact of elastic solids. *J. reine unde agnewandte Mathematik,* vol. 92, 156-171, 1882.
6. Hill, J.M. and Tordesillas, A., The pressure distribution for symmetrical contact of circular elastic cylinders. *Quart. J. Mech. Appl. Math.* vol. 42, 1989.
7. Hill, J.M. and Tordesillas, A., The symmetrical adhesive contact problem for circular elastic cylinders, submitted for publication.
8. Hooke, C.J., The elastohydrodynamic lubrication of heavily loaded contacts. *J. Mech. Engng. Sci.* vol. 19, pp. 149-156, 1977.
9. Johnson, K.L., *Contact Mechanics,* Cambridge University Press, Cambridge, 1985.
10. Loo, T.-T. , Effect of curvature on the Hertz theory for two circular cylinders in contact. *ASME J. Appl. Mech.* vol. 25, pp. 122-124, 1958.
11. Muskhelishvili, N.I., *Some basic problems of the mathematical theory of elasticity* (translation by J.R.M. Radok). Noordhoff, Groningen, 1963.
12. Parish, G.J., Measurements of pressure distribution between metal and rubber covered rollers. *Br. J. Appl. Phys.* vol. 9, pp. 158-161, 1958.
13. Sternberg, E. and Turteltaub, M.J., *Compression of an elastic roller between two rigid plates. Continuum Mechanics and Related Problems of Analysis.* Muskhelishvili Anniversary Volume (Eds. L.I. Sedov et al, Nauka Publishing House, Moscow) pp. 495-515, 1972.

Numerical Modeling of a Scramjet

A. M. WATTS, D. H. SMITH, and L. K. FORBES
Department of Mathematics
The University of Queensland
St. Lucia, Q. 4067, Australia

INTRODUCTION

A scramjet (Supersonic Combustion RAMJET) is a jet engine without moving parts. The external flow, which is at a high Mach number relative to the engine, is contracted by the inlet ducting so that the air is compressed but the flow remains supersonic in the combustion chamber. Fuel such as hydrogen is injected into the hot compressed air and ignites spontaneously so that the pressure in the air is increased further. The duct is then expanded so that the air expands and is accelerated in the downstream directon until the end of the nozzle is reached. A schematic diagram of a scramjet is shown in figure 1.

The scramjet has become of more interest recently because of the potential application in space shuttle vehicles. Its advantage over a rocket engine for flight in a rare atmosphere is that no oxygen has to be carried in the vehicle and there is therefore a considerable weight saving. The weight of hydrogen is small compared to the combined hydrogen and oxygen after combustion.

The aim in the design of a scramjet is to obtain the maximum thrust for a given amount of energy input subject to practical restrictions of weight and strength. It is clear that the increase in total enthalpy through the engine depends only on the amount of heat added.

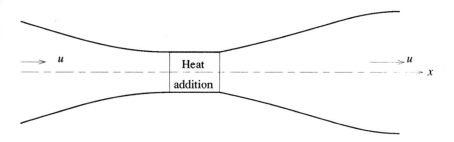

FIGURE 1. Schematic diagram of scramjet

However, the design of the engine determines how the enthalpy is distributed between internal energy, kinetic energy due to transverse velocity and due to longitudinal velocity. The maximum thrust is given with the maximum energy in longitudinal velocity. One of the important considerations in this design is the shape of the internal surfaces and it is largely for this purpose that the numerical method described here is being developed.

A quasi-one dimensional theory of the scramjet indicates that the increase in entropy should be minimised. It is also known from theory of rockets as described in Courant and Friedrichs (1948) that the exit pressure should be the ambient one.

To minimise the increase in entropy, shockwaves and other dissipative modes of flow should be avoided as far as possible and the heat should be added at the highest possible pressure.

In this paper, the plane flow of an inviscid, non-heat conducting fluid is considered where the combustion is represented by a prescribed heat addition to the flow.

GOVERNING EQUATIONS

The continuity, momentum and energy equations for the steady flow of a compressible gas are

$$\nabla.(\rho\mathbf{q}) = 0, \tag{1}$$

$$\nabla.(\rho\mathbf{q}\mathbf{q} + p\mathbf{I}) = \nabla.(\rho\mathbf{q}\mathbf{q}) + \nabla p = 0, \tag{2}$$

$$\nabla.(\rho\mathbf{q}H) = \rho\mathbf{q}.\nabla(h + \frac{1}{2}\mathbf{q}^2) = f. \tag{3}$$

In these equations, ρ is the density, p the pressure, $\mathbf{q} = u\mathbf{i} + v\mathbf{j}$ the velocity, $\mathbf{I}$ the identity tensor, h the enthalpy of the fluid and $H = h + \frac{1}{2}\mathbf{q}^2$ the total entropy, γ the ratio of specific heats, f the energy input per unit volume per unit time, and x the axial coordinate.

We are supposing here that there is no input of mass. The effect of the combustion is represented by an energy source of density f per unit volume per unit time. If the combustion process is to be included in the mathematical model, this density would be given in terms of other gas properties. Also, the relation between the pressure p, density ρ and enthalpy h would be given by rate equations which describe the combustion and chemical processes that occur.

For this study, we assume that the gas is perfect and that there are no chemical changes occurring. The equation of state then becomes

$$h = \frac{\gamma p}{(\gamma - 1)\rho}. \tag{4}$$

FINITE VOLUME METHOD

A finite volume method is used to discretise the equations. This leads to an implicit set of discrete equations which provides stability and allows the grid defined by the volumes to be chosen to conform to the boundary position and to the streamlines. Because the discrete equations are derived directly from the conservation equations, the method is shock capturing, which is of utmost importance in this context. One of the advantages in having one set of grid lines following the streamlines is that, in the future, rate equations describing the chemical and molecular reactions and combustion can be incorporated simply. Another is that the discrete form of the conservation equation becomes simple because there is no convection across the streamwise grid lines.

As is usual in this type of calculation, there is a type of Gibbs phenomenon that occurs when there are discontinuities in the solution of the differential equations. Since there are shock waves in the flows that are being calculated, the Gibbs phenomenon can occur here. To overcome this, an artificial viscosity was introduced with a coefficient whose magnitude depended on the grid size. One of the criticisms that is often levelled at the use of artificial viscosity is that the discontinuities tend to be smeared out with time. A flux-correction method such as that used by Zalesak (1979) could be used but, with the non-linear equations used here, the shocks are not smeared too much. The viscosity leads to shock waves with non-zero thickness which is proportional to the viscosity coefficient and it is found that the thickness can be as small as three or four grid lengths without having excessive oscillations in the solution.

The x axis is in the direction of the inlet flow. The flow region for the plane flow is divided into quadrilaterals, or more particularly trapeziums. The quadrilaterals are then bounded by equally spaced lines in the y direction normal to the x axis and by lines which approximate the streamlines. The corners of the quadrilaterals are at the points $x_i, y_{i,j}$, where $x_{i+1} - x_i = \Delta x$, as shown in Figure 2.

The representation of the flow quantities depends on the order of their derivative with respect to y which is to be approximated. Thus the y velocity component v and the pressure p are approximated by $v_{i,j}$ and $p_{i,j}$ at the corresponding corners of the quadrilaterals and are taken to be continuous, piecewise linear on the edges of the quadrilaterals.

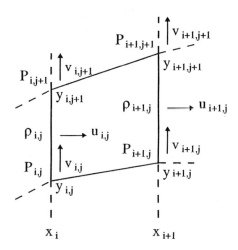

FIGURE 2. Finite volume representation

The density ρ and the x velocity u are piecewise constant functions of y on each of the grid lines $x = x_i$.

The streamlines are lines which satisfy the equation

$$\frac{dy}{dx} = \frac{v}{u}.$$

Since two sides of the quadrilateral are tangential to the streamlines, this equation becomes, in terms of the discrete flow variables,

$$\frac{y_{i+1,j} - y_{i,j}}{x_{i+1} - x_i} = \frac{1}{2}\left(\frac{2v_{i,j}}{u_{i,j-1} + u_{i,j}} + \frac{2v_{i+1,j}}{u_{i+1,j-1} + u_{i+1,j}}\right). \tag{5}$$

DISCRETE FLOW EQUATIONS

The discrete equations are obtained by the application of the divergence theorem to each of the flow equations over a quadrilateral. Conservation of mass, momentum and energy within each quadrilateral are then satisfied exactly in terms of the discrete flow variables, except that as a consequence of equ(5) it is assumed that there is no convection across the streamwise sides of the quadrilaterals.

In discrete form, the continuity, x momentum, y momentum and energy equations become

$$\rho_{i+1,j}u_{i+1,j}[y_{i+1,j+1} - y_{i+1,j}] = \rho_{i,j}u_{i,j}[y_{i,j+1} - y_{i,j}]. \tag{6}$$

$$\rho_{i,j}u_{i,j}(y_{i,j+1} - y_{i,j})u_{i+1,j} + \frac{1}{2}(y_{i+1,j+1} - y_{i,j})p_{i+1,j} + \frac{1}{2}(y_{i,j+1} - y_{i+1,j})p_{i+1,j+1}$$
$$= \rho_{i,j}u_{i,j}(y_{i,j+1} - y_{i,j})u_{i,j} + \frac{1}{2}(y_{i,j+1} - y_{i+1,j})p_{i,j} - \frac{1}{2}(y_{i+1,j+1} - y_{i,j})p_{i,j+1}$$
$$+ \rho_{i,j}u_{i,j}^2(y_{i+1,j+1} - y_{i+1,j}). \tag{7}$$

$$\rho_{i,j}u_{i,j}(y_{i,j+1} - y_{i,j})(v_{i+1,j} + v_{i+1,j+1}) - \Delta x(p_{i+1,j} - p_{i+1,j+1})$$
$$= \rho_{i,j}u_{i,j}(y_{i,j+1} - y_{i,j})(v_{i,j} + v_{i,j+1}) + \Delta x(p_{i,j} - p_{i,j+1}) \tag{8}$$

$$H_{i+1,j} = H_{i,j} + F_{i,j}/[\rho_{i,j}u_{i,j}(y_{i,j+1} - y_{i,j})]. \tag{9}$$

where $F_{i,j}$ is the integral of the heat source density f over the i,j cell. In terms of the other flow variables, $H_{i,j}$ is given by

$$H_{i,j} = \frac{\gamma}{(\gamma - 1)}\frac{p_{i,j} + p_{i,j+1}}{2\rho_{i,j}} + \frac{1}{2}\left(u_{i,j}^2 + \frac{1}{4}(v_{i,j} + v_{i,j+1})^2\right) \tag{10}$$

SOLUTION OF DISCRETE EQUATIONS

Since the difference equations are fully hyperbolic, the solution of the discrete equations at x_{i+1} can be found in terms of the values of the flow variables and y values at x_i. That is, equations (5) to (10) are solved for $\rho_{i+1}, u_{i+1}, v_{i+1}, p_{i+1}, y_{i+1}$. Since they are nonlinear, an iterative method is used. Firstly, we put

$$\rho_{i+1} = \rho_i + \delta\rho_i \tag{11}$$

$$u_{i+1} = u_i + \delta u_i \tag{12}$$

$$v_{i+1} = v_i + \delta v_i \tag{13}$$

$$p_{i+1} = p_i + \delta p_i \tag{14}$$

In equation (6), the $y_{i+1,j}$ and $y_{i+1,j+1}$ are eliminated using equation (5). In the remaining equations, the y's are first estimated using the velocity ratios at x_i, but iteratively corrected after the flow values at x_{i+1} have been calculated.

The equations for the flow variables are solved first ignoring all δ^2 terms and then iterating to include the higher degree terms in δ.

The density of the heat source was taken to be of the form

$$f(x,y) = \frac{A}{K}(1 + \cos\frac{\pi(x - x_c)}{L})(1 + \cos\frac{\pi(y - y_c)}{M})$$

for $x_c - L < x < x_c + L$, $y_c - M < y < y_c + M$ and zero otherwise. Here, $K = 4LM(1 + \pi^{-1})^2$ is a normalising constant.

RESULTS

The method was first tested in a parallel duct without heat addition, where the flow entered the duct at a Mach number of 5.9 at an angle of about 6^0 to the duct. Calculations were done with 100 to 400 cells in the y direction, all the cells being initially square. Without artificial viscosity, there were spurious oscillations in the values of the flow variables. These were eliminated by the addition of viscosity. The level of artificial viscosity did not affect the results noticeably except that the shock thickness increased with viscosity as predicted.

Figure 3 shows the pressure profile at one cross-section of the duct with 400 cells in the y direction. There is clearly a shock wave moving away from the boundary $y = 1$ and an expansion fan centred on the leading edge of the duct at $x = 0, y = 0$. Profiles of the pressure further downstream show that the shock wave moves without change in profile and that the expansion wave spreads in proportion to the distance downstream. With the artificial viscosity at a level which reduces the oscillations to an amplitude which is just visible on a large scale graph, the shock thickness is about 4 grid spacings in the y direction.

Heat addition was then included with an inlet flow that was parallel to the duct walls. On the scale used, the width of the duct was 1, the centre of the heat source was at $x = 0.5, y = 0.5$ and heat was added in an area which extended 0.4 in the x (streamwise) direction and 0.2 in the y direction.

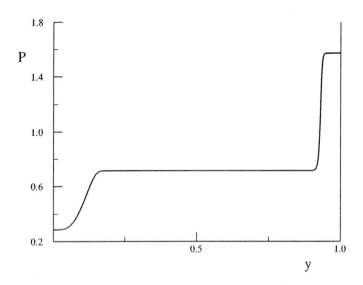

FIGURE 3. Pressure profile for parallel duct with inclined flow

Pressure and density profiles are shown in Figures 4 and 5. The first profiles are a little downstream of the centre of the heat source at $x = 0.6$. The heat source produces a rise in the pressure at that station. Further downstream past the heat source at $x = 1.8$, there is a pressure pulse moving away from the heat source which consists of a shock wave followed by an expansion wave. Because the gas in the centre of the duct is heated, its density there tends to remain low compared with the gas further out. At $x = 2.6$, the pressure pulse is in the process of reflection from the boundaries and at $x = 4.4$, the shock wave at the front of the reflected pressure pulse is coincident with the edge of the heat addition area where there is a contact layer between the unheated gas and the heated gas which is even more clearly shown in the density profile.

REFERENCES

1. Courant, R. and Friedrichs, K.O., *Supersonic Flow and Shock Waves*, Interscience Publishers, New York, 1948 .

2. Zalesak, Steven T., Fully Multidimensional Flux-Corrected Transport Algorithms for Fluids, *J. Comput. Phys.*, vol. 31, pp 335–362, 1979.

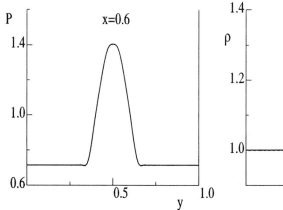

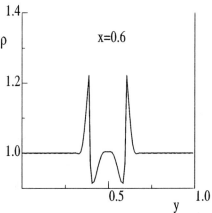

FIGURE 4(a): Pressure profile at x = 0.6   FIGURE 5(a): Density profile at x = 0.6

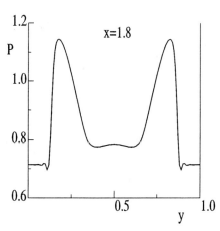

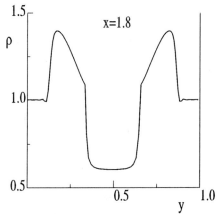

FIGURE 4(b): Pressure profile at x = 1.8   FIGURE 5(b): Density profile at x = 1.8

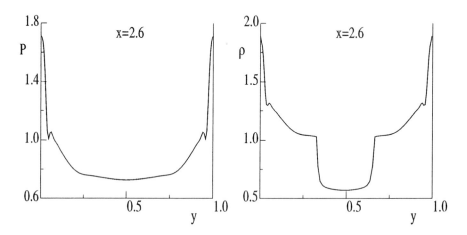

FIGURE 4(c): Pressure profile at x = 2.6   FIGURE 5(c): Density profile at x = 2.6

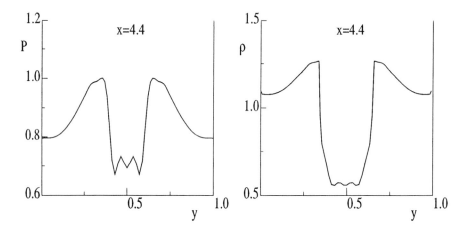

FIGURE 4(d): Pressure profile at x = 4.4   FIGURE 5(d): Density profile at x = 4.4

NETWORKS

Computational Approaches for Solving Some Teletraffic Problems

L. T. M. BERRY
Bond University
Queensland, Australia

INTRODUCTION

Teletraffic science is essentially the study of the interactions of information flows in a variety of types of telecommunications networks. These flows, such as voice, data and video traffic, have stochastic characteristics; thus it is not surprising that applied probability plays a dominant role in such studies. This is especially the case in the performance analysis of communications networks. More generally, mathematics in the wider sense provides many of the tools for modelling and solving teletraffic problems.

The purpose of teletraffic science is to provide concepts, fundamental theories and practical models which will assist with the maintenance of efficient, reliable and cost effective networks.

Since the derivation of Erlang's loss formula, the first major contribution to teletraffic science (Erlang, 1917), computational approaches have featured strongly in the literature. Examination of TELREF, a teletraffic reference database containing 8000 references,revealed that only about 1% of the papers introduced new closed form solutions to problems, some 10% of the papers described a simulation study, whereas more than 70% included computational results of some kind.

The Erlang loss formula

$$E_N(A) = \frac{A^N/N!}{\sum_{j=o}^{N} \frac{A^j}{j!}} \tag{1}$$

gives the time congestion of a Poisson stream of calls offering A erlangs of traffic to a fully available group of N trunks. It also gives the probability that an arriving call in an $M|G|N|N$ queueing system is rejected because no trunk is available (call congestion). Interestingly, some 100 of the papers in the above mentioned database

543

have been devoted to generalizations or properties of the Erlang loss formula, with more than 20 of these being in the last decade. Until the 1970's the value of $E_N(A)$ was usually computed by the recurrence

$$E_0(A) = 1$$

$$E_{N+1}(A) = \frac{AE_N(A)}{N + 1 + AE_N(A)} \tag{2}$$

The numerically superior recurrence

$$E_{N+1}(A)^{-1} = 1 + \frac{(N+1)}{A} E_N(A)^{-1} \tag{3}$$

is now preferred (Jagerman, 1974).

Not only have computational approaches changed for the same model over time but, as always, generalizations have been found, with related computational interests. For example, we consider below Erlang's model generalized to m distinct arrival streams with restrictions which induce a priority on the streams.

A GENERALIZED TRUNK RESERVATION PROBLEM

In the study of a class of loss networks called *non-hierarchical*, the potential for performance degradation when overflow traffic from previous routes mixes with traffic using the route as its first-choice has caused concern. To overcome the possibility of instability, the first-choice traffic is often protected by imposing a reservation number, r, on the other collective traffic. Whenever there are r or fewer *free* trunks, calls other than first-choice are rejected (lost). That is, a restriction is placed on a subset of the calls offered to the trunk.

Consider m Poisson streams with intensities $\lambda_1, \ldots, \lambda_m$ and common holding time μ^{-1}, offered to a group of N trunks. We can identify with each stream k a restriction number r_k such that its calls are accepted only if there are less than $\rho_k = N - r_k$ busy trunks.

Without loss of generality the streams may be ordered by their reservation numbers; $r_j > r_k \Rightarrow j < k$, with $r_m = 0$. A simple analysis of this multi-dimensional birth and death process gives the equilibrium solution:

$$P_k = \frac{A_1^{\rho_1} A_2^{\rho_2 - \rho_1} \ldots A_s^{\rho_s - \rho_{s-1}} A_{s+1}^{k - \rho_s}}{\mu^k k!} P_0 \tag{4}$$

for $\rho_s \leq k < \rho_{s+1}, \rho_0 = 0$ and $k = 0, 1, \ldots, N$,

where $A_k = \lambda_k + \lambda_{k+1} + \ldots + \lambda_m$.

P_0 is readily obtained from the normalization equation. Clearly, one would not compute the different stream blocking probabilities directly from Eq.(4) with appropriate summation over blocking states!

A simple modification of the Erlang recurrence Eq.(3) allows rapid computation of the blocking probabilities $B_k, k = 1, \ldots, m$. See, for example, (Berry, 1988). In outline, we can define the related function $\chi(A; k_0, k_1)$ recursively by

$$\chi(A; k_0, k_0) = 1$$

$$\chi(A; k_0, k_0 + n) = 1 + \frac{(k_0 + n)}{A} \cdot \chi(A; k_0, k_0 + n - 1) \tag{5}$$

for $n = 1, 2, \ldots, (k_1 - k_0)$.

It is straightforward to establish by induction that

$$\chi(A; k_0, k_1) = 1 + \frac{k_1}{A} + \frac{(k_1)_2}{A^2} + \ldots + \frac{(k_1)_{k_1 - k_0}}{A^{k_1 - k_0}} \tag{6}$$

where $(x)_n$ denotes the product $x(x - 1) \ldots (x - n + 1)$.

Generalizing the previous function by continuing the recurrence after the $(k_n - k_{n-1})$th interaction with the argument A_n being replaced by A_{n+1} another inductive proof establishes

$$\chi(A_1, A_2, \ldots, A_m; k_0, k_1, \ldots, k_m) = 1 + \frac{k_m}{A_m} + \frac{(k_m)_2}{A_m^2} + \ldots + \frac{(k_m)_{k_m - k_{m-1}}}{A_m^{k_m - k_{m-1}}}$$

$$+ \frac{(k_m)_{k_m - k_{m-1} + 1}}{A_m^{k_m - k_{m-1}} A_{m-1}} + \ldots + \frac{(k_m)_{k_m - k_0}}{A_m^{k_m - k_{m-1}} A_{m-1}^{k_{m-1} - k_{m-2}} \ldots A_1^{k_1 - k_0}} \tag{7}$$

From Eq.(6) and Eq.(7) it can be shown that

$$\frac{B_k}{B_m} = \chi(A_{k+1}, \ldots, A_m; \rho_k, \rho_{k+1} \ldots, \rho_m) \tag{8}$$

and

$$B_m = \frac{1}{\chi(A_1, \ldots, A_m; \rho_0, \ldots, \rho_m)} \tag{9}$$

Thus the procedure for computing B_m is to begin the Erlang recurrence with starting value 1 and traffic argument A_1, then after ρ_k iterations this is changed to A_{k+1} ($k = 1, 2, \ldots, m - 1$). This produces the denominator in Eq. (9).

During this computation, when step ρ_k is reached an additional parallel computation is started by commencing a new iteration with value 1 and traffic argument A_{k+1}. A simple multiplication by B_m gives the blocking probability B_k.

Sometimes finding an efficient computational method for a repetitive part of a larger calculation is critical in deciding whether or not an algorithmic approach is feasible. We consider next such a requirement that occurred in a major network dimensioning project.

INVERSION OF A DIMENSIONING FORMULA

The term *dimensioning* refers to the process of deciding trunk capacity allocation on links of networks. A minimum cost dimensioning study for the future 1990 Melbourne digital telephone network is described in (Berry, 1985). Our mathematical formulation of this problem required the solution of a large non-linear integer programme.

The alternative routing network had 889 switching nodes and 67,517 distinct origin to destination pairs between them. Some O-D pairs had up to 15 available routes and this resulted in approximately 456,000 routes in total. Since trunks were to be allocated in modules and could be of different types (PCM or VF) an imbedded dynamic programming solution was applied within the main optimisation.

Despite the large number of variables involved, it was demonstrated that the model was computationally feasible and cost savings of at least 5% were possible over conventionally designed networks. We shall focus on a critical part of the modelling which required an interplay between a standard dimensioning formula (Berry, 1971) and its inverse .

The Chain Flow Model

Suppose that we start with a prescribed set of traffic flows on the chains (routes) of a network. In this context the term *flow* refers to a mean value of carried traffic.

Let h_j^k denote the flow on the jth choice route between O-D pair k. Given an arbitrary network topology the chain flow model (Berry, 1971) can be used to compute the required number of trunks, for each link of the network, that would be required to achieve such a network flow pattern. That is, it gives a mapping $\mathbf{h} \to \mathbf{n}$ from the chain flow vector to the trunk vector.

At the link level, an essential formula giving n_i, the number of trunks to allocate to link i, is

$$n_i = x_i + A_i \left[\frac{(M_i - x_i)^2 + v_i}{(M_i - x_i - 1)(M_i - x_i) + v_i} - \frac{(M_i)^2 + V_i}{(M_i - 1)M_i + V_i} \right] \tag{10}$$

where x_i, the total flow on link i, is computed from

$$x_i = \sum_k \sum_j a_{ij}^k h_j^k \tag{11}$$

The variables a_{ij}^k are elements of the link chain incidence matrix for the network. Other variables in Eq.(10) are readily found, see (Berry, 1982), and it will suffice to say that M_i and V_i represent the first two moments of traffic offered to the link i, A_i is a function of these moments and v_i is the variance of traffic overflowing from the link.

The starting point in an iterative solution to this optimal dimensioning problem required specification of trunk allocations to a carefully selected subset of $\mathbf{n}$, called the partial circuit vector $\mathbf{n}_p$. It was then necessary to compute link flows x_i on this subset from Eq.(10). From the detailed model it was possible to determine the chain flow vector, **h**, for the network, subject to performance grades of service bounds, and to then obtain the complete circuit vector **n** corresponding to these chain flows. This constituted one iteration of the algorithm and information gathered in the process was used to make a new selection n_p in such a way that the network cost was reduced whilst maintaining grade of service constraints.

Since at each iteration approximately 27,000 function calls to the inverse dimensioning formula were needed, it was critical to find an accurate and sufficiently fast computational method.

Dropping the subscript, Eq.(10) can be re-arranged to

$$n = x + A\left[\frac{1}{(M - x - 1) + v/(M - x)} - B\right]$$

where $B = M/[M(M - 1) + V]$ \hfill (12)

The overflow variance v is given by

$$v = \frac{(M - x)}{6}\left[3 - (M - x) + \sqrt{(3 - M + x)^2 + 12A(1 - \frac{x}{M})^{0.1}}\right]$$ \hfill (13)

We wish to compute the value of x.

It is tempting to try an iterative approach

$$x_{i+1} = n - A\left[\frac{1}{(M - x_i - 1) + v/(M - x_i)} - B\right]$$

but this fails, even from good starting points; the mapping from x_i to x_{i+1} is not a contraction mapping.

Letting C = n + AB
        D = 5M - 3
        E = -5C - D
        K = CD - 6A

with re-arrangement of Eqs (12) and (13),

$$5x^2 + Ex + K - (x - C)\sqrt{(M - x - 3)^2 + 12A(1 - \frac{x}{m})^{0.1}} = 0$$ \hfill (14)

Newton's method also fails to converge to the the correct root. Although it is not immediately obvious from Eq.(14) the graph of the left-hand side expression has a vertical asymptote with the curve of interest being to its left. Unless the initial iterate is sufficiently close to the correct zero, the tangent to the curve crosses the asymptote and subsequent movement is towards the incorrect root.

A completely reliable method, however, is provided by using midpoint iteration on Eq.(14). A significant increase in efficiency is achieved by including an inverse quadratic interpolation with the midpoint iteration.

Lower and upper bounds bracketing the correct zero were obtained as follows. For overflow traffic $V > M$ thus $C > n$. Noting that $x < n$ we see that $x - C$ is always negative and from Eq.(14)

$$5x^2 + Ex + K < 0$$

Hence

$$x > \frac{-E - \sqrt{E^2 - 20K}}{10}. \tag{15}$$

Thus we arrive at a lower bound for x given by

$$\underline{x} = \max\left(0, \frac{-E - \sqrt{E^2 - 20K}}{10}\right) \tag{16}$$

An upper bound is given by $\min(M, n)$. The subroutine used to compute the link flows by the above method is called FLOW in the network package MINDER (modular interactive network dimensioning and evaluation routines).

As mentioned previously, there are approximately 27,000 calls to FLOW per iteration. This takes less than 25 min CPU time on a Honeywell DPSQ computer.

CONCLUDING REMARKS

For almost a century teletrafficists have solved the practical problems of design and efficient maintenance of large telecommunications networks by building mathematical models and finding particular solutions by computational means. Future problems will become dramatically more complex due to the integration of many new forms of traffic, each stochastically different, on new multiservice networks. Faced with modelling problems orders of magnitude more complex, many teletrafficists are rejecting probabilistic elements and considering deterministic models in order to obtain a simple appreciation of the structural and heavy traffic issues for these network. Some of the simplest structural problems lead to as yet unsolved constrained network problems. Even within the class of tree networks with n nodes, computational time is $0(n^{n-2})$, (Cayley, 1857). Heuristic solutions are appealing, but even with such an approach teletrafficists are now forced to give serious consideration to parallel algorithms and the use of supercomputers.

REFERENCES

1. Berry, L.T.M., A mathematical model for optimizing telephone networks, Ph.D. thesis, The University of Adelaide, 1971.

2. Berry, L.T.M., Optimal dimensioning of circuit switched digital networks, Telecom Report pp. 1-164, 1982.

3. Berry, L.T.M., Harris R.J., Modular design of a large metropolitan telephone network: A case study, *International Teletraffic Congress, Kyoto*, 1985.

4. Berry, L.T.M., Towards an effective stream-control mechanism, *Proceedings of the 3rd Fast Packet Switching Workshop*, Telecom Research Laboratories, 1988.

5. Cayley, A., On the theory of analytical forms called trees,*Phil Mag* Vol 13, pp 172-176,1857.

6. Erlang, A.K., Solution of some problems in the theory of probabilities of significance in automatic telephone exchanges, *Elektroteknikeren*, Vol 13, p 5- , 1917.

7. Jagerman, D.L., Some properties of the Erlang loss function, *B.S.T.J.*, Vol 53, No 3, pp 525-551, 1974.

8. Jagerman, D.L., Methods in traffic calculations, *AT&T Bell Laboratories Technical Journal*,Vol 63, No 7, pp 1283-1310, **1974.**

A Set Partitioning Approach to the Production and Vehicle Routing Problem with Time Windows

P. J. KILBY, J. N. HOLT, and A. M. WATTS
Department of Mathematics
The University of Queensland
St. Lucia, Q. 4067, Australia

INTRODUCTION

The Vehicle Routing Problem (VRP) is concerned with choosing a minimum cost set of routes for a given fleet of vehicles in order to deliver goods to a group of customers with known demands, and without exceeding the capacity restriction of any vehicle. The routes begin and end at a single depot. Bodin et al. (1983) provide an excellent survey of work in this area. The extension to the Vehicle Routing and Scheduling Problem with Time Windows, (VRSPTW), allows for the specification of a time window, between an earliest and latest acceptable delivery time, for each customer. The addition of these temporal considerations adds considerably to the difficulty of the problem. Recent work has included that by Baker and Schaffer (1986),Solomon (1987), Kolen et al. (1987), Solomon et al. (1988)and Desrochers et al. (1988).

We study here an extension to the VRSPTW which occurs in practice, but does not seem to have been considered extensively in the literature. This extension occurs when a production process precedes delivery, so that vehicles must wait for the goods to become available before departing on their routes. Given the existence of time windows for delivery to customers, this has considerable effect on route construction. We must therefore specify the time at which each vehicle is dispatched, as well as the customers it services and the route taken around those customers. We call this problem the Production and Vehicle Routing Problem with Time Windows (PVRPTW). Little work appears to have been done on it.

One example of PVRPTW, which has been studied by Holt and Watts (1988), is the delivery of newspapers from the press to newsagents. In this application, newspapers are loaded straight from the presses into waiting trucks for immediate delivery. Typically, a number of trucks are loaded in parallel at a set of docks fed by conveyors. When a truck is fully loaded and dispatched, another takes its place in the dock. In the newspaper industry, the ability to have a late production start time, and hence include late breaking stories or sports results, can have a significant, beneficial effect on circulation. On the other hand, delaying the start of production increases the cost of delivery to newsagents. The agents have deadlines, totally independent of the production start time, by which they must receive the newspapers if they are to wrap and home deliver by the nominal latest time which the newspaper company considers acceptable. Therefore, an efficient distribution system which seeks to produce an optimal tradeoff between maximimum delay in production start time and minimum distribution costs, is an essential ingredient of the total operation.

The method of solution for PVRPTW which we investigate in this paper is based on Set Partitioning. This model has been used in connection with the VRP by Foster and Ryan (1976), and Ryan and Falkner (1988) have reviewed the properties of the approach to scheduling problems and discussed the application to bus crew scheduling. In this approach, many possible individual vehicle routes are generated first, using *column generation* heuristics. Then the optimal feasible subset of these routes is chosen using a set partitioning zero - one integer programming problem. This Two-Phase approach in effect separates the problem of deciding which vehicle should service which customer from the problem of routing around the customers assigned to a vehicle. The customer-vehicle assignment problem is handled in the mathematical programming phase (Phase II), while for each individual vehicle route produced in Phase I, the routing heuristics have already determined the ordering of customers on the route, and the associated cost of the route.

FORMULATION OF VRSPTW

A set partitioning based formulation of the VRSPTW can be given as follows: Suppose there are n customers, numbered $1 \ldots n$. The demand at customer i is d_i, the waiting time (for unloading, etc.) is w_i, the delivery time window is given by $[e_i, l_i]$, and the time of travel between customers i and j is c_{ij}. Also suppose that there are m vehicles, with m_j of the vehicles having capacity cap_j, $j = 1 \ldots M$. A column vector a_k is used to specify the customers assigned to a potential vehicle route k.

$$a_{jk} = \begin{cases} 1 & \text{if route } k \text{ services customer } j \\ 0 & \text{otherwise} \end{cases}$$

Since this column corresponds to a possible route, we use 'column' and 'route' interchangebly. Note that the entries in the column a_k only specify which customers are serviced by this potential route, not the order in which they are serviced. This must be recorded separately.

In Phase I of the method, a range of column generation heuristics can be used to provide a large set of K possible columns from which the solution will be composed. The cost of column k is given by c_k, and can be calculated using any code for the Travelling Salesman Problem with Time Windows (TSPTW). The column generation procedures will allow only those columns which meet capacity restrictions, and for which an ordering exists such that all time windows can be satisfied. Given this ordering, it is possible to calculate the latest dispatch time, t_k, which allows all customer deadlines to be met.

Once the set of columns is determined, the optimization problem for Phase II of the method can be formulated as an integer programming problem in terms of a set of variables x_k, where

$$x_k = \begin{cases} 1 & \text{if route } k \text{ is used} \\ 0 & \text{otherwise} \end{cases}$$

The formulation is

$$\text{minimise} \sum_{k=1}^{K} c_k x_k \qquad (1)$$

subject to

$$\sum_{k=1}^{K} a_{ik} x_k = 1 \qquad i = 1 \ldots n \tag{2}$$

$$\sum_{k=1}^{K} x_k \leq m \tag{3}$$

$$\sum_{k \in K_i} x_k \leq m - \sum_{j=1}^{i-1} m_j \qquad i = 2 \ldots M \tag{4}$$

$$x_k = 0 \text{ or } 1 \qquad k = 1 \ldots K \tag{5}$$

$$K_i = \{k | \sum_{j=1}^{n} a_{jk} d_j \leq cap_i\} \tag{6}$$

Constraints (2) ensure that each customer is serviced on exactly one route. These are the set partitioning constraints. Constraint (3) restricts the number of routes in the solution to no more than the number of vehicles. The constraints (4), with K_i as defined in (6), enforce the conditions that the set of routes in the solution can be fitted onto the available fleet. If the vehicles are all of the same capacity, constraints (4) may be omitted.

The integer programming problem described by (1) to (6) is generally too large to solve by standard integer programming. Ryan and Falkner (1988) show that it is possible to select the columns for the constraint matrix in scheduling set partitioning problems so that most of the basic feasible solutions of the linear program obtained by relaxing the integrality property have integer components. This reduces the complexity of the problem, allowing feasible solutions to be obtained from the linear program or else relatively few steps of branch and bound to resolve fractionality. The restriction of the set of columns is done using problem structure not already built into the model, and does not significantly reduce the set of useful solutions.

The selection of columns capitilizes on the fact that certain set partitioning matrices are known to produce integer solutions to the linear programming problem. Ryan and Falkner (1988) discuss totally unimodular, balanced and perfect matrices and show that these possess the property. Further, they discuss the strongly integerizing role of constraints involving many columns, even if the set partitioning matrix is not exactly in one of the above classes.

It is possible to generate a totally unimodular matrix in (2) above by generating routes according to the SWEEP method of Gillet and Miller (1974) (See Foster and Ryan (1976) and Ryan and Falkner (1988)). In this method, the customers are numbered radially around the depot, and columns are generated by choosing contiguous groups from this order. More appropriate orderings can be chosen as we shall see later.

FORMULATION OF PVRPTW

In this section, a formulation is presented for a simplified version of PVRPTW. Again, Phase I is concerned with column generation, while Phase II solves a set partitioning problem. The assumptions we will make are that all vehicles have the same capacity cap, and that vehicles are loaded sequentially, rather than in parallel.

Given the constant production rate of goods, $ProdRate$, the total production time required to manufacture all of the required goods is

$$T = \frac{\sum_{i=1}^{n} d_i}{ProdRate}$$

Suppose the total production period T is divided into m equal production periods, each corresponding to the loading time for a vehicle. The starting times of the periods define a discretized approximation to the time axis, corresponding to the starting times for loading vehicles. Using the latest dispatch time, (t_k), for route k to be feasible with respect to its customers' delivery deadlines, we calculate in Phase I the latest period in which this route can be loaded.

Define b_{pk} by

$$b_{pk} = \begin{cases} 1 & \text{if column } k \text{ must be dispatched in period } p \text{ or earlier} \\ 0 & \text{otherwise} \end{cases}$$

Then the new integer programming formulation for Phase II consists of (1), (2), (3), and (5) in the previous section, together with

$$\sum_{k=1}^{K} b_{pk} x_k \leq p \qquad p = 1...m \tag{7}$$

which restricts the number of dispatches up to the end of period p.

While the constraint matrix resulting from this formulation is no longer perfect, the two sections of it corresponding to $[a_{ik}]$ and $[b_{pk}]$ are each unimodular This means that while non-integer solutions can occur in the linear programming relaxation of the problem, they might be expected to be rarer than in an unstructured formulation, and easily reduced to integer solutions by branch and bound (see Foster and Ryan (1976)). Note that in a branch and bound method, branching should be done first on constraints. That is, if one of (3), (4) or (7) has a non-integer left hand side, we branch on it. The aim is to make these constraints active so that they act as dominant side constraints of the form discussed by Ryan and Falkner (1988).

ALGORITHMIC DETAILS FOR PVRPTW
Phase I – Column Generation

In the basic version of the new algorithm, generalized SWEEP-style routes a_k are produced for Phase II in the following way.

In the first instance, an ordering of customers is constructed . This is done by employing a time constrained travelling salesman heuristic on the customers of the problem, using as an objective function

$$\alpha * distance + (1 - \alpha) * slack. \tag{8}$$

The parameter α is in the range [0,1]. slack is the amount of time the route can be delayed without making the route deadline infeasible, and distance is the total distance around the route. By using different values for α, the ordering can reflect different tradeoffs between distance and time. In the present results, we have considered α equal to 1 and 0.75. The results obtained indicate an iterative approach to obtaining the best solution to the problem, as will be seen later.

Given the initial ordering of customers, we generate a set of routes which have the unique subsequence property referred to by Ryan and Falkner (1988) by selecting sequentially from this list. A seed customer is randomly chosen, and a route is initialised using just this customer. The next customer in the ordering is then inserted at an appropriate position into the emerging route using an insertion heuristic, described later. If the total load for the route is greater than some threshhold value (say $cap/2$) the corresponding column a_k is stored. The next customer in the ordered list is then considered, and so on until including the next customer would exceed capacity constraints, or until a time- feasible route cannot be constructed. The cost associated with the column is as for the route, and the particular ordering of the customers within the route is recorded separately for use if this column is used in the final solution. Routes are generated in this way using each of the n customers in turn as seeds.

As an extension of the basic approach, several columns can be constructed for a given set of customers based on each seed customer, by attempting to optimize the objective (8) for different values of α. A column is written out for each distinct route in the collection. Although the columns a_k corresponding to these routes are identical, their costs c_k are not, nor are their latest dispatch periods so that when they are appended by rows arising from (7) they do not lead to identical columns in the integer programming problem.

In order to estimate the best possible route for each choice of α above, several different routes are maintained in parallel using different insertion heuristics. The motivation for this scheme comes from Baker and Schaffer (1986) who found that different algorithms performed better on some data, but not on others. By using a number of different algorithms, the variance in quality of solution caused by this behaviour should be reduced.

Insertion Heuristics

Insertion heuristics are used in this algorithm, as they provide fast sequential procedures for considering the customers in turn, and placing them in the emerging route. This is one concession to speed in the algorithm because of the large number of routes to be generated - we simply include each customer in turn rather than reoptimizing for each. It would be possible to apply improvement heuristics such as 2-Opt or 3-Opt (Lin (1965)) within routes during the construction, or after the column selection phase is completed in order to improve the distance further, but this has not been done at this stage.

The insertion heuristics used here are variations of the two found to be most successful by Baker and Schaffer (1986). These authors define two classes of heuristics, with parameter settings defining instances from the classes. In both classes, a new customer u is inserted between customers i and j (adjacent in the current route) which minimise a cost function. Having established the best insertion position for each uninserted customer, they then select the customer to insert by examining another objective function. In our case, there is no choice of the customer to be inserted. It must be the next one in the initial ordering. We only use the part of the heuristic specifying where it should go on the route.

In the first class, which is the well-known savings heuristic, the cost function depends only on distance:

$$z_1 = c_{iu} + c_{uj} - \mu_1 c_{ij} \tag{9}$$

$\mu_1 = 1.0$ is used in the computations described in the next section.

The second class also includes temporal considerations, in terms of both arrival time and waiting time. Let v_j be the time the vehicle arrives at customer j in the current route, if the vehicle leaves the depot at time 0 (production start time). Then let $v_{j/u}$ be the new time of arrival given that customer u has been inserted before customer j.

Let

$$z_{11} = z_1 \qquad (\text{as in } (9))$$
$$z_{12} = v_{j/u} - v_j$$
$$z_{13} = l_u - v_u$$
$$z_2 = \mu_{11} z_{11} + \mu_{12} z_{12} + \mu_{13} z_{13} \tag{10}$$
$$\mu_{11} + \mu_{12} + \mu_{13} = 1$$
$$\mu_{11} \geq 0, \quad \mu_{12} \geq 0, \quad \mu_{13} \geq 0$$

The parameter settings used for the results in the next section are

μ_{11}	μ_{12}	μ_{13}
0.4000	0.4000	0.2000
0.5000	0.2500	0.2500
0.8000	0.1000	0.1000
0.3333	0.3333	0.3333

This gives a total of 5 heuristic methods used to develop the routes. The values for α used to give the various objectives against which the routes are measured are: 0.0, 1.0 and 0.5.

COMPUTATIONAL RESULTS

There evidently exists no standard set of test problems for the PVRPTW. Baker and Schaffer (1986) describe procedures for generating test problems for the VRSPTW, and we use one of their approaches for these preliminary computational studies, adding various values of two parameters: Production Rate and Production Start. These parameters directly impact the feasibility and cost of solutions. In any application, they are both important economic parameters in the context of the total production, distribution and marketing exercise.

Fifty test problems are considered, each using fifty customers with locations generated randomly using a uniform distribution. Distance between customers is the straight-line distance between them, and time is scaled so that travel time is equal to distance. Time windows are generated following Baker and Schaffer (1986), who construct a Nearest Neighbour tour of the customers, then assign non-overlapping windows with the centre of the window corresponding to the time of arrival using this tour. A certain percentage of customers is chosen randomly, in our case one half. The time windows for these customers are then removed.

Customer loads, in appropriate units, are random in the range 1 - 100. 4 trucks are used in the test problems, each with a capacity of 750 units. This means that, on average, vehicles are 83% loaded. Each of the 50 problems is solved with Production Rates of 2.0, 2.5, 3.0, 3.5, 4.0 (units/unit time) and Starting Time of 0, -30 and -60 (time units offset), giving 750 test problems in all. Further, each of these 750 problems is considered twice, using two different initial orderings of the customers, corresponding to α equal to 1 and 0.75 in the objective (8) used for producing the time constrained travelling salesman route on which the initial orderings are based.

Table I summarizes the results for all 50 sets of customer data for α equal to 1. The zero-one integer programming problem in Phase II of the method was solved using a commercialy available package, AESOP (Murtagh (1986)), which first solves the LP relaxation of the problem, and then uses branch and bound to remove integer infeasibilities if they exist.

TABLE 1. Number of Solved Problems and Average Total Distance for α equal to one.

		Production Rate				
		2.0	2.5	3.0	3.5	4.0
	0	23	28	31	37	39
		1660	1653	1664	1663	1656
Production	-30	25	31	36	39	41
Start		1665	1670	1662	1653	1656
	-60	27	33	39	42	42
		1656	1669	1654	1658	1652

For α equal to 1, the initial ordering of customers is based on distance only, and so it could be anticipated that columns generated from the ordering may not be appropriate when time windows are tight. Due to the random way in which problems were generated, some problems have very tight deadlines, while others are effectively deadline free. This results in some problems having no solution for any value of the parameters, and others having the same solution for all values (indicating a solution not bound by any deadlines). However, all problems demonstrate a non-increasing objective as production rate is increased, and start time decreased, although this is not shown in the table. As the production rate is increased, the vehicles are able to leave the docks earlier, and hence have more time on the run. This means that a problem for which no solution could be found at a lower production rate may now become feasible. Those problems which had feasible solutions for the slow production rates are now able to take advantage of spare time to travel in a more distance-optimal fashion. This results in more problems able to be solved and lower total distances travelled as the production rate increases. Starting production earlier achieves the same result in a more direct way. For α equal to 0.75, the initial ordering takes the time windows into account, and so to some extent chases deadlines. In this case, all problems admit feasible solutions using the columns generated for all values of the production rate and start time. The distances of the resulting solutions are greater than previously, averaging around 1930.

The above discussion suggests an iterative procedure for using α. Suppose we are given a production rate and start time. Using α equal to 1, we could generate an initial ordering, and attempt to solve the problem. If no solution can be found, the time windows are too tight for the ordering. α is reduced by a set amount and the process repeated until a feasible solution is obtained. This has the effect of tailoring the ordering, and hence the columns to the degree of time window influence on the solution. Any further reduction in α would increase the total distance travelled.

The time taken to generate the columns, exclusive of disk I/O, averaged 133 seconds per problem. This remained effectively constant throughout all parameter settings. The computations were carried out on a personal computer with an 80286 processor, and a Norton rating of 10.1. The number of non-integer solutions generated (only 33 problems of the 1500 tested gave rise to non-integer solutions) supports the earlier remarks on the near-perfect structure of the constraint matrix. In 32 of these problems, it was necessary to branch on only one node to produce integer solutions, while the other problem required two more branch and bound steps.

Although the model used here has restricted the fleet to be homogeneous, it is not difficult to see that we could cater for variations in fleet capacities by modifying the constraints (7). Rather than having discrete time intervals corresponding to vehicle loading times, we could take an arbitrarily fine time discretization. The right hand side of (7) would become the total production up to the current time period, and the b_{pk}'s would be the old values multiplied by the load on the kth column. Similarly, we could cater for truck re-use by including columns in the problem which correspond to runs which return to the depot part way through their operation. These aspects will be investigated further.

CONCLUSIONS

The paper has described an idealized version of a vehicle routing problem that occurs in practice, such as the delivery of newspapers, called the Production Vehicle Routing Problem with Time Windows. It has been shown how to formulate the problem using a Set Partitioning model with additional constraints representing the dispatch scheduling. An important aspect of the method of solution is the initial ordering of customers. By selecting customers contiguously from the ordered list to generate routes, near integrality of the solution of the linear programming problem results. It has been shown how to vary this order through the choice of a distance/slack tradeoff, and so create an iterative procedure for estimating the shortest feasible set of routes for the PVRPTW. Tests performed by varying two critical parameters in the production process, namely production rate and production start time, demonstrate the effect these parameters have on the solution.

Another way of using an algorithm such as the one presented here is to fix the production start time and rate and obtain solutions for different numbers of vehicles of different capacities. Economic decisions on fleet composition can then be made. Further work on this topic will aim at removing the restriction that vehicles are of equal capacity, and at catering for truck re-use. In addition, more computational testing needs to be carried out using a range of types of problems, including real world problems.

REFERENCES

1. Baker, E. K. and Schaffer, J. R. , Solution Improvement Heuristics for the Vehicle Routing and Scheduling Problem with Time Window Constraints, *Am. J. Math. Mgmt. Sci.*, vol. 6, pp. 261–300, 1986.
2. Desrochers, M., Lenstra, J. K., Savelsbergh, M. W. P. and Soumis, F. , in *Vehicle Routing: Methods and Studies* , eds. B. L. Golden and A. A. Assad , pp. 65–84, Elsevier Science Publishers , Amsterdam , 1980.
3. Foster, B. A. and Ryan, D. M. , An Integer Programming Approach to the Vehicle Scheduling Problem, *Opl. Res. Q.* , vol. 27, pp. 367–384, 1976.
4. Gillet, B. E. and Miller, L. R. , A Heuristic Algorithm for the Vehicle Dispatch Problem, *Oper. Res.*, vol. 22, pp. 340–349, 1974.
5. Holt, J. N. and Watts, A. M. , in *Vehicle Routing: Methods and Studies* , eds. B. L. Golden and A. A. Assad , pp. 347–358, Elsevier Science Publishers , Amsterdam 1980.
6. Kolen, A., Rinnooy Kan A. and Trienekens, H. , Vehicle Routing with Time Windows , *Oper. Res.*, vol. 35, pp. 266–273, 1987.
7. Lin, S. , Computer Solutions of the Travelling Salesman Problem , *Bell System Tech. J.*, vol. 44, pp. 2245–2269, 1965.
8. Murtagh, B. , AESOP User Manual, 1986.
9. Ryan, D. M. and Falkner, J. C. , On the Integer Properties of Scheduling Set Partitioning Models , *Euro. J. Opl. Res.* , vol. 35, pp. 442–456, 1988.
10. Solomon, M. M. , Algorithms for the Vehicle Routing and Scheduling Problem with Time Window Constraints , *Jour Oper. Res.* , vol. 35, pp. 254–265, 1987.
11. Solomon, M. M. , Baker, E. K. and Schaffer, J. R. , in *Vehicle Routing: Methods and Studies*, eds. B. L. Golden and A. A. Assad , pp. 85–106, Elsevier Science Publishers , Amsterdam 1980.

NUMERICAL METHODS

Choosing Test Matrices
for Numerical Software

A. L. ANDREW
Department of Mathematics
La Trobe University
Bundoora, Victoria 3083, Australia

INTRODUCTION

An important part of the development of mathematical software is systematic testing. Test problems should be chosen to check the performance of the software on a wide range of typical problems and also on any types of problems likely to prove difficult. This paper discusses the selection of test problems. It illustrates some associated difficulties from the author's experience in the search for suitable test matrices to evaluate some routines for the numerical solution of the problem described in Section 2. In particular an important property of one of the numerical examples used by Tan (1986, 1987abc, 1988ab, 1989) and Tan and Andrew (1989) is discussed and implications for the conclusions reached by Tan (1986, 1987abc) are pointed out. However, most of our remarks are relevant for a much wider range of problems, including differential equations.

Often test examples are selected because the solution is known in closed form. Such choices have the obvious advantage that the error in the computed solution is readily determined. Such examples are discussed in Section 3. Software should always be tested on examples with various types of illconditioning and it is particularly convenient when, as often happens, suitable illconditioned examples are available with known closed form solutions. However, when comparing competing methods, it is important to realize that methods that perform best on particular problems do not necessarily perform best for other problems. A major drawback of excessive reliance on problems with known closed form solution is that such examples are available only in rather special cases which may not be representative of the examples on which the routines being tested are likely to be used.

"Real world" applications are another important source of test examples, though it is always important to use a wide variety of examples even if this means that most are "artificial". Even if it is believed that certain difficulties never arise in "real" problems, future applications may be found in which these difficulties do arise and it is always best to know how a routine will perform on as large a range of problems as possible. The difficulty of determining the accuracy of computed solutions for "real" examples is a serious disadvantage of such examples, compared with those with closed form solution, but not usually a fatal one. Comparison of results obtained by independent methods gives a useful indication of accuracy. In problems with differential equations, solutions may be compared with those obtained with a finer grid. Similar methods may be used for matrix problems arising from numerical methods for solving differential equations. The effect of roundoff errors may be estimated by comparing results with different mantissa lengths.

Test matrices are also often generated randomly. Such matrices are the subject of Section 4. It might be thought that the use of randomly generated matrices would ensure the selection of a representative set of test examples. However this depends on the probability distribution used. The apparently natural choice in the default version of the RAND function of the (deservedly) popular package MATLAB turns out to produce examples with rather special properties. Determining the accuracy of computed results also presents much the same difficulties for random examples as for "real" examples.

DERIVATIVES OF EIGENVALUES AND EIGENVECTORS

Sections 3 and 4 are largely concerned with the author's experience with test examples for the following problem. Given a square matrix A which depends smoothly on ρ real parameters $p_1, \ldots, p_\rho$, it is required to compute local numerical values of the partial derivatives of eigenvalues, λ_i, and corresponding eigenvectors, x_i, of A with respect to these parameters. First and second derivatives of eigenvalues are often required when mathematical programming methods are applied to optimal design problems, such as designing aircraft of minimum mass subject to a flutter velocity constraint (Rudisill and Chu, 1975), and computation of second derivatives of λ_i uses derivatives of x_i. Partial derivatives of both eigenvalues and eigenvectors are also used for example in system identification of civil engineering structures (Béliveau, 1987). Several other applications are mentioned by Adelman and Haftka (1986), Andrew (1979), Murthy and Haftka (1988), Rudisill and Chu (1975) and Tan (1988). A special case is that in which $p_1, \ldots, p_\rho$ are simply the matrix elements a_{ij}.

First derivatives of eigenvalues are easily determined by the formula (Adelman and Haftka, 1986, Lancaster; 1964)

$$\lambda_{i,j} = y_i^T A_{,j} x_i / y_i^T x_i \tag{1}$$

where the y_i are left eigenvectors $(y_i^T A = \lambda_i y_i^T)$ and the subscript $,j$ denotes partial differentiation with respect to p_j. A simple and efficient method for computing $x_{i,j}$ is direct solution of the equations resulting from differentiating the equation $Ax_i = \lambda_i x_i$ and a suitable normalizing condition for x_i (Rudisill and Chu, 1975). Failure of some earlier investigators to appreciate the need for a normalizing condition to ensure uniqueness seems to be the reason why some early direct methods (see Rudisill and Chu, 1975; Tan, 1988a for references) are unnecessarily complicated and inefficient, and why numerous attempts (see Murthy and Haftka, 1988; Rudisill and Chu, 1975 for references) have been made to use formulae such as

$$x_{i,j} = \sum_{\substack{r=1 \\ r \neq i}}^{n} \{y_r^T A_{,j} x_i / [(\lambda_i - \lambda_r) y_r^T x_r]\} x_r + \gamma_i x_i \tag{2}$$

for numerical computation of eigenvectors. (See the discussion of Andrew, 1978, 1979 and Tan, 1988a.) Although (2) is important for showing the conditions under which x_i is illconditioned, it is inefficient and numerically unstable for computing $x_{i,j}$ (Andrew, 1979; Tan, 1988a, 1989).

An iterative method for computing $\lambda_{i,j}$ and $x_{i,j}$ was also suggested by Rudisill and Chu (1975) but, although the analysis of Andrew (1978, 1979) led not only to substantial improvements in the method but also to further insight into the problem, the method remained uncompetitive with the best direct methods. At this point my student, Roger Tan, investigated the application of various extrapolation techniques to the iterative method. When he wanted some test examples with known closed form solution, I suggested the examples A_1, A_2 and A_3 described in Section 3. Unfortunately neither of us realized at first that A_3 was defective (though this is noted by Fairweather, 1971) so that, as shown by Andrew (1979), the iterative method could be expected to perform unusually badly. It was not until Roger Tan noted that use of (2) failed completely for A_3 that I realized that defectiveness of A_3 was the reason, and this defectiveness is mentioned in Tan and Andrew (1989). (There is a printer's error in eq.(19) of that paper, where $\neq$ should be $-$.)

Although A_3 is defective, the $x_{i,j}$ computed in Tan (1986, 1987abc, 1988a, 1989) and Tan and Andrew (1989) all corresponded to simple eigenvalues, so that there is no difficulty computing them by direct methods. As all the extrapolation methods tested in Tan (1986, 1987abc) gave highly accurate results for A_1 and A_2, conclusions reached there about the relative accuracy of the methods were based largely on their (much less accurate) results for A_3. Further work is needed to determine the relative merits of these methods for more typical examples, though Tan (1986, 1987abc) clearly showed all of the extrapolation techniques to improve the original iterative method substantially. Also the topological ε–algorithm, first successfully implemented by Tan (1987b), promises to be useful for a wide variety of problems, though it had previously been considered impractical (see Tan, 1988b for references).

EXAMPLES WITH CLOSED FORM SOLUTION

Although there are substantial collections of matrices with known eigenvalues and eigenvectors (Gregory and Karney, 1978), relatively few are suitable for the problem described in Section 2 which requires a matrix which depends on a parameter, p_1, and whose eigenvalues and eigenvectors are known (and differentiable with respect to p_1) for all values of p_1 in some interval. (As noted in Section 2, examples occurring in practice often depend on several parameters, $p_1, \ldots p_\rho$. However, only partial derivatives are computed and, as partial derivatives of a function of several variables are effectively ordinary derivatives of a restriction of that function to a function of one variable, examples depending on a single parameter are adequate for numerical testing.)

The first example chosen was

$$A_1 := \begin{pmatrix} -B_1 + \sigma I_4 & I_4 \\ -B_2 & \sigma I_4 \end{pmatrix}^T \qquad \text{where}$$

$$B_1 := \begin{bmatrix} 3p_1 & 2 & 0 & 0 \\ -(1 + p_1^2 + 2\beta^2) & 0 & 2 & 0 \\ p_1(1 + 2\beta^2) & 0 & 0 & 2 \\ -\beta^2(p_1^2 + \beta^2) & 0 & 0 & 0 \end{bmatrix}, \qquad B_2 := \begin{bmatrix} -1 + 2p_1^2 & 2p_1 & 1 & 0 \\ p_1(1 - p_1^2 - 2\beta^2) & -(p_1^2 + 2\beta^2) & 0 & 1 \\ 2p_1^2\beta^2 & 2p_1\beta^2 & 0 & 0 \\ -p_1\beta^2(p_1^2 + \beta^2) & -\beta^2(p_1^2 + \beta^2) & 0 & 0 \end{bmatrix},$$

I_m is the $m \times m$ identity and $\beta := 1 + p_1$. In the case $\sigma = 0$, the eigenvalues of A_1 are those of the quadratic eigenvalue problem

$$\lambda^2 x + \lambda B_1^T x + B_2^T x = 0 \tag{3}$$

and, if λ_0 is an eigenvalue, x_0 is an eigenvector of (3) if and only if $(\lambda_0 x_0^T, x_0^T)^T$ is an eigenvector of A_1. Problems of the form (3) are important in the theory of damped oscillations. This example was used by Lancaster (1965), Ruhe (1973) and Hayes and Wasserstrom (1976) for testing algorithms for the numerical computation of eigenvalues of matrices depending in a rather general nonlinear manner on the eigenvalue, but the full power of these methods is not needed for this example which is probably best treated by applying standard routines (such as those of EISPACK) to the linearized form $A_1 x = \lambda x$. Various choices of the origin shift σ in A_1 allow different eigenvalues to be made dominant. This flexibility in the choice of dominant eigenvalue enables the methods studied by Tan (1986, 1987abc, 1988a, 1989) and Tan and Andrew (1989) to be tested under a greater variety of conditions.

The eigenvalues of A_1 are listed by Lancaster (1965), Ruhe (1973), Hayes and Wasserstrom (1976) and Tan (1986, 1987ab, 1988a, 1989) and two pairs of (complex conjugate) eigenvectors are given by Tan (1987b, 1988b, 1989) but, as far as the author is aware, the other eigenvectors have not been published. Since A_1 is likely to prove a useful test matrix for other problems, we list the remaining eigenvectors here. They are (i) $[2\beta^2 - i(3\beta^3 + 1), 2\beta^3 + \beta + 1 + i(\beta^2 + 1), -2 + i(\beta^3 + 2\beta + 1), -3\beta - 1 + i(\beta^2 - 3), -3\beta^3 - 1 - 2i\beta^2, \beta^2 + 1 - i(2\beta^3 + \beta + 1), \beta^3 + 2\beta + 1 + 2i, \beta^2 - 3 + i(3\beta + 1)]^T$, (ii) its complex conjugate, (iii) $[-p_1^3(3\beta^2 + p_1^2), -2p_1^2\beta^2, -p_1(1 + 2p_1), 2p_1^2, p_1^2(3\beta^2 + p_1^2), 2p_1\beta^2, 1 + 2p_1, -2p_1]^T$ and (iv) $[0, 0, 0, 0, 0, 0, p_1^2 + \beta^2, 2p_1]^T$, corresponding to the eigenvalues $\sigma + i$, $\sigma - i$, $\sigma - p_1$ and σ respectively. For

563

the problem described in Section 2, these eigenvectors must be suitably normalized. The derivatives of eigenvectors of A_1 are most illconditioned when several of the eigenvalues are close, i.e. when p_1 is close to 0 or -1.

The ease with which the condition of the problem can be changed is one of the advantages of A_1. Its main disadvantage is its small dimensions, especially as iterative methods are used mainly for large matrices. A method used by Tan (1987 abc) to build up somewhat larger matrices can be used quite generally. It reverses a technique described by Andrew (1973) for reducing the eigenvalue problem for certain $2n \times 2n$ matrices to that for two $n \times n$ matrices. It produces matrices of the form

$$A_2 := \begin{bmatrix} B_3 & B_4 P \\ P^{-1} B_4 & P^{-1} B_3 P \end{bmatrix}$$

whose eigenvalues are the same as those of $B_3 + B_4$ and $B_3 - B_4$ and whose eigenvectors are readily computed from those of these matrices. We chose $B_3 + B_4 = A_1, P$ a permutation matrix, and $B_3 - B_4$ a triangular matrix.

In principle this process could be continued, repeatedly doubling the dimensions of the test matrix obtained. This was not done by Tan (1987abc) as I suspected that the matrices produced might be rather atypical, but ironically the other test matrix used by Tan (1987abc) (and also Tan, 1986, 1989 and Tan and Andrew, 1989) was much more atypical. It is the $n \times n$ matrix $A_3 := (a_{3ij})$, studied by Fairweather (1971), where $n > 2$,

$$a_{3ij} := \begin{cases} 0 & \text{if } j > i + 1 \\ 1 & \text{if } j = i + 1 \\ p_1 + \sigma & \text{if } j = i \\ p_1^{1+i-j} & \text{if } j < i \end{cases}$$

and σ is a real scalar. The eigenvalues of A_3 are

$$\lambda_i = \begin{cases} 4p_1 \cos^2(\pi i/(n + 2)) + \sigma & \text{for } i = 1, \ldots, [(n + 1)/2] \\ \sigma & \text{for } i = [(n + 3)/2], \ldots, n \end{cases}$$

where here $[\alpha]$ denotes the largest integer not exceeding α.

As noted by Tan (1986, 1987abc), the eigenvector derivatives of A_3 are more illconditioned when p_1 is very small (making eigenvalues close) or large (making numbers involved in calculations large compared with the solutions), especially when n is large. This illconditioning affects all methods, but iterative methods are especially affected by the defectiveness (not mentioned by Tan (1986, 1987abc)) of the multiple eigenvalue σ, which has geometric multiplicity one. The analysis of Andrew (1979, pp 215–216) shows that this effect is also most marked for large n.

The author is not aware of any such defective matrices arising in real applications where derivatives of eigenvalues and eigenvectors are required. Also if the matrix elements are chosen randomly from any of the usual continuous distributions, the probability of defectiveness is zero. Hence A_3 may be considered very atypical. However A_3 is not really atypical of matrices for which simple closed form derivatives of eigensystems are known. It is the special simple structure of the matrix which makes the closed form solution available. When such simple structure is required, exceptional behaviour, such as defectiveness, is much more likely. That is, matrices required to have known closed form eigenstructure are much more likely than other matrices to have an unusual eigenstructure which may cause algorithms to behave in an unusual way.

The simple structure common in problems with closed form solutions gives rise to atypical examples (and atypical performance of numerical algorithms) for many other problems as well, especially in problems involving differential equations. One with which the author had personal experience concerned the numerical solution of Sturm–Liouville problems. A method for numerical solution of Sturm–Liouville problems was suggested by Leighton (1971) and tested by him on an example with a known closed form solution (and a very simple structure). The method worked much better on this example than on others in two important respects. The extraordinary accuracy of the method for this problem was disguised in Leighton's paper by a mistake in the algorithm suggested and became apparent only when the mistake was corrected by Andrew et al (1981). The reasons for the atypical behaviour of the algorithm for this problem were analysed by Paine and Andrew (1983, Thm 3) and Andrew (1984, pp 846–847).

RANDOM MATRICES

Let $U_n(a,b)$ denote the probability distribution of an $n \times n$ matrix–valued random variable whose elements are independent random variables from the uniform distribution on (a,b). The default version of the MATLAB command $A = \text{rand } (n)$ generates a random matrix from $U_n(0,1)$. Tan and Andrew (1989) proposed a new more stable and efficient iterative method for computing $\lambda_{i,j}$ and $x_{i,j}$ and tested it not only on the matrices A_1, A_2 and A_3 of Section 3 but also on a randomly generated matrix of the form $A_4 + p_j A_5$, where A_4, $A_5 \in U_{80}(0,1)$ were generated using MATLAB.

Matrices in $U_n(0,1)$ are positive and their dominant (Perron) eigenvalue is real, simple and generally close to $n/2$, the other eigenvalues being much smaller. For such matrices, iterative methods perform much better for the dominant eigenvalue and very much worse for all the others than they would for a matrix with a more usual eigenstructure. The reason for this unusual eigenstructure is discussed briefly by Tan and Andrew (1989) and more fully by Andrew (1989). Essentially it is that the expected value of the matrix has rank one and its nonzero eigenvalue is $n/2$. MATLAB may easily be used to generate less unusual random matrices. For example, the command $A = \text{rand } (n) - 0.5*\text{ones}(n)$ generates matrices from $U_n(-\frac{1}{2}, \frac{1}{2})$. Such matrices, having expected value zero, have a much more typical eigenstructure.

It would have been of interest to test the methods of Tan and Andrew (1989) on matrices with several large eigenvalues instead of just one. Random matrices of this type are readily generated by MATLAB simply by adding numbers of $O(n/2)$ to some of the diagonal elements of a matrix from $U_n(0,1)$. For example, it follows from Theorem 1 of Andrew (1989) that adding the number c to one diagonal element of a matrix from $U_n(0,1)$ normally produces a matrix with two eigenvalues close to the solutions of $2\lambda^2 - (n + 2c)\lambda + (n-1)c = 0$ and the other eigenvalues small, while adding instead $n/2$ to two diagonal elements of a matrix from $U_n(0,1)$ produces a matrix with three eigenvalues close to $n/2$, $n/2+(n/2)^{\frac{1}{2}}$ and $n/2-(n/2)^{\frac{1}{2}}$ and the others small.

In the course of this investigation the author noted a strange weakness of the early version of MATLAB (Moler 1980) he was using, when implemented on a VAX. For $n \geq 8$, the programme had difficulty computing the eigenvalues of the matrix obtained by changing one diagonal element of an $n \times n$ matrix of ones, although this problem is well–conditioned. The difficulty appears to be related to the repeated zero eigenvalue (which does not affect the condition of the problem) as it was overcome by a small origin shift.

CONCLUDING REMARKS

Test matrices from some standard sources sometimes perform atypically with numerical algorithms. In particular this is so for examples with known closed form solution, as the simple structure of most such examples makes them more likely to have unusual properties. Even randomly generated examples are very likely to be atypical unless an appropriate probability distribution is used. This does not mean that we should not use examples from such sources. Examples with closed form solutions and randomly generated examples form an important part of a well chosen battery of test examples. Moreover examples chosen from other sources also often show atypical behaviour, and it is not uncommon for methods making their first appearance in the literature to be "supported" by examples on which they perform unusually well. What is important is that we should be cautious drawing too many conclusions from results obtained with a small set of test examples. Most importantly, numerical testing should be complemented to the fullest extent possible by theoretical analysis.

REFERENCES

1. Adelman, H.M. and Haftka, R.T., Sensitivity analysis of discrete structural systems, *AIAA J.*, vol. 24, pp. 823–832, 1986.

2. Andrew, A.L., Eigenvectors of certain matrices, *Linear Algebra Appl.*, vol. 7, pp. 151–162, 1973.

3. Andrew, A.L., Convergence of an iterative method for derivatives of eigensystems, *J. Comput. Phys.*, vol. 26, pp. 107–112, 1978.

4. Andrew, A.L., Iterative computation of derivatives of eigenvalues and eigenvectors, *J. Inst. Math. Appl.*, vol. 24, pp 209–218, 1979.

5. Andrew, A.L., Numerical solution of eigenvalue problems for ordinary differential equations, in: Noye, J. and Fletcher, C., (eds.) *Computational Techniques and Applications: CTAC–83* eds. J. Noye and C. Fletcher, pp. 841–852, North–Holland, Amsterdam, 1984.

6. Andrew, A.L., Eigenvalues and singular values of certain random matrices, Math. Res. paper 89–13, La Trobe University, August 1989.

7. Andrew, A.L., de Hoog, F.R. and Robb, P.J., Leighton's bounds for Sturm–Liouville eigenvalues, *J. Math. Anal. Appl.*, vol. 83, pp. 11–19, 1981.

8. Béliveau, J.G., System identification of civil engineering structures, *Canadian J. Civil Engineering*, vol. 14, pp. 7–18, 1987.

9. Fairweather, G., On the eigenvalues and eigenvectors of a class of Hessenberg matrices, *SIAM Rev.*, vol. 13, pp. 220–221, 1971.

10. Gregory, R.T. and Karney, D.L., *A Collection of Matrices for Testing Computational Algorithms*, Krieger Publishing Coy., Huntington, N.Y., 1978.

11. Hayes, L. and Wasserstrom, E., Solution of non–linear eigenvalue problems by the continuation method, *J. Inst. Math. Appl.*, vol. 17, pp. 5–14, 1976.

12. Lancaster, P., On eigenvalues of matrices dependent on a parameter, *Numer. Math.*, vol. 6, pp. 377–387, 1964.

13. Lancaster, P., *Lambda–Matrices and Vibrating Systems*, Pergaman, Oxford, 1965.

14. Leighton, W., Upper and lower bounds for eigenvalues, *J. Math. Anal. Appl.*, vol. 35, pp. 381–388, 1971.

15. Moler, C., Matlab Users' Guide, Computer Science Dept., University of New Mexico, 1980.

16. Murthy, D.V. and Haftka, R.T., Derivatives of eigenvalues and eigenvectors of a general complex matrix, *Internat. J. Numer. Meth. Engrg.*, vol. 26, pp. 293–311, 1988.

17. Paine, J.W. and Andrew, A.L., Bounds and higher order estimates for Sturm–Liouville eigenvalues, *J. Math. Anal. Appl.*, vol. 96, pp. 388–394, 1983.

18. Rudisill, C.S. and Chu, Y.Y., Numerical methods for evaluating the derivatives of eigenvalues and eigenvectors, *AIAA J.*, vol. 13, pp. 834–837, 1975.

19. Ruhe, A., Algorithms for the nonlinear eigenvalue problem, *SIAM J. Numer. Anal.*, vol. 10, pp. 674–689, 1973.

20. Tan, R.C.E., Accelerating the convergence of an iterative method for derivatives of eigensystems, *J. Comput. Phys.*, vol. 67, pp. 230–235, 1986.

21. Tan, R.C.E., Computing derivatives of eigensystems by the vector ε-algorithm, *IMA J. Numer. Anal.*, vol. 7, pp. 485–494, 1987a.

22. Tan, R.C.E., Computing derivatives of eigensystems by the topological ε-algorithm, *Appl. Numer. Math.*, vol. 3, pp. 539–550, 1987b .

23. Tan, R.C.E., An extrapolation method for computing derivatives of eigensystems, *Internat. J. Comput. Math.*, vol. 22, pp. 63–73, 1987c.

24. Tan, R.C.E., Extrapolation methods and iterative computation of derivatives of eigensystems, Ph.D thesis, La Trobe University, 1988a.

25. Tan, R.C.E., Implementation of the topological ε-algorithm, *SIAM J. Sci. Statist. Comput.*, vol. 9, pp. 839–848, 1988b.

26. Tan, R.C.E., Some acceleration methods for iterative computation of derivatives of eigenvalues and eigenvectors, *Internat. J. Numer. Meth. Engrg.* (In press), vol. 28, 1989.

27. Tan, R.C.E., and Andrew, A.L., Computing derivatives of eigenvalues and eigenvectors by simultaneous iteration, *IMA J. Numer. Anal.*, vol. 9, pp. 111–122, 1989.

The Local Behavior of the Quadratic Hermite-Padé Approximation

R. G. BROOKES
Department of Mathematics
University of Canterbury
Christchurch 1, New Zealand

INTRODUCTION

The subjects of Padé and Taylor approximation to a function $f(x)$ with a valid power series expansion about the origin are well understood. However, in the case of the general Hermite-Padé approximation much remains to be done. Recently (see Brookes and McInnes (to appear)), the existence problem for the quadratic Hermite-Padé approximation was solved, and so it is of interest to compare the performance of this quadratic approximation with those of the more traditional Padé and Taylor approximations. This paper starts to explore this question by examining the particular example of the $(4, 4, 4)$ approximation to $\log(1 + x)$.

Firstly, the definition of Hermite-Padé forms and some notation are introduced:

(i) Let $f(x)$ be a function, analytic in a neighbourhood of the origin, whose power series expansion about the origin is known.

(ii) Let $g_0, g_1, \ldots, g_n$ be functions such that $\forall i \in \{0, \ldots, n\}$, $g_i(f(x))$ is analytic in a neighbourhood of the origin and its power series expansion about the origin is known.

(iii) Let $A_0, A_1, \ldots, A_n \in \mathbf{Z}^+ \cup \{-1\}$ and $N = \sum_{i=0}^{n} A_i$.

(iv) Let $a_0(x), a_1(x), \ldots, a_n(x)$ be polynomials in x with $\deg(a_i(x)) \le A_i \ \forall i \in \{0, \ldots, n\}$, such that

$$\sum_{i=0}^{n} a_i(x) g_i(f(x)) = O\left(x^{N+n}\right) \tag{1}$$

(The only polynomial of degree -1 is, by definition, the zero polynomial.)

Note that such $a_i(x)$, not all zero, must exist since (1) represents a homogeneous system of $N + n$ linear equations in the $N + n + 1$ unknown coefficients of the $a_i(x)$. This system is difficult to solve numerically but using a symbolic manipulation package such as MACSYMA the coefficients of the $a_i(x)$ can easily be found in simple cases.

A set of $a_i(x)$ derived in this way is known as a $(A_n, A_{n-1}, \ldots, A_0)$ Hermite-Padé form for the system $g_i(f(x))$. Such a form will be represented as $\{a_n(x), \ldots, a_0(x)\}$

or as $\sum_{i=0}^{n} a_i(x)g_i(y)$. The set of all $(A_n, A_{n-1}, \ldots, A_0)$ forms for $g_i(f(x))$ forms a linear space for each set of $A_n, \ldots, A_0$.

Set

$$\sum_{i=0}^{n} a_i(x)\, g_i(y(x)) = 0. \tag{2}$$

It is the solution of (2) for $y(x)$ (not necessarily unique) that gives a Hermite-Padé approximation for $f(x)$.

This is the standard way of defining Hermite-Padé approximations. See, for example, Della Dora and Di Crescenzo (1984), Paszkowski (1987) and Baker and Lubinsky (1987) in which there appears an extensive bibliography.

Examples

(i) Setting $g_i(f(x)) = d^i f(x)/dx^i$ gives the differential Hermite-Padé approximants (see Baker and Lubinsky (1987)). The case $n = 1$ is that of the Baker $D-$log approximation (see Baker (1975)).

(ii) Setting $g_i(f(x)) = f(x)^i$ gives the algebraic Hermite-Padé approximants of which there are several important special cases:

1. $n = 1, A_1 = 0$ gives the Taylor approximation.
2. $n = 1$ gives the Padé approximation.
3. $n = 2$ gives the quadratic approximation (again, see Baker and Lubinsky (1987) and the references therein).

In the well-known case of Padé approximation this procedure is followed for $a_1(x)f(x) + a_0(x) = O(x^{A_1+A_0+1})$ which gives $y(x) = -a_0(x)/a_1(x)$. If $a_1(0) \neq 0$ (not a serious restriction) it then follows that $y(x) = f(x) + O(x^{A_0+A_1+1})$. However, in the quadratic case it is not obvious that $a_2(x)y(x)^2 + a_1(x)y(x) + a_0(x) = 0$ yields even an analytic approximation to $f(x)$, still less that it defines a function $y(x)$ such that $y(x) = f(x) + O(x^{A_0+A_1+A_2+2})$.

However, in Brookes and McInnes (to appear) it is shown that given

$$a_2(x)f(x)^2 + a_1(x)f(x) + a_0(x) = O(x^{N+2}), \quad \sum_{i=0}^{2} |a_i(0)| \neq 0$$

then there exists a unique $y(x)$ such that $\sum_{i=0}^{2} a_i(x)y(x)^i = 0$ and $y(x) = f(x) + O(x^K)$ where K is, at worst, $\frac{N}{2} + 1$. This is summarised in Table 1. Note that $D(x) = a_1(x)^2 - 4a_2(x)a_0(x)$.

TABLE 1. Order of approximation

Case	K
$D(0) \neq 0$	$N + 2$
$D(x) = x^{2s}g(x)$ where $\quad g(0) \neq 0$ $2s < N + 1$ $a_0(x) \not\equiv 0$	$N + 2 - s$
$D(0) = 0$ and $\quad a_0(x) \equiv 0$	$\min\{k \in \mathbf{N} : k \geq \frac{N}{2} + 1\}$
$D(x) \equiv 0$	$\min\{k \in \mathbf{N} : k \geq \frac{N}{2} + 1\}$

This immediately raises two questions:

(i) Can the approximation $y(x)$, unique in some neighbourhood of the origin (see Brookes and McInnes (to appear)), be extended to approximate $f(x)$ over some larger region?
(ii) Does this method of approximation give significantly better results than more traditional methods (eg. Padé and Taylor approximations)?

DISCUSSION

It follows from Brookes and McInnes (to appear) that (except for a few special cases) $y(x)$ is of the form

$$y(x) = \frac{-a_1(x) \pm x^s \sqrt{d(x)}}{2a_2(x)} \quad \text{where} \quad x^{2s}d(x) = a_1(x)^2 - 4a_2(x)a_0(x) , \quad d(0) \neq 0.$$

$y(x)$ is analytic everywhere except possibly at the points $x \in \mathbf{C}$ such that $a_2(x) = 0$ (poles) and the points $x \in \mathbf{C}$ such that $d(x) = 0$ (branch points).

Theorem 1. $y(x)$ is single valued and analytic in any simply connected neighbourhood of the origin not containing any of the above points.

Proof. The result follows easily from standard complex variable results. □

It will be assumed that $f(x)$, the function being approximated is single-valued, or at least, that we wish to approximate only one of its Riemann sheets. It is then clearly necessary to restrict the region of approximation, R, so that $y(x)$ is single valued on R. This consideration, with some additional information about $f(x)$ (for instance, the knowledge that $f(x)$ is analytic on some region, or the approximate location of any singularities) turns out to give enough information to accurately approximate some functions over a wide area. It is important at this point to realise that this "additional information" is necessary simply to ensure that the approximation problem is well-defined. The function $f(x)$ is not completely defined by it's power series at the origin, it may have many possible analytic continuations and so more information is needed to decide which possible continuation of the approximation is likely to be closest to $f(x)$.

EXAMPLE : THE $(4, 4, 4)$ APPROXIMATION TO $\log(1 + x)$.

Consider the $(4, 4, 4)$ quadratic approximation to the principal branch of $f(x) = \log(1+x)$ (with a cut taken along $\{x \in \mathbf{R} : x \in (-\infty, -1]\}$). Note that :

(i)
$$\begin{aligned}
& \left(6x^4 - 360x^3 + 180x^2 + 1080x + 540\right) f(x)^2 \\
& + \left(-75x^4 + 1620x^3 + 5310x^2 + 3540x\right) f(x) \\
& + 260x^4 - 4080x^3 - 4080x^2 \qquad\qquad = O\left(x^{14}\right)
\end{aligned}$$

(ii)
$$\begin{aligned}
a_1(x)^2 - 4a_2(x)a_0(x) = {}& -615x^8 + 229320x^7 - 4136580x^6 \\
& + 12612600x^5 + 59667300x^4 \\
& + 64033200x^3 + 21344400x^2 \, .
\end{aligned}$$

Let $x^2 d(x) = a_1(x)^2 - 4a_2(x)a_0(x)$. Using the results of Brookes and McInnes (to appear), the approximation is :

$$y(x) = \frac{-a_1(x) + x\sqrt{d(x)}}{2a_2(x)}$$

where

$$y(x) = f(x) + O\left(x^{13}\right).$$

The roots of $d(x)$ are :

$x = 354.0459$

$x = 10.8301 \pm 0.06444i$

$x = -0.9155 \pm 0.0005i$

$x = -0.9972$

while the roots of $a_2(x)$ are :

$x = 59.4440$

$x = 2.2299$

$x = -0.6904$

$x = -0.9835 \, .$

To ensure that $y(x)$ is single valued, cuts must be taken from the roots of $d(x)$. There are, of course, an infinite number of ways in which this may be done. It turns out that the simplest method gives a very good approximation to $\log(1 + x)$. $x = -0.9972$ is close to the known branch point of $f(x)$ at $x = -1$ and this point is treated simply by by taking $(\infty, -0.9972]$ as a cut. Consider the conjugate pair $x = -0.9155 \pm 0.0005i$. A cut could be taken from each point towards ∞ but in view of the fact that $f(x)$ is analytic on $\mathbf{C} \backslash \{x \in \mathbf{R} : x \in (-\infty, -1]\}$ this choice cannot be expected to give a good approximation. The other alternative is to take a cut between the two points. In fact the behaviour of the two possible values of $y(x)$ on the real axis close to these points (Fig.1) suggests that the best choice is to take the cut $\{w + iy \in \mathbf{C} : w = -0.9155, |y| \leq 0.0005\}$. It is of interest to note that this choice of cuts ensures that none of the roots of $a_2(x)$ are poles of $y(x)$.

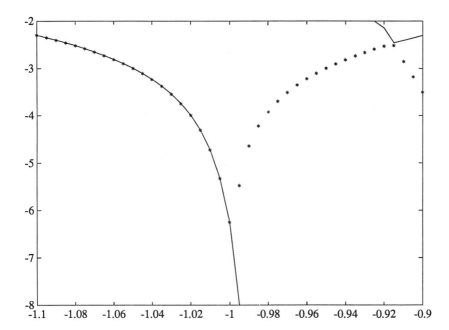

FIGURE 1. Real parts of both possible continuations of $y(x)$.

Fig.1 shows the real parts of both possible continuations of $y(x)$ along the real axis; one, $y_+(x)$, denoted by a solid line, the other $y_-(x)$, by "*". $y_+(x)$, which is the analytic continuation of $y(x)$ along the real axis, has a pole at $x = -0.9835$. The effect of taking the cut between the points $x = -0.9155 \pm 0.0005i$ is to "jump" from $y_+(x)$ to $y_-(x)$ at $x = -0.9155$.

The other conjugate pair of roots is treated similarly, whilst some arbitrary cut is taken from the other single root towards ∞ (this will not concern us here). A region R has now been defined so that $y(x)$ has a unique analytic continuation on R. It remains to calculate $y(x)$ at points in R. Clearly it is not sufficient to just write

$$y(x) = \frac{-a_1(x) + x\sqrt{d(x)}}{2a_2(x)}$$

It is necessary to follow a path inside R "analytically" from the origin to each point. It is easy to formulate an algorithm for this "analytic" procedure to produce numerical values for $y(x)$ on R.

Graphs of the real and imaginary parts of $y(x)$ on the region $\{w + iy \epsilon \, C : |w|, |y| \leq 2\}$ are given in Figures 2 and 3. These graphs are plotted using PC-Matlab with a mesh spacing of 0.1 . The lower left corner corresponds to the point $-2 - 2i$.

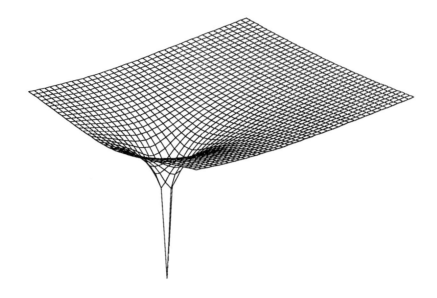

FIGURE 2. Real$(y(x))$.

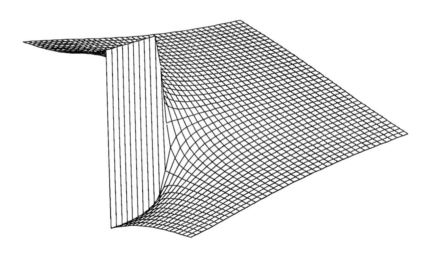

FIGURE 3. Imag$(y(x))$.

The surfaces shown in Figures 2 and 3 are very similar to those corresponding to $f(x)$. Note how well the branch point structure at $x = -1$ is modelled by $y(x)$, behaviour which is impossible to model with rational and polynomial approximations. If one calculates the $(6,6)$ Padé approximation (which also matches $f(x)$ up to $O(x^{13})$) then comparison of the two error functions shows that $y(x)$ gives approximately two more significant figures of accuracy than the Padé approximation on this region. The twelfth degree Taylor approximation is worse still.

REMARKS

Since the existence problem for quadratic approximations has been solved Brookes and McInnes (to appear) and a recurrence algorithm for quadratic forms exists Brookes and McInnes (1989), the next logical question to be addressed is that of the actual calculation of the approximation on some region. This is not a trivial problem because of the different types of singularities present in the approximation and, in particular, its many different possible analytic continuations. We have demonstrated these problems, and possible practical solutions, using the example of the $(4,4,4)$ approximation to $\log(1 + x)$. This approximation was shown to accurately represent the branch point structure of $\log(1+x)$ in a way not possible by more usual methods of approximation. In fact, closer investigation of the uniform error shows that $y(x)$ is significantly more accurate than Padé and Taylor approximation over *much* larger regions. Although this paper has studied only one example, it should be stressed that similar results have been observed with approximations to other functions such as $\cos(x), e^{-x}, \sqrt[3]{(1+x)}$. Although space does not permit it to be shown here, there is much similar evidence to suggest that the quadratic approximation has significant advantages over the Padé and Taylor approximations, both in terms of accuracy and the types of behaviour that it is able to model.

REFERENCES

1. Baker, G.A., *Essentials of Padé Approximants*, Academic Press, New York, 1975.

2. Baker, G.A., and Lubinsky, D.S., Convergence theorems for rows of differential and algebraic Hermite-Padé approximations, *J. Comput. Appl. Math.*, vol. 18, pp. 29–52, 1987.

3. Brookes, R.G., and McInnes, A.W., The Existence and Local Behaviour of the Quadratic Function Approximation, *J. Approx. Th.* (to appear).

4. Brookes, R.G., and McInnes, A.W., A Recurrence Algorithm for Quadratic Hermite-Padé Forms. Research Report No. 48, Department of Mathematics, University of Canterbury, March 1989.

5. Della Dora, J., and di Crescenzo, C., Approximations de Padé-Hermite. *Numer. Math.* vol. 43, pp. 23–57, 1984.

6. Paszkowski, S., Recurrence Relations in Padé-Hermite Approximation. *J. Comput. Appl. Math.* vol. 19, pp. 99–107, 1987.

Superconvergence and Post-Processing Formulas

G. F. CAREY
The University of Texas at Austin
Austin, Texas 78712, USA

R. J. MACKINNON
E. G. & G., Idaho, Inc.
P.O. Box 1625
Idaho Falls, Idaho 83415-2403, USA

INTRODUCTION

In both finite difference and finite element methods, analysts have observed that there are certain points at which the computed solution or its derivatives are exceptionally accurate (Barlow (1976), Carey and Oden (1985), Hinton and Campbell (1974), and Zienkiewicz (1977)). At these pre-determined superconvergence points, the local rate of convergence with mesh refinement is also superior to the global rate (Douglas and Dupont (1974), Dupont (1976) and Zlamal(1978)). For instance, the values at the inter-element "knots" and the derivatives at the interior Gauss-points exhibit superconvergence for certain classes of one-dimensional problems. In addition to direct superconvergence results for the computed solution there are post-processing strategies that can be introduced to produce results with superior accuracy and higher rates of convergence. For example, averaging derivatives from adjacent elements may lead to error cancellation and a more accurate result (Wheeler and Whiteman (1987)). Using auxiliary information (such as the differential equation) may also lead to higher-order post-processing formulas (Wheeler (1974)). Here we briefly summarize some superconvergence point results and a convenient method of analysis, based on Taylor-series concepts, that provides an easy interpretation of the main development. We give some related post-processing formulas and enhancement procedures together with supporting numerical studies. More details are given in our recent work (MacKinnon and Carey (1988, 1989a, 1989b)).

DISCUSSION

As a simple illustrative example for derivative superconvergence, let us consider the simple problem of linear interpolation of grid point values u_i at x_i, $i = 1, 2, \ldots, N$. The derivative of the linear finite element interpolant on interval i is $u_h'(\bar{x}) = (u_{i+1} - u_i)/h$, $h = x_{i+1} - x_i$ for $\bar{x} \in (x_i, x_{i+1})$. By elementary Taylor series expansion, we have

$$u'(\bar{x}) = (u_{i+1} - u_i)/h + \frac{(\delta_{i+1} - \delta_i)}{2}\bar{u}'' + \frac{(\delta_{i+1} + \delta_i)^3}{3!h}\bar{u}''' + \ldots \tag{1}$$

where $\delta_{i+1} = x_{i+1} - \bar{x}$, $\delta_i = \bar{x} - x_i$. Evidently, if $\bar{x}$ is the mid-point of the interval, then $\delta_{i+1} = \delta_i$ and the chord slope is an $O(h^2)$ accurate approximation to u' at this special superconvergence point. Further, the error grows linearly with $\delta_{i+1} - \delta_i$ towards the interval end points.

This approach can be extended to higher degree elements and to higher dimensions. For example, for a quadratic element basis $\{\psi_j\}$ in one-dimension, we get

$$u'(\bar{x}) = u'_h(\bar{x}) - \sum_{j=1}^{3} \psi'_j(\bar{x})(\delta_j^3/3!)u'''(\bar{x}) + O(h^3) \tag{2}$$

and the leading term in the error vanishes at the two interior Gauss points for the element. For the triangle and quadrilateral, this analysis shows that the tangential derivatives are superconvergent at the corresponding Gauss points of the sides. In fact, the results can be deduced directly from the one-dimensional case for this interpolation problem. We can also proceed more formally to seek interior superconvergence points. Following the same procedure, we find that the interior points are not superconvergent for the triangle, but for the special case of the rectangle, derivatives normal to the side bisectors are superconvergent. In particular, for a rectangle aligned with the x, y axes both u_x and u_y are superconvergent at the centre. A similar result holds for the mixed Galerkin finite element method.

Of course, the preceding results apply for interpolation and we are interested in the approximate solution of a boundary-value problem. For a quasi-uniform mesh of bilinear rectangles, nodal superconvergence of the Galerkin finite element solution can be shown. This result can then be applied to demonstrate derivative superconvergence of the Galerkin finite element approximation (MacKinnon and Carey (1989a)).

Post Processing - Auxiliary Data

Other strategies such as averaging may be useful in devising special post-processing formulas that give more accurate results. For example, consider a pair of right isoscelese triangles with common slant side. The tangential derivative along the slant side is continuous and superconvergent at the Gauss points on that side, as noted previously. However, the normal derivative is discontinuous across this side and the individual element values converge at the (lower) global rate. Averaging the normal derivative values returns a post-processed result that is superconvergent. Hence averaging in this case may be viewed as a post-processing strategy. This procedure has been studied in MacKinnon and Carey (1989a).

A more interesting post-processing result can be developed from the Taylor series analysis using the differential equation as auxiliary information. Let us consider, for example, the two-point boundary value problem

$$au'' + bu' + cu = f \tag{3}$$

with a, b, c, f specified functions of position and with appropriate end conditions. Using (3) for u'' in the truncation error term of (1), with $a \neq 0$ in (α, β)

TABLE 1: Convergence of derivative error e_{hx} and superconvergent result e_{hx}^* at interior node $\bar{x} = 3/4$ for
linear and quadratic elements.

h	LINEAR		QUADRATIC	
	e_{hx}	e_{hx}^*	e_{hx}	e_{hx}^*
1/4	1.62	0.35	0.29	0.07
1/8	0.64	0.07	0.06	0.008
1/16	0.29	0.02	0.01	0.0009
	$\sim O(h)$	$\sim O(h^2)$	$\sim O(h^2)$	$\sim O(h^3)$

$$u'(\bar{x}) = (u_{i+1} - u_i)/h + \frac{(\delta_{i+1} - \delta_i)}{2a}(f - bu' - cu)_{x=\bar{x}} + O(h^2) \tag{4}$$

so that

$$\left\{ 1 + \frac{b}{a}(\delta_{i+1} - \delta_i) \right\} u'(\bar{x}) = (u_{i+1} - u_i)/h + \frac{(\delta_{i+1} - \delta_i)}{2a}(f - c\bar{u}) + O(h^2) \tag{5}$$

where $\bar{u} = u(\bar{x})$ is interpolated.

Similarly, under appropriate smoothness assumptions, we can differentiate (3) to obtain an expression for $u'''(\bar{x})$ in (2) and a flux formula that is $O(h^2)$ accurate at *any* interior point $\bar{x}$. Numerical results for a test problem using linear and quadratic elements and the finite element solution are given in Table 1 for $-u'' + u' + 2u = -e^{-3x}$, $u(0) = u(1) = 0$.

The technique can be extended to two dimensions. Consider, for instance, an equilateral triangle and second order elliptic PDE. The differential equation can be introduced to replace one of the high derivatives (eg. u_{yy}). The terms involving u_{xx} and u_{xy} will still remain. Hence, a second-order accurate formula will result if the coefficients of u_{xx} and u_{xy} vanish. This will occur only at special points $(\bar{x}, \bar{y})$. For the Laplacian we find MacKinnon and Carey (1989a) that the locus of these points is a circle of radius $h/2\sqrt{3}$ with center at the centroid and where h is the side length.

Error Enhancement

Estimates of the actual nodal error in the solution are of considerable interest, since this would indicate the adequacy of the grid. For linear problems, accurate extraction formulas can be constructed for computing approximations to the nodal error. Let the differential equation be written compactly as

$$Lu = f \text{ in } \Omega \text{ with } u = g \text{ on } \partial\Omega, \tag{6}$$

where Ω is a bounded domain with smooth boundary $\partial\Omega$. A finite difference or finite element approximate form of (6) can be set up in the usual way. Then the nodal error on a quasi-uniform mesh of bilinear rectangles can be represented as

$$e(x_i, y_i) = (\Delta\xi)^2 w(x_i, y_i) + (\Delta\xi)^2 Q(x_i, y_i, \Delta\xi) \tag{7}$$

where w is the solution to the auxiliary problem

$$Lw = -r_2(x, y, u) \text{ in } \Omega \tag{8}$$

Here, r_2 the leading truncation error contribution from the discrete approximation for u, and Q is uniformly bounded in x_i, y_i and $\Delta\xi$; ($\Delta\xi$, $\Delta\eta$ are the mesh scales with $\Delta\eta = \gamma\Delta\xi$).

Since the operators in (6) and (8) are identical, the discrete problem for (6) is first constructed, factored and solved; then the discrete problem for (8) requires simply substitution sweeps using the previous matrix factors. The error correction can be added to the approximate solution to obtain an enhanced solution with higher order convergence properties with respect to h. The residual or truncation error forcing term on the right in (8) may be obtained by use of the differential equation as before and differencing the approximate solution. Hence the approach is actually equivalent to using the differential equation directly in the truncation error for (6) to generate a corresponding higher-order formula (Bramble and Hubbard (1964), MacKinnon and Carey (1989a)).

Weak Statement/Flux Formulas

In previous studies we have discussed the use of the weak Green-Gauss formulas (eg. see Carey (1982), Carey et al. (1985), Levine (1985), and Lazarov et al. (1989)). Again this corresponds to using the differential equation as auxiliary data as we now see. Consider the simple integration by parts form of the Green-Gauss formula in 1-D on the interval $(0,1)$,

$$\int_0^1 (u'v)' dx = \int_0^1 u''v dx + \int_0^1 u'v' dx \tag{9}$$

for arbitrary admissible u, v. This implies

$$u'v \bigg|_0^1 = \int_0^1 u''v dx + \int_0^1 u'v' dx \tag{10}$$

In particular, by selecting smooth v with $v(1) = 1$ and $v(0) = 0$, we have an extraction formula for $u'(1)$. That is, we can pick, for instance $v = x$ or $v = \phi_n$ (the extreme basis function at $x_n = 1$ in a finite element grid) to obtain the formula. Note that if we use linear elements, the contribution from $\int_0^1 u'v' dx$ on setting $u = u_h$ and $v = \phi_n$ is simply the difference approximation $(u_n - u_{n-1})/h$. If, in addition, we have a differential equation, say $u'' + u = f$, then we can replace u'' to get

$$u'(1) = \int_0^1 (f - u)v dx + \int_0^1 u'v' dx \tag{11}$$

The expression on the right corresponds to the integral in the weak statement and, on introducing the computed approximation u_h for u and ϕ_n for v, the superconvergent flux post-processing formula is obtained

$$u'_*(1) = \int_0^1 (u'_h \phi'_n - u_h \phi_n + f\phi_n) dx \tag{12}$$

In two dimensions, the corresponding Green-Gauss formula for the Laplacian leads to the flux formula using $\Delta u + u = f$,

$$\int_{\partial\Omega} \sigma_h \phi_i ds = \int_{\Omega} (\nabla u_h \cdot \nabla \phi_i - u_h \phi_i + f \phi_i) dx \, dy \tag{13}$$

where σ_h is the normal flux approximation and ϕ_i is the finite element basis function associated with node i on the boundary.

By using Lobatto quadrature as reduced integration (lumping) an explicit nodal projection formula for the flux follows from (13). Numerical results for these flux post-processing formulas are given elsewhere (Carey (1982) and Carey et al. (1985)). We have also recently extended many of these ideas to a class of monotone nonlinear problems (Chow et al. (1989)).

An interesting application of the result in (13) can be constructed for the stream function – vorticity formulation in viscous flow computations. Here the vorticity transport and stream function equations are iteratively decoupled and the difficulty is constructing Dirichlet boundary conditions for vorticity. However, if we consider the stream function equation $-\Delta\psi = \omega$, with stream function ψ and vorticity ω then the procedure for (13) now implies, on replacing f by ω, σ_h by $\partial\psi/\partial n$, and rearranging terms,

$$\int_{\Omega} \omega \phi_i dx \, dy = \int_{\Omega} \nabla \phi_h \cdot \nabla \phi_i dx \, dy - \int_{\partial\Omega} \frac{\partial\psi}{\partial n} \phi_i ds \tag{14}$$

At a solid boundary $\partial\psi/\partial n = 0$ and since $\phi_i(x_j, y_j) = \delta_{ij}$, using the Lobatto (node point) quadrature will provide an explicit superconvergent extraction formula for vorticity ω_i at boundary node i.

ACKNOWLEDGMENTS

This research has been supported in part by the Texas Advanced Technology Research Program and by the Office of Naval Research.

REFERENCES

1. Andreev, A. B. and Lazarov, R. D., Superconvergence of the Gradient for Quadratic Triangular Finite Elements, *Num. Meth.* PDE (to appear), 1987.

2. Barlow, J., Optimal Stress Locations in Finite Element Models, *Int. J. Numer. Meth. Eng.*, Vol. 10, pp. 243-251, 1976.

3. Bramble, J. H. and Hubbard, B. E., Approximation of Derivatives by Finite Difference Methods in Elliptic Boundary Value Problems, Contributions to Differential Equations, Vol. III, No. 4, pp. 399-410, 1964.

4. Carey, G. F., Derivative Calculation from Finite Element Solutions, *J. Comp. Meth. Appl. Mech. Eng.*, Vol. 35, pp. 1-14, 1982.

5. Carey, G. F., Chow, S. S., and Seager, M., Approximate Boundary Flux Calculation, *J. Comp. Meth. Appl. Mech. Eng.*, Vol. 50, pp. 107-120, 1985.

6. Carey, G. F. and Oden, J. T. *Finite Elements: A Second Course*, Prentice-Hall, Englewood Cliffs, NJ, 1983.

7. Carey, G. F. and Wheeler, M. F., C°-Collocation-Galerkin Methods, in *Codes for Boundary Value Problems in Ordinary Differential Equations*, Lecture Notes in Computer Science, Springer-Verlag, pp. 250-256, 1979.

8. Chow, S. S., Carey, G. F., and Lazarov, R.D., Natural and Post-Processed Superconvergence in Semilinear Problems, *Numer. Meth.* P.D.E's, 1989, in press.

9. Douglas, Jr., and Dupont, T., Galerkin Approximations for the Two-Point Boundary-Value Problem Using Continuous, Piece-Wise Polynomial Spaces, *Numer. Math.*, Vol. 22, pp. 99-109, 1974.

10. Dupont, T., A Unified Theory of Superconvergence for Galerkin Methods for Two-Point Boundary Problems, *SIAM J. Numer. Anal.*, Vol. 13, No. 3, pp. 362-368, 1976.

11. Hinton, E. and Campbell, J. S., Local and Global Smoothing of Discontinuous Element Functions Using a Least Squares Method, *Int. J. Numer. Meth. Eng.*, Vol. 8, pp. 461-480, 1974.

12. Lazarov, R. D., Chow, S. S., Pehlivanov, A. I. and Carey, G. F., Superconvergence Analysis of Approximate Boundary-Flux Calculations, in press, 1989.

13. Levine, N., Superconvergent Recovery of the Gradient From Piecewise Linear Finite-Element Approximations, *IMA J. Numer. Anal.*, Vol. 5, pp. 407-427, 1985.

14. MacKinnon, R. J. and Carey, G. F., Analysis of Material Interface Discontinuities and Superconvergent Fluxes in Finite Difference Theory, *J. Comp. Phys.*, Vol. 75, 1, pp. 151-167, 1988.

15. MacKinnon, R. J. and Carey, G. F., Superconvergent Derivatives: A Taylor Series Analysis, *Int. J. Num. Meths. Eng.*, 28, 489-509, 1989.

16. MacKinnon, R. J. and Carey, G. F., Nodal Superconvergence and Solution Enhancement for a Class of Finite Element and Finite Difference Methods, *SIAM J. Sci. Stat. Comp.*, in press, 1989.

17. Wheeler, M. F., A Galerkin Procedure for Estimating the Flux for Two-Point Problems, *SIAM J. Numer. Anal.*, Vol. 11, pp. 764-768, 1974.

18. Wheeler, M. F., and Whiteman, J., Superconvergent Recovery of Gradients on Subdomains from Piecewise Linear Finite Element Approximations, *J. Num. Meth.*, P.D.E.'s, Vol. 3, No. 1, pp. 357-374, 1987.

19. Zienkiewicz, O. C., *The Finite Element Method*, 3rd. Ed., McGraw-Hill, London, 1977.

20. Zlamal, M., Superconvergence and Reduced Integration in the Finite Element Method, *Math. Comput.* Vol. 32, pp. 663-685, 1978.

Iterative Zero-Finding Revisited

TIEN CHI CHEN
Department of Computer Science
The Chinese University of Hong Kong
Shatin, HK

INTRODUCTION

Functional iteration formulas of the Newton type

$$F(z;g) \equiv z - gf(z)/[d\ f(z)/dz] = z - g/s$$

are known to exhibit superlinear convergence to g-fold zeros, if effects of other zeros are negligible. Surprisingly, with dynamic adjustment of g, they exhibit a "global convergence" behavior, approaching polynomial zeros from almost anywhere in the complex plane. See Chen (1988).

The present work provides a systematic framework to examine global convergence behavior. The conventional Newton method with integer g only give linear global convergence, and the Laguerre method, adjusted for multiple zeros, may fail for clusters of high degree, as shown in numerical experiments. A new cluster-adapted formula is introduced; it converges to a zero of the symmetric cluster in one iteration, and exhibits cubic classical convergence in general.

THE SINGLE SYMMETRIC CLUSTER AS TEST VEHICLE

We shall study iterations for zeros $\{\alpha_i\}$ of the Nth degree polynomial:

$$f(z) = C \prod_{i=1}^{n} (z - \alpha_i)^{m(i)} = \sum_{j=0}^{N} a_j z^j$$

using $s \equiv [df(z)/dz]/f(z)$; $\mu \equiv \{1 - f(z)[d^2 f(z)/dz^2]/[df(z)/dz]^2\}^{-1}$

as auxiliary quantities. Classically, μ estimates the multiplicity of the zero being approached. Yet in the present study all the zeros contribute to μ, sometimes equally, and μ is often complex. We shall use as the apparent multiplicity $M \equiv$ Floor $[0.5 + \text{Re}(\mu)]$; it is considered qualified if and only if $N - 1 \geq M \geq 1$ and $M \geq 2|\text{Im}(\mu)|$.

Quantities referring to the kth iteration shall be subscripted by k, which may be omitted where no misunderstanding occurs. The (k + 1)st iteration is often indicated by a prime.

The symmetric cluster

We shall now develop a simple theory to examine the candidate formulas for global convergence, based on test cases which involve a "symmetric cluster function" with $n \geq 2$ zeros, all of multiplicity m, placed symmetrically on a circle. With complex constants C, α, β, we have

$$f(z) = C[(z - \alpha)^n - \beta^n]^m = C \prod_{j=1}^{n} [z - \alpha - \beta \exp(2\pi i j/n)]^m$$

defining $t \equiv (z - \alpha)/\beta$
then $s = N/\beta t(1 - 1/t^n)$
         $\mu = N/[1 + (n - 1)/t^n]$
and $t'= [F(z;g) - \alpha]/\beta = t(1 - g/N + g/Nt^n)$

A necessary condition for global convergence is, iterates from the two extreme regions should enter the transitional region in a finite number of iterations.

Consider $n = 4$, $\alpha = 0 + 0i$, $\beta = 1 + 0i$. From the guess $z_0 = 100 + i$, the four single zeros appear as one 4-fold zero; the observed μ is 3.99999988, hence M is 4. The classical Newton iteration with $g = M = 4$ gives $z = 1 \times 10^{-8} + 0i$, strikingly close to the centroid, offering the hope that global convergence is indeed possible. But there the effects of zeros cancel, and $M = 0 \approx \mu$. Even the minimum integer $g = 1$ moves the iterate towards infinity, to $2.5 \times 10^{17} + 0i$. Thus the method may not be a good candidate for global convergence in general.

The symmetric cluster trichotomizes the complex plane into regions:

a. The asymptotic region: $|t| \gg 1$, say $|t| \geq 10$, $(M \approx N)$;
b. The transitional region: $|t| \approx 1$, say $0.1 < |t| < 10$;
c. The centroid region: $|t| \ll 1$, say $|t| \leq 0.1$, $(M \approx 0)$.

We saw that the Newton iterations may shuttle between the extreme regions, but tends to miss the transitional region in between.

Criteria for global convergence

A systematic study requires a theoretical measure for global convergence towards the perimeter of the cluster disk, in terms of which a zero is as inaccessible from the centroid as from infinity. A good choice is

$\tau \equiv Max(|t|, 1/|t|)$
Thus $\tau'< \tau$,   if $1 \leq |t'|^{\pm 1} < |t|^{\pm 1}$

allowing all sign combinations. Dependent on $\tau'> \tau$, $\tau'= \tau$ or $\tau'< \tau$, a guess can be repelled, unaffected or attracted respectively by the zeros. We seek a development parallel to classical theory, which uses the measure $|\delta| \in [0,1)$, δ being the distance from the guess to its intended zero. Using a guess with $\tau \gg 1$, the following convergence patterns with $q > 1$ are identified:

a. Sublinear: $\tau' = O(\tau^q)$, $|t'| = O(|t|^{\pm q})$.   Analog: $\delta' = O(\delta^{1/q})$
b. Linear: $\tau' = O(\tau)$, $|t'| = O(|t|^{\pm 1})$.   Analog: $\delta' = O(\delta)$
c. Superlinear: $\tau' = O(\tau^{1/q})$, $|t'| = O(|t|^{\pm 1/q})$. Analog: $\delta' = O(\delta^q)$
        (Quadratic or cubic if $q = 2$ or 3)
d. Immediate: $\tau' = 1$, $|t'| = 1$.   Analog: $\delta' = 0$

Global convergence is essentially inoperative if the behavior is sublinear; haphazard or slow if linear. Superlinear global convergence is clearly desirable. Surprising as it may seem, one could reach a zero of a symmetric cluster in one pass.

After arriving at the transitional region, the iterates should aim towards a component zero of f(z). One could switch to another formula with improved classical convergence, but this may not be necessary.

THE INADEQUACY OF CLASSICAL NEWTON FORMULAS

The classical Newton formula uses real, integer g. We have

$$t' = [F(z;g) - \alpha]/\beta = t - t(1 - 1/t^n)g/N$$
$$= t^{1-n}, \qquad\qquad \text{if } g = N \text{ for all } t,$$
$$= t (1 - g/N) + O(t^{1-n}), \text{ if } g \neq N, \text{ for } |t| \gg 1$$
$$= gt^{1-n}/N + O(t), \qquad \text{if } g \neq N, \text{ for } |t| \ll 1$$

Starting from an extreme region, $\tau \gg 1$, the global convergence behavior with an integer g is at best linear; it is usually sublinear, but could <u>never</u> be superlinear. We conclude that this formula is truly useful only in the classical context, when the guess is already in the transitional region ($\tau \approx 1$); even from there ultimate convergence is not assured. Another Newton formula $F(z;\mu)$, classically second order for arbitrary multiplicity, has similar deficiencies despite the explicit use of μ.

THE GENERALIZED LAGUERRE FORMULA

The generalized Laguerre formula corresponds to the generalized Newton formula where g is an irrational function of z:

$$L(z;N,p) \equiv z - N/s\{1 \pm [(N/p - 1)(N/\mu - 1)]^{1/m}\}$$

the sign is chosen to make the real part of the square-root positive. It possesses cubic classical convergence for p-fold zeros; for a recent study with any real N > |p| see Hansen and Patrick (1976, 1977).

We shall call $L(z;N,p)$ <u>degenerate</u>, if it is trivially reducible to one of the classical Newton formulas; these are already known to be deficient, and should be avoided. Degeneracy occurs for p = 0, N or μ, but not for p = M if the latter is a qualified multiplicty estimate.

Choice of parameters

We have for p < N,

$$t' = [L(z;N,p) - \alpha]/\beta = t - t(1 - 1/t^n)/\{1 \pm [(N/p - 1)(n - 1)/t^n]^{1/m}\}$$
$$= 1 \text{ exactly}, \qquad\qquad\qquad\qquad \text{if } N = 2, \text{ for all } t$$
$$= [(N/p - 1)(n - 1)]^{1/2}t^{1-n/2} + O(t^{1-n}), \text{ if } N \neq 2, |t| \gg 1$$
$$= [(N/p - 1)(n - 1)]^{-1/2}t^{1-n/2} + O(t), \quad \text{if } N \neq 2, |t| \ll 1$$

The choice for p clearly should reflect the multiplicity, yet at the same time one must avoid degeneracy and cases of cyclic behavior. Also, if $|Re(\mu)| \ll |Im(\mu)|$, M cannot serve as a valid multiplicity estimate. For low-order polynomials (N < 5), the single-zero formula $L(z;N,1)$ performs well before the onset of classical convergence; but then the

convergence is very slow for multiple zeros. For our working formula we
have chosen p = P as defined below:

Let W = M if M is qualified (i.e., N - 1 $\geq$ M $\geq$ 1 and M $\geq$ 2 $|Im(\mu)|$)
       = 1 otherwise

then P = W   if k mod 3 = 1
       = Min(W, P_{k-1}) if k mod 3 = 2
       = 1   if k mod 3 = 0

The use of W sidesteps cases of undesirable values of M. The dependency
on k breaks cyclic iterate sequences, by keeping P small and by invoking
the reliable L(z;N,1) at least once every three iterations.

Shortcomings in Laguerre's method

A detailed analysis of Laguerre iterations shows that the "fan-out"
parameter n greatly affects the global convergence behavior, which is

immediate if n = 2; quadratic if n = 3; linear if n = 4;
yet <u>sublinear</u> for n $\geq$ 5 where $\tau' = O(\tau^{n/2-1})$.

Thus the formula, while highly effective for polynomials of low symmetry
and/or low degree, nevertheless could fail for symmetric cluster
functions of degree 5 or higher. This observation is confirmed by
numerical experiments below.

A NEW "CLUSTER-ADAPTED" FORMULA

We now propose a new "<u>cluster-adapted</u>" formula

$C(z;N,p) \equiv z - N(Q^{p/N} - 1)/(Q - 1)s$ with $Q \equiv (N/\mu - 1)/(N/p - 1)$

where p is an estimate of the multiplicity m, and the sign is chosen to
make the real part of the root positive. More computation effort is
required here than either the Newton or the Laguerre formulas, but the
merit is believed to outweigh the extra evaluation cost. To be avoided
are the following degenerate cases: $C(z;n;p) = F(z;N)$, $F(z;0)$ and $F(z;\mu)$
if p = 0, N, and μ, when Q = 0, ∞ and 1 respectively.

Global convergence behavior

C(z;N,p) with $|N/p| > 1$ has the following desirable properties:

t '= $[C(z;N,p) - \alpha]/\beta$
   = 1,   if p = m, for all t
   = $O(t^{1-p/m})$,   for all p if either $|t| \gg 1$ or $|t| \ll 1$

Thus C(z;N,m) converges to (α + β), an m-fold zero of the single
symmetric cluster, in one iteration <u>regardless of guess position</u>, except
for roundoff error. By setting 2m > p > 0, superlinear convergence is
assured from the extreme regions; a safe strategy is to set p = 1.

586

Classical convergence properties

The classical convergence behavior of $C(z;N,p)$ is cubic for p-fold zeros, and (p/N + 1)st order to to N-fold zeros. Thus the formula continues to be useful after reaching the transitional region.

Estimating the multiplicity

The immediate convergence of $C(z;N,p)$ requires setting p = m, assuming that f(z) is a symmetric cluster with m-fold zeros. By eliminating t between μ and s, we have, after some manipulations:

$$m = N/n = (N/\{1 - (N/\mu - 1)/[N/(z - \alpha)s - 1]\}$$

where α can be evaluated by $(-a_{N-1}/Na_N)$. The quantity Re(m) is deemed qualified as an estimate if and only if the computation incurs neither overflow nor division by 0, further $N - 1 \geq Re(m) \geq 1$, $Re(m) > 2|Im(m)|$. Re(m) is most effective during the first iteration; afterwards cubical classical convergence probably already begins to take hold.

A practical scheme to exploit immediate convergence

The following scheme with p = R has been found practical:

Let V = Re(m) if Re(m) is qualified
       = 1 otherwise

Then R = V if k = 1
       = W otherwise, where W has been defined for $L(z;N,P)$

REALISTIC SITUATIONS

A good global convergence formula should be able to handle the symmetric cluster, by directing the iterates towards the transitional region where the zeros are. Such an ability probably goes a long way towards global convergence for general polynomials as well.

Zero distributions of lower symmetry

A cluster with lower symmetry should be easier to handle than its symmetric counterpart. The iterate from the asymptotic region, should deviate noticeably from the centroid; and that inside the convex hull formed by the zeros, should stop far short of infinity, as effects of component zeros do not cancel. In both cases the iterate should reach the transitional region sooner, and the latter should also be broader because of the lower symmetry.

Nevertheless, complications arise when the zeros contain an embedded symmetric cluster; near its centroid, the effects of the local zeros may cancel, yet s and μ may stay finite due to external zeros. Should the centroid of one cluster lie in the asymptotic region of another, iterates could conceivably cycle nonstop among several controid regions. The problem clearly merits further study.

Homing in on a zero

After reaching the transitional region, formulas can still fail. For
instance, μ often carries a sizable imaginary part, disqualifying M as
multiplicity estimate. Fortunately, both the Laguerre and the cluster-
adapted formulas seem to be quite capable of converging to a zero,
though the processes are far from understood. The problem of steering
towards for particular zeros of interest is also elusive, and no
solution can be offered as yet.

Exceptional situations

The evaluation of $N/\mu - 1$, required in both $L(z;N,P)$ and $C(z;N,R)$, lead
to cancellation errors if τ is large, creating degeneracy and causing
extra iterations, but seldom nonconvergence. The computation truly
fails if $s = 0$; then z can be assumed to be near α, and β is
conveniently estimated as $[-f(z)/a_N]^{1/N}$.

$C(z;n;R)$ leads to $Q = 1$ hence $0/0$ if p happens to match μ; the resulting
degenerate formula with the singularity removed is $z' = z - \mu/s$.

EXAMPLES USING $L(z;N,P)$ and $C(z;N,R)$

We have employed both the Laguerre formula and the cluster-adapted
formula to find zeros for a number of polynomials, on an IBM PC using
the STSC Pocket APL package, in floating-point arithmetic with a 52-bit
fraction (15.6 decimal digits). When more than one zero is sought,
deflation was employed m times after an m-fold zero was found. There
was no purification after deflation.

    Global convergence was considered reached after the kth iteration
if $|f(z_k)| < 10^{-10}$. A stricter criterion would certaintly enhance
accuracy upon deflation, but further convergence belongs to the realm of
classical convergence and further is affected by machine roundoff. We
observed that usually upon convergence $|f(z_k)| < 10^{-13}$, but errors of
such size was not always reachable via more iterations.

Convergence to a polynomial zero

Table 1 shows the number of iterations needed, for zero distributions of
varying symmetry, using guesses spanning 16 orders of magnitude.
Surprisingly, the global convergence was often faster than many classical
iterations with the zero already localized.

$C(z;N,R)$ was outstanding for symmetric clusters, though slowed by
cancellation if $|z_0|$ was large. It worked well for other cases too,
usually taking fewer than 10 iterations; yet in Case 4, with $z_0 = 0.0001$
+ 0.0001i, though $f(z_1)$ was already correct to 6 decimals, 21 iterations
was required for full accuracy, meanwhile the iterates ping-ponged
between two sub-clusters in a manner not well understood as yet.

The Laguerre formula was found to be excellent in most situations. As
predicted, it fared poorly for the cluster with five symmetrically
placed zeros (Case 2), though surprisingly it still managed to converge
for most initial guesses.

TABLE 1. Global Convergence with L(z;N,P) and C(z;N,R)
The number of iterations required are shown for each initial guess.
L(z): L(z;N,P). C(z): C(z;N,R).

Case	N	Zeros	$10^{-10}+10^{-10}i$		$10^{-3}+10^{-3}i$		Initial guesses $(z0)$ $2+2i$		10^3+10^3i		10^6+10^6i	
			L(z)	C(z)	L(z)	C(z)	L(z)	C(z)	L(z)	C(z)	L(z)	C(z)
1a	4	$(Z1)_2$	10	1	9	1	4	1	10	2	7	3
1b	8	$(Z1)_2^2$	10	1	8	1	5	1	8	2	7	8
1c	12	$(Z1)^3$	10	1	7	1	7	1	7	2	7	12
2a	5	$(Z2)_2$	6	1	>30	1	5	1	23	3	10	3
2b	10	$(Z2)_2^2$	8	1	>30	1	5	1	>30	6	5	10
2c	15	$(Z2)^3$	(OV)	1	>30	1	5	1	>30	9	7	14
3a	4	$(Z3)_2$	6	5	6	5	4	5	6	7	6	6
3b	8	$(Z3)^2$	7	4	7	4	5	4	7	8	7	7
4	8	(Z4)	8	6	8	6	19	8	9	21	9	5
5	16	(Z5)	16	9	10	5	8	9	10	5	8	19

Notations: (OV): Exponent overflow.

(Z1): ($\pm 1 \pm i$), forming corners of a square.
(Z2): (1 + 0i, 0.309 016 994 4 $\pm$ 0.951 056 516 3i, −0.809 016 994 4
       $\pm$ 0.587 785 252 3i), forming vertices of a regular pentagon of
       radius 1 centered at the origin, symmetrical about the X-axis.
(Z3): ($\pm 4 \pm i$), forming corners of a narrow rectangle.
(Z4): ($\pm 1 \pm .01 \pm .01i$), forming supercluster with two square clusters.
(Z5): ($\pm 1 \pm .01 \pm .01i$, $\pm .01 \pm i \pm .01i$), embedding four square clusters.

TABLE 2. Zeros of a polynomial discussed by Ralston and Rabinowitz

$f(z) = (z - 1 - i)(z - 4 + 3i)(z - 4 - 3i)(z - 3.999 - 3i)$
All starting guesses are set at 10 + 0i. $d \equiv 1E\text{-}12 = 1\times10^{-12}$

			L(z;N,P)			C(z;N,R)	
N	k	$|f(z_k)|$	Computed z	k	$|f(z_k)|$	Computed z	
5	4	1E-12	[4 − 3i +(0 + 0i)d]	6	4E-12	[3.999+3i + (6 + 15)d]	
4	7	2E-13	[4 + 3i +(−7+18i)d]	4	1E-12	[1 + i +(200+1195i)d]^2	
3	2	8E-15	[3.999+3i+(7-18i)d]	0	1E-12	---	
2	1	1E-14	[1 + i +(0 + 0i)d]	2	9E-16	[4 + 3i +(446 + 26i)d]	
1	0	0	[1 + i +(0 + 0i)d]	0	0	[4 − 3i + (14 + 28i)d]	

Σ:14 (42 evaluations)   Σ:12 (36 evaluations)

589

Finding all zeros of a polynomial

The 5th degree polynomial used in Table 2 has been discussed by Ralston and Rabinowitz (1978) whose Newton-based trial-and-error method used a total of 150 function evaluations to find all its zeros with worst-case absolute error $|\delta| \leq 8 \times 10^{-9}$; it ran twice as fast as the more accurate ($|\delta| \leq 5 \times 10^{-11}$) method of Jenkins and Traub with 83 evaluations. Our algorithms consumed 42 and 36 evaluations, yielding $|\delta| \leq 2 \times 10^{-11}$ and 1×10^{-9} respectively. Notwithstanding, it must be stated that our iterations used a different convergence criterion and each had required nontrivial computations besides three evaluations.

SUMMARY AND CONCLUSIONS

It is believed that the present work, while in a preliminary stage, has already demonstrated that for polynomials, global convergence formulas could become a feasible alternative to classical iterative zero-finding, and one may be free to start the iterations from almost anywhere.

Our theory, based on the symmetric cluster, has shown that the conventional Newton method and the Laguerre method, both adjusted for multiple zeros, are deficient. The proposed cluster-adapted method converges in one iteration for symmetric clusters, and exhibits cubic classical convergence in general. It could be the formula of choice, especially in combination with the Laguerre formula for added speed.

The steering towards particular zeros clearly merits further study. Also, the choices for multiplicity parameters P and R are probably not optimum; fine-tuning could enhance efficiency and reduce pitfalls.

ACKNOWLEDGMENT

The author acknowledges the stimulation received from Dr. Ralph A. Willoughby, IBM Research Center, Yorktown Heights, USA, and the late Prof. Peter Henrici, ETH, Zürich, Switzerland. He is also grateful to the referees for valuable comments. Part of this study was made while the author was with IBM Corporation, USA, and also IBM Heidelberg Scientific Center, the Federal Republic of Germany.

REFERENCES

1. Chen, T. C., Global iterative convergence to zeros in the presence of clusters, *Proc. International Symposium, Taipei, Taiwan,* pp. 270-276, 1988.

2. Hansen, E., and Patrick, M., Estimating the multiplicity of a root, *Numer. Math.* vol. 27, pp. 121-131, 1976.

3. Hansen, E., and Patrick, M., A family of root finding methods, *Numer. Math.* vol. 27, pp. 257-269, 1977.

4. Ralston, A. , and Rabinowitz, P., *A First Course in Numerical Analysis,* 2d Ed., pp.391, 395, McGraw-Hill, New York, 1978.

On Linear Matrix Equations

KING-WAH ERIC CHU
Mathematics Department
Monash University
Clayton, Victoria 3168, Australia

INTRODUCTION

We shall consider the solution of linear matrix equations (LME)

$$\sum_{j=1}^{N} A_{ij} X_j B_{ij} = C_i , \quad i = 1, \cdots, M .\tag{1}$$

Since the appearance of the important survey paper (Lancaster 1970) on (1), more results have been discovered. For a long time, expanding (1) using the Kronecker product $\otimes$ into the more standard form

$$\sum_{j=1}^{N} G_{ij} v(X_j) = v(C_i) , \quad G_{ij} \equiv A_{ij} \otimes B_{ij}^{T} , \quad i = 1, \cdots, M;\tag{2}$$

where $v(Z)$ is the column vector constructed from lining up the rows of the matrix Z, and then solving (2) by Gaussian elimination or the like, seemed to be the only available solution process. The approach was inefficient as the dimension of the matrix A is an order greater than those of A_{ij} and B_{ij}. Also, any structures in A_{ij} and B_{ij} cannot be utilized effectively. In (Bartels & Stewart 1972), a direct method for the solution of the Sylvester equation $AX - XB = C$ was proposed, using QR decomposition on the square matrices A and B (see details in next section). It was the first time the numerical stable solution of a nontrivial LME could be carried out directly without using the Kronecker product formulation. Also, the well-known solvability conditions could be deduced from the solution method. Similar approaches using matrix decompositions have become popular for the solution as well as the analysis of LME's. The scope of study has also been widened, e.g. relationship of controllability/observability and solutions of LME's, invertibility or rank of solutions, solutions with structures (e.g. symmetry), inertia theorems, least squares solutions, etc., plus the ongoing search for more efficient and robust numerical methods (e.g. methods based on Hessenberg forms) as well as more worthwhile applications (see the reference list for details).

In this paper, we shall present some recent results by the author on the solvability of some LME's, their symmetric solutions and least squares solutions. Various numerical solution processes have also been suggested, either explicitly or implicitly from the analysis of solvability conditions, and are mostly based on stable matrix decompositions like QR, Schur, generalized Schur, singular value and generalized singular value decompositions.

We shall only consider direct methods here. Information on iterative methods can be found in (Hoskins et. al. 1979),(Hoskins & Walton 1979). Methods based on techniques in complex or functional analysis, e.g. Taylor's formula for matrix functions, contour integration or spectral theory, can be found in (Lancaster 1970),(Wimmer & Ziebur 1972). Much work have also been done on the relationship between LME's in arbitrary fields, matrix polynomials and resultants (see (Hartwig 1974)). The reference list also includes some early literature on the topic (see also Note E15 on pp. 787–788 in (Nashed 1976)) and some standard references on matrix theory and numerical linear algebra. The thesis (Kolka 1984) contains a fine account of the state of the topic up to 1984, some interesting applications in control system design, and some useful and new results.

SYLVESTER EQUATION

Consider the Sylvester equation

$$AX - XB = C , \quad A \in \mathcal{R}^{m \times m} , \quad B \in \mathcal{R}^{n \times n}. \tag{3}$$

The Liapunov equation, with $B = A^T$, is a special case. The Sylvester equation has applications in many different areas, ranging from stability in control system designs to perturbation theory for matrix eigenvalue problems.

In (Bartels & Stewart 1972) (which influences our work heavily), the matrices A and B have been decomposed into lower- and upper-triangular forms, using QR or Schur decomposition. Eqt. (3) can then be written into the equivalent form

$$\begin{bmatrix} a_{11} & 0 & \cdots & \cdots & 0 \\ a_{21} & a_{22} & \ddots & & \vdots \\ \vdots & & \ddots & \ddots & \vdots \\ \vdots & & & \ddots & 0 \\ a_{m1} & \cdots & \cdots & \cdots & a_{mm} \end{bmatrix} \tilde{X} - \tilde{X} \begin{bmatrix} b_{11} & b_{12} & \cdots & \cdots & b_{1n} \\ 0 & b_{22} & & & \vdots \\ \vdots & \ddots & \ddots & & \vdots \\ \vdots & & \ddots & \ddots & \vdots \\ 0 & \cdots & \cdots & 0 & b_{nn} \end{bmatrix} = \tilde{C}, \tag{4}$$

where x_{ij} and c_{ij} denote the ij-th elements of $\tilde{X}$ and $\tilde{C}$ respectively. The solution of (4) can be found explicitly in a recursive manner : for $i = 1, \cdots, m$; and $j = 1, \cdots, n$;

$$T_{ij}(x_{ij}) = c_{ij} - \sum_{k<i} a_{ik} x_{kj} + \sum_{k<j} x_{ik} b_{kj} , \quad T_{ij}(z) \equiv a_{ii} z - z b_{jj}. \tag{5}$$

Note that $T_{ij}(z)$ is not written as $(a_{ii} - b_{jj}) z$ so that (5) is still valid for nonscalar a_{ii} or b_{jj} (see the real arithmetic case below).

If the elements x_{ij} are solved row-wise (or column-wise), all the terms on the RHS of (5) are known. From the form of the operators T_{ij}, the Sylvester equation will then be uniquely solvable if and only if $a_{ii} \neq b_{jj}$ for all possible values of i and j. Note that a_{ii} and b_{jj} are the eigenvalues of A and B respectively.

Instead we can use Real Schur forms in forming (4), and a_{ii} and b_{jj} in (5) can be scalars or 2×2 blocks, corresponding to complex eigenvalues. The solution process then involves the solution of Sylvester equations with operators T_{ij}, i.e. linear systems of dimensions at most 4×4 using the Kronecker product formulation. Only real arithmetic will then be required.

In a different technique in (Golub et. al. 1979), only one of the matrices (say A if its dimension is greater than that of B) is decomposed into Schur or Real Schur form, with the other matrix (B) in Hessenberg form. Then (3) is transformed into

$$\begin{bmatrix} a_{11} & 0 & \cdots & \cdots & 0 \\ a_{21} & a_{22} & \ddots & & \vdots \\ \vdots & & \ddots & \ddots & \vdots \\ \vdots & & & \ddots & 0 \\ a_{m1} & \cdots & \cdots & \cdots & a_{mm} \end{bmatrix} \tilde{X} - \tilde{X} \tilde{B} = \tilde{C}, \tag{6}$$

where x_i^T, the i-th row of $\tilde{X}$, can be calculated recursively by solving : for $i = 1, \cdots, m$;

$$T_i\left(x_i^T\right) = c_i^T - \sum_{k<i} a_{ik} x_k^T \ , \quad T_i\left(z^T\right) \equiv a_{ii} z^T - z^T \tilde{B}. \tag{7}$$

Note that $H = a_{ii} \otimes I - I \otimes \tilde{B}^T$, the matrix representation of the operator T_i, is Hessenberg and thus the inversions of T_i in (7) can be carried out efficiently. Solvability conditions can be derived from (7). Only real arithmetics will be involved if the Real Schur form is used to decompose A.

In the above methods, $\tilde{X}$ can be transformed back to X in the original co-ordinates orthogonally.

GENERALIZED SYLVESTER EQUATION

In (Chu 1987a), the following theorem on the solvability of the generalized Sylvester equation

$$AXB - CXD = E \ , \quad A, C \in \mathcal{R}^{m \times m} \ , \quad D, B \in \mathcal{R}^{n \times n}; \tag{8}$$

has been proved :

THEOREM 1 (Chu 1987a) *The matrix equation (8) has a unique solution if and only if (i) $(A - \lambda C)$ and $(D - \lambda B)$ are regular matrix pencils, and (ii) their generalized spectra have empty intersection.*

The theorem is more general than similar results available later elsewhere which consider only solvability conditions for Eqt. (8) with regular matrix pencils.

The approach in (Bartels & Stewart 1972) has been generalized to deal with (8). The matrix pencils $(A - \lambda C)$ and $(D - \lambda B)$ are transformed, using the QZ algorithm, to lower- and upper-triangular Real Schur forms respectively, and (8) becomes

$$\begin{bmatrix} a_{11} & & 0 \\ \vdots & \ddots & \\ a_{p1} & \cdots & a_{pp} \end{bmatrix} \tilde{X} \begin{bmatrix} b_{11} & \cdots & b_{1q} \\ & \ddots & \vdots \\ 0 & & b_{qq} \end{bmatrix} - \begin{bmatrix} c_{11} & & 0 \\ \vdots & \ddots & \\ c_{p1} & \cdots & c_{pp} \end{bmatrix} \tilde{X} \begin{bmatrix} d_{11} & \cdots & d_{1q} \\ & \ddots & \vdots \\ 0 & & d_{qq} \end{bmatrix} = \tilde{E}. \tag{9}$$

The diagonal blocks a_{ii}, b_{ii}, c_{ii} and d_{ii} can be scalar or 2×2 as in the previous section.

The solution can then be found recursively : for $i = 1, \cdots, p$; $j = 1, \cdots, q$;

$$T_{ij}\left(x_{ij}\right) = e_{ij} - \sum_{kl} T_{ij}\left(x_{kl}\right) \ , \quad T_{il}\left(x_{jk}\right) \equiv a_{ij} x_{jk} b_{kl} - c_{ij} x_{jk} d_{kl}, \tag{10}$$

with $\tilde{X}$ and $\tilde{E}$ partitioned appropriately in (9), and the summation $\sum_{kl}$ in (10) carried out for all $k = 1, \cdots, i$; $l = 1, \cdots, j$; and $(k, l) \neq (i, j)$.

The equations in (10) can easily be solved, with the RHS's known if the x_{ij}'s are solved row-wise (or column-wise). The equations involved at most 4×4 linear systems after any possible applications of Kronecker products. Then X can be transformed back to the original co-ordinates orthogonally.

An alternative algorithm in (Epton 1980) decomposes only one matrix pencil into a triangular Schur form, and the other into a triangular Hessenberg form, similar to the approach in (Golub et. al. 1979) for Sylvester equations.

In (Chu 1987a), the set of simultaneous equations in Y and Z, $(YA - DZ, YC - BZ) = (E, F)$, has been proved to be adjoint to Eqt. (8), and numerical algorithms analogous to those in (10) and (Epton 1980) can be devised for its solution. It is easy to generalize the solvability results and numerical algorithms for the more general equation $\sum_i^p f_{1i}(A) X f_{2i}(B) + \sum_i^q f_{3i}(C) X f_{4i}(D) = E$, for well behaved matrix functions f_{ji}.

593

SOLUTION OF $AXB + CYD = E$

The following equation is the true special case of (1) for $M = 1, N = 2$:

$$AXB + CYD = E, \qquad (11)$$

where the matrices A, B, C, D and E do not have to be square.

The consistency conditions of (11), and ultimately its solution, rely upon the novel application of the new and potentially powerful tool in numerical linear algebra, the Generalized Singular Value Decomposition (GSVD), invented in (Paige & Saunders 1981) .

The GSVD, a generalization of the Singular Value Decomposition (SVD), can be described as follows : given two matrices $A \in \mathcal{R}^{m \times n}$ and $B \in \mathcal{R}^{p \times n}$ with the same number of columns, there exists orthogonal matrices U and V and nonsingular matrix X such that

$$A = U\Sigma_A X , \quad B = V\Sigma_B X, \qquad (12)$$

where Σ_A and Σ_B are the same sizes as A and B respectively, $k = \mathrm{rank}(C) = \mathrm{rank}\begin{bmatrix} A \\ B \end{bmatrix}$,

and

$$\Sigma_A = \left[\begin{array}{ccc|c} I_A & & & \\ & S_A & & 0 \\ & & 0_A & \end{array}\right] , \quad \Sigma_B = \left[\begin{array}{ccc|c} 0_B & & & \\ & S_B & & 0 \\ & & I_B & \end{array}\right]. \qquad (13)$$

The matrices Σ_A and Σ_B are partitioned similarly into four blocks of columns in (13), of sizes $r, s, (k - r - s)$ and $(n - k)$ respectively. I_A and I_B are identity matrices, 0_A and 0_B are zero matrices, and

$$S_A = \mathrm{diag}(\alpha_1, \cdots, \alpha_s) , \quad S_B = \mathrm{diag}(\beta_1, \cdots, \beta_s), \qquad (14)$$

with $1 > \alpha_1 \geq \cdots \geq \alpha_s > 0$, $0 < \beta_1 \leq \cdots \leq \beta_s < 1$, and $\alpha_i^2 + \beta_i^2 = 1, i = 1, \cdots, s$. The ordered pairs (α_i, β_i) are called the generalized singular values of the matrix pair (A, B), and are closely related to the CS-decomposition in (Stewart 1982) . Some submatrices in (13) can vanish, depending on the structures of the matrices A and B. The matrix X in (12) will be ill-conditioned when the matrix C has a small but non-zero singular value, i.e. the rank determination of $C^T \equiv [A^T, B^T]$ is not straight-forward. The matrix X has to be inverted in some of the numerical algorithms considered later.

Decomposing the matrix-pairs (A^T, C^T) and (B, D) using the GSVD, (11) is equivalent to

$$X_1^T \Sigma_A^T U_1^T \cdot X \cdot U_2 \Sigma_B X_2 + X_1^T \Sigma_C^T V_1^T \cdot Y \cdot V_2 \Sigma_D X_2 = E, \qquad (15)$$

where the matrices U_i and V_i are orthogonal, and X_i are nonsingular (c.f. (12)).

For $\tilde{X} \equiv U_1^T X U_2$, $\tilde{Y} \equiv V_1^T Y V_2$ and $\tilde{E} \equiv X_1^{-T} E X_2^{-1}$, (15) becomes

$$\left[\begin{array}{c|c} \begin{array}{ccc} I_A & & \\ & S_A & \\ & & 0_A^T \\ \hline & 0 & \end{array} \end{array}\right] \tilde{X} \left[\begin{array}{ccc|c} I_B & & & \\ & S_B & & 0 \\ & & 0_B & \end{array}\right] + \left[\begin{array}{c|c} \begin{array}{ccc} 0_C^T & & \\ & S_C & \\ & & I_C \\ \hline & 0 & \end{array} \end{array}\right] \tilde{Y} \left[\begin{array}{ccc|c} 0_D & & & \\ & S_D & & 0 \\ & & I_D & \end{array}\right] = \tilde{E} \ (16)$$

With $\tilde{X}, \tilde{Y}$ and $\tilde{E}$ partitioned up appropriately, (16) is then equivalent to

$$\left[\begin{array}{ccc|c} x_{11} & x_{12}S_B & 0 & \\ S_A x_{21} & S_A x_{22}S_B + S_C y_{22}S_D & S_C y_{23} & 0 \\ 0 & y_{32}S_D & y_{33} & \\ \hline & 0 & & 0 \end{array}\right] = \left[\begin{array}{ccc|c} e_{11} & e_{12} & e_{13} & \\ e_{21} & e_{22} & e_{23} & e_{\cdot 4} \\ e_{31} & e_{32} & e_{33} & \\ \hline & e_{4\cdot} & & e_{44} \end{array}\right]. \qquad (17)$$

Then the following theorem can be deduced from (17) :

THEOREM 2 (Chu 1987b) *Eqt. (11) is consistent if and only if the following submatrices of $\tilde{E}$ vanish :* $e_{13}, e_{31}, e_{4.}, e_{.4}, e_{44}$.

For consistent equations, the submatrices x_{ij} and y_{kl}, with $i, j = 3$ or $k, l = 1$, are arbitrary. Additional degrees of freedom exist in x_{22} and y_{22}, with $x_{22} = S_A^{-1}(e_{22} - S_C y_{22} S_D) S_B^{-1}$ when y_{22} is arbitrary, or $y_{22} = S_C^{-1}(e_{22} - S_A x_{22} S_B) S_D^{-1}$ when x_{22} is arbitrary.

Least squares solutions can be found for consistent equations (it is more complicated for inconsistent equations, as $\tilde{E} = X_1^{-T} E X_2^{-1}$ and X_i's are not orthogonal). Let the ij-th components of x_{22}, y_{22} and e_{22}, be denoted by x_{2ij}, y_{2ij} and e_{2ij} respectively. From Theorem 2, we have $\begin{bmatrix} \alpha_i \beta_j & \gamma_i \delta_j \end{bmatrix} \begin{bmatrix} x_{2ij} & y_{2ij} \end{bmatrix}^T = e_{2ij}$, where α_i, β_i, γ_i and δ_i denote the diagonal elements of S_A, S_B, S_C and S_D respectively. The over-determined equations can be solved using generalize inverses, QR or SVD.

Then the solution X and Y of (11) can be obtained from $\tilde{X}$ and $\tilde{Y}$ through the orthogonal transformations U_i and V_i.

Recall that the matrices X_i will be ill-conditioned if there exists a small singular value for $[A, C]$ or $[B^T, D^T]$. If such a situation arises, (11) should be re-written with modified matrices on the LHS, with the difference corresponding to these small singular values shifted to the RHS. Then iterative refinement or perturbation techniques can be used to find the solution to the ill-posed equation.

In (Chu 1987b) , some special cases of (11) have also been considered. The equations $AX + YD = E$ and $AXB + Y = E$ can be dealt with satisfactorily using the SVD's or QR decompositions of A, B or D. Similarly, the equation $AXB = E$ can be handled and some known results in (Baksalary & Kala 1980),(Kolka 1984) can be proved in equivalent and much simpler forms. Numerical solutions can be found through the SVD or GSVD.

It has also been proved that the equation in Z, $(AZB, CZD) = (G, H)$, is adjoint to Eqt. (11) and similar analysis and numerical algorithms can be applied to the adjoint equation.

SYMMETRIC SOLUTIONS

Consider the solution of
$$AX = B , X = X^T , A \in \mathcal{R}^{m \times n} , B \in \mathcal{R}^{p \times n}, \tag{18}$$
using the SVD and GSVD (Chu 1989) .

Symmetric Solution of $AX = B$ by SVD

Let A be in its SVD, $A = UDV^T = \begin{bmatrix} U_1 & U_2 \end{bmatrix} \begin{bmatrix} \Sigma & 0 \\ 0 & 0 \end{bmatrix} \begin{bmatrix} V_1^T \\ V_2^T \end{bmatrix}$, (18) is then equivalent to $D \cdot V^T X V = U^T B V$, or

$$\begin{bmatrix} \Sigma & 0 \\ 0 & 0 \end{bmatrix} \begin{bmatrix} x_{11} & x_{12} \\ x_{12}^T & x_{22} \end{bmatrix} = \begin{bmatrix} b_{11} & b_{12} \\ b_{21} & b_{22} \end{bmatrix} \tag{19}$$

with $V^T X V$ and $U^T B V$ partitioned appropriately.

Then the symmetric solution of (18) can be read from (19), and

$$X = V \begin{bmatrix} \Sigma^{-1} b_{11} & \Sigma^{-1} b_{12} \\ b_{12}^T \Sigma^{-1} & Z \end{bmatrix} V^T , \quad Z \text{ arbitrary and symmetric}, \tag{20}$$

provided that b_{21} and b_{22} vanish and $\Sigma^{-1}b_{11}$ is symmetric. The consistency conditions are equivalent to those in (Don 1987) : $(I - AA^+)B = 0$, $AB^T = BA^T$, but are in the transformed co-ordinates. Since the SVD is the tool for the numerical evaluation of generalized inverses, the method in (19) is numerically sound, and more natural and direct.

The symmetric solution in the least squares sense for inconsistent equations can also be obtained from (19) and (20) : $X = V \begin{bmatrix} \text{sym}\,(\Sigma^{-1}b_{11}) & \Sigma^{-1}b_{12} \\ b_{12}^T\Sigma^{-1} & 0 \end{bmatrix} V^T$, with $\text{sym}(P) \equiv (P + P^T)/2$, $\text{asym}(P) \equiv (P - P^T)/2$, and the residual equals to $U \begin{bmatrix} \Sigma\,\text{asym}\,(\Sigma^{-1}b_{11}) & 0 \\ b_{21} & b_{22} \end{bmatrix} V^T$.

Symmetric Solution of $AX = B$ by GSVD

From the last subsection, the consistency of (18) reduces to checking whether or not $U_2^T B$ and $\text{asym}\left(\Sigma^{-1}U_1^T BV_1\right)$ are zero. Numerically, it would be difficult to decide whether a number is negligible or small but nonzero (c.f. the problem of rank determination for matrices). The difficulty can be shifted partially by the application of GSVD instead of the SVD to the problem. Note that the following discussion follows the convention for GSVD's in (12), while a different convention has been used in (Chu 1989).

The matrices A^T and B^T have the same number of columns and can be expanded in their GSVD's, i.e. for orthogonal matrices W_A and W_B and nonsingular M, $A = M\Sigma_A^T W_A$, $B = M\Sigma_B^T W_B$, with Σ_A and Σ_B as in (13). Note that $\text{rank}(A) = r + s$ and $k = \text{rank}(A, B)$. The matrices M, W_A and W_B have been used in place of X^T, U^T and V^T in (12) respectively.

Eqt. (18) is then equivalent to $M\Sigma_A W_A \cdot X = M\Sigma_B W_B$ or

$$\begin{bmatrix} I_A & & \\ & S_A & \\ & & 0_A^T \\ \hline & 0 & \end{bmatrix} \tilde{X} = \begin{bmatrix} 0_B^T & \\ S_B & \\ & I_B \\ \hline 0 & \end{bmatrix} \tilde{W}, \tag{21}$$

with $\tilde{X} \equiv W_A X W_A^T$ and $\tilde{W} \equiv W_B W_A^T$.

Judging from (21), the submatrix I_B on the RHS has to vanish for consistency of the equation, implying $k = r + s$ or $\text{rank}(A, B) = \text{rank}(A)$, a well-known consistency result. In other words, with $\alpha_i = \sin\theta_i$ and $\beta_i = \cos\theta_i$ in (14), the canonical angles θ_i have to be all non-zero. Thus a consistent equation can be written in the simplified form :

$$\text{diag}\,\{I_A, S_A, 0\} \begin{bmatrix} x_{11} & x_{12} & x_{13} \\ x_{12}^T & x_{22} & x_{23} \\ x_{13}^T & x_{23}^T & x_{33} \end{bmatrix} = \text{diag}\,\{0_B^T, S_B, 0\} \begin{bmatrix} w_{11} & w_{12} & w_{13} \\ w_{21} & w_{22} & w_{23} \\ w_{31} & w_{32} & w_{33} \end{bmatrix},$$

which implies the following results :

THEOREM 3 (Chu 1989) *(1) The system $AX = B$ is consistent if and only if the canonical angles θ_i for the GSVD of (A^T, B^T) are all non-zero.*

(2) For symmetric solution $X = X^T$, $w_{21} = 0$ and $x_{22} = S_A^{-1}S_B w_{22}$ has to be symmetric.

(3) The symmetric solution is characterized by $x_{11}, x_{12}, x_{13} = 0$, $x_{22} = S_A^{-1}S_B w_{22}$, $x_{23} = S_A^{-1}S_B w_{23}$, where x_{33} is an arbitrary symmetric matrix.

Other LME's have also been considered in (Chu 1989) : $A^T X B = C$, $A^T X + X^T A = C$, $A^T X B + B^T X^T A = C$, $(AX, CX) = (B, D)$. Solutions to other LME's with $M, N > 2$ are still open problems.

REFERENCES

1. Baksalary, J. K., and Kala, R., The Matrix Equation $AX - YB = C$, *Lin. Alg. Applic.*, vol. 25, pp. 41–43, 1979.

2. Baksalary, J. K., and Kala, R., The Matrix Equation $AXB - CYD = E$, *Lin. Alg. Applic.*, vol. 30, pp. 141–147, 1980.

3. Bartels, R. H., and Stewart, G. W., Solution of the Equation $AX + XB = C$, *Comm. ACM*, vol. 15, pp. 820–826, 1972.

4. Carlson, D., Datta, B. N., and Schneider, H., On the Controllability of Matrix Pairs (A, K) with K Positive Semidefinite, *SIAM J. Alg. Disc. Meth.*, vol. 5, pp. 346–350, 1984.

5. Carlson, D. and Hill, R. D., Controllability and Inertia Theory for Functions of a Matrix, *J. Math. Anal. Applic.*, vol. 59, pp. 260–266, 1977.

6. Chu, K.-w. E., The Solution of the Matrix Equations $AXB - CXD = E$ and $(YA - DZ, YC - BZ) = (E, F)$, *Lin. Alg. Applic.*, vol. 93, pp. 93–105, 1987.

7. Chu, K.-w. E., Singular Value and Generalized Singular Value Decompositions, and the Solution of Linear Matrix Equations, *Lin. Alg. Applic.*, vol. 88/89, pp. 83–98, 1987.

8. Chu, K.-w. E., Symmetric Solutions of Linear Matrix Equations by Matrix Decompositions, *Lin. Alg. Applic.* (forthcoming).

9. Datta, K., The Matrix Equation $XA - BX = R$ and Its Applications, *Lin. Alg. Applic.*, vol. 109, pp. 91–105, 1988.

10. De Souza, E., and Bhattacharyya, S. P., Controllability, Observability and the Solution of $AX - XB = C$, *Lin. Alg. Applic.*, vol. 39, pp. 167–188, 1981.

11. Don, F. J. H., On the Symmetric Solutions of a Linear Matrix Equation, *Lin. Alg. Applic.*, vol. 93, pp. 1–7, 1987.

12. Epton, M. A., Methods of the Solution of $AXD - BXC = E$ and its Applications in the Numerical Solution of Implicit Ordinary Differential Equations, *BIT*, vol. 20, pp. 341–345, 1980.

13. Gantmacher, F. R., *The Theory of Matrices, II*, Chelsea, New York, 1960.

14. Golub, G. H., and Van Loan, C., *Matrix Computations*, Johns Hopkins University Press, Baltimore, 1983.

15. Golub, G. H., Nash, S., and Van Loan, C., A Hessenberg-Schur Method for the Matrix Problem $AX + XB = C$, *IEEE Trans. Autom. Contr.*, vol. AC-24, pp. 909–913, 1979.

16. Hartwig, R. E., $AX - XB = C$, Resultants and Generalized Inverses, *SIAM J. Appl. Maths.*, vol. 28, pp. 154–183, 1974.

17. Hearon, J. Z., Nonsingular Solutions of $TA - BT = C$, *Lin. Alg. Applic.*, vol. 16, pp. 57–63, 1977.

18. Hoskins, W. D., Meek, D. S., and Walton, D. J., High-Order Iterative Methods for the Solution of the Matrix Equation $XA + AY = F$, *Lin. Alg. Applic.*, vol. 23, pp. 121–139, 1979.

597

19. Hoskins, W. D., and Walton, D. J., Methods for Solving the Matrix Equation $TB+CT = -A$ and its Generalizations, *Lin. Alg. Applic.*, vol. 23, pp. 217–225, 1979.

20. Kolka, G. K. G., Linear Matrix Equations and Pole Assignment, Ph. D. Thesis, Dept. of Mathematics and Computer Sciences, Univ. of Salford, U. K., 1984.

21. Lancaster, P., Explicit Solution of Linear Matrix Equations, *SIAM Rev.*, vol. 12, pp. 544–566, 1970.

22. Lancaster, P., and Timenetsky, M., *The Theory of Matrices*, 2nd Edition, Academic Press, Orlando, 1985.

23. Laub, A. J., Numerical Linear Algebra Aspects of Control Design Computations, IEEE Trans. Autom. Contr., vol. AC-30, pp. 97–108, 1985.

24. Laub, A. J., and Linnemann, A., Hessenberg and Hessenberg/Triangular Forms in Linear System Theory, *Int. J. Contr.*, vol. 44, pp. 1523–1547, 1986.

25. Magnus, J. R., L-Structured Matrices and Linear Matrix Equations, *Lin. & Multilinear Alg.*, vol. 14, 67–88, 1983.

26. Mitra, S. K., Common Solutions to a Pair of Linear Matrix Equations $A_1 X B_1 = C_1$ and $A_2 X B_2 = C_2$, *Math. Proc. Cambridge Philos. Soc.*, vol. 74, pp. 213–216, 1973.

27. Nashed, M. Z., (Ed.), *Generalized Inverses and Applications*, Academic Press, New York, 1976.

28. Paige, C. C., and Saunders, M. A., Towards a Generalized Singular Value Decomposition, *SIAM J. Numer. Anal.*, vol. 18, pp. 398–405, 1981.

29. Roth, W. E., The Equations $AX - YB = C$ and $AX - XB = C$ in Matrices, *Proc. Amer. Math. Soc.*, VOL. 3, pp. 392–396, 1952.

30. Stewart, G. W., Computing the CS-Decomposition of a Partitioned Orthogonal Matrix, *Numer. Math.*, vol. 40, pp. 297–306, 1982.

31. Vetter, W. J., Vector Structures and Solutions of Linear Matrix Equations, *Lin. Alg. Applic.*, vol. 10, pp. 181–188, 1975.

32. Wimmer, H., and Ziebur, A. D., Solving the Matrix Equations $\sum_{p=1}^{r} f_p(A) X g_p(B) = C$, *SIAM Rev.*, vol. 14, pp. 318–323, 1972.

33. Wimmer, H. K., Inertia Theorems for Matrices, Controllability, and Linear Vibrations, *Lin. Alg. Applic.*, vol. 8, pp. 337–343, 1974.

34. Wimmer, H. K., The Matrix Equation $X - AXB = C$ and an Analogue of Roth's Theorem, *Lin. Alg. Applic.*, vol. 109, pp. 145–147, 1988.

35. Wimmer, H. K., Linear Matrix Equations, Controllability and Observability, and the Rank of Solutions, *SIAM J. Matrix Anal. Appl.*, vol. 9, pp. 570–579, 1988.

36. Zietak, K., The l_p-Solution of the Matrix Equation $AX + YB = C$, *Computing*, vol. 32, pp. 153–162, 1984.

37. Zietak, K., The Chebyshev Solution of the Linear Matrix Equation $AX + YB = C$, *Numer. Math.*, vol. 46, pp. 455–478, 1985.

Direct vs Iterative Cross Plane Solvers for Three-Dimensional Marching Algorithms

R. COHEN
Department of Mechanical Engineering
University of Sydney
Sydney NSW 2006, Australia

INTRODUCTION

The Finite Difference Method (FDM), or indeed almost any numerical method for solving systems of Partial Differential Equations (PDE's), produces large systems of linear algebraic equations which require solution. These equations are of the form

$$[A]\,\Phi = (R) \tag{1}$$

where $[A]$ is a matrix of coefficients – due to the linearization and discretization of the governing system. Φ is a vector of unknowns to be solved for, and (R) is the right hand side vector, containing all known terms, and those not included implicitly on the left hand side. Solution to this system forms the basis to the overall solution procedure. In the general case where the governing equations are nonlinear, the matrix $[A]$ has a dependence on Φ. The iterative system is of the form

$$\left[A\left(\Phi^{n-1}\right)\right]\Phi^n = (R) \tag{2}$$

where superscript n refers to the current iteration level. This system must be repeatedly solved for Φ^n, until suitable convergence of Φ^n with respect to Φ^{n-1} is obtained.

The above assumes that the linear system is solved directly at each iteration. This means that Φ^n is an *exact* solution to the linear system (2). Direct solution of linear systems, is computationally very expensive. For a full matrix of m equations, Gaussian elimination requires $O\left(m^3\right)$ operations. For many applications it is more efficient to split matrix $[A]$ in such a way as it can be solved iteratively. Alternatively, it may be possible to make an approximation to $[A]$, which allows it to be solved with less computational effort. The solution to the approximate system is then modified to produce a solution to the actual system of eq.(1).

GOVERNING EQUATIONS

Consider the incompressible steady state three dimensional Reduced Navier Stokes (RNS) equations. These equations govern the subsonic viscous fluid flow about a configuration in which there is a predominant (but not exclusive) streamwise direction.

continuity

$$\frac{\partial u}{\partial x} + \frac{\partial v}{\partial y} + \frac{\partial w}{\partial z} = 0$$

x momentum

$$\frac{1}{\varrho}\frac{\partial p}{\partial x} + u\frac{\partial u}{\partial x} + v\frac{\partial u}{\partial y} + w\frac{\partial u}{\partial z} = \mu\left[\frac{\partial^2 u}{\partial y^2} + \frac{\partial^2 u}{\partial z^2}\right] \tag{3}$$

599

y momentum

$$\frac{1}{\varrho}\frac{\partial p}{\partial y} + u\frac{\partial v}{\partial x} + v\frac{\partial v}{\partial y} + w\frac{\partial v}{\partial z} = \mu\left[\frac{\partial^2 v}{\partial y^2} + \frac{\partial^2 v}{\partial z^2}\right] \tag{4}$$

z momentum

$$\frac{1}{\varrho}\frac{\partial p}{\partial z} + u\frac{\partial w}{\partial x} + v\frac{\partial w}{\partial y} + w\frac{\partial w}{\partial z} = \mu\left[\frac{\partial^2 w}{\partial y^2} + \frac{\partial^2 w}{\partial z^2}\right] \tag{5}$$

In these equations, u, v, and w are the velocity components in the x, y, and z directions respectively, and the pressure is p. The equations are implemented as a spatial marching scheme in the x direction. An implicit set of linear equations is set up on a cross plane determined by the y and z directions. This equation set is solved, then marched on to the next cross plane. To account for upstream interaction, this marching procedure is embedded within a global relaxation procedure. Details of this procedure can be found in Rubin and Lin (1980), Cohen and Khosla (1987) and Cohen and Khosla (in pub.). It is in context of this scheme, that various types of solver will be discussed.

The complete scheme presents three stages, each of which forms an intermediate in the overall solution procedure.

- global relaxation

- cross plane relaxation

- nonlinear iteration

On the lowest level is the nonlinear iteration. This is due to the nonlinear terms in the momentum equations. Any numerical solution technique for the RNS equations will require some form of nonlinear iteration. The other levels of iteration are specific to the marching technique. Plane relaxation is the phase at which a cross plane solution is obtained. Before marching can proceed, a solution on the current cross plane is required. Global relaxation requires repeated marching sequences. These begin upstream, and march to the downstream boundary. They then begin marching from the upstream cross plane again, using a pressure field updated from the previous marching pass. It is via global iteration, that information propagates upstream in the flowfield.

It is important how the cross plane relaxation is performed. This is the phase that generates a linear system of the form of eq.(2). The algebraic equations that result from the RNS equations at a gridpoint i,j,k are of the following form

$$a\phi_{i,j-1,k} + b\phi_{i,j,k} + c\phi_{i,j+1,k} + d\phi_{i,j,k-1} + e\phi_{i,j,k+1} + qq\phi_{i,j-1,k-1}$$

$$+rr\phi_{i,j+1,k-1} + ss\phi_{i,j-1,k+1} + tt\phi_{i,j+1,k+1} = r_{i,j,k} \tag{6}$$

$a, b, c, d, qq, rr, ss, tt$ are coefficients of the discretised and linearised equations at gridpoints corresponding to a nine point implicit cell. Each of these coefficients are 4×4 submatrices. $\phi_{i,j,k}$ is a vector of the unknowns, at grid location i,j,k. At this stage, the system can be solved directly, or it can be split into a further iterative process. Usually iterative solution would be preferable. Iterative techniques are computationally less intensive than direct techniques. Common iterative techniques include Alternating Direction Implicit (ADI) schemes, the Conjugate Gradient (CG) Method, as well as many others. The main limitation of iterative techniques is that they are highly dependent on the structure and content of the

600

coefficient matrix. For high Reynolds number flows, these matrices are non diagonal dominant, and thus sensitive to many iterative methods. For this reason, direct solution was considered. Due to the structure of coefficient matrix $[A]$, the Yale Sparse Matrix Package (YSMP) was employed. It is compared to an iterative technique – the Coupled Strongly Implicit (CSIP) Scheme.

ITERATIVE SOLUTION

When implementing an iterative technique, the general approach is to split the coefficient matrix of eq.(1) into two parts. Rewriting $[A]$ as $[M + N]$. The system is then of the form:

$$([M] + [N]) \, \Phi = (R) \tag{7}$$

This is then put into iterative form:

$$[M] \, \Phi^n = (R) - [N] \, \Phi^{n-1} \tag{8}$$

and solved, until $|\Phi^n - \Phi^{n-1}|$ is sufficiently small. The iterative scheme depends on the choice of $[M]$ and $[N]$. The usual choice is such that

$$[M] \, \Phi = RHS.$$

is easily solvable.

Coupled Strongly Implicit Procedure

As an example of an iterative solution technique as applied to eq.(1), the Coupled Strongly Implicit Procedure (CSIP) is considered. Developed by Stone (1960), this procedure is based on a modification of the structure of the original coefficient matrix – so that it can be easily factorised. It is then solved iteratively, until the solution of the modified system approaches the solution to the original. The CSIP has been employed successfully in many applications (Khosla and Rubin (1986) and Rubin and Khosla (1981)), where various two dimensional flows have been calculated. The CSIP has also been used for three dimensional flows by Raven and Hoekstra (1985), and Gordnier (1989).

The standard CSIP procedure operates on a five point implicit star. All corner points $i,j\pm 1,k\pm 1$ are treated explicitly. They are lagged from the previous iteration. The form of the equations is;

$$a\phi_{i,j-1,k} + b\phi_{i,j,k} + c\phi_{i,j+1,k} + d\phi_{i,j,k-1} + e\phi_{i,j,k+1} = g_{i,j,k} \tag{9}$$

where

$$g_{i,j,k} = r_{i,j,k} - \left(qq\phi_{i,j-1,k-1} + rr\phi_{i,j+1,k-1} + ss\phi_{i,j-1,k+1} + tt\phi_{i,j+1,k+1} \right) \tag{10}$$

for solution, let

$$\phi_{i,j,k}^n = gm_{i,j,k} + et_{i,j,k}\phi_{i,j,k}^n + f_{i,j-1,k}\phi_{i,j,k-1}^n \tag{11}$$

where

$$gm_{i,j,k} = \beta_{i,j,k}^1 \left[g_{i,j,k} + c_{i,j,k} \left(gm_{i,j+1,k} + f_{i,j+1,k}\phi_{i,j+1,k-1}^{n-1} \right) \right.$$

$$+e_{i,j,k}\left(gm_{i,j,k+1} + et_{i,j,k+1}\phi_{I,j-1,k+1}^{n-1}\right)\right] \tag{12}$$

$$et_{i,j,k} = -\beta_{i,j,k}^{-1}a_{i,j,k} \tag{13}$$

$$f_{i,j,k} = -\beta_{i,j,k}^{-1}d_{i,j,k} \tag{14}$$

and:

$$\beta_{i,j,k} = b_{i,j,k} + c_{i,j,k}et_{i,j+1,k} + e_{i,j,k}f_{i,j,k} \tag{15}$$

The scheme involves a two step process. An initial forward sweep is performed in which coefficients $et_{i,j,k}$, $f_{i,j,k}$ and $gm_{i,j,k}$ are generated. The second sweep involves backsubstitution for $\phi_{i,j,k}$. The scheme is iterative, because corner points are introduced in the calculation of the coefficients in the forward sweep. All points outside the five point implicit star are lagged – which itself requires iteration. Modifications to the CSIP can be made – whereby seven, or even nine points are treated implicitly. This Modified Strongly Implicit Procedure (MSIP) is described in the appendix of Anderson et.al. (1984).

When solutions were generated using the CSIP, it was found that their stability was highly dependent on the initial conditions. They tended to diverge on local iteration, unless severe marching stepsize limitations were enforced. Raven and Hoekstra (1985) also mention encountering convergence problems with the algorithm. They recommend enhancing the implicitness of the CSIP process. It it these reasons, that have motivated the implementation of a direct solver in ongoing work.

DIRECT SOLUTION

Direct solution involves factorizing matrix $[A]$ into lower and upper triangular matrices $[L]$ and $[U]$ respectively, such that:

$$[A] = [L][U] \tag{16}$$

Then the following triangular systems:

$$[L]Y = R \tag{17}$$

and

$$[U]\Phi = Y \tag{18}$$

must be solved for Y and Φ respectively. There are many methods for performing the LU decomposition of $[A]$. These are usually variants of Gaussian elimination, which have been optimised for specific applications. For matrices generated over the cross plane, which take the form of eq.(6), the most important features which can be exploited are that $[A]$ is banded, and sparse. Direct solution must then focus upon bandsolvers, or sparse solvers. The bandwidth of $[A]$ depends on the extent of cross plane. The distance between coefficient groups qq, d, rr and ss, e, tt can be minimised to about twice the smaller gridpoint dimension in the cross plane, multiplied by the number of equations at each point. For a system of 4 equations, on a 20×20 grid in the cross plane, the bandwidth is about 160. The overall order of the matrix is 1600. Less than 3% of the elements in this matrix would be nonzero.

$$[A] = \begin{bmatrix} 1 & 0 & 0 & 2 & 0 & 0 \\ 3 & 4 & 0 & 0 & 0 & 0 \\ 0 & 0 & 5 & 0 & 0 & 0 \\ 0 & 6 & 0 & 7 & 0 & 0 \\ 0 & 0 & 0 & 0 & 8 & 9 \\ 10 & 0 & 0 & 0 & 0 & 11 \end{bmatrix}$$

	1	2	3	4	5	6	7	8	9	10	11
A	1	2	3	4	5	6	7	8	9	10	11
JA	1	4	1	2	3	2	4	5	6	1	6
IA	1	3	5	6	8	10	12 end of matrix				

FIGURE 1. Sparse Storage Scheme for 6 × 6 matrix.

Sparse Matrix Solvers

For the present scheme, the Yale Sparse Matrix Package has been implemented. The YSMP is a set of Fortran routines developed by Eisenstadt et.al. (1977). The YSMP has been used successfully by Bender and Khosla (1987), and Vanka and Leaf (1973), for a wide range of two dimensional flow problems. The package is an efficient set of subroutines for performing an LU decomposition of coefficient matrix $[A]$. Only non-zero terms of the decomposition, and the original matrix are stored. A brief description of the YSMP follows.

The nonzero coefficients of matrix $[A]$ are stored as a vector in row wise order. An integer vector JA stores the column location of each element of $[A]$. Another integer vector IA contains the starting position of each row in $[A]$. The end of matrix is indicated, by the last element of IA being set equal to the number of nonzero terms plus one. An example of the storage scheme used by the YSMP appears in fig.1. The solver requires that each term on the main diagonal of $[A]$ be non-zero. It does not perform any partial pivoting, or conditioning to achieve diagonal dominance. There are several drivers that can be used when performing the LU decomposition, and backsubstitution during the linear solution process. These are specifically for cases where either storage must be minimised, computational effort must be minimised, or the LU decomposition is retained for multiple right hand side vectors. These options are described more fully in Eisenstadt et.al. (1977), which also gives relative performance figures for them. The YSMP keeps track of all terms generated in the LU decomposition. They are stored in the same form as $[A]$. When the LU decomposition is performed, non-zero terms will be generated. In the worst case the non-zeros generated would fill the entire matrix $[A]$, however bandsolvers will limit this fill in to the diagonal band. The amount of fill in is dependent on the ordering of the rows in $[A]$. A *Minimum Degree Algorithm* has been included with the YSMP to minimise this fill in. This algorithm reorders $[A]$ such that this criteria is met. The algorithm is heuristic in nature, so the reordering is not rigorous for absolute minimum possible fill in. The minimum degree algorithm is described in George (1981).

In the current implementation, fig.2 shows that the size of the LU decomposition is about five times the number of nonzero terms in the original system. For a system of m equations, the performance of a direct solver will be proportional to $m^{(1+\Delta)}$. Gaussian elimination has $\Delta = 2$. For the present system, fig.3 shows Δ to be less then 1.

603

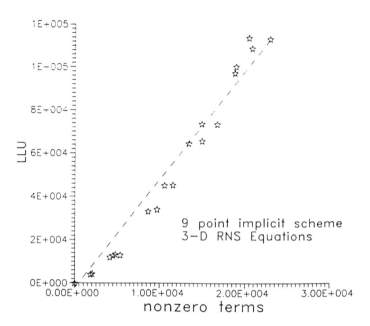

FIGURE 2. Fill in vs. Number of nonzero terms.

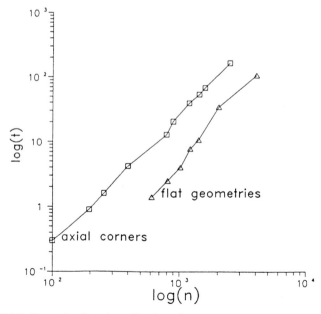

FIGURE 3. Execution Speed vs. Number of nonzeroes.

COMPUTATIONAL RESULTS

This scheme has been implemented for a number of different configurations, including the flow over a finite edge plate, axial corners and a three dimensional bump. For detail of these, and results see Cohen and Khosla (1987), Cohen and Khosla (in pub.) and Cohen (in prep.). It has been highly successful in a wide range of conditions. The sensitive stability characteristics of iterative solvers have not been experienced. In the worst case, of three dimensional separated flows, a pseudo time term had to be included in the momentum equations. This is required to boost the diagonal term in the coefficient matrix. It was found that this term was only needed for the initial few iterations, after which it could be switched off. Other than this small modification to the coefficient matrices, no preconditioning, or special techniques were required to ensure that a stable solution would be obtained. Although slower than iterative techniques, significant problems have been solved on mini and micro computers. The direct solver makes the solution easily attainable. Several strategies for improving the speed of algorithms using direct methods are discussed by Himansu et.al. (1989).

CONCLUSION

Direct solution of large linear systems – although not always computationally efficient, is balanced by it's robustness. Where algorithms are being developed, or new problems are being solved, it is recommended that the solution technique be as insensitive as possible to the matrices generated by a specific application. This may even be at the price of overall efficiency. Once solutions to a class of problems can be obtained, then attention can be focused on improving the solver. When embedded within an overall iterative scheme, the direct solver has proved a successful tool in the solution of three dimensional RNS marching problems. In this context, the overall scheme still retains the advantages of an iterative method globally, whilst at a lower level the benefits of a direct solution technique can be exploited.

REFERENCES

1. D.A. Anderson, J.C. Tannehill, and R.H. Pletcher, *Computational Fluid Mechanics and Heat Transfer.* McGraw-Hill. 1984.

2. E.E. Bender and P.K. Khosla, Solution of the Two Dimensional Navier-Stokes Equations using Sparse Matrix Solvers. *AIAA paper no 86-0603.* Presented at the 25th Aerospace Sciences Meeting. *Reno, Nevada, USA.* Jan. 1987.

3. R. Cohen and P.K. Khosla, Solution to the Reduced Navier-Stokes Equations in Three Dimensions, Using a Global Relaxation Procedure. Proceedings of CTAC-87/ISCFD. *Sydney Australia,* August 1987.

4. R. Cohen and P.K. Khosla, Three Dimensional Reduced Navier Stokes Solutions for Subsonic Separated and Non-Separated Flows, Using a Global Pressure Relaxation Procedure. *International Journal for Numerical Methods in Fluids.* Accepted for Publication.

5. R. Cohen, Ph.D Thesis. The University of Sydney. In preparation.

6. S.C. Eisenstadt, M.C. Gursky, M.H. Schultz and A.H. Sherman, Yale Sparse Matrix Package II. The Nonsymmetric Codes. Yale University. Department of Computer Science. Research Report #114. 1977.

7. A. George, in *Sparse Matrices and their Uses.* ed. I.S. Duff, pp. 290-293, Academic Press. 1981.

8. R. Gordnier, Ph.D Dissertation. University of Cincinnati. 1989.

9. A. Himansu, P.K. Khosla and S.G. Rubin, Relaxation/Sparse-matrix solvers for

Three-Dimensional Recirculating Flows. AIAA Paper AIAA-89-0552. Presented at *27th Aerospace Sciences Meeting.* January 9-12, 1989. Reno, Nevada, USA.

10. P.K. Khosla and S.G. Rubin, Consistent Strongly Implicit Procedures. Presented at *10th International Conference on Numerical Methods in Fluid Dynamics.* Beijing, China, June 1986.

11. H.C. Raven and M. Hoekstra, A Parabolised Navier-Stokes Solution for Ship Stern Calculations. MARIN Report no. 50402-I-SR, 1984. Presented at the *Second International Symposium on Ship Viscous Resistance, Göteborg, Sweden.* March 1985.

12. S.G. Rubin and A. Lin, Marching with the PNS Equations. *Israel Journal of Technology,* vol. 18, 1-2 p21-31, 1980.

13. S.G. Rubin and P.K. Khosla, Navier-Stokes Calculations with a Coupled Strongly Implicit Method - I. *Computers and Fluids,* vol. 9, 1981.

14. H.L. Stone, Iterative Solution of Implicit Approximation of Multi-Dimensional Partial Differential Equations. *SIAM J. Num. Anal.,* 5, 1960.

15. S.P. Vanka and G.K. Leaf, Fully Coupled Solution of Pressure-Linked Fluid Flow Equations. Argonne National Laboratory. Argonne, Il., USA. Report ANL-83-73.

Approximate Factorization Explicit Methods for CFD

C. A. J. FLETCHER
Department of Mechanical Engineering
University of Sydney, Sydney NSW 2006, Australia

INTRODUCTION

Fluid flow problems often require the solution of tridiagonal systems of equations after discretisation by a local method, such as the finite difference, finite element or finite volume methods, particularly if some form of approximate factorisation is introduced (Fletcher, 1988, p. 254). Typically new tridiagonal systems must be solved at each step of the global iteration. The Thomas algorithm and related algorithms allow very efficient direct solution of scalar and block tridiagonal systems on serial computers.

However the rapid maturing of parallel–processing technology prompts the search for tridiagonal solvers that can be implemented efficiently on parallel CPUs. Ortega and Voigt (1985) review some algorithms in this area. Recently Evans (1987, 1988) has introduced alternating group explicit (AGE) iterative methods for the efficient solution of tridiagonal systems of equations on parallel processors. A key feature of the AGE family of algorithms is that they sequentially restructure the overall tridiagonal matrix into independent 2×2 sub–blocks. In principle each sub–block can be solved in parallel on separate processors.

A typical AGE algorithm is less efficient than a single application of a direct tridiagonal solver on a serial computer. However with sufficiently fine granularity, and embedded in a global iterative framework, AGE algorithms are potentially very efficient for computational fluid dynamic (CFD) applications. Preliminary computational experiments with both compressible, low Reynolds number flows and incompressible high Reynolds number flows are encouraging (Satofuka, 1988).

In this paper we develop an approximate factorisation explicit (AFE) algorithm that can be interpreted as an extension of the early AGE algorithm of Evans (1987). Tridiagonal systems produced by high Reynolds number fluid flows with three–point central differencing usually lack diagonal dominance. The problem is exacerbated as the Reynolds number is increased. This is expected to have more impact on the efficiency of an iterative, e.g. AFE, solver than on a direct solver of a tridiagonal system of equations. The sensitivity of the AFE algorithm to increased Reynolds number is examined here in relation to a model convection diffusion problem. Subsequently we apply the AFE algorithm, with and without multigrid, to the two–dimensional Burgers equations to assess the efficiency of the AFE algorithm for an equation system representative of CFD applications.

APPROXIMATE FACTORISATION EXPLICIT (AFE) METHOD

Evans (1987) has introduced the alternating group explicit (AGE) iterative method for solving linear tridiagonal systems of equations, $Au = B$, by splitting A into $G_1 + G_2$ where G_1 and G_2 contain zero elements except for independent (2×2) sub–blocks along the diagonal. Since each sub–block can be inverted independently, the overall algorithm is amenable to efficient computer execution using vector and/or parallel processors.

Here we use a similar strategy to solve $R(u) = B - Au = 0$ by introducing a discrete pseudotransient construction (Fletcher, 1988, p. 208),

$$\Delta u^{n+1}/\Delta t - \{\beta R(u^{n+1}) + (1 - \beta) R(u^n)\} = 0 , \tag{1}$$

where u^{n+1} is the solution at the $n + 1$ iteration (pseudotime) level.

By expanding $R(u^{n+1})$ as a Taylor series about $R(u^n)$, eq. (1) can be approximated by

$$\{(1/\Delta t) I + \beta(G_1 + G_2)\} \Delta u^{n+1} = R(u^n) . \tag{2}$$

For the choice $r = 1/\Delta t$ and $\beta = 1$, an ADI implementation of eq. (2) leads to the basic Evans (1987) algorithm, eq. (35). However we prefer to adopt an approximate factorisation procedure (Fletcher, 1988, p. 254) since it leads to fewer evaluations of $R(u^n)$. This is usually a major contribution to the execution time for CFD governing equations. With $r = 1/\Delta t$ and $\beta = 1$ eq. (2) is updated with a two–stage algorithm,

$$(rI + G_1) \Delta u^{n+\frac{1}{2}} = R(u^n) \tag{3a}$$

$$(rI + G_2) \Delta u^{n+1} = \gamma r \, \Delta u^{n+\frac{1}{2}} \tag{3b}$$

and $u^{n+1} = u^n + \Delta u^{n+1}$. $\tag{3c}$

Expressed in terms of u this becomes
$$\tilde{u} = (G_2 - rI) \, u^n \tag{4a}$$

$$\tilde{G}_1 \, u^{n+\frac{1}{2}} = B - \tilde{u} \tag{4b}$$

$$\tilde{G}_2 \, u^{n+1} = \tilde{u} + r \, \{u^n(2 - \gamma) + \gamma u^{n+\frac{1}{2}}\} , \tag{4c}$$

where $\tilde{G}_1 = G_1 + rI$, etc.

The choice $\gamma = 2$ in eq. (4c) is interesting because it allows the same memory allocation for u^n, $2ru^{n+\frac{1}{2}} + \tilde{u}$ and u^{n+1}, i.e. the solution can be upgraded in situ. It may be noted that, with the choice $\gamma = 2 - \omega$, eq. (4) coincides with the generalised Evans (1987) algorithm, eq. (44).

The extension of the above approximate factorisation to multidimensions is direct. As an example, a finite difference discretisation of the two–dimensional Laplace equation would produce the following in place of eq. (1),

$$\Delta u^{n+1}/\Delta t - (L_{xx} + L_{yy})\{\beta u^{n+1} + (1 - \beta) u^n\} = 0 , \tag{5}$$

where $L_{xx} = [\partial^2/\partial x^2]_d$ etc. With $r = 1/\Delta t$ and $\beta = 1.0$, the equivalent of eq. (2) is

$$\{rI + (G_1^{xx} + G_2^{xx} + G_1^{yy} + G_2^{yy})\} \Delta u^{n+1} = R(u^n) = (L_{xx} + L_{yy})u^n \tag{6}$$

where G_1^{xx}, G_2^{xx} are the (2×2) sub–block decompositions of $-L_{xx}$; and similarly for G_1^{yy}, G_2^{yy}. Equation (6) is updated via a four–stage approximate factorisation,

$$\tilde{G}_1^{xx} \Delta u^{n+\frac{1}{4}} = R(u^n) \tag{7a}$$

$$\tilde{G}_2^{xx} \Delta u^{n+\frac{1}{2}} = \gamma_x r \Delta u^{n+\frac{1}{4}} \tag{7b}$$

$$\tilde{G}_1^{yy} \Delta u^{n+3/4} = \delta r \Delta u^{n+\frac{1}{2}} \tag{7c}$$

$$\tilde{G}_2^{yy} \Delta u^{n+1} = \gamma_y r \Delta u^{n+3/4} \tag{7d}$$

where $\tilde{G}_1^{xx} = G_1^{xx} + rI$ etc.

Each of the pairs of equations, (7a), (7b) and (7c), (7d) can be solved by the equivalent of eqs. (4a) to (4c). Also the use of δ on the right–hand side of eq. (7c) is available even if a conventional scalar tridiagonal solver (Fletcher, 1988, p. 183) is used in place of the AFE algorithm, eqs. (3a) to (3c).

Since the introduction of approximate factorisation in either eqs. (3) or (7) implies an iterative solution it is appropriate to seek ways of accelerating the iterative process. Here we make use of a V–cycle multigrid strategy (Fletcher, 1988, p. 203). On the mth grid a single relaxation process consists of ℓ iterations of eqs. (7) for a two– dimensional problem, where ℓ is $0(10)$. The relaxation process is carried out twice when moving to a coarser grid (reducing m) and once when moving to a finer grid. The relaxation process is represented symbolically by

$$W^{m+1,\upsilon} = \text{RELAX } (W^{m+1}, \text{ eq. (7)}, R^{m+1}), \tag{8}$$

where ν is the number of times the relaxation process is undertaken. On the finest grid W is the current approximation to u; on coarser grids W^{m+1} is an approximation to Δu^{m+1}. The term R^{m+1} is either the equation residual, R, on the finest grid or ΔR on coarser grids. Using $W^{m+1,\upsilon}$ from eq. (8), the residual on the next coarser grid, m, is obtained from

$$R^m = I_{m+1}^m R^{m+1,\upsilon}, \tag{9}$$

where I_{m+1}^m is the restriction operator; typically single, five or nine point operators are used. On each coarser grid the current solution correction, W^m, is set to zero before initiating eq. (8). The combination of eq. (8) and (9) is repeated until the coarsest grid is reached.

The solution correction, W^m, on the coarsest grid is relaxed using eq. (8) and then interpolated onto the next finer grid, m+1, i.e.

$$W^{m+1} = I_m^{m+1} W^m, \tag{10}$$

where I_m^{m+1} denotes the prolongation (interpolation) operator. Here linear and bilinear interpolation are used.

The application of eq. (8) once and eq. (10) is repeated on each finer grid until the finest grid is reached. The process of starting on the fine grid, relaxing twice and restricting until the coarsest grid is reached, and relaxing once and prolonging until the finest grid is again reached constitutes a single (2,1) MG V–cycle. For a smooth problem perhaps 10 V cycles would be necessary for convergence.

ONE–DIMENSIONAL CONVECTION DIFFUSION EQUATION

This problem is governed by the equation

$$u \, \partial T / \partial x - \alpha \, \partial^2 T / \partial x^2 = 0 , \tag{11}$$

with boundary conditions, $T = 0$ at $x = 0$ and $T = 1$ at $x = 1$. The solution, $T(x)$, is uniform and equal to zero in the interior and rises to $T = 1$ in a boundary layer adjacent to $x = 1$. As u/α increases the boundary layer becomes thinner. Consequently the equation is a useful model for studying convection dominated phenomena (Fletcher, 1988, p. 293).

With three–point centered difference discretisation this problem produces a linear tridiagonal system of equations associated with the interior points, $j = 2, NX - 1$,

$$- (1 + 0.5 \, R_{cell}) \, T_{j-1} + 2T_j - (1 - 0.5 \, R_{cell}) \, T_{j-1} = 0 , \tag{12}$$

where $R_{cell} = u \Delta x / \alpha$.

Boundary conditions are implemented as $T_1 = 0$ and $T_{NX} = 1$. As R_{cell} increases beyond 2, eq. (12) becomes increasingly less diagonally dominant and its solution becomes increasingly oscillatory. Although the solution of eq. (12), for $R_{cell} > 2$, is not an accurate solution of eq. (11), the problem described by eq. (12) for increasing R_{cell} is a useful test of the convergence properties of the AFE method, eq. (4).

To test the sensitivity to increasing R_{cell}, solutions have been obtained with $NX = 41$ using eq. (4). The values of r and γ to give the fastest convergence at each value of R_{cell} are indicated in Table 1. Convergence is assumed when the rms error of the current solution compared with the exact solution of eq. (12) is less than 1.0×10^{-5}.

As indicated in Table 1 the number of iterations to convergence (n_{con}) using eq. (4) increases as R_{cell} increases. Using a geometric cycle for r does not improve the convergence rate. Each evaluation of eq. (4) requires $9(A + M)(NX - 2)$ operations, where A denotes addition or subtraction and M multiplication or division. The choice $r = 0.5$, $\gamma = 2.0$ simplifies eq. (4c) and requires $8(A + M)(NX - 2)$ operations. Thus this choice is almost as efficient as the optimum r, γ choice.

A direct solution of eq. (12) requires $5(A + M)(NX - 2)$ operations so the present algorithm is 40 times slower on a scalar machine under the most favourable conditions, $R_{cell} = 2$. The present problem, eq. (12), is particularly convenient for direct solution, but with fine enough granularity, i.e. one processor per interior grid point, the AFE algorithm, eq. (4), would be as efficient.

TABLE 1. Convergence Sensitivity to Increasing R_{cell}

R_{cell}	r	γ	n_{con}	n_{con} $r=0.5, \ \gamma=2.0$
1	0.67	2.38	29	38
2	0.67	2.13	22	32
4	0.67	1.96	36	44
8	0.67	2.17	134	210

TABLE 2. R_{cell} Sensitivity for Multigrid AFE

R_{cell}	r	γ	MG iter.	MG equiv. work	AFE eq. (4)
1	0.67	2.0	6	38	44
2	0.67	2.17	9	56	34
4	0.75	2.27	12	75	52
8	0.85	2.17	21	132	165

Convergence can be accelerated using multigrid. Corresponding optimum choices of r and γ with a (2,1) MG V–cycle are shown in Table 2. The solutions were obtained with NX = 65 and four levels of grid coarsening. A smaller number of multigrid cycles are required for convergence than fine–grid iterations using eq. (4) directly. The results produced by AFE, eq. (4) use the optimum (r, γ) combination indicated in Table 1. However each multigrid cycle requires three relaxations, eq. (8) on each grid and a single MG cycle requires 6.27 as many operations as a single fine grid relaxation in one dimension. Consequently for this problem multigrid gives no practical improvement over simple AFE relaxation on an equivalent work basis, and the previous remarks about computational efficiency of the AFE algorithm still hold.

TWO–DIMENSIONAL BURGERS EQUATION

This problem is an appropriate test case because the equation structure is similar to that of the incompressible fluid flow momentum equations. Consequently the AFE algorithm has to be embedded in a global iterative structure which provides a more realistic indication of its performance.

The governing equations are written in divergence vector form as

$$R(\bar{q}) = \partial \bar{F}/\partial x + \partial \bar{G}/\partial y - (1/Re)\left\{\partial^2 \bar{q}/\partial x^2 + \partial^2 \bar{q}/\partial y^2\right\} - \bar{S} = 0 , \qquad (13)$$

where $\bar{q} = \{u, v\}$, $\bar{F} = \{u^2, uv\}$, $\bar{G} = \{uv, v^2\}$

and $\bar{S} = \{0.5\ Re\ u(u^2 + v^2),\ 0.5\ Re\ v(u^2 + v^2)\}$.

The steady–state solution is of interest; a term, $\partial \bar{q}/\partial t$, is introduced to facilitate a pseudotransient implementation, as in eq. (1). Three–point centered difference formulae are used for discretisation and a linearisation is introduced (Fletcher, 1988, p. 363) to generate two scalar equations similar to eq. (6),

$$\{I/\Delta t + \beta[L_x u + L_y v - (1/Re)\{L_{xx} + L_{yy}\} - Re(u^2 + v^2)]\}\ \Delta \bar{q}^{n+1} = R(\bar{q}^n) . \qquad (14)$$

Equations (14) can be solved efficiently using eqs. (7) with $r = 1/\Delta t$, $\beta = 1.0$ and with the use of the following sub–block decomposition,

$$G_1^{xx} + G_2^{xx} = L_x u - (1/Re)L_{xx} - 0.5\ Re(u^2 + v^2) , \qquad (15)$$

and similarly for $G_1^{yy} + G_2^{yy}$. This decomposition is available since the right–hand side of eq. (15) is tridiagonal. In the results shown below eq. (7) has been used to

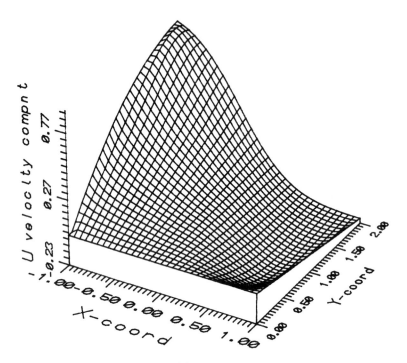

FIGURE 1. Converged solution (u) of eq. (13)

iteratively converge the solution both directly and as the relaxation process, eq. (8), in the multigrid method.

Dirichlet boundary conditions are provided by an exact solution (Fletcher, 1988, p. 361), at Re = 5. This solution in the region, $-1 \le x \le 1$, $0 \le y \le y_{max}$, provides the starting solution for the test problem which is equivalent except that Re = 20. A typical converged solution for u, satisfying eq. (13), is shown in Fig. 1. The solution is characterised by a peak in u and v close to $x = -1$ and $y = y_{max}$. The magnitude of u falls towards $x = +1$ and $y = 0$. The value of v (not shown) falls towards $y = 0$ and reverses in magnitude towards $x = +1$.

The convergence history when eq. (7) is applied on a 33 × 33 uniform grid to eqs. (14) and (15) is shown in Fig. 2. Equation (7) is applied for IF iterations (see Fig. 2) in the form of eq. (4) with $R(\bar{q}^n)$ in eq. (14) frozen. More iterations (larger IF) at each iteration step n produces a performance closer to the conventional approximate factorisation leading to the direct (Fig. 2) inversion of tridiagonal systems, i.e. (7a) and (7b) together and (7c) and (7d) together. However although this produces faster convergence it requires more operations. If IF < 3 the iterative process is divergent. For each case optimal values of r, γ_x, δ and γ_y in eq. (7) are used.

When eq. (7) is applied to eqs. (14) and (15) with a four–level (2, 1) multigrid V cycle the corresponding convergence behaviour is shown in Fig. 3. As in Fig. 2 optimal values of the parameters in eq. (7) have been chosen to maximise the rate of convergence. The behaviour of the iterative solver, eq. (7), as the relaxation process, eq. (8), is similar to its behaviour with simple (fine–grid) approximate

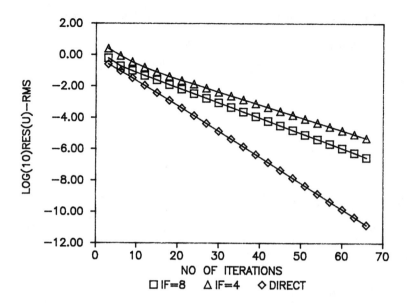

FIGURE 2. Convergence behaviour for approximate factorisation iteration

factorisation iteration (Fig. 2). That is increasing IF gives a faster rate of convergence that requires more operations per multigrid iteration.

Table 3 provides an operation count for the different methods of solving eq. (7) corresponding to Fig. 2. The entries in Table 3 are the number of floating point operations with all four basic operations assumed to require the same CPU time.

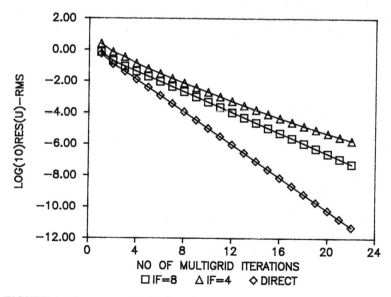

FIGURE 3. Convergence behaviour for multigrid iteration

TABLE 3. Contributions to the Single Iteration Grid Point Operation Count

Solution of eq. (7)	$R(\bar{q}^n)$ eq. (14)	LHS eq. (14)	Solution of (14), using (7)	Total
AFE, IF = 8	43	70	640	753
AFE, IF = 4	43	70	320	433
AF, DIRECT	43	70	32	145

The approximate factorisation iteration results shown in Fig. 2 require the evaluation of the right–hand side of eq. (14), the evaluation of the left–hand side coefficients in eq. (7) based on the left–hand side of eq. (14) and the subsequent solution of eq. (14) using eq. (7). Only the third stage introduces differing operation counts for the three cases shown in Fig. 2. A single evaluation of eq. (7) using the sub–block decomposition is about 2.5 times as expensive as a direct tridiagonal solver. For four passes through eq. (7) (IF = 4) a tenfold operation count multiplier (Table 3) occurs. However for the overall algorithm this corresponds to a threefold operation count penalty. Similar ratios arise with the multigrid results (Fig. 3) since the fine grid iteration dominates the operation count.

From Fig. 2 the sub–block decomposition with IF = 4 requires twice as many iter–ations to reach a comparable stage of convergence as the use of the direct tridiagonal solver. This implies an overall execution time factor of 6 compared with the direct tridiagonal solver. For IF = 8 the overall execution time factor is about 8.

CONCLUDING REMARKS

If more than six parallel processors are available the above results imply that it would be more efficient to use the sub–block decomposition and distribute this part of the algorithm, and the other operations indicated in Table 3, equally between all processors. However for a three–dimensional problem on an n^3 grid the direct tridiagonal solver can be implemented in parallel in one dimension at a time. Therefore if $6(n-2)$ parallel processors are available the AFE algorithm will be more efficient, ignoring overhead effects associated with the parallel implementation and operation count contributions common to both formulations.

REFERENCES

1. Evans, D.J., The Alternating Group Explicit (AGE) Matrix Iterative Method, *Appl. Meth. Modelling*, Vol. 11, pp. 256–263, 1987

2. Evans, D.J., The Modified Alternating Group Explicit (MAGE) Method, *Appl. Meth. Modelling*, Vol. 12, pp. 262–267, 1988

3. Fletcher, C.A.J., *Computational Techniques for Fluid Dynamics*, Vol. 1, Springer– Verlag, Heidelberg, 1988

4. Ortega, J.M. and Voigt, R.G., Parallel Algorithms for Partial Differential Equations, *SIAM Review*, Vol. 27, pp. 147–240, 1985

5. Satofuka, N., Group Explicit Methods for the Solution of Fluid Dynamic Equations, *Proc. ISCFD–Sydney*, eds. G. de Vahl Davis and C.A.J. Fletcher, pp. 117–134, North–Holland, Amsterdam, 1988

Computation of the Collocating Quadratic Hermite-Padé Approximation

A. W. MCINNES
Department of Mathematics
University of Canterbury
Christchurch, New Zealand

INTRODUCTION

The idea of applying Padé approximation to problems in mathematical physics was put forward by Baker and Gammel (1961). Padé approximants, the rational analogue of the Taylor polynomial approximant, could be expected to represent a wider range of behaviour than simple polynomial approximation. A variety of generalisations of this approach have been suggested and these are often collected in the class of what are called Hermite-Padé approximants (see for example Baker and Lubinsky (1987)). In particular, some authors (e.g. Short (1979)) have used quadratic approximants, which have branch points, to approximate multi-valued functions.

A more recent and interesting application of the use of approximations by algebraic functions occurs in the field of computational geometry. This involves the use of algebraic curves and surfaces to approximate and model objects in computational geometry and computer modelling (see for example Bajaj and Kim (1989), Hoffman (1988)). While the approach used by these authors is generally quite different from that of the generalised Padé approximants, the problems are similar, and further study will undoubtedly produce a unified mathematical structure.

In this paper, we study the approximation by a quadratic algebraic function. While the approach of a generalised form of the Padé approximation requires that all the information about the given function, $f(x)$, be at a single point (usually taken to be the origin), such a requirement may be unrealistic in the case of computer modelling. We should also consider the case where the information about the given function is available at a number of distinct points. This is a generalisation of the Newton-Padé, or 'rational interpolation' problem which was first discussed by the author in McInnes (1984).

The two separate aspects of the quadratic approximation problem are the properties of the approximation and the computation of the approximation. The next

---

This paper was written while the author was visiting the Center for Applied Mathematics, Purdue University, West Lafayette, IN 47907

section describes a general formulation of the problem that includes both the distinct and confluent point cases mentioned above, and notes some properties of the approximating function defined in this way. In the final section an algorithm is derived, which computes the collocating quadratic approximants in a Newton form, analogous to the Newton form in polynomial approximation.

FORMULATION OF THE QUADRATIC APPROXIMANT

<u>Definition 1.</u> Let $n = (n_0, n_1, n_2)$, where $n_i \geq -1$ are integers for $i = 0, 1, 2$.

An n *quadratic function* is a function, $Q(x)$, which satisfies the equation

$$P_n(Q, x) = a_0(x) + a_1(x)Q(x) + a_2(x)Q(x)^2 = 0, \tag{1}$$

where the $a_i(x)$ are algebraic polynomials with $\deg(a_i(x)) \leq n_i$ for $i = 0, 1, 2$. By convention the polynomial of degree -1 is identically zero.

If $n = (n_0, n_1)$, the eq. (1) defines a rational function, and if, in addition, $n_1 = 0$ the eq. (1) trivially defines a polynomial function. Let $A_{N-1} = \{x_k, k = 0(1)N - 1\}$ be the set of *distinct* node points corresponding to the given set $\{L_i, i = 0(1)N - 1\}$ of point evaluation functionals defined by $L_i g(x) = g(x_i)$, where $N + 1 = \sum_{i=0}^{2}(n_i + 1)$.

<u>Definition 2.</u> For a given function $f(x)$, the function

$$P_n(f, x) = a_0(x) + a_1(x)f(x) + a_2(x)f(x)^2 = O(w_N(x)) \tag{2}$$

will be called an n *quadratic form*, where the $a_i(x)$ are algebraic polynomials with $\deg(a_i(x)) \leq n_i$, for $i = 0, 1, 2$, $w_N(x) = \prod_{k=0}^{N-1}(x - x_k)$. The order notation is to be interpreted as meaning bounded in the neighbourhood of the points of the collocation node set A_{N-1}. Generally the subscript n on P will be dropped where the context makes this obvious.

Note that $P(f, x)$ may also be written as a function $r(x)$. This function $r(x)$ may be approximated by a polynomial of degree $< N$ by collocation on the points of A_{N-1}. As is shown in elementary numerical analysis texts, the error of this approximation may be written as $O(w_N(x))$. This is the rationale for the notation in (2). Further, if the given linear functionals were $L_k = D^k$ where $D^k r(x) = r^{(k)}(0)/k!$ for $k = 0(1)N - 1$, then this definition becomes the n quadratic form which generalises the form used for Padé approximation in the rational case ($i = 1$) and Taylor polynomial approximation in the polynomial case. (See McInnes (1989).)

As shown in McInnes (1989) for the case of functionals D^k, it is important to carefully distinguish between the approximation of the quadratic form and the approximating properties of the corresponding quadratic function, in order to fully analyse the latter. The details for the case of the point evaluation functionals L_k are analogous and will only be sketched here.

<u>Theorem 3.</u> (Existence) There always exists an n quadratic form for the given linearly independent functionals L_k and given function $f(x)$.

<u>Proof</u>: The quadratic form is defined by the coefficient polynomials $a_i(x)$. The existence of the $a_i(x)$ follows since the application of the linear functionals $L_k, k = 0(1)N - 1$, to (2) leads to a system of N homogeneous linear equations for the $N + 1$ unknown coefficients of the $a_i(x)$. Hence a non-trivial solution, with $a_i(x)$ not all zero, exists. Further, if $a_i(x) \equiv 0$ for $i = 1, 2$, then $a_0(x) \equiv 0$, and hence we must have $a_1(x), a_2(x)$ not both identically zero.

The uniqueness of this quadratic form requires a more careful argument. The matrix form of the system of linear equations represented by applying L_k to (2) has the coefficient matrix:

$$F = [F_{n_0} : F_{n_1} : F_{n_2}]$$

where

$$F_{n_k} = \begin{bmatrix} f_0^k & x_0 f_0^k & \cdots & x_0^{n_k} f_0^k \\ f_1^k & x_1 f_1^k & \cdots & x_1^{n_k} f_1^k \\ \cdots & \cdots & \cdots & \cdots \\ f_{N-1}^k & x_{N-1} f_{N-1}^k & \cdots & x_{N-1}^{n_k} f_{N-1}^k \end{bmatrix}$$

The matrix F has dimensions $N \times (N + 1)$ and hence has a solution space of dimension at least 1. If the rank of the matrix F is N, then any constant multiple of the coefficient polynomials will also be a solution. These solutions may be called *essentially unique* following Baker and Lubinsky (1987). A unique representative of this class may be defined by choosing a suitable normalization.

In this paper, the assumption will be made that rank $F = N$, and hence there is an essentially unique solution $P^*(f, x)$. Once $P^*(f, x)$ has been obtained, we can define a quadratic function, $Q(x)$ satisfying

$$P^*(Q, x) = a_0^*(x) + a_1^*(x)Q(x) + a_2^*(x)Q(x)^2 = 0. \tag{3}$$

Since $P^*(f, x)$ is determined by applying the functionals, L_k, to the quadratic form involving $f(x)$, this function, $Q(x)$, represents an approximation to $f(x)$. Since (3) *normally* has two distinct branches, if the roots of (3) are distinct at the collocation points, we require that the roots of $P^*(Q, x)$, as a polynomial in Q, have distinct roots for normal behaviour.

<u>Definition 4.</u> The function $f(x)$ and the functionals L_k defining the quadratic form $P^*(f, x)$ are called *normal* if $\partial P^*(f, x) / \partial f |_{x \in A_{N-1}} \neq 0$, where A_{N-1} is the set of collocation points associated with the functionals $\{L_k, k = 0(1)N - 1\}$.

In the case of rational approximation, $\partial P^*(f, x) / \partial f = a_1(x)$. Thus this definition corresponds to the usual conditions required in ordinary and multi-point Padé approximation in Baker and Graves-Morris (1981).

If $\partial P^*(f, x) / \partial f = 0$ for some point in the collocation set A_{N-1}, this leads to the concept of unattainable points in rational approximation. The non-normal case for the functionals D^k is discussed in McInnes (1989), but this topic will not be pursued here. Assume hereafter that the quadratic form is normal.

Since $P^*(Q(x_0), x_0) = 0 = P^*(f(x_0), x_0)$, only one of the distinct branches of $Q(x)$ will have $Q(x_0) = f(x_0)$. Hence there is a unique branch of the function $Q(x)$ which passes through the point $(x_0, f(x_0))$, and we consider the approximation properties of this branch as an approximation to $f(x)$.

Theorem 5. Let $P^*(f, x)$ be the normal quadratic form defined by equation (2). Then the quadratic function defined by (3), subject to the condition $Q(x_0) = f(x_0)$, is an approximation to $f(x)$ satisfying

$$Q(x) = f(x) + O(w_N(x)).$$

The proof of this theorem is analogous to that of Theorem 9 in McInnes (1989). A detailed proof will be given in a forthcoming report.

This quadratic function approximation to $f(x)$ may be represented by considering the N equations obtained by applying the functionals, L_k, to (2), plus eq. (1), to give a homogeneous system of $N+1$ linear equations in the $N+1$ unknown coefficients. This system has a non-trivial solution which is expressed by the eliminant of these equations, i.e. $\det B = 0$.

COMPUTATION OF THE QUADRATIC FORM

Although the mathematical solution of the problem of determining the $\underset{\sim}{n}$ quadratic form is easily expressed by the equation $\det B = 0$, which is the eliminant of the system of linear equations (2), the computational solution of this problem may be another matter. If the coefficient polynomials, $a_i^*(x)$, are expressed in terms of the usual basis functions, x^k, numerical problems are not surprising in view of the near dependence of the basis functions. For example, the $(8, 8, 8)$ quadratic forms for collocation at the single point, $x_0 = 0$, to the functions $\log(1 + x)$ and $\exp(x)$ have matrices with singular values 10^{-18} and 10^{-25} respectively. The $(4, 4, 4)$ quadratic forms for collocation at the distinct point set $\{x_i = 0(0.1)1.3\}$, to the same functions have matrices with singular values of order 10^{-16}.

We investigate a Newton formulation to this problem, in an analogous form to the way the Newton basis acts in the special case of polynomial approximation.

In order to motivate the argument, consider the polynomial case. The $\underset{\sim}{n}$ polynomial form is given by

$$(a_0(x) + f(x)) = O(w_N(x)), \qquad (4)$$

where the symbols are as defined for (2), and $N + 1 = n_0 + 1 + 0 + 1$.

By applying the sequence of linear functionals, L_i, $i = 0(1)N - 1$, to equation (4), and adding the equation for the $\underset{\sim}{n}$ polynomial function, $Q(x)$,

$$a_0(x) + Q(x) = 0, \qquad (5)$$

we obtain a homogeneous system of $N + 1$ linear equations in the $N + 1$ unknown coefficients. As noted in the previous section, this system has a nontrivial solution which is expressed by the eliminant of these equations, $\det B = 0$, where the matrix B is given by:

$$B = \begin{bmatrix} 1 & x_0 & \cdots & x_0^{n_0} & f_0 \\ 1 & x_1 & & x_1^{n_0} & f_1 \\ \cdots & \cdots & & & \cdots \\ 1 & x_{n_0} & \cdots & x_{n_0}^{n_0} & f_{n_0} \\ 1 & x & \cdots & x^{n_0} & Q(x) \end{bmatrix} \qquad \text{where } f_k = f(x_k), \quad k = 0(1)n_0.$$

If we apply column operations to this matrix to reduce the matrix to lower triangular form, in an exactly analogous way in which the row operations of Gaussian Elimination would reduce the matrix to an upper triangular form, we obtain an equivalent matrix whose last row is

$$[w_0(x), w_1(x), \ldots, w_{N-1}(x), Q(x) - p_{n_0}(x)]$$

where $w_0(x) = 1, w_j(x) = \prod\limits_{k=0}^{j-1}(x - x_k), j = 1(1)N - 1$ are the Newton basis functions,

$p_{n_0}(x) = \sum\limits_{j=0}^{n_0} f_{01\ldots j} w_j(x)$ is the collocating polynomial expressed in the Newton form and $f_{01\ldots j}$ denotes the jth divided difference of f on the collocating or node point set A_{N-1}.

By carrying out this process, we have in effect changed the basis of the polynomial function in (5) from $\sum\limits_{j=0}^{n_0} a_{0j} x^j + Q(x) = 0$ to $\sum\limits_{j=0}^{n_0} b_j w_j(x) + [Q(x) - p_{n_0}(x)] = 0$.

The solution is now trivial and $\det B = 0$ becomes $\sum\limits_{\substack{i,j=0 \\ i>j}}^{n_0}(x_i - x_j)[Q(x) - p_{n_0}(x)] = 0$.

This gives the desired solution since the set A_{N-1} was assumed to consist of distinct points.

Consider now the $\underset{\sim}{n}$ quadratic function, where, to avoid complications we assume the form is *normal*. Since the $n_0 = n_1 = n_2$ type of quadratic function is the type most commonly sought, this case will be considered, with $n_i = n$. The $\underset{\sim}{n}$ quadratic form is given as in (2), by

$$P(f, x) = a_0(x) + a_1(x)f(x) + a_2(x)f(x)^2 = O(w_N(x)). \tag{6}$$

Applying the sequence of linear functionals, $L_i, i = 0(1)N - 1$, to equation (6), and adding the equation for the $\underset{\sim}{n}$ quadratic function, $Q(x)$,

$$P(Q, x) = a_0(x) + a_1(x)Q(x) + a_2(x)Q(x)^2 = 0, \tag{7}$$

we obtain a homogeneous system of $N + 1$ linear equations in the $N + 1$ unknown coefficients. As before, this system has a non-trivial solution which is expressed by the eliminant of these equations, $\det B = 0$, where B is the $N \times (N + 1)$ matrix, F, augmented by the $(N + 1)$st row

$$[1, x, \ldots x^n, Q(x), xQ(x), \ldots, x^n Q(x), Q(x)^2, xQ(x)^2, \ldots, x^n Q(x)^2].$$

619

The same column operations are applied as before in order to effect a change of basis. The details are tedious but routine algebra. What is interesting is the pattern that emerges in the new basis functions so produced. Since the collocation nodes, A_{N-1}, are all distinct, and the transformed matrix is lower triangular, the solution to $\det B = 0$ is, as before, $\varphi_N(x) = 0$ where $\varphi_N(x)$ is the final basis element. The basis elements satisfy the relations:

$$\varphi_0(x) = 1,$$
$$\varphi_{k+1}(x) = \varphi_k(x)(x - x_k), \qquad k = 0(1)n - 1, \quad \text{and}$$
$$\varphi_{k+n+1}(x) = \varphi_k(x)Q(x) - \sum_{j=0}^{n} R_{jk}\varphi_{k+j}(x), \qquad k = 0(1)2n + 1.$$

The constants R_{jk} are obtained from the divided difference table by the following algorithm:

Coefficient Algorithm:

Step 1: Form the first $n + 1$ columns of the divided difference table in such a way that the first n rows are formed in the standard way by differencing from their neighbouring elements, but the remaining rows are differenced from the nth row. Thus the difference table becomes:

$$
\begin{array}{lllll}
f_0 & & & \\
f_1 & f_{01} & & \\
\cdot & \cdot & & \\
f_n & f_{n-1,n} & \cdots & f_{01\ldots n} & \\
f_{n+1} & f_{n,n+1} & \cdots & f_{12\ldots n,n+1} & f_{01\ldots n,n+1} \\
f_{n+2} & f_{n,n+2} & \cdots & f_{12\ldots n,n+2} & f_{01\ldots n,n+2} \\
\cdot & \cdot & & \cdot & \cdot \\
f_{N-1} & f_{n,N-1} & \cdots & f_{12\ldots n,N-1} & f_{01\ldots n,N-1}
\end{array}
\qquad (8)
$$

Step 2:
For $j = n + 1(1)N - 1$ do
   For $k = 1(1)n$ do
      $F_{k,k+1,\ldots n,j} \leftarrow f_{k,k+1\ldots n,j}/f_{01\ldots n,j}$

Step 3:
For $j = n + 1(1)N - 2$ do
   For $k = j + 1(1)N - 1$ do
      $K \leftarrow F_{j-n,j-n+1,\ldots j-1,k} - F_{j-n,j-n+1,\ldots,j-1,j}$
      For $i = 1(1)n - 1$ do
         $F_{j-n+i,\ldots,j-1,j,k} \leftarrow (F_{j-n+i,\ldots,j-1,k} - F_{j-n+i,\ldots,j-1,j})/K$
      $F_{jk} \leftarrow (f_k - f_j)/K.$

This algorithm produces the table:

f_0

f_1 f_{01}

.   .

f_n $f_{n-1,n}$ $\cdots$ $f_{01\ldots n}$

f_{n+1} $F_{n,n+1}$ $\cdots$ $F_{12\ldots,n+1}$

f_{n+2} $F_{n+1,n+2}$ $\cdots$ $F_{23\ldots,n+2}$

.   .

$$f_{N-1} \quad F_{N-2,N-1} \quad \cdots \quad F_{N-1-n,\ldots,N-1} \tag{9}$$

The constants Rjk are the elements along the kth diagonal of this table. If the successive values of $\varphi_k(x)$ are substituted into the equation $\varphi_N(x) = 0$, and the result rearranged in the form (7), the leading coefficient of $a_2(x)$ is normalized to 1.

<u>Example.</u> Consider the $(1,1,1)$ quadratic approximation to $f(x) = e^x$ with the collocation nodes $A_4 = \{0,1,2,3,4\}$. Since $\underset{\sim}{n} = (1,1,1)$, only the first two columns of the divided difference table are needed and the tables (8) and (9) become

1			and	1		
e	$e-1$			e	$e-1$	
e^2	e^2-e	$(e^2-2e+1)/2$		e^2	$2e/(e-1)$	
e^3	$(e^3-e)/2$	$(e^3-3e+2)/6$		e^3	$e(e-1)(e+2)$	
e^4	$(e^4-e)/3$	$(e^4-4e+3)/12$		e^4	$2e^2/(e^2-1)$	

Hence the result is $\varphi_5(x) = 0$ where

$\varphi_0(x) = 1$
$\varphi_1(x) = \varphi_0(x)(x-0) = x$
$\varphi_2(x) = \varphi_0(x)Q(x) - 1\varphi_0(x) - (e-1)\varphi_1(x)$
$\varphi_3(x) = \varphi_1(x)Q(x) - e\varphi_1(x) - 2e/(e-1)\varphi_2(x)$
$\varphi_4(x) = \varphi_2(x)Q(x) - e^2\varphi_2(x) - e(e-1)(e+2)\varphi_3(x)$
$\varphi_5(x) = \varphi_3(x)Q(x) - e^3\varphi_3(x) - 2e^2/(e^2-1)\varphi_4(x)$

Making the substitutions in $\varphi_5(x) = 0$, and collecting terms, we may obtain (after rationalising terms):

$[2e^5 + 4e^4 + e^4(e^2-1)x] + [-2e^5 - 4e^4 + 4e^2 + 2e + (e^5 + 2e^4 - 2e^2 - e)x]Q(x) +$
$[-4e^2 - 2e + (e^2-1)x]Q(x)^2 = 0,$

which is the form that may be obtained from the direct solution of $\det B = 0$.

<u>Theorem 6.</u> For $k > 0$, the functions $\varphi_k(x)$ vanish for $x = x_j, \quad j = 0(1)k-1$.

Proof: This property of the Newton basis functions in polynomial approximation is shared by these new basis functions. It is obviously true for the basis functions $\varphi_k(x), k = 1(1)n$, since this is just the usual Newton polynomial basis. The proof is by induction. For $\varphi_j(x)$ with $j = k + n + 1$, this function vanishes at $x = x_j, j = 0(1)k-1$ by the induction hypothesis. If we consider the column elimination process referred to above, the constants R_{jk} are essentially the multipliers used to eliminate the elements in the $(k + n + 1)$th column of the matrix, compounded together. That is, since $\varphi_k(x)$ vanishes at $x = x_j, j = 0(1)k - 1$, choose R_{0k} so that $\varphi_k(x)Q(x) - R_{0k}\varphi_k(x)$ vanishes also at $x = x_k$. And in general, since $\varphi_{k+i}(x)$ and $\varphi_k(x)Q(x) - \sum_{j=0}^{i} R_{jk}\varphi_{k+j}(x)$ both vanish for $x = x_j, j = 0(1)k + i - 1$, choose R_{jk} so that the latter expression also vanishes for $x = x_{k+i}$. Since this holds for $i = 0(1)n$, we have constructed $\varphi_{k+n+1}(x)$ which vanishes for $x = x_j, j = 0(1)k+n$, as required.

<u>Theorem 7.</u> The functions $\varphi_k(x)$, $k = 0(1)N - 1$, are linearly independent.

<u>Proof:</u> Let $\sum_{k=0}^{N-1} b_k\varphi_k(x) = 0$. Setting $x = x_k, k = 0(1)N - 1$, in turn and using the result of the previous theorem we obtain $b_k = 0, k = 0(1)N - 1$ since the collocation nodes are distinct.

The interesting feature about these results is that we have obtained a set of basis functions that have similar properties to the Newton polynomial basis functions, including a simple recursive definition. Further, the constants in the recursion relationship are relatively easily derived from the familiar divided difference table.

However, several features deserve further examination. Additional investigations are being carried out to determine if similar patterns can be deduced for the general n quadratic approximation. If the collocation node points are all distinct and the function values $\tilde{f}_k, k = n+1(1)N-1$ are also distinct, the algorithm for the recursion coefficients appears to work satisfactorily. However if at least two of these function values are equal the algorithm will fail and this situation also needs to be addressed. The numerical behaviour of this formulation is also being investigated.

REFERENCES

1. Bajaj, C. L., and Kim, M., Generation of Configuration Space Obstacles: The case of moving algebraic curves, *Algorithmica*, vol. 4, pp. 157-172, 1989.

2. Baker, G. A., and Gammel, J. K., The Padé Approximant, *J. Math. Anal. Appl.*, vol. 2, pp. 21-30, 1961.

3. Baker, G. A., and Graves–Morris, P. R., *Padé Approximants, Parts I and II.*, Encyclopedia of Mathematics and Its Applications, vols. 13 and 14, Addison-Wesley, Reading, MA, 1981.

4. Baker, G. A., and Lubinsky, D. S., Convergence theorems for rows of differential and algebraic Hermite–Padé approximants, *J. Comput. Appl. Math.*, vol. 18, pp. 29-52, 1987.

5. Hoffman, C. M., Algebraic Curves, in *Mathematical Aspects of Scientific Software*, ed. J. R. Rice, Springer Verlag, New York, 1988.

6. McInnes, A. W., Collocation with quadratic approximation, *Math. Chron.*, vol. 13, p. 97, 1984.

7. McInnes, A. W., The existence of approximations by algebraic functions, Technical Report 96, Center for Applied Mathematics, Purdue University, 1989.

8. Short, L., The evaluation of Feynman integrals in the physical region using multi–valued approximants, *J. Phys. G.*, vol. 5, pp. 167-198, 1979.

The Propagation of Rounding Errors in Pivoting Techniques for Toeplitz Matrix Solvers

D. R. SWEET and S. C. BUI
Department of Computer Science
James Cook University of North Queensland
Townsville Q. 4811, Australia

INTRODUCTION

A *Toeplitz matrix* is one whose entries are constant along each diagonal. Toeplitz matrices have many applications in mathematics and engineering - major areas include signal processing, control theory and Padé approximation. There are several "classical" algorithms which solve Toeplitz systems in $O(n^2)$ operations, such as those of Trench(1964) and Bareiss(1969).

It has been shown by Cybenko(1980) that these algorithms are stable in the positive-definite case. There are some applications where the matrix is indefinite, e.g. computation of Padé approximants and the design of eigenfilters (Makhoul(1981)). Sweet (1982) showed that in this case, the classical algorithms can suffer a serious loss of accuracy when one or more leading principal submatrices are ill-conditioned.

This problem can be addressed by incorporating some form of pivoting into the solvers. A pivoting technique based on the algorithm of Bareiss(1969) was proposed by Sweet(1982), and experimental results show a marked improvement over the original algorithm.

It is shown in this paper that there are some cases where the method does not terminate normally; a method of *augmenting* the matrix is proposed which overcomes this problem. An error analysis for the pivoted Bareiss algorithm is then given. It predicts a much smaller error bound than for the Bareiss algorithm and this conclusion is supported by experimental results from a variety of matrices.

THE BAREISS ALGORITHM - ORIGINAL AND PIVOTED VERSIONS

We first present the Bareiss algorithm, and recall a result which shows that ill-conditioned leading principal (LP) submatrices cause small pivots which lead to a loss of accuracy. We then show how alternative pivots can be selected to counter this loss of accuracy.

Bareiss's algorithm

Bareiss's algorithm solves the Toeplitz system

$$T\mathbf{x} = \mathbf{b} \tag{1}$$

by transforming it successively into the pairs of systems $T^{(-0)}x=b^{(-0)}$, $T^{(+0)}x=b^{(+0)}$, ..., $T^{(1-n)}x=b^{(1-n)}$, $T^{(n-1)}x=b^{(n-1)}$, where $T^{(-j)}$ has zeroes along the j sub-diagonals, and $T^{(j)}$ has zeroes along the j super-diagonals, as illustrated in Fig.1. Here, areas bounded by solid lines are of Toeplitz form, areas bounded by dotted lines are of general form, and other areas contain zeroes.

In one cycle of the algorithm, a/c times rows 1 to $n-j-1$ of $T^{(j)}$ are subtracted from rows $j+2$ to n of $T^{(-j)}$, thus eliminating diagonal "a" in the new iterate $T^{(-j-1)}$; then, d/b times rows $j+2$ to n of $T^{(-j-1)}$ are subtracted from rows 1 to $n-j-1$ of $T^{(j)}$, eliminating diagonal "d". Only $O(n)$ distinct operations are required in each cycle, and $O(n^2)$ overall.

Occurrence of small pivots in Bareiss algorithm

It can be shown (Sweet(1982)) that if one or more adjacent LP submatrices are ill-conditioned, then several diagonals adjacent to the diagonal zero-band, including diagonal "b" (in Fig.1) will be small in some *negative-index* iterate, say $T^{(-j)}$. This will cause a large multiplier and (as will be shown later) a loss of accuracy.

Incorporation of pivoting into the Bareiss algorithm

At each major step of the Bareiss algorithm, there are *different* choices which also allow diagonals to be eliminated. Thus at the first half of a Bareiss step, we could use diagonal "d" (in Fig.1) instead of diagonal "c" as pivot; and at the second half of a Bareiss step, we could use diagonal "a" instead of diagonal "b". There is also a choice of whether to eliminate in iterate $(-j)$ first (as in done in the Bareiss algorithm) or to eliminate in iterate (j) first. The basic idea in pivoting is to use these choices judiciously to avoid small pivots.

In the general case, both diagonals "a" and "b", and several adjacent ones can be small. The essence of the Pivoted Bareiss Algorithm (PBA) is to first "move" the zero-band until a large pivot adjoins it, then use this large pivot. The zero-band can be moved p places up (or down) as follows: *Backtrack* by p Bareiss cycles, then perform p steps, selecting the pivots in such a way as to move the zero-band up (or down) at each cycle.

The Backtrack Procedure. Suppose we wish to move the zero-band of Fig.1 *up* by p places so that the diagonal "f" is accessible for use as pivot. We recover Bareiss iterates $(p-j)$ and $(j-p)$ and then perform the following steps p times:

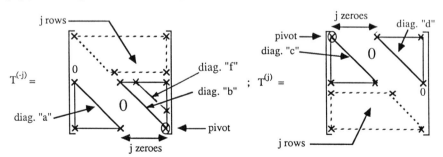

FIGURE 1. Form of Bareiss iterates after cycle j.

624

A-cycle:
1. With diagonal "d" as pivot, eliminate diagonal "b"
2. With the new diagonal "b" as pivot, eliminate diagonal "d"
It can readily be verified that each A-cycle produces an iterate with the zero-band moved up one place from its previous position.

The zero-band can be moved down as follows:
C-cycle:
1. With diagonal "a" as pivot, eliminate diagonal "c"
2. With the new diagonal "c" as pivot, eliminate diagonal "a"

Elimination of Small Diagonals. After the backtrack procedure, the small diagonals all appear on one side of the zero-band. They can be eliminated using normal Bareiss cycles.

Restoration of Bareiss Form (RBF). The zero-band should be moved back to its "normal" position, in which the Toeplitz parts have the shapes illustrated in Fig.1. If the zero-band is displaced *upwards* by p places, it should be moved back to its normal position by p C-cycles. If the band is displaced *downward* p places, it should be moved back up using p A-cycles.

NON-TERMINATION OF PBA : THE AUGMENTATION PROCEDURE

There are certain pivot paths in which PBA cannot terminate normally. This problem can be overcome by *augmenting* the Toeplitz matrix with r rows and columns and extending the right-hand-side such that the first n components of the solution of the augmented system are the same as those of the original system.

Non-termination of PBA.

For PBA to terminate correctly, the Toeplitz part of the iterates must be in Bareiss form. There are some pivot paths in which *restoration of Bareiss form* is not complete at cycle $n-1$, and the zero-band "displacement" (ZBD) is not zero, as illustrated in Fig.2.

Condition for non-zero final ZBD. Let k be cycle the cycle number before the restoration is carried out, and let d_z be the zero-band displacement at this stage. Then the condition for non-termination is

$$k+|d_z| > n-1 \qquad (2)$$

Non-termination can be predicted before backtracking. Suppose a backtrack is needed at step $(-j)$. Let there be p small diagonals above the zero-band and q below. Let d be the number of cycles to backtrack. It can be shown using (2) that PBA will not terminate if

$T^{(1-n)}$ (Toeplitz Part) $T^{(n-1)}$ (Toeplitz Part)

(a)   0 ——————— 0 X   X 0 ——————— 0

(b)   XXX 0——————— 0   XXXX 0———————0

FIGURE 2. Toeplitz parts of $T^{(1-n)}$ and $T^{(n-1)}$ (a) with ZBD=0   (b) with ZBD=3

$$j+p+q+d > n-1 \tag{3}$$

The Augmentation Procedure.

If (3) is satisfied, PBA does not terminate because T is not "large enough" for RBF to finish. If T is augmented by the addition of r rows and columns so that the new matrix is still Toeplitz, then from (3) RBF will terminate provided

$$r = j+p+q+d-n+1$$

The right-hand-side $\mathbf{b}$ of (1) must also be augmented with r entries. Denote the augmented system as

$$\begin{bmatrix} T & U \\ V & W \end{bmatrix} \begin{bmatrix} \mathbf{x}_1 \\ \mathbf{x}_2 \end{bmatrix} = \begin{bmatrix} \mathbf{b} \\ \mathbf{c} \end{bmatrix} \tag{4}$$

If $\mathbf{c}$ is chosen such that $\mathbf{x}_2 = 0$, then from Eq.(4), $\mathbf{x}_1 = \mathbf{x}$.

Computation of $\mathbf{c}$ in the augmented system.

It can be verified (Sweet(1982)) that *column-permuted* upper and lower triangles can be extracted from the PBA iterates. The purpose of augmentation is to allow RBF to finish, so PBA must end in $|d_z|$ A or C-cycles. Consider the former case. It turns out from the triangle extraction procedure that the last $|d_z|$ rows are upper-triangular. It can be shown that if T is well-conditioned, then $|d_z| \geq r$, and the last r PBA-cycles are all A-cycles.

If $\mathbf{c}$ were selected such that $\mathbf{x}_2$ is zero, then the modified right-hand-side $\mathbf{b}'$ corresponding to the permuted upper-triangle, say U', must have zero in its last r components. Since the last r cycles are A-cycles, it can easily be checked from the extraction procedure that

$$u_i{}' = t_L^{(1-i)} \, , \, i = n+1, \ldots, n+r \tag{5}$$

where $i \cdot$ denotes row i and L denotes the last row of the *Toeplitz* part of the matrix.

Let the *multiplier matrices* $M^{(\pm j)}$ be such that $M^{(-j)}T = T^{(-j)}$ and $M^{(j)}T = T^{(j)}$. The $M^{(-j)}$ and $M^{(j)}$ are computed by setting $M^{(0)} = I$, and applying the same row operations as were applied in the corresponding steps of PBA. It is easily checked that $M^{(-j)}$ and $M^{(j)}$ have j Toeplitz rows, each with only $j+1$ non-zero diagonals.

Define M' by $M'T = U'$. Then

$$M' \begin{bmatrix} \mathbf{b} \\ \mathbf{c} \end{bmatrix} = \mathbf{b'} \tag{6}$$

Since the Toeplitz parts of $M^{(-j)}$ and $M^{(j)}$ correspond to the Toeplitz parts of $T^{(-j)}$ and $T^{(j)}$, equations (5) and (6) imply that

$$m_i{}' = m_L^{(1-i)} , \, i = n+1, \ldots, n+r \tag{7}$$

From (6) and (7), we obtain

$$m_{L,1:n}^{(1-i)} \mathbf{b} + m_{L,n+1:n+r}^{(1-i)} \mathbf{c} = 0 , \, i = n+1, \ldots, n+r \tag{8}$$

where $L,p:q$ denotes elements p to q of the last row of the Toeplitz part. The first term in each of equations (8) is known, so we need to invert an $r \times r$ system to obtain $\mathbf{c}$. This still requires $O(r^3)$ operations. An $O(nr)$ algorithm to compute c is developed below.

$O(nr)$ solution of eq.(8). Consider the case where r is even. The case where r is odd can be treated by a simple modification. For A-cycles, the following can be shown

$$\mathbf{m}_{;i}^{(-j)} = \mathbf{m}_{;i}^{(1-j)} - m_{-j}\mathbf{m}_{;i-1}^{(j-1)}, \quad \mathbf{m}_{;i}^{(j)} = \mathbf{m}_{;i}^{(j-1)} - m_j\mathbf{m}_{;i}^{(-j)} \tag{9a,b}$$

where $;i$ denotes the i-th row of the Toeplitz part, and m_{-i} and m_i are the *multipliers* used in cycle i (a/c and d/b for a B-cycle). Define $\mathbf{b}_{;i}^{(\pm j)} = \mathbf{m}_{;i}^{(\pm j)}\tilde{\mathbf{b}}$ where $\tilde{\mathbf{b}} = [\mathbf{b}^T, \mathbf{c}^T]^T$. Now from eq.(8), $\mathbf{b}_{;n+r-i}^{(-i)} = 0$, $i = n, \cdots, n+r-1$. Using this fact and eqs(9), we can show by downward induction on the cycle number that

$$\mathbf{b}_{;i}^{(n-1+r/2)} = \mathbf{m}_{;i}^{(n-1+r/2)}\tilde{\mathbf{b}} = 0, \quad i = 1, \ldots, r/2 \tag{10}$$

$$\mathbf{b}_{;i}^{(1-n-r/2)} = \mathbf{m}_{;i}^{(1-n-r/2)}\tilde{\mathbf{b}} = 0, \quad i = 2, \ldots, 1+r/2 \tag{11}$$

Eqs.(10) and (11) constitute two Toeplitz sets of equations of the form given in Fig.3. By repeatedly using diagonal "f" to eliminate diagonal "g", then the new diagonal "g" to eliminate diagonal "f", this system can be reduced in $O(nr)$ operations to two matrices which can be interlaced to form an r-row set with zeroes above the diagonal passing through the bottom-right corner. Only the last r variables (i.e. $\mathbf{c}$) in this set are unknown, and can be solved from this reduced set. An analogous method can be applied when the reduction finishes with C-cycles.

Results. The modified algorithm was run on a matrix with $n=6$ and the order 3, 4 and 5 LP submatrices ill-conditioned. The modified PBA augmented the matrix to order 7 and gave a satisfactory error performance. The original PBA did not terminate.

ROUNDING ERROR ANALYSIS OF THE PIVOTED BAREISS ALGORITHM

In this section, we show that the *computed* value of $\mathbf{x}$ is the *exact* solution of a system where T and $\mathbf{b}$ are slightly perturbed. The approach is similar to that of the analysis of the Bareiss algorithm (Sweet (1982)). However, some generalization of the analysis is required to cater for the pivoting. We first give an outline of Sweet's method for extracting column-permuted upper and lower triangles

Procedure 1 (extraction of permuted upper and lower triangles from $T^{(\pm i)}$).
For $i \leftarrow 0$ to $n-1$ do
    Determine row nos. $r(\pm i)$ and cycle nos. $k(\pm i)$ according to type of cycle i (A,B or C)
    Extract $\mathbf{u}_{i+1}$. from row $r(-i)$ of $T^{k(-i)}$
    Extract $\mathbf{l}_{n-i}$. from row $r(i)$ of $T^{k(i)}$; divide $\mathbf{l}_{i}$. by factor s_{+i} which depends only on the $m_{\pm j}$

L and U then satisfy the following:

$$T = L^S D^{-S} U \tag{12}$$

where ^{S} denotes transpose about the secondary diagonal, and D is a diagonal matrix of the permuted diagonal elements of L.

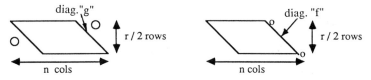

FIGURE 3. Form of co-efficients of equations (10) and (11).

The Error Analysis. The results are embodied in Theorems 1 and 2. Theorem 1 gives the perturbations in terms of difference matrices F^+ and F^-, and Theorem 2 gives recursive expressions for $F^\pm$ in terms of the multipliers $m_{\pm i}$ used during the reduction.

Theorem 1.
Let $\mathbf{x}^\pm$ be the exact solutions of the two systems

$$T^\pm \mathbf{x}^\pm = \mathbf{b}^\pm \tag{13}$$

where T^+ and T^- are the *computed* values of L and U as given in Procedure 1 and $\mathbf{b}^\pm$ are the corresponding right-hand-sides. Then $\mathbf{x}^\pm$ are the exact solutions of the following perturbed systems

$$(T + (L^S D^{-S} + \Delta_L)F^-)\mathbf{x}^- \ = \ \mathbf{b}^- + (L^S D^{-S} + \Delta_L)\mathbf{g}^- \tag{14}$$

$$(T + (U^S + \Delta_U)F^+)\mathbf{x}^+ \ = \ \mathbf{b}^+ + (U^S + \Delta_U)\mathbf{g}^+ \tag{15}$$

where Δ_L and Δ_U are $0(\varepsilon)$ (ε is machine precision) and $F^\pm$ and $\mathbf{g}^\pm$ are defined by

$$F^\pm = T^\pm - \hat{T}^\pm \quad \text{and} \quad \mathbf{g}^\pm = \mathbf{b}^\pm - \hat{\mathbf{b}}^\pm \tag{16a,b}$$

where the *hatted* quantities are the result of applying PBA without rounding error, but with the $m_{\pm i}$ replaced by their computed values.

Proof. Define $\hat{M}_\pm$ by $\hat{M}_\pm T = \hat{T}^\pm$. Substituting (16a,b) into (13) we get

$$(T + (\hat{M}_\pm)^{-1} F^\pm)\mathbf{x}^\pm = \hat{\mathbf{b}}^\pm + (\hat{M}_\pm)^{-1}\mathbf{g}^\pm \tag{17}$$

In *exact* arithmetic, (12) shows that $(\hat{M}_-)^{-1} = L^S D^{-S}$. Define $\Delta_L = L^S D^{-S} - (\hat{M}_-)^{-1}$, and substitute for $\hat{M}_{-1}$ in (17), yielding (14). Clearly $\Delta_L = O(\varepsilon)$. The proof for (15) is similar. QED.

We must now estimate $F^\pm$ and $\mathbf{g}^\pm$. From Procedure 1 and the definition of $T^\pm$ and $F^\pm$ we have

$$\mathbf{f}^\pm_{h(\pm i)\cdot} = \mathbf{t}^\pm_{h(\pm i)\cdot} - \hat{\mathbf{t}}^\pm_{h(\pm i)\cdot} \qquad , i = 0, \ldots, n-1$$

$$= (\mathbf{t}^{k(\pm i)}_{r(\pm i)\cdot} - \hat{\mathbf{t}}^{k(\pm i)}_{r(\pm i)\cdot}) / s_{\pm i} \equiv \mathbf{f}^{k(\pm i)}_{r(\pm i)\cdot} / s_{\pm i} \tag{18}$$

where $h(-i) = i+1$, $h(+i) = n-i$, $s_{-i} = 1$ and the s_{+i} are as in Procedure 1. We now wish to estimate $F^{\pm j}$. We first note that each step in PBA can be written in the following form

$$T^{(\pm i)} = T^{\pm(i-1)} - m_{\pm i} J_{\pm i} T^{\mp(i-v)} \tag{19}$$

where $J_{\pm i}$ is a matrix that *selects* the appropriate rows of $T^{\mp(i-v)}$ and shifts them for subtraction, and v is 0 or 1, depending on sign of the iteration number and cycle type. We next assert the following:

Theorem 2.
Let $E^{(\pm p)}$ be the *local* error incurred in computing $T^{(\pm p)}$. Let $N^{(\pm j)}_{\pm p}$ be the $(\pm p, \pm j)$ *error propagation* matrices, defined by the following recursion:
Procedure 2.

$$N^{(\pm p)}_{\pm p} \leftarrow I$$

For A and B cycles, $N^{(-p)}_p \leftarrow 0; \quad N^{(p)}_{-p} \leftarrow -m_p J_p$

For C cycles, $N^{(p)}_{-p} \leftarrow 0; \quad N^{(-p)}_p \leftarrow -m_{-p} J_{-p}$

$$N^{(\pm j)}_{\pm p} \leftarrow N^{\pm(j-1)}_{\pm p} - m_{\pm j} J_{\pm j} N^{\mp(j-v)}_{\pm p}, \quad j = p+1, \ldots, n-1 \tag{20}$$

where in (20), the signs of p and j are taken independently; *then*

$$F^{(\pm j)} = \sum_{p=1}^{j} (N_p^{(\pm j)}E^{(p)} + N_{-p}^{(\pm j)}E^{(-p)}), \quad j = 1, \ldots, n-1. \tag{21}$$

Proof Outline. The local error incurred in step $\pm p$ is defined by

$$T^{(\pm p)} = T^{\pm(p-1)} - m_{\pm p}J_{\pm p}T^{\mp(p-v)} + E^{(\pm p)} \tag{22}$$

$N_{\pm p}^{(\pm j)}$ is the "factor" by which the error $E^{(\pm p)}$ "propagates" through to a later step $\pm j$. Recalling that $T^{(\pm j)} = \hat{T}^{(\pm j)} + F^{(\pm j)}$, we substitute this in (22), substituting in (21) for the $F^{(\pm j)}$, and equate the co-efficients of $E^{(\pm j)}$, giving Procedure 2.   QED.

Estimation of $N_{\pm p}^{(\pm j)}E^{(\pm p)}$ and $F^{\pm}$. From (22), a straightforward error analysis yields

$$|(N_{\pm p}^{(\pm j)}E_{\pm p})_{kl}| \le 2\varepsilon n^2(\mu_{\pm p} + \mu_{\pm(p-1)})v_{\pm p}^{(\pm j)}t_{max} \tag{23}$$

where $\mu_{\pm p}$, $v_{\pm p}^{(\pm j)}$ and t_{max} are the largest magnitudes in $M^{(\pm p)}$, $N_{\pm p}^{(\pm j)}$, and T respectively. The first two quantities are functions only of the multipliers $m_{\pm j}$, and can be bounded in terms of these; substituting these bounds into (23) and using (21) we get

$$|f_{kl}^{\pm}| \le 8\varepsilon n^3 t_{max} \prod_{i=1-n}^{n-1} (1+m_i)/s_{\pm k} \tag{24}$$

This is a pessimistic bound as it assumes that the multipliers combine in the worst possible way. A similar analysis shows that for the first column of $F^{\pm}$, the product in (24) can be replaced by $P = \prod(1+m_{-i}m_i)$ and usually the bounds for the other elements are less than this. Assuming this to be so, we can say from (24) and (14) that the size of the perturbation matrix is approximately proportional to the product of the multipliers. Since PBA attempts to keep the multipliers small by avoiding small pivots, it would be expected that the error bounds would decrease correspondingly.

EXPERIMENTAL RESULTS

The Bareiss algorithm and various versions of PBA were run on a variety of matrices of orders up to 250. The exact solution in every case was $x = [1,1,\ldots,1]^T$. Measures of expected error used were the product $P = \prod(1+m_{-i}m_i)$ and the "growth factor", the ratio of $max_{i,j,k} |t_{ij}^{(\pm k)}|$ to t_{max}, used in Gauss elimination. Another error measure was the normalized residual in units of machine epsilon ε (the "instability score"). Even for badly-conditioned matrices, a stable algorithm should give a low instability score.

For small well-conditioned matrices (order 6 or 7) when one LP submatrix (Order 4) was ill-conditioned the Bareiss algorithm (BA) gave errors in the solution ranging from 10^{-8} to above 100%. ε was 6×10^{-17}. PBA gave errors of ε to 60ε. Errors were generally proportional to the P-factors. For a somewhat badly-conditioned order-6 matrix, the error in BA was about 10^{-8} while the error in PBA increased proportionally to the condition number, but the instability score was only about 11. One random order-7 matrix caused somewhat poorer results in the original PBA, but a more complex pivoting strategy, which involved a "look-ahead" of multiplier sizes, gave an error of 10ε.

Increasing with matrix size to 13 to 15 caused BA to perform even worse (errors > 100%), and PBA results were slightly degraded (error ~ 200ε), but still good. Increasing the *number* of ill-conditioned submatrices caused a some deterioration in PBA performance (error ~ 5×10^{-13}) when the simple pivot strategy was used, but it was still much better than that of BA. The P-factor was a reasonable rough estimator of error performance, but sometimes it gave overestimates.

Some special matrices were tried, including some ill-conditioned banded ones. These gave very inaccurate results for BA, and rather inaccurate ones for PBA. Using the "look-ahead" technique yielded a PBA instability score of 0.9, however.

For very large, dense, random matrices with some ill-conditioned LP submatrices, BA gave errors over 100%, and PBA's performance was somewhat degraded, but it was still considerably better than for BA. Generally errors were of the order of ε times condition number. Much better results were obtained with the "look-ahead" technique, but an efficient method for "searching" the multipliers has not yet been developed. Further work is required in this area.

CONCLUSIONS

The pivoted Bareiss algorithm has been modified so that it terminates normally on all inputs. An error analysis shows that the expected error is proportional to the multipliers encountered, and the results support this conclusion. Further work will be done to obtain tighter error bounds, possibly in terms of growth factors, and to develop an efficient "look-ahead" technique for optimizing the pivot path.

ACKNOWLEDGEMENT

This project is supported by an Australian Research Council grant.

REFERENCES

1. Bareiss, E. H., Numerical Solution of Linear Equations with Toeplitz and Vector Toeplitz Matrices, *Numer.Math,* vol.13, pp.404-424, 1969.

2. Cybenko, G., The Numerical Stability of the Levinson-Durbin Algorithm for Solving Toeplitz Systems of Equations, *SIAM J.Sci.Stat.Comput.,* vol.6, pp.303-310, 1980.

3. Makhoul, J. On the Eigenvectors of Symmetric Toeplitz Matrices, *IEEE Trans. Acoust., Speech, Signal Processing,* vol.ASSP-29, pp.868-872, 1981.

4. Sweet, D. R., Numerical Methods for Toeplitz Matrices, Ph.D thesis, University of Adelaide, 1982.

5. Trench, W. F., An Algorithm for the Inversion of Finite Toeplitz Matrices, *SIAM J. Appl. Math,* vol.12, pp.515-522, 1964.

OPTIMIZATION

The Integration of Forward Planning LP Models and Shop-Floor Operations

LATIF A. AL-HAKIM
Department of Mechanical and Industrial Engineering
Chrisholm Institute of Technology
Victoria, 3145, Australia

INTRODUCTION

The use of commercial computer packages for production scheduling has for many years been commonplace, as has the application of optimisation techniques such as linear programming, for allocation and budgetary problem investigation. The optimisation models are often used in a forward planning context while the heuristic scheduling algorithms are expected to transform these plans into shop floor reality. Clearly a match between these is unlikely, either through the inbuilt simplifications of the optimisation model or because of shop–floor scheduling restrictions, though the disparity may not in many instances be of a significant magnitude. In batch production organisations however production scheduling factors are often a major consideration and this paper examines, with the aid of a practical application, an iterative approach to tackling such problems.

THE ORGANISATION STUDIED

As an illustration the case of a medium size batch production organisation, based at two main production lines, will be considered. Each line can produce a subset of the company's product range of 60 products and orders are allocated to company from a control marketing organisation (production being linked solely to orders received). All products involve a similar sequence of operations though at certain stages different products require different specialist equipment to do similar jobs. As with many organisations the planning is carried out at two levels – a routine production schedule produced using a commercial package and a 'high level' planning model based upon a linear programme formulation, and it was the uncertain basis for the latter which led to this investigation. It was found that the long range planning model did not appear to allocate orders to company in a cost effective manner (due to non–linearities of the cost functions and the economies of large batch production). In addition the production quantities suggested for the company were often for small batch quantities and, as this was a considerable change–over time these sometimes proved impossible to schedule using any standard scheduling package. Hence there was a requirement to attempt to equate the forward planning optimisation model with the realities of shop floor production scheduling and the obvious economies, in terms of production costs of large batch manufacture.

The disparity between production planning and forward planning will come as little surprise to those with experience of such models. Nevertheless, given a requirement that these two planning methods should present closer solutions, a study of both planning methods was undertaken to highlight areas of conflict.

THE FORWARD PLANNING

The linear programming model to allocate future orders to the company was in use at the start of the study. Although only used partly for actual allocation of orders (there being considerable manual modification), it was in use extensively for facility development planning on the basis of a future anticipated order book. One clear limitation of its usefulness was the difficulty experienced in interpreting the results, especially by senior management. In addition, changes to constraints suggested by management had to be run as a batch job with a resultant slow turnround. As described later, one improvement introduced was the use of an interactive graphics routine which, as well as presenting output, allowed modification of constraints and the running of the resultant model on line.

An examination of the structure of the LP model revealed two major inconsistencies.

(1)   The cost information incorporated in the model was taken from the existing standard costing system which was itself based upon assumed batch sizes. Although limitations were included in the model to limit minimum batch sizes no allowance was (or could easily be) incorporated for the non–linearities in cost associated with various batch sizes.

(2)   In a similar fashion assumed production constraints (by company) did not take into account the often considerable setup times necessary, the LP allocating times on the basis of 'standard' batch sizes.

Both these assumptions might have been acceptable if either fixed batch size for each product has been used or production was on a one–off basis though, even in these cases, the assumption of sufficient available resources would remain.

THE SHOP–FLOOR SCHEDULING

In practice, as stated earlier, it was found necessary to modify by hand the LP allocations before passing orders to the company planning section for scheduling on the shop floor. When using the unmodified LP output the resultants had often been non–feasible and at best gave a gross inbalance between works with a wide gap between aggregate planning horizon and the master schedule planning horizon. Many managers who use the traditional EBQ formula in planning their production have often been disappointed with the results (Bishop 1979) whilst the popular approaches such as MRP have been found ineffective in shop–floor scheduling (Goldratt 1988). In may cases the batch sizes scheduled varied between works, being based primarily upon the experience of the production scheduler involved. Moreover, due to the reputation of JIT and Kanpan, the concepts of dividing batches into subbatches – taking into consideration the limited buffer spaces – and more setups to achieve smooth flows seems almost natural. Thus, not only did the LP cost information assume the 'standard cost batch' but also the actual scheduled batch differed in size from the 'standard' batch.

EXISTING WORK IN THE AREA

Numerous attempts have been made to link the aspects of high and low level decision making but these are generally concerned with the linking of two optimisation models.

In the production scheduling area early work was concerned with optimising schedules (Bowman 1956, 1963) which generally proved infeasible in all but the simplest cases. Various modifications to this approach have been suggested to take account of non–linear cost functions (Carter et.al. 1978) and various penalty functions and mixed heuristic methods have likewise been devised (Mellichamp and Love 1978, Zierer et.al. 1976). The work of Das (1979) has attempted to model a small size production scheduling problem (5 resources, 10 products) as an integer programme but despite the use of a data generator to convert from a network structure to that of an IP the effort and size of model for this scheduling problem would be very great. A recent dynamic approach (Villa and Arcostanzo 1988) employed iterative procedures but quoted the necessity of further work. Several papers have reformulated EBQ formulae (Das and Imo 1983, Mukhopadhyah et.al. 1988) in multi–product multi–stage situations and assessment on capacity utilization. None of these methods, however, tackle the twin problems of batch size determination and minimising the cost of manufacture though several tackle these separately. The approaches to forward planning are likewise only partly linked to production considerations with goal programming (Lee and Moore 1974), decomposition (Christensen and Obel 1978), and non–linear approaches (Lasdon 1970) being just three of many.

Case studies tend to be primarily concerned with the use of scheduling techniques, or with linear programming formulations for forward planning on a single or multi–period basis. Early LP models (Jones and Rope 1964, Smith 1965) tended to be rather small and over–simplified due, in part, to the nature of the available computing technology but later examples (Sanderson 1978, Uhlmann 1988) often present sophisticated multi–period models, which, given the assumed linear relationships do appear to work well. The work of Al–Hakim (1977) has linked LP and PERT in an iterative procedure to model a medium size production scheduling problem. This work is along similar lines to this paper but, although the interactive aspects are similar, the model structure was not well formulated. On a theoretical level the work of Ratcliff (1978) comes close to true optimisation but does require a convex and separable cost function and assumes equal length production periods rather than the order–book situation considered here.

LP FORMULATION

Consider a planning horizon T (see Appendix for the notation), the problem can mathematically be formulated as follows: Given the demand d_i ($i = 1,...,I$), work station capacities in time units C_j ($j = 1,...,J$) and material availabilities M_l ($l = 1,...,L$) in period T, determine x_i, x_{ijk} and y_i to minimize the total costs:

$$\text{Minimize} \sum_i \left[c_i x_i + \sum_k s_{ik} y_i \right] \tag{1}$$

Subject to

$$x_i \geq (d_i - e_i) \qquad\qquad (i = 1,...,I) \tag{2}$$

$$\sum_j \sum_k x_{ijk} = x_i \qquad\qquad (i = 1,...,I) \qquad (3)$$

$$y_i - \frac{x_i}{b_i} \geq 0 \qquad\qquad (i = 1,...,I) \qquad (4)$$

$$\sum_i \sum_k (t_{ijk}x_{ijk} + S_{ik}\, y_i) \leq C_j \qquad\qquad (j = 1,...,J) \qquad (5)$$

$$\sum_i \sum_k m_{ilk}\, x_{ilk} \leq M_l \qquad\qquad (l = 1,...,L) \qquad (6)$$

$$x_i,\ x_{ijk},\ y_i \geq 0$$

The problem is an LP with objective (1) being the total unit and setup costs. Constraints (2) are the demand constraints. Constraints (3) ensure all operations of product i are performed for all $x_i > 0$. Constraints (4) balance number of subbatches y_i, i.e. number of setups, for each batch x_i with (x_i/b_i). Limited workstation capacities are stated in constraints (5) while constraints (6) reflect the availability of materials. Both x_i and y_i are adjusted to integer values. The total setup costs equal $s_{ik}\, y_i$ and the total unit cost is $(c_i + s_{ik}\, y_i/x_i)$. The product mix of product i is given by $(x_i/\underset{i}{\Sigma}x_i)$. However, the formulation can easily be adapted to suit conventional batch production systems by neglecting constraints (3) and all parts of the problem which relate to s_{ik} or S_{ik}.

APPROACHES TO THE PROBLEM

The problems identified — those of batch size determination and the resultant allocation pattern are clearly ones of ideal sized batches within each workstation— thus optimising by works (considering resource utilization) rather than optimising the product—mix. It is clear that the LP model formulated here was intended to act in this manner.

Three basic approaches were suggested in an effort to tackle these problems:

(a)   The allocation of fixed batch sizes by works and by products thus enabling a more realistic (though still not necessarily scheduleable) LP allocation. Or;

(b)   The systematic addition of constraints to the LP model so as to make the resultant schedule feasible. This would require a systematic reduction in LP constraints and adjustment of cost factors but would lead to a sub—optimal, though feasible, solution. This could be conducted in an iterative fashion between LP and PERT. Or;

(c)   Modify only the cost information provided to the LP, this being done on the basis of generated batch sizes and number of setups. This iterative procedure would eventually lead to the input cost information matching that applicable for the output batch sizes. The result of this could then be fed to the PERT package in a multi—product form and using the cost parts of the package a sensitivity analysis carried out around the optimal LP solution.

Clearly none of these approaches would reach an optimal solution with respect to the overall company as distinct from individual production lines. Approach (a) would only optimise on the basis of individual works, approach (b) would tend to be sub—optimal though (c) although more lengthy would lead to a near—

optimal solution – the nearness being dependent upon the skill employed varying batch sizes. The only other approach would be to turn the problem into a true non–linear form with both non–linear constraints and objective function – a procedure which was considered far too complex. It was decided to follow approach (c), but in order to do so it was first decided to make the LP interactive and to provide graphical representation of the results. This was to aid in two ways – rapid modification and solution of the LP and provision of a readily understandable form of output for senior management.

THE INTERACTIVE MODEL

The LP model (480 variables, 240 constraints) was run using the IBM LP package and although initial runs took in excess of 350 cpu. seconds the use of a previous optimal solution as a basis for subsequent runs reduced this generally to around 30 seconds. The output file was read by a 'FORTRAN' program using graphics subroutines and various graphical outputs such as that shown in Figure 1 were produced. This represents (top to bottom) the flow of work for each production line through different processes. The volumes (in various units) are graphically represented using scaled box widths. Following the display of such graphs, modification of constraints and objective function was allowed with constraint values, slacks etc., being displayed for each work. The resultant modified LP could then be re–run and re–displayed, with a total cycle time of typically 40 cpu. seconds.

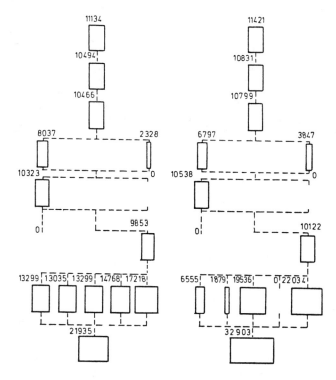

FIGURE 1. Optimised material flows in the two production lines. Numbers represent volume in various units at different manufacturing stages.

THE ITERATIVE APPROACH

Having developed a tool for presentation and rapid modification of results it was then necessary to develop the solution approach. Previous work on the cost model had established fixed and variable cost factors together with overheads for each product and each workstation. (This had been done as part of the development of an incentive payments scheme). Thus it was possible to rapidly determine unit costs and total unit costs, i.e. include setup costs, for each product, given batch sizes and number of setups. A program was developed which when fed with details of batch size, product and works (from the LP) determined unit costs and was able to adjust these details to integer values and to feed them back to an LP formulation. The following procedure is followed.

(a)   The LP planning model with unit costs based upon the "standard cost batch" sizes is run and an allocation determined.

(b)   From this allocation, batch sizes and number of setups are determined and using the cost program the objective function of the LP is modified and the LP re–run. This procedure continues until the data supplied in terms of unit costs nearly equates with the batch sizes produced from the resultant LP.

(c)   The subbatch sizes and requirements are determined and then fed into the PERT package and the cost determined given fixed completion dates. Examination of various allocations about the optimal can be carried out to determine an acceptable schedule for the shop–floor.

RESULTS OF THE APPLICATION OF THE ITERATIVE APPROACH

Clearly, although the solution reached by this procedure will not be optimal the ability to rapidly modify and re–run the planning model, taking account of realistic costs can prove a great advantage. As identified by Sanderson (1978) the iterative control of such a model is a great asset in such situations because it gives greater control over undesirable strategies caused by the inherent characteristics of the LP. When this is aided by graphical output of the results in a form readily interpreted by senior management, the tool becomes even more useful.

The first stage of implementation was to introduce the interactive graphic capability. In order to highlight the possibilities that this introduced, a 3hr. display of its capabilities was undertaken to senior management. This in fact generated considerable interest and as a result co–operation with both production control and the internal accounts functions was greatly aided during the remainder of the project.

The next stage in implementation was to develop the programs which were to automatically adjust the LP input file to take account of the resultant batch size and number of setup changes implicit in each LP solution. As a result of statistical analysis undertaken as part of a productivity incentive scheme, timings and costs were available for set–up and run times for a variety of batch sizes for each item in the product range. The shape of this curve varied significantly from product to product and due allowance for this was incorporated in the costing program. Initially fears had been expressed that such a procedure of cost modification and the re–run of the LP could lead to the solution oscillating, possibly divergently, from one extreme to another but in practice this did not occur and 5 or 6 iterations led to a solution where input cost data, calculated batch sizes and number of setups were within 10% of one another, and from the

batch sizes and number of setups subbatch sizes are calculated and fed to the
'PERT' package.

The method of operation of the LP phase followed closely that of Sanderson
(1978) — an initial monthly run being completed during the third week of the
previous month, to the stage of equating cost input to LP output. Then
undesirable operating strategies are highlighted and the interactive part of the
operation (using constant unit costs) is carried out.

The third stage of operation — the scheduling using the shop–floor scheduling
program is then attempted. Although the LP allocation may still be infeasible it
was found that the interactive modification of the LP together with a realistic
costing approach incorporated in the LP resulted in a schedule which could
generally be accepted with minimal modification to allocations. This may in
some way be due to the varied nature of the order book and concern has been
expressed as to the effect of an order book consisting of a few orders covering
only a small product range. However, it is estimated that the proposed system
will decrease the carrying cost by 65% and will generate a saving of 7·5% of the
total production cost.

CONCLUSIONS

Although only in its exploratory stage it is clear that such an approach, while not
offering any immediate dramatic improvements, does have a number of
advantages. Firstly, unlike many more advanced approaches (two level
programming, goal programming, N.L.P. etc.), it does involve senior management
from the outset and, given an initial LP model results can be rapidly produced at
an early stage. In addition it does immediately question the basis of the LP
model and involves both those at shop–floor level and within the cost accounting
function to justify such assumption. It does require considerable statistical
analysis in order to produce a realistic costing model but in this case such an
analysis had already been conducted. In general it has put forward planning
using the LP on a reliable footing but considerable analysis is still required on
the effect of product mix, specialist equipment etc., on the feasibility of the
allocation for shop–floor scheduling.

APPENDIX

We use the following notations in formulating the problem:

x_i: production quantity, i.e. batch sizes, of product i in period T.

y_i: number of setups to produce one batch of product i.

x_{ijk}:   number of units of product i having their k^{th} operation on workstation (machine) j.

m_{ilk}:   requirement of material l for operation k of product i.

t_{ijk}:   time required for operation k of product i on workstation j.

s_{ik}:   setup cost for operation k or product i.

S_{ik}:   setup time for operation k of product i.

b_i: maximum quantity related to any operation in producing i as a minimum allowable buffer.

c_i: unit cost (without setup costs consideration).

C_j: capacity of workstation j in time units available in period T.

e_i: availability of product i from previous period.

M_l: quantity of material l available in period T.

REFERENCES

1. Al–Hakim, L., Determination of Product Mix Using LP and PERT Techniques, M.Sc. Thesis, University College of Swansea, U.K., 1977.

2. Bishop, J.E., Integrating Critical Elements of Production Planning, *Harvard Business Review,* Sept. – Oct., pp. 154–160, 1979.

3. Bowman, E., Consistence and Optimality in Managerial Decision Making, *Management Science,* vol. 9, no. 2, pp. 310–321, 1963.

4. Bowman, E., Production Scheduling by the Transportation Method of Linear Programming, *Ops. Res.,* vol. 4, no. 1, pp. 100–103, 1956.

5. Carter, P.L., Remus, W.E., and Jenicke, L.O., Production Scheduling Decision Rules in Changing and Non–Linear Environments, *Int. Prod. Res.,* vol. 16, no. 6, pp. 493–496, 1978.

6. Christensen, J., and Obel, B., Simulation of Decentralised Planning in Two Danish Organisations Using Linear Programming Decomposition, *Management Science,* vol. 25, no. 5, pp. 1658–1667, 1978.

7. Das, D., Computer Aided Scheduling, Ph.D. Thesis, University College of Swansea, U.K., 1979.

8. Das, D. and Imo, I., Multi–Stage, Multi–Capacity, Batch Production System with Deterministic Demand Over a Finite Time Horizon, *Int. J. Prod. Res.,* vol. 21, no. 2, pp. 587–596, 1983.

9. Goldratt, E., Computerized Shop Floor Scheduling, *Int. J. Prod. Res.,* vol. 26, no. 3, pp. 443–455, 1988.

10. Jones, W.G. and Rope, C.M., Linear Programming Applied to Production Planning – A Case Study, *Opl. Res. Q.,* vol. 15, no. 4, pp. 1964.

11. Lasdon, L.S., *Optimization Theory for Large Systems,* Macmillan, 1970.

12. Lee, S.M., and Moore, L.J., A Practical Approach to Production Scheduling, *Production and Inventory Management,* vol. 5, no. 1, pp. 79–92, 1974.

13. Mellichamp, J.M., and Love, R.M., Production Switching Heuristics for the Aggregate Planning Problem, *Management Science,* vol. 24, no. 12, pp. 1242–1250, 1978.

14. Mukhopadhyay, S.K., Malik, K.P., and Paul, M., Determination of Batch Sizes in Multi–Product, Multi–Stage Situation and Assessment on Capacity Utilization, *Int. J. Prod. Res.,* vol. 26, no. 7, pp. 1259–1280, 1988.

15. Ratcliff, H.D., Network Models for Production Scheduling Problems with Convex Cost and Batch Processing, *AIIE Transactions,* vol. 10, no. 1, pp. 104–8, 1978.

16. Sanderson, I.W., An Interactive Production Planning System in the Chemical Industry, *Opl. Res.,* vol. 29, no. 8, pp. 731–739, 1978.

17. Smith, S.B., An Input–Output Model for Production and Inventory Planning, *J. Engineering,* vol. 16, no. 1, pp. 64–71, 1965.

18. Uhlmann, A., Linear Programming on a Micro Computer: An Application in Refinery Modelling, *European J. Opl. Res.,* vol. 35, pp. 321–327, 1988.

19. Villa, A., and Arcostanzo, M., Dopp – Dynamically Optimized Production Planning, *Int. J. Prod. Res.,* vol. 26, no. 10, pp. 1637–1650, 1988.

20. Zierer, T.K., Mitchell, W.A., and White, T.R., Practical Applications of Linear Programming to Shell's Distribution Problems, *Interfaces,* vol. 6, no. 4, pp. 17–26, 1976.

Optimal Location of Detectors on a Measurement Surface

N. G. BARTON and D. R. JENKINS
CSIRO Divison of Mathematics and Statistics
P.O. Box 218
Lindfield, NSW 2070, Australia

INTRODUCTION

We consider the problem of placing detectors on a measurement surface in order to locate the source of a signal. This problem occurs widely in medicine and physical science, for example, in electroencephalography or in the location of mineral orebodies using electromagnetic methods. In many applications, it is necessary to gather a lot of data to solve such inverse problems, and hence the number of detectors might be quite large. In some applications, however, mutual interference occurs and so the detectors need to be placed as far from each other as possible. This leads to an optimisation problem which is solved in this paper by two different algorithms - one based on simulated annealing, and the other based on an analogy with particle dynamics.

The present work is motivated by the following problem. Suppose one wished to "map" the normal component of magnetic field due to a source adjacent to a bounded measurement surface. One way to do this would be to place a large number of current loops at various locations on the surface. The current measured in each loop is proportional to the flux of magnetic field through the area enclosed by the loop, and hence the normal component of magnetic field, if the loops are placed tangential to the surface. However, current flowing in a loop creates its own magnetic field which can affect the measurement of the field in a nearby loop. This effect is known as mutual inductance. The aim therefore is to obtain the maximum possible separation between loops in order to minimise the effect of mutual inductance.

It appears that very little work has been done on this type of problem. If the surface is unbounded, then the problem becomes analogous to a sphere packing problem, as described by Coxeter (1969). However, for N around 100 the effects of the boundaries on the solution are significant, and the unbounded solutions are no longer applicable.

Two separate approaches are described here. In each case the finite size of the detectors is ignored, and they are assumed to be points. Firstly, a cost function based on the distance to the nearest neighbour or the boundary is constructed, and an algorithm which moves the points in an attempt to minimise the cost is used. The algorithm we have used is *simulated annealing*. It has recently attracted significant attention for combinatorial problems, particularly in VLSI circuit design, where it is used to optimise the arrangement of circuit elements on a substrate. The algorithm is based on an analogy with the way that metals cool to form crystalline structures. The rate of cooling determines whether or not the metal attains a pure crystalline state, which represents the minimum energy of the system. Simulated annealing uses this analogy to attempt to obtain a global minimum of the specified cost function, rather than a local minimum. The ability to achieve the global minimum is determined by careful choice of a set of parameters, known collectively as the *cooling schedule*. Some experience in constructing an appropriate cooling schedule for particular problems has been gained in recent years (van Laarhoven & Aarts, 1987; Rutenbar, 1989), but it is generally necessary to obtain the cooling schedule empirically. The difficulty in obtaining an appropriate cost function and corresponding cooling schedule has led to limited results for the present problem.

The second approach is to adopt a particle dynamics model. Each point is considered to be a particle which moves due to repulsive forces from each of the other particles, as well as repulsion from the boundary. Their motion is damped so that they eventually reach local equilibrium. This formulation results in a set of 4N first order *o.d.e.*'s which depend upon the nature of the repulsive force and a single damping parameter. The equations tend to be stiff, and hence difficult to solve numerically, and there is no guarantee that the equilibrium obtained is a global one.

We present results obtained using the particle dynamics approach, for various values of N. The surfaces considered so far include a bounded plane (the unit square) and a hemisphere. Extensions to other surface shapes, to finite detector geometry and more complicated interaction laws are discussed.

SIMULATED ANNEALING

Consider the set of N points, $S = \{x_1, x_2, ..., x_N\}$, located on a bounded surface defined by

$$f(\mathbf{x}) = 0 \tag{1}$$

Our aim is to relocate the points so as to maximise the function

$$D(S) = \min_i \left\{ \min \left\{ \min_j |\mathbf{x}_i - \mathbf{x}_j|, e_i \right\} \right\}$$

where the distance norm is Euclidean and e_i is the distance from point $\mathbf{x}_i$ to the nearest edge. The simulated annealing algorithm is designed to minimise a cost function C, which is appropriately defined in this case to be

$$C(S) = \frac{1}{D^2(S)} \tag{2}$$

The algorithm proceeds as follows:

1. Choose an initial arbitrary configuration of points.

2. Set the "temperature" parameter T to an initial value T_0.

repeat

3. Perform a series of M "transitions" of the current configuration, accepting or rejecting each transition according to the Boltzmann distribution, $\exp(-\Delta C/T)$, where ΔC is the change in C from a previously accepted transition. Clearly, if $\Delta C < 0$ then the transition is automatically accepted.

4. Decrease T from its current value T_i to T_{i+1} according to a prescribed rule.

until a *stopping criterion* is satisfied.

The appropriate choice of T_0, M, the rule for decreasing T and the stopping criterion constitute the *cooling schedule*. It is desirable to choose the cooling schedule that minimises the amount of computation required for the solution of the optimisation problem. The following cooling schedule is based upon suggestions made in Chapter 5 of van Laarhoven & Aarts (1987).

The initial temperature is obtained by increasing T_0 until a specified percentage (80%) of transitions are accepted. It is desirable to accept almost all transitions initially. Also, several values of M, the Markov chain length were used, and the cooling rate took the form

$$T_{i+1} = \theta T_i \tag{3}$$

with a typical value of θ being 0.8. The algorithm was stopped if no transitions had been accepted in the previous 3 cooling steps.

A transition is a strategy for changing the current configuration at a given value of T. There is an infinite variety of ways of defining a transition, and no obvious theory that indicates which is optimal. It is obviously important to choose a definition which leads to solutions in a reasonable time, and this aspect is the major difficulty associated with the development of the algorithm. In the present work, a transition is defined as moving a random subset of k points a distance ϵ along the surface in a random direction. Various values of k and ϵ were used.

Solutions have been obtained for small values of N up to about 15, using this method, but it became difficult to decide on the most appropriate values of the various adjustable parameters in the cooling schedule with larger values of N. Thus it was difficult to obtain solutions for large N (i.e. up to about 100) in a reasonable time.

An alternative simulated annealing approach may also be useful (A.J. Baddeley, private communication). The approach is based on the definition of a transition as

1. randomly delete one point x_i with equal probability for each point.

2. randomly add one point according to the probability density

$$g(\mathbf{x}) = V \prod_{j \neq i} |\mathbf{x} - \mathbf{x}_j|^T$$

where V is a normalising constant.

With appropriate choices of the cooling schedule, this approach should lead to the solution of the optimisation problem. Further development of this algorithm is necessary, in order to include effects of the boundaries, and this will form the basis of future work.

PARTICLE DYNAMICS APPROACH

Imagine that each of the N points in S represents the location of a particle having mass m, which moves due to an interparticle repulsive force and a damping force. Then the motion of particle i is determined from an equation of conservation of momentum,

$$\ddot{\mathbf{x}}_i = A \sum_{j \neq i}^{N} \frac{\mathbf{r}_{ij}}{r_{ij}^{n+1}} + \mathbf{F}_{ib} - k\dot{\mathbf{x}}_i + \lambda(\mathbf{x}_i)\mathbf{n}_i \tag{4}$$

where n, A and k are positive constants and the dot denotes differentiation with respect to time. The vector $\mathbf{r}_{ij} = \mathbf{x}_i - \mathbf{x}_j$ and $r_{ij} = |\mathbf{r}_{ij}|$. The first term in equation (4) is due to the interparticle repulsive force, the second term is a repulsive force between particle i and the boundary, the third term is a damping force proportional to the velocity of the particle and the last term is a restoring force, in the direction of the local normal,

$$\mathbf{n}_i = \nabla f(\mathbf{x}_i)$$

The restoring force is necessary to constrain the particle motion to the surface (Becker, 1954). With appropriate rescaling, equation (4) can be reduced to

$$\ddot{\mathbf{x}}_i = \alpha \mathbf{F}_i - \dot{\mathbf{x}}_i + \lambda \nabla f(\mathbf{x}_i) \tag{5}$$

where $\mathbf{F}_i$ represents the total repulsive force on particle i (i.e. the interparticle forces plus the boundary repulsion). The unknown function λ can be obtained by differentiating (1) twice with respect to time and substituting the result into (5) to eliminate the second derivatives, yielding

$$\lambda = - \frac{\dot{\mathbf{x}}_i.(\dot{\nabla f}) + \alpha \mathbf{F}_i.\nabla f}{|\nabla f|^2} \tag{6}$$

Then equations (1),(5) and (6) can be solved to determine the location of particle i at time t, for some specified initial condition. These equations constitute three 2nd order o.d.e.'s and one algebraic equation, for the three components of the particle. One of these equations can be eliminated, and it is most sensible to eliminate one of the o.d.e.'s. Thus the main computational effort in this approach is the solution of $4N$ first-order o.d.e.'s for some given initial condition. The integration of these equations is continued until an equilibrium, which corresponds to an energy minimum, is reached. The equilibrium solution should be close to the optimum locations. The initial conditions chosen are

$$\mathbf{x}_i = \mathbf{R}_i \quad ; \quad \dot{\mathbf{x}}_i = 0 \tag{7}$$

where $\{\mathbf{R}_i, i = 1..N\}$ is a set of randomly chosen positions. The advantage of choosing random initial conditions is that different equilibria can be found for each "simulation", thus giving several "solutions" for any value of N.

There are only two controlling parameters in this model, n and α. The parameter α is a ratio between the repulsive and damping forces. Small α corresponds to strong damping, and local equilibrium will be rapidly achieved. Large α corresponds to minimal damping, so an equilibrium state will only occur after a long time. Large α causes increased stiffness of the *o.d.e.*'s, because a particle path may change rapidly when it gets near another particle. Thus it is desirable to choose a small value of α, although it should be large enough that the repulsive force is sufficient to accelerate the particles towards an equilibrium within a reasonable time.

The value of n affects the range of the interparticle force. Small n means a particle is influenced by a large number of neighbouring particles, while large n means it is influenced by its immediate neighbours only. Thus large n is appropriate to the solution of the placement problem. However, large n increases the stiffness of the *o.d.e.*'s, which implies that it is computationally desirable for n to be small. For a direct analogy with electrostatics, $n = 2$. Experience shows that a suitable intermediate value is $n = 3$.

Numerical solution of the above initial value problem was achieved using the public domain subroutine EPSODE, which is a version of Gear's method for stiff *o.d.e.*'s (Byrne & Hindmarsh, 1975). The integration of the equations was continued until a test for equilibrium was satisfied, namely that the distance moved by any point in the previous time step was less than a specified value, d_{min}, which was set to 10^{-5}.

SAMPLE RESULTS

All the results presented here were obtained using the particle dynamics model. Similar results are possible using the simulated annealing algorithm, although it is more difficult to obtain results for large N due to the complexity of the procedure.

Fig 1 shows results for points within the unit square, for various values of N. In this case $\lambda = 0$. Note that each value of N used is a perfect square, so as to illustrate the effect of the edges of the boundary upon the solution. For small N, as in fig 1(a,b), the solution consists of the points forming an essentially square lattice, while for larger N, as in fig 1 (c-f), the points form an approximately hexagonal lattice near the centre of the square, with an adjustment to a rectangular lattice near the boundary. The figure contains two different equilibria for each of $N = 49$ and $N = 100$, illustrating the difficulty in obtaining the global energy minimum. In each case, the two equilibria are very similar, a fact which is most easily noticed by observing the figure from a very shallow angle.

Fig 2 shows results for points constrained to move on the surface of a hemisphere, with various values of N, again all perfect squares. In this case, the equation of the surface in Cartesian coordinates is

$$f(x, y, z) = z - \sqrt{1 - x^2 - y^2} = 0$$

and consequently

$$\lambda(x_i, y_i, z_i) = -(\dot{x}_i^2 + \dot{y}_i^2 + \dot{z}_i^2) - \alpha(x_i F_{ix} + y_i F_{iy} + z_i F_{iz})$$

The results are shown in the form of a stereographic projection (Fisher *et al.*, 1987) which removes the difficulty of plotting points near the equator. For ease of interpretation, dotted lines on each graph represent lines of latitude in multiples of $30°$.

The figure shows that the solutions consist of points evenly distributed on lines of equal latitude for small and moderate values of N. However, for larger N (as in fig 2(e,f)) the

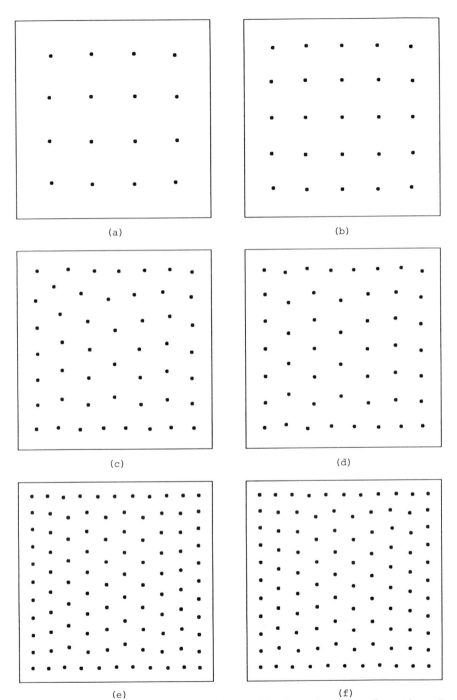

FIGURE 1. Equilibrium solutions for the particle dynamics approach on the unit square, with (a) $N=16$ (b) $N=25$ (c) $N=49$ (d) $N=49$ (e) $N=100$ and (f) $N=100$.

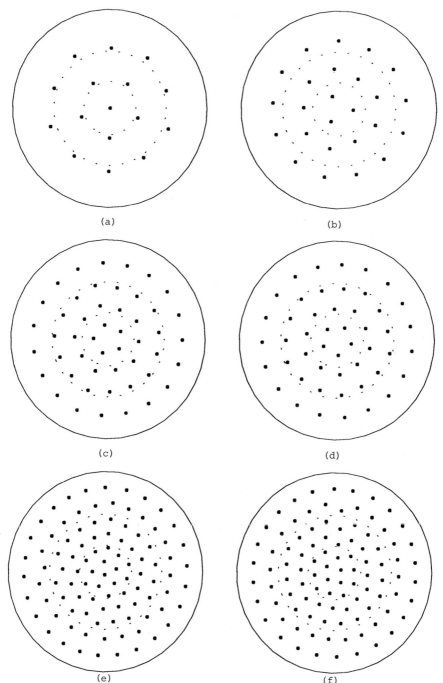

FIGURE 2. Equilibrium solutions for the particle dynamics approach on the surface of a hemisphere, with (a) $N=16$ (b) $N=25$ (c) $N=49$ (d) $N=49$ (e) $N=100$ and (f) $N=100$.

647

ringed formation exists near the equator, but breaks down near the centre. Obviously the ringed formation is due to the edge effect, which does not extend across the whole surface when N is large. Again, the problem of non-uniqueness of the solution is evident in the results shown for $N = 49$ and $N = 100$.

DISCUSSION

The results presented here, using the particle dynamics analogy, are satisfactory for the problem as stated. The algorithm is quite straightforward, and can be easily modified to determine solutions for a different shaped surface, as long as an analytical form for the surface is available. The question arises as to how well this approach can be applied to generalisations of the problem. For example, in the magnetic measurement application, the detectors are of finite size and of varying shape. A possible approach to obtaining a solution for circular detectors with radius ϵ is to include an exclusion zone in the interparticle force $\mathbf{F}_{ij}$, which would then take the form

$$\mathbf{F}_{ij} = \frac{\hat{\mathbf{r}}_{ij}}{(r_{ij} - \epsilon)^n}$$

For a polygonal shaped detector, a possible solution would be to locate a point charge at each vertex of the polygon, and then solve the equations of conservation of both linear and angular momentum for the centre of mass of the polygon.

Future developments may include the inclusion of random forces in the particle dynamics approach, in order to attempt to achieve a global minimum, and alternative simulated annealing algorithms. The aim is to obtain a more efficient algorithm, in terms of computation time, for the solution of the present application.

The authors wish to acknowledge the assistance of Mr H. Butler in the development of the simulated annealing algorithm.

REFERENCES

1. Becker R.A., *Introduction to Theoretical Mechanics*, §11-7, McGraw-Hill, New York, 1954.

2. Byrne G.D. and Hindmarsh A.C., A polyalgorithm for the numerical solution of ordinary differential equations, *ACM Trans. Math. Soft.*, vol. 1, pp. 71-96, 1975.

3. Coxeter H.S.M., *Introduction to Geometry*, 2nd ed., Wiley, New York, 1969.

4. Fisher N.I., Lewis T. and Embleton B.J.J., *Statistical Analysis of Spherical Data*, §3.3.1, Cambridge UP, Cambridge, 1987.

5. van Laarhoven P.J.M. and Aarts E.H.L., *Simulated Annealing: Theory and Applications*, D. Reidel, Dordrecht, 1987.

6. Rutenbar R.A., Simulated annealing algorithms: an overview, *IEEE Circuits and Devices Mag.*, vol. 5(1), pp. 19-26, 1989.

PLANGEN—A Mine Plan Generating Package

M. A. FORBES
OPCOM Pty Limited
P.O. Box 149
Toowong, Q. 4066, Australia

J. N. HOLT and A. M. WATTS
Department of Mathematics
The University of Queensland
St. Lucia, Q. 4067, Australia

INTRODUCTION

In this paper we describe some of the algorithmic details of PLANGEN, an interactive, mine plan generating software package combining mathematical optimization techniques with a high level of user control to obtain efficient and practical mine plans for two dimensional mining problems. The PLANGEN package was developed by OPCOM Pty Limited for COMALCO Aluminium Limited for their Weipa bauxite mining operation. We will not attempt to describe the implementational details such as the user interface here, although these have been taken very seriously in developing the package as they are crucial to the success of the software. Forbes (1987) provides full details of these facilities.

The purpose of mine planning is to determine the schedule of mining and related activities to be carried out during a specified planning period so as to maximize the profitability of the mine while meeting a variety of constraints imposed on the operation throughout the period. Typically, mine plans are required which specify the goals for the next several years. The activities related to the actual mining which are also determined by the mine plan include the clearing of mining regions, the stripping of overburden to reveal the ore, and later the revegetation of mined areas.

Although the problem is conceptually straightforward, near optimal execution of the planning process without the aid of optimizing algorithms is extremely difficult, due to the magnitude and combinatorial nature of the problem.

In the next section, we formulate the general mine planning problem amenable to solution by PLANGEN and briefly mention some unsuccessful solution techniques which have been considered by others in the past. In the following section, we describe the algorithm used by PLANGEN to obtain solutions to the mine planning problem.

FORMULATION OF THE PROBLEM

The problem we consider is a two dimensional one as encountered in surface mining applications. The mining regions are composed of a large number of mine blocks for which ore quality, ore tonnage and mining cost data are available. Each potential mining area is made up of many mine blocks. The final mine plan is specified by stating which blocks (or fractions of blocks) are mined and when they are mined. Although usually square, the mine blocks may be of arbitrary shape and this must be allowed for in the model. Mine faces define which blocks are currently accessible for mining.

649

The mining cost for a block should be the actual total cost per tonne of mining and shipping the ore. It includes fixed, capital, processing and regeneration costs, as well as clearing, stripping and so on. Wet season mine blocks may be specified. That is, some mine blocks may be inaccessible for some part of the year, and this needs to be catered for by the solution algorithm.

Orders are received for shipping parcels of ore. These may be required to be of variable tonnage and quality. The quality is specified by placing bounds on the concentration of basic "blend components" of the ore. In the case of bauxite, the principal ones are silica and alumina although other components of interest include iron and titanium. Critical blend constraints may apply to each parcel separately. The mine planning exercise must take into account all outstanding orders and make some assumptions concerning the extrapolation of orders beyond those currently on hand. Mine plans extending over several time periods (e.g. five year plans) are typically required.

Access constraints apply to each mine block. In practice, it is a requirement that a block be on the mine face when it is mined. This means that the solution technique must ensure that a path from the initial face to the mine block exists before allowing it to be mined. Each block in such a path must be fully mined before the mining of the next block can commence.

With the aid of the following notation we can derive a mathematical formulation of the problem:

TP = number of parcels of ore to be mined.

$T0_i$ = tonnage of ore required for parcel i, $i = 1, ..., TP$.

BC = number of blend constraints for each parcel.

B_{ij} = percentage of blend component j required in parcel i
$\qquad j = 1, ..., BC$; $i = 1, ..., TP$.

Tol_{ij} = permitted tolerance of blend component j in parcel i
$\qquad j = 1, ..., BC$; $i = 1, ..., TP$.

N = number of mine blocks.

T_k = tonnage of ore available in block k, $k = 1, ..., N$.

C_k = cost per tonne of mining the ore in block k,
$\qquad k = 1, ..., N$.

Q_{kj} = percentage of blend component j in block k,
$\qquad j = 1, ..., BC$; $k = 1, ..., N$.

X_{ki} = fraction of block k used in parcel i,
$\qquad k = 1, ..., N$; $i = 1, ..., TP$.

W_k = total fraction of block k used in all parcels
$\qquad = X_{k1} + X_{k2} + ... + X_{k,TP}$, $k = 1, ..., N$.

Y_i = actual tonnes of ore mined in parcel i
$\qquad = X_{1i} * T_1 + X_{2i} * T_2 + ... + X_{Ni} * T_N$,
$\qquad i = 1, ..., TP$.

Z_{ij} = actual tonnage of blend component j in parcel i
$\qquad = X_{1i} * T_1 * Q_{1j} + ... + X_{Ni} * T_N * Q_{Nj}$
$\qquad i = 1, ..., TP$; $j = 1, ..., BC$.

The problem can then be stated as -

Minimize $W_1 * C_1 * T_1 + ... + W_N * C_N * T_N$

Subject to :

(1) Blend constraints

$$B_{ij} - Tol_{ij} \leq Z_{ij}/Y_i \leq B_{ij} + Tol_{ij} \qquad\qquad i = 1,...,TP; \quad j = 1,...,BC.$$

(2) Tonnage constraints

$$Y_i \geq T0_i \qquad\qquad i = 1,...,TP.$$

(3) Block usage constraints

$$0 \leq X_{ki} \leq 1 \qquad\qquad k = 1,...,N; \quad i = 1,...,TP.$$
$$0 \leq W_k \leq 1 \qquad\qquad k = 1,...,N.$$

The procedure used in PLANGEN further constrains the solution to have the practical property that a block is used in at most two consecutive parcels.

(4) Access constraints

These constraints, referred to in the previous section, cannot be represented by linear constraints in continuous variables, and hence place the problem outside the realm of linear programming. If linear programming is used to solve the above model, ignoring the access constraints, infeasible solutions result which require the mining of blocks before access to them has been obtained by earlier mining.

It is possible to formulate the access constraints by introducing integer 0-1 indicator variables. This leads to an integer programming problem in which the number of integer variables is so large that existing branch and bound methods are impractical.

PLANGEN restricts consideration to "access feasible" solutions throughout. In order to decide whether a mine block may be mined in a given time period, PLANGEN stores information about access to each mine block. For each block, a list of potential access blocks is stored. A block may be mined in a given time period if one of its potential access blocks is mined in the same or an earlier time period. At any stage, each block has a specific predecessor so that a unique path back to a mine face is defined for it.

We are able to specify the allowed directions of mining in PLANGEN by editing the access data. For example, we can automatically select one of three global access types :- all neighbours, neighbours closer to face or neighbours not further from face. The access data for individual blocks can be modified to include arbitrary neighbouring blocks.

Constraints (2) in the model reveal that the size of the ore parcels considered determines the unit of time used. For example, a 5 year mine plan with requirements of 6 million tonnes of ore per year could be formulated as 30 parcels of 1 million tonnes or as 5 parcels of 6 million tonnes. The former formulation would be preferred to ensure feasiblity of blend over 1 million tonne parcels, whereas the second formulation would be preferred for a quick approximate solution in answer to "What if ... ?" type questions.

As well as modelling the obvious silica and alumina blend constraints, wet season mining can be controlled by the blend constraints (1). To model the inaccessibility of some blocks during the wet season, a blend component (D) is intoduced, with blocks that can only be mined during the dry season having $Q_{k,D} = 100$ and other blocks having $Q_{k,D} = 0$. $B_{i,D} = 50$ and $Tol_{i,D} = 50$ are used for parcels i which will be mined in the dry season, while $B_{i,D} = 0$ and $Tol_{i,D} = 0$ are used for parcels which will be mined in the wet season.

The mining cost C_k in the objective function, as mentioned earlier, should be the cost per tonne of mining the ore. It should not include any measure of the quality of the ore unless the quality directly affects a cost (eg beneficiation cost).

THE PLANGEN OPTIMIZER

PLANGEN produces cost-effective mine plans by applying a two stage heuristic optimization procedure to the problem described above. In Stage I, an approximate solution is found by simulated annealing, and then linear programming is used in Stage II in an attempt to improve this solution by considering a perturbation problem. We outline the main features of these stages.

Stage I

A starting solution is generated using the following procedure. For each parcel of ore in the plan, repeat the following until the tonnage of the parcel is greater than or equal to the required tonnage. Randomly choose an accessible block. If adding this block to the parcel decreases the infeasibility of the blend constraints, add the block to the parcel. Otherwise add the block to the parcel with probability α. α is increased from time to time through the process. This initial solution will satisfy the tonnage requirements and the access constraints, but will not in general satisfy the blend constraints.

The method of simulated annealing, Kirkpatrick et al (1983), is then applied to the approximate solution obtained. In this technique, random changes to the current solution are generated and are accepted if they lead to a decrease in the objective function. If they lead to an increase , they are accepted with probability $exp(-increase/temperature)$, where temperature is a parameter which is gradually reduced. The parameter "starting temperature" is used as the initial temperature and is reduced by a "reduction ratio" at regular intervals. The method derives from an analogy between combinatorial optimization problems and the behaviour of condensed matter physical systems. It simulates the annealing of a solid to identify its low temperature state. The objective function is thought of as the energy of the system, and the random changes correspond to the changes in state experienced by a system in thermal equilibrium at a certain temperature. The key to the success of the method is that it allows transitions to states of higher energy with some non-zero probability. In the context of optimization, this means that we have a mechanism for escaping from local minima. The following is a skeleton of the algorithm.

```
initialize:
        system state
        starting temperature
while .not.stopping-criterion
        do i=1, maxreps-at-current-temp
                generate perturbed system state
                calculate change in energy
                call Accept
                if (Accept) then
                        change system state
                endif
        enddo
        update temperature
endwhile
```

The objective we use in the annealing algorithm is a composite one with weighted contributions from

(1) infeasibilities   the sum of the extent by which each blend component falls outside the acceptable range

(2) cost   the sum of the mining costs of the material used in the current solution

(3) deficient tonnage   the sum of shortfalls in required tonnage

(4) excess tonnage   the sum of excesses in required tonnage

(5) jumps   a count of the number of times neighbouring mine blocks are used in different parcels

The random changes which produce the perturbed system state are obtained as follows. Choose two blocks which are not neighbours. For each of these two blocks carry out the following. If the block is shared, that is mined in two parcels, adjust the fraction mined in the first parcel by adding a random fraction δ distributed Uniform (-0.5,0.5), truncating to 0 or 1 as necessary. If the block is only mined in one parcel, randomly choose to share it with the previous or next parcel. If this sharing is possible (see (a) and (b) below), set the amount mined in the earlier parcel to be a random fraction distributed Uniform (0,1). Otherwise, choose a replacement block and repeat. Calculate the change in the objective function components. (This takes $O(BC)$ time.) To calculate the objective function completely takes time of $O((N + TP)BC)$. This is done periodically to reduce rounding errors.

(a) A block can be shared with the previous parcel if the mining of its predecessor is completed in or before the previous parcel.

(b) A block can be shared with the next parcel as long as it is not the predecessor of a block mined in the current parcel.

Where possible, we update the predecessor information to satisfy (a) and (b).

Although it is sufficient to consider changes to one block at a time, it has been found in practice that considering two blocks increases the number of changes which improve the objective function, and hence improves the efficiency of simulated annealing.

While PLANGEN is calculating the Stage I solution, various pieces of status information are displayed. This allows the progress of the solution to be monitored. The most important piece of information for most users is COST, which is the cost of the current mine plan as measured by the data supplied.

At regular intervals during the solution process, PLANGEN prompts with the message "Continue solution ? ". If the user answers "yes" to this question, PLANGEN will continue with its approximate solution procedure. In practice, the answer should be based on the progress of the COST figures which are displayed. While these figures continue to decrease steadily and significantly the solution process should be continued.

Stage II

After obtaining an approximate solution, the user will usually want to tighten the solution so that the amount of ore in each parcel is exactly the required amount, the blend constraints are met exactly and the cost is decreased as far as possible. PLANGEN provides a facility to tighten the solution in this way. A "perturbation problem" is constructed which is solved by the AESOP linear programming package, Murtagh (1986). The solution to the perturbation problem is then used to update the current mine plan and thus reduce the total cost. When the perturbation problem has been solved, the user is prompted with the message "Continue tightening ?". If the answer is "yes" to this question, another perturbation problem will be generated and solved. This process should probably continue while the cost is decreasing significantly between perturbation problems.

The "perturbation problem" is generated in the following way. We consider a change to the mining of each mine block in the problem. The magnitudes of the changes are the variables of the linear programming problem described below. The rules for determining the type of these changes are the same as those used in the simulated annealing stage, except that (i) no changes to predecessor information are allowed, and (ii) if a block cannot be shared in its randomly chosen direction, (previous or next parcel), we attempt to share it in the other direction.

The linear program problem has, as its objective to be maximized, the sum of (i) the cost of mining all ore not mined in the current solution, and (ii) the excess ore mined in each parcel multiplied by the index of the parcel. The effect of (i) is to minimize the total cost of mining. The effect of (ii) is to push excess tonnage into later parcels. After a sequence of perturbation problems, the excess tonnage is removed from the plan.

The constraints for the linear programming problem are the blend constraints and the minimum tonnage constraints (1) and (2).

DISCUSSION

PLANGEN can be used to solve short term or long term mine planning problems. The sizes of the blocks and the parcels can be controlled by the user to give the desired resolution. Facilities are provided for aggregating the basic blocks for which information is stored. In practice it is found that the average block size should not exceed about one tenth of the average parcel size, and the blocks should be of similar size, as should the parcels.

The package can also be used as a planning tool to investigate such diverse problems as the effects of different market demand scenarios, the placement of new mine faces and haul roads and the ratio of ore to take from different deposits which may be separated in both distance and mineral composition.

The package has been used to solve problems with as many as 2000 mine blocks, 20 parcels, 20 mine faces and five blend constraints. Stage I optimization takes approximately 10 minutes of CPU time on a Microvax II, and each iteration of Stage II takes approximately 5 minutes. The time for Stage I depends on the number of blocks and the weightings in the objective function. The time for Stage II depends on the number of blocks and parcels. On difficult problems, Stage II may not be able to find a feasible solution, in which case the Stage I process should be repeated before further passes of Stage II.

CONCLUSION

We have described a fast, effective heuristic algorithm for the solution of practical mine planning problems. The PLANGEN system was developed for COMALCO Aluminium, and has been in operation since mid-1987 in the mine planning department at the mine site . It also featured as a Research Project of the Week at the World EXPO-88 in Brisbane. Although the method as described is designed for two dimensional mining applications, its extension to three dimensional problems such as encountered in open cut coal mines is straightforward through suitable redefinition of the access paths for mine blocks. Extra classes of constraints such as those related to dragline movements can be incorporated. The system could be further enhanced by a graphics interface.

REFERENCES

1. Forbes, M. A., PLANGEN User Manual, OPCOM Pty. Limited, 1987.

2. Murtagh, B. A., AESOP User Manual, 1986.

3. Kirkpatrick, S., Gelatt, C. D. Jr. and Vecchi, M.P., Optimization by Simulated Annealing, *Science*, vol. 220, pp. 671-680, 1983.

The L_∞ Exact Penalty Function in Semi-Infinite Programming

C. J. PRICE and I. D. COOPE
Department of Mathematics
University of Canterbury
Christchurch, New Zealand

INTRODUCTION.

The convergence properties of an algorithm for Semi-Infinite Programming are examined. The algorithm requires only first derivatives, and uses standard Sequential Quadratic Programming techniques, together with an L_∞ exact penalty function.

It is shown that convergence of the algorithm is not dependent on the applicability of an implicit function theorem. A weaker assumption concerning the finiteness of the number of local maximizers of the semi-infinite constraints is sufficient.

The algorithm generates a sequence of iterates in the following manner. At each iterate the SIP is replaced by a finite dimensional Non-Linear Programming Problem equivalent to the SIP in some neighborhood of the iterate. SQP techniques are used to obtain the next iterate, where a sufficient reduction of the penalty function is required in the process. The sequence of iterates so generated converges to a set of critical points of the penalty function. If a convergence point of the set of iterates is also a feasible point of the SIP, and suitable values for parameters appearing in the penalty function are used, then that convergence point will also be a stationary point of the SIP.

The semi-infinite programming problem is as follows:

$$\begin{aligned}
minimize \ & f(x) \quad \text{over} \quad x \in R^n, \\
\text{subject to} \quad & g(x,t) \leq 0 \quad \forall t \in T, \quad \text{where} \quad T \subset R^p.
\end{aligned} \tag{1}$$

The objective function $f(x)$ is continuously differentiable, mapping $R^n \to R$. The region T is a Cartesian product of bounded closed intervals of the form

$$\{t_i \in R \ : \ \alpha_i \leq t_i \leq \beta_i\} \quad \text{for} \quad i = 1, ..., p.$$

The function $g(x,t)$ maps $R^n \times R^p \to R$ and is continuously differentiable in all arguments.

In what follows the term "local maximizer" refers to local maximizer of $g(x,t)$ with respect to t, where $t \in T$ and x is some point in R^n, unless stated otherwise. The exact identity of x will be clear from the context. All unmarked norms are 2–norms and iteration numbers are given as bracketed superscripts.

SUFFICIENT CONDITIONS FOR LOCAL EQUIVALENCE TO AN NLPP.

In this section the local equivalence of the SIP to an NLPP is discussed. This equivalence depends on the infinity norm base of the penalty function, where for the purposes of this section the penalty function ϕ is taken to be of the form

$$\phi(x) = f(x) + \mathfrak{p}(\theta), \qquad \text{where} \qquad \theta = max_{t \in T}\, [g(x,t)]_+ \, .$$

Here $\mathfrak{p}(\theta)$ is a strictly monotonically increasing differentiable function of θ satisfying $\mathfrak{p}(0) = 0$, and θ is the infinity norm of the constraint violations. This permits a precise definition of the local equivalence referred to above.

Definition 1.

An NLPP is regarded as locally equivalent to the SIP on a set S iff on S it has the same objective function, the same set of feasible points, and an identical L_∞ exact penalty function, as the SIP (1). $\square$

Given identical objective functions, this local equivalence can be shown by establishing the local equality of the infinity norm of the constraint violations for the SIP, and NLPP. This is clearly the case if the NLPP constraint functions are locally equal to the local maximal values (and hence also the global maximal values) of $g(x,t)$ as functions of x. Each local maximizer at an iterate x_0 gives rise to at least one constraint of the NLPP. In light of this, the following assumption is needed.

Assumption 2.

At all points $x \in R^n$ the number of local maximizers of $g(x,t)$ in T is finite. $\square$

Actually, for the purposes of this discussion, it is only required that the number of local maximizers above a threshold $v(x)$ strictly less than the global maximal value of $g(x,t)$ be finite for all x in some neighborhood of each iterate, and in some neighborhood of each limit point of the sequence of iterates (Price and Coope, 1989).

Assumption 2 implies each local maximizer is both isolated, and strict. Consequently, the local maximizers are locally continuous, as follows.

Proposition 3. Continuity of the Local Maximizers of g(x,t).

Let t_0 be a local maximizer of $g(x_0,t)$. Then $\forall \epsilon > 0$, $\exists \eta > 0$ such that $\forall x \in R^n$ satisfying $\| x - x_0 \| < \eta$, there is at least one local maximizer $t \in T$ of g satisfying $\| t - t_0 \| < \epsilon$. If, for all sufficiently small ϵ there is exactly one such local maximizer, then it is expressible as a continuous function of x, $\forall x \in R^n \; : \; \| x - x_0 \| < \eta$.

Proof: Define

$$S(\epsilon, \epsilon_0) = \{t \in T \; : \; \epsilon \leq \| t - t_0 \| \leq \epsilon_0\},$$

where ϵ_0 is sufficiently small to ensure there is no local maximizer within ϵ_0 of t_0, other than t_0 itself. $S(\epsilon, \epsilon_0)$ is a clearly compact, and so the supremum of $g(x_0,t)$ on $S(\epsilon, \epsilon_0)$ is achieved. Using $g(x_0,t_0) = g_0$ let

$$m(\epsilon) = g_0 - max_{t \in S(\epsilon, \epsilon_0)}\, g(x_0,t).$$

Now the continuity of $\nabla_x g$ with respect to all arguments, and the compactness of T imply g is equicontinuous over T with respect to x. Therefore

$$\forall \epsilon > 0, \quad \exists \eta(\epsilon) > 0, \quad \text{such that} \quad \forall x, \quad \forall t \in T,$$

$$\| x - x_0 \| < \eta \quad \Rightarrow \quad |g(x,t) - g(x_0,t)| < \frac{1}{4}m(\epsilon).$$

Hence, for all x sufficiently close to x_0

$$g(x, t_0) > g_0 - \frac{1}{4}m(\epsilon); \qquad \text{and also}$$

$$\forall t \in S(\epsilon, \epsilon_0), \qquad g(x, t) < g_0 - \frac{3}{4}m(\epsilon).$$

So there must be at least one local maximizer inside the intersection of T, and the open sphere of radius ϵ, centred on t_0. As this holds for all small positive ϵ, the continuity of the local maximizer t_0 at x_0 follows. This continuity can be extended to all such x satisfying $\| x - x_0 \| < \eta$ by applying the above argument to each such x. $\square$

This proposition implies each local maximizer at x_0 has at least one continuation over some neighborhood of x_0, where a continuation of a local maximizer t_0 is a continuous function $t(x)$ on some neighborhood of x_0, mapping x into the set of local maximizers of $g(x, t)$, and satisfying $t(x_0) = t_0$. If the behaviour of each of the local maximizers t_i at x_0 with respect to variations in x is such that each t_i has a unique continuation, then these may be eliminated from the inequalities

$$g(x, t_i) \leq 0 \qquad i = 1, ..., h$$

near x_0 to yield a finite NLPP:

$$min \ f(x) \qquad \text{subject to} \qquad c_i(x) \leq 0 \qquad i = 1, ..., h,$$

where $c_i(x) = g(x, t_i(x))$ for $i = 1, ..., h$. It is possible for a local maximizer at x_0 to have several continuations in some neighborhood of x_0. In this case each continuation $t(x)$ yields one ordinary constraint on being substituted into $g(x, t) \leq 0$.

This NLPP construction tacitly assumes that each local maximizer for x near x_0 is obtainable from a continuation of some local maximizer at x_0. This is not always the case. It is possible for local maximizers to appear out of stationary points at x_0 which are not local maximizers. These stationary points may be defined precisely as follows.

Definition 4. Nascent Local Maximizer.

A point t_0 is a nascent local maximizer of $g(x_0, t)$ if:

1. t_0 is not a local maximizer of $g(x_0, t)$.
2. $\forall \epsilon, \eta > 0 \qquad \exists x : \| x - x_0 \| < \eta$ for which there exists a local maximizer t of $g(x, t)$ satisfying $\| t - t_0 \| < \epsilon$. $\square$

Actually, the NLPP constructed above is locally equivalent to the SIP irrespective of any nascent local maximizers, as is shown next.

Proposition 5. Continuity of the Global Maximizers.

For all x_0 in R^n, $\exists \eta > 0$ such that the global maximizers τ_i, $i = 1, ..., l$ of $g(x_0, t)$ at x_0 are continuous functions of x, $\forall x : \| x - x_0 \| < \eta$, and for all such x all global maximizers of $g(x, t)$ are in the set $G(x) = \{\tau_i(x) : i = 1, ..., l\}$.

Proof: By proposition 3 the positions of all local maximizers are continuous with respect to x at x_0, apart from local maximizers nascent at x_0. Let L be the set of all local maximizers of $g(x_0, t)$ excluding members of the set G. Whence, as the number of local maximizers is finite, and by the continuity of g with respect to all arguments

$$\exists \delta > 0 \quad \text{such that} \quad \forall x : \| x - x_0 \| < \delta, \quad \forall t \in L, \quad g(x, t(x)) < g_0,$$

where g_0 is the global maximal value of $g(x_0, t)$ over T. This will yield the required result if it can be shown that $\exists \epsilon > 0$ such that each local maximizer nascent at x_0 lies below $g_0 - \epsilon$ as x tends to x_0, for values of x at which the maximizer exists.

Let there exist a function $t(x)$ specifying a local maximizer near x_0 which is necessarily discontinuous at x_0, so

$$\exists \epsilon > 0 \quad \forall \delta > 0 \quad \exists x \; : \| x - x_0 \| < \delta, \quad \text{and} \quad \| t(x) - t_0 \| \geq \epsilon \quad \forall t_0 \in L \bigcup G.$$

Therefore there is an infinite sequence $\{z_n\} \to x_0$ such that each $g(z_n, t)$ has a local maximizer τ_n with

$$\tau_n \in T_0, \quad T_0 = T - \bigcup_{t_0 \in L \bigcup G} \{t \in T : \| t - t_0 \| < \epsilon\}$$

for some $\epsilon > 0$. T_0 is compact, so $\{\tau_n\}$ has a limit point τ_∞ in T_0, and τ_∞ is a nascent local maximizer. There is a subsequence $\{z_{*n}\}$ such that $\{\tau_{*n}\}$ converges to τ_∞. Now, τ_∞ may lie on some of the bounding constraints of T. As $\tau_{*n} \to \tau_\infty$, in the limit $n \to \infty$ the active set for τ_{*n} must be a subset of the active set for τ_∞. So, as each τ_{*n} satisfies the first order Karesh Kuhn Tucker conditions for $g(z_{*n}, t)$ on T, and by the C^1 continuity of g, τ_∞ is a KKT point of $g(x_0, t)$. Whence τ_∞ is a stationary point, and not a local maximizer.

As T_0 is compact, and as $g(x_0, t)$ is continuous, the set of function values corresponding to nascent local maximizers is closed, so this set achieves its supremal value which must be strictly less than the global maximum. By the argument given at the beginning of the proof, the result holds. $\square$

The importance of this proposition is that all global maximizers at x are obtainable from continuations of those at x_0. The penalty function is based on the infinity norm, which depends only on the global maximizers. Hence the L_∞ penalty function is continuous everywhere. In contrast the corresponding L_1 penalty function may be discontinuous at points outside the feasible region as it depends on the local maximizers (Tanaka et. al., 1988). The continuity of the global maximizers permits a finite NLPP to be defined which locally has an identical penalty function to the SIP.

Theorem 6. Local Equivalence of the SIP, and NLPP.

At any point $x_0 \in R^n$, $\exists \eta > 0$ such that over the region $\{x \in R^n \; : \| x - x_0 \| < \eta\}$, the SIP is locally equivalent to the NLPP

$$min_x \; f(x) \quad \text{subject to} \quad c_i(x) \leq 0, \quad \forall i = 1, ..., r(x_0),$$

where $c_i(x) = g(x, \tau_i(x))$ for each $i = 1, ..., r$, and where τ_i are the continuations of the local maximizers of $g(x_0, t)$ over $\{x \in R^n \; : \| x - x_0 \| < \eta\}$.

Proof: By proposition 5, $\exists \eta > 0$ where $\forall x \; : \| x - x_0 \| < \eta$ the global maximizers of $g(x, t)$ are obtainable by the continuations of the global (and hence local) maximizers of $g(x_0, t)$. So, using $c(x) = [c_1(x), ..., c_r(x)]^T$, gives

$$\| [c(x)]_+ \|_\infty = max_i \; [c_i(x)]_+ = [max_i \; c_i(x)]_+ = max_{t \in T} \; [g(x, t)]_+.$$

Hence the SIP, and NLPP have identical infinity norms of their respective constraint violations in $\{x \in R^n \; : \| x - x_0 \| < \eta\}$. Consequently the feasible regions are equal inside $\{x \in R^n \; : \| x - x_0 \| < \eta\}$, and the result follows. $\square$

This theorem implies that the SIP can, to some extent, be treated as an NLPP with its constraints defined in an unusual way. Due to the SQP base of the algorithm, it is required that the NLPP constraints be continuously differentiable.

Proposition 7. C^1 continuity of the NLPP constraints.

If $t(x)$ is a continuous function specifying the position of a local maximizer of $g(x,t)$ for $x : \| x - x_0 \| < \eta$ for some $\eta > 0$ then the constraint $c(x) = g(x, t(x))$ is continuously differentiable, and $\nabla_x c(x) = \nabla_x g(x, t)$ at $t = t(x)$.

Proof: See (Price and Coope, 1989). $\square$

The local equivalence of the SIP to an NLPP with C^1 constraints permits optimality conditions to be derived easily.

Theorem 8.

Let x^* be an optimal point of (1), at which the following regularity assumption holds:

$$\exists h \in R^n \quad \text{such that} \quad g(x^*, t) + h^T \nabla g(x^*, t) < 0 \quad \forall t \in T.$$

The first order necessary conditions for optimality are: there exists a finite number of global maximizers t_i^* of $g(x^*, t)$ each with an associated Lagrange multiplier λ_i^* satisfying

$$\nabla f + \sum_{i=1}^{m} \lambda_i^* \nabla_x g(x^*, t_i^*) = 0 \quad \text{with} \quad m \leq n,$$

where $g(x^*, t_i^*) = 0$, and $\lambda_i^* \geq 0$, $\forall i : i = 1, ..., m$.

Here the NLPP constraints have been written as $g(x, t_i(x)) \leq 0$, where $t_1(x), ..., t_r(x)$ are the continuations of the local maximizers of $g(x^*, t)$ about x^*.

Proof: By theorem 1, page 249 of (Luenberger, 1969). $\square$

In the next section an L_∞ penalty function algorithm exploiting the local equivalence to an NLPP with C^1 constraints is described.

AN L_∞ ALGORITHM FOR SIP.

An exact non-differentiable penalty function based on the infinity norm is used. The penalty function is constructed so that any local minimum of the SIP (1) is also a local minimum of the penalty function ϕ. Specifically

$$\phi(\mu, \nu; x) = f(x) + \mu\theta + \frac{1}{2}\nu\theta^2, \quad \text{where} \quad \theta = max_{t \in T} \; [g(x, t)]_+, \tag{2}$$

is used as the penalty function, and $[x]_+$ denotes the maximum of x and zero. The two variables μ, and ν serve as penalty parameters, where $\mu > 0$, and $\nu \geq 0$ are required. The quadratic term in θ has been included in order to prevent μ from being set at an unreasonably high value in the first few iterations of the algorithm. If this does occur it can have a detrimental effect on the algorithm's performance (Coope, 1985).

Rather than searching for a local minimizer of the SIP directly the algorithm seeks a local minimizer of ϕ with suitable values of μ and ν. This is achieved by generating a sequence of iterates, the choice of each iterate being based on an L_∞ Inequality Quadratic Programme approximation to (2) at the previous iterate, such that the sequence of penalty function values so obtained is monotonically decreasing.

The L_∞ quadratic approximation to ϕ at $x^{(k)}$, denoted by ψ, is as follows:

$$\psi^{(k)}(s) = f\left(x^{(k)}\right) + h^T s + \frac{1}{2}s^T H^{(k)} s + \mu \, \| \ell_+(s) \|_\infty + \frac{1}{2}\nu \, \| \ell_+(s) \|_\infty^2,$$

where $\psi\left(x^{(k)}; \mu, \nu; s\right)$ has been shortened to $\psi^{(k)}(s)$. $\tag{3}$

Here $H^{(k)}$ is a positive definite matrix, $h = \nabla f(x^{(k)})$, and

$$\ell_+(s) = [\ell(s)]_+, \qquad \text{where} \qquad \ell(s) = c\left(x^{(k)}\right) + s^T \nabla c\left(x^{(k)}\right)$$

is the linearization of the constraint vector about $x^{(k)}$, and evaluated at the point $x^{(k)} + s$. The problem of minimizing $\psi^{(k)}(s)$ may be solved as an IQP; its solution $s^{(k)}$ is used to form the search direction at the current iterate. The algorithm searches either along the line $x^{(k)} + \alpha s^{(k)}$, or along the quadratic arc $x^{(k)} + \alpha s^{(k)} + \alpha^2 \sigma^{(k)}$, where $\sigma^{(k)}$ is a correction vector chosen to prevent the Maratos effect (Mayne and Polak, 1982). In either case a modified Armijo search is conducted. The first such α satisfying the sufficient descent condition

$$\phi\left(x^{(k)}\right) - \phi\left(x^{(k)} + q^{(k)}(\alpha)\right) \geq \rho\left[\psi\left(x^{(k)}; 0\right) - \psi\left(x^{(k)}; \alpha s^{(k)}\right)\right] \tag{4}$$

is used, where $q^{(k)}(\alpha)$ is either the line, or arc step as specified above, and $0 < \rho < 1$. The next iterate is obtained by augmenting the current iterate with $q^{(k)}(\alpha^{(k)})$.

This completes the outline of the algorithm apart from the method used to update the penalty parameters. These are adjusted in order to obtain equivalence of the stationary points of the SIP, and the feasible critical points of the penalty function to which the algorithm is converging. This equivalence occurs if the penalty parameters satisfy the conditions given in theorem 10 (below), which are $\mu \geq \| \lambda^* \|_1$, and $\nu \geq 0$, where λ^* are the optimal Lagrange multipliers. In practice estimates λ^*_{est} of the optimal Lagrange multipliers are used. To ensure an acceptable value for μ is eventually achieved these estimates must be lower semi-continuous with respect to x. They may be formed using information from the solution of the L_∞IQP (3), or by other methods (Gill and Murray, 1979).

Algorithm summary.

An iteration of the algorithm passes through the following steps:

1. All the local maximizers of g above a (strictly negative) threshold are found by first finding approximations to them using a grid search, and refining these approximations using a Quasi-Newton method.
2. Form the approximating L_∞ quadratic penalty function $\psi^{(k)}$ and find its solution $s^{(k)}$; if necessary increasing the penalty parameters in order to ensure that near feasibility is obtained, if this is possible.
3. Compare the predicted, and the actual descent via the inequality (4). If the actual descent obtained is insufficient then either do a line search if $x^{(k)}$ is highly infeasible (ie θ exceeds a strictly positive fixed parameter τ_1), otherwise do an arc search. If $\| \sigma^{(k)} \| > \| s^{(k)} \|$ then a line search is used in any case.
4. Estimate the Lagrange multipliers at the new iterate. If $\mu \leq \frac{5}{4} \| \lambda^*_{est} \|_1$ then one, or both penalty parameters are increased. If the optimal Lagrange multiplier estimates are expected to be reasonably accurate, μ is increased sufficiently to yield $\mu > \frac{3}{2} \| \lambda^*_{est} \|_1$. The estimates of the optimal Lagrange multipliers are assumed not to be accurate in the first few iterations, and at other iterations in which θ exceeds a strictly positive fixed parameter τ_2.
5. Update the matrix $H^{(k)}$ by a Quasi-Newton scheme. The scheme used is to maintain positive definiteness.
6. If sufficient accuracy has not been attained, begin another iteration.

CONVERGENCE OF THE ALGORITHM.

The purpose of this section is to examine the convergence properties of the algorithm. For convenience, the algorithm uses only first derivatives. Accordingly, it is desirable that the algorithm be capable of solving C^1 problems, and so only first order optimality conditions are available. Therefore any stationary point of (1) is regarded as an acceptable convergence point for the purposes of this discussion. Similarly, any point x for which the penalty function has a non-negative Gateaux derivative in every direction will be regarded as a valid convergence point for the penalty function. Such points will be referred to as critical points of the penalty function. Firstly, the following definition is made.

Definition 9. Critical Point

A point x is a critical point of the penalty function ϕ iff

$$\forall s \in R^n : \| s \| = 1, \qquad lim_{\epsilon \to 0+} \frac{\phi(x + \epsilon s) - \phi(x)}{\epsilon} \geq 0. \quad \square$$

The Gateaux differentiability of ϕ is an immediate consequence of the C^1 nature of $f(x)$ and $c(x)$, and the Gateaux differentiability of $\| [.]_+ \|_\infty$.

Theorem 10. Acceptable values for the Penalty Parameters.

A point x^* is a stationary point of (1), with an associated Lagrange multiplier vector λ^*, iff for all μ, and ν satisfying

$$\mu \geq \| \lambda^* \|_1 \qquad \text{and} \qquad \nu \geq 0, \tag{5}$$

x^* is a feasible critical point of ϕ. Furthermore, if $\mu > \| \lambda^* \|_1$, then a feasible point x^* is a local minimum of (1) iff it is a local minimum of ϕ.

Proof: See, for example (Price and Coope, 1989). $\square$

Theorem 11. Convergence.

Given:
1. All iterates generated by the algorithm lie in a bounded region of R^n.
2. For all $x \in R^n$ the number of local maximizers of $g(x, t)$ in T is finite.
3. The sequence of matrices $H^{(k)}$ is bounded above in the 2–norm, and each such matrix is positive definite.
4. The parameters μ and ν are only altered a finite number of times.

Then firstly, the algorithm converges to a set $\mathcal{K}$ of critical points of $\phi(\mu, \nu; x)$, where μ and ν are the final values of these parameters; and secondly, the penalty function ϕ is constant on the set $\mathcal{K}$.

Furthermore, if the sequence of 2–norms of the matrices $H^{(k)}$ also has a strictly positive lower bound then the set $\mathcal{K}$ is connected, and either $\mathcal{K}$ is a subset of the feasible region, or $\mathcal{K}$ is a subset of the infeasible region. Consequently, if the set of points to which the algorithm converges is feasible, the penalty parameters satisfy the conditions (5), and the stationary points of the SIP are isolated, then the sequence of iterates converges to a single point, and that convergence point is a stationary point of the SIP (1).

Proof: See (Price and Coope, 1989). $\square$

CONCLUSION.

It has been shown that the sequence of iterates generated by the algorithm converges to a connected set of critical points of the penalty function given the SIP is locally equivalent to an NLPP in the region of interest. If one such point is feasible, then all limit points of the sequence of iterates are stationary points of the SIP.

The fact that the algorithm may converge to a connected set of points rather than a single point is of no great consequence. Each of the points in this set has the same penalty function value. If one such limit point is feasible, then all such points are feasible, and they all have the same value for the objective function. Provided the objective function is a sensible measure of the relative merits of each feasible point, it matters not which limit point the final iterate generated best approximates, as each limit point is equally good.

Unlike similar algorithms based on an exact L_1 penalty function, this algorithm does not depend on the validity of an implicit function theorem to ensure the local equivalence of the SIP, and NLPP. A finite number of local maximizers for each of the semi-infinite constraints is sufficient to yield this equivalence. Given the inability of any computer to deal with an infinite number of local maximizers, this restriction is quite mild.

Obviously, there are many possible ways of implementing the algorithm. In particular it is desirable that the implementation used converge superlinearly on problems possessing the required degree of continuity. These points, along with numerical results, will be discussed in (Price, forthcoming).

REFERENCES.

1. Coope, I. D., The Maratos effect in sequential quadratic programming algorithms using the exact L_1 penalty function, Report CS-85-32, Computer Science Department, University of Waterloo, August 1985.
2. Coope, I. D., and Watson, G. A., A projected Lagrangian algorithm for semi-infinite programming, *Mathematical Programming*, vol. 32, no. 3, pp. 337–356, 1985.
3. Fiacco, A. V., and Kortanek, K. O., Semi-Infinite Programming, and Applications, *Proceedings of an International Symposium*, Springer-Verlag, Berlin, 1983.
4. Gill, P. E., and Murray, W., The computation of Lagrange multiplier estimates for constrained optimization, *Mathematical Programming*, vol. 17, no. 1, pp. 32–60, 1979.
5. Gustafson, S. A., in *Semi-Infinite Programming, and Applications*, eds. Fiacco and Kortanek, pp. 138–157, Springer-Verlag, Berlin, 1983.
6. Luenberger, D. G., *Optimization by Vector Space Methods*, pp. 249–250, J. Wiley & Sons, 1969.
7. Mayne, D. Q., and Polak, E., A superlinearly convergent algorithm for constrained optimization problems, *Mathematical Programming Study*, vol. 16, pp. 45–61, 1982.
8. Price, C. J., PhD thesis, forthcoming.
9. Price, C. J., and Coope, I. D., The L_∞ penalty function in Semi-infinite Programming, Departmental Report, Mathematics Department, University of Canterbury, Christchurch, New Zealand, 1989.
10. Tanaka, Y., Fukushima, M., and Ibaraki, T., A globally convergent SQP method for semi-infinite non-linear optimization, *Journal of Computational and Applied Mathematics*, vol. 23, no. 2, pp. 141–153, 1988.
11. Watson G. A., Globally convergent methods for semi-infinite programming, *BIT*, vol. 21, no. 2, pp. 362–373, 1981.

Generating Information about Constrained Optima in the Design Decision Support Process

J. D. YANG, D. W. KELLY, and C. PATTERSON
School of Mechanical and Industrial Engineering
University of New South Wales
Kensington, NSW 2033, Australia

INTRODUCTION

Modern engineering design for manufacture is frequently concerned with producing near optimal designs for systems having multiple objectives which are not easy to characterize mathematically. For example, ease of manufacture and serviceability can be over-riding considerations. In addition the designer may deliberately choose not to approach a constraint which implies a bifurcation failure of the system such as buckling. In this environment the design engineer requires information in an interactive decision support process rather than fully automated optimization of the system.

Finite element software is used almost universally to provide information about the strength, stiffness and dynamic properties of the design. However this software has the capability of providing substantially more information. Sensitivity derivatives can be determined from the solution equations which are normally in the form of the linear algebraic equations

$$Ku = P. \tag{1}$$

Taking derivatives

$$K\frac{\partial u}{\partial x} = \frac{\partial P}{\partial x} - \frac{\partial K}{\partial x}u \tag{2}$$

and if $\sigma = DBu$, then

$$\frac{\partial \sigma}{\partial x} = DB\frac{\partial u}{\partial x} + \frac{\partial(BD)}{\partial x}u \tag{3}$$

where σ is the stress vector and u the displacement vector. These stress and displacement derivatives have long been used in optimization algorithms to guide the redesign process. In the application envisaged here they could be presented interactively to the design engineer to guide redesign of the structure to meet design objectives.

665

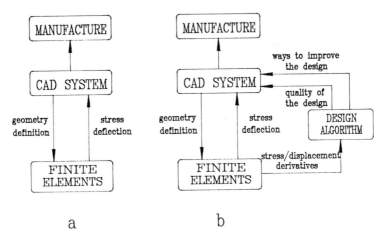

FIGURE 1: Linking of CAD and FEM

To extend further the information which can be made available we propose in this paper to view the design process as a convex optimization problem which requires the determination of the saddle point of the Lagrangean function. Primal and dual problems can then be identified such that feasible primal and dual points will bound the optimum of a nominated objective function. The Lagrange multipliers defining a non-optimal dual point indicate a possible constraint set which could satisfy the Kuhn-Tucker conditions of optimality.

These procedures were originally proposed by Kelly et.al.(1975 and 1977), but are now receiving attention because of the proliferation of interactive CAD/finite element systems such as AutoCAD/ANSYS and AutoCAD/STRAND5 implemented on microcomputers. In some cases the finite element analysis depicted in Eq.(1) can require several hours of central processing time and little consideration has been given to fully automating the design process. Here the design exercise is depicted in Fig.1a. The new procedures would be those depicted in Fig.1b providing the maximum decision support to the design engineer.

In the next section the procedures for obtaining information about the constrained optimum are defined. A redesign strategy is then suggested and applied to a simple problem. A practical application is considered briefly in the section Applications in Structural Design.

LAGRANGEAN FORMULATION

A well-posed optimal structural design problem may be stated as:

$$Minimize: \ f(x_1, x_2, \ldots, x_n) = f(\mathbf{x}) \tag{4}$$

subject to the satisfaction of a set of constraints which may consist of inequality constraints

$$g_i(\mathbf{x}) \leq 0, \quad i = 1, \ldots, l, \tag{5}$$

and equality constraints

$$g_i(\mathbf{x}) = 0, \quad i = l+1, \ldots, m. \tag{6}$$

The design variables are also usually restricted to be nonnegative:

$$x_i \geq 0, \quad i = 1, \ldots, n, \quad \mathbf{x} \in E^n \tag{7}$$

but these conditions can, if necessary, be included directly in the inequality constraint set (5) through additional constraints (such as $-x_i \leq 0$, i=1,...,n). Any design $\mathbf{x}$ which satisfies the constraints is said to be a feasible design, and a feasible design $\mathbf{x}^*$ which makes f($\mathbf{x}$) an absolute minimum is the optimum design.

It is well known that the optimization problem can be replaced by the problem of finding the stationary point of the Lagrangean function

$$L(\mathbf{x}, \lambda) = f(\mathbf{x}) + \sum_{i=1}^{m} \lambda_i g_i(\mathbf{x}), \quad \lambda_i \geq 0, \quad i = 1, \ldots, l. \tag{8}$$

which satisfies the Kuhn-Tucker conditions:

$$\frac{\partial L}{\partial x_j} = 0 \quad for \ x_j \in X \Longrightarrow \frac{\partial f}{\partial x_j} + \sum_{i=1}^{m} \lambda_i \frac{\partial g_i}{\partial x_j} = 0 \tag{9}$$

$$\frac{\partial L}{\partial \lambda_i} = 0 \quad for \ \lambda_i \in \lambda \Longrightarrow \lambda_i g_i(\mathbf{x}^*) = 0, \quad i = 1, \ldots, l, \tag{10}$$

where

$$\lambda = \{[\lambda_1, \ldots, \lambda_m]^t | \lambda_i \geq 0, \ i = 1, \ldots, l\}, \quad X = \{\mathbf{x} | x_j > 0, \ \mathbf{x} \in E^n\}.$$

Bounds on the optimum value of $f(\mathbf{x})$ follow by defining the primal and dual problems. Let us define the primal function to be

$$\bar{L}(\mathbf{x}) = \max_{\lambda_i \in \lambda} [L(\mathbf{x}, \lambda)] \tag{11}$$

and the dual function to be

$$\hat{L}(\lambda) = \min_{\mathbf{x}} [L(\mathbf{x}, \lambda)]. \tag{12}$$

It can be shown, for example Beightler et.al.(1979), that these two problems con verge to the Lagrangean saddle point

$$\hat{L}(\lambda) \leq L(\mathbf{x}^o, \lambda^o) = f(\mathbf{x}^o) \leq \bar{L}(\mathbf{x}) \tag{13}$$

where $\mathbf{x}^0$ defines the optimum value of the design variables.

Thus, for any $(\mathbf{x}^A, \lambda^A)$ which is feasible for the dual problem, the dual function $\hat{L}(\lambda^A)$ always provides a lower bound for the primal optimum

$$\hat{L}(\lambda^A) = f(\mathbf{x}^A) + \sum_{i=1}^{l} \lambda_i^A g_i(\mathbf{x}^A) \leq \bar{L}(\mathbf{x}^o) = f(\mathbf{x}^o). \tag{14}$$

Based on Eq.(14), the method of generating a lower bound can be established. For a given design $\mathbf{x}$, the Lagrange multipliers λ are determined by solving the linear equation (9)

$$\nabla g(\mathbf{x})\lambda = -\nabla f(\mathbf{x}). \tag{15}$$

The lower bound on the global minimum value of the merit function can then be found from the dual function

$$\hat{L}(\lambda) = f(\mathbf{x}) + \sum_{i=1}^{l} \lambda_i g_i(\mathbf{x}). \tag{16}$$

The number of variables n however does not always equal the number of constraints (including equality and inequality) m. In this case, the matrix $\nabla g(\mathbf{x})$ in Eq.(15) is not $m \times m$, and the equation can't be solved in a normal sense. The following linear programming problem is established to find the lower bound

$$\max_{\lambda} [L(\lambda)] = f(\mathbf{x}) + \sum_{i=1}^{l} \lambda_i g_i(\mathbf{x}) \tag{17}$$

s.t.

$$\nabla f(\mathbf{x}) + \nabla g(\mathbf{x})\lambda = 0 \quad \text{and} \quad \lambda \geq 0.$$

A feasible dual solution (all $\lambda_i \geq 0$) should always be possible if size limitations are placed on all the design variables.

A REDESIGN STRATEGY

The lower bound which is got from the dual analysis indicates the merit of a design. It is therefore useful in deciding when to terminate a redesign process. The bound will contain the global minimum and will therefore indicate whether the optimum design has been achieved. If not, the user must try to improved the design using the information which is available to him.

For a given design $\mathbf{x}$, the Lagrange multipliers λ which are found from Eq.(15) reflect the activity levels of the constraints at this stage. They provide an indication of a possible set of active constraints at the optimum with the linear programming algorithm eliminating non-active constraints by defining the associated λ_i to be zero. From Eq.(14), the difference of the primal function $f(\mathbf{x})$ and the lower bound $L(\mathbf{x}, \lambda)$ is $\sum \lambda_i g_i$. Therefore an attempt at reducing the absolute value of active constraints $|g_i|$ should be made in the redesign procedure, especially for the higher activity levels (larger λ). The constraint derivatives $\partial g_i/\partial x_j$ can be used as the guide for reducing $|g_i|$.

The following design procedure can be proposed:
(1) Define the initial design.
(2) Execute a finite element analysis to get deflection, stress and their derivatives.
(3) Generate the lower bound.
(4) Quality assessment. If the current design is close to the bound then stop.
(5) Determine the active constraints. Change the design to make the active constraints $g_i \rightarrow 0$. Go to (2).

This design procedure is suitable for commercial use in structural design. It is anticipated that the times required to generate the bound and apply the active set strategy will be of the same order as that for the initial finite element analysis.

A table designed to represent the relationships between variables and performance levels can be set up. This table has the following form:

TABLE 1: Rule-based Sensitivity Table (RST)

			x_1	x_2	x_3
g_1	λ_1	$\lambda_1 g_1$	$\partial g_1/\partial x_1$	$\partial g_1/\partial x_2$	$\partial g_1/\partial x_3$
g_2	λ_2	$\lambda_2 g_2$	$\partial g_2/\partial x_1$	$\partial g_2/\partial x_2$	$\partial g_2/\partial x_3$
g_3	λ_3	$\lambda_3 g_3$	$\partial g_3/\partial x_1$	$\partial g_3/\partial x_2$	$\partial g_3/\partial x_3$
g_4	λ_4	$\lambda_4 g_4$	$\partial g_4/\partial x_1$	$\partial g_4/\partial x_2$	$\partial g_4/\partial x_3$
$f(\mathbf{x})$	$L(\mathbf{x}, \lambda)$	$\sum \lambda_i g_i$	$\partial f/\partial x_1$	$\partial f/\partial x_2$	$\partial f/\partial x_3$

EXAMPLE. A THREE BAR TRUSS

The redesign strategy based on the rule based sensitivity table was used to execute the minimum mass design for the simple truss shown in Fig.2. The ends of the bars are pinned so that they carry only axial stress σ. The values of these stresses are to be limited to being less than the yield stress, and for ease of fabrication, the cross-sectional area A_i of each bar must not be less than 1 cm^2.

For simplicity the stress constraints which have the form $\sigma_i \leq \sigma_{lim}$ have been rewritten in the form $\sigma_i/\sigma_{lim} = \bar{\sigma}_i \leq 1$. The problem can therefore be posed as

$Minimize \quad f = mass$

subject to

$$g_1 = \bar{\sigma}_1 - 1.0 \leq 0, \quad g_2 = \bar{\sigma}_2 - 1.0 \leq 0,$$
$$g_3 = 1.0 - A_1 \leq 0, \quad g_4 = 1.0 - A_2 \leq 0.$$

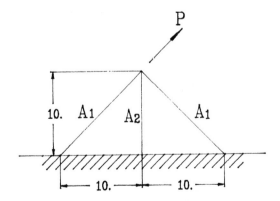

FIGURE 2: Design details for 3-bar truss

The structural mass, stresses, displacements and their derivatives are calculated by the finite element software STRAND5. The initial design is $A_1 = 8.0$, $A_2 = 8.0$, and the associated sensitivity table is given in Table 2. For this design the merit function has a value of 30.63 and the bound 7.14. Therefore redesign is required. The active constraints are g_1 and g_4. The redesign procedure is to reduce the value of $\sum \lambda_i g_i$ by forcing active constraints g_i to zero. From Table 2 this could be done by concentrating on $\lambda_1 g_1$ and reducing A_1 and A_2 in proportion to the gradients in the last two columns. With insight however A_2 can be reduced more rapidly because it is the only variable contributing to g_4.

After several iterations a near optimum design with a mass of 17.03 was achieved. The details of each iteration are given in Table 3. The corresponding lower bounds are also given to assess the merit of each design.

TABLE 2: Rule-based Sensitivity Table for 3-bar

	Initial design		$A_1 = 8.0$	$A_2 = 8.0$
g_i	λ_i	$\lambda_i g_i$	$\partial g_i/\partial A_1$	$\partial g_i/\partial A_2$
G1=-4.299E-01	4.788E+01	-2.058E+01	-5.916E-02	-1.222E-02
G2=-6.660E-01	0.000E+00	0.000E+00	-1.729E-02	-2.444E-02
G3= 7.000E+00	0.000E+00	0.000E+00	-1.001E+00	0.000E+00
G4=-7.000E+00	4.158E-01	-2.910E+00	0.000E+00	-1.001E+00
$f(A)$	$L(A, \lambda)$	$\sum \lambda_i g_i$	$\partial f/\partial A_1$	$\partial f/\partial A_2$
F= 3.063E+01	7.134E+00	-2.349E+01	2.832E+00	1.001E+00

	Redesign 1		$A_1 = 6.0$	$A_2 = 4.0$
g_i	λ_i	$\lambda_i g_i$	$\partial g_i/\partial A_1$	$\partial g_i/\partial A_2$
G1=-1.858E-01	2.499E+01	-4.645E+00	-1.133E-01	-3.338E-02
G2=-4.467E-01	0.000E+00	0.000E+00	-4.739E-02	-6.706E-02
G3=-5.000E+00	0.000E+00	0.000E+00	-1.001E+00	0.000E+00
G4=-3.000E+00	1.671E-01	-5.012E-01	0.000E+00	-1.000E+00
$f(A)$	$L(A, \lambda)$	$\sum \lambda_i g_i$	$\partial f/\partial A_1$	$\partial f/\partial A_2$
F= 2.097E+01	1.582E+01	-5.146E+00	2.832E+00	1.001E+00

TABLE 3: Designs for the 3-bar truss

Design	$A_1(cm^2)$	$A_2(cm^2)$	Mass(kg)	Bound
Initial	8.00	8.00	30.63	7.14
Redesign 1	6.00	4.00	20.97	15.82
Redesign 2	5.50	2.80	18.36	16.75
Redesign 3	5.10	2.80	17.22	16.95
Optimum	5.06	2.72	17.03	16.98

APPLICATIONS IN STRUCTURAL DESIGN

As a more practical example the machine casting shown in Fig.3 was also considered. The structure was drawn using AutoCAD and analyzed using the STRAND5. The sequence of designs is detailed in Table 4. The design variables are T_1, T_2 and T_3, the thickness of plate in the frame head, the base and ribs in the base respectively.

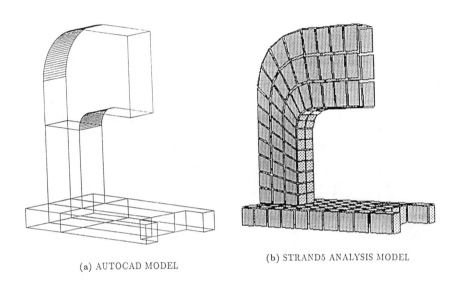

(a) AUTOCAD MODEL

(b) STRAND5 ANALYSIS MODEL

FIGURE 3: Design of a milling machine frame

TABLE 4: Design for the milling machine frame

Design	$T_1(cm)$	$T_2(cm)$	$T_3(cm)$	Mass(kg)	Bound
Initial	1.5	1.5	1.5	338.4	146.2
redesign 1	1.2	1.2	1.2	271.5	185.5
redesign 2	1.1	0.2	2.5	227.4	225.7

CONCLUDING DISCUSSION

Modern engineering design uses computer based software for drafting and analysis. In this paper we have shown that a interactive design process can be supported by decision support algorithms which do not simply present data on the current design, such as stress and displacement, but also provide information about the quality of the design and guidance on how to improve the design. These procedures could provide valuable support for a design engineer in those general engineering applications which do not warrant a fully automated optimisation process.

REFERENCES

1. AutoCAD, *AutoCAD Reference manual.* AUTODESK Australia Pty Ltd.

2. Beightler, C.S., Phillips, D.T. and Wilde, D.J., *Foundations of Optimization,* Prentice-Hall, 1979.

3. Kelly, D. W., A dual formulation for generating information about constrained optima in automated design, *Computer Methods in Applied Mechanics and Engineering.* no. 5, pp. 339-352, 1975.

4. Kelly, D. W., Morris, A. J., Bartholomew, P., A review of techniques for automated structural design, *Computer Methods in Applied Mechanics and Engineering.* no. 12, pp. 219-242, 1977.

5. STRAND5, *User's manual,* G+D Computing, 3 Smail St., Sydney, Australia.

GENERAL

Numerical Analysis of Some Three-Dimensional Problems in Jointed Rock Masses

H. ALEHOSSEIN, J. P. CARTER, and J. C. SMALL
Centre for Geotechnical Research
University of Sydney
Sydney, NSW 2006, Australia

INTRODUCTION

Natural rock masses are usually composed of blocks of intact material separated by planar joints or discontinuities. The mechanical behaviour of these rock masses is governed not only by the properties of the intact rock but also by the deformational characteristics of the discontinuity planes. In attempting to model this behaviour at least two approaches are possible: either the joints are included explicitly in any mathematical model of the mass, in which case the discontinuous nature of the rock mass is addressed directly, or their effects are incorporated implicitly in the choice of constitutive relations used to represent the mass as an equivalent continuum. The latter approach has the major attraction that it is computationally more efficient than the explicit approach. Its validity will of course depend on the scale of the problem, e.g. how widely spaced the joints are compared to the size of the loaded region. Furthermore, many problems in rock mechanics require a full three–dimensional treatment and there are many reasons why this may be necessary, e.g. the loading may lack symmetry or the rock mass may have important spatial variations of its characteristic properties. In such cases the numerical analyses required to solve practical problems can be very costly, particularly if the effects of jointing are to be included explicitly.

However, there is a significant number of applications where it is possible to make simplifying assumptions about the nature of the rock mass or the nature of the loading, so that simpler and less costly forms of analysis may be employed. Typical examples of this include the use of the assumptions of plane strain or axial symmetry and the implicit allowance for the effects of jointing, as described above. A special class of problems where simpli– fying assumptions are possible is addressed in this paper, and a convenient method of analysis, employing Fourier integral transforms and a finite element procedure, is presented.

PROBLEM DEFINITION

The problem addressed is that of a uniform vertical pressure, q, applied over a rectangle of dimensions BxL on the surface of a regularly jointed rock mass (see the inset to Fig. 2). Solutions are obtained for the settlement of the surface of the rock mass, which may contain either one, two or three sets of continuous joints. Of particular interest is the effect of jointing on the predicted settlements. All joints in any given set are assumed to be parallel and planar and they have sufficiently similar mechanical properties and spacing such that they can be represented by only two material constants, as described in detail in the following section. Where more than one joint set is present in the rock mass they are orthogonal. In the most general case considered here the joint planes may be inclined at oblique angles to the surface and the axes of the loaded area, but the individual joints remain parallel or orthogonal to each other. Furthermore, it is assumed that the blocks of intact rock that are defined by the intersecting joint planes are composed of isotropic, linear elastic material.

Although the basic components of the geomechanical model are relatively simple, the composite behaviour of the rock mass (intact blocks plus discontinuity planes or joints) will be linear but anisotropic. Details of this anisotropy are given below. This model is considered suitable for the prediction of the movements of foundations within the working load range of behaviour. Typical values of the material constants required in the model that are applicable to real rock masses have been suggested by a number of authors, e.g. Kulhawy (1978), Bandis et al (1983).

EQUIVALENT CONTINUUM MODEL

Consider an ideal rock mass where the intact material is isotropic and elastic, with Young's modulus E_r and Poisson's ratio ν_r. Initially, consider the case where the blocks of intact material are separated by a single set of parallel discontinuities or joints, as depicted schematically in Fig. 1. All joints have the same orientation and their spacing, S, is constant. The mechanical behaviour of each joint is characterized by an elastic shear stiffness K_s and a normal stiffness K_n, i.e. the shear and normal modes of joint behaviour are both linear and uncoupled.

A number of authors have described the overall mechanical behaviour of this type of discontinuous medium by a set of anisotropic compliance relations. A basic assumption of this formulation is that the vector of

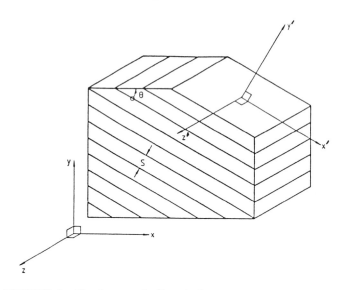

FIGURE 1. Single set of aligned discontinuities

strain components in the rock mass, ϵ' (measured with respect to a local coordinate system, x'y'z', which is attached to the joint plane), is given by the summation of two component vectors, i.e.

$$\epsilon' = \epsilon_c' + \epsilon_j' \tag{1}$$

where ϵ_c' and ϵ_j' are the 'strains' in the intact rock and the set of joints, respectively. ϵ_c' respresents the actual continuum strains in the rock blocks and the components of ϵ_j' are related to the shear and normal displacements on any joint plane divided by the uniform joint spacing, S. ϵ' may be related to the stresses in the local coordinate system, σ', via the compliance matrix, C', i.e.

$$\epsilon' = C'\sigma' \tag{2}$$

In a three–dimensional description, C' may be written as,

$$C' = (1/E_r) \begin{bmatrix} 1 & -\nu_r & -\nu_r & 0 & 0 & 0 \\ & 1+R_n & -\nu_r & 0 & 0 & 0 \\ & & 1 & 0 & 0 & 0 \\ & & & 2(1+\nu_r)+R_s & 0 & 0 \\ & & & & 2(1+\nu_r)+R_s & 0 \\ \text{symmetric} & & & & & 2(1+\nu_r) \end{bmatrix} \tag{3a}$$

in which: $R_s = E_r/(SK_s)$ (3b)

$R_n = E_r/(SK_n)$ (3c)

where: $\epsilon' = (\epsilon_{x'}, \epsilon_{y'}, \epsilon_{z'}, \gamma_{x'y'}, \gamma_{y'z'}, \gamma_{z'x'})^T$ (3d)

$\sigma' = (\sigma_{x'}, \sigma_{y'}, \sigma_{z'}, \tau_{x'y'}, \tau_{y'z'}, \tau_{z'x'})^T$. (3e)

The rigidity matrix for this material, D', is simply the inverse of the compliance matrix C'. As can be seen from equations (3), the joint relative stiffnesses, R_s and R_n, and the elastic properties of the intact blocks, E_r and ν_r, are the only material parameters required to define this geomechanical model.

In most problems it will be convenient to express the compliance relations in terms of a global reference frame (x,y,z), viz.

$$\epsilon = C \sigma \tag{4}$$

The matrix C may be obtained by a transformation of C', i.e.

$$C = TC'T^T \tag{5}$$

where T is the appropriate matrix for coordinate transformation. Explicit expressions for T for general and specific conditions can be found in standard texts (e.g. see Goodman, 1980, page 350).

It is not difficult to extend the formulation for a single set of parallel joint planes to include cases where multiple joint sets are present in the rock mass. Because the material response is linear, the overall strains can be regarded as the superposition of the approriate component parts corresponding to each of the joint sets and the intact blocks. Accordingly, it is possible to show that the compliance matrix for a rock mass with n different sets of joints may be expressed as,

$$C = C_c' + \sum_{i=1}^{n} T_i C_{ji}' T_i^T \qquad (6)$$

where T_i is the appropriate rotation matrix for joint set i having the form implied by equation (5), C_{ji}' represents the compliance of joint set i, C_c' is the compliance of the rock blocks only and n is the total number of joint sets in the rock mass.

FOURIER TRANSFORMS

There are many problems in geomechanics which are three dimensional in nature but for which a full three dimensional analysis is not necessary. An example of problems of this type are those for which the geological profile does not vary in one or two coordinate directions. Fourier integral transformation techniques may be applied to these problems (Small and Wong, 1988), and the finite element technique can be used to approximate the transformed field quantities in a plane containing two of the coordinate axes. Any important field quantity, e.g. $\delta(x,y,z)$, may be transformed by a Fourier integral to $\Delta(x,y,\alpha)$, where

$$\Delta(x,y,\alpha) = [1/(2\pi)] \int_{-\infty}^{+\infty} \delta(x,y,z) e^{-i\alpha z} dz, \qquad (7)$$

provided the material properties and problem geometry do not change in the z direction. Equation (7) may be written for all displacement and stress components to obtain their transforms. Assuming that the solution for a quantity in transform space is Δ, the actual solution δ may be obtained from the inversion formula,

$$\delta = \int_{-\infty}^{+\infty} \Delta \, e^{+i\alpha z} d\alpha, \qquad (8)$$

using numerical integration.

SURFACE DEFLECTION OF A LOADED RECTANGULAR AREA

The problem of a finite layer of jointed rock of depth H, subjected to a uniform vertical load acting on a rectangular area of length L and breadth B has been considered. A number of different types of rock mass were analysed and these include cases with one, two and three joint sets, a variety of joint orientations and a range of joint properties. Rectangles with three different aspect ratios, viz. L/B = 7, 2 and 1 were specifically investigated.

The results of this parametric study are presented in Figs 2 to 6, which show vertical deflections of the rock surface plotted against either the distance from the centre of the footing or the aspect ratio of the loaded rectangle. Carter and Alehossein (1989) have shown that by normalising the surface deflections of a jointed rock mass by the corresponding deflections for an isotropic layer with elastic properties the same as the intact rock (i.e. E_r and ν_r), the effect of layer depth (H) is only of minor significance in many practical problems. Hence, although the results presented in Figs 2 to 6 were obtained for the specific case where H/B = 5 they should have much wider application.

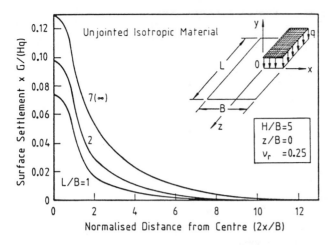

FIGURE 2. Rectangular loading on an unjointed layer.

Isotropic Rock Layer

For completeness, the surface settlements due to rectangular loadings applied to a homogeneous, isotropic (unjointed) elastic layer are included in Fig. 2. The vertical axis shows the surface settlement normalised in this case by the elastic shear modulus of the medium, G, the depth of the elastic layer, H, and the intensity of the surface loading, q. The settlements along a centreline ($z = 0$) have been plotted against the non-dimensional distance from the centre of the footing ($2x/B$), and the effect of footing shape is indicated by curves corresponding to different aspect ratios (L/B). Note that there are no distinguishable differences between the predictions for $L/B = 7$ and ∞ (the plane strain case) at this scale of plotting.

Effect of Joint Orientation

A study was made of the influence of one, two and three joint sets on the settlements of the rectangular footing. Three different joint sets were considered and these are designated as sets A, B and C. Each of the joints in set A is perpendicular to the y axis (i.e. the joint planes are horizontal), each joint in set B is perpendicular to the x axis (i.e. the joints are vertical and run parallel to the longer side of the rectangular footing), while all joints in set C are perpendicular to the z axis (i.e. these joints are also vertical but strike parallel to the shorter axis of the footing). All joint sets A, B and C have the same mechanical properties, characterized by relative stiffnesses $R_s = R_n = 1$. These values are reasonably typical for many rock masses encountered in practice (e.g. see Kulhawy, 1978)

The influence of one joint set on the surface settlements is indicated in Fig. 3. This shows the central settlement for a jointed rock mass normalised by the central settlement of an isotropic layer with elastic properties E_r and ν_r, (i.e. a rock mass with no joints) plotted against the aspect ratio of the footing, L/B. Results are given for three different rock masses, with each containing only one of the three sets A, B or C. It can be seen that the central settlement ratio is almost insensitive to the aspect ratio of the footing. Furthermore, the introduction of either of the vertical joint sets

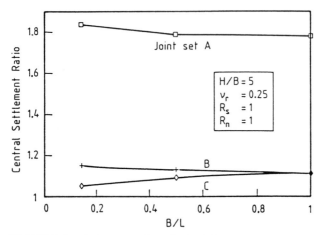

FIGURE 3. Influence of a single joint set

(i.e. B or C) with $R_n = R_s = 1$ into the rock mass causes at most an increase in footing settlement of only 15%. On the other hand, the introduction of horizontal joints with the same relative stiffness causes the settlements to increase by approximately 75 to 85%. This indicates that the vertical deflections of footings on jointed rock masses will be influenced largely by the normal stiffness of the joints and the greatest effects will occur when the jointing is horizontal or near horizontal.

The predicted settlements for cases in which two joint sets are present in the rock mass are plotted in Fig. 4. As for the case of a single joint set, the central settlement ratio is not very sensitive to the footing shape. Furthermore, the horizontal joint set has much more influence on the settlement ratio than either of the vertical joint sets.

Fig. 5 shows the predicted settlement ratios for rock masses containing three orthogonal joint sets, where each set is characterized by the same relative shear and normal stiffness. Curves have been plotted for the case

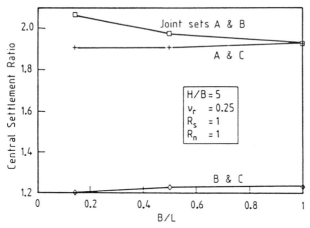

FIGURE 4. Influence of two joint sets.

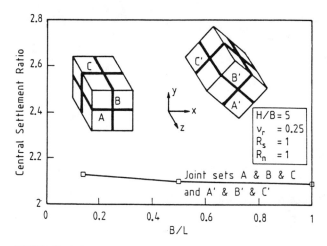

FIGURE 5. Influence of three joint sets

in which the three sets of orthogonal joints are parallel to the coordinate axes, i.e. one set (A) is horizontal and the other two (B and C) are vertical and parallel to the sides of the rectangular footing. Also plotted on Fig. 5 are predictions for the case where three joint sets A', B' and C' are included in the rock mass. These joints have the same properties as sets A, B and C, i.e. $R_s = R_n = 1$. Each of A' and B' is inclined at an angle of 45° to the x and y axes. C' is vertical and perpendicular to the z axis, i.e. the same as set C. Again the settlement ratio shows almost no dependence on the footing shape. As demonstrated in this figure, the orientation of the orthogonal sets A' and B' has no influence on the settlement for this particular class of problem. It can be shown that this result is true for any pair of orthogonal joint sets for which the joint stiffnesses $R_s = R_n$ are the same for each set, no matter what their orientation in the x–y plane (Carter and Alehossein, 1989).

Joint Stiffness

A limited study was also made of the influence of the joint stiffness on the foundation settlement. Specifically, square footings (L/B = 1) on rock masses in which the joints are characterized by $R_s = R_n = 1$ and $R_s = R_n = 10$ were examined and the results are given in Fig. 6. The three lower curves in this figure correspond to cases where $R_s = R_n = 1$ and the upper three curves to rock masses with softer joints, i.e. $R_s = R_n = 10$. In this figure the ratio plotted as ordinate is given by the settlement for the jointed rock mass divided by the central settlement of a square footing on an unjointed isotropic layer with elastic properties E_r and ν_r. These settlement ratios have been plotted against distance from the footing centre in the direction parallel to its shorter side. A number of interesting features are evident in this figure. It is clear that the relative stiffness of the joints has an important effect on the footing settlement, e.g. for cases in which $R_s = R_n = 10$ the presence of jointing causes the settlements near the centre of the footing to be increased by factors on the order of 10 above the settlements for an unjointed rock mass. The corresponding factor for rock masses with $R_s = R_n = 1$ is about 2. The greater significance of horizontal joints (set A) compared with vertical joints (sets B and C) is also evident.

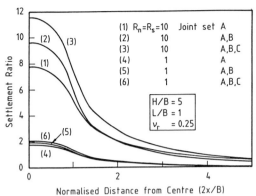

FIGURE 6. Influence of joint stiffnesses.

REMARKS

Three major assumptions were made in this work. Firstly, linear elastic behaviour was assumed for both the intact rock and the joints. Secondly, uncoupled spring models were adopted for the joints under shear and normal loading. Finally, in the present analysis there is no distinction between the normal stiffness of a joint in tension or one in compression. These are major simplifications of reality, necessary to permit tractable solutions to be found for some important boundary loading problems. It is well known that real joints may exhibit non-linear behaviour and a coupling of the shear and normal modes of behaviour (particularly dilatancy), especially at relatively high stress levels. However, the simplifications made here are consistent with the levels of data that are most often available to describe real joint systems and they are likely to provide a reasonable description of the rock mass response at working load levels in many instances. In an earlier study (Carter and Alehossein, 1989) the lack of distinction between the response of the joints in tension and compression was found to be relatively unimportant in this class of problems.

REFERENCES

1. Bandis, S.C., Lumsden, A.C., and Barton, N.R., Fundamentals of Rock Joint Deformation, *Int. J. Rock. Mechanics Mining Sciences & Geomech. Abstr,* vol. 20, no. 6, pp. 249–268, 1983.

2. Carter, J.P., and Alehossein, H., Settlement of Strip Foundations on Regularly Jointed Rock Masses, Foundation Engineering Congress, ASCE, Chicago, 1989.

3. Goodman, R.E., *Introduction to Rock Mechanics,* John Wiley and Sons, New York, 1980.

4. Kulhawy, F., Geomechanical Model for Rock Foundation Settlement, *J. Geot. Eng. (ASCE),* vol. 104, no. 2, pp. 211–227, 1978.

5. Small, J.C., and Wong, H.K.W., The Use of Integral Transforms in Solving Three Dimensional Problems in Geomechanics, *Computers and Geotechnics,* vol. 6, pp. 199–216, 1988.

The Influence of Modal Characteristics on the Dynamic Response of Compliant Cylinders in Waves

N. HARITOS
Department of Civil and Agricultural Engineering
University of Melbourne
Parkville, Victoria 3052, Australia

INTRODUCTION

The study of the dynamic response of a single compliant vertical cylinder subjected to hydrodynamic loading from uni-directional ocean waves is invaluable to engineers concerned with the design of offshore structures.

Such structures are observed to respond dynamically to the influence of hydrodynamic loading effects principally in their fundamental (or first mode) of vibration.

For a large class of offshore structures, Morison's equation (or its modification) can be used to evaluate the alongwave hydrodynamic loading (Sarpkaya & Isaacson, 1981). When the structure response amplitudes to this loading become significant, the modified form of Morison's equation leads to additional non-linearity due to the enhancement in the relative motion of the structure and the fluid produced by the vibration of the structure itself.

This paper seeks to identify the influence that modal characteristics (mode shape, natural frequency, stiffness, etc.) would have on the character of the dynamic response of a compliant vertical cylinder in uni-directional waves conforming to Pierson-Moskowitz spectra through the conduct of a range of simulation studies chosen to highlight these influences.

MODEL FOR THE DYNAMIC RESPONSE

A single-degree-of-freedom oscillator model can be used to describe the response of the vertical cylinder of diameter D in depth of water h, depicted in Fig.1, in its fundamental mode characterized by natural circular frequency ω_o, 'effective' stiffness k, damping ratio ζ and mode shape, $\psi(z)$.

The mode shape and natural frequency for such a cylinder can be evaluated from knowledge of how mass and stiffness are distributed along the structure and the manner in which the structure is supported. Classical solutions are possible in situations where these properties are relatively simple (e.g. a uniform cantilever) else some form of numerical approximation using any of the standard structural analysis computer packages possessing dynamics analysis capabilities may be used. Allowances here in the model for ω_o and $\psi(z)$ would need to be made for 'added mass' in the submerged portion of the cylinder and for the effects of buoyancy and the so-called 'P/δ' effect on the stiffness of the cylinder at the mean water level (Haritos[1], 1989).

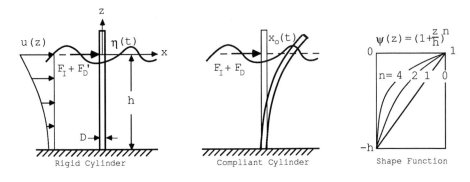

FIGURE 1.   Schematic of Vertical Surface-Piercing Cylinder and Function $\psi(z)$

When a Morison description for the hydrodynamic loading of this single-degree-of-freedom oscillator is adopted, the equation of motion for the time varying response x_o at the mean water level (MWL) can be written as

$$\ddot{x}_o + 2\omega_o \zeta_o \dot{x}_o + \omega_o^2 x_o = \frac{F_I + F_D}{M_o + M'} \tag{1}$$

where ω_o is the natural circular frequency of the cylinder in water, ζ_o is the basic value of 'effective' damping (assumed to be viscous) in water, M_o is the 'effective' mass of the cylinder at the MWL, and M' is the cylinder effective 'added mass' also at the MWL and

$$F_I \approx \int_{-h}^{0} \frac{\pi}{4} \rho \, C_M \, D^2 \, \dot{u}(z) \, \psi(z) \, dz \tag{2}$$

$$F_D \approx \int_{-h}^{0} \frac{1}{2} \rho \, C_D \, D \, (u(z) - \dot{x}_o \, \psi(z)) \, |u(z) - \dot{x}_o \, \psi(z)| \, \psi(z) \, dz \tag{3}$$

$$M' \approx \int_{-h}^{0} \frac{\pi}{4} \rho \, (C_M - 1) \, D^2 \, \psi^2(z) \, dz \tag{4}$$

in which $u(z)$ and $\dot{u}(z)$ represent the time varying alongwave water particle velocity and acceleration at level z respectively, for the uni-directional sea state under consideration.

F_I and F_D are the Morison equation representations for inertia force and drag, with C_M and C_D being, respectively, the inertia and drag coefficients for the cylinder, where 'relative velocity' has been considered responsible for the forcing (Sarpkaya & Isaacson, 1981). (The integrations in Eqs.(2) to (4) have been taken to the MWL in lieu of η, the instantaneous value of waveheight, as an approximation).

It is apparent from simple observation of the above equations that the form of $\psi(z)$ will directly influence added mass M' (and hence structure natural circular frequency ω_o) and will also directly influence the scaling for the loading (Eq.(2) and Eq.(3)) and indirectly influence it via M' in Eq.(1).

It is also apparent from inspection of Eq.(3) that the drag force is non-

684

linear and coupled with the response velocity $\dot{x}_o$.

Lipsett (1986) used several linearisation techniques on this drag term contribution in a solution of a form of Eq.(1) in which a unit length of cylinder was considered so that there was no variation in z for mode shape $\psi(z)$ or in the water particle kinematics, $u(z)$ and $\dot{u}(z)$, in his treatment. (Some of the approximations used would allow for direct simulation of the response in the frequency domain negating the need for a time step solution of the equation of motion). As may be expected, the results indicated that linearisation of the drag term alters the character of the modeled response and this effect is most dramatic in situations where the relative magnitude of drag to inertia force loading becomes increasingly significant (ie for high values of Keulegan-Carpenter number, K_c).

In situations where the drag force is non-negligible, therefore, the solution to Eq.(1) in the context used here can only be obtained numerically for any given condition of sea state. Standard time-stepping procedures such as Houbolt's method, Wilson θ and Newmark β methods, all of which have been traditionally used in earthquake engineering applications, can be utilized to solve for x_o in Eq.(1).

DESCRIPTION OF THE WAVE KINEMATICS

A uni-directional irregular sea state can be considered to be composed of a Fourier series of Airy wavelets with random phases and amplitudes that satisfy the requirements of a nominated spectral description. Here, the Pierson-Moskowitz form of wave spectrum is assumed viz

$$S_\eta(f) = \frac{0.0005}{f^5} \; e^{-5/4(f_p/f)4} \tag{5}$$

where f_p is the frequency at the peak in the spectrum and $f_p = 1.38/V$ where V is the value of characteristic mean wind speed responsible for producing the 'fully-developed' sea state.

Now, an Airy wavelet of amplitude A_i, circular frequency ω_i wave number κ_i and phase ϕ_i is given by

$$\eta_i = A_i \cos (\omega_i t - \phi_i) \tag{6}$$

and satisfies the dispersion relationship

$$\omega_i^2 = g \; \kappa_i \tanh \kappa_i h \tag{7}$$

For this Airy wave, the corresponding wave kinematics are given by

$$u_i(z) = \frac{\omega_i \cosh \kappa_i(z+h)}{\sinh \kappa_i h} \; \eta_i \tag{8}$$

$$\dot{u}_i(z) = \frac{\omega_i^2 \cosh \kappa_i(z+h)}{\sinh \kappa_i h} \; \eta_{i,\frac{\pi}{2}} \tag{9}$$

where $\eta_{i,\pi/2}$ is also of sinusoidal form but lags wavelet η_i by $\pi/2$.

The amplitude of the wavelets, A_i, is chosen to satisfy

$$A_i = \sqrt{2 \; S_\eta(f) \, df} \tag{10}$$

For a regular frequency increment of df = 1/T where T is the total time
length of record to be simulated, the number of wavelets used, say N/2, will
produce a total of N equally spaced ordinates of time trace via an Inverse
Fast Fourier Transform (IFFT) technique which would be Gaussian by virtue of
the Central Limit Theorem (Newland, 1983).

Equation (8), and an IFFT procedure similar to that used for the waveheight
trace can also be adopted to simulate the time series for the corresponding
water particle velocity traces at selected depth locations, z, viz u(z).

EVALUATION OF THE INERTIA FORCING

The intertia force F_I of Eq.(2) itself can also be simulated using an IFFT
procedure similar to that described above. A closed form solution to the
amplitudes of the 'wavelets' that apply to the simulation of the inertia
force trace can only be obtained for simple analytical representations of
mode shape $\psi(z)$.

Here, for the purposes of generality, $\psi(z)$ is considered to conform to a
'power law' profile with exponent n and is given by

$$\psi(z) = (1 + \frac{z}{h})^n \tag{11}$$

For n a zero or positive-valued integer, the amplitude of the inertia force
wavelet, F_{Ii}, can be obtained from the recursive relationship

$$F_{Ii} = \frac{\pi}{4} \rho g \ C_M \ D^2 \ I_n(\kappa_i h) \ \eta_i, \frac{\pi}{2} \tag{12}$$

where

$$I_n(\kappa_i h) = I_0(\kappa_i h) - \frac{n}{\kappa_i h} (1 - \frac{n-1}{\kappa_i h} I_{n-2}(\kappa_i h)) \qquad\qquad ,n \geq 2 \tag{13}$$

$$I_0(\kappa_i h) = \tanh \kappa_i h \qquad\qquad ,n = 0 \tag{14}$$

$$I_1(\kappa_i h) = I_0(\kappa_i h) - \frac{1}{\kappa_i h} (1 - \frac{1}{\cosh \kappa_i h}) \qquad\qquad ,n = 1 \tag{15}$$

Figure 2 presents plots of $I_n(\kappa h)$ for a selected range of values of n.

It is clear from these plots that the effect of high order values of n is to
increasingly attenuate, or filter out, the energy that would otherwise be

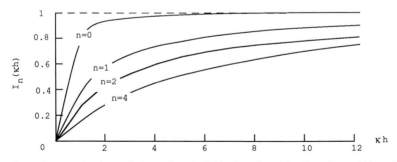

FIGURE 2. Variation of Function $I_n(\kappa h)$ for Inertia Forcing with values of n

available to promote this forcing and that this attenuation appears to affect the lower frequency wavelets more than those at higher frequencies. Further, for n greater than zero, the so-called 'deep water' limit (as far as inertia forcing is concerned) would be significantly greater than the traditional value for wave kinematics (viz, $\kappa h = \pi$).

The entire time length of trace for F_I can be simulated from an IFFT procedure on the wavelets with scaling as given by Eq.(12).

EFFECT ON ADDED MASS

The shape function for $\psi(z)$ specified by Eq.(11) can easily be integrated in Eq.(4) to obtain

$$\frac{M'}{M_w} = \frac{C_M - 1}{2n+1} \tag{16}$$

in which C_M has been assumed to be constant-valued and M_w represents the mass of fluid displaced by the cylinder in depth of water, h.

It is obvious from inspection of Eq.(16) that M' decreases monotonically with increasing values of n.

EFFECT OF THE DRAG FORCING

As discussed previously, the non-linear nature and coupling observed in the expression for drag force necessitates a numerical time step procedure for the solution to the response in which the drag force itself needs to be determined at each time step from the current value of $\dot{x}_o$ and from traces for $u(z_j)$ obtained at the relevant positions z_j, e.g.

$$F_D \approx \frac{1}{2} \rho \, C_D \, D \sum_{j=1}^{m} (u(z_j) - \dot{x}_o \psi(z_j)) |u(z_j) - \dot{x}_o \psi(z_j)| \, \psi(z_j) \, dz_j \tag{17}$$

where

$$h = \sum_{j=1}^{m} dz_j \tag{18}$$

and where the segment interval dz_j at level z_j and the number of segments m can be suitably chosen in order to 'optimise' on the numerical approximation to the integration involved in Eq.(2). Alternatively other acceptable forms of numerical integration can be used (e.g. Simpson's rule for m odd and dz_j evenly spaced).

EVALUATION OF THE RESPONSE

Equation (1) can now be solved using any suitable numerical integration scheme to obtain a time series for x_o at a regular time step corresponding to that used in the simulation for the forcing.

Details of the geometry (D, h, $\psi(z)$, ω_o, ζ_o, M_o) and level of sea state need be specified as well as initial conditions for the response.

Here, Newmark's β method has been used for a cylinder of diameter D = 2m, in depth of water h = 40m with C_M =2.0 and C_D =1.0, and for a range of Pierson-

Moskowitz sea states, (V = 8, 12, 16, 24 and 32 m/s) and for n = 0, 1, 2 and 4 in Eq.(11) for the shape of $\psi(z)$, by way of an example. A time step of 0.1 secs and a total time length of record of 819.2 secs, representing a total of 8192 points have been used in this simulation procedure for x_0.

In order to be able to demonstrate the influence mode shape would have in the drag forcing term of Eq.(1), a corresponding set of solutions to this response have also been obtained for the case where the 'relative velocity' assumption has been removed in the specification for the drag force, viz

$$F_D' \approx \int_{-h}^{0} \frac{1}{2} \rho \, C_D \, D \, u(z) \, |u(z)| \, \psi(z) \, dz \qquad (19)$$

The model for forcing based upon $F_I + F_D'$ is interpreted as the equivalent forcing on a 'rigid' cylinder or the 'unmodified' form of Morison's equation (UFM). Comparison of the response trace to such a forcing model with that obtained from the model retaining the relative velocity assumption as in Eq.(3) (i.e. the modified form of Morison's equation, (MFM)), would be indicative of the influence of structure/fluid interaction and the role that $\psi(z)$ would have in this interaction.

In order to highlight the influence of structure/fluid interaction and to also indicate the effects that the modal characteristics ω_0 and $\psi(z)$ (through the choice of exponent n in Eq.(11)) would have on the character of the resultant response, it is convenient to define a so-called peak 'wave factor', R, viz

$$R = \frac{k \, \bar{x}_{max}}{\sigma_I} \qquad (20)$$

where $\qquad \sigma_I = \frac{\pi}{4} \, \rho g \, C_M \, D^2 \, \sigma_\eta \qquad (21)$

and $\bar{x}_{max}$ represents the expected peak response in a one-hour record, k is the equivalent stiffness at the MWL in the model of Eq.(1) so that $k \, \bar{x}_{max}$ in effect is the expected value of peak spring force in this model. σ_I is interpreted as the 'deep water' limit to the standard deviation in the inertia force component of the forcing as obtained from Eqs.(2) and (12) where $I_n(\kappa h)$ has effectively been set equal to 1.

Quantity R would therefore be akin to the 'peak factor' approach of 'second moment' methods and would be representative of an aspect of the response characteristics of particular interest to a design engineer: the extremes.

Here x_{max} has been determined from an upcrossing analysis on the last 600 seconds of each 819.2 second simulation record for the response providing a generous allowance for any influence of 'transient' behaviour associated with the initial conditions in the earlier part of each record. (A Monte Carlo approach for ten separate sets of simulations for x_{max} all at the same conditions have in fact been conducted in order to estimate $\bar{x}_{max}$ and obtain an indication of the variability in this estimate).

A model in the form of a Weibull relationship has been fitted to these up-crossing observations and extrapolated to predict the level of x_{max} for a one hour period. Now if N_0 is the count of upcrossing at the mean response level (here taken as zero-valued) and N_x is the corresponding count for level x, then

$$\frac{N_x}{N_o} = e^{-\alpha (x/\sigma_x)^\beta} \tag{22}$$

where α and β are constants fitted using the method of least squares to observations of N_x/N_o for 100 regular increments in x in the range $\sigma_x \leq x \leq x_{max}$ and where σ_x is the standard deviation in response, x_o. (Values of $\alpha = 1/2$, $\beta = 2$ would be indicative of a Rayleigh variation for upcrossings and a Normally distributed x_o).

Figures 3(a) and 3(b) consider the separate influences of mass ratio, $M_r = (M_o + M')/M_w$, and of exponent n in the shape function $\psi(z)$, respectively, on the value for peak wave factor, R, when other modal characteristics and level of storm activity are held fixed at the values depicted on these figures. It is clear that the 'relative velocity' assumption in the MFM results in a reduction in response amplitudes and in the expected peak response in a one-hour storm, from that obtained from the corresponding UFM. The reduction is largest for 'milder' variations in shape function (low values of n) than for 'sharper' variations in this shape (high values of n). Conversely, this reduction appears largest for higher values of the mass ratio, M_r, than for lower values of this ratio.

Figures 4(a) and 4(b) consider the separate influences of storm intensity (wind speed V in the Pierson-Moskowitz spectrum, for fixed modal characteristics) and of frequency ratio f_o/f_p (where structure stiffness k is varied for otherwise fixed modal characteristics and constant level of storm intensity), respectively, upon the value of peak wave factor, R. (Parameters chosen for these representative results are as depicted on the figures).

Figure 4(a) indicates that the peak wave factor R decreases with storm intensity and a fixed value for f_o/f_p ratio more rapidly for the MFM than for the UFM. The variation in Fig.4(b) for the MFM indicates that R is sensitive to f_o/f_p ratio and differs significantly from the corresponding result for the UFM. These differences are largest at low values of f_o/f_p (between 0.75 and 2.0) and can be attributed to the influence of 'hydrodynamic damping' in the structure/fluid interaction (Haritos[1], 1989).

In all cases, the MFM produces values of peak wave factor R, lower than those corresponding to the UFM for the same conditions of simulation.

The values of α and β are observed to depart significantly from the Rayleigh values at the higher ratios of f_o/f_p in both the UFM and MFM approaches

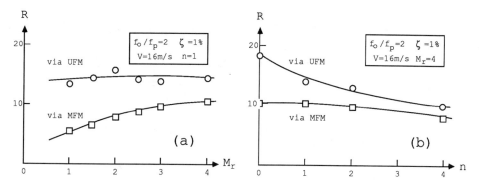

FIGURE 3.   The Variations of Peak Wave Factor with Mass Ratio and Exponent n

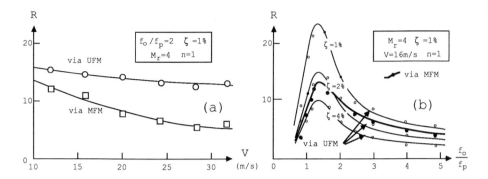

FIGURE 4.   The Variations of Peak Wave Factor with V and Frequency Ratio.

(because of the largely non-Gaussian nature of the drag force component) but approach these values closely for low ratios of f_o/f_p (between 0.75 and 2.0) when the 'resonant' contribution of the overall response approaches a 'narrow band' condition and becomes more Gaussian in character irrespective of the highly non-Gaussian nature of the forcing itself (Haritos[2], 1989).

CONCLUSIONS

Modal characteristics (ω_o, $\psi(z)$, M_r, ζ) significantly influence the character of the resultant dynamic response of compliant cylinders in irregular uni-directional Pierson-Moskowitz sea states.

The relative velocity assumption in the modified form of Morison's equation produces a form of 'hydrodynamic damping' due to structure-fluid interaction that plays a large role in determining peak response values for such a cylinder.   The nature of this role has been evidenced from simulation records and a comparison with the corresponding results for the unmodified form of Morison's equation.

REFERENCES

1. Haritos[1], N., Hydrodynamic Damping of Compliant Vertical Cylinders in Waves, forthcoming *Proc. Conf. Simulation Society Australia*, Canberra, Sept., 1989.

2. Haritos[2], N., The Distribution of Peak Wave Forces on Rigid and Compliant Cylinders in 2 and 3-Dimensional Sea States, forthcoming *Proc. 9th. Aust. Conf. on Coastal & Ocean Engineering*, Adelaide, December, 1989.

3. Lipsett, A.W., A Perturbation Method for Nonlinear Structural Response to Oscillatory Flow, *Applied Ocean Research*, vol. 8, no. 4, pp 183-189, 1986.

4. Newland, D.E., *An Introduction to Random Vibrations and Spectral Analysis*, Longman, London, 1983.

5. Sarpkaya, T., and Isaacson, M., *Mechanics of Wave Forces on Offshore Structures*, Van Nostrand Reinhold, New York, 1981.

Computational Approach to the Prediction of Erosion in Economiser Tubebanks

A. N. KITCHEN and C. A. J. FLETCHER
University of Sydney
Sydney, NSW 2006, Australia

INTRODUCTION

Flyash erosion of power station boiler tubes is a world-wide problem. A high percentage of boiler tube failures may be directly attributed to the thinning of tube walls caused by impaction by highly abrasive flyash. The consequent loss of availability and reduction of boiler life makes the understanding of the process and its alleviation a major economic concern. This paper describes a computational approach to the prediction of erosion within economiser tubebanks.

The computational model has to synthesize the flyash distribution and density upstream of the boiler tubebank; predict the trajectories of individual flyash particles through the tube bank, and predict and accumulate the erosion caused by each particle impaction. The computational model, in the form of a computer program, is to be used as 'black-box' design tool to identify local areas of very high erosion and to test strategies, such as introducing baffles, to reduce the erosion. The repeated running of the computer program implies that it must be very economical. The 'black-box' nature of its eventual usage requires that it should be very robust. These two requirements guide the choice of methods for the individual subtasks in the program.

DESCRIPTION OF PROGRAM

An interactive Fortran computer program has been developed as a design tool for the Electricity Commission of NSW. The program has been written to be user-friendly, achieved by a menu-driven design and prompts where applicable.

The approach has been to model the inlet ash burden conditions as discrete numbers of fundamental particles of selected sizes at discrete locations with appropriate initial velocities. Each of the fundamental particles are 'released' and its trajectory through the economiser is calculated. The overall erosion is then calculated by summing the effects of the individual particle/tube collisions, and scaling them appropriately as determined by the inlet distribution. This Lagrangian prescription is preferred to the Eulerian approach (Crowe, 1982) in order to reduce the CPU time. The program consists of several sections described below.

Calculation of dust burden. The inlet dust burden data is available as either actual measured values or calculated from knowledge of the ash content of the coal. A size distribution of the particles (typically in the range from 10 to 150 microns) must also be known or assumed.

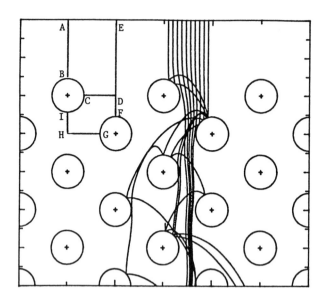

FIGURE 1. Typical particle trajectories; particle size = 50 microns, step size = 0.0001
$U_\infty = 10ms^{-1}$, tube size = $50mm$, tube spacing $150mm \times 100mm$,
inlet gas temperature = $543°C$, inlet water temperature = $254°C$.

Other parameters required are the gas flow velocity and the particle density. A number of discrete particle sizes (nps) is selected, and the number of inlet locations (npr) also specified. This latter parameter has a great influence on the accuracy of the overall calculation and will be considered below. The overall calculation time is approximately proportional to the product of the number of particle sizes and the number of release locations. Finally, the number of particles for each size/location combination is calculated as determined by the inlet distribution. Typically these numbers are of the order of 10^7.

Gas flow field. The underlying gas flow field determines to a large extent the severity and location of the erosion damage. The fluid velocities are calculated over the whole geometry and are stored for use by the trajectory calculation routines. Initially, the flow field has been solved and results obtained assuming inviscid flow. Further research to obtain the turbulent flow field is presently being undertaken.

Because of the inviscid flow assumption, only two simple geometries need to be considered to represent the entire flow. The first geometry is the half-width above the first row of tubes (ABCDE, Fig. 1). The second geometry is the fundamental internal domain (CDFGHI, Fig. 1). All other regions within the flow are reflections or rotations of these two. A non-uniform grid is constructed to ensure that grid lines are tangential to the tube surface at the zero and ninety degree angles around the cylinder. This provides a starting point for the velocity calculation near the surface described below. The Laplace equation for the stream function is discretised using three-point central finite difference formulae and the discrete equations are solved by an SOR scheme. The velocities at the grid points are obtained by central difference differentiation of the stream function values. Close to the curved boundary, mass flow conservation is successively applied to the grid cells whilst forcing tangential boundary velocities at the tube surface. This enables the grid point velocities in grid cells adjacent to the tube surface to be determined so that the correct particle trajectories are obtained. Because of the severe particle deceleration close to the tube surface (eq. (1) and Fig. 1), a failure to make this correction leads to inaccurate particle trajectories, which in turn leads to inaccurate predictions of the erosion (eq. (4)).

The stream function solution obtained from the entry region (ABCDE, Fig. 1) is applied as an inlet condition to the region within the tube bank (CDFGHI, Fig. 1). The velocities are obtained in the internal domain in a similar manner to that described above. These solutions may be written to external files, enabling a number of different flowfield solutions for different geometries to be stored and employed as required, without necessitating the computationally expensive process of solving the gas flow field every time a trajectory/erosion prediction is required.

The viscous flow solutions (velocity, pressure) will be obtained by a finite volume numerical solution of the incompressible turbulent Navier-Stokes equations. The influence of cylinder wakes, and turbulent boundary layer effects is expected to make a significant difference to particle trajectories, but not to the procedures to calculate the particle/tube collisions and the tube erosion. This will be reported in a later paper.

Particle trajectory calculation. The particle trajectories are calculated by integrating numerically the first order differential equations of motion. Non-dimensionalised with respect to particle diameter and free stream velocity, these are :

$$\frac{du_p}{dt} = \frac{3}{4}\frac{C_D}{s}\frac{1}{d_p}(u_g - u_p)V_s \tag{1}$$

$$\frac{dv_p}{dt} = \frac{3}{4}\frac{C_D}{s}\frac{1}{d_p}(v_g - v_p)V_s + \frac{ga}{U_\infty^2}(\frac{1}{s} - 1) \tag{2}$$

where the right hand sides indicate the component of the aerodynamic drag associated with the slip velocity V_s, given by :

$$V_s = ((u_g - u_p)^2 + (v_g - v_p)^2)^{0.5} \tag{3}$$

The subscripts p and g denote particle and gas phases respectively, and s is the ratio of particle and gas densities. The drag coefficient is determined from an empirical relation of the form $C_D = 24/Ref(Re)$ (Bauver et al, 1984). The equations are integrated by a fourth order Runge-Kutta scheme using velocities interpolated from the previously calculated grid point solutions. The interpolation routine sets a flag if the cylinder boundary has been crossed. The trajectory is then recalculated from the previous location using a smaller time step until the boundary is crossed again. The exact collision point may then be determined by finding the intersection between the boundary and the line of the last two trajectory points.

Presently, rebound behaviour is determined by empirical restitution relationships dependant upon the incident angle and particle velocity developed by Vittal and Tabbakof (1982) . It is anticipated that actual restitution data for the flyash in use will be employed later as experimental data becomes available. The particle trajectories are quite sensitive to changes in the rebound conditions, so accurate knowledge of these relationships is important. Fig. 1 shows some typical particle trajectories through the tube bank. The rebound behaviour and the tendency for the particles to migrate to a central channel in which no further tube collisions will occur is also what is observed in practice.

Erosion model The erosion model has been developed at the University of Sydney by incorporating empirical relations and experimental data available in the literature, e.g. (Levy et al, 1986). It was desired to find a model that was as universal as possible, requiring only modifications to various constants to account for the different erosion behaviour observed in practice. To this end, the model arrived at has identified the principal factors influencing the erosion and incorporated these effects in individual non-coupled terms. We have chosen the significant factors to be particle velocity, metal temperature, particle size, impact angle and the particle/target material combination. The form of the erosion model is:

$$e = K \cdot f(T/T_m) \cdot g(d_p/180) \cdot h(\beta/90) \cdot (V/V_{ref})^n \tag{4}$$

where
e	=	erosion rate (g/g)
K	=	constant
T	=	tube metal temperature $(^\circ C)$
T_m	=	reference temperature (e.g. melting temperature)
d_p	=	particle diameter (μm)
β	=	angular location around tube from the front stagnation point
V	=	particle velocity (ms^{-1})
V_{ref}	=	reference velocity (e.g. 20 ms^{-1})
n	=	velocity exponent ($n = 2.36$ here.)

For flyash impacting 304 stainless steel, the terms in eq. (4) are indicated below :

$$f(T/T_m) = 1.00 - 0.5\theta + 6.5\theta^2 \tag{5}$$

where $\theta = T/T_m$ and $T_m = 1430^\circ C$

$$g(d_p/180) = 1.0085529(1.0 - \exp(-4.77(d_p/180))) \tag{6}$$

$$h(\beta/90) = 2.29716\hat{\beta} - 1.605212\hat{\beta}^2 + 0.318942\hat{\beta}^3 - 0.010891\hat{\beta}^4 \tag{7}$$

where $\hat{\beta} = \beta/90$. The angle of maximum erosion, $\beta_{max} = 30.79^\circ$, and the ratio of the erosion at normal impact to that at angle β_{max} is 0.17. The value of K for this combination is 3.944×10^{-6}. For some effects, such as particle shape, there is insufficient data to establish a formal correlation. Thus K can be interpreted as containing all these secondary influences.

It can be seen from eqs (4) to (7) that the primary effect of an error in predicting the particle's trajectory will be experienced through the $h(\beta/90)$ and $(V/V_{ref})^n$ functions. Of these two functions, the sensitivity to (V/V_{ref}) is greater since $n = 2.36$. Additionally, an error in the trajectory will be magnified in the location of subsequent impacts. However, this factor may be compensated for by considering a large number of trajectories.

Erosion data handling. When a particle strikes a tube, the cylinder number, collision co-ordinates and particle velocity components are stored in an array. This array is referenced by a second array acting as a pointer. In this way, efficient data storage is achieved. After all the trajectory runs have been completed, each collision entry is used to calculate the eroded mass using eq. (4). Each cylinder is divided into 27 windward sectors comprising two 15 degree sectors, ten 10 degree sectors, and the remainder 5 degrees. The sector number and row number is determined from the collision coordinate data and the local tube metal temperature is calculated from an overall heat transfer balance. Together with the particle size and knowledge of the tube and particle materials, sufficient information is available for determining the erosion in the relevant tube sector, using eq. (4).

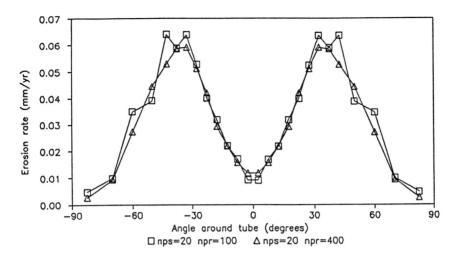

FIGURE 2. Influence of *npr* on the tube erosion map.

Erosion map production. After the erosion contributions to each sector have been calculated, an overall tube erosion map may be built up. As the released particles for the erosion
calculation were released only from one upstream sector (AE in Fig. 1), all the eroded sectors
in one row may be mapped onto one representative tube for each row to account for particle
migration from other inlet locations. In addition the results are also reflected to account for
contributions from the other side.

Each of the contributory eroded masses are scaled up by the number of particles determined
initially in the ash burden calculations and summated. An overall map can then be produced
showing the total erosion in each tube sector (see Figs. 2 and 3). A row-wise plot of maximum tube erosion is also available to highlight potential problem areas (Figs. 4 and 5).

RESULTS AND DISCUSSION

There are two principal parameters controlling the accuracy of the erosion calculations; the
number of particle release points (*npr*), and the number of discrete particle sizes (*nps*) on
AE in Fig. 1. The former will influence the spread of erosion damage on the typical tube – a
low number producing an erosion map featuring a few discrete peak damage areas, whereas
a larger number will produce a more uniform pattern, approaching that achieved in practice.
However as program execution time is approximately proportional to the product of these
two parameters, it is desirable to find the minimum values which will produce acceptable
predictions.

Fig. 2 shows the effect of the number of particle release points (*npr*) on the erosion pattern
on a tube in row 1. Increasing smoothness of the erosion pattern with increasing particle
release number may be seen here. The pattern for *npr* = 400 is a much smoother curve than
that produced for *npr* = 100, although the magnitudes of the erosion peaks are similar. The
smoothness of the curve indicates that a sufficient density of particles are being released to
represent the ash burden incident on row 1. If the value of *npr* is too small, a continuous
erosion pattern cannot be obtained, and the overall magnitude of the erosion is inaccurately

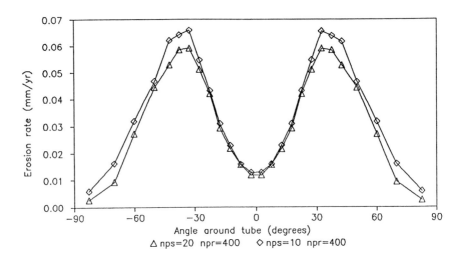

FIGURE 3.   Influence of *nps* on the tube erosion map.

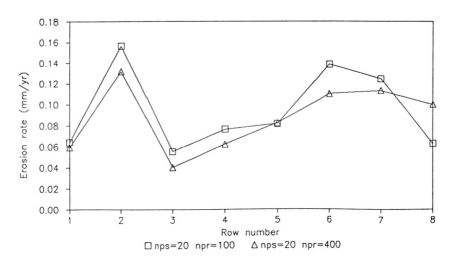

FIGURE 4.   Influence of *npr* on the row-wise erosion map.

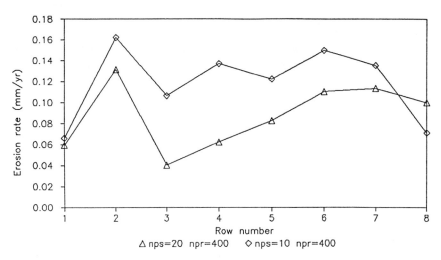

FIGURE 5.   Influence of *nps* on the row-wise erosion map.

predicted as a relatively large fraction of the dust burden finds its way between the tubes , producing no contribution to the final erosion values.

This distribution (Fig. 2), indicates a peak erosion at about 35° to 45° from the front stagnation point, a lower level of erosion in the front stagnation point region, and a fall-off in erosion as β approaches 90°. This distribution is similar to that found experimentally for isolated bodies (Bauver et al, 1984).

The effect of the choice of the number of particle sizes (*nps*) is shown in Fig. 3. The program chooses each particle size to be the mean of each of the particle size bands specified. The ash distribution used in this work is based on a sieving analysis obtained from samples taken at the inlet to the economiser. The ash particles used in the analysis are in the range from 1 to 200 microns. This means that as more particle sizes are included, additional smaller and larger particles are introduced into the inlet distribution. The particle size has several influences. The primary influence is on the the particle trajectory followed, and secondly, the erosion damage caused at impact. From eqs. (1) and (2) it can be seen that large particles, having higher inertia, are unable to follow curved streamlines Thus, large particles are involved in more collisions and, through eq. (4), cause more damage at each collision. Particles under 10 microns in size have been found to cause negligible erosion. However, as the erosion map is an integration of all erosion caused on all tubes lying in a certain row, the discrete trajectory effect tends to be minimised as long as sufficient release points (Fig. 2) are used.

For a given ash distribution at the entrance to the economiser, fewer particle size bands (*nps*) will imply more particles will be assigned to bands associated with larger particles. These follow trajectories that produce more collisions in the 35° to 45° sectors. Consequently, if *nps* is too small, the maximum erosion rate (which is the critical parameter to be estimated) will be overpredicted (Fig. 3).

The maximum erosion for each row is shown by Figs. 4 and 5. These graphs follow similar trends to their corresponding tube erosion plots Figs. 2 and 3. In all cases, the maximum erosion is found to have occurred in the second row. This is attributed to the second row's unique situation; receiving both primary impacts and a high percentage of secondary impacts from the first row. Intuitively, we would expect the erosion rate to decrease further into the tube bank as an increasing number of particles find their way into the channel between the tubes and no longer contribute to erosion (Fig. 1).

However, the larger particles that remain follow trajectories that impact mainly in the 35° to 45° sectors. These sectors are the ones in which maximum erosion occurs (Figs. 2 and 3, and eq. (4)). Consequently, the maximum erosion rate grows slightly between tuberows 3 and 6 or 7 (Figs. 4 and 5). Due to the loss of kinetic energy at each impact, these larger particles eventually end up in the inter-tube channel. This corresponds to the reduction in maximum erosion rate beyond tuberow 7 (Figs. 4 and 5).

CONCLUSIONS

A computer program has been developed to predict the location and magnitude of flyash erosion in an economiser tubebank. A Lagrangian approach to the prediction of erosion has been outlined which will form the basis of a powerful diagnostic tool. The present results show qualitative agreement with observed erosion damage. Further tuning is required to match observed data closely. The gas velocity field will be subsequently determined for turbulent flow in a variety of geometries, e.g. non-staggered tube banks and close to the boiler wall.

ACKNOWLEDGEMENTS

The authors are grateful to the Electricity Commission of NSW for their continuing support of the Boiler Tube Erosion Project, to which the above results are contributing.

REFERENCES

1. Crowe C. T., Review - Numerical Models for Dilute Gas-Particle Flows, *Journal of Fluids Engineering*, vol. 104, pp. 297-302, 1982.

2. Bauver W.P., Bianca J. D., Fishburn J. D., and McGowan J. G., Characterization of Erosion of Heat Transfer Tubes in Coal Fired Power Plants, Joint Power Generation Conference, Toronto, 1984.

3. Vittal B. V. R. and Tabakoff W., Two Phase Flow Around a Two-Dimensional Cylinder, *A.I.A.A. Journal*, vol. 25, no. 5, 1986.

4. Levy A.V., Yan J., and Patterson J., Elevated Temperature Erosion of Steels, *Wear*, vol. 108, pp. 43-60, 1986.

Nonlinear Dynamic Analysis of Shallow Shells by the Method of Spline Collocation

J. L. MEEK and LIN WENJING
Department of Civil Engineering
University of Queensland
St. Lucia, Q. 4067, Australia

XU CIDA
Department of Engineering Mechanics
Tongji University
PRC

INTRODUCTION

The geometric nonlinear dynamic analysis of shells is of considerable interest in many areas of engineering. Typical examples of such problems include computation of the dynamic response to loads due to explosive forces, earthquake loads, splashing waves and industrial dynamic loads in mechanical, aeronautical and marine engineering, etc. The complexity of the nonlinear problem requires that simple, efficient solution techniques be devised.

This paper suggests the use of the collocation method to analyze geometric nonlinear dynamic response of shells, using the cubic B-spline function as the trial function in the time domain. The equation of nonlinear motion of shells is derived using the Lagrangian equation. Setting residuals of the equation at the spline time nodes to zero, a recurrent cubic algebraic equation with a single variable is obtained. Solving the equation recurrently, successive constants at progressive time nodes can be found. As a result, the successive mode displacement, velocity and acceleration of a shell at any node may be obtained. Since the B-spline function has good approximation properties and is a compact expression, it becomes very convenient and easily programmable to solve problems of nonlinear dynamic response of shells by means of the method presented.

DERIVATION OF THE GOVERNING NONLINEAR EQUATION OF MOTION

In the problems of the geometric nonlinear dynamic response of shells, the magnitude of the deflection of the shells is of the same order of magnitude as the thickness of the shell. For the analysis of nonlinear dynamic response the influence of the deformation of the middle surface should not be ignored. Therefore, the relationship between the deformation of the middle surface and displacement should be (1. Timoshenko & Woinowsky 1959)

$$\varepsilon_{ij} = \frac{1}{2} \left(\frac{\partial u_i}{\partial x_j} + \frac{\partial u_j}{\partial x_i} - k_{ij}u_3 + \frac{\partial u_3}{\partial x_j} \frac{\partial u_3}{\partial x_i} \right)$$

$$\chi_{ij} = -\frac{\partial u_3^2}{\partial x_i \partial x_j} \tag{1}$$

By the Kirchhoff assumption, the strain at a point of a distance z from the middle surface is given by

$$\varepsilon'_{ij} = \varepsilon_{ij} + z\chi_{ij} \qquad (2)$$

The general constitutive equations of shells are

$$\sigma'_{ij} = C_{ijlm}\varepsilon'_{lm} \qquad (3)$$

The strain energy of shells is

$$U = \frac{1}{2}\int_v \{\sigma'\}^T\{\varepsilon'\}dv \qquad (4)$$

The analysis assumes that the shell is shallow and so the distortion of a thin shell loaded transversely can be described in terms of the curvatures of the middle surface. These curvatures are second derivatives of the transverse deflection u_3. Therefore, the displacements u_1 and u_2 in the plane can be neglected. Substituting eqtns (1)-(3) into (4), taking notice of u_3 being a function of x_1 and x_2 and being independent from x_3, the strain energy of shell can be written as

$$U = \frac{K}{2}\int_F\int\{[1-\frac{1}{2}[(\frac{\partial u_3}{\partial x_1})^2 + (\frac{\partial u_3}{\partial x_2})^2]-(k_1+k_2)u_3]^2 - 2(1-\mu)$$

$$[(\frac{1}{2}(\frac{\partial u_3}{\partial x_1})^2 - k_1 u_3)(\frac{1}{2}(\frac{\partial u_3}{\partial x_2})^2 - k_2 u_3) - \frac{1}{4}(\frac{\partial u_3}{\partial x_1}\frac{\partial u_3}{\partial x_2}$$

$$-2k_{12}u_3)^2]\}dx_1 dx_2 + \frac{D}{2}\int_F\int\{(\frac{\partial^2 u_3}{\partial x^2}+\frac{\partial^2 u_3}{\partial x^2})^2$$

$$-2(1-\mu)[\frac{\partial^2 u_3}{\partial x_1^2}\frac{\partial^2 u_3}{\partial x_2^2}-(\frac{\partial^2 u_3}{\partial x_1 \partial x_2})^2]\}dx_1 dx_2 \qquad (5)$$

in which

$$K = \frac{Eh}{1-\mu^2} \qquad\qquad D = \frac{Eh^3}{12(1-\mu^2)} \qquad (6)$$

For simply supported shells, the deflection of function is given as 7. Xu (1985)

$$u_3 = \sum_m \sum_n A_{mn}(t)\sin\alpha_m x_1 \sin\beta_n x_2 \qquad (7)$$

By using the standard expressions for the Lagrangian equation

$$\frac{d}{dt}\frac{\partial T}{\partial \dot{q}_i} - \frac{\partial T}{\partial q_i} + \frac{\partial U}{\partial q_i} - \frac{\partial W_d}{\partial q_i} = \frac{\partial W_e}{\partial q_i} \qquad (8)$$

the equations of nonlinear motion of the shells are derived

$$\ddot{A}_{mn}(t) + B_1\dot{A}_{mn}(t) + B_2 A_{mn}(t) + B_3 A_{mn}^2(t) + B_4 A_{mn}^3(t) = p_{mn}(t) \qquad (9)$$

$$(m, n = 1,3,5,\ldots\ldots)$$

700

in which

$$B_1 = \frac{C^*}{\bar{m}}$$

$$B_2 = \frac{1}{\bar{m}} \{ K[k_1^2 + k_2^2 + 2\mu k_1 k_2 + 2(1-\mu)k_{12}^2] + D\pi^4 (\frac{m^2}{a^2} + \frac{n^2}{b^2})^2 \}$$

$$B_3 = -\frac{4K}{3\bar{m}mn} [\frac{m^2}{a^2}(k_1 + \mu k_2) + \frac{n^2}{b^2}(k_2 + \mu k_1)] (\cos m\pi - 1)(\cos n\pi - 1)$$

$$B_4 = -\frac{K\pi^4}{16\bar{m}} [\frac{9}{2}(\frac{m^4}{a^4} + \frac{n^4}{b^4}) + \frac{m^2 n^2}{a^2 b^2}]$$

$$P_{mn}(t) = \frac{4qf(t)}{mn\pi^2 \bar{m}} (\cos m\pi - 1)(\cos n\pi - 1)$$

THE SOLUTION OF THE EQUATION OF MOTION USING SPLINE COLLOCATION

Let $A_{mn}(t)$ in eqtn (9) be represented by a cubic B-spline function (3 Li and Qi (1979))

$$A_{mn}(t) = \sum_{j=-1}^{r+1} c_j \, \Omega_3(\frac{t - t_j}{\Delta t}) \tag{10}$$

in which r is the number of the spline nodes, t_j the time at the spline node j such that $t_j = t_0 + j\Delta t$, and c_j (j $=-1,0,1,...,$ r+1) are coefficients to be determined. If t_i is also at the spline node i, $t_i = t_0 + i\Delta t$,then $(t_j - t_i)/\Delta t = $ i-j, t_0 being the initial time.

Therefore, eqtn (10) and its 1st and 2nd derivatives are

$$A_{mn}(t) = \sum_{j=-1}^{r+1} c_j \, \Omega_3(i-j)$$

$$\dot{A}_{mn}(t) = \sum_{j=-1}^{r+1} c_j \, \Omega_3'(i-j)/\Delta t \tag{11}$$

$$\ddot{A}_{mn}(t) = \sum_{j=-1}^{r+1} c_j \, \Omega_3''(i-j)/\Delta t^2$$

Substituting eqtn(11) into eqtn(9), the residual equation of motion is produced

$$R_I = \frac{1}{\Delta t^2} \sum_{j=-1}^{r+1} c_j \, \Omega_3''(i-j) + B_1 \frac{1}{\Delta t} \sum_{j=-1}^{r+1} c_j \, \Omega_3'(i-j) + B_2 \sum_{j=-1}^{r+1} c_j$$

$$\Omega_3(i-j) + B_3 [\sum_{j=-1}^{r+1} c_j \, \Omega_3(i-j)]^2 + B_4 [\sum_{j=-1}^{r+1} c_j \Omega_3(i-j)]^3 - P_{mn}(t) \tag{12}$$

According to the properties of the spline function, the values of the spline function and its first and second derivatives are defined at spline nodes with i-j =-1,0, +1. Equation (10) can be expressed as

$$A_{mn}(t) = \frac{1}{6}(c_{i-1} + 4c_i + c_{i+1})$$

$$\dot{A}_{mn}(t) = \frac{1}{2\Delta t}(-c_{i-1} + c_{i+1})$$ (13)

$$\ddot{A}_{mn}(t) = \frac{1}{\Delta t^2}(c_{i-1} - 2c_i + c_{i+1})$$

Setting the residuals of eqtn (12) at the spline nodes to zero, a recurrent cubic algebraic equation of single variable c_{i+1} is obtained

$$a_1 c_{i+1}^3 + a_2 c_{i+1}^2 + a_3 c_{i+1} + a_4 = 0$$ (14)

in which a_1, a_2 a_3 and a_4 are expressions of c_{i-1}, c_i, $P_{mn}(t)$, Δt and B_i. Substituting the initial conditions $A_{mn}(t_0)$, $\dot{A}_{mn}(t_0)$ and $\ddot{A}_{mn}(t_0)$ into eqtn(14), c_{-1}, c_0 and c_1 can be obtained. Then substituting these coefficients into eqtn(18) and solving it recurrently, successive constants c_{-1}, c_0, c_1, ... c_r,c_{r+1} at progressive time nodes can be produced. Making use of eqtn (13), the successive mode displacement, velocity and acceleration of the shell at any time node may be obtained. Superimposing the modal displacements, the dynamic displacement of the shell is produced.

NUMERICAL EXAMPLES

In order to show the simplicity, accuracy and efficiency of the method presented, three numerical examples have been selected for the large displacement dynamic response of shells.

Example 1. A simply supported hyperboloid shallow shell is shown in Fig. 1. The geometry is given

$a = b = 35m$ $h = 0.08m$ $R_1 = R_2 = 45.5m$

$E = 3*10^4 MPa$ $\mu = 0$ $\rho = 2.5^T/_{m^3}$

$\xi = 0.2$ $k_1 = k_2 = \frac{1}{R_1} = \frac{1}{R_2} = 0.021978^1/_m$ $k_{12} = 0$

the mass of unit area of shell $\bar{m} = 0.020408^T/_{m^2}$, the frequency of free vibration for simply supported hyperboloid shallow shell is

$$\omega_{mn} = \frac{1}{\bar{m}}\{K[k_1^2 + k_2^2 + 2\mu k_1 k_2 + 2(1-\mu)k_{12}^2] + D\pi^4(\frac{m^2}{a^2} + \frac{n^2}{b^2})^2]\}$$

and the damping coefficient $C^* = 2\bar{m} \xi\omega_{mn}$.

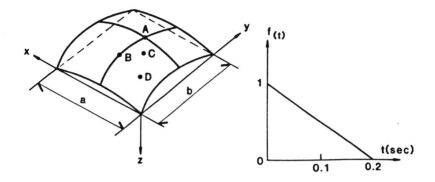

FIGURE 1. Hyperboloid shallow shell   FIGURE 2. Load as a function of t

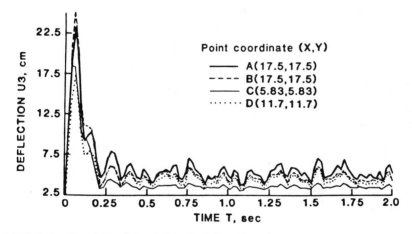

FIGURE 3. The deflection of the shell Vs. time t

The shell is subjected to uniform dead load , $q = 8^{T}/_{m}2$ and uniform impact load f(t) as shown in Fig. 2, the duration of which is 0.2sec. The initial conditions are $A_{mn}(t_0) = \dot{A}_{mn}(t_0) = 0$.

Superimposing the calculated mode displacement for m,n = 1,3,5, the dynamic displacement of the shell at different points are shown in Fig 3. From Fig. 3, it is seen that the amplitude decays quickly with the disappearing of impact force. This is the general observation for most examples.

Different boundary conditions can be incorporated by changing eqtn (7). As a result, the dynamic response of shells with different boundary conditions can be analyzed. For a concentrated load, q is replaced by

703

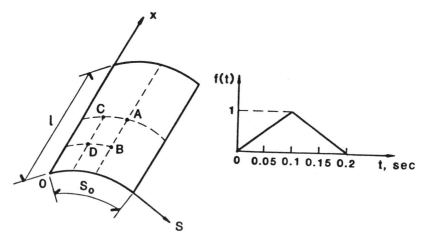

FIGURE 4. Cylindric shallow shell　　FIGURE 5. Load as a function of t

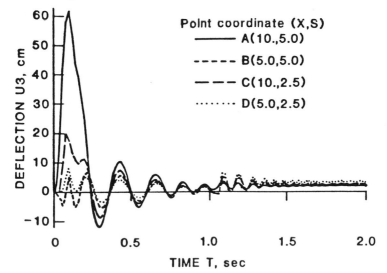

FIGURE 6. The deflection of the shell Vs. time t under the uniform load $p\delta(x-x_o)\delta(y-y_o)$. Here $\delta(...)$ is Dirac delta function and (x_o,y_o) is the point of the load applied.

Example 2. A simply supported open cylindric shallow shell is shown in Fig.4. The geometry is given

$$L = 20m \quad S = 10m \quad h = 0.08m \quad R = 50.m \quad \bar{m} = 0.002^T/_{m^2}$$
$$\zeta = 0.2 \quad k_1 = 0 \quad k_2 = \frac{1}{R} = 0.002^1/_m \quad k_{12} = 0$$
$$E = 3*10^4 MPa \quad \mu = 0 \quad p = 2.5^T/_{m^3}$$

704

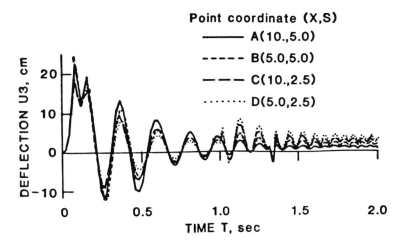

FIGURE 7. The deflection of the shell Vs. t under the concentrated load

In this example, the deflections of the shell under the uniform dynamic load are calculated. The load is a function of time as shown in Fig.5. In addition, the deflections of the shell under a concentrated load p applied at the central point of the shell is calculated (where p=q*L*S). The dynamic deflections of the shell at different points are shown in Figs.6 and 7. It is seen that the amplitude of the deflection at the point of the concentrated load is much larger than that of the uniform load when the instantaneous load applied. This is in accord with the general known solutions.

CONCLUSION

In the study presented, a recurrent cubic algebraic equation with a single variable is obtained making use of the cubic B-spline function as the trial function of mode displacements in the time domain. Putting the residuals of the differential equation of motion at spline nodes to zero and solving the equation recurrently, successive constants at progressive time nodes can be obtained. Successive mode displacements, velocities and accelerations of the shell at any time node can be calculated. Due to the good approximation properties and the compact expression of the B-spline function, the use of the method presented in this paper in solving the nonlinear dynamic response of shallow shells is convenient and easily programmable.

REFERENCES

1.   Timoshenko S., and Woinowsky S., *Theory of Plates and Shells*, McGraw-HillBook Company, New York, 1959.

2.   Xu Cida, "Dynamic Response of Plates and Shells by the Method of Spline Collocation", *Numerical Structural Mechanics and its Applications*, vol. 2 No. 1, pp72-76 1985. (in Chinese)

3.   Li Yuesheng and Qi Dongxu, *Methods of Spline Function*, Science Publishing House 1979. (in Chinese)

A Computationally Efficient Solution Technique to Laplacian Flow Problems with Mixed Boundary Conditions

W. W. READ
Department of Mathematics
James Cook University of North Queensland
Townville Q. 4811, Australia

R. E. VOLKER
Australian Centre for Tropical Freshwater Research
James Cook University of North Queensland
Townville Q. 4811, Australia

INTRODUCTION

Solutions to Laplace's equation are required for a wide range of fluid flow problems, where the assumption of a potential function is appropriate. A good example is creeping flow in porous media, where

$$\nabla^2\phi(x,y) = 0 \text{ within the saturated region } \Omega \tag{1}$$

$$\text{sub. to } a(x,y)\phi(x,y)+b(x,y)\frac{\partial\phi(x,y)}{\partial m} = c(x,y) \text{ on the boundary } \partial\Omega \tag{2}$$

(if $\partial/\partial m$ denotes differentiation normal to the boundary). Problems in this category include sloping hillslopes under irrigation, where knowledge of flow patterns and seepage face lengths is important in any hydrological management plan. The flow region Ω can be roughly described as a long, shallow permeable layer with an impermeable bottom boundary. The impermeable side boundaries may be almost vertical, but the upper and lower boundaries may have variable slope (Fig.1). Estimates of the potential function can be obtained using a variety of techniques, including conformal mapping, purely numerical schemes (such as finite element approximations or the boundary integral technique), and orthogonal function theory.

Analytical solutions to a variety of ground water flow problems have been obtained by conformal transformations, in the case of plane flow (Polubarinova-Kochina,1962). Solutions for sloping hillslopes with infinitely deep bottom boundaries (Warrick,1970) have been obtained by this method. However, this approach is unsuitable for all but the simplest plane geometries. Numerical solutions can be obtained by the finite element method (Choi,1978;Neiber&Walter,1981), but potential function values are not immediately available at non-nodal points. A closely related approach, the boundary integral method (BIEM), has also been applied to the hillslope seepage problem (Volker&Read,1987), but again without producing detailed information about the interior region.

Analytical solutions can also be obtained by expanding the

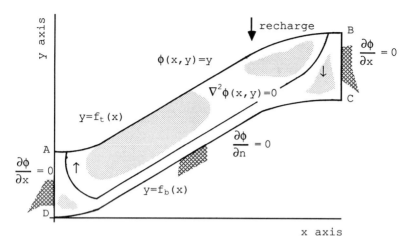

FIGURE 1. Hillside seepage flow region Ω and boundaries $\partial\Omega$.

velocity potential in terms of appropriate basis functions
(Powers et.al.,1967). Such series solutions to the hillslope
seepage problem have been obtained for a variety of surface
configurations and horizontal impermeable bottom boundaries at
finite and infinite depths. The solutions obtained give a
complete hydrological description of the entire flow region,
including the boundaries. This paper outlines an efficient
solution technique closely related to this approach and
classical least squares, applied to the hillside seepage
problem with arbitrary upper and lower boundaries. In the
following sections, the problem is formulated mathematically,
the new solution technique detailed and applied to four soil
geometries, with appropriate discussion and conclusions.

MATHEMATICAL PROBLEM

An example of the geometry of the permeable soil layer is
depicted in Fig. 1. Assuming the soil is homogeneous and
isotropic, so that seepage is governed by Darcy's law, the
velocities U, V in the x, y directions are given by:

$$U = -K\frac{\partial\phi(x,y)}{\partial x} \quad , \quad V = -K\frac{\partial\phi(x,y)}{\partial y} \qquad (3)$$

where $\phi(x,y)$ is the hydraulic potential, and K is the
(constant) hydraulic conductivity. Invoking the continuity
condition for incompressible fluids leads to Laplace's
equation: i.e.,

$$\frac{\partial^2\phi}{\partial x^2} + \frac{\partial^2\phi}{\partial y^2} = 0 \qquad (4)$$

Along the impermeable boundaries (AD,BC,DC), the normal
velocity is zero: i.e.

$$\frac{\partial\phi}{\partial x} = 0 \qquad (x=0,s) \qquad (5)$$

$$\frac{\partial\phi}{\partial m} = 0 \qquad (y=f_b(x)) \qquad (6)$$

The upper permeable boundary (AB) is subject to an applied
recharge, so the hydraulic potential is equal to the elevation
of the surface above an arbitrary datum: i.e.

$$\phi(x,y) = y = f_t(x) \tag{7}$$

Noting (5), the classical method of separation of variables therefore leads to a series solution :

$$\phi(x,y) = \sum_{n=1}^{\infty} u_n(x,y)A_n + \sum_{n=1}^{\infty} v_n(x,y)B_n + C_0 \tag{8}$$

where $u_n(x,y) = \cos\dfrac{n\pi x}{s}e^{\frac{n\pi y}{s}}$ and $v_n(x,y) = \cos\dfrac{n\pi x}{s}e^{\frac{-n\pi y}{s}}$ (9)

Now, adopting the approximate truncated form:

$$\phi(x,y) = \sum_{n=1}^{N} u_n(x,y)A_n + \sum_{n=1}^{N} v_n(x,y)B_n + C_0 \tag{10}$$

The constants C_0, A_n, B_n (n=1,...,N) can be evaluated using the upper and lower boundary conditions. The upper boundary condition (Eq.7) becomes

$$\phi(x,f_t(x)) = \phi_t(x) = \sum_{n=1}^{N} u_n(x,f_t(x))A_n + \sum_{n=1}^{N} v_n(x,f_t(x))B_n + C_0 \tag{11}$$

$$= \sum_{n=1}^{N} u_n^t(x)A_n + \sum_{n=1}^{N} v_n^t(x)B_n + C_0 \tag{12}$$

The bottom boundary condition (Eq.6) may be written

$$\frac{\partial\phi(x,y)}{\partial y} - f'_b(x)\frac{\partial\phi(x,y)}{\partial x} = 0 \tag{13}$$

where $f'_b(x)$ is the gradient of the bottom boundary, so

$$\sum_{n=1}^{N} u_n^b(x)A_n + \sum_{n=1}^{N} v_n^b(x)B_n + 0.C_0 = 0 \tag{14}$$

where

$$u_n^b(x) = \frac{\pi}{s}n(\cos\frac{n\pi x}{s} + f'_b(x)\sin\frac{n\pi x}{s})e^{\frac{n\pi f'_b(x)x}{s}} \tag{15}$$

$$v_n^b(x) = -\frac{\pi}{s}n(\cos\frac{n\pi x}{s} - f'_b(x)\sin\frac{n\pi x}{s})e^{\frac{-n\pi f'_b(x)x}{s}} \tag{16}$$

The bottom boundary condition can alternatively be expressed in terms of the conjugate (or _stream_) function $\psi(x,y)$ satisfying the Cauchy equations:

$$\frac{\partial\phi}{\partial x} = -\frac{\partial\psi}{\partial y} \, , \; \frac{\partial\phi}{\partial y} = \frac{\partial\psi}{\partial x} \tag{17}$$

Hence

$$\psi(x,y) = \sum_{n=1}^{N} us_n(x,y)A_n + \sum_{n=1}^{N} vs_n(x,y)B_n + C_1 \tag{18}$$

where $us(x,y) = \sin\dfrac{n\pi x}{s}e^{\frac{n\pi y}{s}}$, $vs(x,y) = -\sin\dfrac{n\pi x}{s}e^{\frac{-n\pi y}{s}}$ (19)

Along the impermeable bottom and side boundaries, the zero normal velocity condition (Eqs. 5,6) is equivalent to a constant stream function value. Choosing a value of zero along the bottom boundary, and taking the arbitrary constant C1 as zero, the bottom boundary condition can be expressed in terms of the stream function as

$$\psi(x,f_b(x)) = \psi_b(x) = \sum_{n=1}^{N} us_n^b(x)A_n + \sum_{n=1}^{N} vs_n^b(x)B_n + 0.C_0 = 0 \tag{20}$$

where $us_n^b(x) = us_n(x,f_b(x))$, $vs_n^b(x) = vs_n(x,f_b(x))$ (21)

SOLUTION TECHNIQUE

Previous researchers have exploited Gram-Schmidt orthonormalisation, when the bottom boundary is horizontal. In this case, a simple relationship exists between the coefficients (A_n, B_n) of the two sets of basis functions, and the problem is reduced to fitting the upper boundary conditions using one set of basis functions. For an arbitrary bottom boundary shape however, there is no longer any simple relationship between the coefficients.

Classical least squares involves forming and minimising the squared error of the upper and lower boundary approximations, and this produces a system of $(2N+1)\times(2N+1)$ linear equations:

$$E_t = \int_0^s (f_t(x) - \{\sum_{n=1}^N u_n^t(x)A_n + \sum_{n=1}^N v_n^t(x)B_n + C_0\})^2 dx \qquad (22)$$

$$E_b = \int_0^s (\sum_{n=1}^N u_n^b(x)A_n + \sum_{n=1}^N v_n^b(x)B_n)^2 dx \qquad (23)$$

Ignoring the constant term C_0 for the moment, minimisation w.r.t. A_j leads to a set of $2N$ equations:

$$\sum_{n=1}^N <u_j^t, u_n^t>A_n + \sum_{n=1}^N <u_j^t, v_n^t>B_n + <u_j^t, 1>C_0 = <u_j^t, f_t> , \quad j=1,\cdots,N \qquad (24)$$

$$\sum_{n=1}^N <u_j^b, u_n^b>A_n + \sum_{n=1}^N <u_j^b, v_n^b>B_n + 0.C_0 = 0 , \quad j=1,\cdots,N \qquad (25)$$

where $<a,b>$ denotes an inner product:

$$<a,b> = \int_0^s a(x)b(x)dx \qquad (26)$$

A similar set of equations results when the minimisation is performed w.r.t. B_j: viz.

$$\sum_{n=1}^N <v_j^t, u_n^t>A_n + \sum_{n=1}^N <v_j^t, v_n^t>B_n + <v_j^t, 1>C_0 = <v_j^t, f_t> , \quad j=1,\cdots,N \qquad (27)$$

$$\sum_{n=1}^N <v_j^b, u_n^b>A_n + \sum_{n=1}^N <v_j^b, v_n^b>B_n + 0.C_0 = 0 , \quad j=1,\cdots,N \qquad (28)$$

However, it is not clear whether the minimisation should be carried out w.r.t. A_j or B_j, or a combination of both.

The solution technique detailed in this paper hinges on developing a set of equations, based on the form of Eqs. 24,25,27,28. Note that $\cos(j\pi x/s)$ is a common term in all of them, whereas the exponential terms vary in the sign of the exponent. Also, the cosine term does not involve the equations for the upper and lower boundaries. This suggests that the least squares solution can be approximated by replacing the first basis function in all the terms of the form $<a,b>$ with $r_j(x)=\cos(j\pi x/s)$, resulting in a single set of $2N$ equations, consistent with both Eqs. 24,25 and Eqs. 27,28. The equation for the constant term C_0 is generated by setting j to zero in Eq.24 or Eq.27. The matrix equation generated can then be solved for the coefficients A_j and B_j. In matrix form, the system of equations becomes:

$$U\mathbf{g} = \mathbf{h} \qquad (29)$$

where
$$\mathbf{g}_i = A_i , \quad i=1,\ldots,N$$
$$= B_{i-N} , \quad i=N+1,\ldots,2N \qquad (30)$$
$$= C_0 , \quad i=2N+1$$

$$\mathbf{h}_i = <r_i, f_t> , \quad i=1,\ldots,N$$
$$= 0 , \quad i=N+1,\ldots,2N \qquad (31)$$
$$= <r_0, f_t> , \quad i=2N+1$$

for i=1,···,N , $U_{ij} = <r_i, u_j^t>$, j=1,···,N

$= <r_i, v_{j-N}^t>$, j=N+1,···,2N

$= <r_i, 1>$, j=2N+1

for i=N+1,···,2N , $U_{ij} = <r_{i-N}, u_j^b>$, j=1,···,N

$= <r_{i-N}, v_{j-N}^b>$, j=N+1,···,2N $\qquad$ (32)

$= 0$, j=2N+1

for i=2N+1 , $U_{ij} = <r_0, u_j^t>$, j=1,···,N

$= <r_0, v_{j-N}^t>$, j=N+1,···,2N

$= <r_0, 1>$, j=2N+1

When expressed in terms of the stream function, minimising the error in the bottom boundary condition leads to a set of 2N equations similar to Eqs. 25,28:

$$\sum_{n=1}^{N} <us_j^b, us_n^b>A_n + \sum_{n=1}^{N} <us_j^b, vs_n^b>B_n + 0.C_0 = 0 , \quad j=1,···,N \qquad (33)$$

$$\sum_{n=1}^{N} <vs_j^b, us_n^b>A_n + \sum_{n=1}^{N} <vs_j^b, vs_n^b>B_n + 0.C_0 = 0 , \quad j=1,···,N \qquad (34)$$

Inspection of each of the terms in Eqs. 33,34 reveals that $rs_j=\sin(j\pi x/s)$ is a common factor, so the solution process can be modified by replacing the leading basis function in all the bracketed terms in the expression with rs_j. The vectors **g,h** remain the same, with the matrix U_{ij} (Eq.32) replaced by Us_{ij}, where Us_{ij} is given by:

for i=1,...,N , $Us_{ij} = U_{ij}$, j=1,...,2N+1

for i=N+1,···,2N , $Us_{ij} = <rs_{i-n}, us_j^b>$, j=1,···,N

$= <rs_{i-n}, vs_{j-N}^b>$, j=N+1,···,2N $\qquad$ (35)

$= 0$, j=2N+1

for i=2N+1 , $Us_{ij} = U_{ij}$, j=1,...,2N+1

RESULTS

The solution technique detailed previously involves evaluating and solving a matrix equation for the series coefficients C_0, A_i, B_i, i=1,...,N in Eq.10. In order to evaluate the matrix equation, the upper and lower boundaries $f_t(x)$ and $f_b(x)$ must be specified. Four representative hillslope geometries have been selected to demonstrate the solution technique (c.f. Tab.1). Potential solutions have been obtained, using both the velocity and stream function lower boundary conditions. Most of the terms in the matrix equation involve integrals, which have been evaluated numerically using the IMSL library routines, to the precision of the machine. The upper limit N to the summation in Eq.10 must first be prescribed. Some experimentation revealed that N=10 gave very good results (larger values did not significantly improve the accuracy of the approximation), so all the computations involved 21 terms (to approximate the velocity potential). The resulting matrices proved to be well conditioned, and the corresponding system of equations were solved using IMSL routines, once again to the precision of the machine.

In order to compare the results using both types of bottom boundary conditions, the minimum recharge R_m, stagnation point coordinates (X_s, Y_s), and the root mean square (rms) errors e_p^t, e_v^b, e_s^b of the upper and lower boundary approximations have been calculated. The mean square error e^2 of an approximation of g(x) by h(x) can be defined as

711

TABLE 1.   Lower and upper flow region boundary equations

Region		Lower boundary	Upper boundary
1	$f_b(x)$	$=2.5\times10^{-2}x^2, \quad 0\le x<10$ $=0.05x-0.25, \quad 10\le x<30$ $= 1.25\times10^{-3}x^2+0.125x-1.375,$ $\quad\quad 30\le x\le50$	$f_t(x)=f_b(x)+1, \quad 0\le x\le50$
2	$f_b(x)$	$=0.05x, \quad\quad\quad 0\le x<50$	$f_t(x) \begin{cases}=2.5\times10^{-2}x^2+1, 0\le x<10\\ =0.05x-0.25, 10\le x<30\\ =1.25\times10^{-3}x^2+0.125x\\ \quad -1.375, \quad 30\le x\le50\end{cases}$
3	$f_b(x)$	$=-4.0\times10^{-4}x^2+0.05x, 0\le x<50$	$f_t(x)\begin{cases}= 5.0\times10^{-3}x^2, 0\le x<5\\ =0.05x-0.875, 5\le x<45\\ =-5.0\times10^{-3}x^2+0.5x\\ \quad -9.25, \quad 45\le x\le50\end{cases}$
4	$f_b(x)$	$=1.0\times10^{-3}x^2 \quad ,0\le x<50$	same as for Region 3

$$e^2 = <(g-h), (g-h)>/<1,1> \tag{36}$$

The rms errors (obtained by taking the square root of Eq.36) correspond to the upper boundary potential approximation (e^t_p), the lower boundary zero velocity condition (e^b_v), and the lower boundary zero stream function condition (e^b_s). The minimum recharge necessary to keep the slope saturated will be the free surface fluid velocity at the highest elevation. This occurs at x=s, where the velocity is vertical. i.e.,

$$R_m = -(\partial\phi(x,y)\setminus\partial y)|_{x=s,y=ft(s)} \tag{37}$$

The stagnation point ψ_s (i.e., (X_s, Y_s)) occurs at the minimum value of the stream function on the upper boundary. ψ_s was determined numerically, by dividing the free surface into a grid of 1000 points and selecting the smallest value of the stream function $\psi(x, f_t(x))$.

Table 2 contains estimates of R_m, X_s, Y_s, e^t_p, e^b_v and e^b_s for each of the four flow regions. Surprisingly, the bottom velocity boundary condition can lead to smaller errors than the stream function condition. (In the case of regions 1 and 2, note that e^b_v and e^b_s are roughly half the size, whereas the errors for regions 3 and 4 are of similar size.). However, all the approximations seem fairly good, with reasonable agreement between the sets of values for R_m, X_s, Y_s. Regions 1 and 2 require roughly 10% ~ 20% of the recharge necessary to keep regions 3 and 4 saturated, possibly due to their shallower depth and smaller flow volume. A similar line of reasoning can

TABLE 2. R_m, X_s, Y_s, e^t_p, e^b_v and e^b_s for flow regions 1,2,3,4

	Region 1		Region 2		Region 3		Region 4	
	str.	vel.	str.	vel.	str.	vel.	str.	vel.
$R_m \times10^{-2}$K	3.78	3.86	1.79	1.15	30.9	30.9	18.3	18.3
X_s	13.1	13.3	13.9	13.6	42.5	42.5	23.5	23.4
Y_s	1.40	1.41	1.45	1.43	3.00	3.00	2.05	2.04
$e^t_p\times10^{-4}$	5.74	5.74	6.00	6.00	12.4	12.4	12.1	12.1
$e^b_v\times10^{-5}$	0.256	0.170	1.09	0.482	1.11	1.09	2.87	2.89
$e^b_s\times10^{-4}$	0.691	0.330	3.37	1.34	2.19	2.11	5.96	5.99

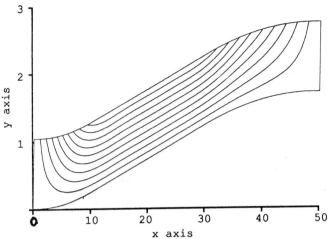

FIGURE 2. Normalised stream function plot for Region 1

be used to explain the difference between the recharge rates of
region 1 compared with region 2, and region 3 compared with
region 4. The surface geometry of regions 1 and 2 is the same,
and the marked similarity of the results for the stagnation
point indicates that this is a principal consideration in
determining seepage surfaces for shallow, rolling hillslopes.

Figures 2 and 3 show plots of normalised streamlines for
flow regions 1 and 3. The contours correspond to 10%
increments of the volume of fluid seeping through the region.
The streamlines were calculated numerically using IMSL
routines, by solving an implicit equation for the normalised
stream function:

$$\Psi_n(x,y) = p/100 = (\Psi(x,y) - \Psi_{min})/(\Psi_{max} - \Psi_{min}) \ , \ 0 \leq p \leq 100 \qquad (38)$$

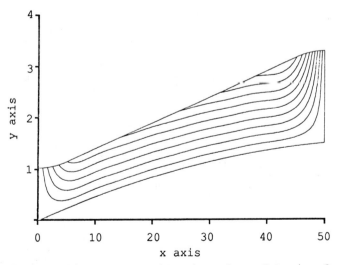

FIGURE 3. Normalised stream function plot of Region 3

713

The minimum value of the stream function is ψ_s; the (constant) maximum occurs on the impermeable boundary, and can be estimated by $\psi(0, f_t(1))$. Inspection of Figs. 2,3 reveals the location of possible soil erosion and salt trapping. Streamlines that are relatively far apart indicate low fluid velocities, and hence an area for solute deposition. Alternatively, relatively adjacent streamlines indicate high velocities and hence maximal erosive pressure.

DISCUSSION AND SUMMARY

The least squares approach caters for arbitrary bottom boundary shape, by using the bottom boundary conditions to evaluate a numerical relationship between the basis functions. This entails generating and solving a system of equations, to evaluate the coefficients for both sets of basis functions. The bottom boundary conditions can be expressed in terms of a velocity condition or a stream function condition, when the base is impermeable. The velocity condition appears to produces slightly more accurate results.

This solution technique provides a detailed solution to the problem, through the velocity potential. The solution technique is computationally efficient, compared to numerical schemes such as the finite element method and the boundary integral equation method (BIEM). For example, BIEM solutions for long, shallow sloping hillslopes require the generation and solution of a system of equations typically involving 100 or more variables, with the number of variables increasing in proportion to the length of slope. On the other hand, the least squares approach requires between 20 and 30 variables, regardless of the length of slope. Essentially, the complexity of the problem is reduced by an order of magnitude (i.e., from two dimensional to one dimensional). The use of a velocity potential in this technique also simplifies the calculation of equipotential lines and streamlines, by providing an analytical solution for the velocity potential.

REFERENCES

1. Choi,E.C.C., A Finite Element Study of Steady State Flow in an Unconfined Aquifer Resting on a Sloping Bed, *Water Resour.Res.,* Vol. 14, No.3, pp.391-394,1978.

2. Niebler, J.C. and Walter, M.F., Two - Dimensional Soil Moisture Flow in a Sloping Rectangular Region: Experimental and Numerical Studies, *Water Resour. Res.,* Vol. 17, No. 6, pp. 1722-1730, 1981.

3. Polubarinova-Kochina, P. YA., *Theory of ground water movement* Translated by J.M.R.De Wiest, Princeton University Press, 1962.

4. Powers, W.L. Kirkham, D. and Snowden, G., Orthonormal function tables and the seepage of steady rain through soil bedding. *J. of Geophys. Res.,* Vol. 72, No. 24, pp. 6225-6237, 1967.

5. Volker R.E. and Read W.W., Numerical and Analytical Solutions for Seepage Face Lengths on Hillslopes Under Irrigation, *Proc. Conf. on Hydraulics in Civil Engineering,* I.E. Aust., pp. 116-121, 1987.

6. Warrick, A.W., A mathematical solution to a hillside seepage problem. *Soil Sci. Soc. Amer. Proc.,* Vol. 34, pp. 849-853, 1970.

Low Dimensionality—Approximations in Mechanics

A. J. ROBERTS
Department of Applied Mathematics
The University of Adelaide
G.P.O. Box 498
Adelaide 5001, South Australia

INTRODUCTION

In many physical problems it is recognised that the physical system evolves quickly to some characteristic state of balance, and then evolves relatively slowly. To put it another way, we often divide the behaviour of a complex physical system into two parts: exponentially decaying transients which are of no interest, and the remaining long-time relatively-slow evolution. The division between the two parts is arbitrary, although in some problems it may be clear-cut. More mathematically, the system settles exponentially quickly onto some curving hyper-surface in the solution space on which it evolves relatively slowly. Such a hyper-surface is called an invariant manifold and contains all the interesting long-time behaviour of the physical system. In particular, such an invariant manifold contains the attractor at large time, whether it is simple or strange.

Invariant manifolds are not only relevant to problems of the above dissipative nature, they are implicitly assumed in a wide variety of common approximations in mechanics. Muncaster (1983[5]) shows how invariant manifolds arise in the derivation of "coarse grain" theories from "fine grain" theories. For example, the derivation of the equations of continuum mechanics from the particle equations of a monatomic gas, and the derivation of full body dynamics from the equations for an elastic continuum. Another important occurrence is in quasi-geostrophic models of the atmosphere, as discussed by Vautard & Legras (1986) and Lorenz (1986), where the relatively fast gravity-wave oscillations of the atmosphere are filtered out of the governing equations leaving a set of equations of lower dimensionality which describes the evolution on an invariant manifold. Other occurrences of invariant manifolds in applied mathematics are described by Roberts (1989[9]).

Of course one of the most common situations where we reduce the dimension of the problem at hand is in *numerical approximations*. For example, if we are interested in the evolution of a continuum then we end up having to replace the infinite number of degrees-of-freedom by a numerical approximation with only a finite number of degrees-of-freedom. Looking at numerical approximations from an invariant manifold viewpoint we see that such an approximation may be divided into two parts. The first part is the reduction of the dimensionality where we "ignore", "throw away" or "average over" exponentially decaying transients (or rapid oscillations, depending upon the nature of the physical approximation) by only considering solutions with a finite number of degrees-of-freedom. The set of these exact solutions form a finite dimensional invariant manifold of the full system. The second part of the numerical

---

This work ws supported by a University of Adelaide Research Grant.

approximation is the calculation of the invariant manifold, rarely can it be found exactly. Indeed, a common class of numerical approximations, in effect, represent the invariant manifold by its tangent plane at the origin.

Here I describe a formal procedure which calculates approximations to invariant manifolds based upon a fixed point of the dynamical system. It is based on centre manifold theory, see Carr (1981) and Roberts (1985,1988), for which rigorous results are known. Once a description of such an invariant manifold is constructed, the long-time asymptotic behaviour of the system can be immediately written down. In addition, given a specific initial condition and forcing of the full problem, this new geometric view provides a method for calculating the best initial condition and the best forcing for the low-dimensional approximation. To be definite I apply the techniques to a specific partial differential equation.

THE KURAMOTO–SIVASHINSKY EQUATION

The particular problem addressed in this paper is the Kuramoto-Sivashinsky equation

$$u_t + uu_x + u_{xx} + \frac{1}{\Lambda^2}u_{xxxx} = 0 , \tag{1}$$

to be solved for real $u(x,t)$ subject to the boundary conditions

$$u = u_{xx} = 0 \quad \text{on} \quad x = 0, \pi , \tag{2}$$

where the subscripts x or t denote differentiation with respect to the subscript. This equation (1) combines a simple nonlinearity uu_x with a destabilizing diffusion term u_{xx} and a stabilizing fourth-order dissipation u_{xxxx}/Λ^2. This last term is the only stabilizing term in the equation and so we consider Λ to be analogous to a Reynold's or Rayleigh number (alternatively we may consider $1/\Lambda$ to be a type of viscosity). The Kuramoto-Sivashinsky equation arises as an approximation in a number of different physical contexts, for example: the Belouzov-Zabotinskii reaction, instabilities of flame fronts, and Poiseuille flow of a film layer. More detailed references to these applications may be found in the articles by Hyman & Nicolaenko (1986) or Frisch et al (1986). The parameter Λ may also be considered as being proportional to the length of a box in which the ($\Lambda = 1$)-version of (1) applies. Whichever way it is considered, the larger the value of Λ, the richer in structure the solutions to (1) become.

This may be seen in the linearised version of (1). A complete set of eigenmodes of the linearised (1) which satisfy the boundary condition (2) is simply $\sin(kx)$ for wavenumbers $k = 1, 2, \ldots$. The linear growth-rate of the kth mode, or decay-rate if negative, is

$$\lambda_k = k^2 \left(1 - k^2/\Lambda^2\right) . \tag{3}$$

Thus small wavenumber modes with $0 < k < \Lambda$ are linearly unstable, while high wavenumber modes with $k > \Lambda$ are linearly stable — the larger Λ is the more unstable modes there are. Stable finite-amplitude solutions occur because the non-linearity in the equation transfers "energy" from low wavenumbers to high wavenumbers where it can be dissipated. This suggests that at any given Λ the interesting dynamically-active modes are those of low wavenumber, while high wavenumber modes are just driven by the low wavenumber modes and so behave relatively simply. If this occurs then we say that the high wavenumber modes are slaved to the low wavenumber modes. The practical upshot of this is that the interesting behaviour of solutions to (1) may be governed by a set of ordinary differential equations for the low wavenumber modes, which are called *master modes*, in which the strongly-dissipative dynamics of the high wavenumber modes, called *slave modes*, have been implicitly taken into account. Furthermore, we expect that this state of quasi-balance will be achieved quickly as the initial "energy" in the slave modes decays exponentially.

ITERATIVE CONSTRUCTION OF AN INVARIANT MANIFOLD

The clearest way to construct an invariant manifold of (1) is to take the write down the Fourier sine-transform of u

$$u(x,t) = \sum_{k=1}^{\infty} a_k(t) \sin(kx) , \qquad (4)$$

and then consider the evolution of the Fourier coefficients $a_k(t)$ for $k = 1, 2, 3, \dots$. Upon substituting (4) into the Kuramoto-Sivashinsky equation (1) we find that

$$\dot{a}_k = \lambda_k a_k - \frac{1}{2} \sum_{\ell=1}^{\infty} \ell a_\ell \left[a_{k+\ell} + \mathrm{sgn}(k - \ell) a_{|k-\ell|} \right] , \qquad (5)$$

where ($\dot{}$) denotes the time derivative d/dt, and where λ_k can be recognised as the linear growth/decay-rate of the kth Fourier mode given by (3). This transformation is equivalent to making a linear change of basis to the problem (for simplicity, both a_k and $\sin(kx)$ are henceforth called the kth mode). Its utility is that it puts the governing equations (5) in the very special form that the linearised problem is diagonal. This allows the calculation and the visualisation of an invariant manifold to be in the simplest possible form.

For the moment, just consider the linearised version of (5). According to (3) the Fourier modes with wavenumber $k > \Lambda$ will decay to zero, while the modes with wavenumber $k < \Lambda$ will grow and dominate the solution. Thus the long time evolution of the system is dictated by the evolution of the small wavenumber modes $k \leq \Lambda$. Returning to the complete version of (5) we may argue that the nonlinear coupling terms just modify this simple picture (see Carr (1981), Chapter 1). The modification is that the small wavenumber modes (perhaps including some of the slowly decaying modes as well) will still dominate the solution at large time, but all the remaining modes will quickly tend to some hyper-surface in the solution space (an invariant manifold) which is a function of the dominant modes and is approximately zero near the origin. Thus for some range of the instability parameter Λ it is appropriate to select $a_1, \dots, a_M$ (denoted by **a**) as M master modes which are dominant at large time, because for $\Lambda \leq M$ these are either the linearly growing modes or the modes of slowest decay. The remaining quickly-decaying slave modes, $a_{M+1}, a_{M+2}, \dots$, are assumed to have no great impact on the long time evolution and are thus assumed to be a function of the master modes. The ansatz is then to pose that, however the master modes $a_1, \dots, a_M$ evolve, the system stays on the invariant manifold described by

$$a_s = h_s(\mathbf{a}) \quad \text{for} \quad s = M + 1, M + 2, \dots , \qquad (6)$$

for some functions h_s to be found.

To derive an equation for the invariant manifold $h_s(\mathbf{a})$ it is convenient to rewrite the governing equation (5) in the general form

$$\dot{a}_k = \lambda_k a_k + \mathcal{N}_k(\mathbf{a}, \mathbf{h}) , \qquad (7)$$

717

where $\mathbf{h}$ denotes all of the slave modes and where $\mathcal{N}_k$ contains all the nonlinear terms of the equations so that in this problem

$$\mathcal{N}_k(\mathbf{a}, \mathbf{h}) = -\frac{1}{2} \sum_{\ell=1}^{\infty} \ell h_\ell \left[h_{k+\ell} + \operatorname{sgn}(k - \ell) h_{|k-\ell|} \right] \ , \tag{8}$$

in the right-hand side of which h_m for $m = 1, \ldots, M$ is an alias for a_m. Now, if the invariant manifold $h_s(\mathbf{a})$ were known then (7) for $k = 1, \ldots, M$ would describe the evolution of the master modes, and hence gives the evolution on the invariant manifold. If the system is to evolve so that it stays on the manifold then the natural evolution of the slave modes as given by (7) must be matched by the evolution on the curving manifold, thus

$$\lambda_s h_s + \mathcal{N}_s(\mathbf{a}, \mathbf{h}) = \dot{a}_s = \sum_m \frac{\partial h_s}{\partial a_m} \dot{a}_m = \sum_m \frac{\partial h_s}{\partial a_m} \left[\lambda_m a_m + \mathcal{N}_m(\mathbf{a}, \mathbf{h}) \right] \ . \tag{9}$$

(Note that, unless otherwise indicated, the convention adopted is that the subscript m for master will implicitly range over $m = 1, \ldots, M$, and the subscript s for slave will implicitly range over $s = M + 1, M + 2, \ldots$.) Rearranging the above equation gives

$$- \lambda_s h_s + \sum_m \frac{\partial h_s}{\partial a_m} \lambda_m a_m = \mathcal{N}_s(\mathbf{a}, \mathbf{h}) - \sum_m \frac{\partial h_s}{\partial a_m} \mathcal{N}_m(\mathbf{a}, \mathbf{h}) \ , \tag{10}$$

as the fundamental set of equations to be solved to find the invariant manifold h_s.

In practise, exact solutions to an equation for an invariant manifold are impossible to find in general. The only practical course is to find an approximate solution for an invariant manifold. Here this is easy to do via iteration, although to calculate the invariant manifold to high-order it is better to substitute a formal asymptotic expansion. Let $h_s^{(r)}$ denote the rth iterative approximation to the invariant manifold. Starting with the initial approximation that $h_s^{(1)}(\mathbf{a}) = 0$, as this is appropriate near the origin, we find successive iterates through (10) by solving

$$- \lambda_s h_s^{(r+1)} + \sum_m \frac{\partial h_s^{(r+1)}}{\partial a_m} \lambda_m a_m = \mathcal{N}_s \left(\mathbf{a}, \mathbf{h}^{(r)} \right) - \sum_m \frac{\partial h_s^{(r)}}{\partial a_m} \mathcal{N}_m \left(\mathbf{a}, \mathbf{h}^{(r)} \right) \ . \tag{11}$$

This is a first-order non-homogeneous partial differential equation for $h_s^{(r+1)}$. Since the non-linearity (8) in the equation is a multinomial in the various modes it easily follows that each iterate is a multinomial in the master modes a_m.

In summary, the above arguments allow us to find an approximation to an invariant manifold $h_s(\mathbf{a})$ such that, no-matter what the initial conditions are, the solutions of the Kuramoto-Sivashinsky equation tend exponentially quickly to a state described by

$$u(x, t) = \sum_m a_m \sin(mx) + \sum_s h_s(\mathbf{a}) \sin(sx) \ , \tag{12}$$

where the master amplitudes a_m evolve in time according to

$$\dot{a}_m = \lambda_m a_m + \mathcal{N}_m(\mathbf{a}, \mathbf{h}) \ . \tag{13}$$

Such a description will be valid over some range of the instability parameter Λ depending on the number of master modes M.

TWO LOW DIMENSIONAL INVARIANT MANIFOLDS

We now construct approximate descriptions of the 2-dimensional and the 4-dimensional invariant manifolds of the Kuramoto-Sivashinsky equation. In all the invariant manifold descriptions the fixed point at the origin and the nearby time dependent behaviour is always reproduced accurately as the descriptions of the invariant manifolds are based at this point. So, as a nontrivial comparison, a nontrivial fixed point away from the origin was calculated (usually stable) and comparisons made by looking at predictions of the time dependent behaviour near the fixed point. We limit the instability parameter to the range $\Lambda \in [1,4]$ as for $\Lambda < 1$ the origin is the only fixed point (and is stable), and $\Lambda < 4$ is the approximate range of validity of the 4-dimensional invariant manifold.

The 2-dimensional invariant manifold

One of the simpler examples of the application of an invariant manifold to the asymptotic description of a dynamical system is when the invariant manifold is two-dimensional; that is, there are precisely two master modes. Applying the procedure outlined in the previous section to the Kuramoto-Sivashinsky equation with the slave modes being considered functions of a_1 and a_2 we start the iteration with the approximation

$$h_s^{(1)} = 0 \quad \text{for} \quad s = 3, 4, 5, \dots ; \tag{14}$$

that is, the invariant manifold is "flat" near the origin. This approximates the manifolds by their tangent-plane at the origin, this is a fair approximation for small Λ.

To find a more accurate approximation to the invariant manifold we substitute (14) into the right-hand side of (11) and solve it to obtain

$$h_s^{(2)} = \begin{cases} -3\Lambda^2 a_1 a_2 / \left[8 \left(16 - \Lambda^2 \right) \right] & \text{if } s = 3 \\ -\Lambda^2 a_2^2 / \left[8 \left(28 - \Lambda^2 \right) \right] & \text{if } s = 4 \\ 0 & \text{otherwise} \end{cases} \tag{15}$$

This gives the quadratic shape of the invariant manifold near the origin. A further refinement can be found by carrying out a further step of the iteration to obtain

$$h_s^{(3)} = \begin{cases} -3d_{16} a_1 a_2/8 - d_{13} d_{16} a_1{}^3/32 + (d_{16} - d_{28}) \Lambda^2 a_1 a_2{}^2/256 & \text{if } s = 3 \\ -d_{28} a_2{}^2/8 + 3 \left(6d_{16} - d_{28} \right) d_{119/5} a_1{}^2 a_2/80 & \text{if } s = 4 \\ 5 \left(3d_{16} + d_{28} \right) d_{37} a_1 a_2{}^2/256 & \text{if } s = 5 \\ d_{28} d_{52} a_2{}^3/64 & \text{if } s = 6 \\ 0 & \text{otherwise} \end{cases} \tag{16}$$

where the coefficients $d_\alpha = 1/ \left(\alpha/\Lambda^2 - 1 \right)$. As many further iterations as desired and are practical may be carried out to provide more and more details about the shape of the invariant manifold.

The system tends to the above manifold exponentially quickly (at least near the origin where the rate of approach is given by the eigenvalue of the dominant slave mode). Once on the manifold the evolution of the system is given by (13), although we can only approximate the evolution as the manifold is only known approximately, given by (14), (15) or (16) for example. The very first approximation to the invariant manifold (14), that it is zero, is just a modal numerical model of the system truncated to $M = 2$ modes. This is because the evolution equation (13) becomes

$$\dot{a}_1 = \lambda_1 a_1 + a_1 a_2/2 \,, \tag{17}$$
$$\dot{a}_2 = \lambda_2 a_2 - a_1{}^2/2 \,,$$

which is just the original modal equations (5) with all modes $k > 2$ ignored.

The second approximation to the evolution on the invariant manifold is given by substituting (15) into (13) and just modifies (17) to

$$\dot{a}_1 = \lambda_1 a_1 + a_1 a_2/2 - 3d_{16} a_1 a_2{}^2/16 \,, \tag{18}$$
$$\dot{a}_2 = \lambda_2 a_2 - a_1{}^2/2 - 3d_{16} a_1{}^2 a_2/8 - d_{28} a_2{}^3/8 \,.$$

These evolution equations take account of the quadratic curvature of the invariant manifold near the origin. The third approximation to the evolution equations is obtained by substituting (16) into (13) and gives further and higher-order corrections to (18).

The conventional modal approximation (17), while quantitatively accurate for $\Lambda < 1.3$, is only qualitatively correct for $\Lambda < 2$. In contrast, the modified approximation (18) is quantitatively accurate for $\Lambda < 2.4$ when the $\sin(3x)$ mode begins to be significant (it becomes linearly unstable at $\Lambda = 3$). As a description of the evolution of the system this simple two-dimensional invariant manifold, taking the curving nature of the invariant manifold into account, performs well. Further, it seems that the linear growth rate (about the origin) is a fair predictor of when a model of a given dimensionality will fail.

The 4-dimensional invariant manifold

Here a_1, a_2, a_3 and a_4 are taken to be master modes and the rest to be slaves; as before the invariant manifold is found by iteration. We just describe the predictions of the first (modal) approximation and the second approximation without going into the details.

There is little difference between these two approximations for small instability parameter $\Lambda \leq 2.3$. However, the second approximation does much better over the rest of the range $\Lambda \leq 4$, in fact the second approximation is so good as to be virtually indistinguishable from numerically exact solutions. In contrast the first (modal) approximation is only *quantitatively* satisfactory for $\Lambda \leq 2.3$. For example, the first approximation predicts that the discovered fixed point undergoes a Hopf bifurcation at $\Lambda \approx 3.2$; in fact, nothing of the sort happens—the fixed point actually undergoes a pitchfork bifurcation at about $\Lambda = 3.6$. There is nothing in the full picture of the time dependent behaviour which is not also in the second of the four-dimensional invariant manifold approximations.

APPROPRIATE INITIAL CONDITIONS

This view of low-dimensional approximations now suggests how correct initial conditions can be found for a given approximation in terms of the given initial conditions for the full system. The rationale is that if the manifold is exponentially attracting then there exists a solution on the manifold which is approached exponentially quickly by the exact solution — the location of the invariant manifold solution at the initial time is the correct initial condition for the low-dimensional approximation. This argument has been developed by Roberts (1989[7]) and we find that provided the given initial condition is not too far from the invariant manifold then it should be projected onto the manifold by a location dependent projection.

The usual numerical approach is simply to neglect those components which are not explicitly in the model. For example, if the initial condition for the Kuramoto-Sivashinsky equation is $a_k(0) = a_{0k}$ for $k = 1, 2, \ldots$ then the usual approach is to start the low-dimensional model with $a_m(0) = a_{0m}$ for $m = 1, 2, \ldots, M$. Geometrically this is a projection of the actual initial point $\mathbf{a}_0$ onto a point $(\mathbf{a}(0), \mathbf{h}(\mathbf{a}(0)))$ on the manifold such that the projection is orthogonal to the set of M (infinite-dimensional) vectors $\mathbf{z}_m = (\delta_{mk})$. That is, we find $\mathbf{a}(0)$ such that

$$\mathbf{z}_m \cdot [(\mathbf{a}(0), \mathbf{h}(\mathbf{a}(0))) - \mathbf{a}_0] = 0 \quad \text{for } m = 1, 2, \ldots, M . \tag{19}$$

The correct approach is precisely the same except that the vectors $\mathbf{z}_m$, which define the projection, now depend upon the location $\mathbf{a}(0)$ instead of being fixed.

Following the analysis in Roberts (1989[7]) in the case of the Kuramoto-Sivashinsky equation leads to an equation for the $\mathbf{z}_m$ which can be solved iteratively. The starting guess for the iteration is the usual naive approximation $\mathbf{z}_m^{(1)} = (\delta_{mk})$. For the two-dimensional invariant manifold described earlier I find that the first correction to this guess is

$$\mathbf{z}_1^{(2)} = (1, 0, d_8 a_2/24, 0, 0, \ldots) \quad \text{and} \quad \mathbf{z}_2^{(2)} = (0, 1, d_{11} a_1/6, d_{16} a_2/16, 0, \ldots) \tag{20}$$

Using these approximations to $\mathbf{z}_m$ in (19) we appreciate that these correct projections make a quadratic correction to the initial condition of the low-dimensional model. Higher-order approximations to $\mathbf{z}_m$ may be systematically calculated and used to refine the initial condition and consequently improve the long-term predictions of the model.

APPROPRIATE FORCING

The Kuramoto-Sivashinsky equation (1) is a homogeneous differential equation. However, in physical problems there is often an inhomogeneous forcing of the equations. For example, if the right-hand-side of (1) is replaced by the forcing $f(x, t)$ then the right-hand-side of (5) would include its sine-transform $f_k(t)$. Once again the usual numerical approach would be to simply use the forcing $f_m(t)$ of the master modes and ignore the forcing $f_s(t)$ of the slave modes. However, this can now be shown to be only the first approximation to the correct forcing of the low-dimensional model.

Roberts (1989[7],§7) argued that a forcing may be approximated by a series of discrete δ-function impulses. During each force free interval between impulses the system tends exponentially quickly to the invariant manifold; an impulse then pushes it away from the invariant manifold, only for the system to again tend back exponentially quickly. The earlier work on initial conditions suggests that at the start of each force free interval the solution on the invariant manifold should be started at a location which best matches where the full system would have been kicked to. This starting location is generally different from the location before the impulse occurred, and so corresponds to a specific impulse of the evolution equations (13) on the invariant manifold. The consequence of this argument is that correct forcing of (13), $f_m^{\mathcal{M}}(t)$ say, is given by the solution of the system of linear equations

$$\sum_{m'} \left[\mathbf{z}_m \cdot \frac{\partial(\mathbf{a}, \mathbf{h}(\mathbf{a}))}{\partial a_{m'}} \right] f_{m'}^{\mathcal{M}} = \mathbf{z}_m \cdot \mathbf{f} \tag{21}$$

at each time. For example, for the Kuramoto-Sivashinsky equation as we have analysed it, we find that the matrix of coefficients from the left-hand-side of (21) is the identity matrix with quadratic and higher order corrections. Thus, for the two-dimensional invariant manifolds described earlier the correct forcing of the evolution equations, (17) or (18), is

$$f_1^{\mathcal{M}} = f_1 + d_8 a_2 f_3/24 + O(a^2) \quad \text{and} \quad f_2^{\mathcal{M}} = f_2 + d_{11} a_1 f_3/6 + d_{16} a_2 f_4/16 + O(a^2) \ . \tag{22}$$

These corrections may have a significant effect depending upon the circumstances. For example, if all the forcing is zero except for f_3 then, upon including the appropriate forcing in (17) with $\Lambda = 1$, we predict that the fixed point at the origin splits into three fixed points at $(a_1, a_2) = (0,0), (d_{11} f_3/3, 0)$ and $-(d_8 f_3/12)(1, (d_8 + 4d_{11}) f_3/24)$. The usual approach would be to ignore the presence of f_3 and hence to ignore this particular phenomena.

CONCLUSION

This invariant manifold view of low-dimensional approximations is very appealing. In particular, it shows that conventional numerical models can be viewed as the first of a sequence of more refined and accurate asymptotic models. Also we can now systematically calculate the appropriate initial conditions and forcing for such a numerical model, and show how the conventional practises should be modified in order to give exponential accuracy in the long term.

REFERENCES

1. Carr, J.,, Applications of centre manifold theory, in *Applied Math. Sci.*, Springer-Verlag, 1981.

2. Frisch, U., She, Z. S., and Thual, O.,, Visco-elastic behaviour of cellular solutions to the Kuramoto-Sivashinsky model, *J. Fluid Mech.*, vol. 168, pp. 221–240, 1986.

3. Hyman, J. M., and Nicolaenko, B.,, The Kuramoto-Sivashinsky equation: a bridge between PDEs and dynamical systems, *Physica D*, vol. 18, pp. 113–126, 1986.

4. Lorenz, E. N.,, On the existence of a slow manifold, *J. Atmos. Sci.*, vol. 43, pp. 1547–1557, 1986.

5. Muncaster, R. G.,, Invariant manifolds in mechanics I: the general construction of coarse theories from fine theories, *Arch. Rat. Mech. & Anal.*, vol. 84, pp. 353–373, 1983.

6. Roberts, A. J.,, The application of centre manifold theory to the evolution of systems which vary slowly in space, *J. Austral. Math. Soc. Series B*, vol. 29, pp. 480–500, 1988.

7. Roberts, A. J.,, Appropriate initial conditions for asymptotic descriptions of the long term evolution of dynamical systems, *J. Austral. Math. Soc. Series B*, vol. 31, pp. 48–75, 1989.

8. Roberts, A. J.,, Simple examples of the derivation of amplitude equations possessing bifurcations, *J. Austral. Math. Soc. Series B*, vol. 27, pp. 48–65, 1985.

9. Roberts, A. J.,, The utility of an invariant manifold description of the evolution of a dynamical system, *SIAM J. Math. Anal.* forthcoming, 1989.

10. Vautard, R., and Legras, B.,, Invariant manifolds, quasi-geostrophy and initialisation, *J. Atmos. Sci.*, vol. 43, pp. 565–584, 1986.

Analysis of a Reinforced Arbitrary Shaped Block

A. G. THOMPSON
CSIRO Division of Geomechanics
Nedlands, Western Australia

INTRODUCTION

A three-dimensional analysis technique has been developed for the design of reinforcement of arbitrary shaped rock blocks with a combination of free faces and faces which interact with the adjacent rock mass. The rock block can have three translational and three rotational modes of displacement. This differs from the previously published work of a number of workers (e.g. [1] and [2]) in which translations only were considered. Reinforcing elements are modelled with non-linear load-displacement characteristics. The solution technique for block translations and rotations is unique in that joint interactions are modelled as constraints on block displacements.

THE DESIGN PROBLEM

A common safety and productivity problem experienced in mining and civil engineering is the design of reinforcement for unstable blocks. The types of problems that may be analysed using the writer's method are shown in Fig. 1. The blocks may be of arbitrary shape and reinforcing elements may be located anywhere on the block with any orientation.

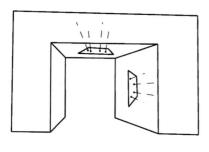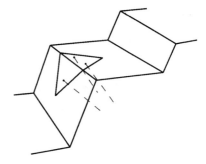

FIGURE 1. Typical problems which may be analysed.

ASSUMPTIONS OF THE ANALYSIS METHOD

The major assumptions made in developing the theory are:
1. The analysis is applicable to a single rigid block.
2. Block displacements are small.
3. The driving force on the block is due solely to gravity.
4. The passive forces acting on the block are due to the reinforcement and interactions across the joints forming the block.

DESCRIPTION OF THE ANALYSIS METHOD

The solutions for reinforcement loads and joint reactions in problems such as those shown in Fig. 1 are generally statically indeterminate. However, with the assumptions detailed previously, it is possible to obtain a unique solution by satisfying force equilibrium and displacement compatibility. Also, because of the non-linear nature of the problem, an iterative procedure using manipulation of matrix equations needs to be used to solve for the increments in displacements of the block, reinforcement forces and joint reactions. In the following equations, upper case letters represent total displacements and forces whilst lower case letters represent increments in these quantities. Global coordinates are indicated by capital letters. Coordinates in lower case letters are relative to a block corner or a reinforcing element.

The equilibrium of a reinforced block shown in Fig. 2, considering the applied centroidal forces $[P_0]$ and the summations of passive reaction forces for the N_s reinforcing elements and N_c corner reactions can be expressed in matrix form as

$$[P_0] = \sum_{i=1}^{N_s} [C_i]^T [S_{g\,i}] + \sum_{j=1}^{N_c} [C_j]^T [R_{g\,j}] \tag{1}$$

where $[S_{g\,i}]$ are the components of support forces in the global coordinate directions

$[R_{g\,j}]$ are the components of corner reactions in the global coordinate directions

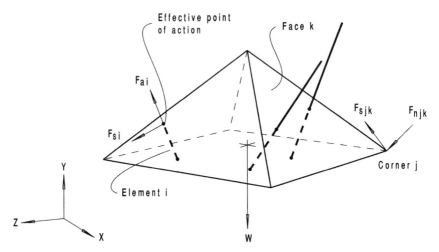

FIGURE 2. Forces acting on reinforced block.

724

$$[C] = \begin{bmatrix} 1 & 0 & 0 & -Y_1 & 0 & Z \\ 0 & 1 & 0 & X_1 & -Z_1 & 0 \\ 0 & 0 & 1 & 0 & Y_1 & -X_1 \end{bmatrix} \qquad (2)$$

where X_1, Y_1, Z_1 are the local coordinates at any point on the block. The six equilibrium equations require the solution for $3N_s + 3N_c$ unknown forces. In general, the total number of unknowns will be in excess of 6. In order to solve for a unique set of forces it is necessary to reduce the number of unknowns to six by using conditions of displacement compatibility.

BLOCK DISPLACEMENTS

With the assumptions of a rigid block and small displacements, the global displacements $[U_g]$ at any point on the block can be expressed in terms of the centroidal displacements $[D_c]$ as

$$[U_g] = [C] [D_c] \qquad (3)$$

where $[D_c] = [U_c \ V_c \ W_c \ \Theta_c \ \Phi_c \ \Psi_c]^T$ are the translations of the centroid in the X,Y and Z directions and the clockwise rotations about the Z, X and Y axes respectively.

In order to evaluate the local displacements applicable to each reinforcing element or corner of a face, local coordinate axes are defined. The local displacements $[U_{1i}]$ are related to the global displacements by

$$[U_{1i}] = [T_i] [U_{gi}] \qquad (4)$$

where $[T_i]$ is a transformation matrix consisting of the direction cosines for each of the local axes.

REINFORCEMENT CHARACTERISTICS

Reinforcing elements are assumed to act at the point where they exit from the block and penetrate the intact rock. The response of the reinforcement is assumed to be resolvable into axial and shear components and that each component may be approximated by a segmental non-linear curve such as the one shown in Fig. 3.

If a reinforcing element has a locally displaced position with shear displacement D_s and axial displacement D_a, the defined characteristic curves may be used to define the forces and the stiffnesses of response to further displacement from the current displaced condition.

$$F_s = F_s' + K_s \ d_s \qquad (5a)$$
$$F_a = F_a' + K_a \ d_a \qquad (5b)$$

where F_a, F_s = unknown axial and shear forces

F_a', F_s' = known axial and shear forces at displacements D_a and D_s

In matrix form Eqs.(5a) and (5b) become

$$\begin{bmatrix} F_{sx} \\ F_{sy} \\ F_{sz} \end{bmatrix} = \begin{bmatrix} F_{sx}' \\ F_{sy}' \\ F_a' \end{bmatrix} + \begin{bmatrix} K_s & 0 & 0 \\ 0 & K_s & 0 \\ 0 & 0 & K_a \end{bmatrix} \begin{bmatrix} d_{sx} \\ d_{sy} \\ d_{sz} \end{bmatrix} \qquad (6)$$

725

or

$$[F_{1i}] = [F'_{1i}] + [K_i] [T_i] [C_i] [d_c] \tag{7}$$

The forces in the reinforcement local coordinate axes are converted to the global coordinate system through

$$[S_{gi}] = [T_i]^T [F_{1i}] \tag{8}$$

By combining eqs. (1) and (8) and (7), the equilibrium equations (considering only the block weight and the effect of all reinforcing elements) become

$$[f_{k-1}] = \sum_{i=1}^{N_s} [C_i]^T [T_i]^T [K_i] [T_i] [C_i] [d_{ck}] \tag{9}$$

Further, Eq.(9) can be rewritten as

$$[f_{k-1}] = [K_{k-1}] [d_{ck}] \tag{10}$$

where $[d_{ck}]$ is the unknown centroidal displacement increments for the kth iteration

$[K_{k-1}]$ is the tangent stiffness matrix for all supports
$[f_{k-1}]$ is the out-of-balance forces after the previous iteration

and $[f_{k-1}] = [P_0] - \sum_{i=1}^{N_s} [C_i]^T [T_i]^T [F'_{1i}] \tag{11}$

JOINT INTERACTIONS

Joint interactions may be treated in much the same way as the reinforcing elements by defining a relationship between normal reaction force and joint closure. This relationship will be of the general form shown in Fig. 4. This method is used by the multiple discrete element codes initiated by [3] and continued by a number of other workers. The concept of joint stiffness is artificial in some respects and is difficult to relate to measurable quantities.

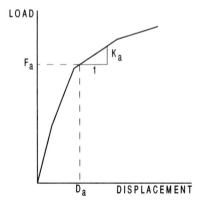
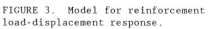

FIGURE 3. Model for reinforcement load-displacement response.

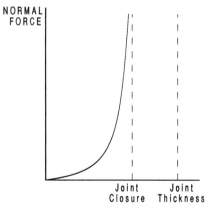

FIGURE 4. Assumed model for joint closure under compression.

726

The method used here assumes that a joint can be characterised by a maximum normal closure which imposes a constraint normal to the joint surface. These constraints are used to modify the equilibrium equations so that they contain a mixture of known forces with corresponding unknown displacements and unknown forces with corresponding known displacements. Any number of corners, edges or faces may impose restrictions on block movement, but only a maximum of 6 (the total number of block centroid displacements and rotations) of these constraints can be independent. No constraint occurs when the joint is open. The joint interactions are assumed to be concentrated at the corners of the block faces where there is an associated shear resistance.

Each face may have one, two or three constraints corresponding to a normal constraint and up to two rotational constraints as shown in Fig. 5. Constraints are detected during iterations by accumulated closures in excess of the defined maximum joint closure D_0. The calculated normal displacement D_{zjk} for corner j of face k and D_0 are used to define the increment in normal displacement for the next increment for the corner d_{zjk} according to

$$d_{zjk} = [l_{3jk} \quad m_{3jk} \quad n_{3jk}] [C_j] [d_c] \qquad (12)$$

where $d_{zjk} = -(D_{zjk} - D_0)$

The second constraint for the same face is stated as imposing a rotational increment ψ_y of opposite sense to the local rotation Ψ_y

$$\psi_y = -\Psi_y = [0 \quad 0 \quad 0 \quad n_{2jk} \quad l_{2jk} \quad m_{2jk}] [d_c] \qquad (13)$$

The third constraint on the same face is stated as imposing a rotation increment ϕ_x of opposite sense to the local rotation Φ_x and means the face is in intimate contact over the entire face and

$$\phi_x = -\Phi_x = [0 \quad 0 \quad 0 \quad n_{1jk} \quad l_{1jk} \quad m_{1jk}] [d_c] \qquad (14)$$

Subsequent corners for the same face will not change the number of constraints possible for the face. However, they are considered in the determination of the force distribution for the face.

The shear reactions are evaluated from the normal reaction R_n for the corner. The limiting value of the shear force $R_{s\ max}$ is given by

$$R_{s\ max} = R_n \tan \phi_k \quad \text{where} \quad \phi_k \text{ is the friction angle for face k} \qquad (15)$$

It is assumed that maximum shear resistance occurs at a defined amount of shear displacement D_{sm} as shown in Fig. 6. This allows for cases where full mobilisation of the shear resistance is not necessary for block stability. The mobilised shear resistance R_s at shear displacement D_s is given by

$$R_s = R_{s\ max} (D_s/D_{sm}) \quad \text{subject to } R_s \leq R_{s\ max} \qquad (16)$$

Cohesion is considered only for full-face contact and is distributed to each corner of the face. The local corner reactions $[F_{ljk}]$ are converted to global forces through

$$[S_{gj}] = [T_{jk}]^T [F_{ljk}] \qquad (17)$$

where the matrix T_{jk} represents the direction cosines of the orthogonal axes for corner j of face k.

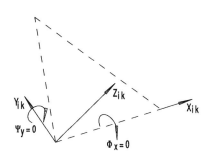
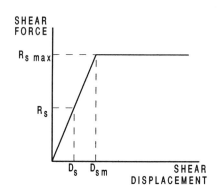

FIGURE 5. Definition of rotation   FIGURE 6. Joint shear mobilisation
constraints on a face.   model.

The known corner constraint displacement and rotation increments $[d']$
and some of the unknown centroidal displacement and rotational increments
are related to the unknown increments of centroidal translation and
rotation increments through

$$[d'] = [R]^{-1} [d_c] \qquad (18)$$

Equation (18) is comprised of the conditions described in Eqs.(12)-(14).

SOLUTION OF EQUILIBRIUM EQUATIONS

The original equilibrium Eq.(10) is modified to reflect the mixture of
known forces and displacements and unknown forces and displacements. This
is achieved by replacing $[d_c]$ with $[d']$ in the original equations and
pre-multiplying both sides of the equation by $[R]^T$.

$$[R]^T [f] = [R]^T [K] [R] [d'] \qquad (19)$$

or

$$[f'] = [K'] [d'] \qquad (20)$$

To solve for the unknown block displacements and joint normal reactions,
the modified equilibrium Eq.(20) is partitioned in the following way

$$\begin{bmatrix} f'_1 \\ f'_2 \end{bmatrix} = \begin{bmatrix} K'_{11} & K'_{12} \\ K'_{21} & K'_{22} \end{bmatrix} \begin{bmatrix} d'_1 \\ d'_2 \end{bmatrix} \qquad (21)$$

where $[d'_1]$ are the known correction displacement increments

   $[d'_2]$ are the unknown centroidal displacement increments

   $[f'_1]$ are the unknown corner forces increments

   $[f'_2]$ are the known out-of-balance forces

Equation 21 is solved by considering the following two equations.

$$[f'_1] = [K'_{11}] [d'_1] + [K'_{12}] [d'_2] \qquad (22)$$

$$[f'_2] = [K'_{21}] [d'_1] + [K'_{22}] [d'_2] \qquad (23)$$

728

Equation (23) is used to solve for the independent block displacements

$$[d_2'] = [K_{22}']^{-1} \{ [f_2'] - [K_{21}'] [d_1'] \} \tag{24}$$

The increments in all centroidal translations and rotations are then obtained by using Eq.(18) such that

$$[d_c] = [R] [d'] \tag{25}$$

The accumulated centroidal displacements are then given by

$$[D_c] = \sum_{k=1}^{N_i} [d_{ck}] \tag{26}$$

where d_{ck} are the incremental displacements at the kth iteration.
N_i is the total number of iterations

NORMAL REACTIONS AT CORNERS

The increments in normal forces corresponding to the constraints are obtained by substituting the block displacements and constraint displacements into Eq.(22) so that

$$[f_1''] = [K_{11}'] [d_1'] + [K_{12}'] [d_2'] - [f_1'] \tag{27}$$

where f_1'' is the increment in the force matrix due to changes in all the normal forces corresponding to the constraints. The duplicate constraints complicate how the individual corner forces are related to the force vector defined by Eq.(27). Since the block is assumed rigid, force distributions are not critical so long as they satisfy force and moment equilibrium of the block.

The solution of redundant forces is obtained in a similar way to the procedure used for the solution of the entire design problem. Using Eq.(12), the normal displacement w_{nj} at any corner j may be defined. Assuming for the present that normal force f_{nj} is proportional to normal displacement allows

$$f_{nj} = K w_{nj} \tag{28}$$

where K is a fictitious normal stiffness

and $$w_{nj} = [T_{3j}] [R_{11}] [d_1'] \tag{29}$$

where $[R_{11}]$ is the partitioned component of the R matrix corresponding to the displacement constraints

The global forces and moments of the normal force are then given by

$$[f_{1j}'] = [C_j]^T [T_{3j}]^T [f_{nj}]$$

and $$[f_{1j}''] = [R_{11}]^T [f_{1j}'] \tag{30}$$

For all corner constraints then, by comparison with Eqs.(10) and (16),

$$[f_1''] = [S_{11}] [K d_1'] \tag{31}$$

where $$[S_{11}] = \sum_{j=1}^{N_r} [R_{11}]^T [C_j]^T [T_{3j}]^T [T_{3j}] [C_j] [R_{11}] \tag{32}$$

and $[S_{11}]$ is a square matrix of order corresponding to the number of independent constraints.

N_r is the total number of corner contacts

By retaining the product $[K d_1']$ with units of force, specific values of normal stiffness are not required. Equation (32) may then be solved as

$$[K d_1'] = [S_{11}']^{-1} [f_1''] \tag{33}$$

Individual corner force increments are determined from

$$f_{nj} = [T_{3j}] [C_j] [R_{11}] [K d_1'] \tag{34}$$

CONVERGENCE CRITERIA

The calculations described in the sections above are repeated until the out-of-balance forces indicated by Eq.(11) and the increments in blocks displacements described by Eq.(25) are negligible.

COMPUTER PROGRAM

A computer program based on the above theory has been written for a personal computer. The block is defined in terms of vertex and face numbers. The coordinates of the vertices are used to calculate the face areas, volume and centre of mass of the block. Reinforcing elements are specified by the lengths of anchorage in the block and the stable rock beyond the block boundaries. Each reinforcing element has an individual load-displacement characteristic which depends on these anchor lengths. Each face may have different values of joint closure, cohesion and friction. The solution provides estimates of the displaced position of the block, reinforcement axial and shear loads and the joint reactions. The software enables engineers to quickly evaluate different reinforcing schemes and to produce suitable reinforcement designs for unstable blocks.

SUMMARY

A method has been developed for the analysis of reinforcement of unstable, arbitrary shaped blocks. The method is complementary to rigid block analyses developed by a number of other workers. The analysis method is capable of incorporating non-linear load-displacement relationships which are characteristic of reinforcing systems. Personal computer compatible software based on the method has been written.

ACKNOWLEDGEMENT

The financial and collaborative contributions of the mining industry through the Australian Mineral Industries Research Association Limited (AMIRA) are gratefully acknowledged.

REFERENCES

1. Warburton, P.M., Vector stability analysis of an arbitrary polyhedral block with any number of free faces. *Int. J. Rock Mech. Min. Sci. & Geomech. Abstr.*, vol. 18, pp. 415-427, 1981.

2. Goodman, R.E. and Gen-Hua Shi, *Block theory and its application to rock engineering*, pp. 1-338, Prentice-Hall, New Jersey, 1985

3. Cundall, P.A., A computer model for simulating progressive, large scale movements in blocky rock systems, *Proc. International Symposium on Rock Fracture*, Nancy, France, vol. 2, Paper 8, 1971.

A Numerical Method for Transient Temperature Change in Buildings

R. VAN KEER
Faculty of Applied Sciences, R.U.G.
St.-Pietersnieuwstraat 39, 9000 Ghent, Belgium

INTRODUCTION

a. An important topic in heat transmission in buildings concerns the effect of changes in the outdoor temperature on the temperature of the air, θ_i , inside an enclosure, related to the thermal properties of the external wall and the internal mass, as well as to the rate of heat added to the air, q, as well as to the rate of ventilation, v.

As a model problem of practical relevance, see Pratt (1981), consider the situation illustrated in Fig. 1, reprensented by the following set (P) of equations for the smooth triple $\Theta(x,t) = (\theta(x,t), \theta_i(t), \theta_{is}(t))$, $0 \leq x \leq \ell, t \geq 0$.

$$K.\frac{\partial^2\theta(x,t)}{\partial x^2} = \frac{\partial\theta(x,t)}{\partial t}, \qquad 0 < x < \ell, t > 0 \tag{1}$$

$$k.\frac{\partial\theta(x,t)}{\partial x} = h_0.\theta(x,t), \qquad x = 0, t > 0 \tag{2}$$

$$k.\frac{\partial\theta(x,t)}{\partial x} = h_i[\theta_i(t) - \theta(x,t)], \qquad x = \ell, t > 0 \tag{3}$$

$$q = h_i[\theta_i(t) - \theta(x,t)] + v.\theta_i(t) - h_s.[\theta_{is}(t) - \theta_i(t)], \qquad x = \ell, t > 0 \tag{4}$$

$$c.\frac{d\theta_{is}(t)}{dt} = -h_s.[\theta_{is}(t) - \theta_i(t)], \qquad t > 0 \tag{5}$$

$$\theta(x,0) = \theta_0 + \frac{(1/h_0 + x/k)}{1 + v.a}.(q - v.\theta_0), \qquad 0 < x < \ell \tag{6}$$

$$\theta_{is}(0) = \theta_0 + \frac{a}{1 + v.a}.(q - v.\theta_0) = \theta_i(0) \tag{7}$$

$$a = \ell/k + 1/h_0 + 1/h_i$$

(P)

Here the I.C's (6) - (7) represent the steady state regime holding until the sudden drop over θ_0 of the outdoor temperature. The capacity of the air and the resistance of the internal mass (with capacity c) are assumed to be zero. The external wall is a large slab, thickness ℓ, of isotropic material, with conductivity k and diffusivity K, conducting heat only in the x direction, perpendicular to the surfaces. k, K as well as the three heat transfer coefficients h_0, h_i, h_s are (positive) constants, which is also the case with v and q.

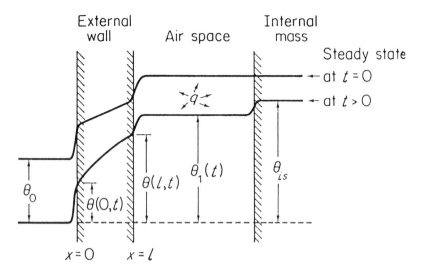

FIGURE 1. Schematic representation of temperature change in a heated enclosure, Pratt (1981)

b. Eliminating $\theta_i(t)$ and $\theta_{is}(t)$ between (4) - (5) and (7), $\theta_i(t)$ is seen to obey a linear Volterra integral equation of the convolution type in terms of $\theta(1,t)$, which may easily be solved. Hence the B.C. (3) may be rewritten as

$$\frac{\partial \theta(x,t)}{\partial x} = F(\theta(x,t), \int_0^t dr.g(t-r).\theta(x,r)) \quad , \quad x = \ell, t > 0 \tag{8}$$

where $F : \mathbf{R}_+^2 \to \mathbf{R}$ is a linear function, and the kernel $g : \mathbf{R}_+ \to \mathbf{R}$ is also known. Thus $\theta(x,t)$ obeys a nonlinear parabolic problem with an integral B.C. at $x = \ell$. Rewritten in a suitable variational form this problem is seen to be well posed, see e.g. Slodicka (1989). The analytical solution seems not to be known.

(8) suggests a Laplace transformation method to be applied, as in the image function space the B.C. at $x = \ell$ becomes an ordinary Robin condition. An exact, but very complicated expression for $L[\theta_i(t)](p)$ can be obtained, the inversion of which requires a careful analysis and suitable approximations, see Pratt (1981) and Davies (1984). Alternatively the (well posed) elliptic problem in the image space can be solved numerically by a standard finite element method, from which $L[\theta_i(t)](p)$ is seen to be a rational function of the transformation parameter p, which may be obtained exactly by symbolic computation (e.g. by REDUCE). The inversion can be performed either semi-analytical by numerical factorization of the denominator or by a numerical L^{-1} transformation-algorithm, see Honig and Hirdes (1984).

c. In this paper we discuss a general numerical method for the evaluation of $\Theta(x,t)$, viz. a finite difference-finite element method. Although, for brevity, we deal with the problem (P), the method can easily be adapted to more involved cases, covering e.g. a multi-layer external wall, or thermal properties which are space or time dependent. Before putting the problem in a variational framework, we invoke the linear superposition principle to simplify (P) considerably

AN EQUIVALENT FORM

By linear superposition we have

$$\Theta(x,t) = \Theta^{(\theta_0)}(x,t) + \Theta^{(q)}(x), \qquad 0 \le x \le \ell, t \ge 0$$

where $\Theta^{(\theta_0)}$ is the solution of (P) when $q = 0$, while $\Theta^{(q)}$ corresponds to the case $\theta_0 = 0$. The latter is independent of time and is

$$\Theta^{(q)}(x) = (\theta(x,0), \theta_i(0), \theta_{is}(0)), \qquad 0 \le x \le \ell$$

defined by (6) - (7) where $\theta_0 = 0$.

Hence we may restrict ourselves to the case $q = 0$. For notational simplicity we rewrite $\Theta^{(\theta_0)}$ as Θ. Next introducing

$$U(x,t) = (u(x,t), u_i(t), u_{is}(t));$$
$$u(x,t) = \theta(x,t) - \theta(x,0), u_i(t) = \theta_i(t) - \theta_i(0), u_{is}(t) = \theta_{is}(t) - \theta_{is}(0)$$

we are left with the problem for U, $0 \le x \le \ell, t \ge 0$,

$$K.\frac{\partial^2 u(x,t)}{\partial x^2} = \frac{\partial u(x,t)}{\partial t}, \qquad 0 < x < \ell, t > 0 \tag{9}$$

$$k.\frac{\partial u(x,t)}{\partial x} = h_0.[u(x,t) + \theta_0], \qquad x = 0, t > 0 \tag{10}$$

$$k.\frac{\partial u(x,t)}{\partial x} = h_i[u_i(t) - u(x,t)], \qquad x = \ell, t > 0 \tag{11}$$

$$0 = h_i[u_i(t) - u(x,t)] + v.u_i(t) - h_s.[u_{is}(t) - u_i(t)], \qquad x = \ell, t > 0 \tag{12}$$

$$c.\frac{du_{is}(t)}{dt} = -h_s.[u_{is}(t) - u_i(t)], \qquad t > 0 \tag{13}$$

$$u(x,0) = 0, \qquad 0 < x < \ell ; u_{is}(0) = 0 = u_i(0) \tag{14 - 15}$$

(P')

Here the inhomogenities are transferred to the B.C. at $x = 0$. This will be advantageous in what follows.

SEMI DISCRETE AND FULL DISCRETE GALERKIN APPROXIMATIONS

Considering, for the time being, $u_i(t)$ as a known function, the problem (9) - (11) for $u(x,t)$ may be formulated in a weak variational form by a standard technique, see e.g. Wait and Mitchell (1985) for a classical reference or Dautray-Lions (1985) for a more general, abstract approach.

Introduce $V = H^1(0,\ell)$, the first order Sobolev space on $(0,\ell)$. Consider the problem of finding $u(x,t)$ so that

$$u(x,t) \in V \qquad \text{and} \qquad \frac{\partial u}{\partial t}(x,t) \in L_2(0,\ell) \quad , \quad t > 0$$

$$\frac{k}{K} \int_0^\ell \frac{\partial u}{\partial t}.v.dx + B(u,v) = f(v), \qquad \forall v \in V, \quad t > 0$$

where

$$B(u, v) = k. \int_0^\ell \frac{\partial u}{\partial x} . v'.dx + h_0.u(0,t).v(0) + h_i.u(\ell,t).v(\ell)$$

$$f(v) = h_i.u_i(t).v(\ell) - h_0.\theta_0.v(0)$$

[Recall that $H_1(0, \ell) \hookrightarrow C^0([0, \ell])$ by the Sobolev embeddings theorem, so that $v(0), v(\ell), \cdots$, are well defined].

In fact, this statement must be made more precise as far as the depending of u on t is concerned by the introduction of abstract function spaces such as $L_2(0, T; V^*), V^*$ being the dual space of V and $T > 0$ arbitrary. The bilinear form B on $V \times V$ is easily seen to be bounded and (strongly) coercive, while f defines a linear bounded functional on V. Hence the problem just mentioned has a unique, stable solution, which moreover may shown to be 'smooth' see Dautray-Lions (1985).

We denote by V_h a Lagrange finite element subspace of $H^1(0, \ell)$, corresponding to a partition τ_h of $[0, \ell]$. To fix the ideas, take linear elements. Denote the nodes as $(x_n)_{n=1}^N$, $x_1 = 0$, $x_N = \ell$. Then

$$V_h = \{v \in C^0([0, \ell]) \, ; \, v|_{[x_e, x_{e+1}]} \text{ is a linear polynomial}, \quad e = 1, \ldots, N-1\}$$

We introduce the cardinal basis $(\varphi_m)_{m=1}^N$ of V_h by

$$\varphi_m(x_n) = \delta_{mn}, m \text{ and } n = 1, \ldots, N$$

The semi discrete Galerkin-finite element approximation is defined by

$$u_h(x, t) = \sum_{m=1}^N d_m(t).\varphi_m \in V_h, \qquad t \geq 0 \tag{16}$$

where the column matrix $D(t) = (d_1(t), \ldots, d_N(t))^T$ obeys

$$M.\dot{D}(t) + S.D(t) = F(t) \quad , \quad t > 0 \quad ; \quad D(0) = 0 \tag{17}$$

with

$$M = (M_{mn})_{1 \leq m,n \leq N}, \qquad M_{mn} = \frac{k}{K} . \int_0^\ell \varphi_m(x).\varphi_n(x).dx$$

$$S = (S_{mn})_{1 \leq m,n \leq N}, \qquad S_{mn} = S_{mn}^{(1)} + S_{mn}^{(2)},$$

$$S_{mn}^{(1)} = k \int_0^\ell \varphi_m'(x).\varphi_n'(x)dx \, ; \qquad S_{mn}^{(2)} = h_0.\delta_{m1}.\delta_{n1} + h_i.\delta_{mN}.\delta_{nN}$$

$$F(t) = (F_m(t))_{1 \leq m \leq N}, \qquad F_m(t) = h_i.u_i(t).\delta_{mN} - h_0.\theta_0.\delta_{m1}$$

This corresponds to introducing, in the variational problem above, $v = \varphi_m, 1 \leq m \leq N$, and replacing u by u_h, (16)

To (16) - (17) we add (12) - (13) and (15) where $u(\ell, t)$ is replaced by $u_h(\ell, t)$, i.e. by $d_N(t)$. Denote

$$d_{N+1}(t) = u_i(t), \quad d_{N+2}(t) = u_{is}(t), \quad t > 0 \tag{18}$$

The initial value problem for

$$\tilde{D}(t) = (D(t)^T \quad | \quad d_{N+1}(t), d_{N+2}(t))^T, \qquad t \geq 0. \tag{19}$$

formed by (17) and by

$$\left. \begin{aligned} &-h_i.d_N(t) + (h_i + v + h_s).d_{N+1}(t) - h_s.d_{N+2}(t) = 0 \\ &c.\frac{du_{N+2}(t)}{dt} = -h_s[d_{N+2}(t) - d_{N+1}(t)] \\ &d_{N+1}(0) = d_{N+2}(0) = 0 \end{aligned} \right\} \tag{17'}$$

may be rearranged to

$$\tilde{M}.\dot{\tilde{D}}(t) + \tilde{S}.\tilde{D}(t) = \tilde{F} \quad, \quad t > 0; \qquad \tilde{D}(0) = 0 \tag{20}$$

where we use the block matrices

$$\tilde{M} = \left(\begin{array}{c|cc} M & 0 \\ \hline & 0 & 0 \\ 0 & & \\ & 0 & c \end{array} \right) \quad , \quad \tilde{F} = \left(\begin{array}{c} -h_0.\theta_0 \\ 0 \\ \vdots \\ 0 \\ \hline 0 \\ 0 \end{array} \right)$$

$$\tilde{S} = \tilde{S}^{(1)} + \tilde{S}^{(2)} \quad , \quad \tilde{S}^{(1)} = \left(\begin{array}{c|c} S^{(1)} & 0 \\ \hline 0 & 0 \end{array} \right) \quad , \quad \tilde{S}^{(2)} = \left(\begin{array}{c|ccc} h_0 & & & \\ \hline & h_i & -h_i & 0 \\ & -h_i & \alpha & -h_s \\ 0 & & & \\ & 0 & -h_s & h_s \end{array} \right)$$

$$\alpha = v + h_i + h_s.$$

(19) - (20) define the semi discrete Galerkin finite element approximation of the solution of (P').

The initial value problem (20) will be solved numerically by a standard μ-family of finite difference schemes, $0 \leq \mu \leq 1$. Explicitly, introducing a time step Δt and time points $t_r = r.\Delta t$, $r \in \mathbf{N}$, we define the approximation $G^{(r)} \approx \tilde{D}(t_r)$ by the recurrent scheme of regular linear systems

$$(\tilde{M} + \mu.\Delta t.\tilde{S}).G^{(r)} = (\tilde{M} - (1 - \mu).\Delta t.\tilde{S}).G^{(r-1)} + \Delta t.\tilde{F}, \quad r \in \mathbf{N}_0; \quad G^{(0)} = 0$$

Thus we obtain a full discrete approximation of the solution of (P'),

$$G^{(r)} \simeq \left(u(x_1, t_r), u(x_2, t_r), \ldots, u(x_N, t_r) | u_i(t_r), u_{is}(t_r) \right)^T \quad r \in \mathbf{N}_0$$

For $\mu \in [\frac{1}{2}, 1]$ the method is unconditionally stable. The order is $0((\Delta t)^2)$ if $\mu = \frac{1}{2}$ (Crank-Nicholson scheme) and $0(\Delta t)$ otherwise, the exact solution being smooth.

NUMERICAL RESULT

As an illustrative example we calculate the 'cooling curve' for an enclosure with $\frac{k}{\ell} = 0,75$; $\frac{K}{\ell^2} = 0,053$; $c = 19,2$; $h_i = h_s = 1,43$; $h_0 = 3,33$. (which corresponds to an 'office building'). We consider both the unventilated case, $v = 0$, and the case with normal ventilation, $v = 0,398$.

We take $\mu = 1$ (backward Euler finite difference scheme) with $\Delta t = 0,1h$, and use a linear Lagrange net with 30 equal elements. Recall that from the considerations of §2

$$\frac{u_i(t)}{\theta_0} = \frac{\theta_i(t) - \theta_i(0)}{\theta_0}$$

The result is in good agreement with the one of Pratt (1981) for $v = 0$, but not for $v \neq 0$. See Fig.2 .

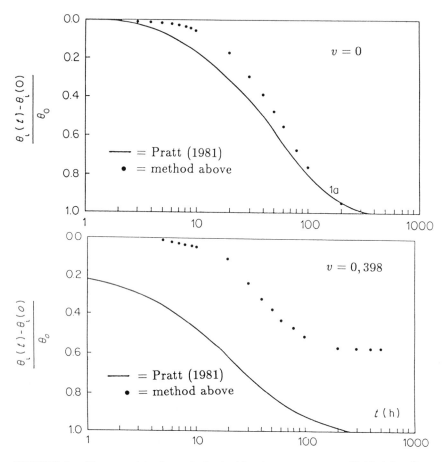

FIGURE 2. Change with time of the inside air temperature, divided by the causal drop in the outdoor temperature. Cases $v = 0$ and $v = 0,398$.

CONCLUDING REMARKS

a. The method above is appropriate to calculate the cooling curves and related physical quantities, such as the time constant, for different building types. This task required a very substantial lot of work in the paper of Pratt (1981). Besides, this author overlooked the simplification, mentioned in §2. The lack of agreement in the ventilated case ($v \neq 0$) between his results and ours is probably due to the fact that he (abusively) uses q instead of ($q - v.\theta_0$) in (6) - (7).

More important the present approach can be extended to more involved physical models. For instance in the case of a multi-layer wall, after a suitable adaptation of the underlying variational problem, the further analysis only requires 'technical' modifications, the construction of the tridiagonal mass matrix M and stifness matrix $S^{(1)}$ being less elementary than in the present problem. This remains true if, in each slab, e.g. the conductivity k is space dependent (requiring a modification of the differential equation (1), of course). A similar remark concerns the time dependency of q in (4), and so on.

b. The hybrid finite element-Laplace transformation method, mentioned in the introduction, consists in the L-transformation of (P'), followed by the application of the (standard) finite element method (the one of §3, say) to the resulting well posed elliptic problem for $L[u(x,t)](p)$, $p > 0$. This method is equivalent to the application of the L-transformation to the initial value problem (20), which itselves results from the semi-discrete finite element method, used for (P'). Both hybrid methods lead to the same regular algebraic system

$$(p.\tilde{M} + \tilde{S}).Y(p) = \frac{1}{p}.\tilde{F}, \qquad p > 0$$

where

$$Y(p) = (y_1(p), \ldots, y_N(p) \mid y_{N+1}(p), y_{N+2}(p))^T; \qquad y_n(p) = L[d_n(t)](p),$$
$$n = 1, \ldots, N + 2$$

$Y(p)$ can be obtained exactly by symbolic computation (e.g. by MACSYMA). Once $det(p.\tilde{M} + \tilde{S})$ has been factorized, the analytical inversion of $y_{N+1}(p)$, which is clearly a rational function of p (the degrees of the numerator and the denominator respectively being N and $N + 2$), gives

$$d_{N+1}(t) = \sum_{\text{poles}} \text{Res}[y_{N+1}(p).e^{p.t}], \qquad t \geq 0 \tag{21}$$

This may also be evaluated by MACSYMA.

As compared to the full discrete method of §3, the present approach directly gives $d_{N+1}(t)$ at any required time point, without iteration with respect to a time step. This feature is especially important in the long time run. Moreover, for not too small values of t only a few terms must be retained in (21), which is clear from the comparison of the poles, all having negative real parts, of course.

Finally we note that $y_{N+1}(p)$ may also be inverted numerically, e.g. by the algorithm of Honig and Hirdes (1984). Here $y_{N+1}(p)$ must be known at a number of points on a suitably chosen vertical line in the complex plane. This technique has recently been discussed by Chen et al. (1987), for simple model problems.

ACKNOWLEDGEMENTS

We are grateful to Ir. K. Audenaert for a fruitful discussion and for providing us with Fig.2. We thank the Belgian National Science Foundation for financial support. We also thank the referee for signalizing us a mistake in a statement in the Concluding Remarks.
Our warmest thanks go to Mrs R. Van Hove and Z. Oost for the painstaking task of typing the manuscript.

REFERENCES

1. Chen H-T., Chen T-M., Chen C-K., Hybrid Laplace transform finite element method for one dimensional heat conduction problems, *Computer Methods in Applied Mechanics and Engineering,* vol 33, pp. 83-95, 1987.

2. Dautray R., Lions J-L., *Analyse mathématique et calcul numérique pour les sciences et les techniques, tome 3,* pp. 615-642, Masson, Paris, 1985.

3. Davies M.G., Comments on Pratt's transient cooling solution, *International Journal of Heat and Mass Transfer,* vol 27, no 7, p.1123, 1984.

4. Honig G., Hirdes U., A method for the Numerical Inversion of Laplace Transforms, *Journal of Computational and Applied Mathematics,* vol 10, pp. 113-132, 1984.

5. Pratt W., *Heat Transmission in Buildings,* pp. 121-157, J. Wiley, Chichester, 1981.

6. Slodicka M., An investigation of convergence and error estimate of the approximate solution for a quasilinear parabolic integro differential equation, forthcoming, *Applikace Mathematiky,* 1989.

7. Wait R. and Mitchell A.R., *Finite Element Analysis and Applications,* pp.146-169 and pp.212-214, J. Wiley, Chichester, 1985.

Numerical Study of Acoustic Scattering and Internal Focussing of Dilatational Waves by Fluid Cylinders

B. C. H. WENDLANDT
DSTO, Materials Research Laboratory
P.O. Box 50
Ascot Vale, 3032, Victoria, Australia

INTRODUCTION

In the present investigation a simple numerical scheme able to study the reflection, focussing and scattering of dilatational waves by elastic and viscoelastic fluid cylinders is presented.   The cylinders are immersed in water and insonified by a plane wave and the time evolution of the scattered waves of the near field around the cylinder is calculated.   Multiple reflections are observed as predicted previously (Davis et al 1978).   The material properties of the cylinders are shown in Table 1. Concentration of acoustic excitation inside the cylinders is observed at the internal focal points as calculated from ray theory.

THEORY

Wave Equation

If the displacement of an acoustic wave is described by a vector $\underset{\sim}{u}$, the vector obeys the wave equation (Sommerfeld 1950)

$$\rho \frac{\partial^2 \underset{\sim}{u}}{\partial t^2} = \nabla\{\lambda(1+\beta \frac{\partial}{\partial t}) \nabla.\underset{\sim}{u}\} \tag{1}$$

when the shear modulus is neglected.   The material parameters are ρ – density of the medium, λ – first Lame constant, β – viscoelastic relaxation constant.

TABLE 1.   Material Properties of Fluids

	Density (kg/m^3) ρ	Speed of Sound (m/s) $\sqrt{\lambda/\rho}$	λ (GPa)	Loss Tangent tan $\delta=\omega\beta$
Cylinder A	1500	1000	1.5	0
Cylinder B	2000	1000	2.0	0
Cylinder C	1500	1000	1.5	0.5
Water	1000	1500	2.25	0

Loss Function

Power is dissipated in a lossy fluid by loss mechanisms which become active during the deformation. The power dissipated by these mechanisms is given by dW/dt where W is the work done by the stresses across the surface of an element of volume $\delta x\ \delta y\ \delta z$.

In the absence of shear it can be shown (Sommerfeld 1950) that

$$dW = (\lambda[1 + \beta \frac{\partial}{\partial t}]\theta)\ d\theta\ dy\ dx\ dz \tag{11}$$

On integration over a cycle $2\pi/\omega$ the loss function L for the dilatational wave of sinusoidal form $\theta = \theta_0 \sin \omega t$, where θ_0 is the amplitude of the dilatation, is

$$L = \frac{\omega^2 \beta \lambda \theta_0^2}{2} \tag{12}$$

In the numerical scheme used in the present work, this loss function L_{ij} at all ij is calculated from

$$L_{ij} = \frac{\omega^2 \lambda_{ij} \beta_{ij} \theta_{0ij}^2}{2} \tag{13}$$

SCATTERING CALCULATIONS

The acoustic response of three lossless and viscoelastic fluid cylinders immersed in water to an incident sine wave packet was calculated. The material properties of Table 1, were assumed in the calculations to enable the echo structure computed to be compared with previously published work (Davis et al 1978) based on Kirchhoff analytical approximation.

Following previously published work the incident wave packet was centered at a dimensionless frequency ka=12, where k is the wave number of the sinusoidal component of the wave packet in water and a is the radius of the cylinder.

RESULTS

The density cross section of the elastomer cylinder is shown without scales in Fig. 1. The acoustic responses of the lossless fluid cylinders are shown in Figs 2-3, where I denotes the incident wavepacket, R the wavepacket or packets reflected by the cylinders and T labels the wavepackets transmitted by the cylinder along the axis of advance of the incident packet, Fig. 1.

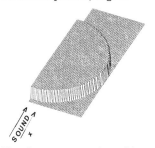

FIGURE 1   Density cross-section of insonified cylinder.

Equation (1) can be expressed in terms of the dilatation θ, which is more amenable to multi-dimensional numerical treatment than the direct evaluation of Eqn. (1). If θ is defined as in (Sommerfeld 1950) by

$$\nabla \cdot \underline{u} \; = \; \theta \tag{2}$$

and the divergence is taken of Eqn. (1), Eqn. (1) can be written as

$$\frac{\partial^2 \theta}{\partial t^2} = \frac{1}{\rho} \nabla^2 \{\lambda(1+\beta \frac{\partial}{\partial t}) \; \theta\} - \frac{(\nabla \rho)}{\rho^2} \cdot \nabla \{\lambda(1+\beta \frac{\partial}{\partial t}) \; \theta\} \tag{3}$$

Numerical Analogue to Wave Equation

Equation (1) and (3) assume that λ and ρ are differentiable with respect to the spatial co-ordinates. Their use in discontinuous media may require special consideration of boundary conditions or some smearing of the discontinuites to be introduced. The scheme presented here will approximate the discontinuites to second order accuracy in space. A somewhat similar scheme has been published for a wave equation able to describe the propagation of torsion waves (Brown 1984). That work considers a wave equation after the assumptions of material continuity have removed the material modulus from the second order spatial differentiation. The approach of (Brown 1984) was unable to successfully model the problem considered in the present paper.

The wave equations (3) can most readily be solved by an explicit finite difference scheme which steps forward in time across the spatial grid. The value of θ is updated at each spatial grid point in turn. When all of the spatial grid has been updated a solution of at that point in time has been calculated for the problem considered. Care must be taken that the velocities of sound, moduli of elasticity, density, and relaxation time parameters are changed at the appropriate nodes in the finite difference representation of Eqn. (3) in order that the acoustic behaviour of a discontinuous medium is modelled adequately.

Using centered differences for the time derivative of θ and limiting the present considerations to two space dimensions (x,y) only

$$\frac{\partial^2 \theta^n}{\partial t^2} \; ij \; = \; \frac{\theta_{ij}^{n+1} - 2 \theta_{ij}^{n} + \theta_{ij}^{n-1}}{(\delta t)^2} \tag{4}$$

If centered differences are also used for the ∇^2 terms in Eqn. (3) and backward differences are used to represent the time derivative describing the viscous Kelvin-Voigt effect (Jaeger 1956) then the first term on the RHS of Eqn. (3), in two space dimensions (x,y) becomes, where δx is the interval between nodes ij and ij+1 or i+1j,

$$\frac{\nabla^2}{\rho}\{\lambda(1 + \beta\frac{\partial}{\partial t})\,\theta\}^n_{ij} \approx \frac{1}{\rho_{ij}(\delta x)^2}\{\lambda_{i+1j}[(1 + \underset{\delta t}{\beta_{i+1j}})\,\theta^n_{i+1j} - \underset{\delta t}{\beta_{i+1j}}\,\theta^{n-1}_{i+1j}]$$

$$- 2\lambda_{ij}\,[(1 + \underset{\delta t}{\beta_{ij}})\,\theta^n_{ij} - \underset{\delta t}{\beta_{ij}}\,\theta^{n-1}_{ij}]\,(1 + [\frac{\delta x}{\delta y}]^2)$$

$$+ \lambda_{i+1j}\,[(1+\underset{\delta t}{\beta_{i-1j}})\,\theta^n_{i-1j} - \underset{\delta t}{\beta_{i-1j}}\,\theta^{n-1}_{i-1j}] + (\frac{\delta x}{\delta y})^2\{\lambda_{ij+1}[(1 + \underset{\delta t}{\beta_{ij+1}})\,\theta^n_{ij+1}$$

$$- \underset{\delta t}{\beta_{ij+1}}\,\theta^{n-1}_{ij+1}] + \lambda_{ij-1}\,[(1+ \underset{\delta t}{\beta_{ij-1}})\,\theta^n_{ij-1} - \underset{\delta t}{\beta_{ij-1}}\,\theta^{n-1}_{ij-1}]\}\} \tag{5}$$

The last term on the right hand side of Eqn. (3) can be approximated similarly, using centered differences to minimize truncation error, to yield

$$(\frac{\nabla\rho}{\rho^2})\cdot\nabla\{\lambda(1 + \underset{\delta t}{\beta\partial})\,\theta\,\}^n_{ij} \approx \frac{(\rho_{ij+1} - \rho_{ij-1})}{4\rho^2_{ij}\,(\delta y)^2}\{\lambda_{ij+1}\,[(1 + \underset{\delta t}{\beta_{ij+1}})\,\theta^n_{ij+1}$$

$$- \underset{\delta t}{\beta_{ij+1}}\,\theta^{n-1}_{ij+1}] - \lambda_{ij-1}\,[(1 + \underset{\delta t}{\beta_{ij-1}})\,\theta^n_{ij-1} - \underset{\delta t}{\beta_{ij-1}}\,\theta^{n-1}_{ij-1}]\}$$

$$+ \frac{(\rho_{i+1j} - \rho_{i-1j})}{4\rho^2_{ij}(\delta x)^2}\{\lambda_{i+1j}\,[(1 + \underset{\delta t}{\beta_{i+1j}})\,\theta^n_{i+1j} - \underset{\delta t}{\beta_{i+1j}}\,\theta^{n-1}_{i+1j}] - \lambda_{i-1j}$$

$$[(1+\underset{\delta t}{\beta_{i-1j}})\,\theta^n_{i-1j} - \underset{\delta t}{\beta_{i-1j}}\,\theta^{n-1}_{i-1j}]\} \tag{6}$$

The dimensional variables of Eqn. (4)–(6) may be transformed into the following dimensionless variables,

$$\alpha_{ij} = \beta_{ij}/t_1, \quad \tau = t/t_1, \quad \xi = x/1, \quad \eta = y/1, \quad \gamma_{ij} = \rho_{ij}/\rho_0, \quad a_{ij} =$$

$$\lambda_{i+1j}/\lambda_{ij}, \quad b_{ij} = \lambda_{i-1j}/\lambda_{ij}, \quad C_{ij} = \lambda_{ij+1}/\lambda_{ij}, \quad B_{ij} = t^2_1\,\lambda_{ij}/1^2\rho_{ij},$$

$$d_{ij} = \lambda_{ij-1}/\lambda_{ij}, \tag{7}$$

where 1, t_1 and ρ_0 are length, time and density characteristic of the calculation of interest.

Eqn. (3) can be written using Eqns. (4–7) as

$$\theta_{ij}^{n+1} = 2\theta_{ij}^{n} - \theta_{ij}^{n-1} + B_{ij} \left(\frac{\delta\tau}{\delta\xi}\right)^2 \{a_{ij}[(1 + \frac{\alpha_{i+1j}}{\delta\tau})\,\theta_{i+1j}^{n} - \frac{\alpha_{i+1j}}{\delta\tau}\,\theta_{i+1j}^{n-1}]$$

$$+ b_{ij}[(1 + \frac{\alpha_{i-1j}}{\delta\tau})\theta_{i-1j}^{n} - \frac{\alpha_{i-1j}}{\delta\tau}\,\theta_{i-1j}^{n-1}] - 2(1 + [\frac{\delta\xi}{\delta\eta}]^2)\,\{[(1 + \frac{\alpha_{ij}}{\delta\tau})\,\theta_{ij}^{n}$$

$$- \frac{\alpha_{ij}}{\delta\tau}\,\theta_{ij}^{n-1}] + \left(\frac{\delta\xi}{\delta\eta}\right)^2 \{C_{ij}[(1 + \frac{\alpha_{ij+1}}{\delta\tau})\,\theta_{ij+1}^{n} - \frac{\alpha_{ij+1}}{\delta\tau}\theta_{ij+1}^{n-1}] + d_{ij} \tag{8}$$

$$[(1 + \frac{\alpha_{ij-1}}{\delta\tau})\theta_{ij}^{n} - \frac{\alpha_{ij-1}}{\delta\tau}\,\theta_{ij}^{n-1}]\} - \frac{(\gamma_{i+1j} - \gamma_{i-1j})}{4\gamma_{ij}\rho_o}\,\{a_{ij}[(1+\frac{\alpha_{i+1j}}{\delta\tau})$$

$$\theta_{i+1j}^{n} - \frac{\alpha_{i+1j}}{\delta t}\,\theta_{i+1j}^{n-1}] - b_{ij}\,[(1+\frac{\alpha_{i-1j}}{\delta\tau})\,\theta_{i-1j}^{n} - \frac{\alpha_{i-1j}}{\delta\tau}\,\theta_{i-1j}^{n-1}]\} -$$

$$\frac{(\gamma_{ij+1} - \gamma_{ij-1})}{4\gamma_{ij}\rho_o}\left(\frac{\delta\xi}{\delta\eta}\right)^2\{C_{ij}[(1+\frac{\alpha_{ij+1}}{\delta\tau})\,\theta_{ij+1}^{n} - \frac{\alpha_{ij+1}}{\delta\tau}\,\theta_{ij+1}^{n-1}] - d_{ij}\,[(1+\frac{\alpha_{ij-1}}{\delta\tau})$$

$$\theta_{ij-1}^{n} - \frac{\alpha_{ij-1}}{\delta t}\,\theta_{ij-1}^{n-1}]\}\}$$

The expected local error of computation is of second order of the spatial intervals and time increments and proportional to the fourth order derivative of the function being approximated (McCracken et al 1969). In practice an error of about five percent was observed in runs using values of 0.1 for the nondimensional spatial intervals.

Stability of the computation is governed by the von Neumann stability condition (McCracken et al 1969) for a wave equation.

$$\frac{2B_{ij}(\delta\tau)^2}{[(\delta\xi)^2 + (\delta\eta)^2]}\,(1 + \frac{\alpha_{ij}}{\delta\tau}) \le 1 \tag{9}$$

When $\frac{\alpha_{ij}}{\delta\tau} \gg 1$, Eqn. (8) becomes a diffusion equation. The stability criterion for Eqn. (8) then becomes

$$\frac{2B_{ij}\,\delta\tau\,\alpha_{ij}}{[(\delta\xi)^2 + (\delta\eta)^2]} \le \frac{1}{2} \tag{10}$$

Evaluation of Eqns. (9,10) shows that the finite differences form of Eqn. (8) obeys a stricter stability criterion than the finite differences form of the wave equations.

Where δt is an increment in time and θ_{ij}^{n+1}, θ_{ij}^{n}, θ_{ij}^{n-1} are dilatations at the three consecutive time instants t^{n+1}, t^{n}, t^{n-1} and $t^{n} = t^{n-1} + \delta t$ at the spacial node ij.

The numerical calculations show that the specular reflection from a fluid cylinder whose acoustic impedance equals that of water, cylinder A in Table 1, is very small, or negligible, in agreement with simple impedance theory, Fig. 2. The specular reflection is labelled R_1 and its computed magnitude falls within the error of computation. The dominant echo from such a cylinder appears to be caused by waves travelling or creeping around the outside of the cylinder and is generated by the cylinder after the specular reflection and is shown R_2 in Fig. 3. The amplitude of this echo agrees to within 10% of previously published results which were based on Kirchhoff approximation theory (Davis et al 1978). The specular reflection from a fluid cylinder whose acoustic impedance is different from that of water (cylinder B in Table 1) is shown in Fig. 4. Again, the amplitude calculated by the present numerical scheme agrees to within 10% with previous work by Davis.

The amplitude of acoustic excitation within a lossy cylinder C is shown in Fig. 5. The power losses are proportional to the square of the amplitude, as given by Eqn. (13). Figure 5 shows amplitude information to bring out details of the acoustic response at the rim. The losses are concentrated at the back of the cylinder where focussing within the cylinder is expected from ray theory Losses also appear at the rim of the cylinder which are due to complex internal reflections, or due to circumferential waves moving at and close to the surface of the cylinder. The present study has not isolated circumferential or creeping waves from multiple internal reflection phenomena. The observed calculated patterns are caused by the five cycle incident wavepacket and should be enhanced and perhaps be somewhat different for long wavetrains, hence the θ_0 scale is arbitrary in Fig. 5.

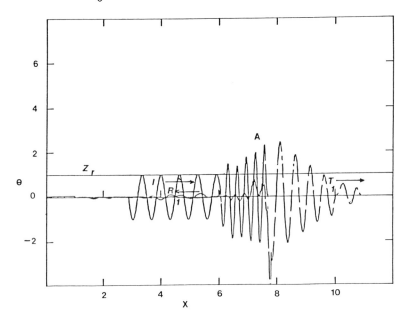

FIGURE 2   Specular echo generated by fluid cylinder A. The echo is normalised with respect to the amplitude of the incident wave packet. The circular cross section is located as shown in Fig. 3, between x = 6 and x = 8.3, where the unit of x is three wavelengths of the incident wave packet. I – incident wave packet, R_1 – first echo or specular echo generated by cylinder, T_1 – first transmitted wavepacket, Z_r – impedance relative to impedance of water.

744

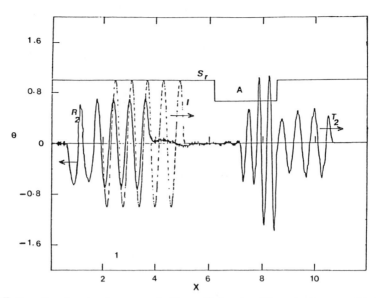

FIGURE 3   Dominant echo generated by cylinder A. The echo is normalised with respect to the amplitude of the incident wave packet. The distance x is normalised to three wavelengths of the incident wave packet. I – incident wave packet (shown for reference only), R_2 – second or dominant echo generated by cylinder, T_2 – second transmitted wave packet, S_r – ratio of speed of sound in cylinder fluid to speed of sound in surrounding water.

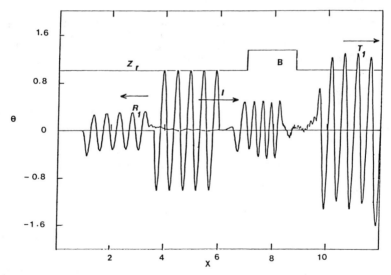

FIGURE 4   Specular echo generated by cylinder B. The echo is normalised with respect to the amplitude of the incident wave packet. The distance x is normalised to three wavelengths of the incident wave packet. I – incident wave, R_1 – first or specular echo generated by cylinder, T_1 – first transmitted wave packet, Z_r – ratio of acoustic impedance of cylinder fluid to acoustic impedance of water.

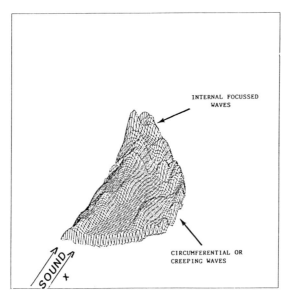

FIGURE 5   Amplitude θ_o of acoustic excitation in cylinder C.   The x and y scales are normalised to the radius of the cylinder.   The θ_o is arbitrary.   Because the acoustic excitation pattern is symmetric about the diameter of the cylinder, only the θ_o in that half of the cross-section of the cylinder shown in Fig. 1 is displayed.

CONCLUSION

The agreement between the echoes calculated by the numerical scheme of Eqn. (8) and previously published work (Davis et al 1978) gives confidence that Eqn. (8) can be used to calculate the acoustic responses including circumferential or creeping waves for more complex elastomeric acoustic systems.

REFERENCES

1.   Brown, D.L., A Note on the Numerical Solution of the Wave Equation with Piecewise Smooth Coefficients, *Mathematics of Computation,* vol. 42, no. 166, pp.369–391, 1984.

2.   Davis, C.M., Dragonette, L.R. and Flau, L., Acoustic Scattering from Silicone Rubber Cylinders and Spheres, *Journal of Acoustical Society of America,* vol. 63, no. 6, pp.1694–1698, 1978.

3.   Jaeger, J.C., *Elasticity, Fracture and Flow,* p.100, Methuen, London, 1956.

4.   McCracken and Dorn, W.S., *Numerical Methods and Fortran Programming,* pp.380–385, Wiley, London, 1964.

5.   Sommerfeld, A., *Mechanics of Deformable Bodies,* p.71, Academic Press, New York, 1950.

6.   Sommerfeld, A., *Mechanics of Deformable Bodies,* p.106, Academic Press, New York, 1950.

Index

Explicit methods (*Cont.*):
    (*See also specific applications,*
       methods, problems)
Explosions, bubble dynamics of, 219–226
Express operating system, 513

Factorization:
    alternating group explicit methods,
       607–614
    computer solutions, 147–148
    NMR modelling and, 173–174
    preconditioners and, 132
    tridiagonal systems, 607–614
Films:
    bearing load capacity and, 383,
       453–460
    non-conforming line contact, 525–532
    point contact problems, 277–284
    slot flow and, 493–500
Filters, adaptive algorithm, 517–524
Finite difference methods (FDM):
    accuracy of, 349–356
    advection-diffusion equations, 327,
       341–348, 349–356
    box method, 351
    building temperature transients, 732
    conjugate gradient and, 164
    dilatational waves, 741
    elastohydrodynamic problems, 279–280
    implicit three-level methods, 341–348
    leapfrog method, 341–348, 351–353,
       356
    mantle convection problems, 161–168
    non-staggered grids, 287–294
    reservoir simulation, 61–62, 121–133
    shock wave model, 366
    submerged gas injection, 471–478
    superconvergence points, 577
    tidal flows, 295–302
    for unsteady convection-diffusion,
       311–317
    wave response parameters, 350
    wind-induced currents, 261–268
    (*See also specific applications,*
       methods, problems)
Finite element model (FEM):
    automatic detail removal, 405–412
    bearing deformation, 383–390,
       453–460
    boundary fitted coordinate transform,
       461–468
    building temperature transients, 732
    engineering design optimization,
       665–672

fracture mechanics, 397–404
free surface flow problems, 243–250,
    461–468
granular solid flows and, 445–452
jointed rock masses, 675–682
machine vision problems, 512
magnetostatic problems, 235
multilevel substructuring, 413–419
resistive plasma stability, 429–436
slot flows, 493–500
solenoidal elements and, 421–428
spectral method, 253–259
superconvergence points, 577
supercomputer simulations, 60
unsteady Navier-Stokes equations,
    253–259
(*See also specific methods, problems*)
Finite strip-element method, 391–396
Finite volume method, scramjet modelling,
    534–536
Flexible manufacturing system, 108–110
Flow problems:
    adaptive grid analysis, 156
    boundary element method and,
       243–250
    flux-split procedure, 437–434
    fractured media, 65–89
    free surfaces, 243–250, 461–468, 476,
       493–500
    Galerkin method for, 261–268
    granular solid flows and, 445–452
    integral equations for, 211–218
    mantle convection problems, 161–168
    mixed boundary problems, 707–714
    narrow extrema, 303–310
    porous media infiltration, 269–276
    reduced Navier-Stokes equations,
       373–381
    sloping hillside seepage, 707–714
    slot flows, 493–500
    solenoidal elements, 421–428
    three-dimensional Euler code, 185–
       192
    tidal flows, 295–302
    tridiagonal systems of equations,
       607–614
Fluorocarbon coolants, 53
Flux-corrected transport, 327–332,
    366–371
Flux-split procedure, 437–444
Forward planning, 633–640
Fourier spectral methods, 253–259
Fractured media, 65–89
Fracture mechanics, 397–404
Fredholm equation, 211–218, 230

750